Digital Wave
Advanced Technology of
Industrial Internet

数字浪潮 工业互联网先进技术 丛书

国家出版基金项目
NATIONAL PUBLICATION FOUNDATION

"十四五"时期国家重点出版物
出版专项规划项目

Digital Wave
Advanced Technology of Industrial Internet

数字浪潮
工业互联网先进技术 丛书

Control of
Autonomous Intelligent Systems

自主智能系统控制

唐漾 金鑫 徐加鹏 黄家豪 著

化学工业出版社
·北京·

内容简介

本书对自主智能系统前沿控制理论问题以及主要方法做了详细阐述，主要从自主智能系统的状态估计、协同控制和网络安全防御三个方面展开，内容涵盖事件触发风险敏感状态估计、具有相对熵约束的事件触发最小最大状态估计、多刚体的有限时间协同控制、多刚体系统的姿态同步、多刚体系统的最优一致性、多智能体系统在网络攻击下的安全控制以及多智能体系统的分布式攻击检测与防御机制等理论研究成果。

本书可供人工智能领域理论研究者或工程技术人员参考，也可供高等院校自动控制专业的本科生和研究生学习。

图书在版编目（CIP）数据

自主智能系统控制 / 唐漾等著. —北京：化学工业出版社，2022.12
（"数字浪潮：工业互联网先进技术"丛书）
ISBN 978-7-122-42289-7

Ⅰ.①自… Ⅱ.①唐… Ⅲ.①人工智能-自动控制系统 Ⅳ.①TP18

中国版本图书馆CIP数据核字（2022）第181263号

责任编辑：宋 辉
文字编辑：李亚楠 陈小滔
责任校对：宋 玮
装帧设计：王晓宇

出版发行：化学工业出版社
（北京市东城区青年湖南街13号 邮政编码100011）
印 装：中煤（北京）印务有限公司
710mm×1000mm 1/16 印张25½ 字数400千字
2023年5月北京第1版第1次印刷

购书咨询：010-64518888
售后服务：010-64518899
网 址：http：//www.cip.com.cn
凡购买本书，如有缺损质量问题，本社销售中心负责调换。

定 价：148.00元

序言
FOREWORD

当前，人类社会来到第四次工业革命的十字路口。数字化、网络化、智能化是新一轮工业革命的核心特征与必然趋势。工业互联网是新一代信息通信技术与工业经济深度融合的新型基础设施、应用模式和工业生态，通过对人、机、物、系统等的全面连接，构建起覆盖全产业链、全价值链的全新制造和服务体系，为工业乃至产业数字化、网络化、智能化发展提供了实现途径，是第四次工业革命的重要基石。目前，我国经济社会发展处于新旧动能转换的关键时期，作为在国民经济中占据绝对主体地位的工业经济同样面临着全新的挑战与机遇。在此背景下，我国将工业互联网纳入新型基础设施建设范畴，相关部门相继出台《“十四五”规划和2035年远景目标纲要》《“十四五”智能制造发展规划》《“十四五”信息化和工业化深度融合发展规划》等一系列与工业互联网紧密相关的政策，希望把握住新一轮的科技革命和产业革命，推进工业领域实体经济数字化、网络化、智能化转型，赋能中国工业经济实现高质量发展，通过全面推进工业互联网的发展和应用来进一步促进我国工业经济规模的增长。

因此，我牵头组织了“数字浪潮：工业互联网先进技术”丛书的编写。本丛书是一套全面、系统、专门研究面向工业互联网新一代信息技术的丛书，是“十四五”时期国家重点出版物出版专项规划项目和国家出版基金项目。丛书从不同的视角出发，兼顾理论、技术与应用的各方面知识需求，构建了全面的、跨层次、跨学科的工业互联网技术知识体系。本套丛书着力创新、注重发展、体现特色，既有基础知识的介绍，更有应用和探索中的新概念、新方法与新技术，可以启迪人们的创新思维，为运用新一代信息技术推动我国工业互联网发展做出重要贡献。

为了确保“数字浪潮：工业互联网先进技术”丛书的前沿性，我邀请杜文莉、侍洪波、顾幸生、牛玉刚、唐漾、严怀成、杨文、和望利、王喆等20余位专家参与编写。丛书编写人员均为工业互联网、自动化、人工智能领域的领军人物，包含多名国家级高层次人才、国家杰出青年基金获得者、国家优秀青年基金获得者，以及各类省部级人才计划入选者。多年来，这些专家对工业互联网关键理论和技术进行了系统深入的研究，取得了丰硕的理论与技术成果，并积累了丰富的实践经验，由他们编写的这套丛书，系统全面、结构严谨、条理清晰、文字流畅，具有较高的理论水平和技术水平。

这套丛书内容非常丰富，涉及工业互联网系统的平台、控制、调度、安全等。丛书不仅面向实际工业场景，如《工业互联网关键技术》《面向工业网络系统的分布式协同控制》《工业互联网信息融合与安全》《工业混杂系统智能调度》《数据驱动的工业过程在线监测与故障诊断》，也介绍了工业互联网相关前沿技术和概念，如《信息物理系统安全控制设计与分析》《网络化系统智能控制与滤波》《自主智能系统控制》和《机器学习关键技术及应用》。通过本套丛书，读者可以了解到信息物理系统、网络化系统、多智能体系统、多刚体系统等常用和新型工业互联网系统的概念表述，也可掌握网络化控制、智能控制、分布式协同控制、信息物理安全控制、安全检测技术、在线监测技术、故障诊断技术、智能调度技术、信息融合技术、机器学习技术以及工业互联网边缘技术等最新方法与技术。丛书立足于国内技术现状，突出新理论、新技术和新应用，提供了国内外最新研究进展和重要研究成果，包含工业互联网相关落地应用，使丛书与同类书籍相比具有较高的学术水平和实际应用价值。本套丛书将工业互联网相关先进技术涉及到的方方面面进行引申和总结，可作为高等院校、科研院所电子信息领域相关专业的研究生教材，也可作为工业互联网相关企业研发人员的参考学习资料。

工业互联网的全面实现是一个长期的过程，当前仅仅是开篇。“数字浪潮：工业互联网先进技术”丛书的编写是一次勇敢的探索，系统论述国内外工业互联网发展现状、工业互联网应用特点、工业互联网基础理论和关键技术，希望本套丛书能够对读者全面了解工业互联网并全面提升科学技术水平起到推进作用，促进我国工业互联网相关理论和技术的发展。也希望有更多的有志之士和一线技术人员投身到工业互联网技术和应用的创新实践中，在工业互联网技术创新和落地应用中发挥重要作用。

钱锋

前言 PREFACE

随着国务院关于《新一代人工智能技术发展规划》的发布，以人工智能技术为核心驱动的自主智能系统技术近年来在学术界和工业界都引起了广泛关注，其相关应用已经渗透工业制造、无人驾驶、航空航天、国防军事等各个领域。自主智能系统是一个具有交叉学科特点的研究领域，其涵盖的理论与技术研究问题十分广泛。其中，自主智能系统的运动控制是最基本也是最重要的问题之一。为了实现自主智能系统的精确控制，对其状态的滤波与估计算法设计十分关键。此外，随着信息技术的不断革新以及自主智能系统网络化程度的不断提高，数据之间的传输和共享面临着极大的安全问题。在此背景下，本书旨在瞄准自主智能系统的控制理论研究前沿问题，从自主智能系统的状态估计、自主控制与系统安全三个角度对理论研究成果做了整理和总结，进一步为我国自主智能系统技术的发展与应用提供理论指导与技术支撑。

本书以智能体系统为研究对象，涵盖了自主智能系统的主要建模与控制方法，主要内容从状态估计、协同控制以及安全防御三个方面展开，对自主智能系统相关问题的理论结果和分析方法做了介绍。全书共分为 11 章，大体可分为四个部分：第一部分包括第 1、2 章，主要介绍自主智能系统的相关背景知识以及主要的建模方法；第二部分包括第 3 ～ 5 章，主要介绍自主智能系统的状态估计方法；第三部分包括第 6 ～ 8 章，主要介绍多刚体系统有限时间协同控制和多刚体系统的姿态同步控制方法，以及多刚体系统的最优一致性控制；第四部分包括第 9 ～ 11 章，主要介绍智能体系统在网络攻击下的安全控制以及多智能体系统的分布式攻击检测与防御机制等内容。

本书由唐漾、金鑫、徐加鹏、黄家豪编写，特别感谢参与本书各章节主要内容整理工作的张丹丹和王赛威等同学。同时感谢在本书的撰写与勘误过程中提出宝贵意见的同行和同学们。

尽管本书在编撰过程中做了大量工作，但由于时间紧张，水平有限，书中疏漏之处在所难免，敬请读者批评指正。

著　者

目录
CONTENTS

符号和缩略语说明

$\mathbb{R}$	实数集
$\mathbb{R}_{\geqslant 0}$	全体非负实数
$\mathbb{R}_{>0}$	全体正实数
$\mathbb{Z}_{\geqslant 0}\mathbb{N}$	全体非负整数
$\mathbb{N}$	整数集合 $\{1, 2, \cdots, n\}$
$\mathbb{R}^n\left(\mathbb{C}^n\right)$	n 维实（复）向量集合
$\left[\boldsymbol{x}_i\right]_{i\in\mathbb{N}}$	一个 n 维实向量
$\boldsymbol{x}\leqslant\boldsymbol{y}$	对于向量$\boldsymbol{x},\boldsymbol{y}\in\mathbb{R}^n$,$\boldsymbol{x}$的所有元素小于等于 $\boldsymbol{y}$ 中相应的元素
$\boldsymbol{x}_{0k}$	向量序列$\left\{\boldsymbol{x}_0,\cdots,\boldsymbol{x}_k\right\}$的简写
$\Vert\boldsymbol{x}\Vert$	向量$\boldsymbol{x}\in\mathbb{R}^n$的欧氏范数
$\Vert\boldsymbol{x}\Vert_{\boldsymbol{Z}}^2$	对于向量$\boldsymbol{x}\in\mathbb{R}^n$和半正定矩阵$\boldsymbol{Z}\in\mathbb{R}^{n\times n}$，定义$\Vert\boldsymbol{x}\Vert_{\boldsymbol{Z}}^2:=\boldsymbol{x}'\boldsymbol{Z}\boldsymbol{x}$
$\mathbb{R}^{m\times n}(\mathbb{C}^{m\times n})$	$m\times n$ 维实（复）矩阵集合
$\boldsymbol{I}$	适维的单位阵
$\mathrm{diag}(\boldsymbol{A}_1,\boldsymbol{A}_2)$	矩阵 $\boldsymbol{A}_1$ 和 $\boldsymbol{A}_2$ 组成的块对角矩阵
$\mathrm{diag}\{\cdot\}$	块对角矩阵
$\mathrm{vec}\{\cdot\}$	向量化操作
$\mathrm{rank}(\cdot)$	矩阵的秩
$\mathrm{tr}(\cdot)$	矩阵的迹
$\otimes$	Kronecker 积
$\mathbb{E}[\cdot], \overline{\mathbb{E}}[\cdot], \widetilde{\mathbb{E}}[\cdot]$	分别为关于概率测度 $\mathbb{P}$、$\overline{\mathbb{P}}$、$\widetilde{\mathbb{P}}$的期望操作算子
$\Pr\{\cdot\}, \overline{\Pr}\{\cdot\}, \widetilde{\Pr}\{\cdot\}$	分别为关于概率测度 $\mathbb{P}$、$\overline{\mathbb{P}}$、$\widetilde{\mathbb{P}}$的一个随机事件的概率
$\boldsymbol{A}\geqslant 0(>0)$	实对称半正定（正定）矩阵 $\boldsymbol{A}$
$\boldsymbol{A}\leqslant 0(<0)$	实对称半负定（负定）矩阵 $\boldsymbol{A}$
$\mathbb{S}_+^n$	$n\times n$ 维实半正定矩阵集合
$(\boldsymbol{A},\boldsymbol{B})$	$\left[\boldsymbol{A}^{\mathrm{T}},\boldsymbol{B}^{\mathrm{T}}\right]^{\mathrm{T}}$

$\det\boldsymbol{A}, \lvert\boldsymbol{A}\rvert$	方阵 $\boldsymbol{A}$ 的行列式
$\rho(\boldsymbol{A})$	方阵 $\boldsymbol{A}$ 或者操作算子的谱半径
$\boldsymbol{A}^{\dagger}$	矩阵 $\boldsymbol{A}$ 的 Moore-Penrose 逆
$\mathrm{sgn}(\bullet)$	符号函数
$\mathrm{sig}(\bullet)^{a}$	$\mathrm{sig}(\bullet)^{a}=\mathrm{sgn}(\bullet)*\lvert\bullet\rvert^{a}$
$\overline{\mathrm{con}}$	闭凸包
$\mathbb{B}(\mathbb{B}^{\circ})$	$\mathbb{R}^{n}$ 中的闭（开）单位球
$\mathcal{C}^{1}$	连续可微函数集
$A \subseteq B$	$\boldsymbol{A}$ 是 $\boldsymbol{B}$ 的子集
$\rightrightarrows$	集值映射
$\lvert x\rvert_{\mathcal{A}}=\inf_{y\in\mathcal{A}}\lvert x-y\rvert$	x 到闭集合 $\mathcal{A}$ 的欧氏距离
$U^{\circ}(x,v)$	函数 U 在 x 处、在方向 v 上的 Clarke 广义方向导数
$\gamma\in\mathcal{K}$	连续的、严格递增的函数$\gamma:\mathbb{R}_{\geqslant 0}\to\mathbb{R}_{\geqslant 0}$，D 属于$\mathcal{K}$类函数，$\gamma(0)=0$
$\mathcal{K}_{\infty}$	无界的$\mathcal{K}$类函数
$\gamma\in\mathcal{KL}$	γ属于$\mathcal{KL}$类函数，则函数$\gamma:\mathbb{R}_{\geqslant 0}^{2}\to\mathbb{R}_{\geqslant 0}$满足$\gamma(\cdot,t)\in\mathcal{K}$，而且$\gamma(s,\cdot)$递减到 0, 其中$t\in\mathbb{R}_{\geqslant 0}$和$s\in\mathbb{R}_{>0}$
$\beta\in\mathcal{KLL}$	γ属于$\mathcal{KLL}$类函数，则函数 $\beta:\mathbb{R}_{\geqslant 0}^{3}\to\mathbb{R}_{\geqslant 0}$满足$\beta(\cdot,\cdot,t)$, $\beta(\cdot,t,\cdot)\in\mathcal{KL}$, 其中$t\in\mathbb{R}_{\geqslant 0}$

Digital Wave
Advanced Technology of
Industrial Internet

Control of Autonomous Intelligent Systems

自主智能系统控制

第 1 章

绪论

1.1
自主智能系统控制介绍

1.1.1 自主智能系统控制的特点及研究现状

自主智能系统是一个新兴的且具有学科交叉特点的研究领域，其内容涵盖了网络通信、自动控制、机器人等多个学科范畴。近年来，随着无线通信技术、人工智能技术的革新，自主智能系统达到了更高的水平。各种类型的自主智能系统在军事国防、航空航天、交通运输、物流仓储等领域相继涌现，比如无人机、无人车辆、无人水下航潜器、空间机器人等。

自主化、网络化、智能化是目前自主智能系统控制的主要特点，其内容包括感知估计、控制决策、网络安全等。

首先，在感知与估计方面，考虑到自主智能系统的目标任务环境往往是具有危险性、高动态性、强对抗性的未知无人环境，如深海、深地、深空以及军事战场等，因此，解决智能自主系统在复杂动态环境下的感知与认知问题是关键，同时也是智能自主系统形成自主控制与决策的基础。目前主要的研究方向有以下几种。

① 未知复杂环境下的自主感知与目标识别　未知复杂环境感知主要实现自主智能系统在复杂环境下的自主定位、导航与目标识别，提高系统在未知环境下的感知认知能力。目前研究的主要挑战在于解决面向夜晚、雨天等复杂环境下的图像特征识别，面向无导航信号下的自主定位以及开放环境下的目标识别与检测等。

② 多传感器下的信息融合与数据滤波　多传感器信息融合指对不同信息源或传感器采集的数据进行融合，以实现对复杂外部环境更好的理解。多传感器融合的优势在于各种不同类型的传感器之间取长补短，优势互补，能获得丰富的环境感知信息，提高时间和空间上的感知范围，增强系统的鲁棒性和灵活性。但是实现难度也较大，主要挑战在于解决

多源异构数据的融合，面向复杂环境下的数据融合滤波问题等。

③ 基于类脑智能的态势感知与认知　类脑智能是借鉴神经机制和认知行为机制，对自主智能系统获取的大量环境数据，包括图像、声音、自然语言等非结构化的数据，进行快速分析、处理和感知，能提升自主智能系统对新环境新任务的自适应能力、对新信息与新技能的自动获取能力，以及在复杂动态环境下进行有效决策的能力。目前基于类脑智能的态势感知与认知的主要研究内容有面向小样本数据场景的态势感知，基于弱监督学习的复杂场景迁移等[1-3]。

其次，在控制与决策方面，主要解决自主智能系统面对复杂环境的不确定性，如何实现智能系统的自主行为与规划，提高自主智能系统的高效性和灵活性。目前主要的研究方向有以下几种。

① 复杂动态环境下的自主智能系统建模　自主智能系统的动力学建模是完成协同控制的基础。然而，由于自主智能平台动力学模型种类繁多，如地面机器人、空中机器人、水下机器人等，其动力学模型存在较大差异性。此外，不同自主智能系统的运动环境也不相同，如在空中机器人动力学模型中需考虑复杂气流变化的影响，水下航行器需考虑急流、阻力等因素，这是自主智能平台动力学建模的主要难题。当前主要的研究内容有：面向异构多智能体系统的动力学建模、基于强化学习的异构多智能体系统建模等。

② 资源受限下的自主智能系统信息传输　自主智能系统的小型化和协同化是近年来的一个主要发展趋势，为此实现智能系统之间信息的高效交互与共享是协同工作的关键基础。然而，由于环境因素以及智能系统本身载荷的限制，智能系统之间的信息传输面临能量资源、通信资源、计算资源受限的难题。该领域主要研究内容有资源受限下的自主智能系统网络拓扑设计与优化，资源受限下的自主智能系统信息传输机制设计问题。

③ 自主智能系统的协同控制与博弈决策　自主智能系统协同控制与博弈决策是实现自主智能系统在复杂场景下的自主协作与群体智能控制以及高效完成任务的核心理论，主要解决的是多智能体在复杂环境下基于自身对外界环境的感知以及与邻居交互的感知信息，基于个体之间的

合作或者竞争关系，完成自主决策和自身动作控制的问题。目前研究的主要挑战在于解决面向复杂场景下的任务规划与动态管理问题、面向复杂异质动力学系统的协同控制问题、开放动态环境下多智能体系统的博弈决策问题[4-6]。

最后，在网络安全方面，考虑到自主智能系统在强对抗环境中往往会受到网络攻击，从而导致自主智能系统无法实现准确的环境感知以及有效的自主控制。因此，为了提高自主智能系统的安全性，需要研究潜在的网络攻击模型，并且提出相应的安全防御方法。目前主要的研究方向有以下几种。

① 面向自主智能系统的网络攻击建模　针对自主智能系统，分析潜在的网络攻击模型以及策略是设计安全防御方法的前提条件。一般而言，网络攻击往往受到攻击能量以及资源的约束。此外，为了屏蔽已有的攻击检测机制，网络攻击的策略设计也需要满足一定的隐秘性条件。因此，其主要研究内容有：资源受限下的网络攻击建模，隐秘性约束下的网络攻击策略设计与优化。

② 自主智能系统的安全防御方法　自主智能系统的安全防御方法可以有效提升网络攻击的检测概率，提升网络攻击下自主智能系统的环境感知精度以及自主控制性能。在检测层面，所提出的攻击检测技术需要在兼顾误报率的基础上，最大化网络攻击的检测率；在感知估计以及控制决策层面，需要最优化估计精度以及控制性能，结合网络攻击的检测机制，提出相应的安全估计以及控制算法。其主要研究内容有：网络攻击的检测机制及其优化，网络攻击下的数据融合以及安全估计算法，网络攻击下的弹性协同控制算法。

③ 面向自主智能系统的安全攻防博弈决策　自主智能系统中攻击者以及防御者往往在动态交互过程中，实时地优化攻击策略以及防御方案，最终实现双方决策的最优均衡。因此，需要在安全攻防博弈的框架下研究自主智能系统的安全防御方法。此外，考虑到在强对抗环境中，攻防双方彼此的状态以及动作信息往往是不完美的，需要采用强化学习方法实现均衡策略的求解。其主要研究内容有：攻防博弈下的安全防御方法设计，基于强化学习的博弈均衡策略求解[7-9]。

1.1.2 多智能体系统的分布式估计和协同控制

1.1.2.1 多智能体系统的分布式估计

多智能体系统的分布式估计问题在无线传感器网络的信息融合与目标跟踪、非合作移动目标下多机器人系统的编队跟踪等问题中具有重要应用价值，因此受到了研究者的极大关注。

2014 年，Yang W 等 [10] 研究了在网络资源受限下，传感器随机激活的分布式滤波问题，得到了一个最优分布式估计器并详细分析了估计器的性能和稳定性，最终减少了传感器能量消耗。2015 年，Das 和 Moura[11] 提出了新息一致的分布式卡尔曼滤波，通过对观测结果进行一致分析，来代替对状态的一致分析，实现了分布式估计。在分布式估计中，每个传感器需要和邻居节点进行通信，使得整个无线传感器网络通信量巨大。因此，事件触发通信在分布式估计中也备受关注。通过在每个传感器中配备一个“send-on-delta”触发条件，当前的估计值与上一被发送的估计值作差，Li 等 [12] 设计了事件触发最优和次优的 Kalman 一致性估计器，并给出了一个充分条件来保证次优估计器的随机稳定性。同样基于“send-on-delta”触发条件思想，Liu 等 [13] 设计了一个分布式估计器，给出了估计误差协方差上界的 Riccati 方程表达式，并迭代地设计了估计器参数来最小化该上界。在系统整体可观条件下（不做局部可观要求），Battistelli 等 [14] 设计了估计误差均方有界的事件触发分布式估计器。文献 [15] 利用线性矩阵不等式的方法设计了稳定的事件触发分布式估计器，且估计器满足一个预定义的估计性能标准。

1.1.2.2 多智能体系统的协同控制

在多智能体系统的协同控制（图 1-1）中，一个基本问题是一致性问题。一致性问题最早产生于计算机科学领域，该研究奠定了分布式计算的理论基础 [16]。首次在控制系统领域中被研究的一致性问题是 Borkar、Varaiya 等人 [17] 和 Tsitsiklis、Bertsekas、Athans 等人 [18] 在分布式决策和并行计算中的异步渐近一致问题。在多智能体系统中，一致性指的是每个智能体的状态趋于一致。在 Fax 和 Murray[19] 的工作中，首次提出了多智能体一致性的理论框架，并且利用图论、拉普拉斯矩阵等数学工具，分析了连续和离散时间下的分布式一致性协议的收敛性能。此后近二十年来，国内外有大量文献对多智能体系统一致性问题进行了深入的研究[20-21]。

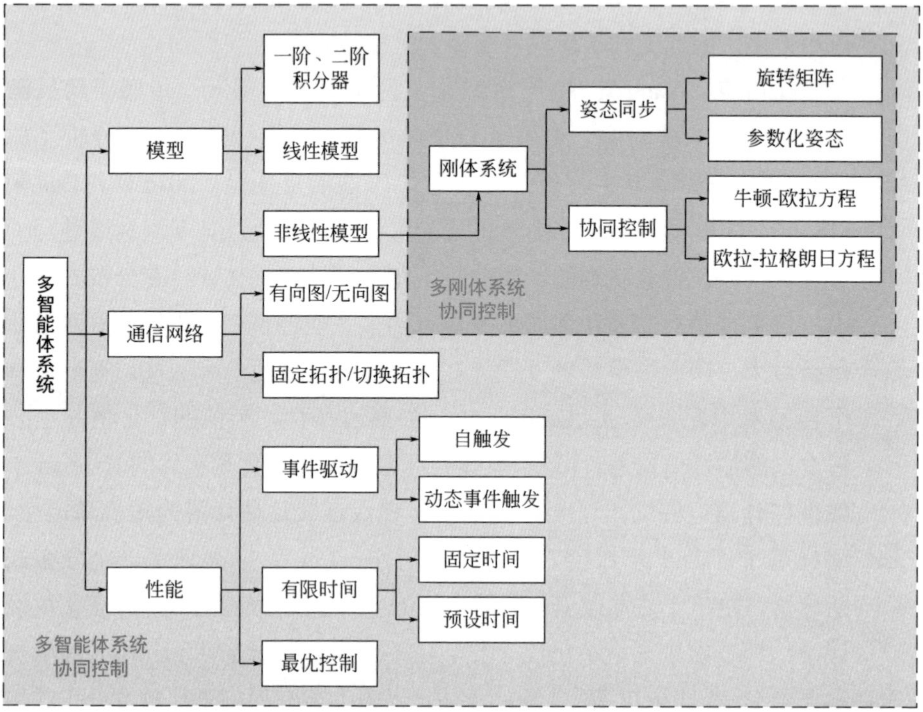

图 1-1　多智能体系统协同控制

下面介绍目前国内外对一致性问题的基本研究情况。一个多智能体系统中，对每个智能体的动态有不同的建模方法，如积分器系统、线性系统、非线性系统等。对相同的一致性协议，不同的系统动态模型使得最终的一致性状态也不尽相同。文献 [22] 研究了当每个智能体的动态模型为二阶积分器时的一致性问题，并且分别考虑了控制输入有界、不需要相对速度测量、参考速度信号对所有智能体已知、参考速度信号对所有智能体未知这四种情况下的一致性算法。对于一般的线性系统模型，文献 [23] 给出了一致性反馈协议存在的充分性条件和相应的一致性协议的生成算法。实际的系统动态特性往往是更加复杂的非线性系统。在文献 [24] 中，作者针对多个耦合的非线性系统研究了两类一致性问题，即状态一致和聚集问题。在通信拓扑是联合连通的条件下，得出了多智能体系统状态一致的充分必要条件。同时，文章中将聚集问题作为状态一致的一种特殊情况，即通信拓扑随状态改变，证明了上述的一致性条件同样成立。

一致性问题是多智能体协同控制的最基本也是最重要的一类问题。根据应用场景的不同，不少文献对其拓展问题，如领航者跟踪问题、包围控制问题、编队控制问题做了深入研究。文献 [25] 研究了多智能体网络中仅有部分智能体已知预设的时变参考轨迹条件下的一致性跟踪问题。进一步，分别讨论了在有向固定拓扑和有向动态切换拓扑下一致性跟踪算法的收敛性。当含有单个领航者时，文献 [26] 研究了线性高阶系统的领航者跟踪问题，领航者的动态由线性方程来描述，每个智能体仅仅依靠自身获得的局部信息实现对领航者的轨迹跟踪。当含有多个领航者时，文献 [27] 首先研究了静态领航者的情况，提出了分布式的包围控制算法，并且给出了实现渐近包围控制的网络拓扑和控制增益的充分条件。进一步，对动态领航者的情况，分别研究了在固定有向拓扑和切换拓扑下的包围控制问题。随着近年来无人飞行器技术的发展，对编队控制问题的研究在多智能体系统领域中一直十分活跃[28-29]。在多智能体系统的编队控制当中，由于有限的感知能力和外界环境的约束，每个智能体只能与它周围的邻居进行信息交互。因此，在现有多智能体协同控制的文献中，运用集中式控制协议的方法并不适用于编队控制。最近几年，不同的分布式编队控制策略，如领航者跟踪控制、包围控制和编队跟踪控制相继被研究。在文献 [30] 中，针对四旋翼蜂群系统提出了基于一致性协议的分布式编队控制协议。此类编队问题考虑的任务是让每个跟随者形成并保持编队的同时与多个领航者的中心保持一个稳定的相对距离。文献 [31] 总结了目前编队控制的三种类型：基于位置的编队控制、基于位移的编队控制以及基于距离的编队控制。基于位置的编队控制需要每个智能体测得自身在全局坐标系中的绝对位置。文献 [32] 研究了基于绝对位置信息的多智能体系统编队控制问题。基于位移的编队控制，则只需要每个智能体获得在全局坐标系中与邻居智能体的相对位置信息，即要求每个智能体需要知道自己的绝对方位信息。在文献 [33] 中，预设的编队队形由在全局坐标系中的相对位置来确定，基于与邻居的相对位移信息的一致性控制协议，整个智能体网络的通信拓扑将保持不变。基于距离的编队控制是要求信息最弱的编队控制方式，每个智能体仅仅需要获取与邻居之间的相对距离即可，不需要建立全局坐标系。基于距离的编队控制是一种完全分布式的控制方法，如文献 [34] 研究了在带有外部干扰的条件下基于距离的多智能体系统编队问题，提出了一种分布式的鲁棒自适应梯度编队控制协议。此外，在多智能体一致性问题的一些实际工程应用中，除了要求每个智能体最终的状态趋于一致，还需要满足一些性能指标，比如控制时间、能量消耗、误差精度等，因此也推动了针对如有限时间一致性 [35]、最优一致性 [36] 等问题的研究。文献 [37] 首次研究了有限时间稳定的概念。针对一般的非线性系统，给出了有限时间稳定的概念，并且给出了李雅普诺夫意义下的有限时间稳定的充分条件和驻留时间的表达式。此后，针对多智能体系统，不少文献也给出了相应的有限时间一致性算法 [38-40]。基于非光滑稳定性分析，文献 [41] 设计了符号梯度一致性控制协议，保证了每个智能体能在有限时间内收敛到一致。还有一些文献提出了基于连续时间的一致性协议 [42]，文献 [42] 针对带有未知非线性动力学的多智能体网络，提出了一类连续的有限时间收敛协议，基于比较定理分析了其有限时间一致收敛性。考虑到有限时间收敛的驻留时间初值对系统状态初值的敏感性，文献 [35] 针对高阶积分器型多智能体系统，提出了固定时间的一致性跟踪算法，证明了一致性的收敛时间不依赖智能体的状态初值。进一步，文献 [43] 研究了更为一般的预设时间收敛问题，使得收敛时间能够不依赖系统的参数任意给定。文献 [44] 针对多智能体系统的领航者跟踪问题，首先设计了预设时间分布式观测器，使得每个跟随者在任意给定时间内能够估计领航者的状态，进一步，结合一个预设时间补偿器最终实现预设时间一致性跟踪。最优一致性通过引入性能成本函数，使得系统的控制性能在某些方面（如能量消耗、控制精度等）达到最优。文献 [45] 基于模型预测控制，研究了一阶和高阶积分器多智能体系统的最优一致性问题。文献 [36] 考虑了一类带有输入饱和线性多智能体系统的最优一致性问题。通过设计强化学习的分布式策略迭代算法，得到雅可比贝尔曼方程的最优解，实现了多智能体系统的最优一致性。

1.2 自主智能系统举例

1.2.1 工业智能控制系统

工业智能控制系统指的是一类无需人干预就能够自主地驱动智能机器，对实际工业环境或过程进行组织、决策和规划，以实现问题求解及其目标的自动控制系统，是使用计算机模拟人类智能的一个重要领域。工业控制系统中的智能控制技术主要有模糊控制、基于知识的专家控制、神经网络控制和集成智能控制等，以及常用的优化算法，如遗传算法、蚁群算法、免疫算法等。具体而言，模糊控制以模糊集合、模糊语言变量、模糊推理作为理论基础，以先验知识和专家经验作为控制规则，基本思想是用机器模拟人对系统的控制，就是在被控对象的模糊模型的基础上，运用模糊控制器近似推理等手段，实现系统控制。专家控制是将专家系统的理论技术和控制理论技术相结合，仿效专家的经验，实现对系统控制的一种智能控制方法。主体由知识库和推理机构成，通过对知识的获取与组织，按某种策略实时选用恰当的规则进行推理，以实现对工业过程的智能控制。神经网络是通过模拟人脑神经元的活动，利用神经元之间的联结与权重值的分布来表示专家的控制规则。基于学习等方法，神经网络的模型可以不断优化神经元之间的权重值，从而实现自校正控制、神经网络模型预测控制等先进的智能控制技术。

近年来，相关学者大力推广智能控制技术的发展，并将其广泛应用于工业控制系统。例如，针对常规 PID 控制难以处理的多时变、有惯性的特点，文献 [46] 采用双模糊控制算法设计了好氧池曝气系统的溶解氧双模糊控制，并通过对双模糊控制系统中的偏差及其微分的论域和模糊子集的选择，改变量化因子的值，以提高系统的性能。针对工业机械臂系统高精度的位置跟踪控制要求，文献 [47] 提出了一种模糊滑模变结构控制器。该控制器将指令信号由滑模变结构控制器产生的滑模等效控制项和由模糊控制器产生的切换控制项相加作为运动指令信号，对机械臂伺服驱动系统进行最佳调整。针对工业锅炉燃烧系统，文献 [48] 建立了模糊控制规则以及控制规律，通过串级控制方案，形成了一种强鲁棒、高控制精度的锅炉燃烧模糊控制算法。为了实现工业 CT 电液位置的精确控制，文献 [49] 基于工业 CT 试验台位置控制系统提出了一种模糊 PID 控制算法，有效地降低了控制系统的稳定误差。文献 [50] 采用模糊 PID 控制算法的工业烤箱温度控制系统，克服了温度对象的纯滞后，减小了系统的超调量，加快了响应速度，取得了良好的控制效果。同样针对工业温度控制系统，文献 [51] 提出了一种模糊自整定 PID 控制方法，在单位阶跃输入下实现了比模糊 PID 控制器更好的控制效果。近年来，随着计算机硬件以及机器学习算法等的大力发展，采用人工智能技术的专家系统来控制工业生产已逐步取代仪表和一般的计算机控制。例如，文献 [52] 将工业以太网与专家控制结合起来，介绍了一种基于以太网的专家控制系统的构建及实时控制过程，从而使得控制系统具备简单灵活、经济可靠、易于实现、维护方便等特点。文献 [53] 将专家控制与传统 PID 控制相结合，设计了基于 LonWorks 的专家 PID 工业控制系统，并分析了该系统的硬件组成和软件实现方案。针对加热炉燃烧过程中热值变化、燃气流量和空气流量变化导致燃烧不稳定的缺点，文献 [54] 提出了一种利用专家控制系统动态改变空燃比的控制方式，使加热炉在各种燃烧工况条件下，燃料在烧嘴内充分混合，从而提高燃烧效率，节约能源，减少氧化烧损。作为人工智能技术之一，基于神经网络的控制方法可以有效地解决工业控制系统中多样性、非线性以及模型不确定性等问题。例如，文献 [55] 提出一种基于神经网络的机械臂自适应参考模型控制方案，从而使得工业机械臂系统能适应由模型不确定性和外界干扰所产生的未知变化。文献 [56] 基于神经网络预测模型，采用“前馈 + 反馈”的控制器架构，研发并投运了镀锌厚度自动控制系统，从而有效地减小了镀层厚度的控制方差及镀层厚度切换过渡时间。文献 [57] 设计了一种基于计算机神经网络的工业电弧炉智能化控制系统，有效地抑制了外部扰动对控制系统性能的影响，实现了炼钢作业效率的提高和优化。针对工业控制系统在各种新型攻击下的运行安全问题，文献 [58] 提出一种改进概率神经网络的工业控制系统安全态势评估方法，从而有效地提升了对攻击类型的分类准确率和精确率。受生物进化过程中“优胜劣汰”的自然选择机制和遗传信息的传递规律的启发，专家学者们设计了进化计算等一系列智能优化算法，并将其用于解决工业控制系统中多优化目标相互耦合牵制的问题。例如，文献 [59] 提出了一种基于差分进化膜计算的 MMC-HVDC 控制参数优化方法，利用差分进化膜计算优化控制系统参数，有效地改善了系统受到短路故障扰动后的动态特性。为了解决传统 PID 控制的实时性和精度较低的问题，文献 [60] 提出了进化计算的 PID 控制参数快速整定算法，并且利用温度控制系统进行了算法验证，证明了该算法具有很强的鲁棒性，可以快速达到系统整定。针对目前双闭环串级比例控制技术中存在的过氧燃烧和缺氧燃烧问题，文献 [61] 提出一种基于遗传算法的工业锅炉变偏置双交叉限幅燃烧控制方法，从而提高了锅炉的燃烧效率，大幅度降低了工业燃煤锅炉污染物的排放。

1.2.2 多机器人协作系统

开发多机器人系统潜力的关键是实现智能协作。机器人应该在任何可能的情况下大力协作，以最大化它们的整体任务表现。现代机器人配备了高带宽通信与各种各样的传感器和执行器阵列，这些资源可以而且应该被开发，以实现群体层面的合作行为。通过共享信息和利用彼此的技能，一组机器人的能力可以真正超越其部件的总和。

机器人的协调控制包括控制一组机器人的位置，使它们执行所需的任务，例如根据每个机器人的测量结果优化目标函数，以及相对于参考点镇定或跟踪所需的位置。从参考轨迹或路径的角度来看，可以将机器人网络的协调控制分为两大类：① 预先已知参考轨迹；② 预先未知参考轨迹。在预先已知参考轨迹中，通常有三种常用的方法：主从式的（leader-following[62]）、行为式的（behavioral[63]）和虚拟结构式的（virtual structure[64]）。关于这几种方法的具体介绍可以参考文献 [65]。大多数涉及机器人协调控制的研究工作，通常会结合这几种方法以集中或分散的方式进行研究。集中式方案（文献 [66-67]）使用单个控制器，进而在工作空间中生成无碰撞轨迹。虽然集中式方案可以保证以一个完整的解决方案来实现机器人的协调控制，但它需要很高的计算能力，而且不具备鲁棒性。相比之下，分散式方案（例如文献 [68]）需要较少的计算工作，并且相对来说更适合团队规模。然而，对临界点的预测和控制是非常困难的。对于预先未知参考轨迹，协调控制通常涉及优化目标函数，进而给出参考轨迹以供网络中的每一个机器人去跟踪此轨迹，包括势场法（the potential field method[62]）和李雅普诺夫直接方法（Lyapunov's direct method[69]），以及基于 Voronoi 分区优化的几何构造方法 [70]。

传统上，计算机网络中的通信使用点对点单播信息传递，也就是说，当计算机 A 想要向计算机 B 发送一条消息时，计算机 A 按名称向计算机 B 指定的地址发送消息，并将消息传递给网络。这种通信模型在机器人系统中存在问题的原因来源于两个方面。首先，与广播消息相比，单播在向多个接收方传递消息时效率非常低。使用广播，则可以发送一次信息，被每个机器人接收，产生一个潜在的巨大的带宽节省，所以这是多机器人系统中一个应该被考虑的重要因素。其次，我们希望被控制的多机器人系统是流动的。机器人可以进出通信范围，通信系统本身可以发送消息，此外，机器人还可能发生许多不同的故障。为了容忍这些动态变化，通信层不应该通过名称来处理机器人。相反，机器人应该通过广播的方式进行匿名交流。

然而，在广播信息的途径中，无线电干扰信号诱发导致的拒绝服务攻击在一定时间内会使信息通信无法成功通过网络通道传输 [71]，也就是说，拒绝服务攻击的主要目的是影响信息交换的及时性，但会导致新数据包因无法及时传输而被丢弃（可以看作是数据包丢失）。网络通道中的数据传输极易受到拒绝服务攻击，机器人会直接偏离预期的路线，因为速度和方向无法受到实时控制 [72]，从而影响机器人协调任务的完成。

另一方面，与仅在运动学水平上进行控制设计相比，尽管考虑动态水平在理论上更具有挑战性，但在实际操作中却是必要的，因为实际控制输入的是用于驱动机器人车轮的电机电流和电压。实际上，从机器人的速度到电压的静态映射可能会显著地影响控制性能。在设计的控制器中，电机的动力学特性往往被忽略，因为电机的动力学响应通常比机器人的动力学响应快得多，所以将应用于机器人车轮上的力矩作为动态水平的控制输入是合理的，并且不应该被忽略。

Digital Wave
Advanced Technology of Industrial Internet

Control of Autonomous Intelligent Systems

自主智能系统控制

第2章

自主智能系统建模方法

2.1
面向状态估计的自主智能系统建模

自主智能系统的状态感知与估计是系统控制的前提与基础。针对自主智能系统的状态估计问题，可以将系统描述为一个线性系统。考虑到传感器中的噪声影响，可以进一步将其描述为线性随机系统。本节将介绍自主智能系统的状态估计系统模型以及相关的基本理论方法。

2.1.1 线性随机系统的 Kalman 滤波方程

对于线性随机系统，Kalman 滤波器是在最小均方估计误差意义下的最优线性状态估计器。考虑如下离散时间线性定常随机系统：

$$\boldsymbol{x}_{k+1} = \boldsymbol{A}\boldsymbol{x}_k + \boldsymbol{w}_k$$

$$\boldsymbol{y}_k = \boldsymbol{C}\boldsymbol{x}_k + \boldsymbol{v}_k \tag{2-1}$$

其中，$\boldsymbol{x}_k \in \mathbb{R}^n$是系统状态，$\boldsymbol{y}_k \in \mathbb{R}^m$是传感器的测量输出。$\boldsymbol{w}_k$和$\boldsymbol{v}_k$是互相独立的零均值白噪声，协方差分别为$\boldsymbol{Q} \geqslant 0$和$\boldsymbol{R} > 0$。初始状态$\boldsymbol{x}_0$均值为$\bar{\boldsymbol{x}}_0$，协方差为$\boldsymbol{\Pi}_0$，且独立于$\boldsymbol{w}_k$和$\boldsymbol{v}_k$。

Kalman 滤波理论可适用于一般的线性时变系统，本节只是以定常系统为例。记$\boldsymbol{y}_{0k} := \{\boldsymbol{y}_0, \cdots, \boldsymbol{y}_k\}$，则定义如下一些常用的符号：

$$\hat{\boldsymbol{x}}_{k|k-1} := \mathbb{E}[\boldsymbol{x}_k | \boldsymbol{y}_{0(k-1)}] \tag{2-2}$$

$$\hat{\boldsymbol{x}}_{k|k} := \mathbb{E}[\boldsymbol{x}_k | \boldsymbol{y}_{0k}] \tag{2-3}$$

$$\boldsymbol{P}_{k|k-1} := \mathbb{E}[(\boldsymbol{x}_k - \hat{\boldsymbol{x}}_{k|k-1})(\boldsymbol{x}_k - \hat{\boldsymbol{x}}_{k|k-1}) y_{0(k-1)}] \tag{2-4}$$

$$\boldsymbol{P}_{k|k} := \mathbb{E}[(\boldsymbol{x}_k - \hat{\boldsymbol{x}}_{k|k})(\boldsymbol{x}_k - \hat{\boldsymbol{x}}_{k|k}) y_{0k}], \tag{2-5}$$

其中，$\hat{\boldsymbol{x}}_{k|k-1}$和$\hat{\boldsymbol{x}}_{k|k}$分别为系统状态的先验和后验估计（也称预测和估计），$\boldsymbol{P}_{k|k-1}$和$\boldsymbol{P}_{k|k}$为相应的估计误差协方差。

Kalman 滤波递推公式如下，它包含两个步骤：

（1）开环预测

$$\hat{\boldsymbol{x}}_{k|k-1} = \boldsymbol{A}\hat{\boldsymbol{x}}_{k-1|k-1} \tag{2-6}$$

$$P_{k|k-1} = AP_{k-1|k-1}A^{\mathrm{T}} + Q \tag{2-7}$$

（2）闭环校正

$$K_k = P_{k|k-1}C^{\mathrm{T}}\left(CP_{k|k-1}C^{\mathrm{T}} + R\right)^{-1} \tag{2-8}$$

$$\hat{x}_{k|k} = \hat{x}_{k|k-1} + K_k\left(y_k - C\hat{x}_{k|k-1}\right) \tag{2-9}$$

$$P_{k|k} = P_{k|k-1} - K_kCP_{k|k-1} \tag{2-10}$$

校正阶段的增益矩阵K_k是权衡预测与校正的最优加权矩阵。很明显，较小意味着预测值较可靠而测量值噪声较大，较大意味着测量较可靠但预测不准。新息$y_k - C\hat{x}_{k|k-1}$，也称为残差，它的作用在于利用更新的测量值y_k来修正状态的预测值$\hat{x}_{k|k-1}$。

虽然前面已经给出了最优滤波增益矩阵和误差协方差的递推关系，但 Kalman 滤波还可以变换为一种称作信息滤波的形式，其开环预测是用协方差矩阵的逆表示，但它本质上还是 Kalman 滤波的等价表示：

$$P_{k|k}^{-1} = P_{k|k-1}^{-1} + CR^{-1}C^{\mathrm{T}}$$

$$K_k = P_{k|k}^{-1}C^{\mathrm{T}}R^{-1} \tag{2-11}$$

由于其在闭环校正阶段的计算方便，已被广泛应用在多传感器融合估计和控制当中。

下面给出 Kalman 滤波一些重要的性质。

① 最优性：当噪声 w_k 和 v_k 服从高斯分布时，Kalman 滤波器不局限于线性最优，为全局最优。

② 无偏性：Kalman 滤波器是无偏的，也就是

$$\mathbb{E}[x_k - \hat{x}_{k|k}] = 0 \tag{2-12}$$

③ 递推性：Kalman 滤波是一种递推算法，即估计值$\hat{x}_{k|k}$的计算只需要上一步估计值$\hat{x}_{k-1|k-1}$和当前步的 y_k，而不需要所有的测量输出 y_{0k}。此外，滤波增益矩阵 K_k 的计算与测量无关，可以离线算出。

2.1.2　线性指数二次高斯状态估计方程

针对线性高斯系统，本小节给出最小化指数型的均方估计误差的最优状态估计器。同样考虑方程 (2-1) 所示的线性系统，不同于上一小节

的是，这里只针对高斯噪声情形，且假设$\boldsymbol{Q}$、$\boldsymbol{R}$、$\boldsymbol{\Pi}_0$均为正定矩阵。

本小节考虑的代价函数具有一个指数形式，也称风险敏感性能指标。首先，定义因果估计器$\{\hat{\boldsymbol{x}}_0,\cdots,\hat{\boldsymbol{x}}_T\}$，使得$k$时刻估计值是基于估计器在$k$时刻的可用测量信息，也就是

$$\hat{\boldsymbol{x}}_k = g_k\left(\boldsymbol{y}_{0k}\right) \tag{2-13}$$

其中，g_k是将可利用信息$\boldsymbol{y}_{0k}$映射到状态估计值$\hat{\boldsymbol{x}}_k$的函数。此时，待最小化的性能指标是对累积估计误差的指数二次函数的期望[73]：

$$J = hat\{x\} - \{0,\bar{1}\}\mathbb{E}[\exp(\theta\boldsymbol{\Psi}_{0,T})] \tag{2-14}$$

其中，$\theta > 0$是个标量，且

$$\boldsymbol{\Psi}_{0,T} = \frac{1}{2}\sum_{k=0}^{T}\|\boldsymbol{x}_k - \hat{\boldsymbol{x}}_k\|_{\boldsymbol{W}}^2$$

这里$\boldsymbol{W}$是一个正定权重矩阵。参数θ可以用来权衡名义（Nominal）模型(2-1)的最优估计与对系统不确定性的鲁棒性，通常θ越大，鲁棒性越好。指数型代价函数的鲁棒性可以这样来解释：根据指数函数$\mathrm{e}^x = \sum_{k=0}^{\infty}\boldsymbol{x}^k / k!$的解析性质，式(2-14)可以展开成

$$J = \mathbb{E}\left[1 + \theta\boldsymbol{\Psi}_{0,T} + \theta^2\boldsymbol{\Psi}_{0,T}^2 + \cdots\right] \tag{2-15}$$

可以看出，指数型代价函数J包含了二次以上的高阶误差，这减少了估计器出现很大估计误差的风险。随着$\theta \to 0$，意味着一阶近似$\mathbb{E}\left[\exp\left(\theta\boldsymbol{\Psi}_{0,T}\right)\right] \to 1 + \mathbb{E}\left[\theta\boldsymbol{\Psi}_{0,T}\right]$恢复到原始的二次性能标准。

最优估计器具有如下类 Kalman 的结构：

$$\hat{\boldsymbol{x}}_{k|k} = \hat{\boldsymbol{x}}_{k|k-1} + \boldsymbol{K}_k\left(\boldsymbol{y}_k - \boldsymbol{C}\hat{\boldsymbol{x}}_{k|k-1}\right) \tag{2-16}$$

$$\hat{\boldsymbol{x}}_{k+1|k} = \boldsymbol{A}\hat{\boldsymbol{x}}_{k|k} \tag{2-17}$$

$$\boldsymbol{P}_{k+1} = \boldsymbol{A}\left(\boldsymbol{\Sigma}_k^{-1} - \theta\boldsymbol{W}\right)^{-1}\boldsymbol{A}^{\mathrm{T}} + \boldsymbol{Q} \tag{2-18}$$

$$\boldsymbol{\Sigma}_k^{-1} = \boldsymbol{P}_k^{-1} + \boldsymbol{C}^{\mathrm{T}}\boldsymbol{R}^{-1}\boldsymbol{C}，\ \boldsymbol{\Sigma}_k^{-1} - \theta\boldsymbol{W} > 0 \tag{2-19}$$

$$\boldsymbol{K}_k = \boldsymbol{\Sigma}_k\boldsymbol{C}^{\mathrm{T}}\boldsymbol{R}^{-1} \tag{2-20}$$

上面递归始于$\hat{\boldsymbol{x}}_{0|-1} = \boldsymbol{\pi}_0$和$\boldsymbol{P}_0 = \boldsymbol{\Pi}_0$。

2.1.3 隐马尔可夫模型的状态估计方程

一个隐马尔可夫模型（Hidden Markov Model, HMM）是一个由测量

过程观测的有限状态 Markov 链[74]，它的状态不能直接得到。HMMs 已经广泛应用到各种领域，比如语音识别、机器监控和功率控制[75]。与线性系统相比，HMMs 具有有限状态这一本质特点。

考虑一个在概率空间$(\Omega,\mathcal{F},\mathbb{P})$上的 HMM，令 $\{\boldsymbol{X}_k\}$ 是一个离散时间、齐次的一阶有限状态的 Markov 过程。假设给定初始状态 $\boldsymbol{X}_0$ 的分布，不失一般性，假定 $\boldsymbol{X}_k$ 的状态空间是一个有限集合$S_X=\{\boldsymbol{e}_1,\boldsymbol{e}_2,\cdots,\boldsymbol{e}_N\}$，其中 $\boldsymbol{e}_i$ 是 $\mathbb{R}^N$ 中第 i 个元素为 1 的单位向量，记由$\boldsymbol{X}_0,\cdots,\boldsymbol{X}_k$生成的$\sigma$- 域为$\mathcal{F}_k^{X,0}:=\sigma\{\boldsymbol{X}_0,\cdots,\boldsymbol{X}_k\}$，并让$\{\mathcal{F}_k^X\}$是由$\mathcal{F}_k^{X,0}$生成的完备虑子（complete filtration），根据 Markov 性质：

$$\Pr\left\{\boldsymbol{X}_{k+1}=\boldsymbol{e}_i \mid \mathcal{F}_k^X\right\}=\Pr\left\{\boldsymbol{X}_{k+1}=\boldsymbol{e}_i \mid \boldsymbol{X}_k\right\} \tag{2-21}$$

令 $a_{ij}:=\Pr\left\{\boldsymbol{X}_{k+1}=\boldsymbol{e}_i \mid \boldsymbol{X}_k=\boldsymbol{e}_j\right\}$，$\boldsymbol{A}:=(a_{ij})\in\mathbb{R}^{N\times N}$，使得$\sum_{i=1}^N a_{ij}=1$。因此，有

$$\mathbb{E}\left[\boldsymbol{X}_{k+1} \mid \mathcal{F}_k^X\right]=\mathbb{E}\left[\boldsymbol{X}_{k+1} \mid \boldsymbol{X}_k\right]=\boldsymbol{A}\boldsymbol{X}_k \tag{2-22}$$

注意，$\boldsymbol{X}_k$不能直接被观测到，但可以得到一个来自传感器测量过程的输出$\boldsymbol{y}_k$：

$$\boldsymbol{y}_k=\boldsymbol{C}\boldsymbol{X}_k+\boldsymbol{v}_k \tag{2-23}$$

其中，$\boldsymbol{v}_k\in\mathbb{R}^m$是具有严格正的概率密度函数$\phi$的白噪声❶。另外，$\boldsymbol{C}:=[\boldsymbol{c}_1,\cdots,\boldsymbol{c}_N]\in\mathbb{R}^{m\times N}$是测量矩阵，其中$\boldsymbol{c}_i\in\mathbb{R}^m,1\leqslant i\leqslant N$。让$\{\mathcal{F}_k^y\}$是由$\sigma\{\boldsymbol{y}_0,\cdots,\boldsymbol{y}_k\}$生成的完备虑子。

注释 2.1 由于 $\boldsymbol{X}_k$ 状态空间的势为 N，任意实函数$f(\boldsymbol{X}_k):\mathbb{R}^N\to\mathbb{R}^m$可以表示为一个线性函数$f(\boldsymbol{X}_k)=\boldsymbol{C}\boldsymbol{X}_k$，这里$\boldsymbol{C}\in\mathbb{R}^{m\times N}$。因此，采用式 (2-23) 中线性表达式 $\boldsymbol{C}\boldsymbol{X}_k$ 是合理的。

下面给出状态 $\boldsymbol{X}_k$ 在给定观测信息 $\boldsymbol{y}_{0k}$ 下的条件期望表达式。首先，定义一个非归一化的条件分布 $\boldsymbol{q}_k$（也称前向变量[76]），它的第 i 个元素为

$$\boldsymbol{q}_k^i=P_{\mathrm{r}}\left(\boldsymbol{y}_0,\cdots,\boldsymbol{y}_k,\boldsymbol{X}_k=\boldsymbol{e}_i\right) \tag{2-24}$$

❶ 因为 ϕ 将会在 Radon-Nikodym [式 (3-6)] 中作为除数，这里要求 ϕ 为正。这不是一个严格的要求且著名的高斯密度函数满足该情况。

那么，$\boldsymbol{q}_k$可由如下线性递归获得[76]

$$\boldsymbol{q}_{k+1}=\operatorname{diag}\left\{\boldsymbol{b}_{k+1}\right\}\boldsymbol{A}\boldsymbol{q}_k,\quad \boldsymbol{b}_{k+1}=\left[\phi\left(\boldsymbol{y}_{k+1}-\boldsymbol{C}\boldsymbol{e}_i\right)\right]_{i\in\mathbb{N}_{1:N}} \tag{2-25}$$

进一步，可得条件期望$\mathbb{E}[\boldsymbol{X}_k|\mathcal{F}_k^y]$，它的第$i$个元素由下式计算得到

$$\mathbb{E}\left[\left\langle\boldsymbol{X}_k,\boldsymbol{e}_i\right\rangle|\mathcal{F}_k^y\right]=\sum_{j=1}^{N}P_{\mathrm{r}}\left(\boldsymbol{X}_k=\boldsymbol{e}_j|\mathcal{F}_k^y\right)\left\langle\boldsymbol{e}_j,\boldsymbol{e}_i\right\rangle=P_{\mathrm{r}}\left(\boldsymbol{X}_k=\boldsymbol{e}_i|\mathcal{F}_k^y\right)=\frac{\boldsymbol{q}_k^i}{\sum_{l=1}^{N}\boldsymbol{q}_k^l} \tag{2-26}$$

最后，基于$\mathbb{E}[\boldsymbol{X}_k|\mathcal{F}_k^y]$，可得最大后验概率估计值或者最小均方误差估计值。

2.2 基于刚体动力学的自主智能系统建模

自主智能系统在大部分场景中表现为机器人、车辆、无人机等刚体对象。因此，针对自主智能系统的控制问题，本节主要对刚体的运动学以及动力学模型做相关介绍。

2.2.1 旋转矩阵的基本知识

三维正交旋转群$\mathbb{SO}(3)$在一点$\boldsymbol{\mathcal{R}}^*$的切空间表示为$\mathbb{T}_{\boldsymbol{\mathcal{R}}^*}\mathbb{SO}(3)=\left\{\boldsymbol{\mathcal{R}}^*\boldsymbol{S},\boldsymbol{S}\in\mathfrak{so}(3)\right\}$，其中$\mathfrak{so}(3)=\left\{\boldsymbol{\Omega}\in\mathbb{R}^{3\times3}:\boldsymbol{\Omega}^{\mathrm{T}}=-\boldsymbol{\Omega}\right\}$。给定$\mathfrak{so}(3)$中的一个元素

$$\boldsymbol{\Omega}=\begin{bmatrix}0&-w_3&w_2\\w_3&0&-w_1\\-w_2&w_1&0\end{bmatrix}\in\mathfrak{so}(3) \tag{2-27}$$

和$\boldsymbol{w}=\left[w_1,w_2,w_3\right]\in\mathbb{R}^3$。算符$(\cdot)^{\vee}$和$(\cdot)^{\wedge}$被定义为从$\mathfrak{so}(3)$到$\mathbb{R}^3$的映射和逆映射。我们将切向量$\boldsymbol{X}_1,\boldsymbol{X}_2\in\mathbb{T}_{\boldsymbol{\mathcal{R}}^*}\mathbb{SO}(3)$认同于向量$\boldsymbol{x}_1,\boldsymbol{x}_2\in\mathbb{R}^3$通过算子$(\cdot)^{\vee}$，在$\mathbb{SO}(3)$上的标准内积由下式给出

$$\left\langle\boldsymbol{X}_1,\boldsymbol{X}_2\right\rangle=\boldsymbol{x}_1^{\mathrm{T}}\boldsymbol{x}_2 \tag{2-28}$$

下面介绍指数映射与对数映射的定义。给定一点$\mathcal{R}_1 \in \mathbb{SO}(3)$与一个切向量$\boldsymbol{V} = \mathcal{R}_1 \boldsymbol{v} \in \mathbb{T}_{\mathcal{R}_1}\mathbb{SO}(3)$，其中$\boldsymbol{v} \in \mathfrak{so}(3)$。指数映射定义为$\exp_{\mathcal{R}_1}: \mathbb{T}_{\mathcal{R}_1}\mathbb{SO}(3) \to \mathbb{SO}(3)$，可由罗德里格斯公式计算

$$\exp_{\mathcal{R}_1}\boldsymbol{V} := \mathcal{R}_1\left(\boldsymbol{I}_3 + \frac{\sin\Theta}{\Theta}\boldsymbol{v}^{\wedge} + \frac{1-\cos\Theta}{\Theta^2}\left(\boldsymbol{v}^{\wedge}\right)^2\right) \tag{2-29}$$

其中，$\Theta = \langle \boldsymbol{v}, \boldsymbol{v} \rangle^{\frac{1}{2}}$是$\boldsymbol{v}$的范数。

给定开集合$\mathcal{D}_{\mathcal{R}_1} \subset \mathbb{T}_{\mathcal{R}_1}\mathbb{SO}(3)$满足$\exp_{\mathcal{R}_1}: \mathcal{D}_{\mathcal{R}_1} \to \mathbb{SO}(3)$是一个微分同胚映射。令$\mathcal{I}_{\mathcal{R}_1} = \exp_{\mathcal{R}_1}\left(\mathcal{D}_{\mathcal{R}_1}\right) \subset \mathbb{SO}(3)$和$\mathcal{R}_2 \in \mathcal{I}_{\mathcal{R}_1}$，则对数映射是指数映射的逆映射，定义为$\log_{\mathcal{R}_1}: \mathcal{I}_{\mathcal{R}_1} \to \mathbb{T}_{\mathcal{R}_1}\mathbb{SO}(3)$，可由下面公式计算：

$$\log_{\mathcal{R}_1}\mathcal{R}_2 := \mathcal{R}_1 \log\left(\mathcal{R}_1^{\mathrm{T}}\mathcal{R}_2\right) = \mathcal{R}_1 \frac{\tilde{\Theta}}{2\sin\tilde{\Theta}}\left(\mathcal{R}_1^{\mathrm{T}}\mathcal{R}_2 - \mathcal{R}_2^{\mathrm{T}}\mathcal{R}_1\right) \tag{2-30}$$

其中，$\tilde{\Theta} = \arccos\dfrac{\mathrm{tr}\left(\mathcal{R}_1^{\mathrm{T}}\mathcal{R}_2\right)-1}{2}$。对于两个旋转矩阵$\mathcal{R}_1, \mathcal{R}_2 \in \mathbb{SO}(3)$的测地线距离定义为$d\left(\mathcal{R}_1, \mathcal{R}_2\right) := \arccos\dfrac{\mathrm{tr}\left(\mathcal{R}_1^{\mathrm{T}}\mathcal{R}_2\right)-1}{2}$。

2.2.2 姿态表征和运动学

令$\mathcal{I}$表示世界坐标系，$\mathcal{F}_i, i = 1, \cdots, N$表示刚体$i$的局部坐标系。每个刚体的姿态相对于世界坐标系$\mathcal{I}$可表示为$\mathcal{R}_i \in \mathbb{SO}(3)$。令$\mathcal{B}_r\left(\boldsymbol{I}_3\right) = \left\{\mathcal{R}_i \in \mathbb{SO}(3): d\left(\mathcal{R}_i, \boldsymbol{I}_3\right) < r\right\}$表示局部覆盖半径为$r$的$\mathbb{SO}(3)$，其中$d\left(\mathcal{R}_i, \boldsymbol{I}_3\right)$表示旋转矩阵$\mathcal{R}_i$和单位矩阵$\boldsymbol{I}_3$在$\mathbb{SO}(3)$上的黎曼距离。开球$\mathcal{B}_{r'}(0) = \left\{\boldsymbol{x}_i \in \mathbb{R}^3 : \left\|\boldsymbol{x}_i\right\| < r'\right\}$表示$\mathbb{R}^3$中的开子集。值得一提的是，如果$r = \pi, \mathcal{B}_r\left(\boldsymbol{I}_3\right)$可几乎全局覆盖$\mathbb{SO}(3)$，且$\mathcal{B}_{\frac{\pi}{2}}\left(\boldsymbol{I}_3\right)$是$\mathbb{SO}(3)$上的最大凸集[77]。任何局部姿态表征可以被定义为一个微分同胚映射$f: \mathcal{B}_r\left(\boldsymbol{I}_3\right) \to \mathcal{B}'_{\pi}\left(\boldsymbol{0}_3\right)$，表示为$f\left(\mathcal{R}_i\right) = g\left(\Theta_i\right)\boldsymbol{u}_i$，其中$g: (-r, r) \to \mathbb{R}$是一个增函数[77]。在集合$\mathcal{B}_{\pi}\left(\boldsymbol{I}_3\right)$中，对应于旋转矩阵$\mathcal{R}_i$的旋转向量$\boldsymbol{u}_i$和旋转角度$\Theta_i$可用对数映射计算

$$\boldsymbol{u}_i^{\wedge} = \frac{1}{2\sin\Theta_i}\left(\mathcal{R}_i - \mathcal{R}_i^{\mathrm{T}}\right) \tag{2-31}$$

$$\cos\Theta_i = \frac{\mathrm{tr}\left(\mathcal{R}_i\right)-1}{2} \tag{2-32}$$

令$\boldsymbol{w}_i$表示机体局部坐标系的角速度向量，并且为需要设计的控制输

入。姿态的运动学使用旋转矩阵为

$$\dot{\mathcal{R}}_i(t)=\mathcal{R}_i(t)\boldsymbol{w}_i^{\wedge}(t),\ \ t\in(0,+\infty) \tag{2-33}$$

其中，$\mathcal{R}_i(t)\boldsymbol{w}_i^{\wedge}(t)\in\mathbb{T}_{\mathcal{R}_i}\mathbb{SO}(3)$。令$\boldsymbol{y}_i(t)=f(\mathcal{R}_i(t))$为局部姿态表征，有

$$\dot{\boldsymbol{y}}_i(t)=\mathrm{D}f(\mathcal{R}_i(t))\dot{\mathcal{R}}_i(t)\Big|_{\mathcal{R}_i(t)=f^{-1}(y_i(t))},\ \ t\in(0,+\infty) \tag{2-34}$$

其中，$\mathrm{D}f(\mathcal{R}_i(t))$是$f(\mathcal{R}_i(t))$相对于$\mathcal{R}_i$的导数。

2.2.3 轴角姿态表征与姿态动力学

令$\boldsymbol{x}_i\in\mathbb{R}^3$表示每个刚体 i 的轴角姿态向量，有

$$\begin{aligned}\boldsymbol{x}_i^{\wedge}&=\log\mathcal{R}_i\\&=\theta_i\boldsymbol{u}_i^{\wedge}\end{aligned} \tag{2-35}$$

其中，log：$\mathbb{SO}(3)\to\mathfrak{so}(3)$是对数映射；$\theta_i$为相对于旋转轴$\boldsymbol{u}_i$的旋转角度。这个向量在部分文献中被称为旋转向量，是因为它可以表示为$\boldsymbol{x}_i=\theta_i\boldsymbol{u}_i$，其中$\boldsymbol{u}_i\in\mathbb{S}^2$是一个单位向量且$\theta_i\in[0,\ \pi)$，$i\in\mathcal{V}$是以$\boldsymbol{u}_i$为旋转轴的旋转角度。旋转向量$\boldsymbol{x}_j$，$j\in\mathcal{N}_i$在局部坐标系$\mathcal{F}_i$中表达为$\boldsymbol{x}_{ij}=\log(\mathcal{R}_i^{\mathrm{T}}\mathcal{R}_j)^{\vee}$。两个旋转$\mathcal{R}_i$和$\mathcal{R}_j$之间的黎曼距离可表示为$d(\mathcal{R}_i,\ \mathcal{R}_j)=\|\boldsymbol{x}_{ij}\|$。旋转矩阵$\mathcal{R}_i$可根据$\boldsymbol{x}_i$通过罗德里格斯公式计算得到[78]，

$$\mathcal{R}_i=\boldsymbol{I}_3+\sin\theta_i\boldsymbol{u}_i^{\wedge}+(1-\cos\theta_i)(\boldsymbol{u}_i^{\wedge})^2 \tag{2-36}$$

基于旋转向量，刚体的动力学方程如下：

$$\begin{aligned}\dot{\boldsymbol{x}}_i&=\boldsymbol{L}_{\boldsymbol{x}_i}\boldsymbol{w}_i\\\boldsymbol{J}_i\dot{\boldsymbol{w}}_i&=\boldsymbol{\tau}_i-S(\boldsymbol{w}_i)\boldsymbol{J}_i\boldsymbol{w}_i\end{aligned} \tag{2-37}$$

其中，$\boldsymbol{w}_i$和$\boldsymbol{\tau}_i$分别是局部坐标系$\mathcal{F}_i$中的瞬时角速度矢量和外部扭矩；$\boldsymbol{J}_i$是对称正定惯性矩阵；$S(\boldsymbol{w}_i)=\boldsymbol{w}_i^{\wedge}$，$\boldsymbol{L}_{\boldsymbol{x}_i}$定义为雅可比矩阵，

$$\boldsymbol{L}_{\boldsymbol{x}_i}=\boldsymbol{I}_3+\frac{\theta_i\boldsymbol{u}_i^{\wedge}}{2}+\left(1-\frac{\theta_i}{2}\cot\frac{\theta_i}{2}\right)(\boldsymbol{u}_i^{\wedge})^2 \tag{2-38}$$

根据式(2-38)，若$\theta_i\in[0,\ \pi]$，可知雅可比矩阵是正定的，即$\boldsymbol{x}_i^{\mathrm{T}}\boldsymbol{J}_{\boldsymbol{x}_i}\boldsymbol{x}_i>0$对任何$\boldsymbol{x}_i\neq\boldsymbol{0}_3$，$\boldsymbol{x}_i\in\mathbb{R}^3$[79]。此外，已知雅可比矩阵具有的几何性质即第二项和第三项垂直于$\boldsymbol{x}_i$，$\boldsymbol{x}_i^{\mathrm{T}}\boldsymbol{J}_{\boldsymbol{x}_i}=\boldsymbol{x}_i^{\mathrm{T}}$。

2.2.4 刚体的欧拉－拉格朗日系统模型

一个刚体系统的动力学模型可由欧拉 - 拉格朗日方程描述如下

$$\boldsymbol{M}(\boldsymbol{q})\ddot{\boldsymbol{q}}+\boldsymbol{C}(\dot{\boldsymbol{q}},\ \boldsymbol{q})\dot{\boldsymbol{q}}+\boldsymbol{G}(\boldsymbol{q})=\boldsymbol{\Gamma} \tag{2-39}$$

其中，$\boldsymbol{q}\in\mathbb{R}^n$表示广义坐标；$\boldsymbol{\Gamma}\in\mathbb{R}^n$表示控制力矩；$\boldsymbol{M}(\boldsymbol{q})\in\mathbb{R}^{n\times n}$表示惯性矩阵；$\boldsymbol{C}(\dot{\boldsymbol{q}},\ \boldsymbol{q})\dot{\boldsymbol{q}}\in\mathbb{R}^n$表示离心力/科式力向量；$\boldsymbol{G}(\boldsymbol{q})\in\mathbb{R}^n$表示重力向量。针对欧拉-拉格朗日模型(2-39)，有下列三个重要的性质：

① 矩阵$\dot{\boldsymbol{M}}(\boldsymbol{q})-2\boldsymbol{C}(\dot{\boldsymbol{q}},\ \boldsymbol{q})$具有反对称性。

② 惯性矩阵 $\boldsymbol{M}(\boldsymbol{q})$ 是对称正定矩阵，且满足$\underline{k}_M\boldsymbol{I}_n\leqslant\boldsymbol{M}(\boldsymbol{q})\leqslant\bar{k}_M\boldsymbol{I}_n$，其中$\underline{k}_M$和$\bar{k}_M$是两个正常数，离心力 / 科氏力矩阵范数满足$\|\boldsymbol{C}(\dot{\boldsymbol{q}},\ \boldsymbol{q})\|\leqslant k_C\|\dot{\boldsymbol{q}}\|$，其中 k_C 是一个正常数。重力向量满足$\|\boldsymbol{G}(\boldsymbol{q})\|\leqslant k_g$，其中 k_g 是正常数。

③ 对于任意向量$\boldsymbol{x},\boldsymbol{y}\in\mathbb{R}^n$，存在一个线性回归矩阵，$\boldsymbol{Y}(\dot{\boldsymbol{q}},\boldsymbol{q},\boldsymbol{x},\boldsymbol{y})$满足$\boldsymbol{M}(\boldsymbol{q})\boldsymbol{x}+\boldsymbol{C}(\dot{\boldsymbol{q}},\boldsymbol{q})\boldsymbol{y}+\boldsymbol{G}(\boldsymbol{q})=\boldsymbol{Y}(\dot{\boldsymbol{q}},\boldsymbol{q},\boldsymbol{x},\boldsymbol{y})\boldsymbol{\Theta}$，其中$\boldsymbol{\Theta}$是描述欧拉 - 拉格朗日系统物理参数的向量。

2.2.5 基于欧拉－拉格朗日系统的四旋翼无人机模型

下面以四旋翼无人机举例，给出其基于欧拉 - 拉格朗日方程的动力学模型。四旋翼无人机可以视为一个六自由度的刚体，因此可以定义广义坐标$\boldsymbol{q}_i=\left[\boldsymbol{\xi}_i^{\mathrm{T}}\ \boldsymbol{\eta}_i^{\mathrm{T}}\right]^{\mathrm{T}}\in\mathbb{R}^6$，其中$\boldsymbol{\xi}_i=\left[x_i\ y_i\ z_i\right]^{\mathrm{T}}$是每架无人机相对于惯性坐标系在三维空间中的位置，$\boldsymbol{\eta}_i=\left[\alpha_i\ \beta_i\ \varphi_i\right]^{\mathrm{T}}$分别表示无人机的航向角、俯仰角以及翻滚角。无人机的平动动能和旋转动能可以分别被计算为$\mathcal{K}_{trans}=\dfrac{1}{2}m_i\dot{\boldsymbol{\xi}}_i^2$和$\mathcal{K}_{rot}=\dfrac{1}{2}\boldsymbol{w}_i^{\mathrm{T}}\boldsymbol{I}_i\boldsymbol{w}_i$，其中 m_i 是质量，$\boldsymbol{I}_i=\mathrm{diag}\left\{I_i^{xx},I_i^{yy},I_i^{zz}\right\}$是惯性矩，$\boldsymbol{w}_i$ 是角速度矢量。角速率$\dot{\boldsymbol{\eta}}_i$和角速度矢量 $\boldsymbol{w}_i$ 的关系可以表示为$\boldsymbol{w}_i=\boldsymbol{\Omega}_i\dot{\boldsymbol{\eta}}_i$，其中

$$\boldsymbol{\Omega}_i=\begin{bmatrix}-\sin\beta_i & 0 & 1\\ \cos\beta_i\sin\varphi_i & \cos\varphi_i & 0\\ \cos\beta_i\cos\varphi_i & -\sin\varphi_i & 0\end{bmatrix} \tag{2-40}$$

四旋翼无人机的重力势能可表示为$\mathcal{V}=mgz_i$。因此，根据欧拉 - 拉格朗日方程 (2-39)，我们可以导出无人机的平动动力学模型和旋转动力学

模型如下：

$$\begin{bmatrix} m_i\ddot{\boldsymbol{\xi}}_i + m_i\boldsymbol{g} & 0 \\ 0 & \boldsymbol{M}_i\left(\boldsymbol{\eta}_i\right)\ddot{\boldsymbol{\eta}}_i + \boldsymbol{C}_i\left(\dot{\boldsymbol{\eta}}_i,\boldsymbol{\eta}_i\right)\dot{\boldsymbol{\eta}}_i \end{bmatrix} = \begin{bmatrix} \boldsymbol{f}_i \\ \boldsymbol{\tau}_i \end{bmatrix} \tag{2-41}$$

其中，$\boldsymbol{g}=[0\ 0\ 9.8]^{\mathrm{T}}$是重力加速度向量；$\boldsymbol{f}_i$和$\boldsymbol{\tau}_i$分别是合外力和合外力矩。惯性矩$\boldsymbol{M}_i\left(\boldsymbol{\eta}_i\right)$和离心力/科氏力矩阵$\boldsymbol{C}_i\left(\dot{\boldsymbol{\eta}}_i,\boldsymbol{\eta}_i\right)$的具体形式如下，为了表示简洁，我们令$s$表示sin，$c$表示cos。

$$\boldsymbol{M}_i\left(\boldsymbol{\eta}_i\right) = \begin{bmatrix} \eta_i^{11} & \eta_i^{12} & \eta_i^{13} \\ \eta_i^{21} & \eta_i^{22} & \eta_i^{23} \\ \eta_i^{31} & \eta_i^{32} & \eta_i^{33} \end{bmatrix} \tag{2-42}$$

其中

$$\begin{aligned} &\eta_i^{11} = I_i^{xx}s_{\beta_i}^2 + I_i^{yy}c_{\beta_i}^2 s_{\varphi_i}^2 + I_i^{zz}c_{\beta_i}^2 c_{\varphi_i}^2，\ \eta_i^{12} = c_{\beta_i}c_{\varphi_i}s_{\varphi_i}\left(I_i^{yy} - I_i^{zz}\right)，\ \eta_i^{13} = -I_i^{xx}s_{\beta_i} \\ &\eta_i^{21} = c_{\beta_i}c_{\varphi_i}s_{\varphi_i}\left(I_i^{yy} - I_i^{zz}\right)，\ \eta_i^{22} = I_i^{yy}c_{\varphi_i}^2 + I_i^{zz}s_{\varphi_i}^2，\ \eta_i^{23} = 0, \\ &\eta_i^{31} = -I_i^{xx}s_{\beta_i}，\ \eta_i^{32} = 0，\ \eta_i^{33} = I_i^{xx} \end{aligned} \tag{2-43}$$

和

$$\boldsymbol{C}_i\left(\dot{\boldsymbol{\eta}}_i,\boldsymbol{\eta}_i\right) = \begin{bmatrix} c_i^{11} & c_i^{12} & c_i^{13} \\ c_i^{21} & c_i^{22} & c_i^{23} \\ c_i^{31} & c_i^{32} & c_i^{33} \end{bmatrix} \tag{2-44}$$

其中

$$\begin{aligned} c_i^{11} = {} & I_i^{xx}\dot{\beta}_i s_{\beta_i}c_{\beta_i} + I_i^{yy}\left(-\dot{\beta}_i s_{\beta_i}c_{\beta_i}s_{\varphi_i}^2 + \dot{\varphi}_i c_{\beta_i}^2 s_{\varphi_i}c_{\varphi_i}\right) \\ & - I_i^{zz}\left(\dot{\beta}_i s_{\beta_i}c_{\beta_i}s_{\varphi_i}^2 + \dot{\varphi}_i c_{\beta_i}^2 s_{\varphi_i}c_{\varphi_i}\right) \\ c_i^{12} = {} & I_i^{xx}\dot{\alpha}_i s_{\beta_i}c_{\beta_i} - I_i^{yy}\left(\dot{\beta}_i s_{\beta_i}s_{\varphi_i}c_{\varphi_i} + \dot{\varphi}_i c_{\beta_i}c_{\varphi_i}^2 - \dot{\varphi}_i c_{\beta_i}c_{\varphi_i}^2\right. \\ & \left. + \dot{\alpha}_i s_{\beta_i}c_{\beta_i}s_{\varphi_i}^2\right) + I_i^{zz}\left(\dot{\varphi}_i c_{\beta_i}s_{\varphi_i}^2 - \dot{\varphi}_i c_{\beta_i}c_{\varphi_i}^2\right. \\ & \left. - \dot{\alpha}_i s_{\beta_i}c_{\beta_i}c_{\varphi_i}^2 + \dot{\beta}_i s_{\beta_i}s_{\varphi_i}c_{\varphi_i}\right) \\ c_i^{13} = {} & -I_i^{xx}\dot{\beta}_i c_{\beta_i} + I_i^{yy}\dot{\alpha}_i c_{\beta_i}^2 s_{\varphi_i}c_{\varphi_i} - I_i^{zz}\dot{y}c_{\beta_i}^2 s_{\varphi_i}c_{\varphi_i} \\ c_i^{21} = {} & -I_i^{xx}\dot{\alpha}_i s_{\beta_i}c_{\beta_i} + I_i^{yy}\dot{\alpha}_i s_{\beta_i}c_{\beta_i}s_{\varphi_i}^2 + I_i^{zz}\dot{\alpha}_i s_{\beta_i}c_{\beta_i}c_{\varphi_i}^2 \\ c_i^{22} = {} & -I_i^{yy}\dot{\varphi}_i s_{\varphi_i}c_{\varphi_i} + I_i^{zz}\dot{\varphi}_i s_{\varphi_i}c_{\varphi_i} \\ c_i^{23} = {} & I_i^{xx}\dot{\alpha}_i c_{\beta_i} + I_i^{yy}\left(-\dot{\beta}_i s_{\varphi_i}c_{\varphi_i} + \dot{\alpha}_i c_{\beta_i}c_{\varphi_i}^2 - \dot{\alpha}_i c_{\beta_i}s_{\varphi_i}^2\right) \end{aligned}$$

$$
\begin{aligned}
&\quad + I_i^{zz}\left(\dot{\alpha}_i c_{\beta_i} s_{\varphi_i}^2 - \dot{\alpha}_i c_{\beta_i} c_{\varphi_i}^2 + \dot{\beta}_i s_{\varphi_i} c_{\varphi_i}\right) \\
&c_i^{31} = -I_i^{yy}\dot{\alpha}_i c_{\beta_i}^2 s_{\varphi_i} c_{\varphi_i} + I_i^{zz}\dot{\alpha}_i c_{\beta_i}^2 s_{\varphi_i} c_{\varphi_i} \\
&c_i^{32} = -I_i^{xx}\dot{\alpha}_i c_{\beta_i} + I_i^{yy}\left(\dot{\beta}_i s_{\varphi_i} c_{\varphi_i} + \dot{\alpha}_i c_{\beta_i} s_{\varphi_i}^2 - \dot{\alpha}_i c_{\beta_i} c^2{}_{\varphi_i}\right) \\
&\quad - I_i^{zz}\left(\dot{\alpha}_i c_{\beta_i} s_{\varphi_i}^2 - \dot{\alpha}_i c_{\beta_i} c_{\varphi_i}^2 + \dot{\beta}_i s_{\varphi_i} c_{\varphi_i}\right) \\
&c_i^{33} = 0
\end{aligned}
\tag{2-45}
$$

2.3 基于事件触发通信机制的自主智能系统建模

自主智能系统的信息交互是完成协同控制任务的关键，然而自主智能系统本身能量、通信资源往往十分受限，因此，建立资源受限下自主智能系统的信息传输模型具有重要意义。本节主要面向自主智能系统的协同控制问题，讨论几类基于事件触发的网络通信模型和基本设计方法，为后续的理论研究奠定基础。

2.3.1 动态事件触发机制的设计原则 I

由于本章节的重点是设计拒绝服务攻击下的事件触发策略，所以我们忽略可能存在的网络延迟。在网络化控制系统中，通信确实经常受到网络延迟的影响，这可能会恶化协调性能 [80-81]，所以关于时滞的解决方法可以按照 Heemels 等 [82] 的思路进行分析。

针对完整状态测量输出，控制输入 $\boldsymbol{u}_\tau$ 在离散的更新时刻$\{t_r\}_{r\in\mathbb{Z}_{\geq 0}}$进行更新，其中$0\leqslant t_0 < t_1 < \cdots < t_r < t_{r+1}$并且$0 < \underline{\Delta} \leqslant t_{r+1} - t_r$。在每一个采样时刻 t_r，采样值$\hat{\mathrm{R}}\mathrm{e}(t_r)$经过网络通道到达控制器端进而更新控制器 $\boldsymbol{u}_\tau$。在$t\in[t_r, t_{r+1}]$期间，零阶保持器使得采样值$\hat{R}e(t_r)$保持不变，即

$$
\dot{\hat{R}}e(t) = 0
$$

基于此，定义$t\in[t_r, t_{r+1}]$期间的采样误差为

$$
e(t) = \hat{R}e(t_r) - Re(t) \tag{2-46}
$$

在每一个更新时刻t_r, $\hat{R}e(t_r)$被重置为$Re(t_r)$，即

$$\hat{R}e\left(t_r^+\right)=Re\left(t_r\right)$$

因此有

$$\begin{cases}\dot{e}(t):=g\left(e(t)\right),\ t\in\left[t_r,\ t_{r+1}\right]\\ e\left(t_{r+1}^+\right)=0\end{cases}\tag{2-47}$$

拒绝服务攻击表示网络通信在攻击期间被阻塞，因此采样值$Re\left(t_k\right)$不能经过网络成功传输。定义一系列时间区间$\{H_n\}_{n\in\mathbb{Z}_{\geqslant 0}}$表示拒绝服务攻击发生的时间段，即$H_n=\{h_n\}\cup\left[h_n,h_n+\tau_n\right)$，其中$\tau_n\in\mathbb{R}_{\geqslant 0}$表示一次拒绝服务攻击的持续时间长度，且满足$h_n\leqslant h_n+\tau_n<h_{n+1}$。令$\mathcal{T}=\bigcup_{n\in\mathbb{Z}_{\geqslant 0}}H_n$表示拒绝服务攻击的总时间，因此拒绝服务攻击的集合可以表示为$\Xi\left(T_1,T_2\right)=\left(\bigcup_{n\in\mathbb{Z}_0}H_n\right)\cap\left[T_1,T_2\right]$，其中任意给定的区间$\left[T_1,T_2\right]$满足$0\leqslant T_1\leqslant T_2$，相应地，$\Theta\left(T_1,T_2\right)=\left[T_1,T_2\right]\backslash\Xi\left(T_1,T_2\right)$表示$\left[T_1,T_2\right]$中没有攻击的时间段。

因此$Re\left(t_r\right)$的更新可以表示为

$$\hat{R}e\left(t_k^+\right)=\begin{cases}Re\left(t_r\right),\quad t_r\notin\mathcal{T}\\ \hat{R}e\left(t_r\right),\quad t_r\in\mathcal{T}\end{cases}$$

其中，$r\in\mathbb{Z}_{\geqslant 0}$。在不引起混淆的情况下，$\Theta\left(T_1,T_2\right)$和$\Xi\left(T_1,T_2\right)$可以分别表示为$\Theta$和$\Xi$。

2.3.1.1 基于阈值的事件触发机制

本章节中采样值 Re 被连续传输给事件驱动策略来产生一系列离散的传输时刻，即一旦事件驱动策略被触发，采样值会经过网络通道传输给控制器：如果有拒绝服务攻击存在，那么采样数据将不能经过网络传输给控制器；如果没有拒绝服务攻击，那么传输将会成功。基于事件驱动的采样通信机制为

$$t_{r+1}=\inf_{r\in\mathbb{Z}_{\geqslant 0}}\left\{t:\ t\geqslant t_r+\tau_{\text{miet}}^m\mid\ \chi(t)\leqslant 0,r\in\mathbb{Z}_{\geqslant 0},\ t_0=0\right\}\tag{2-48}$$

其中，函数$\chi(t)$表示事件触发函数，布尔变量$m\in\{0,1\}$被定义为：①$m=0$表示传输尝试会成功，其中传输尝试时刻t_{r+1}会在自上一次传输时刻t_r再经过时间τ_{miet}^0之后且当$\chi(t)\leqslant 0$的时候到来；②$m=1$表示传输尝试时刻t_{r+1}会在自上一次传输时刻t_r再经过τ_{miet}^1之后且$\chi(t)\leqslant 0$的时候到来，但是数

据不会被成功传输给控制器，因为此时网络通道因拒绝服务攻击而不可用。值得注意的是，τ_{miet}^m表示最小允许传输间隔并满足$0<\tau_{\text{miet}}^1\leqslant\tau_{\text{miet}}^0$，这可以帮助排除奇诺现象。触发时刻方程(2-48)表示事件触发机制只有可能自上次传输时刻再经过τ_{miet}^m之后被触发，而且参数τ_{miet}^m使得在攻击发生期间传输更频繁，使得事件驱动机制尽早知道攻击何时结束。本章节在控制器和事件驱动策略之间采用确认机制，使事件触发机制知道控制器是否接收到采样值，为此我们遵循基于事件触发策略(2-48)的设计机理，这确保了控制系统能够对不可预测的变化（由拒绝服务攻击引起）做出响应，以保持跟踪性能（弹性控制），而正是数字信息和通信技术使我们能够实现这一目标，进而使飞行器姿态控制系统更加智能化。事件驱动函数$\chi(t)$可以设计为

$$\begin{cases}\mathrm{d}\chi(t)=\Psi\big(m(t),\ o(t),\ \chi(t)\big)\mathrm{d}t,\ t\in\left[t_r,\ t_{r+1}\right)\\ \chi^+(t_r)=\begin{cases}\tilde{\chi}(t_r), & t_r\in\Theta_Z\\ \chi(t_r), & t_r\in\Xi_Z\end{cases}\end{cases}\tag{2-49}$$

其中，$o=(e,\tau,\phi)\in\mathbb{R}^e\times\mathbb{R}_{\geqslant 0}\times\left[\lambda,\lambda^{-1}\right]$表示部分变量，$\lambda\in(0,1)$。时间变量$\tau$表示自上次传输尝试之后经过的时间，函数$\phi$将用于推导最小允许传输间隔（将在后面的章节中分析），而函数Ψ和$\tilde{\chi}$将会在具体问题中给出具体表达式。

2.3.1.2 事件触发机制的触发间隔

为此我们首先给出Φ：$\mathbb{R}_{\geqslant 0}\rightarrow\mathbb{R}$，即

$$\frac{\mathrm{d}\phi}{\mathrm{d}\tau}=\begin{cases}-2L\phi-\gamma\left[(m\epsilon+1)\phi^2+\dfrac{m(\tilde{\gamma}-\gamma)+\gamma}{\gamma}\right], & \tau\in\left[0,\ \tau_{\text{miet}}^m\right]\\ 0, & \tau>\tau_{\text{miet}}^m\end{cases}\tag{2-50}$$

其中，$\epsilon\geqslant 0$，$\tilde{\gamma}\geqslant\gamma$，常数$L$和$\gamma$为严格正数，并满足$\lim\limits_{t\to\tau_{\text{miet}}}\phi(t)=\phi(\tau_{\text{miet}})$使得$\phi$在$\tau_{\text{miet}}$处连续。于是，式(2-50)中的$\tau_{\text{miet}}^m$的取值范围为$(0,\tau_{\text{mati}}^m)$，其中$\tau_{\text{mati}}^m$为

$$\tau_{\text{mati}}^{m}=\begin{cases}\dfrac{1}{Lr}\arctan\left(\varGamma\right), & \varLambda>L\\ \tilde{\varGamma}, & \varLambda=L\\ \dfrac{1}{Lr}\operatorname{arctanh}\left(\varGamma\right), & \varLambda<L\end{cases} \tag{2-51}$$

其中，$\lambda\in(0,1)$来自式(2-49)；$r=\sqrt{\left|\dfrac{\varLambda^2}{L^2}-1\right|}$；

$\varLambda=\sqrt{\gamma(1+m\epsilon)\left[m\tilde{\gamma}+\gamma(1-m)\right]}$；$\tilde{\varGamma}=\dfrac{1}{\lambda\gamma(1+m\epsilon)+L}-\dfrac{\lambda}{\lambda L+\gamma(1+m\epsilon)}$；

$$\varGamma=\frac{r(1-\lambda)}{\dfrac{\lambda}{1+\lambda}\left[\dfrac{\gamma(1+m\epsilon)+m\tilde{\gamma}+\gamma(1-m)}{L}-2\right]+1+\lambda}$$

接下来我们给出关于式(2-51)的引理。

引理 2.1[83]

假设τ_{mati}^{m}满足式 (2-51)，那么方程

$$\dot{\tilde{\phi}}=-2L\tilde{\phi}-\gamma\left[(m\epsilon+1)\tilde{\phi}^2+m\frac{\tilde{\gamma}}{\gamma}+1-m\right] \tag{2-52}$$

对应初始值$\tilde{\phi}(0)=\lambda^{-1}$的解满足$\tilde{\phi}(t)\in\left[\lambda,\lambda^{-1}\right]$，其中$t\in\left[0,\tau_{\text{mati}}^{m}\right]$和$\tilde{\phi}\left(\tau_{\text{mati}}^{m}\right)=\lambda$。

基于式 (2-51) 和引理 2.1，τ_{miet}^{1}和τ_{miet}^{0}可以取值为：$\tau_{\text{miet}}^{1}\leqslant\tau_{\text{mati}}^{1}\leqslant\tau_{\text{mati}}^{0}$以及$\tau_{\text{miet}}^{1}\leqslant\tau_{\text{miet}}^{0}\leqslant\tau_{\text{mati}}^{0}$。为方便进一步分析，我们定义$\phi_{\text{miet}}=\tilde{\phi}\left(\tau_{\text{miet}}^{0}\right)$，其中$\tilde{\phi}$来自式 (2-52)，并满足$\tilde{\phi}(0)=\lambda^{-1}$和$\phi_{\text{miet}}\geqslant\lambda$。

注释 2.2

跟 Postoyan 等[84]相比，式 (2-52) 额外考虑两个参数ϵ和$\tilde{\gamma}$，目的是尽早告知事件驱动策略拒绝服务攻击何时结束[83]，因为这两个参数会加快在拒绝服务攻击发生时的传输尝试频率。所以可以根据需要选择这两个参数ϵ和$\tilde{\gamma}$：ϵ或者$\tilde{\gamma}$的值越大，会使得$m=1$时候的传输间隔更小，即拒绝服务攻击发生时的传输尝试频率更高，进而更有利于事件驱动策略察觉并确定拒绝服务攻击在何时结束。

2.3.2 动态事件触发机制的设计原则Ⅱ

在这部分中，我们考虑网络分配给每个节点的传输优先级由传输协议$h\left(\kappa,e\left(t_k\right)\right)$决定，而传输时刻是由中央协调器根据动态系统的当前状态通过实时检测事件触发条件来决定的。所以接下来我们会设计合适的事件触发策略。

根据上面的描述，在每次传输尝试时刻t_k，$k\in\mathbb{Z}_{\geqslant 0}$，$Re$ 的状态更新可以进一步表示为

$$\hat{R}e\left(t_k^+\right)=\begin{cases}Re\left(t_k\right)+h\left(\kappa,e\left(t_k\right)\right), & t_k\notin\mathcal{T}\\ \hat{R}e\left(t_k\right), & t_k\in\mathcal{T}\end{cases}$$

在不引起歧义的情况下，$\varTheta\left(T_1,T_2\right)$和$\varXi\left(T_1,T_2\right)$可以分别简写为$\varTheta$和$\varXi$。相应地，传输误差可以表示为

$$e\left(t_k^+\right)=\begin{cases}h\left(\kappa,\ e\left(t_k\right)\right), & t_k\notin\mathcal{T}\\ e\left(t_k\right), & t_k\in\mathcal{T}\end{cases}$$

下面分别介绍在拒绝服务攻击下基于时钟变量的事件驱动机制，即

$$t_{k+1}=\begin{cases}\inf\left\{t\text{: } t\geqslant t_k+\varDelta_0 \mid m=0,\ \eta\left(t\right)=a_\lambda\lambda^2+b_\lambda \quad k\in\mathbb{Z}_{\geqslant 0}\right\}\\ \inf\left\{t\text{: } t\geqslant t_k+\varDelta_1 \mid m=1,\ k\in\mathbb{Z}_{\geqslant 0}\right\}\end{cases} \tag{2-53}$$

其中，$t_0=0$，函数$\eta(t)$表示事件驱动函数，正参数$b_\lambda\in\left(0,a_\lambda\left(1-\lambda^2\right)\right)$和$a_\lambda\in\left(\lambda,+\infty\right)\left(\lambda\in\left[0,1\right)\right)$，$\max\left\{\varDelta_0,\varDelta_1\right\}\leqslant t_{k+1}-t_k\leqslant T$和$0\leqslant\varDelta_1\leqslant\varDelta_0\leqslant T$ [83,85]。参数$\lambda\in[0,1)$会在具体问题中给出具体分析。本章节使用确认机制，即事件触发机制可以知道控制器端是否成功接收数据包[83]。对于$m\in\{0,1\}$：①$m=0$表示没有拒绝服务攻击的情况，即采样数据可以成功传输，因而有$e\left(t_k^+\right)=h\left(\kappa,e\left(t_k\right)\right)$；②$m=1$表示存在拒绝服务攻击，采样数据不能成功传输，因而有$e\left(t_k^+\right)=e\left(t_k\right)$。另一方面，驱动函数$\eta(t)$满足

$$\begin{cases}\dot{\eta}\left(t\right)=\varPsi\left(m,\ o\left(t\right),\ \eta\left(t\right)\right),\ t\in\left[t_k,\ t_{k+1}\right)\\ \eta^+\left(t_k\right)=\begin{cases}a_\lambda, & t_k\notin\mathcal{T}\\ \eta\left(t_k\right), & t_k\in\mathcal{T}\end{cases}\end{cases}$$

其中，$o(t)$代表部分信息变量；函数$\Psi\left(m,o\left(t\right),\eta\left(t\right)\right)$会在具体问题中给出具体表达式。

如上所示，事件触发策略依赖时钟变量η，即当函数η下降到$a_\lambda\lambda^2+b_\lambda$的时候，事件驱动策略会被触发进而传输采样数据。也就是说，事件触发间隔对应着η从a_λ下降到$a_\lambda\lambda^2+b_\lambda$所经过的时间。更详细地，①如果$t_k\in\Theta\setminus\Xi$（即$m=0$），那么通过实时数据，可以对事件触发策略进行连续正常评估进而正常更新控制器数据；②如果$t_k\in\Xi$（即$m=1$），那么确认机制会通知事件驱动策略进而停止评估，但是会按照式(2-53)中第二个等式产生周期性的通信时刻，直到攻击消失，然后数据成功传输给控制器，然后事件驱动被告知控制器端收到新数据，进而转回正常的驱动机制，即式(2-53)中第一个等式根据系统状态继续产生非周期性的通信时刻。

实际上，可以用其他建模方法来设计拒绝服务攻击下的事件触发策略，如De和Tesi[85]的研究。所以当在拒绝服务攻击下建立系统模型时，可以选择De和Tesi[85]中的一般形式，或者Dolk等[83]研究中的标准混杂形式，其实它们之间没有本质的区别，都是依赖传输误差决定是否传输数据，我们不能评估哪种形式能使事件驱动建模策略更好，但是标准的混杂形式在事件触发调度方面表现出很强的系统性，通过适当选择参数a_λ和b_λ可以避免奇诺现象，而且这些参数也使标准混杂形式下的事件触发机制更加灵活。

2.4 面向网络安全的自主智能系统建模

如今，随着信息技术的不断发展，自主智能系统之间通过信息交互实现互联互通，提高了系统的灵活性和高效性。然而，网络环境的开放性也使得自主智能系统暴露在各种网络攻击之下。针对自主智能系统的网络安全问题，本节主要介绍几种不同类型的网络攻击和相应的检测和防御策略模型。

2.4.1 网络攻击模型

根据文献 [86]，网络攻击一般有两种划分方式，即拒绝服务攻击（DoS 攻击），以及完整性攻击。这两种网络攻击的主要目的分别是破坏无线传输数据的可用性以及完整性。具体而言，DoS 攻击通过阻塞 CPS 的网络层来阻止传感器测量和控制输入等信息的交换，从而实现降低系统性能的目的。在介绍 DoS 攻击的模型之前，我们首先概述无线传感网络中随机数据丢包与信噪比之间的关系。由于信道衰弱以及外界干扰等因素，无线传感网络中存在数据随机丢包的现象。假设传感网络使用了正交幅度调制（QAM），并且存在加性高斯白噪声（AWGN），则符号错误率（SER）和信噪比（SNR）之间的关系如下：

$$\mathrm{SER}=2Q\left(\sqrt{\alpha \mathrm{SNR}}\right),\ Q(x)=\frac{1}{\sqrt{2\phi}}\int_x^{\infty}\exp\left(-\eta^2/2\right)\mathrm{d}\eta \tag{2-54}$$

其中$\alpha>0$是一个参数。在存在DoS干扰攻击者的情况下，文献[87]将通信网络中的SNR重写为$\mathrm{SINR}=\dfrac{p_k}{w_k+\sigma^2}$，其中$p_k$表示传感器在时刻$k$消耗的通信能量，$\sigma^2$表示加性高斯白噪声的功率，$w_k$表示攻击者的干扰功率。我们假设远程估计器可以通过循环冗余校验（CRC）检测符号错误。因此，传感器和远程估计器之间的数据是否传输成功可以用一个伯努利随机过程γ_k，$k\in\mathbb{N}$来刻画：

$$\gamma_k=\begin{cases}1, & \text{如果传输数据在时刻}k\text{成功到达}\\ 0, & \text{否则，视为丢包}\end{cases} \tag{2-55}$$

基于式(2-54)以及上述讨论，我们有

$$\lambda_k=\mathbb{P}\left[\gamma_k=1\right]=1-2Q\left(\sqrt{\alpha\frac{p_k}{w_k+\sigma^2}}\right) \tag{2-56}$$

需要注意的是，SINR不仅取决于传感器使用的传输功率，还受DoS攻击者干扰功率的影响。不同的SINR导致不同的丢失率和远程估计性能。然后由于DoS攻击者总共的干扰能量有限，攻击者需要在每个时刻k最优化地分配其能量，从而最大化地破坏系统性能。这类问题已经被文献[87]等广泛地研究。

与 DoS 攻击不同的是，完整性攻击主要通过修改传输数据的内容来破坏其完整性，它的数据模型一般可以写成$\tilde{\boldsymbol{x}}_k=\boldsymbol{x}_k+\boldsymbol{\Delta}_k$，其中$\boldsymbol{x}_k$表示初始的传输数据，$\boldsymbol{\Delta}_k$表示攻击者注入的错误数据，$\tilde{\boldsymbol{x}}_k$表示被攻击者篡改后的错误数据。根据攻击者攻击模式的不同，完整性攻击可以进一步细分成重放攻击以及欺诈攻击等。根据文献 [88-89] 中的定义，重放攻击可以从时刻 k_0 到时刻$k_0+\tau-1$从无线传感网络中窃听传输的数据，并且将这些数据存储于内存中，然后从时刻$k_0+\tau$到时刻$k_0+2\tau-1$将无线传感网络中的传输数据替换成这些被窃听的数据。因此重放攻击的攻击模型可以写成：

$$\tilde{\boldsymbol{x}}_k=\boldsymbol{x}_{k-\tau} \tag{2-57}$$

欺诈攻击可以在篡改传输数据的同时，屏蔽掉传统的检测器。例如，文献[90-91]提出了一种基于新息的线性中间人攻击，其攻击模型为：

$$\tilde{\boldsymbol{z}}_k=\boldsymbol{T}_k\boldsymbol{z}_k+\boldsymbol{b}_k \tag{2-58}$$

其中，$\tilde{\boldsymbol{z}}_k$是被攻击者篡改的新息；$\boldsymbol{T}_k\in\mathbb{R}^{m\times m}$是任意矩阵；$\boldsymbol{b}_k$是零均值、协方差为$\boldsymbol{\Sigma}_b$的高斯随机变量，与$\boldsymbol{z}_k$不相关。因此$\tilde{\boldsymbol{z}}_k$也遵循零均值、协方差为$\boldsymbol{T}_k\boldsymbol{\Sigma}_z\boldsymbol{T}_k^{\mathrm{T}}+\boldsymbol{\Sigma}_b$的正态分布。由于传统的检测器的机理是基于新息的统计特性，攻击者只要设计其攻击策略满足条件$\boldsymbol{T}_k\boldsymbol{\Sigma}_z\boldsymbol{T}_k^{\mathrm{T}}+\boldsymbol{\Sigma}_b=\boldsymbol{\Sigma}_z$即可保证攻击前后数据的统计特性保持不变，从而实现攻击的隐身。

2.4.2 网络攻击的检测以及识别技术

检测器可以通过一系列数学运算来检验无线传感网络中的传输数据，并且和其门限值做对比，从而实现网络攻击的检测。检测器包含 χ^2 检测器、累计和检测器（CUSUM）、多变量指数加权移动平均检测器（MEWMA）等。其中，χ^2 检测器基于归一化残差$\bar{\boldsymbol{r}}_n\in\mathbb{R}^q$来开发统计检测。因为当不存在攻击的时候，有$\bar{\boldsymbol{r}}_n\sim N(0,I)$，$\chi^2$检测器可以被定义为：

$$\bar{\boldsymbol{r}}_n^{\mathrm{T}}\bar{\boldsymbol{r}}_n < \boldsymbol{\tau}_{\chi^2} \tag{2-59}$$

其中，检测统计量$\bar{\boldsymbol{r}}_n^{\mathrm{T}}\bar{\boldsymbol{r}}_n$在不存在攻击的时候满足$\bar{\boldsymbol{r}}_n^{\mathrm{T}}\bar{\boldsymbol{r}}_n \sim \chi^2(q)$。作为攻击的结果，残差分布的变化可能会改变$\chi^2$分布，但并非所有攻击都是如此。例如，攻击可以将残差向量替换为：$\boldsymbol{r}_n = \boldsymbol{\Sigma}_r^{1/2}\bar{\boldsymbol{r}}_n = \boldsymbol{\Sigma}_r^{1/2}\left[\sqrt{\boldsymbol{c}_n}, 0, \cdots, 0\right]^{\mathrm{T}}$，其中$\boldsymbol{c}_n \sim \chi^2(q)$，从而导致了$\boldsymbol{r}_n^{\mathrm{T}}\boldsymbol{r}_n = \boldsymbol{c}_n \sim \chi^2(q)$。$\chi^2$检测器将无法区分此类攻击，同时增加了观察状态的误差。此外，χ^2检测器是无记忆的，这会使检测残差范数的微小增加变得困难。

CUSUM 检测器通过在检测统计量中引入动态特性，解决当攻击者在归一化残差范数中注入微小但持续的错误数据时，χ^2 检测器难以检测的问题：

$$a_C\boldsymbol{R}_n = \max\left(a_C\boldsymbol{R}_{n-1} + \bar{\boldsymbol{r}}_n^{\mathrm{T}}\bar{\boldsymbol{r}}_n - \gamma,\ 0\right) < \tau_C \tag{2-60}$$

其中，$a_C\boldsymbol{R}_{-1} = 0$并且$\gamma$被称为遗忘因子。为确保检验统计量稳定，需要满足$\gamma > q$。与$\chi^2$检测器类似，CUSUM检测器在没有警报的假设下限制归一化残差的范数：

$$\bar{\boldsymbol{r}}_n^{\mathrm{T}}\bar{\boldsymbol{r}}_n - \gamma \leqslant a_C\boldsymbol{R}_{n-1} + \bar{\boldsymbol{r}}_n^{\mathrm{T}}\bar{\boldsymbol{r}}_n - \gamma < \tau_C \tag{2-61}$$

对于CUSUM检测器，归一化残差范数的持续增加会增加$\bar{\boldsymbol{r}}_n^{\mathrm{T}}\bar{\boldsymbol{r}}_n > \gamma$的可能性。当多个时刻都是如此时，CUSUM检测值会累积增加，从而触发警报。

与 CUSUM 检测器类似，MEWMA 检测器测试统计量也包含动态。MEWMA 检测器使用归一化残差的指数加权移动平均值：

$$\boldsymbol{G}_n = \beta\bar{\boldsymbol{r}}_n + (1-\beta)\boldsymbol{G}_{n-1} \tag{2-62}$$

其中，$\boldsymbol{G}_{-1} = 0$并且参数$\beta \in (0,1]$也被称为遗忘因子。MEWMA检测器被定义为：

$$a_M\boldsymbol{R}_n = \frac{2-\beta}{\beta}\boldsymbol{G}_n^{\mathrm{T}}\boldsymbol{G}_n < \tau_M \tag{2-63}$$

当$\beta = 1$时，其检测统计量等于χ^2检测器的检测统计量。对于较小的β，可以获得与CUSUM检测器类似的效果，因为对于遗忘因子$\beta \in (0,1)$，残差范数的持续增加会导致$\boldsymbol{G}$的较大值，从而导致更高的检测统计值。

对于 LQG 闭环控制系统，文献 [88-89] 考虑了通过在控制信号 $\boldsymbol{u}_k^*$ 中加入水印的方式来提升检测器对重放攻击的检测性能：

$$\boldsymbol{u}_k = \boldsymbol{u}_k^* + \boldsymbol{\zeta}_k \tag{2-64}$$

其中，$\boldsymbol{u}_k^*$表示最优LQG控制信号；$\boldsymbol{\zeta}_k$表示水印信号。假设水印信号$\{\boldsymbol{\zeta}_k\}$是独立于噪声过程$\{\boldsymbol{w}_k\}$以及$\{\boldsymbol{v}_k\}$的p维平稳零均值高斯过程。定义自协方差函数T：$\mathbb{Z} \to \mathbb{R}^{p \times p}$为：

$$T(d) = \mathrm{Cov}(\boldsymbol{\zeta}_0,\ \boldsymbol{\zeta}_d) = \mathbb{E}\boldsymbol{\zeta}_0\boldsymbol{\zeta}_d^{\mathrm{T}} \tag{2-65}$$

假设水印是由隐马尔可夫模型生成的，有

$$\boldsymbol{\xi}_{k+1} = \boldsymbol{A}_h\boldsymbol{\xi}_k + \boldsymbol{\psi}_k,\ \ \boldsymbol{\zeta}_k = \boldsymbol{C}_h\boldsymbol{\xi}_k \tag{2-66}$$

其中，$\boldsymbol{\psi}_k \in \mathbb{R}^{n_h}$；$k \in \mathbb{Z}$是独立同分布的零均值高斯随机序列，其方差为$\boldsymbol{\Psi}$，并且$\boldsymbol{\xi}_k \in \mathbb{R}^{n_h}$是隐状态。为了使$\{\boldsymbol{\zeta}_k\}$成为平稳过程，假设$\boldsymbol{\xi}_0$的协方差是以下李雅普诺夫方程的解：

$$\mathrm{Cov}(\boldsymbol{\xi}_0) = \boldsymbol{A}_h\mathrm{Cov}(\boldsymbol{\xi}_0)\boldsymbol{A}_h^{\mathrm{T}} + \boldsymbol{\Psi} \tag{2-67}$$

通过注入水印的方式，没有攻击时的无线传输数据和当前时刻生成的水印$\boldsymbol{\zeta}_k$呈现出一定的相互关系，然而重放攻击下的无线传输数据则与$\boldsymbol{\zeta}_k$无关。因此，基于水印的检测方式可以实现重放攻击的攻击检测。

2.4.3 抵抗网络攻击的鲁棒滤波

如上所述，完整性攻击可以通过篡改无线传输数据的内容，从而破坏系统的性能。一般而言，被篡改的错误数据偏离原始传输数据越远，攻击者造成的破坏也就越大。因此，为了缓解攻击对系统的影响，有些文献提出了基于饱和新息更新（SIU）的弹性滤波算法。在 SIU 中，每个智能体的目的是融合其邻居的传输数据，从而维护和更新其自身的本地估计 $\boldsymbol{x}_n(t)$。SIU 的每次迭代包括两个步骤：信息传递以及估计更新。为了实现初始化，假设每个智能体 n 设置其本地估计的初始值为 $\boldsymbol{x}_n(0) = 0$。

① 信息传递：在每次迭代中，每个智能体 n 将其当前时刻的估计值 $\boldsymbol{x}_n(t)$ 传输到每个邻居。每个智能体 n 在每次迭代中共传输 d_n 条信息，其中 d_n 表示智能体 n 的邻居个数。

② 估计更新：每个智能体n的状态估计更新为：

$$\boldsymbol{x}_n(t+1)=\boldsymbol{x}_n(t)-\beta_t\sum_{l\in\Omega_n}\left(\boldsymbol{x}_n(t)-\boldsymbol{x}_l(t)\right)+\alpha_t K_n(t)\left(\boldsymbol{y}_n-\boldsymbol{x}_n(t)\right) \tag{2-68}$$

其中，$\alpha_t>0,\beta_t>0$是要在后续中指定的参数序列；$K_n(t)$是本地新息的时变增益，定义为

$$K_n(t)=\begin{cases}1, & \|\boldsymbol{y}_n(t)-\boldsymbol{x}_n(t)\|\leqslant\gamma_t\\ \dfrac{1}{\left\|\boldsymbol{y}_n(t)-\boldsymbol{x}_n(t)\right\|_2}, & 其他\end{cases} \tag{2-69}$$

其中，γ_t是要在后续中指定的参数。$K_n(t)$确保对于所有n和所有t, $K_n(t)\left(\boldsymbol{y}_n-\boldsymbol{x}_n(t)\right)$的2范数的上限为$\gamma_t$。设计SIU算法的挑战是选择自适应的阈值$\gamma_t$，它表示允许传感器测量值偏离本地估计值的程度。如果γ_t太小，则增益$K_n(t)$将限制未受损害的传感器n对其估计精度的补偿效果。另一方面，如果γ_t太大，那么SIU算法将难以有效地抑制攻击对估计精度的影响，从而在攻击下造成更大的本地估计的误差。因此，关键问题是如何选择γ_t来平衡上述两种效应。

Control of
Autonomous Intelligent Systems

自主智能系统控制

第3章

隐马尔可夫模型的事件触发风险敏感状态估计

3.1 概述

与线性高斯系统相比，由于 HMMs 具有有限状态这一本质特点，考虑 HMMs 的事件触发状态估计问题存在一些优势：即使对于确定性事件触发（Deterministic Event-Triggered，DET）条件和丢包情况，仍然可以得到准确的计算可处理的事件触发估计器[92-93]。

另一方面，现有文献大多考虑具有 MMSE 或者最大后验概率（Maximum a Posteriori Probability, MAP）性能指标（称为风险中性指标[94]）。当准确概率模型已知时，这是最优估计方法。然而，考虑到在许多实际工程应用中，不能准确地知道数学模型[95-96]，在这种情况下，或许需要考虑其他估计误差性能指标。风险敏感估计性能指标是个可选择的方案，它考虑的是一个具有指数形式的估计误差函数，能够减少出现较大估计误差的概率，并且函数内部有一个可以调整的参数，称为风险敏感参数。文献 [97] 给出了风险敏感估计的鲁棒性明确理论性解释。因此，本章针对 HMMs 研究一个具有风险敏感性能指标的事件触发状态估计问题。本章所考虑的事件触发条件具有很强的一般性，使得所提出的状态估计算法可以应用到不同的特定事件触发条件当中。通过概率测度变换的方法，该估计问题得到解决。特别地，如果测量噪声服从高斯分布，本章得到了随机性事件触发（Stochastic Event-Triggered, SET）条件和一类 DET 条件下明确的估计器表达式。仿真比较表明了所提出结果的有效性。

3.2 问题描述

在 2.1.3 小节中，已给出 HMMs 的数学描述。现在介绍 HMMs 的风险敏感 MAP（RMAP）的状态估计问题[98-99]。给定$\hat{\boldsymbol{X}}_0,\cdots,\hat{\boldsymbol{X}}_{k-1}$，递归地

定义$\hat{\boldsymbol{X}}_k \in S_X$为 $\boldsymbol{X}_k$ 的 RMAP 估计值，使得

$$\hat{\boldsymbol{X}}_k = \arg\min_{\zeta \in S_X} J_k(\zeta),\ k \in \mathbb{N} \tag{3-1}$$

其中

$$J_k(\zeta) = \mathbb{E}\left[\exp\left(\theta \boldsymbol{\Psi}_{0,k}(\zeta)\right) \mid \mathcal{F}_k^y\right]$$

是风险敏感代价函数。这里，$\theta > 0$是风险敏感参数且

$$\boldsymbol{\Psi}_{0,k}(\zeta) = \hat{\boldsymbol{\Psi}}_{0,k-1} + \mu(\boldsymbol{X}_k,\ \zeta),\ k \in \mathbb{N}$$

其中，$k \geqslant 1$时$\hat{\boldsymbol{\Psi}}_{0,k-1} = \sum_{i=0}^{k-1} \mu\left(\boldsymbol{X}_i,\ \hat{\boldsymbol{X}}_i\right)$，$k=0$时$\hat{\boldsymbol{\Psi}}_{0,k-1}=0$，且

$$\mu(\boldsymbol{u},\ \boldsymbol{v}) = \begin{cases} 0, & 如果\boldsymbol{u}=\boldsymbol{v} \\ 1, & 其他 \end{cases}$$

参数θ可以在名义模型时的最优估计和系统具有不确定性时避免出现大估计误差之间，起到权衡作用。

在本节中，针对 HMMs 考虑带有事件触发测量值的 RMAP 远程状态估计问题（见图 3-1）。

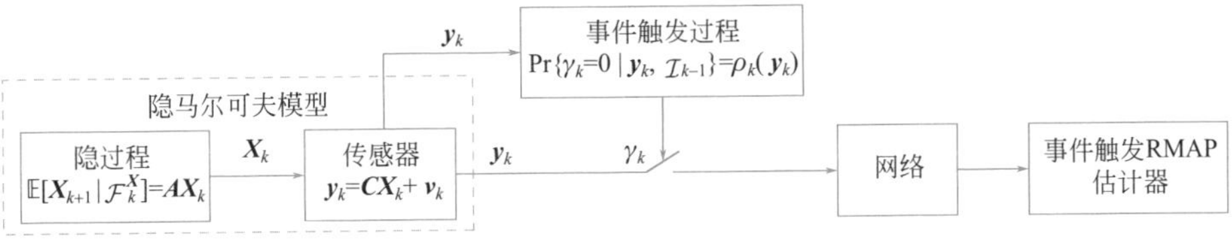

图 3-1 HMMs 的事件触发 RMAP 估计

由于受限的传感器能量和通信资源，我们设计了一个事件触发过程$\{\gamma_k\}$来决定 $\boldsymbol{y}_k$ 是否发送到远程估计器中。如果$\gamma_k=1$，则$\boldsymbol{y}_k$被发送到远端估计器；否则不发送任何信号。定义

$$\mathcal{I}_k := \left\{\gamma_i \middle| i \in \mathbb{N}_{0:k},\ \gamma_i = 0\right\} \cup \left\{\boldsymbol{y}_i \middle| i \in \mathbb{N}_{0:k},\ \gamma_i = 1\right\}$$

且$\mathcal{I}_{-1} := \varnothing$作为$k$时刻估计器的可利用信息集合。这里考虑一个一般性的事件触发方案，也就是文献[100]指出的事件触发调度器：

$$\Pr\left\{\gamma_k = 0 \mid \boldsymbol{y}_k, \mathcal{I}_{k-1}\right\} = \Pr\left\{\gamma_k = 0 \mid \mathcal{F}_k^X \cup \mathcal{F}_k^y \cup \mathcal{F}_{k-1}^\gamma\right\} = \rho_k(\boldsymbol{y}_k) \tag{3-2}$$

其中，$\rho_k(\cdot)$是一个给定$\mathcal{I}_{k-1}$下的已知函数；$\left\{\mathcal{F}_k^\gamma\right\}$是由$\sigma\{\gamma_0,\cdots,\ \gamma_k\}$生成的完备虑子。因此，仅仅需要设计一个在给定$\boldsymbol{y}_k$和过去信息$\mathcal{I}_{k-1}$的传感器休眠概率。事件触发方案(3-2)覆盖了现有大多数的事件触发条件，比

如DET条件[92,101-103]和SET条件[93,104-106]，且也能包含一些潜在的满足定义(3-2)的事件触发条件。注意，当$\gamma_k=0$时，估计器虽然没有接收到精确的$\boldsymbol{y}_k$信息，但仍然可以通过事件触发方案推测到一些关于$\boldsymbol{y}_k$的信息。这是与丢包和离线调度情况的不同之处。

本节目标是：在估计器可利用的事件触发信息集合$\mathcal{I}_k$下，找出隐状态的 RMAP 估计值$\hat{\boldsymbol{X}}_k$。该最优问题可以表述成

$$\hat{\boldsymbol{X}}_k=\arg\min_{\zeta\in S_X}\mathbb{E}\left[\exp\left(\theta\boldsymbol{\Psi}_{0,\ k}(\boldsymbol{\zeta})\right)\mid\mathcal{I}_k\right],\ k\in\mathbb{N} \tag{3-3}$$

上述估计问题在原始概率测度$\mathbb{P}$下描述，下一小节将利用测度变换方法[107]，在一个新的参考概率测度下求解该问题。

3.3 递归估计结果

本节将利用测度变换方法来求解所考虑的事件触发状态估计问题。通过引入一个新的测度，可以获得一个递归非标准化的信息状态，这对处理风险敏感的标准很有帮助。另一方面，在新测度下$\{\boldsymbol{y}_k\}$，$k\in\mathbb{N}$是 i.i.d.（独立同分布）随机变量且独立于过程$\{\boldsymbol{X}_k\}$，这便于处理本节复杂的事件触发估计问题。接下来，首先引入一个新测度并将它与原始测度联系起来，基于此可以得到一个等价的代价指标，该递归的估计问题也能够得到解决。

3.3.1 参考概率测度

现在引入一个新的参考概率测度$\overline{\mathbb{P}}$，在该测度下仍然有

$$\begin{aligned}&\overline{\mathbb{E}}\left[\boldsymbol{X}_{k+1}\mid\mathcal{F}_k^X\right]=\overline{\mathbb{E}}\left[\boldsymbol{X}_{k+1}\mid\boldsymbol{X}_k\right]=\boldsymbol{A}\boldsymbol{X}_k\\&\overline{\Pr}\left\{\gamma_k=0\mid\boldsymbol{y}_k,\ \mathcal{I}_{k-1}\right\}=\overline{\Pr}\left\{\gamma_k=0\mid\mathcal{F}_k^X\cup\mathcal{F}_k^y\cup\mathcal{F}_{k-1}^\gamma\right\}=\rho_k\left(\boldsymbol{y}_k\right)\end{aligned} \tag{3-4}$$

但是此时$\{\boldsymbol{y}_k\}$，$k\in\mathbb{N}$为一个i.i.d.随机变量序列，密度函数是ϕ且满足

$$\overline{\Pr}\left\{\boldsymbol{y}_k\leqslant\boldsymbol{t}\mid\mathcal{F}_k^X\cup\mathcal{F}_{k-1}^y\cup\mathcal{F}_{k-1}^\gamma\right\}=\overline{\Pr}\left\{\boldsymbol{y}_k\leqslant\boldsymbol{t}\right\}=\int_{-\infty}^{t}\phi\left(\boldsymbol{y}_k\right)\mathrm{d}\boldsymbol{y}_k \tag{3-5}$$

其中，$\boldsymbol{t}\in\mathbb{R}^m$。注意到，式(3-5)中第一个等式意味着在$\overline{\mathbb{P}}$下，过程$\{\boldsymbol{y}_k\}$独立于$\{\boldsymbol{X}_k\}$且进一步$\{\gamma_k\}$独立于$\{\boldsymbol{X}_k\}$。

为了概念上简洁，记$\mathcal{G}_k := \mathcal{F}_k^X \cup \mathcal{F}_k^y \cup \mathcal{F}_k^\gamma$。为了将$\overline{\mathbb{P}}$和原始测度 $\mathbb{P}$ 关联起来，在$\mathcal{G}_k$中定义一个 Radon-Nikodym 导数 [108]：

$$\left.\frac{\mathrm{d}\mathbb{P}}{\mathrm{d}\overline{\mathbb{P}}}\right|_{\mathcal{G}_k} := \overline{\boldsymbol{\Lambda}}_k \tag{3-6}$$

其中，$\overline{\boldsymbol{\Lambda}}_k = \prod_{l=0}^k \overline{\boldsymbol{\lambda}}_l$，$\overline{\boldsymbol{\lambda}}_l = \phi\left(\boldsymbol{y}_l - \boldsymbol{C}\boldsymbol{X}_l\right)/\phi\left(\boldsymbol{y}_l\right)$。那么有如下引理。

引理 3.1

如果概率测度$\overline{\mathbb{P}}$下的模型 (3-4) 和 (3-5) 通过 Radon-Nikodym 导数 [式 (3-6)] 被映射到原始测度 $\mathbb{P}$，那么在 $\mathbb{P}$ 下，式 (2-22)、式 (2-23) 和式 (3-2) 成立。

证明：

由$\overline{\boldsymbol{\lambda}}_k$的定义可知，

$$\begin{aligned}
&\overline{\mathbb{E}}\left[\overline{\boldsymbol{\lambda}}_{k+1} \mid \mathcal{F}_k^X \cup \mathcal{F}_k^y\right] \\
&= \overline{\mathbb{E}}\left[\phi\left(\boldsymbol{y}_{k+1} - \boldsymbol{C}\boldsymbol{X}_{k+1}\right)/\phi\left(\boldsymbol{y}_{k+1}\right) \mid \mathcal{F}_k^X \cup \mathcal{F}_k^y\right] \\
&= \sum_{i=1}^N \overline{\mathbb{E}}\left[\frac{\phi\left(\boldsymbol{y}_{k+1} - \boldsymbol{C}\boldsymbol{X}_{k+1}\right)}{\phi\left(\boldsymbol{y}_{k+1}\right)}\left\langle \boldsymbol{X}_{k+1}, \boldsymbol{e}_i\right\rangle \mid \mathcal{F}_k^X \cup \mathcal{F}_k^y\right] \\
&= \sum_{i=1}^N \overline{\mathbb{E}}\left[\frac{\phi\left(\boldsymbol{y}_{k+1} - \boldsymbol{C}\boldsymbol{e}_i\right)}{\phi\left(\boldsymbol{y}_{k+1}\right)}\left\langle \boldsymbol{X}_{k+1}, \boldsymbol{e}_i\right\rangle \mid \mathcal{F}_k^X \cup \mathcal{F}_k^y\right]
\end{aligned}$$

其中，$\left\langle \boldsymbol{X}_{k+1}, \boldsymbol{e}_i\right\rangle = \boldsymbol{X}_{k+1}^{\mathrm{T}}\boldsymbol{e}_i$，第二个等式是由于$\sum_{i=1}^N\left\langle \boldsymbol{X}_{k+1}, \boldsymbol{e}_i\right\rangle = 1$，最后一个等式是因为如果$\boldsymbol{X}_{k+1} \neq \boldsymbol{e}_i$，则$\left\langle \boldsymbol{X}_{k+1}, \boldsymbol{e}_i\right\rangle = 0$。因为在$\overline{\mathbb{P}}$下，$\boldsymbol{y}_{k+1}$独立于$\left\{\boldsymbol{y}_0, \boldsymbol{y}_1, \cdots, \boldsymbol{y}_k\right\}$和过程$\left\{\boldsymbol{X}_k\right\}$，可以进一步得到

$$\begin{aligned}
&\overline{\mathbb{E}}\left[\overline{\boldsymbol{\lambda}}_{k+1} \mid \mathcal{F}_k^X \cup \mathcal{F}_k^y\right] \\
&= \sum_{i=1}^N \overline{\mathbb{E}}\left[\frac{\phi\left(\boldsymbol{y}_{k+1} - \boldsymbol{C}\boldsymbol{e}_i\right)}{\phi\left(\boldsymbol{y}_{k+1}\right)}\right]\overline{\mathbb{E}}\left[\left\langle \boldsymbol{X}_{k+1}, \boldsymbol{e}_i\right\rangle \mid \mathcal{F}_k^X \cup \mathcal{F}_k^y\right] \\
&= \sum_{i=1}^N \overline{\mathbb{E}}\left[\left\langle \boldsymbol{X}_{k+1}, \boldsymbol{e}_i\right\rangle \mid \boldsymbol{X}_k\right] = \sum_{i=1}^N \left\langle \boldsymbol{A}\boldsymbol{X}_k, \boldsymbol{e}_i\right\rangle = 1
\end{aligned} \tag{3-7}$$

类似地，

$$\begin{aligned}&\overline{\mathbb{E}}\left[\overline{\boldsymbol{\lambda}}_k \mid \mathcal{F}_k^{\boldsymbol{X}} \cup \mathcal{F}_{k-1}^{\boldsymbol{y}}\right]\\=&\overline{\mathbb{E}}\left[\phi\left(\boldsymbol{y}_k-\boldsymbol{C}\boldsymbol{X}_k\right)/\phi\left(\boldsymbol{y}_k\right)\mid \mathcal{F}_k^{\boldsymbol{X}} \cup \mathcal{F}_{k-1}^{\boldsymbol{y}}\right]\\=&\overline{\mathbb{E}}\left[\phi\left(\boldsymbol{y}_k-\boldsymbol{C}\boldsymbol{X}_k\right)/\phi\left(\boldsymbol{y}_k\right)\mid \boldsymbol{X}_k\right]=1\end{aligned}\tag{3-8}$$

根据文献[107]第2章的定理3.2，有

$$\mathbb{E}\left[\boldsymbol{X}_{k+1} \mid \mathcal{F}_k^{\boldsymbol{X}} \cup \mathcal{F}_k^{\boldsymbol{y}}\right]=\frac{\overline{\mathbb{E}}\left[\overline{\boldsymbol{\Lambda}}_{k+1}\boldsymbol{X}_{k+1} \mid \mathcal{F}_k^{\boldsymbol{X}} \cup \mathcal{F}_k^{\boldsymbol{y}}\right]}{\overline{\mathbb{E}}\left[\overline{\boldsymbol{\Lambda}}_{k+1} \mid \mathcal{F}_k^{\boldsymbol{X}} \cup \mathcal{F}_k^{\boldsymbol{y}}\right]}$$

然后由式(3-7)及$\boldsymbol{\Lambda}_k$是$\mathcal{F}_k^{\boldsymbol{X}} \cup \mathcal{F}_k^{\boldsymbol{y}}$-可测这一事实，可以得到

$$\begin{aligned}\mathbb{E}\left[\boldsymbol{X}_{k+1} \mid \mathcal{F}_k^{\boldsymbol{X}} \cup \mathcal{F}_k^{\boldsymbol{y}}\right]&=\frac{\overline{\mathbb{E}}\left[\overline{\boldsymbol{\lambda}}_{k+1}\boldsymbol{X}_{k+1} \mid \mathcal{F}_k^{\boldsymbol{X}} \cup \mathcal{F}_k^{\boldsymbol{y}}\right]}{\overline{\mathbb{E}}\left[\overline{\boldsymbol{\lambda}}_{k+1} \mid \mathcal{F}_k^{\boldsymbol{X}} \cup \mathcal{F}_k^{\boldsymbol{y}}\right]}\\&=\overline{\mathbb{E}}\left[\phi\left(\boldsymbol{y}_{k+1}-\boldsymbol{C}\boldsymbol{X}_{k+1}\right)/\phi\left(\boldsymbol{y}_{k+1}\right)\boldsymbol{X}_{k+1} \mid \mathcal{F}_k^{\boldsymbol{X}} \cup \mathcal{F}_k^{\boldsymbol{y}}\right]\end{aligned}$$

根据导出$\overline{\mathbb{E}}\left[\overline{\boldsymbol{\lambda}}_{k+1} \mid \mathcal{F}_k^{\boldsymbol{X}} \cup \mathcal{F}_k^{\boldsymbol{y}}\right]$过程中的类似论据，进一步可得：

$$\begin{aligned}&\mathbb{E}\left[\boldsymbol{X}_{k+1} \mid \mathcal{F}_k^{\boldsymbol{X}} \cup \mathcal{F}_k^{\boldsymbol{y}}\right]\\=&\sum_{i=1}^{N}\overline{\mathbb{E}}\left[\frac{\phi\left(\boldsymbol{y}_{k+1}-\boldsymbol{C}\boldsymbol{e}_i\right)}{\phi\left(\boldsymbol{y}_{k+1}\right)}\boldsymbol{X}_{k+1}\left\langle\boldsymbol{X}_{k+1},\ \boldsymbol{e}_i\right\rangle \mid \mathcal{F}_k^{\boldsymbol{X}} \cup \mathcal{F}_k^{\boldsymbol{y}}\right]\\=&\sum_{i=1}^{N}\overline{\mathbb{E}}\left[\boldsymbol{X}_{k+1}\left\langle\boldsymbol{X}_{k+1},\ \boldsymbol{e}_i\right\rangle \mid \boldsymbol{X}_k\right]\\=&\sum_{i=1}^{N}\boldsymbol{e}_i\overline{\mathbb{E}}\left[\left\langle\boldsymbol{X}_{k+1},\ \boldsymbol{e}_i\right\rangle \mid \boldsymbol{X}_k\right]=\boldsymbol{A}\boldsymbol{X}_k\end{aligned}$$

根据迭代期望法则（文献[108]，引理2.4.8），因为$\mathcal{F}_k^{\boldsymbol{X}} \subset \mathcal{F}_k^{\boldsymbol{X}} \cup \mathcal{F}_k^{\boldsymbol{y}}$，所以有：

$$\mathbb{E}\left[\boldsymbol{X}_{k+1} \mid \mathcal{F}_k^{\boldsymbol{X}}\right]=\mathbb{E}\left[\mathbb{E}\left[\boldsymbol{X}_{k+1} \mid \mathcal{F}_k^{\boldsymbol{X}} \cup \mathcal{F}_k^{\boldsymbol{y}}\right] \mid \mathcal{F}_k^{\boldsymbol{X}}\right]=\boldsymbol{A}\boldsymbol{X}_k\tag{3-9}$$

下面将表明在$\mathbb{P}$下，测量过程

$$\boldsymbol{y}_k=\boldsymbol{C}\boldsymbol{X}_k+\boldsymbol{v}_k\tag{3-10}$$

成立，其中，$\boldsymbol{v}_k$，$k\in\mathbb{N}$是i.i.d.且密度为ϕ的随机变量。按照类似的程序，有

$$\begin{aligned}&\Pr\left\{\boldsymbol{v}_k\leqslant\boldsymbol{t} \mid \mathcal{F}_k^{\boldsymbol{X}} \cup \mathcal{F}_{k-1}^{\boldsymbol{y}}\right\}\\=&\mathbb{E}\left[\boldsymbol{I}\left(\boldsymbol{v}_k\leqslant\boldsymbol{t}\right) \mid \mathcal{F}_k^{\boldsymbol{X}} \cup \mathcal{F}_{k-1}^{\boldsymbol{y}}\right]\end{aligned}$$

$$
\begin{aligned}
&= \frac{\overline{\mathbb{E}}\left[\overline{\boldsymbol{\Lambda}}_k \boldsymbol{I}\left(\boldsymbol{v}_k \leqslant \boldsymbol{t}\right) \mid \mathcal{F}_k^X \cup \mathcal{F}_{k-1}^y\right]}{\overline{\mathbb{E}}\left[\overline{\boldsymbol{\Lambda}}_k \mid \mathcal{F}_k^X \cup \mathcal{F}_{k-1}^y\right]} \\
&= \overline{\mathbb{E}}\left[\frac{\phi\left(\boldsymbol{y}_k - \boldsymbol{C}\boldsymbol{X}_k\right)}{\phi\left(\boldsymbol{y}_k\right)} \boldsymbol{I}\left(\boldsymbol{y}_k - \boldsymbol{C}\boldsymbol{X}_k \leqslant \boldsymbol{t}\right) \mid \mathcal{F}_k^X \cup \mathcal{F}_{k-1}^y\right] \\
&= \int_{\mathbb{R}^m} \phi\left(\boldsymbol{y}_k - \boldsymbol{C}\boldsymbol{X}_k\right) \boldsymbol{I}\left(\boldsymbol{v}_k \leqslant \boldsymbol{t}\right) \mathrm{d}\boldsymbol{y}_k \\
&= \int_{-\infty}^{t} \phi\left(\boldsymbol{v}_k\right) \mathrm{d}\boldsymbol{v}_k = \Pr\left\{\boldsymbol{v}_k \leqslant \boldsymbol{t}\right\}
\end{aligned} \tag{3-11}
$$

其中，第三个等式从式(3-8)及$\overline{\boldsymbol{\Lambda}}_{k-1}$是$\mathcal{F}_k^X \cup \mathcal{F}_{k-1}^y$-可测这一事实得来。上述结果意味着在$\overline{\mathbb{P}}$下，$\boldsymbol{v}_k := \boldsymbol{y}_k - \boldsymbol{C}\boldsymbol{X}_k$，$k \in \mathbb{N}$是一个密度函数为$\phi$的i.i.d.随机变量序列，这表明式(3-10)成立。值得提及的是，式(3-11)指出的$\boldsymbol{v}_k$和$\boldsymbol{X}_k$之间的独立性不与$\boldsymbol{v}_k$的定义起冲突，这是因为只要$\boldsymbol{y}_k$包含$\boldsymbol{C}\boldsymbol{X}_k$，$\boldsymbol{y}_k - \boldsymbol{C}\boldsymbol{X}_k$和$\boldsymbol{X}_k$独立这一情况便存在。最后，

$$
\begin{aligned}
&\Pr\left\{\gamma_k = 1 \mid \mathcal{F}_k^X \cup \mathcal{F}_k^y \cup \mathcal{F}_{k-1}^\gamma\right\} = \mathbb{E}\left[\gamma_k \mid \mathcal{F}_k^X \cup \mathcal{F}_k^y \cup \mathcal{F}_{k-1}^\gamma\right] \\
&= \frac{\overline{\mathbb{E}}\left[\overline{\boldsymbol{\Lambda}}_k \boldsymbol{\gamma}_k \mid \mathcal{F}_k^X \cup \mathcal{F}_k^y \cup \mathcal{F}_{k-1}^\gamma\right]}{\overline{\mathbb{E}}\left[\overline{\boldsymbol{\Lambda}}_k \mid \mathcal{F}_k^X \cup \mathcal{F}_k^y \cup \mathcal{F}_{k-1}^\gamma\right]} = \overline{\mathbb{E}}\left[\gamma_k \mid \mathcal{F}_k^X \cup \mathcal{F}_k^y \cup \mathcal{F}_{k-1}^\gamma\right] \\
&= \overline{\Pr}\left\{\gamma_k = 1 \mid \mathcal{F}_k^X \cup \mathcal{F}_k^y \cup \mathcal{F}_{k-1}^\gamma\right\}
\end{aligned}
$$

其中，第三个等式是由于$\boldsymbol{\Lambda}_k$是$\mathcal{F}_k^X \cup \mathcal{F}_k^y \cup \mathcal{F}_{k-1}^\gamma$-可测这一事实。这意味着

$$
\begin{aligned}
&\Pr\left\{\gamma_k = 0 \mid \mathcal{F}_k^X \cup \mathcal{F}_k^y \cup \mathcal{F}_{k-1}^\gamma\right\} \\
&= \overline{\Pr}\left\{\gamma_k = 0 \mid \mathcal{F}_k^X \cup \mathcal{F}_k^y \cup \mathcal{F}_{k-1}^\gamma\right\} = \rho_k\left(\boldsymbol{y}_k\right)
\end{aligned} \tag{3-12}
$$

证毕。 ■

因此，根据条件的 Bayes 定理 (文献 [107]，第 2 章定理 3.2)，有

$$
\mathbb{E}\left[\exp\left(\theta \boldsymbol{\Psi}_{0,k}(\boldsymbol{\zeta})\right) \mid \mathcal{I}_k\right] = \frac{\overline{\mathbb{E}}\left[\overline{\boldsymbol{\Lambda}}_k \exp\left(\theta \boldsymbol{\Psi}_{0,k}(\boldsymbol{\zeta})\right) \mid \mathcal{I}_k\right]}{\overline{\mathbb{E}}\left[\overline{\boldsymbol{\Lambda}}_k \mid \mathcal{I}_k\right]} \tag{3-13}
$$

这是因为$\mathcal{I}_k \subset \mathcal{G}_k$。上面的等式意味着能够在新测度$\overline{\mathbb{P}}$下求解$\hat{\boldsymbol{X}}$。注意，上面等式右侧的分母不包含变量$\boldsymbol{\zeta}$。所以，原始估计问题(3-3)等价于

$$
\hat{\boldsymbol{X}}_k = \arg\min_{\boldsymbol{\zeta} \in S_X} \overline{\mathbb{E}}\left[\overline{\boldsymbol{\Lambda}}_k \exp\left(\theta \boldsymbol{\Psi}_{0,\ k}(\boldsymbol{\zeta})\right) \mid \mathcal{I}_k\right],\quad k \in \mathbb{N} \tag{3-14}
$$

3.3.2 递归估计

首先，需要定义一个非标准化的信息状态以帮助求解$\hat{\boldsymbol{X}}$。下面将表明，信息状态具有线性递归形式。它与文献 [76] 中的前向变量具有相同的作用，该变量也能够通过递归计算得出并可进一步用于求基于 HMM 的估计值。

定义 3.1

定义$\boldsymbol{\alpha}_k=\left[\boldsymbol{\alpha}_k\left(\boldsymbol{e}_1\right),\cdots,\boldsymbol{\alpha}_k\left(\boldsymbol{e}_N\right)\right]^{\mathrm{T}}$为一个非标准化的信息状态，使得对于$r\in\mathbb{N}_{1:N}$有

$$\boldsymbol{\alpha}_k\left(\boldsymbol{e}_r\right):=\overline{\mathbb{E}}\left[\overline{\boldsymbol{\Lambda}}_k\exp\left(\theta\hat{\boldsymbol{\Psi}}_{0,k-1}\right)\left\langle\boldsymbol{X}_k,\boldsymbol{e}_r\right\rangle\mid\mathcal{I}_k\right],\ k\in\mathbb{N} \tag{3-15}$$

注意，$\boldsymbol{\alpha}_k$不仅包括隐状态$\boldsymbol{X}_k$，还包括部分风险敏感代价（如文献 [97-98]）。我们也许可以用文献 [76] 中的方式来求解估计问题 (3-3)，也就是通过在原始测度$\mathbb{P}$构造前向递归变量帮助求解$\hat{\boldsymbol{X}}_k$。然而，式 (3-3) 的代价函数是一种指数累积误差的形式，并且期望是针对$\boldsymbol{X}_0,\cdots,\boldsymbol{X}_k$，而不仅仅是$\boldsymbol{X}_k$。因此，很难找到这样一个前向变量并导出它在$\mathbb{P}$下的递归表达式。这里，采用参考测度，并导出如下信息状态的递归表达式。

定理 3.1

对于具有事件触发方案 (3-2) 的 HMM［式 (2-22) 和式 (2-23)］，信息状态$\boldsymbol{\alpha}_k$有如下线性递归：

$$\boldsymbol{\alpha}_k=\operatorname{diag}\left\{b_k\right\}\boldsymbol{A}\operatorname{diag}\left\{d_{k-1}\right\}\boldsymbol{\alpha}_{k-1} \tag{3-16}$$

其中，$d_{k-1}=\left[\exp\left(\theta\mu\left(\boldsymbol{e}_i,\hat{\boldsymbol{X}}_{k-1}\right)\right)\right]_{i\in\mathbb{N}_{1:N}}$，且如果$\gamma_k=0$，

$$b_k=1/\overline{\Pr}\left\{\gamma_k=0\mid\mathcal{I}_{k-1}\right\}\left[\int_{\mathbb{R}^m}\phi\left(\boldsymbol{y}_k-\boldsymbol{c}_i\right)\rho_k\left(\boldsymbol{y}_k\right)\mathrm{d}\boldsymbol{y}_k\right]_{i\in\mathbb{N}_{1:N}} \tag{3-17}$$

否则，

$$b_k=\left[\phi\left(\boldsymbol{y}_k-\boldsymbol{c}_i\right)/\phi\left(\boldsymbol{y}_k\right)\right]_{i\in\mathbb{N}_{1:N}} \tag{3-18}$$

证明：

根据$\boldsymbol{\alpha}_k\left(\boldsymbol{e}_r\right)$的定义，有

$$\boldsymbol{\alpha}_k(\boldsymbol{e}_r)=\overline{\mathbb{E}}[\overline{\boldsymbol{\Lambda}}_k\exp(\theta\hat{\boldsymbol{\Psi}}_{0,k-1})\langle\boldsymbol{X}_k,\ \boldsymbol{e}_r\rangle|\mathcal{I}_k]$$
$$=\overline{\mathbb{E}}\left[\phi(\boldsymbol{y}_k-\boldsymbol{c}_r)/\phi(\boldsymbol{y}_k)\exp\left(\theta\mu\left(\boldsymbol{X}_{k-1},\ \hat{\boldsymbol{X}}_{k-1}\right)\right)\overline{\boldsymbol{\Lambda}}_{k-1}\exp\left(\theta\hat{\boldsymbol{\Psi}}_{0,\ k-2}\right)\langle\boldsymbol{X}_k,\ \boldsymbol{e}_r\rangle|\ \mathcal{I}_k\right]$$
$$=\overline{\mathbb{E}}\left[\overline{\mathbb{E}}\left[\phi(\boldsymbol{y}_k-\boldsymbol{c}_r)/\phi(\boldsymbol{y}_k)\exp\left(\theta\mu\left(\boldsymbol{X}_{k-1},\ \hat{\boldsymbol{X}}_{k-1}\right)\right)\overline{\boldsymbol{\Lambda}}_{k-1}\right.\right.$$
$$\left.\left.\times\exp\left(\theta\hat{\boldsymbol{\Psi}}_{0,\ k-2}\right)\langle\boldsymbol{X}_k,\ \boldsymbol{e}_r\rangle|\ \mathcal{F}_{k-1}^{\boldsymbol{X}}\cup\mathcal{F}_k^{\boldsymbol{y}}\cup\mathcal{F}_k^{\gamma}\right]|\ \mathcal{I}_k\right]$$
$$=\overline{\mathbb{E}}\left[\phi(\boldsymbol{y}_k-\boldsymbol{c}_r)/\phi(\boldsymbol{y}_k)\exp\left(\theta\mu\left(\boldsymbol{X}_{k-1},\ \hat{\boldsymbol{X}}_{k-1}\right)\right)\overline{\boldsymbol{\Lambda}}_{k-1}\right.$$
$$\left.\times\exp\left(\theta\hat{\boldsymbol{\Psi}}_{0,\ k-2}\right)\langle\boldsymbol{A}\boldsymbol{X}_{k-1},\ \boldsymbol{e}_r\rangle|\ \mathcal{I}_k\right]$$

其中，第三个等式是由于$\mathcal{I}_k\subset\mathcal{F}_{k-1}^{\boldsymbol{X}}\cup\mathcal{F}_k^{\boldsymbol{y}}\cup\mathcal{F}_k^{\gamma}$和迭代期望法则（文献[108]，引理2.4.8）。利用$\langle\boldsymbol{A}\boldsymbol{X}_{k-1},\boldsymbol{e}_r\rangle=\sum_{i=1}^{N}a_{ri}\langle\boldsymbol{X}_{k-1},\boldsymbol{e}_i\rangle$，

$$\boldsymbol{\alpha}_k(\boldsymbol{e}_r)$$
$$=\sum_{i=1}^{N}a_{ri}\overline{\mathbb{E}}\left[\phi(\boldsymbol{y}_k-\boldsymbol{c}_r)/\phi(\boldsymbol{y}_k)\exp\left(\theta\mu\left(\boldsymbol{e}_i,\ \hat{\boldsymbol{X}}_{k-1}\right)\right)\overline{\boldsymbol{\Lambda}}_{k-1}\exp\left(\theta\hat{\boldsymbol{\Psi}}_{0,\ k-2}\right)\langle\boldsymbol{X}_{k-1},\ \boldsymbol{e}_i\rangle|\ \mathcal{I}_k\right]$$
$$=\overline{\mathbb{E}}\left[\phi(\boldsymbol{y}_k-\boldsymbol{c}_r)/\phi(\boldsymbol{y}_k)|\ \mathcal{I}_k\right]\sum_{i=1}^{N}a_{ri}\exp\left(\theta\mu\left(\boldsymbol{e}_i,\ \hat{\boldsymbol{X}}_{k-1}\right)\right)$$
$$\times\overline{\mathbb{E}}\left[\overline{\boldsymbol{\Lambda}}_{k-1}\exp\left(\theta\hat{\boldsymbol{\Psi}}_{0,\ k-2}\right)\langle\boldsymbol{X}_{k-1},\ \boldsymbol{e}_i\rangle|\ \mathcal{I}_{k-1}\right]$$
$$=\overline{\mathbb{E}}\left[\frac{\phi(\boldsymbol{y}_k-\boldsymbol{c}_r)}{\phi(\boldsymbol{y}_k)}|\ \mathcal{I}_k\right]\sum_{i=1}^{N}a_{ri}\exp\left(\theta\mu\left(\boldsymbol{e}_i,\ \hat{\boldsymbol{X}}_{k-1}\right)\right)\boldsymbol{\alpha}_{k-1}(\boldsymbol{e}_i)$$

其中，第二个等式是因为$\hat{\boldsymbol{X}}_{k-1}$为$\mathcal{I}_{k-1}$-可测，$\{\boldsymbol{y}_k\}$是一个i.i.d.随机变量序列，且在$\overline{\mathbb{P}}$下，$\{\boldsymbol{y}_k\}$和$\{\gamma_k\}$独立于$\{\boldsymbol{X}_k\}$。现在，计算$\boldsymbol{\alpha}_k(\boldsymbol{e}_r)$的关键点在于获得上面最后一个等式中$\overline{\mathbb{E}}\left[\phi(\boldsymbol{y}_k-\boldsymbol{c}_r)/\phi(\boldsymbol{y}_k)|\ \mathcal{I}_k\right]$的具体表达式，这和事件触发方案有关。为此，分别考虑$\gamma_k=0$和$\gamma_k=1$两种情况，并记$\overline{f}(\cdot)$为$\overline{\mathbb{P}}$下的概率密度函数。

① $\gamma_k=0$时：

$$\overline{\mathbb{E}}\left[\phi(\boldsymbol{y}_k-\boldsymbol{c}_r)/\phi(\boldsymbol{y}_k)|\ \mathcal{I}_k\right]=\overline{\mathbb{E}}\left[\phi(\boldsymbol{y}_k-\boldsymbol{c}_r)/\phi(\boldsymbol{y}_k)|\gamma_k=0,\ \mathcal{I}_{k-1}\right]$$
$$=\int_{\mathbb{R}^m}\phi(\boldsymbol{y}_k-\boldsymbol{c}_r)/\phi(\boldsymbol{y}_k)\overline{f}(\boldsymbol{y}_k|\ \gamma_k=0,\ \mathcal{I}_{k-1})\mathrm{d}\boldsymbol{y}_k$$
$$=\int_{\mathbb{R}^m}\frac{\phi(\boldsymbol{y}_k-\boldsymbol{c}_r)}{\phi(\boldsymbol{y}_k)}\frac{\overline{\Pr}\{\gamma_k=0|\ \boldsymbol{y}_k,\ \mathcal{I}_{k-1}\}\overline{f}(\boldsymbol{y}_k|\mathcal{I}_{k-1})}{\overline{\Pr}\{\gamma_k=0|\ \mathcal{I}_{k-1}\}}\mathrm{d}\boldsymbol{y}_k \tag{3-19}$$
$$=1/\overline{\Pr}\{\gamma_k=0|\ \mathcal{I}_{k-1}\}\int_{\mathbb{R}^m}\phi(\boldsymbol{y}_k-\boldsymbol{c}_r)\rho_k(\boldsymbol{y}_k)\mathrm{d}\boldsymbol{y}_k$$

② $\gamma_k = 1$时：

$$
\begin{aligned}
&\bar{\mathbb{E}}\left[\phi\left(\boldsymbol{y}_k - \boldsymbol{c}_r\right) / \phi\left(\boldsymbol{y}_k\right) | \mathcal{I}_k\right] \\
&= \bar{\mathbb{E}}\left[\frac{\phi\left(\boldsymbol{y}_k - \boldsymbol{c}_r\right)}{\phi\left(\boldsymbol{y}_k\right)} | \boldsymbol{y}_k\right] = \phi\left(\boldsymbol{y}_k - \boldsymbol{c}_r\right) / \phi\left(\boldsymbol{y}_k\right)
\end{aligned} \tag{3-20}
$$

那么在矩阵概念下，便可以完成递归表达式(3-16)的证明。证毕。 ■

注释 3.1 假设$\boldsymbol{\pi}_0 \in \mathbb{R}^N$是初始状态$\boldsymbol{X}_0$的概率分布，那么$\boldsymbol{\alpha}_0 = \mathrm{diag}\{b_0\}\boldsymbol{\pi}_0$，其中$b_0$在$\gamma_0 = 0$时对应于式(3-17)，$\gamma_0 = 1$时对应于式(3-18)。该证明可以通过与定理 3.1 中类似的论据来完成。

下面表明，基于$\boldsymbol{\alpha}_k$可以得到估计值$\hat{\boldsymbol{X}}$。

定理 3.2

事件触发 RMAP 估计问题 (3-3) 可求解如下：

$$
\hat{\boldsymbol{X}}_k = \boldsymbol{e}_{i^*}, \quad i^* = \arg\max_{i \in \mathbb{N}_{1:N}} \boldsymbol{\alpha}_k\left(\boldsymbol{e}_i\right), \quad k \in \mathbb{N} \tag{3-21}
$$

证明：

$$
\begin{aligned}
&\bar{\mathbb{E}}\left[\bar{\boldsymbol{\Lambda}}_k \exp\left(\theta \boldsymbol{\Psi}_{0,k}(\boldsymbol{\zeta})\right) | \mathcal{I}_k\right] \\
&= \bar{\mathbb{E}}\left[\exp\left(\theta \mu\left(\boldsymbol{X}_k, \boldsymbol{\zeta}\right)\right) \bar{\boldsymbol{\Lambda}}_k \exp\left(\theta \hat{\boldsymbol{\Psi}}_{0,k-1}\right) | \mathcal{I}_k\right] \\
&= \sum_{j=1}^N \bar{\mathbb{E}}\left[\exp\left(\theta \mu\left(\boldsymbol{e}_j, \boldsymbol{\zeta}\right)\right) \bar{\boldsymbol{\Lambda}}_k \exp\left(\theta \hat{\boldsymbol{\Psi}}_{0,k-1}\right) \left\langle \boldsymbol{X}_k, \boldsymbol{e}_j \right\rangle | \mathcal{I}_k\right] \\
&= \sum_{j=1}^N \exp\left(\theta \mu\left(\boldsymbol{e}_j, \boldsymbol{\zeta}\right)\right) \boldsymbol{\alpha}_k\left(\boldsymbol{e}_j\right)
\end{aligned} \tag{3-22}
$$

令$\boldsymbol{\zeta} = \boldsymbol{e}_i$，$i \in \mathbb{N}_{1:N}$，然后由$\mu$的定义可知，

$$
\begin{aligned}
&\bar{\mathbb{E}}\left[\bar{\boldsymbol{\Lambda}}_k \exp\left(\theta \boldsymbol{\Psi}_{0,k}(\boldsymbol{\zeta})\right) | \mathcal{I}_k\right] \\
&= \exp(\theta) \sum_{j \in \mathbb{N}_{1:N} \setminus i} \boldsymbol{\alpha}_k\left(\boldsymbol{e}_j\right) + \boldsymbol{\alpha}_k\left(\boldsymbol{e}_i\right) \\
&= \exp(\theta) \sum_{j=1}^N \boldsymbol{\alpha}_k\left(\boldsymbol{e}_j\right) + \left(1 - \exp(\theta)\right) \boldsymbol{\alpha}_k\left(\boldsymbol{e}_i\right)
\end{aligned} \tag{3-23}
$$

因为$\theta > 0$，利用式(3-14)可完成该证明。证毕。 ■

定理 3.2 表明事件触发 RMAP 估计值$\hat{\boldsymbol{X}}_k$与通常的 MAP 定义一致：MAP 估计值是在给定测量下出现概率最大的值[109]。

注释 3.2 从定理 3.1 和 3.2 可知，随着 $\theta \to 0$，由式 (3-21) 得到的$\hat{\boldsymbol{X}}_k$降为风险中性情况下的事件触发估计值。该估计值是文献 [92-93] 在可靠信道场景的连续范围测量值版本。注意式 (2-23) 的连续范围测量值模型比文献 [92-93] 的有限状态测量值模型更一般。另一方面，如果传输概率$\rho_k(\boldsymbol{y}_k)$是一个固定值ρ，那么不能获得任何关于$\boldsymbol{y}_k$的信息。此时，由定理 3.1 和 3.2 导出的相应的估计结果与丢包或者离线调度时的情况等价。后面将利用这些估计结果［$\theta \to 0$ 和$\rho_k(\boldsymbol{y}_k) = \rho$两种情况］进行仿真比较，来举例说明所提出事件触发估计器的优势。

3.3.3 测量噪声为高斯时的解

从定理 3.1 和定理 3.2 可知，求解$\hat{\boldsymbol{X}}$的主要难点在于式 (3-17) 中$\int_{\mathbb{R}^m} \phi(\boldsymbol{y}_k - \boldsymbol{c}_i)\rho_k(\boldsymbol{y}_k)\mathrm{d}\boldsymbol{y}_k$的计算，这是由于函数$\phi(\cdot)$和$\rho(\cdot)$的一般性导致$\int_{\mathbb{R}^m} \phi(\boldsymbol{y}_k - \boldsymbol{c}_i)\rho_k(\boldsymbol{y}_k)\mathrm{d}\boldsymbol{y}_k$缺乏明确表达式以及存在数值积分。本小节将在测量噪声 $\boldsymbol{v}_k$ 是高斯时，推导出 SET 条件下的封闭解和一类 DET 条件下的明确解。具体来说，假定 $\boldsymbol{v}_k$ 是零均值高斯，具有正定协方差矩阵 $\boldsymbol{R}$。

首先，考虑 SET 条件情况，该条件由文献 [104] 首次提出。下面，考虑如下一般性的 SET 条件：

$$\gamma_k = \begin{cases} 0, & \text{如果}\, \boldsymbol{\tau}_k \leqslant \exp\left(-\dfrac{1}{2} \| \boldsymbol{y}_k - \boldsymbol{\beta}_k \|_{\boldsymbol{Y}}^2\right) \\ 1, & \text{其他} \end{cases} \tag{3-24}$$

其中，$\boldsymbol{\tau}_k$是均匀分布在[0,1]上的i.i.d.随机变量；$\boldsymbol{Y} \in \mathbb{R}^{m\times m}$是一个正定权重矩阵；$\boldsymbol{\beta}_k \in \mathbb{R}^m$是一个估计器基于可利用信息集合$\mathcal{I}_{k-1}$可以得知的量，$\boldsymbol{\beta}_k$不需要传输至估计器。注意到如果$\boldsymbol{\beta}_k$是上一被发送的测量值，条件(3-24)降为“send-on-delta”触发策略[110]的随机版本。

定理 3.3

对于一般性 SET 条件 (3-24)，事件触发 RMAP 估计问题 (3-3) 的解有如下封闭形式：

$$\hat{\boldsymbol{X}}_k=\boldsymbol{e}_{i^*},\quad i^*=\arg\max_{i\in\mathbb{N}_{1:N}}\tilde{\boldsymbol{\alpha}}_k\left(\boldsymbol{e}_i\right),\quad k\in\mathbb{N} \tag{3-25}$$

其中，$\tilde{\boldsymbol{\alpha}}_k\left(\boldsymbol{e}_i\right)$是$\tilde{\boldsymbol{\alpha}}_k:=\left[\tilde{\boldsymbol{\alpha}}_k\left(\boldsymbol{e}_i\right)\right]_{i\in\mathbb{N}_{1:N}}$的第$i$个元素，且$\tilde{\boldsymbol{\alpha}}_k$由如下递归给出

$$\tilde{\boldsymbol{\alpha}}_k=\operatorname{diag}\left\{\tilde{b}_k\right\}\boldsymbol{A}\operatorname{diag}\left\{d_{k-1}\right\}\tilde{\boldsymbol{\alpha}}_{k-1},\quad \tilde{\boldsymbol{\alpha}}_0=\operatorname{diag}\left\{\tilde{b}_0\right\}\boldsymbol{\pi}_0 \tag{3-26}$$

其中，$d_{k-1}=\left[\exp\left(\theta\mu\left(\boldsymbol{e}_i,\ \hat{\boldsymbol{X}}_{k-1}\right)\right)\right]_{i\in\mathbb{N}_{1:N}}$，且如果$\gamma_k=0$则

$$\tilde{b}_k=\left[\exp\left(-\frac{1}{2}\|\boldsymbol{c}_i-\boldsymbol{\beta}_k\|^2_{\left(\boldsymbol{R}+\boldsymbol{Y}^{-1}\right)^{-1}}\right)\right]_{i\in\mathbb{N}_{1:N}} \tag{3-27}$$

否则，

$$\tilde{b}_k=\left[\exp\left(-\frac{1}{2}\|\boldsymbol{y}_k-\boldsymbol{c}_i\|^2_{\boldsymbol{R}^{-1}}\right)\right]_{i\in\mathbb{N}_{1:N}} \tag{3-28}$$

证明：

从式 (3-4) 的第二个等式和式 (3-24) 可知，

$$\begin{aligned}\rho_k\left(\boldsymbol{y}_k\right)&=\overline{\Pr}\left\{\gamma_k=0|\ \boldsymbol{y}_k,\ \mathcal{I}_{k-1}\right\}\\&=\overline{\Pr}\left\{\exp\left(-\frac{1}{2}\|\boldsymbol{y}_k-\boldsymbol{\beta}_k\|^2_{\boldsymbol{Y}}\right)\geqslant\boldsymbol{\tau}_k|\ \boldsymbol{y}_k,\ \mathcal{I}_{k-1}\right\}\\&=\exp\left(-\frac{1}{2}\|\boldsymbol{y}_k-\boldsymbol{\beta}_k\|^2_{\boldsymbol{Y}}\right)\end{aligned}$$

然后考虑$\gamma_k=0$的情况，得到

$$\begin{aligned}&\int_{\mathbb{R}^m}\phi\left(\boldsymbol{y}_k-\boldsymbol{c}_i\right)\rho_k\left(\boldsymbol{y}_k\right)\mathrm{d}\boldsymbol{y}_k\\&=\frac{1}{\sqrt{(2\pi)^m|\boldsymbol{R}|}}\int_{\mathbb{R}^m}\exp\left(-\frac{1}{2}\left(\|\boldsymbol{y}_k-\boldsymbol{c}_i\|^2_{\boldsymbol{R}^{-1}}+\|\boldsymbol{y}_k-\boldsymbol{\beta}_k\|^2_{\boldsymbol{Y}}\right)\right)\mathrm{d}\boldsymbol{y}_k\\&=\boldsymbol{\eta}_k\exp\left(-\frac{1}{2}\|\boldsymbol{c}_i-\boldsymbol{\beta}_k\|^2_{\left(\boldsymbol{R}+\boldsymbol{Y}^{-1}\right)^{-1}}\right),\quad i\in\mathbb{N}_{1:N}\end{aligned} \tag{3-29}$$

其中，$\boldsymbol{\eta}_k$是与$\boldsymbol{c}_i$和$\boldsymbol{\beta}_k$无关的量。从定理3.2可知，$\boldsymbol{\alpha}_k$的缩放不改变所产生的估计值$\hat{\boldsymbol{X}}_k$。因此，在求解$\hat{\boldsymbol{X}}_k$时，可以忽略式(3-17)中的$\overline{\Pr}\left\{\gamma_k=0|\ \mathcal{I}_{k-1}\right\}$、式(3-18)中的$\phi(\boldsymbol{y}_k)$以及式(3-29)中的$\boldsymbol{\eta}_k$。所以，得到结果式(3-27)和式(3-28)。式(3-25)的结果和$\tilde{\boldsymbol{\alpha}}_k$的递归分别与定理3.2和定理3.1一样。证毕。 ■

下面，考虑如下一类 DET 条件：

$$\gamma_k = \begin{cases} 0, & \text{如果} -\boldsymbol{\delta} \leqslant \boldsymbol{y}_k - \boldsymbol{\xi}_k \leqslant \boldsymbol{\delta} \\ 1, & \text{其他} \end{cases} \tag{3-30}$$

其中，$\boldsymbol{\delta} = [\delta_1, \cdots, \delta_m] \geqslant 0$是个设定门限；$\boldsymbol{\xi}_k \in \mathbb{R}^m$与$\boldsymbol{\beta}_k$一样，被估计器所知。令$\boldsymbol{\xi}_k = \boldsymbol{CA}\hat{\boldsymbol{X}}_{k-1}$且$\delta_1 = \cdots = \delta_m$，那么式(3-30)降为基于新息无穷范数的触发条件[102]。定义集合$\Xi_k \in \mathbb{R}^m$为

$$\Xi_k := \left\{ \boldsymbol{y}_k \in \mathbb{R}^m : -\boldsymbol{\delta} \leqslant \boldsymbol{y}_k - \boldsymbol{\xi}_k \leqslant \boldsymbol{\delta} \right\}$$

则有如下结果：

定理 3.4

对于式 (3-30) 这一类 DET 条件，事件触发 RMAP 估计问题 (3-3) 的解为：

$$\hat{\boldsymbol{X}}_k = \boldsymbol{e}_{i^*}, \quad i^* = \arg\max_{i \in \mathbb{N}_{1:N}} \tilde{\boldsymbol{\alpha}}_k(\boldsymbol{e}_i), \quad k \in \mathbb{N}, \tag{3-31}$$

其中，$\tilde{\boldsymbol{\alpha}}_k(\boldsymbol{e}_i)$是$\tilde{\boldsymbol{\alpha}}_k := \left[\tilde{\boldsymbol{\alpha}}_k(\boldsymbol{e}_i)\right]_{i \in \mathbb{N}_{1:N}}$的第$i$个元素，且$\tilde{\boldsymbol{\alpha}}_k$由如下公式递归地给出

$$\tilde{\boldsymbol{\alpha}}_k = \mathrm{diag}\left\{\tilde{b}_k\right\} \boldsymbol{A} \mathrm{diag}\left\{d_{k-1}\right\} \tilde{\boldsymbol{\alpha}}_{k-1}, \quad \tilde{\boldsymbol{\alpha}}_0 = \mathrm{diag}\left\{\tilde{b}_0\right\} \boldsymbol{\pi}_0 \tag{3-32}$$

其中，$d_{k-1} = \left[\exp\left(\theta\mu\left(\boldsymbol{e}_i, \hat{\boldsymbol{X}}_{k-1}\right)\right)\right]_{i \in \mathbb{N}_{1:N}}$，且如果$\gamma_k = 0$，则

$$\tilde{b}_k = \left[\frac{1}{\sqrt{(2\pi)^m |\boldsymbol{R}|}} \int_{\xi_k - \delta - c_i}^{\xi_k + \delta - c_i} \exp\left(-\frac{1}{2} \| \boldsymbol{z} \|_{\boldsymbol{R}^{-1}}^2\right) \mathrm{d}\boldsymbol{z}\right]_{i \in \mathbb{N}_{1:N}} \tag{3-33}$$

否则，

$$\tilde{b}_k = \left[\exp\left(-\frac{1}{2} \| \boldsymbol{y}_k - \boldsymbol{c}_i \|_{\boldsymbol{R}^{-1}}^2\right)\right]_{i \in \mathbb{N}_{1:N}} \tag{3-34}$$

证明：

从式 (3-4) 中可以推出

$$\rho_k(\boldsymbol{y}_k) = \overline{\Pr}\left\{\gamma_k = 0 | \boldsymbol{y}_k, \mathcal{I}_{k-1}\right\} = \begin{cases} 1, & \text{如果} \boldsymbol{y}_k \in \Xi_k \\ 0, & \text{其他} \end{cases}$$

因此，考虑$\gamma_k = 0$的情况，得到

$$
\begin{aligned}
&\int_{\mathbb{R}^m} \phi(\boldsymbol{y}_k - \boldsymbol{c}_i)\rho_k(\boldsymbol{y}_k)\mathrm{d}\boldsymbol{y}_k \\
&= \frac{1}{\sqrt{(2\pi)^m |\boldsymbol{R}|}} \int_{\Xi_k} \exp\left(-\frac{1}{2}\| \boldsymbol{y}_k - \boldsymbol{c}_i \|^2_{\boldsymbol{R}^{-1}}\right)\mathrm{d}\boldsymbol{y}_k \\
&= \frac{1}{\sqrt{(2\pi)^m |\boldsymbol{R}|}} \int_{\boldsymbol{\xi}_k - \boldsymbol{\delta} - \boldsymbol{c}_i}^{\boldsymbol{\xi}_k + \boldsymbol{\delta} - \boldsymbol{c}_i} \exp\left(-\frac{1}{2}\| \boldsymbol{z} \|^2_{\boldsymbol{R}^{-1}}\right)\mathrm{d}\boldsymbol{z}
\end{aligned}
$$

其余的证明可以通过遵循与定理3.3证明相似的过程来完成。证毕。

■

在定理 3.4 中，如果$\gamma_k = 0$，则$\tilde{\boldsymbol{\alpha}}_k$的计算主要需要计算 m 维超矩形的多元正态累积概率，这很容易在数值上实现。

注释 3.3

我们也可以考虑其他类别的 DET 条件来获得与定理 3.4 中相似的结果。例如，可以使用$\|\boldsymbol{y}_k - \boldsymbol{\xi}_k\|_1 \leqslant \sigma$或$\|\boldsymbol{y}_k - \boldsymbol{\xi}_k\|^2 \leqslant \sigma$，$\sigma > 0$代替式 (3-30) 中$-\boldsymbol{\delta} \leqslant \boldsymbol{y}_k - \boldsymbol{\xi}_k \leqslant \boldsymbol{\delta}$，此时，需要为所考虑区域内的多元正态累积概率开发有效的数值计算方法。

注释 3.4

为了分析事件触发条件下估计性能和参数之间的关系，可以分别定义$\mathbb{E}\left[\left\langle \boldsymbol{X}_k, \hat{\boldsymbol{X}}_k \right\rangle\right]$和$\mathbb{E}\left[\left\langle \boldsymbol{X}_k, \hat{\boldsymbol{X}}_k \right\rangle | \mathcal{I}_k\right]$作为无条件和有条件的估计性能，表示 k 时刻的预期估计正确率。但是，如文献 [104] 中所示，很难获得无条件估计性能的解析解，因为涉及对触发序列 $\{\gamma_k\}$ 的数学期望操作。

对于条件估计性能，可以得到以下结果：

$$
\begin{aligned}
&\mathbb{E}\left[\left\langle \boldsymbol{X}_k,\ \hat{\boldsymbol{X}}_k \right\rangle | \mathcal{I}_k\right] = \mathbb{E}\left[\left\langle \boldsymbol{X}_k,\ \boldsymbol{e}_{i^*} \right\rangle | \mathcal{I}_k\right] \\
&= \frac{\bar{\mathbb{E}}\left[\bar{\boldsymbol{\Lambda}}_k \left\langle \boldsymbol{X}_k,\ \boldsymbol{e}_{i^*} \right\rangle | \mathcal{I}_k\right]}{\bar{\mathbb{E}}\left[\bar{\boldsymbol{\Lambda}}_k | \mathcal{I}_k\right]} = \frac{\bar{\mathbb{E}}\left[\bar{\boldsymbol{\Lambda}}_k \left\langle \boldsymbol{X}_k,\ \boldsymbol{e}_{i^*} \right\rangle | \mathcal{I}_k\right]}{\sum_{j=1}^{N} \bar{\mathbb{E}}\left[\bar{\boldsymbol{\Lambda}}_k \left\langle \boldsymbol{X}_k,\ \boldsymbol{e}_j \right\rangle | \mathcal{I}_k\right]} \\
&= \lim_{\theta \to 0} \tilde{\boldsymbol{\alpha}}_k\left(\boldsymbol{e}_{i^*}\right) / \sum_{j=1}^{N} \lim_{\theta \to 0} \tilde{\boldsymbol{\alpha}}_k\left(\boldsymbol{e}_j\right)
\end{aligned}
\tag{3-35}
$$

其中，$i^* = \arg\max_{i \in \mathbb{N}_{1:N}} \tilde{\boldsymbol{\alpha}}_k(\boldsymbol{e}_i)$是由式 (3-25) 或式 (3-31) 给出。值得注意的是，结果 (3-35) 是针对潜在的不确定系统，而不仅仅是名义系统。

3.4 仿真示例

下面，将用一个带有测量高斯噪声的建模不准确的 HMM 来举例说明所提出的事件触发 RMAP 估计器。考虑一个 $N=5$ 的 Markov 链 $\{\boldsymbol{X}_k\}$，表示状态为 $\{0,1,2,3,4\}$ 的标量实值过程 $\{\boldsymbol{x}_k\}$, 其用单位向量 $\{\boldsymbol{e}_1,\boldsymbol{e}_2,\boldsymbol{e}_3,\boldsymbol{e}_4,\boldsymbol{e}_5\}$ 来表示。构造转移概率矩阵 $\boldsymbol{A}$ 中的元素为

$$a_{ij} = C_{N-i}^{j-i}(1-p)^{N-j}p^{j-i}，0 \leqslant p \leqslant 1 \tag{3-36}$$

其中，如果$0 \leqslant n \leqslant l$，则$C_l^n$表示为$l$个对象中的$n$个对象$(l, n \in \mathbb{N})$的组合数，否则$C_l^n = 0$。令初始概率分布为$\boldsymbol{\pi}_0 = [\boldsymbol{a}_{i1}]_{i \in \mathbb{N}_{1:N}}$。此模型可以表示故障检测场景，其中过程$\boldsymbol{x}_k$表示故障设施的数量，此时$p$是每个设备发生故障的概率。有关式(3-36)的详细推导，请参见文献[111]。传感器的测量过程是

$$\boldsymbol{y}_k = [0.30 \ \ 0.33 \ \ 0.50 \ \ 0.85 \ \ 0.84]\boldsymbol{X}_k + \boldsymbol{v}_k \tag{3-37}$$

其中，$\boldsymbol{v}_k$是零均值高斯，且协方差为$R = 0.01$。

首先，为了显示 RMAP 估计器的鲁棒性，考虑所有测量值都可被估计器使用的场景。考虑如下情况：p 的建模值为$p_{\text{nominal}} = 0.01$，而实际值为$p_{\text{true}} \in [0.0001, 0.1]$。选择风险敏感参数$\theta = 0.04$。仿真时间范围为 100，并且执行一个 10000 次运行的 Monte Carlo 仿真。图 3-2 呈现了与 MAP 估算器相比，RMAP 估算器的估计性能❶。如预期的那样，RMAP 估计器在不确定性域内具有可接受的平均估计正确率❷，而 MAP 估计器存在较大的误差。

❶ 在图3-2中，似乎RMAP和MAP的最大平均精度为$p_{\text{true}}=p_{\text{nominal}}=0.01$，但是通过观察实际数据，这不是真实的。直观地，平均估计正确率确实不必要地在$p_{\text{true}}=p_{\text{nominal}}$处最大。

❷ 图3-2和图3-3中的平均估计正确率是通过计算在10000次仿真下的正确估计值的平均百分比来计算的。

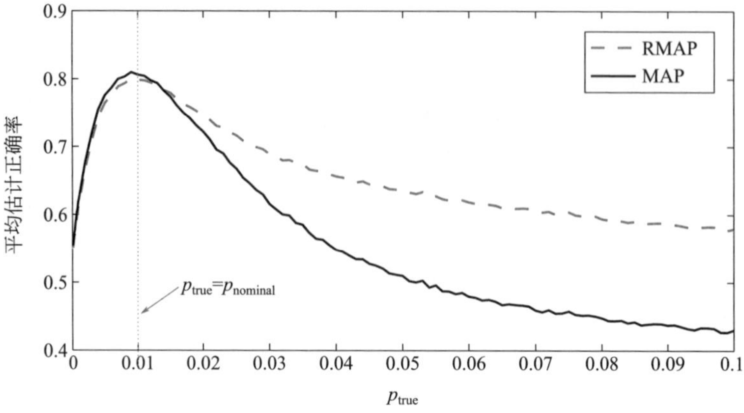

图 3-2 RMAP 和 MAP 估算器之间的性能比较

现在，分别展示定理 3.3 和 3.4 所提出的随机性事件触发 RMAP（Stochastic Event-Triggered RMAP，SET-RMAP）和确定性事件触发 RMAP（Deterministic Event-Triggered MAP, DET-RMAP）估计器的优点。此时，令 $p_{true}=0.03$，其他模型参数与上述相同。对于所考虑的 SET 条件 (3-24) 和 DET 条件 (3-30)，考虑"send-on-delta"触发策略[110]，即式 (3-24) 的 $\boldsymbol{\beta}_k$ 和式 (3-30) 的 $\boldsymbol{\xi}_k$ 是上一被传输的测量值。为了进行性能比较，同时考虑随机离线 RMAP（Random Offline RMAP, RO-RMAP）、随机性事件触发 MAP（Stochastic Event-Triggered MAP, SET-MAP）及确定性的事件触发 MAP（Deterministic Event-Triggered MAP, DET-MAP）估计器，这些估计器可以基于注释 3.2 的分析得到。对于 RO-RMAP 估计器，在每个时刻，传感器按照一个相等的固定概率来决定是否发送测量值。通过 Monte Carlo 仿真，图 3-3 说明了在不同的平均通信率（通过分别在 SET 和 DET 估计器中设置不同 $\boldsymbol{Y}$ 和 $\boldsymbol{\delta}$ 的值）下，所提出的 SET-RMAP 和 DET-RMAP 估计器以及其他三个估计器的性能比较情况。显然，在平均估计正确率方面，所提出的两个事件触发的 RMAP 估计器均优于其他三个。此外，对于相同的性能标准，还显示 DET 优于 SET，这与使用线性高斯模型[112]在特定实验设置获得的结果一致。此外，可以观察到，当通信率处于一个较低水平时，与 RO-RMAP 相比，DET-RMAP 和 SET-RMAP 仍然可以保持不错的性能。图 3-4 显示了 DET-RMAP 和

RO-RMAP 的一个运行实例，其中平均通信率为 0.0692，相应的平均估计正确率分别为 63.28% 和 49.12%。图 3-5 给出了一个使用相同触发时刻的 DET-RMAP 和 DET-MAP 的实例，其中平均通信率为 0.3295，相应的平均估计正确率分别为 67.26% 和 59.37%。

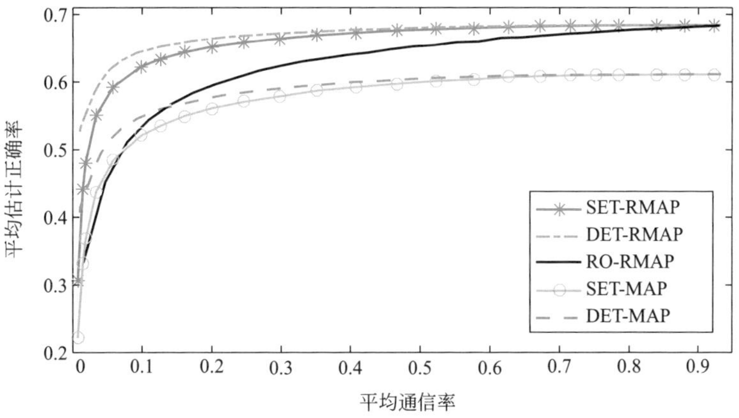

图 3-3 $p_{true}=0.03$ 时不同平均通信率下的性能比较

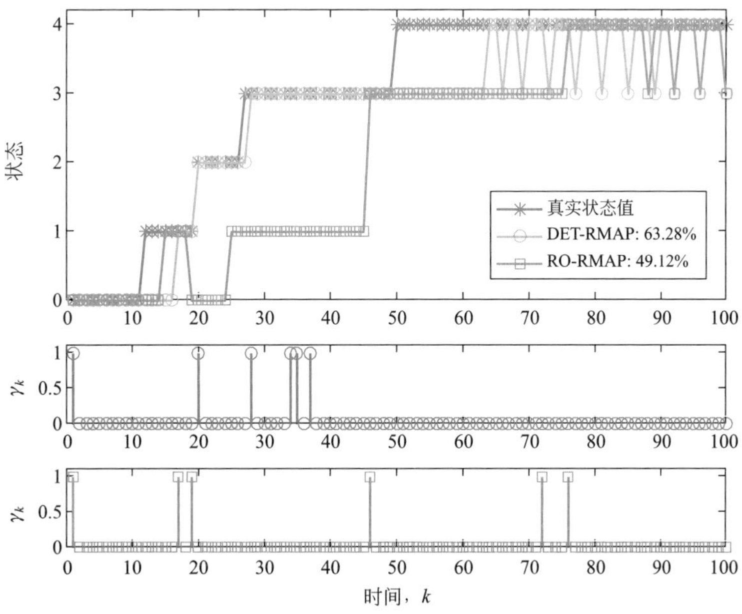

图 3-4 $p_{true}=0.03$ 时同样平均通信率下的一个 DET-RMAP 和 RO-RMAP 估计器运行实例

通过以上实验，可以得出结论：与离线风险敏感估计器相比，所提出的事件触发风险敏感估计器具有提升的性能，并且具有比风险中性估计器更好的鲁棒性。

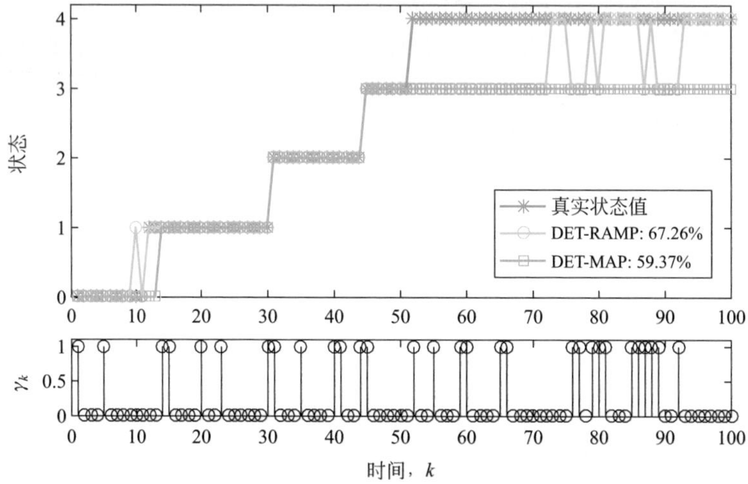

图 3-5 p_{true}=0.03 时同样触发时刻下的一个 DET-RMAP 和 DET-MAP 估计器运行实例

Digital Wave
Advanced Technology of Industrial Internet

Control of Autonomous Intelligent Systems

自主智能系统控制

第4章

具有相对熵约束的事件触发最小最大状态估计

4.1 概述

在上一章中，所考虑的状态变量是离散的（有限的状态），且估计器的鲁棒性没有针对特定类型的不确定性系统。本章则将考虑具有连续范围的状态变量的系统，研究具有相对熵约束这一类型的随机不确定系统的事件触发 minimax 估计问题。相对熵是一种衡量两个概率模型之间不匹配程度的标准，自然地，它可以用来描述系统的不确定性程度 [113-114]。具有相对熵约束的鲁棒状态估计和控制是鲁棒控制系统领域中一个重要的针对不确定系统的处理方法。该方法的主要特点是，将系统的不确定性建模成分布未知的加性噪声，但噪声形成的概率测度满足某个相对熵约束，此时所得的控制与估计结果可视为 LQG 控制器与 Kalman 滤波器在不确定系统下的扩展。

首先，本章将所研究估计问题转化成一个等价的线性二次高斯指数估计问题，然后利用信息状态方法和动态规划（Dynamic Programming, DP）理论给出了该问题的一个解决方案。值得指出的是，经典的处理涉及指数代价的估计和控制问题的信息状态方法（如文献 [94,107]）是通过使用测度变换方法 [107]，在一个新的概率测度下定义信息状态。本章则是在原始测度下定义信息状态，在数学上更直截了当。进一步，由于该最优估计问题没有解析解，本章考虑了一个一步形式的次优问题并引入后验相对熵约束，最终得到了一步次优问题的解析解。此外，本章采用序贯式融合方法将所得到的一步形式的估计结果推广到集中式传感器网络场景。仿真比较呈现了所提出的一步形式的事件触发鲁棒估计结果的有效性。

4.2
问题描述

4.2.1 不确定系统

考虑如下在有限时域 T 下的离散时间线性时变随机不确定系统：

$$\begin{aligned} \boldsymbol{x}_{k+1} &= \boldsymbol{A}_k \boldsymbol{x}_k + \tilde{\boldsymbol{w}}_k \\ \boldsymbol{y}_k &= \boldsymbol{C}_k \boldsymbol{x}_k + \tilde{\boldsymbol{v}}_k \\ \boldsymbol{z}_k &= \boldsymbol{L}_k \boldsymbol{x}_k \end{aligned} \tag{4-1}$$

其中，$\boldsymbol{x}_k \in \mathbb{R}^n$、$\boldsymbol{y}_k \in \mathbb{R}^m$和$\boldsymbol{z}_k \in \mathbb{R}^p$分别是系统的状态、测量和待估计的变量；$\boldsymbol{A}_k$、$\boldsymbol{C}_k$和$\boldsymbol{L}_k$是已知的矩阵，但噪声$\tilde{\boldsymbol{w}}_k$和$\tilde{\boldsymbol{v}}_k$以及初始状态$\tilde{\boldsymbol{x}}_0$是分布未知的随机变量。

为了刻画可容许（admissible）扰动噪声过程，有必要指定名义（nominal）噪声过程和初始状态，表示为 $\{\boldsymbol{w}_k\}$、$\{\boldsymbol{v}_k\}$ 和 $\boldsymbol{x}_0$。考虑这样一个广泛被采用的高斯假设：$\boldsymbol{w}_k$ 和 $\boldsymbol{v}_k$ 是密度函数分别为$\psi_k(\boldsymbol{w}_k) := \mathcal{N}(\boldsymbol{w}_k; 0, \boldsymbol{Q}_k)$和 $\phi_k(\boldsymbol{v}_k) := \mathcal{N}(\boldsymbol{v}_k; 0, \boldsymbol{R}_k)$的相互独立高斯白噪声，$\boldsymbol{x}_0$ 是密度函数为 $\mu(\boldsymbol{x}_0) = \mathcal{N}(\boldsymbol{x}_0; \boldsymbol{\pi}_0, \boldsymbol{\Pi}_0)$的高斯变量，且独立于 $\boldsymbol{w}_k$ 和 $\boldsymbol{v}_k$。这里，$\boldsymbol{Q}_k$、$\boldsymbol{R}_k$、$\boldsymbol{\Pi}_0$ 是半正定矩阵。由于系统 (4-1) 的动态演变实际上依赖于$\tilde{\boldsymbol{x}}_0$和噪声过程$\{\tilde{\boldsymbol{w}}_k\}$、$\{\tilde{\boldsymbol{v}}_k\}$，系统 (4-1) 可以被认为是定义在未知的噪声空间 $\{\tilde{\boldsymbol{x}}_0, \tilde{\boldsymbol{w}}_{0T}, \tilde{\boldsymbol{v}}_{0T}\}$，该空间的概率测度记为$\tilde{\mathbb{P}}(\tilde{\boldsymbol{x}}_0, \tilde{\boldsymbol{w}}_{0T}, \tilde{\boldsymbol{v}}_{0T})$，是扰动的概率测度。相应地，有名义概率测度$\mathbb{P}(\boldsymbol{x}_0, \boldsymbol{w}_{0T}, \boldsymbol{v}_{0T})$。根据 $\boldsymbol{x}_0$、$\boldsymbol{w}_k$ 和 $\boldsymbol{v}_k$ 三者之间的独立性质，有

$$\begin{aligned} & \mathrm{d}\mathbb{P}(\boldsymbol{x}_0, \boldsymbol{w}_{0T}, \boldsymbol{v}_{0T}) \\ &= \mathrm{d}\mathbb{P}(\boldsymbol{x}_0) \prod_{k=0}^{T} \mathrm{d}\mathbb{P}(\boldsymbol{w}_k, \boldsymbol{v}_k | \boldsymbol{x}_0, \boldsymbol{w}_{0k-1}, \boldsymbol{v}_{0k-1}) \\ &= \mu(\boldsymbol{x}_0) \mathrm{d}\boldsymbol{x}_0 \prod_{k=0}^{T} \psi_k(\boldsymbol{w}_k) \mathrm{d}\boldsymbol{w}_k \prod_{k=0}^{T} \phi_k(\boldsymbol{v}_k) \mathrm{d}\boldsymbol{v}_k \end{aligned} \tag{4-2}$$

这里d$\mathbb{P}$表示概率测度$\mathbb{P}$的微分，与勒贝格积分相关。

进一步采用相对熵来衡量概率测度$\tilde{\mathbb{P}}$与 $\mathbb{P}$ 之间的差异[113-114]，相对熵 $h(\tilde{\mathbb{P}}\|\mathbb{P})$定义如下：

$$h\left(\tilde{\mathbb{P}}\|\mathbb{P}\right):=\begin{cases}\tilde{\mathbb{E}}[\log\dfrac{\mathrm{d}\tilde{\mathbb{P}}}{\mathrm{d}\mathbb{P}}], & \tilde{\mathbb{P}}\ll\mathbb{P}\text{且}\log\dfrac{\mathrm{d}\tilde{\mathbb{P}}}{\mathrm{d}\mathbb{P}}\in L_1(\tilde{\mathbb{P}})\\ +\infty, & \text{其他}\end{cases}$$

其中，$\tilde{\mathbb{P}}\ll\mathbb{P}$表示扰动测度$\tilde{\mathbb{P}}$与$\mathbb{P}$绝对连续；$L_1$指的是$L_1$空间，使得 $\log\dfrac{\mathrm{d}\tilde{\mathbb{P}}}{\mathrm{d}\mathbb{P}}\in L_1\left(\tilde{\mathbb{P}}\right)$，意味着$\int\left|\log\dfrac{\mathrm{d}\tilde{\mathbb{P}}}{\mathrm{d}\mathbb{P}}\right|\mathrm{d}\tilde{\mathbb{P}}<\infty$。记$\mathcal{P}$为所有满足$h\left(\tilde{\mathbb{P}}\|\mathbb{P}\right)<\infty$的测度$\tilde{\mathbb{P}}$的集合。根据相对熵的链规则（chain rule）（文献[115]，引理7.9），可得

$$\begin{aligned}&h\left(\tilde{\mathbb{P}}\left(\tilde{\boldsymbol{x}}_0,\tilde{\boldsymbol{w}}_{0T},\tilde{\boldsymbol{v}}_{0T}\right)\|\mathbb{P}\left(\boldsymbol{x}_0,\boldsymbol{w}_{0T},\boldsymbol{v}_{0T}\right)\right)\\ =&h\left(\tilde{\mathbb{P}}\left(\tilde{\boldsymbol{x}}_0\right)\|\mathbb{P}\left(\boldsymbol{x}_0\right)\right)+\sum_{k=0}^{T}h\left(\tilde{\mathbb{P}}\left(\tilde{\boldsymbol{w}}_k,\tilde{\boldsymbol{v}}_k\mid\tilde{\boldsymbol{x}}_0,\tilde{\boldsymbol{w}}_{0k-1},\tilde{\boldsymbol{v}}_{0k-1}\right)\|\mathbb{P}\left(\boldsymbol{w}_k,\boldsymbol{v}_k\mid\boldsymbol{x}_0,\boldsymbol{w}_{0k-1},\boldsymbol{v}_{0k-1}\right)\right)\end{aligned}\tag{4-3}$$

下面给出不确定系统 (4-1) 的相对熵约束定义。

定义 4.1

给定一个常数$d>0$，如果一个概率测度$\tilde{\mathbb{P}}\in\mathcal{P}$满足约束

$$h\left(\tilde{\mathbb{P}}\|\mathbb{P}\right)\leqslant\tilde{\mathbb{E}}\left[V_u\right]+d,\quad V_u:=\frac{1}{2}\sum_{k=0}^{T}\|\boldsymbol{D}_k\boldsymbol{x}_k\|^2\tag{4-4}$$

其中，$\boldsymbol{D}_k$是不确定性矩阵，则称$\tilde{\mathbb{P}}$为可容许测度。所有可容许$\tilde{\mathbb{P}}$组成的集合记为$\varXi$。

定义 4.1 给出的相对熵约束可以涵盖很多有用的不确定性描述。例如，对于不确定性系统 (4-1)，可使$\tilde{\boldsymbol{x}}_0:=\boldsymbol{x}_0+\boldsymbol{\xi},\tilde{\boldsymbol{w}}_k:=\varDelta_1\left(\boldsymbol{x}_k\right)+\boldsymbol{w}_k$以及$\tilde{\boldsymbol{v}}_k:=\varDelta_2\left(\boldsymbol{x}_k\right)+\boldsymbol{v}_k$。这里，$\boldsymbol{\xi}$是一个未知的常数矩阵并与$\{\boldsymbol{w}_k\}$、$\{\boldsymbol{v}_k\}$独立，$\varDelta_1(\boldsymbol{x}_k)$和$\varDelta_2(\boldsymbol{x}_k)$是未知但有界的量。那么根据式 (4-3)，此时相对熵约束 (4-4) 可以明确地表达为如下二次和约束

$$\frac{1}{2}\|\boldsymbol{\xi}\|_{\varPi_0^{-1}}+\frac{1}{2}\sum_{k=0}^{T}\tilde{\mathbb{E}}\left[\|\varDelta_1\left(\boldsymbol{x}_k\right)\|_{\boldsymbol{Q}_k^{-1}}^2+\|\varDelta_2\left(\boldsymbol{x}_k\right)\|_{\boldsymbol{R}_k^{-1}}^2-\sum_{k=0}^{T}\|\boldsymbol{D}_k\boldsymbol{x}_k\|^2\right]\leqslant d\tag{4-5}$$

4.2.2 事件触发策略

鉴于有限的传感器能量和通信带宽，引入一个事件触发策略来决定

是否将 $\boldsymbol{y}_k$ 发送至远程估计器中（如图 4-1 所示）。如果$\gamma_k=1$，传感器发送 $\boldsymbol{y}_k$；否则不发送。令$\mathbb{S}_k:=\{i\in\mathbb{N}_{0:k}|\ \gamma_i=0\}$并定义

$$\mathcal{I}_k:=\{\gamma_i|\ i\in\mathbb{S}_k\}\cup\{\boldsymbol{y}_i|i\in\mathbb{N}_{0:k}\setminus\mathbb{S}_k\},\ \mathcal{I}_{-1}:=\varnothing$$

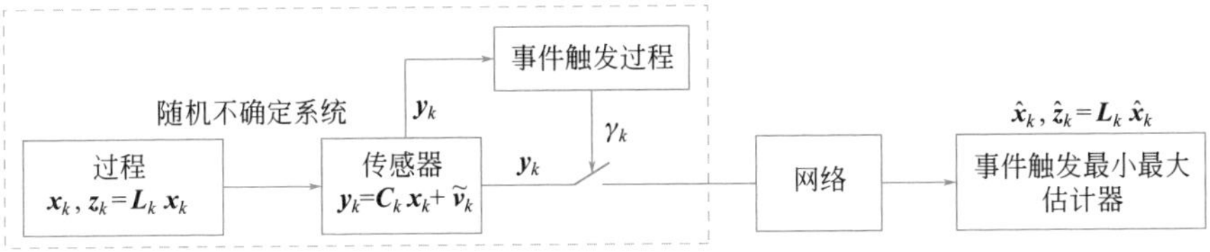

图 4-1　针对随机不确定系统的事件触发最小最大状态估计

作为k时刻估计器可利用的信息集。一个随机性事件触发方案描述如下❶[116]

$$\begin{aligned}\widetilde{\Pr}\{\gamma_k=0|\ \boldsymbol{y}_k,\mathcal{I}_{k-1}\}&=\widetilde{\Pr}\{\gamma_k=0|\ \boldsymbol{y}_{0k},\boldsymbol{x}_{0k},\gamma_{0k-1}\}\\&=\exp\left(-\frac{1}{2}\|\ \boldsymbol{y}_k-\boldsymbol{\beta}_k\|_{\boldsymbol{Y}_k}^2\right)\end{aligned}\tag{4-6}$$

其中，$\boldsymbol{Y}_k\in\mathbb{R}^{m\times m}$是一个正定的权重矩阵；$\boldsymbol{\beta}_k\in\mathbb{R}^m$是一个估计器基于$I_{k-1}$可以得知的量。矩阵参数$\boldsymbol{Y}_k$扮演着调节传输率的角色：一般地，$\boldsymbol{Y}_k$取值越小意味着一个更低的传输率。特别地，$\boldsymbol{Y}_k\to\infty$时，传输率趋于100%，即测量值每个时刻都被发送。$\boldsymbol{\beta}_k$的选择依赖于传感器实际所拥有的资源，常见的选择有$\boldsymbol{\beta}_k$为上一被发送的测量值和$\boldsymbol{y}_k$的一步预测值[117]。式(4-6)的第一个等式指的是k时刻的传输概率仅仅依赖于当前的测量$\boldsymbol{y}_k$和过去的信息集$\mathcal{I}_{k-1}$。对于名义系统(4-2)，事件触发过程(4-6)变成

$$\begin{aligned}\Pr\{\gamma_k=0|\ \boldsymbol{y}_k,\mathcal{I}_{k-1}\}&=\Pr\{\gamma_k=0|\ \boldsymbol{y}_{0k},\boldsymbol{x}_{0k},\gamma_{0k-1}\}\\&=\exp\left(-\frac{1}{2}\|\ \boldsymbol{y}_k-\boldsymbol{\beta}_k\|_{\boldsymbol{Y}_k}^2\right)\end{aligned}\tag{4-7}$$

---

❶ 在每个时刻k，式(4-6)中的概率$\widetilde{\Pr}\{\gamma_k=0|\ \boldsymbol{y}_k,\mathcal{I}_{k-1}\}$可以通过引入一个独立同分布的随机变量$\xi_k$来实施。$\xi_k$在 [0.1] 区间均匀分布，如果$\zeta_k<\exp\left(-\frac{1}{2}\|\ \boldsymbol{y}_k-\boldsymbol{\beta}_k\|_{\boldsymbol{Y}_k}^2\right)$，则$\gamma_k=0$；否则$\gamma_k=1$。

注释 4.1

注意，严格地讲，事件触发器中的随机过程 $\{\zeta_k\}$ 也应该定义在概率测度$\tilde{\mathbb{P}}$和 $\mathbb{P}$ 上。因为 $\{\zeta_k\}$ 不存在不确定性，使得

$$h\left(\tilde{\mathbb{P}}\left(\tilde{\boldsymbol{x}}_0,\tilde{\boldsymbol{w}}_{0T},\tilde{\boldsymbol{v}}_{0T},\boldsymbol{\zeta}_{0T}\right)\|\mathbb{P}\left(\boldsymbol{x}_0,\boldsymbol{w}_{0T},\boldsymbol{v}_{0T},\boldsymbol{\zeta}_{0T}\right)\right)$$
$$=h\left(\tilde{\mathbb{P}}\left(\tilde{\boldsymbol{x}}_0,\tilde{\boldsymbol{w}}_{0T},\tilde{\boldsymbol{v}}_{0T}\right)\|\mathbb{P}\left(\boldsymbol{x}_0,\boldsymbol{w}_{0T},\boldsymbol{v}_{0T}\right)\right)$$

所以这种差异对本章考虑的问题没有影响。

4.2.3 事件触发最小最大最优估计问题

下面，给出本节所研究的事件触发最小最大最优估计问题。考虑一个因果估计器$\{\hat{z}_0,\cdots,\hat{z}_T\}$，也就是估计值是基于估计器当前时刻可利用的信息得到：

$$\hat{\boldsymbol{z}}_k=\hat{\boldsymbol{z}}_k\left(\mathcal{I}_k\right)=\boldsymbol{L}_k\hat{\boldsymbol{x}}_k\left(\mathcal{I}_k\right)=\boldsymbol{L}_k\hat{\boldsymbol{x}}_k \tag{4-8}$$

为了评估估计器性能，定义估计误差代价为

$$V_e:=\frac{1}{2}\sum_{k=0}^{T}\|\boldsymbol{z}_k-\hat{\boldsymbol{z}}_k\|^2=\frac{1}{2}\sum_{k=0}^{T}\|\boldsymbol{x}_k-\hat{\boldsymbol{x}}_k\|_{\boldsymbol{W}_k}^2 \tag{4-9}$$

其中，$\boldsymbol{W}_k=\boldsymbol{L}_k^{\mathrm{T}}\boldsymbol{L}_k$。那么，在定义4.1所给出的不确定性约束下，事件触发最小最大最优估计问题可以被描述为

$$(P_0):\qquad \inf_{\hat{\boldsymbol{x}}_{0T}}\sup_{\tilde{\mathbb{P}}\in\Xi}\widetilde{\mathbb{E}}[V_e] \tag{4-10}$$

注释 4.2

虽然上述的最小最大问题和传统时间触发通信（估计器每个时刻都能收到测量值）下的相应问题 [114] 具有相同的形式，但它们的代价函数 V_e 不同。这里的估计值$\hat{\boldsymbol{x}}_k$依赖于事件触发过程 $\{\gamma_k\}$，这使得估计器的设计和分析更为复杂。

使用拉格朗日乘子法，可以将有约束的最小最大问题 (4-10) 转化为一个无约束的形式。

引理 4.1

（文献 [113]，定理 3.1）考虑如下无约束的最小最大问题：

$$J_\tau := \inf_{\hat{x}_{0T}} \sup_{\tilde{\mathbb{P}}\in\mathcal{P}} \left\{ \tilde{\mathbb{E}}\left[V_e\right] - \tau\left[h\left(\tilde{\mathbb{P}} \| \mathbb{P}\right) - \tilde{\mathbb{E}}\left[V_u\right] - d\right]\right\}$$
$$= \tau\left\{\inf_{\hat{x}_{0T}} \sup_{\tilde{\mathbb{P}}\in\mathcal{P}} \left[\tilde{\mathbb{E}}\left[\tau^{-1}V_\tau\right] - h\left(\tilde{\mathbb{P}} \| \mathbb{P}\right)\right] + d\right\} \tag{4-11}$$

其中

$$V_\tau = V_e + \tau V_u = \sum_{k=0}^{T} \frac{1}{2}\left(\| \boldsymbol{x}_k - \hat{\boldsymbol{x}}_k \|_{\boldsymbol{W}_k}^2 + \tau \| \boldsymbol{D}_k \boldsymbol{x}_k \|^2\right)$$

定义$\Gamma := \{\tau : \tau > 0且J_\tau < \infty\}$，那么有约束的最小最大问题(4-10)取值有限当且仅当$\Gamma \neq \varnothing$。此外，如果$\Gamma \neq \varnothing$，则

$$\inf_{\hat{x}_{0T}} \sup_{\tilde{\mathbb{P}}\in\Xi} \tilde{\mathbb{E}}\left[V_e\right] = \inf_{\tau\in\Gamma} J_\tau \tag{4-12}$$

由于相对熵$h\left(\tilde{\mathbb{P}} \| \mathbb{P}\right)$不具有明确的表达式，很难直接得到参数化随机博弈问题 (4-11) 的解。一个可行的方式是利用自由能量和相对熵之间的对偶性关系 [118]，将该问题转换为一个等价的事件触发线性指数二次高斯（Linear Exponential Quadratic Gaussian, LEQG）估计问题。在这样一个对偶关系下，对于无约束问题 (4-11)，有

$$J_\tau = \tau\left\{\left(\inf_{\hat{x}_{0T}} \log J^L(\tau)\right) + d\right\} \tag{4-13}$$

其中

$$J^L(\tau) = \mathbb{E}\left[\exp\left(\tau^{-1}V_\tau\right)\right] \tag{4-14}$$

是事件触发LEQG代价。因此，最小最大问题(P_0)等价为

$$(P_0'):\quad \inf_{\tau\in\Gamma} \tau\left\{\left(\inf_{\hat{x}_{0T}} \log J^L(\tau)\right) + d\right\} \tag{4-15}$$

从(P_0')中可以得知，要解决所考虑的事件触发的最小最大最优估计问题(P_0)，首先要解决如下的事件触发LEQG问题

$$\inf_{\hat{x}_{0T}} J^L(\tau) \tag{4-16}$$

然后由$\tau^* = \arg\inf_{\tau\in\Gamma} J_\tau$找出最优的$\tau$。所以，通过导出LEQG估计器并令$\tau = \tau^*$，可以得到事件触发最小最大最优估计器。

4.3
事件触发 LEQG 估计

现有文献已给出传统时间触发的最优 LEQG 估计问题的解[73]，该解是一个类 Kalman 估计器。本节将调查事件触发 LEQG 估计问题，结果表明此问题更为复杂且没有解析解。

因为在 LEQG 代价式 (4-14) 中，数学期望是对于名义概率测度 $\mathbb{P}$ 而言的，所以本节不需要考虑扰动测度 $\tilde{\mathbb{P}}$。名义概率测度 $\mathbb{P}$ 有具体的概率分布，这方便了问题的分析和求解。接下来，给出求解该问题的方案：首先定义一个信息状态变量，并得到了其递归表达式，然后基于该信息状态，利用 DP 理论求解该问题。

4.3.1 递归的信息状态

本小节将给出新定义的信息状态的递归封闭形式表达式。构建一个信息状态的准则是它必须包含关于系统状态所有可利用的信息（文献 [119], 79 页）。因此，常见的 $\boldsymbol{x}_k$ 条件密度函数是一个信息状态，它满足信息状态的准则要求。然而，条件密度不适合用来处理指数代价。一个更合适的信息状态应该包含部分 LEQG 代价[97,120]。为此，将 LEQG 代价 [式 (4-14)] 分解为

$$J^L(\tau)=\mathbb{E}\left[\exp\left(\sum_{k=0}^{T}\overline{\boldsymbol{\varPsi}}_k\left(\boldsymbol{x}_k,\hat{\boldsymbol{x}}_k\right)\right)\right] \tag{4-17}$$

其中

$$\overline{\boldsymbol{\varPsi}}_k\left(\boldsymbol{x}_k,\hat{\boldsymbol{x}}_k\right)=\frac{1}{2}\left(\tau^{-1}\|\boldsymbol{x}_k-\hat{\boldsymbol{x}}_k\|_{\boldsymbol{W}_k}^2+\|\boldsymbol{D}_k\boldsymbol{x}_k\|^2\right) \tag{4-18}$$

然后构建如下一个信息状态。

定义 4.2

定义信息状态 $\boldsymbol{\alpha}_k\left(\boldsymbol{x}_k\right)$ 为

$$\boldsymbol{\alpha}_k\left(\boldsymbol{x}_k\right):=\int_{\mathbb{R}^n}\cdots\int_{\mathbb{R}^n}\exp\left(\boldsymbol{\varPsi}_{0,k-1}\right)f\left(\boldsymbol{x}_{0k},\mathcal{I}_k\right)\mathrm{d}\boldsymbol{x}_{0k-1},\quad k=1,2,\cdots,T \tag{4-19}$$

其中，$f\left(\boldsymbol{x}_{0k},\mathcal{I}_k\right)$表示$\boldsymbol{x}_{0k}$和$\mathcal{I}_k$的联合分布，并且$\boldsymbol{\Psi}_{0,k-1}=\sum_{i=0}^{k-1}\overline{\boldsymbol{\Psi}}_i\left(\boldsymbol{x}_i,\hat{\boldsymbol{x}}_i\right)$。信息状态的初始值为$\boldsymbol{\alpha}_0\left(\boldsymbol{x}_0\right):=f\left(\boldsymbol{x}_0,\mathcal{I}_0\right)$。

注释 4.3

值得指出的是，本节定义的信息状态在数学上比经典的处理涉及指数代价的估计和控制问题的信息状态方法（如文献[94,107,120-122]）更直接。在这些文献中，通过使用测度变换方法[107]，在一个新的概率测度下定义信息状态，这有些复杂。具体地，此方法通过定义一个 Radon-Nikodym 导数[107-108]引入一个新的测度$\overline{\mathbb{P}}$，该 Radon-Nikodym 导数在新测度$\overline{\mathbb{P}}$和原始测度$\mathbb{P}$之间建立起一个联系。然后可以证明在$\overline{\mathbb{P}}$下，$\{\boldsymbol{y}_k\}$是密度函数为ϕ_k且独立于$\{\boldsymbol{x}_k\}$的一序列i.i.d.随机变量。$\overline{\mathbb{P}}$下的信息状态可以定义为

$$\boldsymbol{\alpha}_k\left(x\right)\mathrm{d}x=\overline{\mathbb{E}}\left[\overline{\boldsymbol{\Lambda}}_k\exp\left(\boldsymbol{\Psi}_{0,k-1}\right)\boldsymbol{I}\left(\boldsymbol{x}_k\in\mathrm{d}x\right)\mid\mathcal{I}_k\right]\tag{4-20}$$

其中$\overline{\boldsymbol{\Lambda}}_k=\prod_{l=0}^{k}\phi_l\left(\boldsymbol{y}_l-\boldsymbol{C}_l\boldsymbol{x}_l\right)/\phi_l\left(\boldsymbol{y}_l\right)$是被定义的 Radon-Nikodym 导数$\mathrm{d}\mathbb{P}/\mathrm{d}\overline{\mathbb{P}}$。$\boldsymbol{I}\left(\boldsymbol{x}_k\in\mathrm{d}x\right)$是一个指示器函数。$\boldsymbol{x}_k\in\mathrm{d}x$时，$\boldsymbol{I}\left(\boldsymbol{x}_k\in\mathrm{d}x\right)=\boldsymbol{I}$，否则$\boldsymbol{I}\left(\boldsymbol{x}_k\in\mathrm{d}x\right)=0$。引入测度变换方法的优点是$\{\boldsymbol{y}_k\}$具有独立性质，这也许可以简化推导过程。但是，本节表明，仍然可以在原始测度$\mathbb{P}$下定义一个合适的信息状态来处理带有指数代价的问题，从而可以使用所熟知的Bayes推论。

下面，利用 Bayes 推论给出$\boldsymbol{\alpha}_k\left(\boldsymbol{x}_k\right)$的递归表达式。

引理 4.2

对于带有事件触发方案(4-7)的名义系统，由式(4-19)定义的信息状态$\boldsymbol{\alpha}_k\left(\boldsymbol{x}_k\right)$具有如下递归形式：

① 如果$\gamma_k=0$，

$$\begin{aligned}&\boldsymbol{\alpha}_k\left(\boldsymbol{x}_k\right)\\&=\int_{\mathbb{R}^n}\exp\left(\overline{\boldsymbol{\Psi}}_{k-1}\left(\boldsymbol{z},\hat{\boldsymbol{x}}_{k-1}\right)\right)\psi_{k-1}\left(\boldsymbol{x}_k-\boldsymbol{A}_{k-1}\boldsymbol{z}\right)\boldsymbol{\alpha}_{k-1}\left(\boldsymbol{z}\right)\mathrm{d}\boldsymbol{z}\\&\times\int_{\mathbb{R}^m}\Pr\left\{\gamma_k=0\mid\boldsymbol{y}_k,\mathcal{I}_{k-1}\right\}\phi_k\left(\boldsymbol{y}_k-\boldsymbol{C}_k\boldsymbol{x}_k\right)\mathrm{d}\boldsymbol{y}_k\end{aligned}\tag{4-21}$$

② 如果$\gamma_k = 1$，

$$\boldsymbol{\alpha}_k(\boldsymbol{x}_k) = \phi_k(\boldsymbol{y}_k - \boldsymbol{C}_k\boldsymbol{x}_k)\int_{\mathbb{R}^n}\exp\left(\overline{\boldsymbol{\Psi}}_{k-1}(\boldsymbol{z},\hat{\boldsymbol{x}}_{k-1})\right) \times\psi_{k-1}(\boldsymbol{x}_k - \boldsymbol{A}_{k-1}\boldsymbol{z})\boldsymbol{\alpha}_{k-1}(\boldsymbol{z})\mathrm{d}\boldsymbol{z} \tag{4-22}$$

当$\gamma_0 = 0$时，上述递归始于

$$\boldsymbol{\alpha}_0(\boldsymbol{x}_0) = \mu(\boldsymbol{x}_0)\int_{\mathbb{R}^m}\Pr\{\gamma_0 = 0 | \boldsymbol{y}_0\}\phi_0(\boldsymbol{y}_0 - \boldsymbol{C}_0\boldsymbol{x}_0)\mathrm{d}\boldsymbol{y}_0 \tag{4-23}$$

当$\gamma_0 = 1$时，则

$$\boldsymbol{\alpha}_0(\boldsymbol{x}_0) = \phi_0(\boldsymbol{y}_0 - \boldsymbol{C}_0\boldsymbol{x}_0)\mu(\boldsymbol{x}_0). \tag{4-24}$$

证明：

根据式 (4-19) 中$\boldsymbol{\alpha}_k(\boldsymbol{x}_k)$的定义，如果$\gamma_k = 0$，可得

$$\begin{aligned}
&\boldsymbol{\alpha}_k(\boldsymbol{x}_k)\\
&= \int_{\mathbb{R}^n}\cdots\int_{\mathbb{R}^n}\exp\left(\boldsymbol{\Psi}_{0,k-1}\right)f(\boldsymbol{x}_{0k},\gamma_k = 0,\mathcal{I}_{k-1})\mathrm{d}\boldsymbol{x}_{0k-1}\\
&= \int_{\mathbb{R}^n}\cdots\int_{\mathbb{R}^n}\exp\left(\boldsymbol{\Psi}_{0,k-1}\right)\Pr\{\gamma_k = 0 | \boldsymbol{x}_{0k},\mathcal{I}_{k-1}\}f(\boldsymbol{x}_k | \boldsymbol{x}_{k-1})f(\boldsymbol{x}_{0k-1},\mathcal{I}_{k-1})\mathrm{d}\boldsymbol{x}_{0k-1}
\end{aligned} \tag{4-25}$$

其中，最后一个等式来自Bayes法则和噪声的独立性质。因为

$$\begin{aligned}
&\Pr\{\gamma_k = 0 | \boldsymbol{x}_{0k},\mathcal{I}_{k-1}\}\\
&= \int_{\mathbb{R}^m}\Pr\{\gamma_k = 0 | \boldsymbol{x}_{0k},\boldsymbol{y}_k,\mathcal{I}_{k-1}\}f(\boldsymbol{y}_k | \boldsymbol{x}_{0k},\mathcal{I}_{k-1})\mathrm{d}\boldsymbol{y}_k\\
&= \int_{\mathbb{R}^m}\Pr\{\gamma_k = 0 | \boldsymbol{y}_k,\mathcal{I}_{k-1}\}\phi_k(\boldsymbol{y}_k - \boldsymbol{C}_k\boldsymbol{x}_k)\mathrm{d}\boldsymbol{y}_k
\end{aligned} \tag{4-26}$$

进一步可得

$$\begin{aligned}
&\boldsymbol{\alpha}_k(\boldsymbol{x}_k)\\
&= \int_{\mathbb{R}^n}\ldots\int_{\mathbb{R}^n}\exp\left(\overline{\boldsymbol{\Psi}}_{k-1}(\boldsymbol{x}_{k-1},\hat{\boldsymbol{x}}_{k-1})\right)\exp\left(\boldsymbol{\Psi}_{0,k-2}\right)f(\boldsymbol{x}_k | \boldsymbol{x}_{k-1})f(\boldsymbol{x}_{0k-1},\mathcal{I}_{k-1})\mathrm{d}\boldsymbol{x}_{0k-1}\\
&\times\int_{\mathbb{R}^m}\Pr\{\gamma_k = 0 | \boldsymbol{y}_k,\mathcal{I}_{k-1}\}\phi_k(\boldsymbol{y}_k - \boldsymbol{C}_k\boldsymbol{x}_k)\mathrm{d}\boldsymbol{y}_k\\
&= \int_{\mathbb{R}^n}\exp\left(\overline{\boldsymbol{\Psi}}_{k-1}(\boldsymbol{z},\hat{\boldsymbol{x}}_{k-1})\right)\psi_{k-1}(\boldsymbol{x}_k - \boldsymbol{A}_{k-1}\boldsymbol{z})\boldsymbol{\alpha}_{k-1}(\boldsymbol{z})\mathrm{d}\boldsymbol{z}\\
&\quad\times\int_{\mathbb{R}^m}\Pr\{\gamma_k = 0 | \boldsymbol{y}_k,\mathcal{I}_{k-1}\}\phi_k(\boldsymbol{y}_k - \boldsymbol{C}_k\boldsymbol{x}_k)\mathrm{d}\boldsymbol{y}_k
\end{aligned} \tag{4-27}$$

其中，最后一个等式来自$\boldsymbol{\alpha}_{k-1}\left(\boldsymbol{x}_{k-1}\right)$的定义。

对于$\gamma_k=1$，类似地，通过 Bayes 法则有

$$\begin{aligned}&f\left(\boldsymbol{x}_{0k},\boldsymbol{y}_k,\mathcal{I}_{k-1}\right)\\&=f\left(\boldsymbol{x}_k,\boldsymbol{y}_k\mid\boldsymbol{x}_{0k-1},\mathcal{I}_{k-1}\right)f\left(\boldsymbol{x}_{0k-1},\mathcal{I}_{k-1}\right)\\&=f\left(\boldsymbol{y}_k\mid\boldsymbol{x}_k\right)f\left(\boldsymbol{x}_k\mid\boldsymbol{x}_{k-1}\right)f\left(\boldsymbol{x}_{0k-1},\mathcal{I}_{k-1}\right)\end{aligned}\tag{4-28}$$

并且进一步得到

$$\begin{aligned}&\boldsymbol{\alpha}_k\left(\boldsymbol{x}_k\right)\\&=\int_{\mathbb{R}^n}\cdots\int_{\mathbb{R}^n}\exp\left(\boldsymbol{\Psi}_{0,k-1}\right)f\left(\boldsymbol{x}_{0k},\boldsymbol{y}_k,\mathcal{I}_{k-1}\right)\mathrm{d}\boldsymbol{x}_{0k-1}\\&=\int_{\mathbb{R}^n}\cdots\int_{\mathbb{R}^n}\exp\left(\overline{\boldsymbol{\Psi}}_{k-1}\left(\boldsymbol{x}_{k-1},\hat{\boldsymbol{x}}_{k-1}\right)\right)\exp\left(\boldsymbol{\Psi}_{0,k-2}\right)\\&\times f\left(\boldsymbol{y}_k\mid\boldsymbol{x}_k\right)f\left(\boldsymbol{x}_k\mid\boldsymbol{x}_{k-1}\right)f\left(\boldsymbol{x}_{0k-1},\mathcal{I}_{k-1}\right)\mathrm{d}\boldsymbol{x}_{0k-1}\\&=\phi_k\left(\boldsymbol{y}_k-\boldsymbol{C}_k\boldsymbol{x}_k\right)\int_{\mathbb{R}^n}\exp\left(\overline{\boldsymbol{\Psi}}_{k-1}\left(\boldsymbol{z},\hat{\boldsymbol{x}}_{k-1}\right)\right)\psi_{k-1}\left(\boldsymbol{x}_k-\boldsymbol{A}_{k-1}\boldsymbol{z}\right)\boldsymbol{\alpha}_{k-1}\left(\boldsymbol{z}\right)\mathrm{d}\boldsymbol{z}\end{aligned}\tag{4-29}$$

证毕。 ■

从引理 4.2 可以观察到，要得到$\boldsymbol{\alpha}_k\left(\boldsymbol{x}_k\right)$的明确表达式，需要高斯密度函数相乘的计算。为此，下面给出一个一般化的完全平方技术。

引理 4.3

给定$\boldsymbol{x}\in\mathbb{R}^n$，一组矩阵$\left\{\boldsymbol{G}_i\right\}\in\mathbb{R}^{m_i\times n}$，一组对称矩阵$\left\{\boldsymbol{X}_i\right\}\in\mathbb{R}^{m_i\times m_i}$，以及一组向量$\left\{\boldsymbol{b}_i\right\}\in\mathbb{R}^{m_i},1\leqslant i\leqslant N$，定义$\boldsymbol{G}_{1\mathrm{N}}:=\sum_{i=1}^{N}\boldsymbol{G}_i^{\mathrm{T}}\boldsymbol{X}_i\boldsymbol{G}_i$和$G_{\mathrm{b}}:=\sum_{i=1}^{N}\boldsymbol{G}_i^{\mathrm{T}}\boldsymbol{X}_i\boldsymbol{b}_i$。如果$\boldsymbol{G}_{1\mathrm{N}}$是非奇异的，那么

$$\sum_{i=1}^{N}\|\boldsymbol{G}_i\boldsymbol{x}+\boldsymbol{b}_i\|_{\boldsymbol{X}_i}^2=\|\boldsymbol{x}+\boldsymbol{G}_{1\mathrm{N}}^{-1}\boldsymbol{G}_{\mathrm{b}}\|_{\boldsymbol{G}_{1\mathrm{N}}}^2+\sum_{i=1}^{N}\|\boldsymbol{b}_i\|_{\boldsymbol{G}_i}^2-\|\boldsymbol{G}_{\mathrm{b}}\|_{\boldsymbol{G}_{1\mathrm{N}}^{-1}}^2\tag{4-30}$$

证明：

利用$N=2$时的完全平方结果（如文献 [116] 中引理 15），可以用数学归纳法证明该引理。 ■

基于引理 4.2 和 4.3，可以得到如下关于$\boldsymbol{\alpha}_k\left(\boldsymbol{x}_k\right)$的明确递归结果。

定理 4.1

对于带有事件触发方案 (4-7) 的名义系统，信息状态$\boldsymbol{\alpha}_k\left(\boldsymbol{x}_k\right)$是一个

非标准化的高斯密度，给出如下：

$$\boldsymbol{\alpha}_k\left(\boldsymbol{x}_k\right)=\boldsymbol{S}_k\exp\left(-\frac{1}{2}\|\boldsymbol{x}_k-\overline{\boldsymbol{x}}_k\|_{\boldsymbol{\Sigma}_k^{-1}}^2\right) \tag{4-31}$$

这里，$\boldsymbol{\Sigma}_k^{-1}$满足如下递归方程：

$$\begin{cases}\boldsymbol{\Sigma}_k^{-1}=\boldsymbol{C}_k^{\mathrm{T}}\left[\boldsymbol{R}_k+(1-\gamma_k)\boldsymbol{Y}_k^{-1}\right]^{-1}\boldsymbol{C}_k+\boldsymbol{Q}_{k-1}^{-1}-\boldsymbol{Q}_{k-1}^{-1}\boldsymbol{A}_{k-1}\overline{\boldsymbol{\Sigma}}_k\boldsymbol{A}_{k-1}^{\mathrm{T}}\boldsymbol{Q}_{k-1}^{-1}\\ \overline{\boldsymbol{\Sigma}}_k^{-1}=\boldsymbol{A}_{k-1}^{\mathrm{T}}\boldsymbol{Q}_{k-1}^{-1}\boldsymbol{A}_{k-1}+\boldsymbol{\Sigma}_{k-1}^{-1}-\tau^{-1}\boldsymbol{W}_{k-1}-\boldsymbol{D}_{k-1}^{\mathrm{T}}\boldsymbol{D}_{k-1}\end{cases} \tag{4-32}$$

上面的递归成立，当且仅当对于所有$k<T$，$\boldsymbol{A}_k^{\mathrm{T}}\boldsymbol{Q}_k^{-1}\boldsymbol{A}_k+\boldsymbol{\Sigma}_k^{-1}-\tau^{-1}\boldsymbol{W}_k-\boldsymbol{D}_k^{\mathrm{T}}\boldsymbol{D}_k>0$成立。$\overline{\boldsymbol{x}}_k$和$\boldsymbol{S}_k$由如下递归给出：

$$\begin{cases}\boldsymbol{\Sigma}_k^{-1}\overline{\boldsymbol{x}}_k=\boldsymbol{Q}_{k-1}^{-1}\boldsymbol{A}_{k-1}\overline{\boldsymbol{\Sigma}}_k\left(\boldsymbol{\Sigma}_{k-1}^{-1}\overline{\boldsymbol{x}}_{k-1}-\tau^{-1}\boldsymbol{W}_{k-1}\hat{\boldsymbol{x}}_{k-1}\right)\\ \qquad+\boldsymbol{C}_k^{\mathrm{T}}\left[\boldsymbol{R}_k+\left(1-\gamma_k\right)\boldsymbol{Y}_k^{-1}\right]^{-1}\left[\gamma_k\boldsymbol{y}_k+\left(1-\gamma_k\right)\boldsymbol{\beta}_k\right]\\ \boldsymbol{S}_k=\boldsymbol{S}_{k-1}\left(2\pi\right)^{\frac{-m\gamma_k}{2}}|\boldsymbol{R}_k|^{-\frac{1}{2}}|\boldsymbol{R}_k^{-1}+\boldsymbol{Y}_k|^{\frac{\gamma_k-1}{2}}|\boldsymbol{Q}_{k-1}|^{-\frac{1}{2}}|\overline{\boldsymbol{\Sigma}}_k|^{\frac{1}{2}}\exp\left\{\frac{1}{2}\left[\|\hat{\boldsymbol{x}}_{k-1}\|_{\tau^{-1}\boldsymbol{W}_{k-1}}^2\right.\right.\\ \qquad-\|\overline{\boldsymbol{x}}_{k-1}\|_{\boldsymbol{\Sigma}_{k-1}^{-1}}^2-\|\gamma_k\boldsymbol{y}_k+\left(1-\gamma_k\right)\boldsymbol{\beta}_k\|_{\left[\boldsymbol{R}_k+(1-\gamma_k)\boldsymbol{Y}_k^{-1}\right]^{-1}}^2\\ \qquad\left.\left.+\|\boldsymbol{\Sigma}_{k-1}^{-1}\overline{\boldsymbol{x}}_{k-1}-\tau^{-1}\boldsymbol{W}_{k-1}\hat{\boldsymbol{x}}_{k-1}\|_{\overline{\boldsymbol{\Sigma}}_k}^2+\|\overline{\boldsymbol{x}}_k\|_{\boldsymbol{\Sigma}_k^{-1}}^2\right]\right\}\end{cases} \tag{4-33}$$

上面递归始于

$$\begin{cases}\boldsymbol{\Sigma}_0^{-1}=\boldsymbol{C}_0^{\mathrm{T}}\left[\boldsymbol{R}_0+\left(1-\gamma_0\right)\boldsymbol{Y}_0^{-1}\right]^{-1}\boldsymbol{C}_0+\boldsymbol{\Pi}_0^{-1}\\ \boldsymbol{\Sigma}_0^{-1}\overline{\boldsymbol{x}}_0=\boldsymbol{\Pi}_0^{-1}\boldsymbol{\pi}_0+\boldsymbol{C}_0\left[\boldsymbol{R}_0+\left(1-\gamma_0\right)\boldsymbol{Y}_0^{-1}\right]^{-1}\left[\gamma_0\boldsymbol{y}_0+\left(1-\gamma_0\right)\boldsymbol{\beta}_0\right]\\ \boldsymbol{S}_0=\left(2\pi\right)^{-\frac{n+\gamma_0 m}{2}}\left|\boldsymbol{R}_0\right|^{-\frac{1}{2}}\left|\boldsymbol{R}_0^{-1}+\boldsymbol{Y}_0\right|^{-\frac{(1-\gamma_k)}{2}}\left|\boldsymbol{\Pi}_0\right|^{-\frac{1}{2}}\exp\left\{\frac{1}{2}\left[\|\overline{\boldsymbol{x}}_0\|_{\boldsymbol{\Sigma}_0^{-1}}^2-\|\boldsymbol{\pi}_0\|_{\boldsymbol{\Pi}_0^{-1}}^2\right.\right.\\ \qquad\left.\left.-\|\gamma_0\boldsymbol{y}_0+\left(1-\gamma_0\right)\boldsymbol{\beta}_0\|_{\left[\boldsymbol{R}_0+(1-\gamma_0)\boldsymbol{Y}_0^{-1}\right]^{-1}}^2\right]\right\}\end{cases} \tag{4-34}$$

证明：

采用数学归纳法证明该定理。同引理 4.2，分别考虑$\gamma_k=0$和$\gamma_k=1$两种情况。因为$\boldsymbol{\alpha}_0\left(\boldsymbol{x}_0\right)$的推导与一般情况$\boldsymbol{\alpha}_k\left(\boldsymbol{x}_k\right)$的推导类似，$\boldsymbol{\alpha}_0\left(\boldsymbol{x}_0\right)$的推导过程被省略。假设在 $k-1$ 时刻，

$$\boldsymbol{\alpha}_{k-1}\left(\boldsymbol{x}_{k-1}\right)=\boldsymbol{S}_{k-1}\exp\left(-\frac{1}{2}\|\boldsymbol{x}_{k-1}-\overline{\boldsymbol{x}}_{k-1}\|^2_{\boldsymbol{\Sigma}_{k-1}^{-1}}\right) \tag{4-35}$$

然后证明对于$\boldsymbol{\alpha}_k\left(\boldsymbol{x}_k\right)$，式 (4-31) 成立。为了简洁，将$\boldsymbol{x}_k$、$\boldsymbol{y}_k$简化成$\boldsymbol{x}$和$\boldsymbol{y}$。根据引理4.2，对于$\gamma_k=0$，可以得到

$$\boldsymbol{\alpha}_k\left(\boldsymbol{x}\right)=\boldsymbol{S}_{k-1}\left(2\pi\right)^{-\frac{m+n}{2}}\left|\boldsymbol{R}_k\right|^{-\frac{1}{2}}\left|\boldsymbol{Q}_{k-1}\right|^{-\frac{1}{2}}\int_{\mathbb{R}^m}\exp\left(-\boldsymbol{s}_1/2\right)\mathrm{d}\boldsymbol{y}\int_{\mathbb{R}^n}\exp\left(-\boldsymbol{s}_2/2\right)\mathrm{d}\boldsymbol{z} \tag{4-36}$$

其中，

$$\boldsymbol{s}_1=\|\boldsymbol{y}-\boldsymbol{C}_k\boldsymbol{x}\|^2_{\boldsymbol{R}_k^{-1}}+\|\boldsymbol{y}-\boldsymbol{\beta}_k\|^2_{\boldsymbol{Y}_k}$$

以及

$$\boldsymbol{s}_2=\|\boldsymbol{z}-\overline{\boldsymbol{x}}_{k-1}\|^2_{\boldsymbol{\Sigma}_{k-1}^{-1}}-\|\boldsymbol{z}-\hat{\boldsymbol{x}}_{k-1}\|^2_{\tau^{-1}\boldsymbol{W}_{k-1}}-\|\boldsymbol{D}_{k-1}\boldsymbol{z}\|^2+\|\boldsymbol{A}_{k-1}\boldsymbol{z}-\boldsymbol{x}\|^2_{\boldsymbol{Q}_{k-1}}$$

通过利用引理4.3的完全平方技术和矩阵求逆引理，得

$$\begin{aligned}\boldsymbol{s}_1&=\|\boldsymbol{y}-\boldsymbol{\beta}_k-\left(\boldsymbol{C}_k\boldsymbol{x}-\boldsymbol{\beta}_k\right)\|^2_{\boldsymbol{R}_k^{-1}}+\|\boldsymbol{y}-\boldsymbol{\beta}_k\|^2_{\boldsymbol{Y}_k}\\&=\|\boldsymbol{y}-\boldsymbol{\beta}_k-\left(\boldsymbol{R}_k+\boldsymbol{Y}_k^{-1}\right)^{-1}\left(\boldsymbol{C}_k\boldsymbol{x}-\boldsymbol{\beta}_k\right)\|^2_{\boldsymbol{R}_k+\boldsymbol{Y}_k^{-1}}+\|\boldsymbol{C}_k\boldsymbol{x}-\boldsymbol{\beta}_k\|^2_{\left(\boldsymbol{R}_k+\boldsymbol{Y}_k^{-1}\right)^{-1}}\end{aligned} \tag{4-37}$$

再次使用引理4.3，

$$\begin{aligned}\boldsymbol{s}_2=&\|\boldsymbol{z}\|_{\overline{\boldsymbol{\Sigma}}_k}-\|\hat{\boldsymbol{x}}_{k-1}\|^2_{\tau^{-1}\boldsymbol{W}_{k-1}}+\|\overline{\boldsymbol{x}}_{k-1}\|^2_{\boldsymbol{\Sigma}_{k-1}^{-1}}+\|\boldsymbol{x}\|^2_{\boldsymbol{Q}_{k-1}^{-1}}\\&-\|\boldsymbol{A}_{k-1}^{\mathrm{T}}\boldsymbol{Q}_{k-1}^{-1}\boldsymbol{x}+\boldsymbol{\Sigma}_{k-1}^{-1}\overline{\boldsymbol{x}}_{k-1}-\tau^{-1}\boldsymbol{W}_{k-1}\hat{\boldsymbol{x}}_{k-1}\|^2_{\overline{\boldsymbol{\Sigma}}_k}\end{aligned} \tag{4-38}$$

其中，$\overline{\boldsymbol{\Sigma}}_k$由式(4-32)给出。将式(4-37)和式(4-38)代入到式(4-36)，并利用高斯密度函数在其整个定义域的积分为1这一事实，可以得到

$$\begin{aligned}\boldsymbol{\alpha}_k\left(\boldsymbol{x}\right)=&\boldsymbol{S}_{k-1}|\boldsymbol{R}_k|^{-\frac{1}{2}}|\boldsymbol{R}_k^{-1}+\boldsymbol{Y}_k|^{-\frac{1}{2}}|\boldsymbol{Q}_{k-1}|^{-\frac{1}{2}}|\overline{\boldsymbol{\Sigma}}_k|^{\frac{1}{2}}\exp\left(\frac{1}{2}\left(\|\hat{\boldsymbol{x}}_{k-1}\|^2_{\tau^{-1}\boldsymbol{W}_{k-1}}-\|\overline{\boldsymbol{x}}_{k-1}\|^2_{\boldsymbol{\Sigma}_{k-1}^{-1}}\right)\right)\\&\times\exp\left(-\frac{1}{2}\left(\|\boldsymbol{C}_k\boldsymbol{x}-\boldsymbol{\beta}_k\|^2_{(\boldsymbol{R}_k+\boldsymbol{Y}_k^{-1})^{-1}}+\|\boldsymbol{x}\|^2_{\boldsymbol{Q}_{k-1}^{-1}}\right.\right.\\&\left.\left.-\|\boldsymbol{A}_{k-1}^{\mathrm{T}}\boldsymbol{Q}_{k-1}^{-1}\boldsymbol{x}+\boldsymbol{\Sigma}_{k-1}^{-1}\overline{\boldsymbol{x}}_{k-1}-\tau^{-1}\boldsymbol{W}_{k-1}\hat{\boldsymbol{x}}_{k-1}\|^2_{\overline{\boldsymbol{\Sigma}}_k}\right)\right)\end{aligned} \tag{4-39}$$

注意，式(4-36)中对于z的积分取值有限当且仅当$\boldsymbol{A}_k^{\mathrm{T}}\boldsymbol{Q}_k^{-1}\boldsymbol{A}_k+\boldsymbol{\Sigma}_k^{-1}-\tau^{-1}\boldsymbol{W}_k-\boldsymbol{D}_k^{\mathrm{T}}\boldsymbol{D}_k>0$。再次使用引理4.3，得到

$$\boldsymbol{\alpha}_k\left(\boldsymbol{x}\right)=\boldsymbol{S}_k\exp\left(-\frac{1}{2}\|\boldsymbol{x}-\overline{\boldsymbol{x}}_k\|^2_{\boldsymbol{\Sigma}_k^{-1}}\right) \tag{4-40}$$

这里，当$\gamma_k=0$时，$\boldsymbol{\Sigma}_k$、$\overline{\boldsymbol{x}}_k$和$\boldsymbol{S}_k$的表达式由式(4-32)和式(4-33)给出。类似地，可以证明对于$\gamma_k=1$，式(4-40)成立。证毕。■

4.3.2 动态规划

利用迭代期望法则（文献[108]，引理2.4.8）和$\boldsymbol{\alpha}_k(\boldsymbol{x}_k)$的定义，LEQG代价[式(4-17)]可以重写为

$$\begin{aligned}J^L(\tau)&=\mathbb{E}\left[\mathbb{E}\left[\exp\left(\sum_{k=0}^{T}\overline{\boldsymbol{\Psi}}_k(\boldsymbol{x}_k,\hat{\boldsymbol{x}}_k)\right)\mid\mathcal{I}_T\right]\right]\\&=\mathbb{E}\left[\frac{\int_{\mathbb{R}^n}\boldsymbol{\alpha}_T(\boldsymbol{x}_T)\exp\left(\overline{\boldsymbol{\Psi}}_T(\boldsymbol{x}_T,\hat{\boldsymbol{x}}_T)\right)\mathrm{d}\boldsymbol{x}_T}{f(\mathcal{I}_T)}\right]\end{aligned}\tag{4-41}$$

这表明LEQG代价是一个关于信息状态$\boldsymbol{\alpha}_T(\boldsymbol{x}_T)$的函数。下面，将基于定理4.1中$\boldsymbol{\alpha}_k(\boldsymbol{x}_k)$的递归结果，通过解决一个DP问题来得到事件触发LEQG估计器。

定理 4.2

最小化事件触发LEQG代价$\mathbb{E}\left[\exp\left(\tau^{-1}V_\tau\right)\right]$的$\boldsymbol{x}_{0T}$最优估计值$\hat{\boldsymbol{x}}_{0|0},\cdots,\hat{\boldsymbol{x}}_{T|T}$可以通过求解如下DP方程获得：

$$J_T\left(\overline{\boldsymbol{x}}_T\right)=\inf_{\hat{\boldsymbol{x}}_T}\int_{\mathbb{R}^n}\boldsymbol{\alpha}_T(\boldsymbol{x}_T)\exp\left(\overline{\boldsymbol{\Psi}}_T(\boldsymbol{x}_T,\hat{\boldsymbol{x}}_T)\right)\mathrm{d}\boldsymbol{x}_T/f(\mathcal{I}_T)\tag{4-42}$$

$$J_k\left(\overline{\boldsymbol{x}}_k\right)=\inf_{\hat{\boldsymbol{x}}_k}\mathbb{E}\left[J_{k+1}\left(\overline{\boldsymbol{x}}_{k+1}\right)\mid\mathcal{I}_k\right],\ k=T-1,\cdots,0\tag{4-43}$$

其中，$J_k\left(\overline{\boldsymbol{x}}_k\right)$是$k$时刻的最优值函数，且数学期望是针对$\boldsymbol{y}_{k+1}$和$\boldsymbol{\gamma}_{k+1}$。此外，最优的LEQG代价为

$$\inf_{\hat{\boldsymbol{x}}_{0T}}J^L(\tau)=\mathbb{E}\left[J_0\left(\overline{\boldsymbol{x}}_0\right)\right]\tag{4-44}$$

这里是对$\boldsymbol{y}_0$和γ_0求期望。

证明：

记

$$\overline{J}:=\frac{\int_{\mathbb{R}^n}\boldsymbol{\alpha}_T(\boldsymbol{x}_T)\exp\left(\overline{\boldsymbol{\Psi}}_T(\boldsymbol{x}_T,\hat{\boldsymbol{x}}_T)\right)\mathrm{d}\boldsymbol{x}_T}{f(\mathcal{I}_T)}\tag{4-45}$$

令$J_k^*\left(\bar{\boldsymbol{x}}_k\right)$表示式(4-16)的子问题（时间$k$到$T$）的最优代价：

$$J_k^*\left(\bar{\boldsymbol{x}}_k\right) := \inf_{\hat{x}_{kT}} \mathbb{E}\left[\bar{J} \mid \mathcal{I}_k\right] \tag{4-46}$$

下面将会通过数学归纳法证明

$$J_k^*\left(\bar{\boldsymbol{x}}_k\right) = J_k\left(\bar{\boldsymbol{x}}_k\right),\quad k = 0,1,\cdots,T \tag{4-47}$$

这使得在$k = 0$时，证明了

$$\inf_{\hat{x}_{0T}} \mathbb{E}\left[\bar{J} \mid \mathcal{I}_0\right] = J_0^*\left(\bar{\boldsymbol{x}}_0\right) = J_0\left(\bar{\boldsymbol{x}}_0\right) \tag{4-48}$$

成立，且证明了由式(4-42)和式(4-43)生成的$\hat{\boldsymbol{x}}_{0|0},\cdots,\hat{\boldsymbol{x}}_{T|T}$的确最小化了LEQG代价。

首先，对于$k = T$，因为$\bar{J}$是关于$\mathcal{I}_T$的一个函数，所以根据式(4-46)可知

$$J_T^*\left(\bar{\boldsymbol{x}}_T\right) = \inf_{\hat{x}_T} \bar{J} \tag{4-49}$$

因此，$J_T^*\left(\bar{\boldsymbol{x}}_T\right) = J_T\left(\bar{\boldsymbol{x}}_T\right)$。假设在$k+1$时刻有$J_{k+1}\left(\bar{\boldsymbol{x}}_{k+1}\right) = J_{k+1}^*\left(\bar{\boldsymbol{x}}_{k+1}\right)$，那么在$k$时刻，

$$\begin{aligned}
J_k^*\left(\bar{\boldsymbol{x}}_k\right) &= \inf_{\hat{x}_{kT}} \mathbb{E}\left[\bar{J} \mid \mathcal{I}_k\right] \\
&= \inf_{\hat{x}_k} \mathbb{E}\left[\inf_{\hat{x}_{k+1T}} \mathbb{E}\left[\bar{J} \mid \mathcal{I}_{k+1}\right] \mid \mathcal{I}_k\right] = \inf_{\hat{x}_k} \mathbb{E}\left[J_{k+1}^*\left(\bar{\boldsymbol{x}}_{k+1}\right) \mid \mathcal{I}_k\right] \\
&= \inf_{\hat{x}_k} \mathbb{E}\left[J_{k+1}\left(\bar{\boldsymbol{x}}_{k+1}\right) \mid \mathcal{I}_k\right] = J_k\left(\bar{\boldsymbol{x}}_k\right)
\end{aligned} \tag{4-50}$$

其中，第二个等式由迭代期望法则和最优性原则[123]得到；第三个等式可根据$J_{k+1}^*\left(\bar{\boldsymbol{x}}_{k+1}\right)$的定义得知。进一步可得

$$\inf_{\hat{x}_{0T}} J^L(\tau) = \inf_{\hat{x}_{0T}} \mathbb{E}\left[\bar{J}\right] = \inf_{\hat{x}_{0T}} \mathbb{E}\left[\mathbb{E}\left[\bar{J} \mid \mathcal{I}_0\right]\right] = \mathbb{E}\left[J_0\left(\bar{\boldsymbol{x}}_0\right)\right] \tag{4-51}$$

其中，第一个等式来自式(4-41)和式(4-45)；第三个等式来源于式(4-48)。

证毕 ■

在DP方程(4-42)和(4-43)中，$\boldsymbol{\Sigma}_k^{-1}\bar{\boldsymbol{x}}_k$和$\boldsymbol{\Sigma}_k^{-1}$是状态变量，它们是根据定理4.1中的前向递归表达式进行动态演变，$\hat{\boldsymbol{x}}_k$是控制变量，最后$\gamma_{k+1} \in \{0,1\}$和$\boldsymbol{y}_{k+1} \in \mathbb{R}^m$是扰动变量。虽然定理4.2给出最优估计值的计算表达式，然而该表达式不是一个解析的形式。即使在一些特例中，比如在事件触发条件中令$\boldsymbol{\beta}_k = 0$和$\boldsymbol{Y}_k = a\boldsymbol{R}_k$($0 < a < \infty$是个标量因子)，也无法得到解析解。需要借助DP算法的数值执行方式来近似地求解该DP方

程，这对计算方面提出了挑战。具体地，首先需要离散化$\hat{\boldsymbol{x}}_k$和$\boldsymbol{y}_{k+1}$，假定在它们的每个分量上离散化c个点，分别得到c^n和c^m个点。然后根据DP问题的状态方程，也就是式(4-32)和式(4-33)，可知$\boldsymbol{\Sigma}_k^{-1}$状态空间的维数是$2^{k+1}$，且记$\boldsymbol{\Sigma}_k^{-1}\bar{\boldsymbol{x}}_k$的状态空间维数为$\mathbb{D}_k$，可以递归地计算出：

$$\mathbb{D}_k=\left(c^m+1\right)c^n\mathbb{D}_{k-1},\ \ \mathbb{D}_0=c^m+1 \tag{4-52}$$

它随着k的增长呈指数增长。注意，为了方便计算$\mathbb{D}_k$，已令β_k是个常数。因此，对于有限时域T，k阶段时的DP算法计算操作数是$c^n c^m\mathbb{D}_k=c^{(k+1)n}\left(c^m+1\right)^{k+1}c^m$（文献[123]，第6章），这表明计算复杂度与所考虑的系统模型的维数和时域T呈指数关系。所以，DP方法的计算复杂度为$O\left(\sum_{k=0}^{T-1}c^{(k+1)n}\left(c^m+1\right)^{k+1}c^m\right)$。

注释 4.4 文献[104]中显示，对于不存在不确定性的线性高斯系统，随机性事件触发条件下可以得到MMSE估计器的封闭形式表达式。然而，在具有相对熵约束的不确定系统情况下随机性事件触发条件下最优的LQEG估计值$\hat{\boldsymbol{x}}_{0|0},\cdots,\hat{\boldsymbol{x}}_{T|T}$和最优的LEQG代价$\mathbb{E}\left[J_0\left(\bar{\boldsymbol{x}}_0\right)\right]$没有解析的形式。这是由于DP、指数二次代价和事件触发器引入的二位决策变量γ_k一起的影响。具体地说，DP方程(4-43)中的期望是关于所有$k\in\mathbb{N}_{0:T-1}$时刻的γ_{k+1}和y_{k+1}，这导致在每个时刻k出现了多个不同的关于决策变量$\hat{\boldsymbol{x}}_k$的指数二次函数。因此，$\hat{\boldsymbol{x}}_k$的最优解不能解析地表达为$\bar{\boldsymbol{x}}_k$的函数。注意到这与时间触发的LEQG状态估计问题[73,114]不同。因为对于时间触发的情形，每个时刻k只有一个关于$\hat{\boldsymbol{x}}_k$指数二次函数，所以$\hat{\boldsymbol{x}}_k$是$\bar{\boldsymbol{x}}_k$的线性函数。另外值得提及的是，与传统Bayes推论相比，由于文献[94,107,120-122]使用的概率测度变换方法不会导致不同的结果，因此不能指望通过概率测度变换方法得到一个解析的结果。

考虑实际的实施性，接下来将进一步探索一个更简单也更有趣的一步的事件触发最小最大估计问题。下一节结果表明，此问题描述使得可以导出一个具有封闭形式解的一步事件触发估计器。

4.4
一步事件触发最小最大估计

一步的估计问题可以这样来描述：在每个时刻 k，仅仅将$\hat{\boldsymbol{x}}_k$视作决策变量并假设 k 时刻之前的所有估计值已经被确定。这意味着估计误差代价 V_e[式 (4-9)] 被松弛为

$$V_{e,k}=\frac{1}{2}\sum_{i=0}^{k-1}\|\boldsymbol{x}_i-\hat{\boldsymbol{x}}_{i|i}\|_{\boldsymbol{W}_i}^2+\frac{1}{2}\|\boldsymbol{x}_k-\hat{\boldsymbol{x}}_k\|_{\boldsymbol{W}_k}^2 \tag{4-53}$$

其中，估计值$\hat{\boldsymbol{x}}_{i|i}, i<k$已知，$\hat{\boldsymbol{x}}_{k|k}$有待于被确定。当然，这里要求$\hat{\boldsymbol{x}}_{k|k}$是关于$\mathcal{I}_k$的函数。在现有文献中，一步的策略已经被广泛应用在估计问题中[94,97,124]。

此时，原始的最优问题变成了一个前向递归的估计问题，这使得在 k 时刻设计最小最大估计器时，信息集合 $\mathcal{I}_k$ 可以得到利用。为此，在给定的当前信息集合 $\mathcal{I}_k$ 下，定义一个条件的扰动测度$\tilde{\mathbb{P}}(\cdot|\mathcal{I}_k)$和条件的名义测度$\tilde{\mathbb{P}}(\cdot|\mathcal{I}_k)$之间的相对熵：

$$h_0\left(\tilde{\mathbb{P}}(\cdot|\mathcal{I}_k)\|\mathbb{P}(\cdot|\mathcal{I}_k)\right):=\begin{cases}\tilde{\mathbb{E}}\left[\log\dfrac{\mathrm{d}\tilde{\mathbb{P}}(\cdot|\mathcal{I}_k)}{\mathrm{d}\mathbb{P}(\cdot|\mathcal{I}_k)}\middle|\mathcal{I}_k\right], & \tilde{\mathbb{P}}(\cdot|\mathcal{I}_k)\ll\mathbb{P}(\cdot|\mathcal{I}_k)\\ +\infty, & \text{其他}\end{cases} \tag{4-54}$$

所有满足$h_0\left(\tilde{\mathbb{P}}(\cdot|\mathcal{I}_k)\|\mathbb{P}(\cdot|\mathcal{I}_k)\right)<+\infty$的$\tilde{\mathbb{P}}(\cdot|\mathcal{I}_k)$的测度表示为集合$\mathcal{P}_k$。与之前定义的相对熵$h(\tilde{\mathbb{P}}\|\mathbb{P})$比较，$h_0\left(\tilde{\mathbb{P}}(\cdot|\mathcal{I}_k)\|\mathbb{P}(\cdot|\mathcal{I}_k)\right)$可以被视为一个后验相对熵。在HMMs中，已有这样的一个后验相对熵被刻画[125]。此时，定义4.1中的相对熵应改成相应的后验形式：

$$h_0\left(\tilde{\mathbb{P}}(\cdot|\mathcal{I}_k)\|\mathbb{P}(\cdot|\mathcal{I}_k)\right)\leqslant\tilde{\mathbb{E}}\left[V_{u,k}|\mathcal{I}_k\right]+d,\quad V_{u,k}:=\frac{1}{2}\sum_{i=0}^{k}\|\boldsymbol{D}_i\boldsymbol{x}_i\|^2 \tag{4-55}$$

另外，表示所有可容许测度$\tilde{P}(\cdot|\mathcal{I}_k)$为集合$\Xi_k$。那么最优估计问题$(P_0)$被松弛为

$$(P_1):\quad \inf_{\hat{x}_k}\sup_{\tilde{\mathbb{P}}(\cdot|\mathcal{I}_k)\in\Xi_k}\tilde{\mathbb{E}}\left[V_{e,k}|\mathcal{I}_k\right] \tag{4-56}$$

注释 4.5　注意到，原始的事件触发最小最大最优估计问题(P_0')是在有限时域T范围内，但一步形式的问题可以考虑无穷时域范围。所以，对于一步问题，时间k不再局限于有限时域T上。

犹如在4.2节中原始的最优问题描述，通过拉格朗日乘子法（引理4.1），(P_1)可以被变换为如下无约束问题：

$$J_{\tau_k} := \tau_k\left\{\inf_{\hat{x}_k}\sup_{\tilde{\mathbb{P}}\in\mathcal{P}_k}\left[\tilde{\mathbb{E}}\left[\hat{\boldsymbol{\Psi}}_{0,k-1}+\overline{\boldsymbol{\Psi}}_k\left(\boldsymbol{x}_k,\hat{\boldsymbol{x}}_k\right)|\mathcal{I}_k\right]-h_0\left(\tilde{\mathbb{P}}(\cdot|\mathcal{I}_k)\|\mathbb{P}(\cdot|\mathcal{I}_k)\right)\right]+d\right\}$$

$$\inf_{\hat{x}_k}\sup_{\tilde{\mathbb{P}}(\cdot|\Gamma_k)\in\Xi_k}\tilde{\mathbb{E}}\left[V_{e,k}|\mathcal{I}_k\right]=\inf_{\tau_k\in\Gamma_k}J_{\tau_k} \tag{4-57}$$

其中，函数$\overline{\boldsymbol{\Psi}}_k$由式(4-18)定义，但$\tau$为时变，使得

$$\overline{\boldsymbol{\Psi}}_k\left(\boldsymbol{x}_k,\hat{\boldsymbol{x}}_k\right)=\frac{1}{2}\left(\tau_k^{-1}\|\boldsymbol{x}_k-\hat{\boldsymbol{x}}_k\|_{\boldsymbol{W}_k}^2+\|\boldsymbol{D}_k\boldsymbol{x}_k\|^2\right)$$

以及$k\geqslant 1$时，$\hat{\boldsymbol{\Psi}}_{0,k-1}=\sum_{i=0}^{k-1}\frac{1}{2}\left(\tau_k^{-1}\|\boldsymbol{x}_i-\hat{\boldsymbol{x}}_{i|i}\|_{\boldsymbol{W}_k}^2+\|\boldsymbol{D}_i\boldsymbol{x}_i\|^2\right), k=0$时有$\hat{\boldsymbol{\Psi}}_{0,-1}=0$。这里，$\Gamma_k:=\left\{\tau_k:\tau_k>0且J_{\tau_k}<\infty\right\}$。利用对偶性关系[118]，$(P_1)$可以进一步转换为如下等价问题：

$$(P_1'):\inf_{\tau_k\in\Gamma_k}\tau_k\left\{\left(\inf_{\hat{x}_k}\log J^L\left(\tau_k\right)\right)+d\right\} \tag{4-58}$$

其中，

$$J^L\left(\tau_k\right)=\mathbb{E}\left[\exp\left(\hat{\Psi}_{0,k-1}+\overline{\Psi}_k\left(x_k,\hat{x}_k\right)\right)|\mathcal{I}_k\right] \tag{4-59}$$

是一步事件触发LEQG代价。因此，通过导出一步LEQG估计器并得到

$$\tau_k^*=\arg\inf_{\tau_k\in\Gamma_k}\tau_k\left(\log\inf_{\hat{x}_k}J^L\left(\tau_k\right)+d\right)$$

可以获得一步事件触发最小最大估计器及相应的最小最大代价[式(4-58)]。

4.4.1 一步事件触发 LEQG 估计

根据 LEQG 代价 [式 (4-59)]，得到如下一步事件触发 LEQG 估计问题：

$$\hat{\boldsymbol{x}}_{k|k} = \arg\min_{\hat{\boldsymbol{x}}_k} \mathbb{E}\left[\exp\left(\hat{\boldsymbol{\Psi}}_{0,k-1} + \overline{\boldsymbol{\Psi}}_k\left(\boldsymbol{x}_k, \hat{\boldsymbol{x}}_k\right)\right) | \mathcal{I}_k\right] \tag{4-60}$$

对于该问题，需要在一步的情况下重新定义信息状态$\boldsymbol{\alpha}_k\left(\boldsymbol{x}_k\right)$，使得

$$\boldsymbol{\alpha}_k\left(\boldsymbol{x}_k\right) := \int_{\mathbb{R}^n} \dots \int_{\mathbb{R}^n} \exp\left(\hat{\boldsymbol{\Psi}}_{0,k-1}\right) f\left(\boldsymbol{x}_{0k}, \mathcal{I}_k\right) \mathrm{d}\boldsymbol{x}_{0k-1} \tag{4-61}$$

且初始值仍然是$\boldsymbol{\alpha}_0\left(\boldsymbol{x}_0\right) := f\left(\boldsymbol{x}_0, \mathcal{I}_0\right)$。同样，式(4-33)中$\boldsymbol{\alpha}_k\left(\boldsymbol{x}_k\right)$的递归需要做些改变来符合重新定义的$\boldsymbol{\alpha}_k\left(\boldsymbol{x}_k\right)$。

基于该重新定义的信息状态，可以给出如下结果。

定理 4.3

对于一步事件触发 LEQG 估计问题 (4-60)，最优估计值$\hat{\boldsymbol{x}}_{k|k}$为

$$\hat{\boldsymbol{x}}_{k|k} = \boldsymbol{K}_k \overline{\boldsymbol{x}}_k,\ \boldsymbol{K}_k = \boldsymbol{I}_n + \left(\boldsymbol{\Sigma}_k^{-1} - \boldsymbol{D}_k^{\mathrm{T}} \boldsymbol{D}_k\right)^{-1} \boldsymbol{D}_k^{\mathrm{T}} \boldsymbol{D}_k \tag{4-62}$$

其中$\overline{\boldsymbol{x}}_k$和$\boldsymbol{\Sigma}_k$由如下递归方程给出：

$$\begin{cases} \boldsymbol{\Sigma}_t^{-1} \overline{\boldsymbol{x}}_t = \boldsymbol{Q}_{t-1}^{-1} \boldsymbol{A}_{t-1} \overline{\boldsymbol{\Sigma}}_t \left(\boldsymbol{\Sigma}_{t-1}^{-1} \overline{\boldsymbol{x}}_{t-1} - \tau_k^{-1} \boldsymbol{W}_{t-1} \hat{\boldsymbol{x}}_{t-1|t-1}\right) \\ \qquad + \boldsymbol{C}_t^{\mathrm{T}} \left[\boldsymbol{R}_t + \left(1-\gamma_t\right) \boldsymbol{Y}_t^{-1}\right]^{-1} \left[\gamma_t \boldsymbol{y}_t + \left(1-\gamma_t\right) \boldsymbol{\beta}_t\right] \\ \boldsymbol{\Sigma}_t^{-1} = \boldsymbol{C}_t^{\mathrm{T}} \left[\boldsymbol{R}_t + \left(1-\gamma_t\right) \boldsymbol{Y}_t^{-1}\right]^{-1} \boldsymbol{C}_t + \boldsymbol{Q}_{t-1}^{-1} - \boldsymbol{Q}_{t-1}^{-1} \boldsymbol{A}_{t-1} \overline{\boldsymbol{\Sigma}}_t \boldsymbol{A}_{t-1}^{\mathrm{T}} \boldsymbol{Q}_{t-1}^{-1} \\ \overline{\boldsymbol{\Sigma}}_t^{-1} = \boldsymbol{A}_{t-1}^{\mathrm{T}} \boldsymbol{Q}_{t-1}^{-1} \boldsymbol{A}_{t-1} + \boldsymbol{\Sigma}_{t-1}^{-1} - \tau_k^{-1} \boldsymbol{W}_{t-1} - \boldsymbol{D}_{t-1}^{\mathrm{T}} \boldsymbol{D}_{t-1},\ 0 < t \leqslant k \end{cases} \tag{4-63}$$

且初始值$\overline{x}_0$和Σ_0与式(4-34)相同。最优估计器[式(4-62)]具有有限的LEQG代价$J^L\left(\tau_k\right)$［式(4-59)］当且仅当下面不等式成立：

$$\boldsymbol{\Sigma}_t^{-1} - \tau_k^{-1} \boldsymbol{W}_t - \boldsymbol{D}_t^{\mathrm{T}} \boldsymbol{D}_t > 0,\ t = k \tag{4-64}$$

$$\boldsymbol{A}_t^{\mathrm{T}} \boldsymbol{Q}_t^{-1} \boldsymbol{A}_t + \boldsymbol{\Sigma}_t^{-1} - \tau_k^{-1} \boldsymbol{W}_t - \boldsymbol{D}_t^{\mathrm{T}} \boldsymbol{D}_t > 0,\ t < k \tag{4-65}$$

此外，相应的最小LEQG代价为

$$\inf_{\hat{\boldsymbol{x}}_k} J^L\left(\tau_k\right) = \frac{(2\pi)^{\frac{n}{2}} \boldsymbol{S}_k}{f\left(\mathcal{I}_k\right)} \left|\boldsymbol{\Sigma}_k^{-1} - \tau_k^{-1} \boldsymbol{W}_k - \boldsymbol{D}_k^{\mathrm{T}} \boldsymbol{D}_k\right|^{-\frac{1}{2}} \exp\left(\frac{1}{2} \| \overline{\boldsymbol{x}}_k \|^2_{\boldsymbol{\Sigma}_k^{-1} \boldsymbol{K}_k - \boldsymbol{\Sigma}_k^{-1}}\right) \tag{4-66}$$

其中，$\boldsymbol{S}_k$由如下递归取得

$$\boldsymbol{S}_t=\boldsymbol{S}_{t-1}(2\pi)^{\frac{-m\gamma_t}{2}}\left|\boldsymbol{R}_t\right|^{-\frac{1}{2}}\left|\boldsymbol{R}_t^{-1}+\boldsymbol{Y}_t\right|^{\frac{\gamma_t-1}{2}}\left|\boldsymbol{Q}_{t-1}\right|^{-\frac{1}{2}}\left|\overline{\boldsymbol{\Sigma}}_t\right|^{\frac{1}{2}}\exp\left\{\frac{1}{2}\left[\|\hat{\boldsymbol{x}}_{t-1|t-1}\|^2_{\tau_k^{-1}\boldsymbol{W}_{t-1}}-\|\overline{\boldsymbol{x}}_{t-1}\|^2_{\boldsymbol{\Sigma}_{t-1}^{-1}}\right.\right.$$
$$\left.\left.-\|\gamma_t\boldsymbol{y}_t+(1-\gamma_t)\boldsymbol{\beta}_t\|^2_{\left[\boldsymbol{R}_t+(1-\gamma_t)\boldsymbol{Y}_t^{-1}\right]^{-1}}+\|\boldsymbol{\Sigma}_{t-1}^{-1}\overline{\boldsymbol{x}}_{t-1}-\tau_k^{-1}\boldsymbol{W}_{t-1}\hat{\boldsymbol{x}}_{t-1|t-1}\|^2_{\overline{\boldsymbol{\Sigma}}_t}+\|\overline{\boldsymbol{x}}_t\|^2_{\boldsymbol{\Sigma}_t^{-1}}\right]\right\}$$
$$0<t\leqslant k \tag{4-67}$$

证明：

该证明主要是基于引理4.2中$\overline{\boldsymbol{x}}_k$、$\boldsymbol{\Sigma}_k$和$\boldsymbol{S}_k$的递归结果。对于式(4-60)中的代价函数，有

$$\mathbb{E}\left[\exp\left(\hat{\boldsymbol{\Psi}}_{0,k-1}+\overline{\boldsymbol{\Psi}}_k(\boldsymbol{x}_k,\hat{\boldsymbol{x}}_k)\right)|\ \mathcal{I}_k\right]$$
$$=\int_{\mathbb{R}^n}\boldsymbol{\alpha}_k(\boldsymbol{x}_k)\exp\left(\overline{\boldsymbol{\Psi}}_k(\boldsymbol{x}_k,\hat{\boldsymbol{x}}_k)\right)\mathrm{d}\boldsymbol{x}_k\ /\ f(\mathcal{I}_k)$$
$$=\frac{\boldsymbol{S}_k}{f(\mathcal{I}_k)}\int_{\mathbb{R}^n}\exp\left(-\frac{1}{2}\left(\|\boldsymbol{x}_k-\overline{\boldsymbol{x}}_k\|^2_{\boldsymbol{\Sigma}_k^{-1}}-\|\boldsymbol{x}_k-\hat{\boldsymbol{x}}_k\|^2_{\tau_k^{-1}\boldsymbol{W}_k}-\|\boldsymbol{D}_k\boldsymbol{x}_k\|^2\right)\right)\mathrm{d}\boldsymbol{x}_k$$
$$=(2\pi)^{\frac{n}{2}}\left|\boldsymbol{\Sigma}_k^{-1}-\tau_k^{-1}\boldsymbol{W}_k-\boldsymbol{D}_k^{\mathrm{T}}\boldsymbol{D}_k\right|^{-\frac{1}{2}}\boldsymbol{S}_k\ /\ f(\mathcal{I}_k)\exp(\boldsymbol{q}_k/2) \tag{4-68}$$

其中，

$$\boldsymbol{q}_k=\|\hat{x}_k\|^2_{\tau_k^{-1}\boldsymbol{W}_k}-\|\overline{\boldsymbol{x}}_k\|^2_{\boldsymbol{\Sigma}_k^{-1}}+\|\boldsymbol{\Sigma}_k^{-1}\overline{\boldsymbol{x}}_k-\tau_k^{-1}\boldsymbol{W}_k\hat{\boldsymbol{x}}_k\|^2_{\left(\boldsymbol{\Sigma}_k^{-1}-\tau_k^{-1}\boldsymbol{W}_k-\boldsymbol{D}_k^{\mathrm{T}}\boldsymbol{D}_k\right)^{-1}}$$

且上式中积分存在当且仅当

$$\boldsymbol{\Sigma}_k^{-1}-\tau_k^{-1}\boldsymbol{W}_k-\boldsymbol{D}_k^{\mathrm{T}}\boldsymbol{D}_k>0 \tag{4-69}$$

注意到，不等式(4-69)自身保证了$\boldsymbol{q}_k$在$\hat{\boldsymbol{x}}_k$上为正。那么通过求解$\partial\boldsymbol{q}_k/\partial\hat{\boldsymbol{x}}_k=0$，可得到一步决策情形下的最优估计值$\hat{\boldsymbol{x}}_{k|k}$：

$$\hat{\boldsymbol{x}}_{k|k}=\boldsymbol{K}_k\overline{\boldsymbol{x}}_k,\boldsymbol{K}_k=\boldsymbol{I}_n+\left(\boldsymbol{\Sigma}_k^{-1}-\boldsymbol{D}_k^{\mathrm{T}}\boldsymbol{D}_k\right)^{-1}\boldsymbol{D}_k^{\mathrm{T}}\boldsymbol{D}_k \tag{4-70}$$

这里$\overline{\boldsymbol{x}}_k$和$\boldsymbol{\Sigma}_k$由递归方程(4-63)得到，且根据定理4.1可知，对于所有的$t<k$，要求满足如下条件

$$\boldsymbol{A}_t^{\mathrm{T}}\boldsymbol{Q}_t^{-1}\boldsymbol{A}_t+\boldsymbol{\Sigma}_t^{-1}-\tau_k^{-1}\boldsymbol{W}_t-\boldsymbol{D}_t^{\mathrm{T}}\boldsymbol{D}_t>0 \tag{4-71}$$

将式(4-70)代入式(4-68)中，可以得到式(4-66)。证毕。 ■

4.4.2 一步事件触发最小最大估计器

由问题(P_1')和重新描述的最小LEQG代价[式(4-66)]可知，需要

$f\left(\mathcal{I}_k\right)$的一个明确表达式来确定最优的拉格朗日乘子$\boldsymbol{\tau}_k^*$。接下来，定义一个信息状态$\eta_k\left(\boldsymbol{x}_k\right)$来帮助求解$f\left(\mathcal{I}_k\right)$。

定义 4.3

定义信息状态$\boldsymbol{\eta}_k\left(\boldsymbol{x}_k\right)$为

$$\boldsymbol{\eta}_k\left(\boldsymbol{x}_k\right):=f\left(\boldsymbol{x}_k,\mathcal{I}_k\right) \tag{4-72}$$

其中，$f\left(\boldsymbol{x}_k,\mathcal{I}_k\right)$代表$\boldsymbol{x}_k$和$\mathcal{I}_k$的联合分布。

引理 4.4

信息状态$\boldsymbol{\eta}_k\left(\boldsymbol{x}_k\right)$是一个非标准化的高斯密度，给出如下：

$$\boldsymbol{\eta}_k\left(\boldsymbol{x}_k\right)=\boldsymbol{Z}_k\exp\left(-\frac{1}{2}\|\boldsymbol{x}_k-\tilde{\boldsymbol{x}}_k\|_{\boldsymbol{\Lambda}_k^{-1}}^2\right) \tag{4-73}$$

其中，$\tilde{\boldsymbol{x}}_k$、$\boldsymbol{\Lambda}_k$和$\boldsymbol{Z}_k$满足如下递归方程：

$$\begin{cases}\tilde{\boldsymbol{x}}_k=\boldsymbol{A}_{k-1}\tilde{\boldsymbol{x}}_{k-1}+\boldsymbol{\Lambda}_k\boldsymbol{C}_k^{\mathrm{T}}\left[\boldsymbol{R}_k+\left(1-\gamma_k\right)\boldsymbol{Y}_k^{-1}\right]^{-1}\left[\gamma_k\boldsymbol{y}_k+\left(1-\gamma_k\right)\boldsymbol{\beta}_k-\boldsymbol{C}_k\boldsymbol{A}_{k-1}\tilde{\boldsymbol{x}}_{k-1}\right]\\ \boldsymbol{\Lambda}_k^{-1}=\left[\boldsymbol{A}_{k-1}\boldsymbol{\Lambda}_{k-1}\boldsymbol{A}_{k-1}^{\mathrm{T}}+\boldsymbol{Q}_{k-1}\right]^{-1}+\boldsymbol{C}_k^{\mathrm{T}}\left[\boldsymbol{R}_k+\left(1-\gamma_k\right)\boldsymbol{Y}_k^{-1}\right]^{-1}\boldsymbol{C}_k\\ \boldsymbol{Z}_k=\boldsymbol{Z}_{k-1}\left(2\pi\right)^{\frac{-m\gamma_k}{2}}\left|\boldsymbol{R}_k\right|^{-\frac{1}{2}}\left|\boldsymbol{R}_k^{-1}+\boldsymbol{Y}_k\right|^{\frac{\gamma_k-1}{2}}\left|\boldsymbol{Q}_{k-1}\right|^{-\frac{1}{2}}\left|\bar{\boldsymbol{\Lambda}}_k\right|^{\frac{1}{2}}\exp\left\{\frac{1}{2}\left[\|\tilde{\boldsymbol{x}}_{k-1}\|_{\boldsymbol{\Lambda}_{k-1}^{-1}\bar{\boldsymbol{\Lambda}}_k\boldsymbol{\Lambda}_{k-1}^{-1}-\boldsymbol{\Lambda}_{k-1}^{-1}}^2\right.\right.\\ \qquad\left.\left.-\|\gamma_k\boldsymbol{y}_k+\left(1-\gamma_k\right)\boldsymbol{\beta}_k\|_{\left[\boldsymbol{R}_k+\left(1-\gamma_k\right)\boldsymbol{Y}_k^{-1}\right]^{-1}}^2+\|\tilde{\boldsymbol{x}}_k\|_{\boldsymbol{\Sigma}_k^{-1}}^2\right]\right\}\\ \bar{\boldsymbol{\Lambda}}_k^{-1}=\boldsymbol{A}_{k-1}^{\mathrm{T}}\boldsymbol{Q}_{k-1}^{-1}\boldsymbol{A}_{k-1}+\boldsymbol{\Lambda}_{k-1}^{-1}\end{cases} \tag{4-74}$$

$\tilde{\boldsymbol{x}}_k$、$\boldsymbol{\Lambda}_k$、$\boldsymbol{Z}_k$的初始值分别与式(4-34)中的$\bar{\boldsymbol{x}}_k$、$\boldsymbol{\Sigma}_k$、$\boldsymbol{S}_k$相等。

证明：

根据定义 4.2 可知，如果$\tau^{-1}\to 0$和$\boldsymbol{D}_k=0,\boldsymbol{\alpha}_k\left(\boldsymbol{x}_k\right)$则降为$\boldsymbol{\eta}_k\left(\boldsymbol{x}_k\right)$。那么，通过让$\tau^{-1}\to 0$和$\boldsymbol{D}_k=0,\boldsymbol{\eta}_k\left(\boldsymbol{x}_k\right)$的递归结果可由定理 4.1 直接得出。证毕。■

根据$\boldsymbol{\eta}_k\left(\boldsymbol{x}_k\right)$的定义和引理 4.4，可以得到

$$f\left(\mathcal{I}_k\right)=\int_{\mathbb{R}^n}\boldsymbol{\eta}_k\left(\boldsymbol{x}_k\right)\mathrm{d}\boldsymbol{x}_k=\left(2\pi\right)^{\frac{n}{2}}\left|\boldsymbol{\Lambda}_k\right|^{\frac{1}{2}}\boldsymbol{Z}_k \tag{4-75}$$

将定理4.3、引理4.4以及式(4-75)应用到最小最大问题(P_1')，可以得到如下命题。

命题 4.1

对于一步事件触发最小最大估计问题(P_1')，最小最大估计器由相应的 LEQG 估计器（定理 4.3）给出，且

$$\tau_k^* = \arg\inf_{\tau_k \in \Gamma_k} \tau_k \left\{ \left(\inf_{\hat{x}_k} \log J^L(\tau_k) \right) + d \right\} \tag{4-76}$$

其中，

$$\inf_{\hat{x}_k} \log J^L(\tau_k) = \log \boldsymbol{S}_k - \log \boldsymbol{Z}_k + 0.5 \Big(\| \overline{\boldsymbol{x}}_k \|^2_{\boldsymbol{\Sigma}_k^{-1}\boldsymbol{K}_k - \boldsymbol{\Sigma}_k^{-1}}$$
$$- \log \left(\left| \boldsymbol{\Sigma}_k^{-1} - \tau_k^{-1} \boldsymbol{W}_k - \boldsymbol{D}_k^{\mathrm{T}} \boldsymbol{D}_k \| \boldsymbol{\Lambda}_k \right| \right) \Big).$$

此外，生成的一步最小最大代价是$\tau_k^* \left(\log \inf_{\hat{x}_k} J^L(\tau_k^*) + d \right)$。

注释 4.6　注意到，$\tau_k \left(\log \inf_{\hat{x}_k} J^L(\tau_k) + d \right) < \infty$可以由约束$\tau_k \in \Gamma_k$来保证，该约束与条件 (4-64) 和 (4-65) 等价。虽然不容易从式 (4-76) 中得到τ_k^*的明确表达式，但是可以在$\tau_k \in [\tau_{k,\mathrm{b}}, \infty)$数值中找到$\tau_k^*$，这里$\tau_{k,b}$是个关键值，小于它则条件 (4-64) 和 (4-65) 不成立。然后，可以通过τ_k^*得到每个时刻的一步最小最大代价 [式 (4-58)]，并且该最小最大代价保证了估计性能。另外，从式(4-64) 和式 (4-65) 可知，如果不确定性矩阵 $\boldsymbol{D}_k$ 非常大，$\tau_{k,\mathrm{b}}$将不会存在。因此，$\boldsymbol{D}_k$ 存在一个上界，这是利用相对熵约束方法处理不确定系统估计和控制问题的一个限制 [113-114]。

注释 4.7　记$\theta_k = \tau_k^{-1}$，可以注意到所提出的一步最小最大估计器和事件触发的风险敏感估计器 [122] 有着类似的形式。参数τ_k^{-1}扮演着与风险敏感参数类似的角色。这是由于事实上一步最小最大估计问题也涉及一个指数二次代价，与风险敏感代价一致。不同的是式 (4-60) 的指数上的二次函数包含$\| \boldsymbol{D}_k \boldsymbol{x}_k \|^2$项，该项由相对熵约束 [式 (4-55)] 引入。加之，本节所得的估计算法涉及 τ_k 的优化，然而文献 [122] 中没有有效的理论方法找出最优风险敏感参数。仿真部分将会提供与风险敏感估计器的比较，将表明 $\boldsymbol{D}_k$ 的存在能够提升估计性能的鲁棒性。

4.5 多传感器场景

本节将之前的估计结果采用序贯融合的方式扩展到多传感器系统场景。考虑名义系统中有 M 个传感器：

$$\boldsymbol{y}_k^i = \boldsymbol{C}_k^i \boldsymbol{x}_k + \boldsymbol{v}_k^i, i = 1, 2, \cdots, M \tag{4-77}$$

其中，$\boldsymbol{y}_k^i \in \mathbb{R}^{m_i}$是第$i$个传感器测量值；$\boldsymbol{v}_k^i, i \in \mathbb{N}_{1:M}$是互相独立的高斯噪声，密度函数为$\phi_k^i \sim \mathcal{N}\left(0, \boldsymbol{R}_k^i\right)$且独立于$\boldsymbol{w}_k$和$\boldsymbol{x}_0$。那么，名义概率测度$\mathbb{P}$被定义在噪声空间$\left\{\boldsymbol{x}_0, \boldsymbol{w}_{0T}, \boldsymbol{v}_{0T}^1, \cdots, \boldsymbol{v}_{0T}^M\right\}$。

现在考虑每个传感器 i 配备一个事件触发器，由此来决定是否将$\boldsymbol{y}_k^i$发送至远程估计器中。同样，令$\gamma_k^i \in \{0,1\}$表示$\boldsymbol{y}_k^i$的传输决策变量。定义$\mathbb{I}_k^l := \left\{i \in \mathbb{N}_{1:l} \mid \gamma_i = 0\right\}, 1 \leqslant l \leqslant M$，进一步定义

$$\mathbb{I}_k^l := \bigcup_{i \in \mathbb{N}_{0:k-1}} \left\{\left\{\gamma_i^j \mid j \in \mathbb{I}_k^M\right\} \cup \left\{\boldsymbol{y}_i^j \mid j \in \mathbb{N}_{1:M} \setminus \mathbb{I}_k^M\right\}\right\}$$

$$\cup \left\{\left\{\gamma_k^j \mid j \in \mathbb{I}_k^l\right\} \cup \left\{\boldsymbol{y}_k^j \mid j \in \mathbb{N}_{1:l} \setminus \mathbb{I}_k^l\right\}\right\}$$

为估计器在k时刻可利用的信息集（在传感器$l+1$之前），这里$\mathcal{I}_0^0 := \varnothing$。记$\mathcal{I}_k^0 := \mathcal{I}_{k-1}^M$。然后对于每个传感器，一个一般性的随机事件触发方案可以被设计为

$$\Pr\left\{\gamma_k^i = 0 \mid \boldsymbol{y}_k^i, I_k^{i-1}\right\} = \exp\left(-\frac{1}{2} \| \boldsymbol{y}_k^i - \boldsymbol{\beta}_k^i \|_{\boldsymbol{Y}_k^i}^2\right) \tag{4-78}$$

其中，$\boldsymbol{Y}_k^i \in \mathbb{R}^{m_i \times m_i}$是正定矩阵；$\boldsymbol{\beta}_k^i \in \mathbb{R}^{m_i}$是一个估计器基于$\mathcal{I}_k^{i-1}$可以得知的量。

同单传感器情况，使用相对熵约束来描述系统不确定性。同样，对于多传感器情况，不能得到事件触发最优估计器的解析解。因此，考虑如$\left(P_1'\right)$的一步问题：

$$\left(P_2'\right): \inf_{\tau_{k,i} \in \Gamma_{k,i}} \tau_{k,i} \left\{\left(\inf_{\hat{\boldsymbol{x}}_k} \log J^{L,i}\left(\tau_{k,i}\right)\right) + d\right\} \tag{4-79}$$

其中，

$$J^{L,i}\left(\tau_{k,i}\right)=\mathbb{E}\left[\exp\left(\hat{\boldsymbol{\Psi}}_{0,k-1}+\overline{\boldsymbol{\Psi}}_k\left(\boldsymbol{x}_k,\hat{\boldsymbol{x}}_k\right)\right)\mid \mathcal{I}_k^i\right] \tag{4-80}$$

函数$\overline{\boldsymbol{\Psi}}_k$与$\hat{\boldsymbol{\Psi}}_{0,k-1}$和单传感器情况相同。同样有如下LEQG估计问题：

$$\hat{\boldsymbol{x}}_{k|k}^i=\arg\min_{\hat{\boldsymbol{x}}_k}\mathbb{E}\left[\exp\left(\hat{\boldsymbol{\Psi}}_{0,k-1}+\overline{\boldsymbol{\Psi}}_k\left(\boldsymbol{x}_k,\hat{\boldsymbol{x}}_k\right)\right)\mid \mathcal{I}_k^i\right] \tag{4-81}$$

并定义多传感器场景下的信息状态：

$$\boldsymbol{\alpha}_k^i\left(\boldsymbol{x}_k\right):=\int_{\mathbb{R}^n}\cdots\int_{\mathbb{R}^n}\exp\left(\hat{\boldsymbol{\Psi}}_{0,k-1}\right)f\left(\boldsymbol{x}_{0k},\mathbb{I}_k^i\right)\mathrm{d}\boldsymbol{x}_{0k-1} \tag{4-82}$$

初始值为$\boldsymbol{\alpha}_0^i\left(\boldsymbol{x}_0\right):=f\left(\boldsymbol{x}_0,\mathcal{I}_0^i\right)$。那么对于多传感器系统，有如下一步事件触发最小最大估计结果。

定理 4.4

对于每个传感器配备事件触发方案 (4-78) 的多传感器系统 (4-77)，一步事件触发最小最大估计值$\hat{\boldsymbol{x}}_{k|k}^i$为

$$\hat{\boldsymbol{x}}_{k|k}^i=\boldsymbol{K}_k^i\overline{\boldsymbol{x}}_k^i,\boldsymbol{K}_k^i=\boldsymbol{I}+\left[\left(\sum\nolimits_k^i\right)^{-1}-\boldsymbol{D}_k^{\mathrm{T}}\boldsymbol{D}_k\right]^{-1}\boldsymbol{D}_k^T\boldsymbol{D}_k,i=1,2,\cdots,M \tag{4-83}$$

其中，$\overline{\boldsymbol{x}}_k^i$和$\boldsymbol{\Sigma}_k^i$由下式在$0<t\leqslant k$上的递归求得：

① 对于$1<i\leqslant M$，

$$\begin{cases}\left(\boldsymbol{\Sigma}_t^i\right)^{-1}\overline{\boldsymbol{x}}_t^i=\left(\boldsymbol{C}_t^i\right)^{\mathrm{T}}\left[\boldsymbol{R}_t^i+\left(1-\gamma_t^i\right)\left(\boldsymbol{Y}_t^i\right)^{-1}\right]^{-1}\left[\gamma_t^i\boldsymbol{y}_t^i+\left(1-\gamma_t^i\right)\boldsymbol{\beta}_t^i\right]+\left(\Sigma_t^{i-1}\right)^{-1}\overline{\boldsymbol{x}}_t^{i-1}\\ \left(\boldsymbol{\Sigma}_t^i\right)^{-1}=\left(\boldsymbol{C}_t^i\right)^{\mathrm{T}}\left[\boldsymbol{R}_t^i+\left(1-\gamma_t^i\right)\left(\boldsymbol{Y}_t^i\right)^{-1}\right]^{-1}\boldsymbol{C}_t^i+\left(\boldsymbol{\Sigma}_t^{i-1}\right)^{-1}\end{cases} \tag{4-84}$$

② 对于$i=1$，

$$\begin{cases}\left(\boldsymbol{\Sigma}_t^i\right)^{-1}\overline{\boldsymbol{x}}_t^i=\boldsymbol{Q}_{t-1}^{-1}\boldsymbol{A}_{t-1}\overline{\boldsymbol{\Sigma}}_t\left[\left(\boldsymbol{\Sigma}_{t-1}^M\right)^{-1}\overline{\boldsymbol{x}}_{t-1}^M-\tau_{k,i}^{-1}\boldsymbol{W}_{t-1}\hat{\boldsymbol{x}}_{t-1|t-1}^M\right]\\ \qquad+\left[\boldsymbol{R}_t^i+\left(1-\gamma_t^i\right)\left(\boldsymbol{Y}_t^i\right)^{-1}\right]^{-1}\left[\gamma_t^i\boldsymbol{y}_t^i+\left(1-\gamma_t^i\right)\boldsymbol{\beta}_t^i\right]\\ \left(\boldsymbol{\Sigma}_t^i\right)^{-1}=\left(\boldsymbol{C}_t^i\right)^{\mathrm{T}}\left[\boldsymbol{R}_t^i+\left(1-\gamma_t^i\right)\left(\boldsymbol{Y}_t^i\right)^{-1}\right]^{-1}\boldsymbol{C}_t^i+\boldsymbol{Q}_{t-1}^{-1}-\boldsymbol{Q}_{t-1}^{-1}\boldsymbol{A}_{t-1}\overline{\boldsymbol{\Sigma}}_t\boldsymbol{A}_{t-1}^{\mathrm{T}}\boldsymbol{Q}_{t-1}^{-1}\\ \overline{\boldsymbol{\Sigma}}_t^{-1}=\boldsymbol{A}_{t-1}^{\mathrm{T}}\boldsymbol{Q}_{t-1}^{-1}\boldsymbol{A}_{t-1}+\left(\boldsymbol{\Sigma}_{t-1}^M\right)^{-1}-\tau_{k,i}^{-1}\boldsymbol{W}_{t-1}-\boldsymbol{D}_{t-1}^{\mathrm{T}}\boldsymbol{D}_{t-1}\end{cases} \tag{4-85}$$

最优估计器 [式 (4-83)] 存在当且仅当

$$\left(\boldsymbol{\Sigma}_t^i\right)^{-1}-\tau_{k,i}^{-1}\boldsymbol{W}_t-\boldsymbol{D}_t^{\mathrm{T}}\boldsymbol{D}_t>0,\quad t=k \tag{4-86}$$

$$\boldsymbol{A}_t^{\mathrm{T}}\boldsymbol{Q}_t^{-1}\boldsymbol{A}_t+\left(\boldsymbol{\Sigma}_t^M\right)^{-1}-\tau_{k,i}^{-1}\boldsymbol{W}_t-\boldsymbol{D}_t^{\mathrm{T}}\boldsymbol{D}_t>0,\quad t<k \tag{4-87}$$

此外，$\tau_{k,i}^*$由下式给出

$$\tau_{k,i}^*=\arg\inf_{\tau_{k,i}\in\varGamma_{k,i}}\tau_{k,i}\left\{\left(\inf_{\hat{\boldsymbol{x}}_k}\log J^{L,i}\left(\tau_{k,i}\right)\right)+d\right\} \tag{4-88}$$

其中，

$$\begin{aligned}&\inf_{\tilde{\boldsymbol{x}}_k}\log J^{L,i}\left(\tau_{k,i}\right)\\&=\log\tilde{\boldsymbol{S}}_k^i-\log\tilde{\boldsymbol{Z}}_k^i+0.5\left(\|\overline{\boldsymbol{x}}_k^i\|^2_{\left(\boldsymbol{\Sigma}_k^i\right)^{-1}\boldsymbol{K}_k^i-\left(\boldsymbol{\Sigma}_k^i\right)^{-1}}\log\left(\left|\left(\boldsymbol{\Sigma}_k^i\right)^{-1}-\tau_{k,i}^{-1}\boldsymbol{W}_k-\boldsymbol{D}_k^{\mathrm{T}}\boldsymbol{D}_k\|\boldsymbol{\Lambda}_k^i\right|\right)\right)\end{aligned}$$

这里，对于所有$0<t\leqslant k$，如果$1<i<M$，则$\tilde{\boldsymbol{S}}_k^i$可由下式得到

$$\begin{aligned}\tilde{\boldsymbol{S}}_k^i=&\tilde{\boldsymbol{S}}_k^{i-1}\left(2\pi\right)^{\frac{-m_i\gamma_k^i}{2}}\left|\boldsymbol{R}_k^i\right|^{-\frac{1}{2}}\left|\left(\boldsymbol{R}_k^i\right)^{-1}+\boldsymbol{Y}_k^i\right|^{\frac{\gamma_k^i-1}{2}}\\&\times\exp\left(-\frac{1}{2}\|\gamma_k^i\boldsymbol{y}_k^i+\left(1-\gamma_k^i\right)\boldsymbol{\beta}_k^i\|^2_{\left[\boldsymbol{R}_k^i+\left(1-\gamma_k^i\right)\left(\boldsymbol{Y}_k^i\right)^{-1}\right]^{-1}}\right),\end{aligned}$$

如果$i=1$，则$\tilde{\boldsymbol{S}}_k^i$具有与式(4-67)中$\boldsymbol{S}_k$同样的形式。如果$\tau_{k,i}^{-1}\to 0$且$\boldsymbol{D}_k=0$，则$\tilde{\boldsymbol{S}}_k^i$和$\boldsymbol{\Sigma}_k^i$分别降低至$\tilde{\boldsymbol{Z}}_k^i$和$\boldsymbol{\Lambda}_k^i$。

证明：

该证明与单传感器情况类似，故省略。

注释 4.8

本节聚焦在一个一般性的随机事件触发方案，一般性指的是每个传感器 i 在 k 时刻的传输概率与前一时刻信息集合$\mathcal{I}_k^{i-1}$相关。当选择$\boldsymbol{\beta}_k^i=\boldsymbol{C}_k^i\hat{\boldsymbol{x}}_k^{i-1}$时，成立。因此，$\overline{\boldsymbol{x}}_{k|k}$和$\hat{\boldsymbol{x}}_{k|k}^i$的计算要序贯地更新测量信息，犹如文献 [103] 那样。然而，如果事件触发方案是$\Pr\left\{\gamma_k^i=0|\ \boldsymbol{y}_k^i\right\}$[105,116]的形式，触发概率独立于前一时刻的信息集，那么能够采用将 k 时刻所有测量信息同时更新并堆叠在一起的方式来计算$\overline{\boldsymbol{x}}_{k|k}$。

4.6 仿真示例

下面用两个示例来说明本节提出的估计结果：单传感器系统场景和多传感器系统场景。

（1）单传感器系统

考虑如下二阶不确定系统

$$\begin{aligned} \boldsymbol{x}_{k+1} &= \begin{bmatrix} 0.6 & 0.2 \\ 0 & -0.6 \end{bmatrix} \boldsymbol{x}_k + \tilde{\boldsymbol{w}}_k \\ \boldsymbol{y}_k &= \begin{bmatrix} 1 & 1 \end{bmatrix} \boldsymbol{x}_k + \tilde{\boldsymbol{v}}_k \\ \boldsymbol{z}_k &= \begin{bmatrix} 1 & 0 \\ 0 & 1 \end{bmatrix} \boldsymbol{x}_k \end{aligned} \tag{4-89}$$

假设名义噪声信号$\boldsymbol{w}_k$和$\boldsymbol{v}_k$服从高斯分配，其协方差分别是$\boldsymbol{Q}_k = \boldsymbol{I}_2$和$R_k = 1$。另外，假设初始条件$x_0$服从高斯分布$\mathcal{N}(0, \boldsymbol{I})$。接下来使用命题4.1中的一步事件触发技术来处理带有相应的相对熵约束[式(4-55)]的不确定系统(4-89)。这里，不确定性矩阵$\boldsymbol{D}_k$和常数d由下式给出：

$$\boldsymbol{D}_k = 0.38\boldsymbol{I}，\ d = 10^{-6} \tag{4-90}$$

考虑不确定系统带有参数化的扰动情况，让扰动的过程噪声为

$$\tilde{\boldsymbol{w}}_k = \delta \boldsymbol{I}_2 \boldsymbol{x}_k + \boldsymbol{w}_k，\ \delta \in [-0.38, 0.38] \tag{4-91}$$

假设真实的测量噪声$\tilde{\boldsymbol{v}}_k$和初始条件$\tilde{\boldsymbol{x}}_0$与名义情况相同，那么根据式(4-5)，条件名义测度$\mathbb{P}$和条件扰动测度$\tilde{\mathbb{P}}$之间的相对熵为

$$h_0\left(\tilde{\mathbb{P}}_k(\cdot|\mathcal{I}_k) \| \mathbb{P}_k(\cdot|\mathcal{I}_k)\right) = \frac{1}{2}\sum_{i=0}^{k}\tilde{\mathbb{E}}\left[\|\boldsymbol{x}_i\|^2_{\delta^2\boldsymbol{Q}_i^{-1}} \mid \mathcal{I}_k\right] < \frac{1}{2}\sum_{i=0}^{k}\tilde{\mathbb{E}}\left[\|\boldsymbol{D}_i\boldsymbol{x}_i\|^2 \mid \mathcal{I}_k\right] + d$$

这严格满足后验相对熵约束[式(4-55)]。

为了呈现一步事件触发最小最大估计器的鲁棒性，考虑针对名义系统设计的事件触发 Kalman 滤波器 [104] 和事件触发风险敏感估计器，这三个估计器均在随机性事件触发条件 (4-7) 上使用“send-on-delta”触发方式 [110]，也就是式 (4-7) 中的 $\boldsymbol{\beta}_k$ 是上一被传输的测量值。选择触发参数

$Y_k = 0.5$。因此，这三个估计器的触发时刻相同。首先，数值搜索出最优的拉格朗日乘子τ_k^*，来获得一步事件触发最小最大估计器，图 4-2 描绘了在一个实现的信息集合下，$\delta = 0.38$时的$k = 1, \cdots, 300$时刻的τ_k^*。在此鲁棒性比较中，时域范围是$T = 100$，并且执行了一个 10000 次试验的 Monte Carlo 仿真来计算估计误差代价$\sum_{k=1}^{T} \tilde{\mathbb{E}}\left[\| \boldsymbol{x}_k - \hat{\boldsymbol{x}}_{k|k} \|_{\boldsymbol{L}_k^{\mathrm{T}} \boldsymbol{L}_k}^2 /2\right]$。由于对于风险敏感估计器，还没有一个有效的方法来指导如何选择最优的风险敏感参数 θ，这里固定$\delta = 0.38$并用 Monte Carlo 方法来找出一个相对好的 θ 取值。图 4-3 表明$\theta = 0.37$时代价最小，说明 $\theta = 0.37$ 是最好的选择。然后，图 4-4 描绘了在整个δ范围下的三个估计器的估计误差代价❶。从图 4-4 中可以观察到，与事件触发风险敏感估计器和事件触发 Kalman 滤波器比较，一步最小最大估计器对扰动更不敏感。这验证了注释 4.7 的分析：相对熵约束中 $\boldsymbol{D}_k$ 的存在提升了估计鲁棒性。

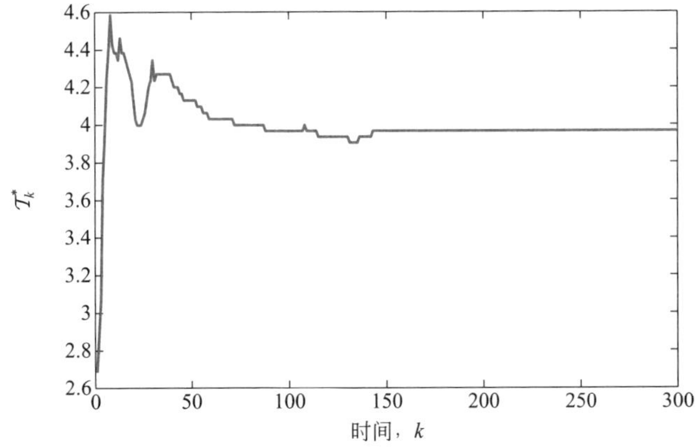

图 4-2 单传感器系统：$\delta = 0.38$ 时最优拉格朗日乘子 τ_k^*

（2）多传感器系统

在该示例中，对定理 4.4 中一步事件触发最小最大估计器和时间触发最优估计器 [114] 进行比较，来说明当通信资源有限时所提出的估计器的有效性：考虑由如下三个传感器来测量系统的状态。

$$\boldsymbol{y}_k^1 = \begin{bmatrix} 1 & 1 \end{bmatrix} \boldsymbol{x}_k + \boldsymbol{v}_k^1$$

❶ 由于不同的δ值会导致不同的真实系统，即使采用相同的Y_k，不同δ值下的传感器到估计器的传输率也不是相同的。对于图 4-4，平均传输率在$\delta \in [-0.38, 0.38]$范围内从0.7985递减至0.4792。

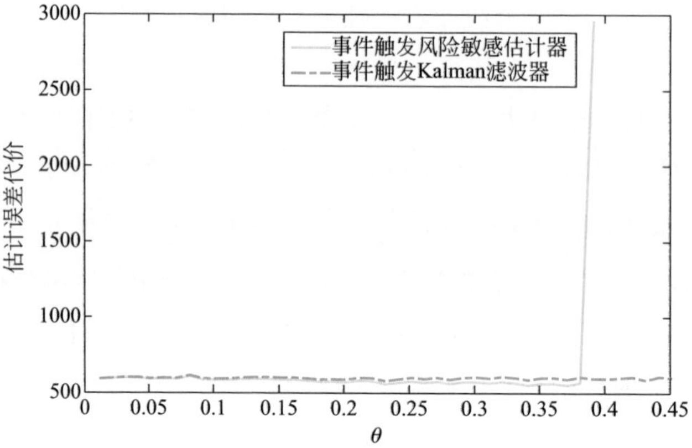

图 4-3　$\delta=0.38$ 时风险敏感参数 θ 不同取值下的估计误差代价

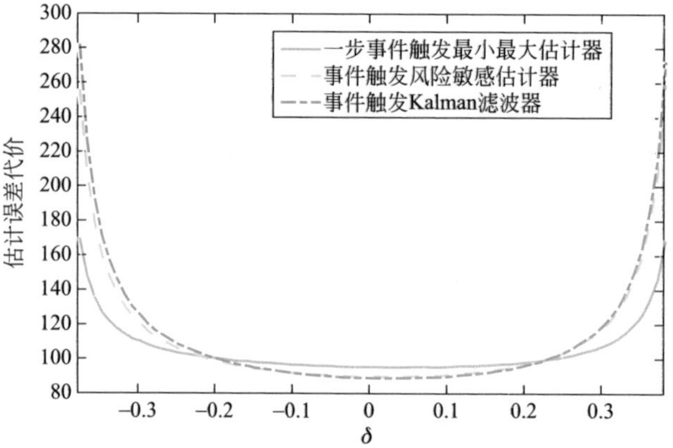

图 4-4　扰动系统下估计误差代价比较

$$\begin{aligned} \boldsymbol{y}_k^2 &= \begin{bmatrix} 1 & 0 \end{bmatrix} \boldsymbol{x}_k + \boldsymbol{v}_k^2 \\ \boldsymbol{y}_k^3 &= \begin{bmatrix} 0 & 1 \end{bmatrix} \boldsymbol{x}_k + \boldsymbol{v}_k^3 \end{aligned} \tag{4-92}$$

其中，噪声协方差为$R_k^1=1, R_k^2=0.1, R_k^3=0.1$。同样，使用“send-on-delta”触发策略并令

$$Y_k^1=1,\quad Y_k^2=4,\quad Y_k^3=3 \tag{4-93}$$

在这种情况下，考虑系统扰动参数为$\delta=0.38$。对于时间触发最小最大估计器，得到最优的τ为 0.1681。这两个估计器的估计性能如图4-5所示。这里，传感器 1、2、3的平均传输率分别为 0.5688、0.5938、0.6625，且

图4-5的γ_k由$\frac{1}{3}\sum_{i=1}^{3}\gamma_i$计算得出。从图4-5中可以观察到，尽管每个传感器的传输率不是很高，与时间触发最优估计器相比，一步事件触发最小最大估计器的估计性能没有退化太多。另一方面，该估计性能比较表明所采纳的一步思想有着合理的近似。

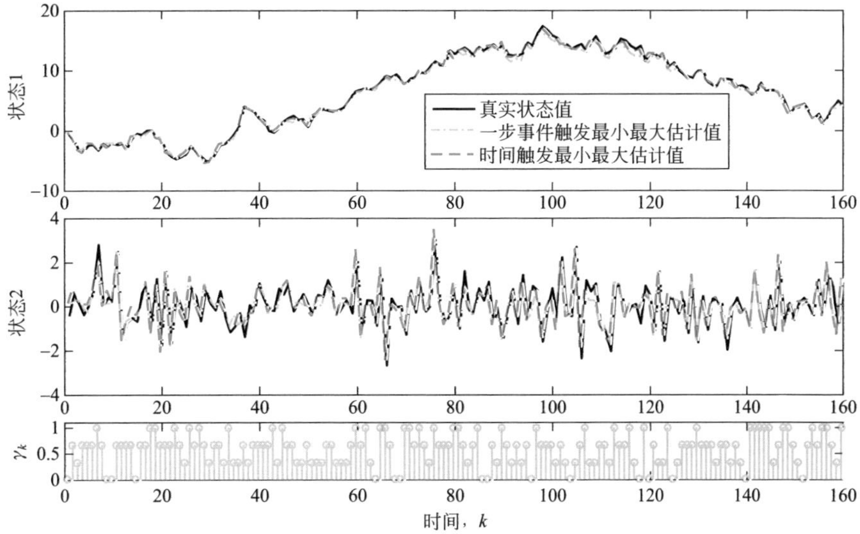

图 4-5　多传感器系统：$\delta = 0.38$ 时一步事件触发最小最大估计器的估计性能

Control of
Autonomous Intelligent Systems

自主智能系统控制

第5章

具有随机丢包的最优状态估计

5.1 概述

具有随机丢包的状态估计问题是控制领域近二十年来的一个研究热点，已取得了丰富的理论成果。一个先驱的成果是由 Sinopoli 等在 2004 年发表的具有 i.i.d. 随机观测丢失的 Kalman 滤波研究[126]。文献 [127-128] 进一步研究了 Markovian 丢包下的 Kalman 滤波稳定性问题。Markovian 丢包刻画了网络环境的可能相关性，它采用一个二位状态的 Markov 链来建模随机丢包。本章同样研究 i.i.d. 和 Markovian 丢包信道下的状态估计问题，但和现有文献研究的具体问题不同。

5.2 节研究了一个具有风险敏感性能指标的 i.i.d. 随机丢包最优状态估计问题。不同于现有文献研究的二次型估计误差代价函数，风险敏感性估计问题针对的是一个指数型二次估计误差代价函数。采用信息状态的方法，本节导出了最优的估计器及最优的代价，并给出了最优估计器存在的充要条件。进一步，本节讨论了所得的间歇性估计器和现有的间歇性 Kalman 滤波器、间歇性 H_∞估计器之间的联系。仿真比较结果表明了所提出间歇性估计器比间歇性 Kalman 滤波器更具有鲁棒性。

5.3 节研究了多个 Markovian 丢包信道下最优稳态状态估计问题。不同于时变 Kalman 滤波器[126-132]，稳态估计器的增益是个常值，可以离线计算，进而减轻估计器实时计算负担。虽然文献 [133-134] 已研究了单个 Markovian 丢包信道下的稳态估计问题，但多个 Markovian 丢包信道情形却没有得到足够的重视，且文献 [133-134] 没有充分研究稳态估计器的稳定性问题。首先，采用 Markovian 跳变线性系统（Markovian Jump Linear Systems，MJLSs）的方法，本节得到了一个最优的稳态估计器。然后，深入研究了估计器的稳定性问题：耦合的代数 Riccati 方程均方镇定解的存在性和系统均方可检测性。最后，提出了一个复杂度低的局部最优估计器，并对被解耦的子系统均方可检测性，给出了一些解析的条件。仿真示例说明了均方镇定解与均方可检测结果以及所得到最优与局部最优稳态估计器的性能。

5.2
独立同分布丢包信道下最优风险敏感状态估计

5.2.1　最优估计问题

考虑在时间范围 T 内的如下离散时间线性时变 Gauss-Markov 系统：

$$\boldsymbol{x}_{k+1} = \boldsymbol{A}_k \boldsymbol{x}_k + \boldsymbol{w}_k$$
$$\boldsymbol{y}_k = \boldsymbol{C}_k \boldsymbol{x}_k + \boldsymbol{v}_k \tag{5-1}$$

其中，$\boldsymbol{x}_k \in \mathbb{R}^n$是系统状态；$\boldsymbol{y}_k \in \mathbb{R}^m$是传感器测量输出；$\boldsymbol{w}_k$和$\boldsymbol{v}_k$是密度函数分别为$\psi_k(\boldsymbol{w}_k) = \mathcal{N}(\boldsymbol{w}_k; 0, \boldsymbol{Q}_k)$和$\phi_k(\boldsymbol{v}_k) = \mathcal{N}(\boldsymbol{v}_k; 0, \boldsymbol{R}_k)$的互相独立高斯白噪声。初始条件$\boldsymbol{x}_0$为高斯分布且其密度函数为$\mu(\boldsymbol{x}_0) = \mathcal{N}(\boldsymbol{x}_0; \boldsymbol{\pi}_0, \boldsymbol{\Pi}_0)$，并与$\boldsymbol{w}_k$和$\boldsymbol{v}_k$独立。这里，假设$\boldsymbol{Q}_k$、$\boldsymbol{R}_k$、$\boldsymbol{\Pi}_0$是正定矩阵。

本节考虑 2.1.2 节所描述的最优线性指数二次高斯（Linear Exponential Quadratic Gaussian, LEQG）/ 风险敏感状态估计问题，其中，观察值 $\boldsymbol{y}_k$ 通过丢包信道传输到估计器（见图 5-1）。k 时刻 $\boldsymbol{y}_k$ 的到达由一个二位随机变量 γ_k 来刻画，且$\Pr\{\gamma_k = 1\} = p$。此外，$\{\gamma_k\}$ 是一个 i.i.d. 随机过程。这里，$\gamma_k = 1$ 代表通信信道可靠且估计器能够收到 $\boldsymbol{y}_k$；否则认为丢包发生。定义 $\mathbb{S}_k := \{i \in \mathbb{N}_{0:k} \mid \gamma_i = 0\}$，那么在 k 时刻，估计器可以利用如下信息集合

$$\mathcal{I}_k := \{\gamma_i \mid i \in \mathbb{S}_k\} \cup \{\boldsymbol{y}_i \mid i \in \mathbb{N}_{0:k} \setminus \mathbb{S}_k\}$$

且$\mathcal{I}_{-1} := \varnothing$。

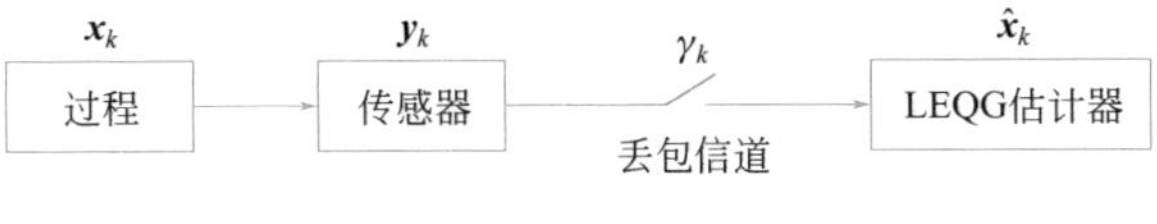

图 5-1　具有随机测量丢包的 LEQG 估计示意图

本节聚焦于因果估计器$\{\hat{\boldsymbol{x}}_0, \cdots, \hat{\boldsymbol{x}}_T\}$，使得 k 时刻估计值是基于估计器在 k 时刻的可用信息，也就是

$$\hat{\boldsymbol{x}}_k = g_k(\mathcal{I}_k) \tag{5-2}$$

其中，g_k是将可利用信息$\mathcal{I}_k$映射到状态估计值$\hat{\boldsymbol{x}}_k$的函数。正如文献[73]中

一样，寻求待最小化的LEQG性能指标是对累积估计误差的指数二次函数的期望：

$$J = \inf_{\hat{x}_{0T}} \mathbb{E}\left[\exp\left(\theta \boldsymbol{\Psi}_{0,T}\right)\right] \tag{5-3}$$

其中，$\theta > 0$是个标量，且

$$\boldsymbol{\Psi}_{0,T} = \frac{1}{2}\sum_{k=0}^{T} \| \boldsymbol{x}_k - \hat{\boldsymbol{x}}_k \|_{\boldsymbol{W}_k}^2$$

这里$\boldsymbol{W}_k$是一个正定权重矩阵。参数θ可以用来权衡名义模型(5-1)的最优估计与对系统不确定性的鲁棒性，通常θ越大，对越大的不确定性，鲁棒性越好。注意，式(5-3)中的期望不仅针对随机向量$\boldsymbol{x}_0$、噪声序列$\{\boldsymbol{w}_k\}$和$\{\boldsymbol{v}_k\}$，还针对到达序列$\{\gamma_k\}$。

5.2.2 最优 LEQG 估计

5.2.2.1 递归的信息状态

定义 5.1

定义$\boldsymbol{\alpha}_k(\boldsymbol{x}_k)$为一个信息状态，使得

$$\boldsymbol{\alpha}_k(\boldsymbol{x}_k) := \int_{\mathbb{R}^n} \cdots \int_{\mathbb{R}^n} \exp\left(\theta \boldsymbol{\Psi}_{0,k-1}\right) f\left(\boldsymbol{x}_{0k}, \mathcal{I}_k\right) \mathrm{d}\boldsymbol{x}_{0k-1},\ k = 1,2,\cdots,T \tag{5-4}$$

其中，$f(\boldsymbol{x}_{0k}, \mathcal{I}_k)$代表$\boldsymbol{x}_{0k}$和$\mathcal{I}_k$的联合分布，其初始值为$\boldsymbol{\alpha}_0(\boldsymbol{x}_0) := f(\boldsymbol{x}_0, \mathcal{I}_0)$。

注释 5.1　如文献 [119] 中所述，信息状态的概念为：信息状态是所有可利用的观测信息的函数。在本节中，为了处理指数代价，信息状态$\boldsymbol{\alpha}_k(\boldsymbol{x}_k)$不仅包括当前可用信息，还包括一部分 LEQG 代价。

下面，给出$\boldsymbol{\alpha}_k(\boldsymbol{x}_k)$的明确递归表达式。

定理 5.1

在$k \in [0,T]$时域内，信息状态$\boldsymbol{\alpha}_k(\boldsymbol{x}_k)$是一个非标准化的高斯密度：

$$\boldsymbol{\alpha}_k(\boldsymbol{x}_k) = \boldsymbol{S}_k \exp\left(-\frac{1}{2} \| \boldsymbol{x}_k - \bar{\boldsymbol{x}}_k \|_{\boldsymbol{\Sigma}_k^{-1}}^2\right) \tag{5-5}$$

这里，$\boldsymbol{\Sigma}_k$根据如下递归演化：

$$\begin{cases}\boldsymbol{\Sigma}_k^{-1}=\boldsymbol{Q}_{k-1}^{-1}-\boldsymbol{Q}_{k-1}^{-1}\boldsymbol{A}_{k-1}\overline{\boldsymbol{\Sigma}}_k\boldsymbol{A}_{k-1}^{\mathrm{T}}\boldsymbol{Q}_{k-1}^{-1}+\gamma_k\boldsymbol{C}_k^{\mathrm{T}}\boldsymbol{R}_k^{-1}\boldsymbol{C}_k\\ \overline{\boldsymbol{\Sigma}}_k^{-1}=\boldsymbol{A}_{k-1}^{\mathrm{T}}\boldsymbol{Q}_{k-1}^{-1}\boldsymbol{A}_{k-1}+\boldsymbol{\Sigma}_{k-1}^{-1}-\theta\boldsymbol{W}_{k-1}\end{cases} \tag{5-6}$$

并且上面递归成立，当且仅当有如下不等式成立：

$$\boldsymbol{A}_t^{\mathrm{T}}\boldsymbol{Q}_t^{-1}\boldsymbol{A}_t+\boldsymbol{\Sigma}_t^{-1}-\theta\boldsymbol{W}_t>0,\quad t=0,\cdots,k-1 \tag{5-7}$$

式(5-5)的$\overline{\boldsymbol{x}}_k$和$\boldsymbol{S}_k$按照如下递归演化：

$$\begin{cases}\boldsymbol{\Sigma}_k^{-1}\overline{\boldsymbol{x}}_k=\boldsymbol{Q}_{k-1}^{-1}\boldsymbol{A}_{k-1}\overline{\boldsymbol{\Sigma}}_k\left(\boldsymbol{\Sigma}_{k-1}^{-1}\overline{\boldsymbol{x}}_{k-1}-\theta\boldsymbol{W}_{k-1}\hat{\boldsymbol{x}}_{k-1}\right)+\gamma_k\boldsymbol{C}_k^{\mathrm{T}}\boldsymbol{R}_k^{-1}\boldsymbol{y}_k\\ \boldsymbol{S}_k=\boldsymbol{S}_{k-1}(2\pi)^{\frac{-m\gamma_k}{2}}(1-p)^{1-\gamma_k}\left|\boldsymbol{R}_k\right|^{-\frac{\gamma_k}{2}}\left|\boldsymbol{Q}_{k-1}\right|^{-\frac{1}{2}}\left|\overline{\boldsymbol{\Sigma}}_k\right|^{\frac{1}{2}}\\ \qquad\times\exp\left\{\frac{1}{2}\left[\|\hat{\boldsymbol{x}}_{k-1}\|_{\theta\boldsymbol{W}_{k-1}}^2-\|\overline{\boldsymbol{x}}_{k-1}\|_{\boldsymbol{\Sigma}_{k-1}^{-1}}^2-\gamma_k\|\boldsymbol{y}_k\|_{\boldsymbol{R}_k^{-1}}^2\right.\right.\\ \qquad\left.\left.+\|\boldsymbol{\Sigma}_{k-1}^{-1}\overline{\boldsymbol{x}}_{k-1}-\theta\boldsymbol{W}_{k-1}\hat{\boldsymbol{x}}_{k-1}\|_{\overline{\boldsymbol{\Sigma}}_k}^2+\|\overline{\boldsymbol{x}}_k\|_{\boldsymbol{\Sigma}_k^{-1}}^2\right]\right\}\end{cases} \tag{5-8}$$

上述递归始于

$$\begin{cases}\boldsymbol{\Sigma}_0^{-1}=\gamma_0\boldsymbol{C}_0^{\mathrm{T}}\boldsymbol{R}_0^{-1}\boldsymbol{C}_0+\boldsymbol{\Pi}_0^{-1}\\ \boldsymbol{\Sigma}_0^{-1}\overline{\boldsymbol{x}}_0=\boldsymbol{\Pi}_0^{-1}\boldsymbol{\pi}_0+\gamma_0\boldsymbol{C}_0^{\mathrm{T}}R_0^{-1}\boldsymbol{y}_0\\ \boldsymbol{S}_0=(2\pi)^{-\frac{n+\gamma_0 m}{2}}(1-p)^{1-\gamma_k}\left|\boldsymbol{R}_0\right|^{-\frac{1}{2}}\left|\boldsymbol{\Pi}_0\right|^{-\frac{1}{2}}\\ \qquad\times\exp\left\{\frac{1}{2}\left[\|\overline{\boldsymbol{x}}_0\|_{\boldsymbol{\Sigma}_0^{-1}}^2-\|\boldsymbol{\pi}_0\|_{\boldsymbol{\Pi}_0^{-1}}^2-\gamma_0\|\boldsymbol{y}_0\|_{\boldsymbol{R}_0^{-1}}^2\right]\right\}\end{cases} \tag{5-9}$$

证明：

该证明主要基于 Bayes 推论和完全平方技术。首先，根据式 (5-4) 中$\boldsymbol{\alpha}_k\left(\boldsymbol{x}_k\right)$的定义，如果$\gamma_k=0$，可以得到

$$\begin{aligned}&\boldsymbol{\alpha}_k\left(\boldsymbol{x}_k\right)\\ &=\int_{\mathbb{R}^n}\cdots\int_{\mathbb{R}^n}\exp\left(\theta\boldsymbol{\Psi}_{0,k-1}\right)f\left(\boldsymbol{x}_{0k},\gamma_k=0,\mathcal{I}_{k-1}\right)\mathrm{d}\boldsymbol{x}_{0k-1}\\ &=\int_{\mathbb{R}^n}\cdots\int_{\mathbb{R}^n}\exp\left(\theta\boldsymbol{\Psi}_{0,k-1}\right)\Pr\left\{\gamma_k=0|\ \boldsymbol{x}_{0k},\mathcal{I}_{k-1}\right\}f\left(\boldsymbol{x}_k|\ \boldsymbol{x}_{k-1}\right)f\left(\boldsymbol{x}_{0k-1},\mathcal{I}_{k-1}\right)\mathrm{d}\boldsymbol{x}_{0k-1}\\ &=\left(1-p\right)\int_{\mathbb{R}^n}\cdots\int_{\mathbb{R}^n}\exp\left(\theta\boldsymbol{\Psi}_{0,k-1}\right)f\left(\boldsymbol{x}_k|\ \boldsymbol{x}_{k-1}\right)f\left(\boldsymbol{x}_{0k-1},\mathcal{I}_{k-1}\right)\mathrm{d}\boldsymbol{x}_{0k-1}\end{aligned}$$

其中，第二个等式来自Bayes法则和噪声的独立性质，最后一个等式是

基于如下事实：$\{\gamma_k\}$是独立于动态系统(5-1)的i.i.d.Bernoulli过程，其中$\Pr\{\gamma_k=1\}=p$。还有：

$$
\begin{aligned}
&\boldsymbol{\alpha}_k(\boldsymbol{x}_k)\\
&=(1-p)\int_{\mathbb{R}^n}\cdots\int_{\mathbb{R}^n}\exp\left(\frac{\theta}{2}\|\boldsymbol{x}_k-\hat{\boldsymbol{x}}_k\|_{\boldsymbol{W}_k}^2\right)\exp\left(\theta\boldsymbol{\Psi}_{0,k-2}\right)\\
&\quad\times f(\boldsymbol{x}_k|\ \boldsymbol{x}_{k-1})f(\boldsymbol{x}_{0k-1},\mathcal{I}_{k-1})\mathrm{d}\boldsymbol{x}_{0k-1}\\
&=(1-p)\int_{\mathbb{R}^n}\exp\left(\frac{\theta}{2}\|\boldsymbol{x}_k-\hat{\boldsymbol{x}}_k\|_{\boldsymbol{W}_k}^2\right)\psi_{k-1}(\boldsymbol{x}_k-\boldsymbol{A}_{k-1}\boldsymbol{z})\boldsymbol{\alpha}_{k-1}(\boldsymbol{z})\mathrm{d}\boldsymbol{z}
\end{aligned}
\tag{5-10}
$$

其中，用到$\boldsymbol{\alpha}_{k-1}(\boldsymbol{x}_{k-1})$的定义。另外，根据$\boldsymbol{\alpha}_0(\boldsymbol{x}_0)$的定义，可以很容易地获得$\boldsymbol{\alpha}_0(\boldsymbol{x}_0)=(1-p)\mu(\boldsymbol{x}_0)$。

对于$\gamma_k=1$，根据 Bayes 法则类似地有

$$
\begin{aligned}
f(\boldsymbol{x}_{0k},\boldsymbol{y}_k,\mathcal{I}_{k-1})&=f(\boldsymbol{x}_k,\boldsymbol{y}_k|\ \boldsymbol{x}_{0k-1},\mathcal{I}_{k-1})f(\boldsymbol{x}_{0k-1},\mathcal{I}_{k-1})\\
&=f(\boldsymbol{y}_k|\ \boldsymbol{x}_k)f(\boldsymbol{x}_k|\ \boldsymbol{x}_{k-1})f(\boldsymbol{x}_{0k-1},\mathcal{I}_{k-1})
\end{aligned}
$$

且进一步得到

$$
\begin{aligned}
\boldsymbol{\alpha}_k(\boldsymbol{x}_k)&=\int_{\mathbb{R}^n}\cdots\int_{\mathbb{R}^n}\exp\left(\theta\boldsymbol{\Psi}_{0,k-1}\right)f(\boldsymbol{x}_{0k},\boldsymbol{y}_k,\mathcal{I}_{k-1})\mathrm{d}\boldsymbol{x}_{0k-1}\\
&=\int_{\mathbb{R}^n}\cdots\int_{\mathbb{R}^n}\exp\left(\frac{\theta}{2}\|\boldsymbol{x}_k-\hat{\boldsymbol{x}}_k\|_{\boldsymbol{W}_k}^2\right)\exp\left(\theta\boldsymbol{\Psi}_{0,k-2}\right)\\
&\quad\times f(\boldsymbol{y}_k|\ \boldsymbol{x}_k)f(\boldsymbol{x}_k|\ \boldsymbol{x}_{k-1})f(\boldsymbol{x}_{0k-1},\mathcal{I}_{k-1})\mathrm{d}\boldsymbol{x}_{0k-1}\\
&=\phi_k(\boldsymbol{y}_k-\boldsymbol{C}_k\boldsymbol{x}_k)\int_{\mathbb{R}^n}\exp\left(\frac{\theta}{2}\|\boldsymbol{x}_k-\hat{\boldsymbol{x}}_k\|_{\boldsymbol{W}_k}^2\right)\psi_{k-1}(\boldsymbol{x}_k-\boldsymbol{A}_{k-1}\boldsymbol{z})\boldsymbol{\alpha}_{k-1}(\boldsymbol{z})\mathrm{d}\boldsymbol{z}
\end{aligned}
\tag{5-11}
$$

对于初始值，有$\boldsymbol{\alpha}_0(\boldsymbol{x}_0)=\phi_0(\boldsymbol{y}_0-\boldsymbol{C}_0\boldsymbol{x}_0)\mu(\boldsymbol{x}_0)$。

然后将通过数学归纳法进行此证明。这里仅给出$\gamma_k=1$的情况，因为$\gamma_k=0$时$\boldsymbol{\alpha}_k(\boldsymbol{x}_k)$的证明可以根据前一种情况的论据来完成。显然对于初始时刻，式 (5-5) 和式 (5-9) 成立。假设在$k-1$时刻，

$$
\boldsymbol{\alpha}_{k-1}(\boldsymbol{x}_{k-1})=\boldsymbol{S}_{k-1}\exp\left(-1/2\|\boldsymbol{x}_{k-1}-\overline{\boldsymbol{x}}_{k-1}\|_{\boldsymbol{\Sigma}_{k-1}^{-1}}^2\right)
$$

然后证明对于$\boldsymbol{\alpha}_k(\boldsymbol{x}_k)$，式(5-5)成立。根据式(5-11)，$\gamma_k=1$时有

$$
\begin{aligned}
\boldsymbol{\alpha}_k(\boldsymbol{x})&=\boldsymbol{S}_{k-1}(2\pi)^{-\frac{n}{2}}|\boldsymbol{R}_k|^{-\frac{1}{2}}|\boldsymbol{Q}_{k-1}|^{-\frac{1}{2}}\\
&\quad\times\exp\left(-\frac{1}{2}\|\boldsymbol{y}_k-\boldsymbol{C}_k\boldsymbol{x}\|_{\boldsymbol{R}_k^{-1}}^2\right)\int_{\mathbb{R}^n}\exp(-\boldsymbol{s}/2)\mathrm{d}\boldsymbol{z}
\end{aligned}
\tag{5-12}
$$

其中，

$$\boldsymbol{s}=\|\boldsymbol{z}-\overline{\boldsymbol{x}}_{k-1}\|_{\boldsymbol{\Sigma}_{k-1}^{-1}}^2-\|\boldsymbol{z}-\hat{\boldsymbol{x}}_{k-1}\|_{\theta\boldsymbol{W}_{k-1}}^2+\|\boldsymbol{A}_{k-1}\boldsymbol{z}-\boldsymbol{x}\|_{\boldsymbol{Q}_{k-1}}^2$$

通过利用向量的完全平方技术，$\boldsymbol{s}$可以重写为

$$\begin{aligned}\boldsymbol{s}=&\|\boldsymbol{z}\|_{\overline{\boldsymbol{\Sigma}}_k}^2-\|\hat{\boldsymbol{x}}_{k-1}\|_{\theta\boldsymbol{W}_{k-1}}^2+\|\overline{\boldsymbol{x}}_{k-1}\|_{\boldsymbol{\Sigma}_{k-1}^{-1}}^2+\|\boldsymbol{x}\|_{\boldsymbol{Q}_{k-1}^{-1}}^2\\&-\|\boldsymbol{A}_{k-1}^{\mathrm{T}}\boldsymbol{Q}_{k-1}^{-1}\boldsymbol{x}+\boldsymbol{\Sigma}_{k-1}^{-1}\overline{\boldsymbol{x}}_{k-1}-\theta\boldsymbol{W}_{k-1}\hat{\boldsymbol{x}}_{k-1}\|_{\overline{\boldsymbol{\Sigma}}_k}^2\end{aligned}$$

其中，$\overline{\boldsymbol{\Sigma}}_k$由式(5-6)给出。由于高斯密度函数在整个空间上的积分等于1，因此有

$$\begin{aligned}\int_{\mathbb{R}^n}\exp\left(-\boldsymbol{s}/2\right)\mathrm{d}\boldsymbol{z}=&\left|\overline{\boldsymbol{\Sigma}}_k\right|^{\frac{1}{2}}\exp\left(\frac{1}{2}\left(\|\hat{\boldsymbol{x}}_{k-1}\|_{\theta\boldsymbol{W}_{k-1}}^2-\|\overline{\boldsymbol{x}}_{k-1}\|_{\boldsymbol{\Sigma}_{k-1}^{-1}}^2\right)\right)\\&\times\exp\left(-\frac{1}{2}\left(\|\boldsymbol{x}\|_{\boldsymbol{Q}_{k-1}^{-1}}^2-\|\boldsymbol{A}_{k-1}^{\mathrm{T}}\boldsymbol{Q}_{k-1}^{-1}\boldsymbol{x}+\boldsymbol{\Sigma}_{k-1}^{-1}\overline{\boldsymbol{x}}_{k-1}-\theta\boldsymbol{W}_{k-1}\hat{\boldsymbol{x}}_{k-1}\|_{\overline{\boldsymbol{\Sigma}}_k}^2\right)\right)\end{aligned}\tag{5-13}$$

当且仅当不等式条件(5-7)成立时上面积分取得有限值。将式(5-13)代入式(5-12)并再次使用完全平方的方法，可得

$$\boldsymbol{\alpha}_k\left(\boldsymbol{x}\right)=\boldsymbol{S}_k\exp\left(-1/2\|\boldsymbol{x}-\overline{\boldsymbol{x}}_k\|_{\boldsymbol{\Sigma}_k^{-1}}^2\right)\tag{5-14}$$

其中，当$\gamma_k=1$时，$\boldsymbol{\Sigma}_k$、$\overline{\boldsymbol{x}}_k$和$\boldsymbol{S}_k$由式(5-6)给出。类似地，对于$\gamma_k=0$可以证明式(5-14)成立。证毕。 ■

5.2.2.2 动态规划

为了利用由定理5.1给出的信息状态$\boldsymbol{\alpha}_k\left(\boldsymbol{x}_k\right)$的递归，将式(5-3)中LEGQ代价重写为一个关于$\boldsymbol{\alpha}_T\left(\boldsymbol{x}_T\right)$的函数❶：

$$\begin{aligned}\mathbb{E}\left[\exp\left(\theta\boldsymbol{\Psi}_{0,T}\right)\right]&=\mathbb{E}\left[\mathbb{E}\left[\exp\left(\sum_{k=0}^{T}\frac{\theta}{2}\|\boldsymbol{x}_k-\hat{\boldsymbol{x}}_k\|_{\boldsymbol{W}_k}^2\right)|\ \mathcal{I}_T\right]\right]\\&=\mathbb{E}\left[\int_{\mathbb{R}^n}\cdots\int_{\mathbb{R}^n}\exp\left(\sum_{k=0}^{T}\frac{\theta}{2}\|\boldsymbol{x}_k-\hat{\boldsymbol{x}}_k\|_{\boldsymbol{W}_k}^2\right)f\left(\boldsymbol{x}_{0T}|\ \mathcal{I}_T\right)\mathrm{d}\boldsymbol{x}_{0T}\right]\\&=\mathbb{E}\left[\int_{\mathbb{R}^n}\boldsymbol{\alpha}_T\left(\boldsymbol{x}_T\right)\exp\left(\frac{\theta}{2}\|\boldsymbol{x}_T-\hat{\boldsymbol{x}}_T\|_{\boldsymbol{W}_T}^2\right)\mathrm{d}\boldsymbol{x}_T/f\left(\mathcal{I}_T\right)\right]\end{aligned}\tag{5-15}$$

---

❶ 通过定义另外一个信息状态$\boldsymbol{\eta}_k(\boldsymbol{x}_k):=f(x_k,\mathcal{I}_k)$，$k=0,\cdots T$，可以证明式（5-15）和式（5-22）中的除数$f(\mathcal{I}_T)$和$f(\mathcal{I}_k)$是大于0的数。

其中，第一个等式来自迭代期望法则（文献[108]，引理2.4.8）。

定义最优值函数$J_k\left(\overline{\boldsymbol{x}}_k\right)$为

$$J_k\left(\overline{\boldsymbol{x}}_k\right)=\inf_{\hat{\boldsymbol{x}}_{kT}}\mathbb{E}\left[\int_{\mathbb{R}^n}\boldsymbol{\alpha}_T\left(\boldsymbol{x}_T\right)\exp\left(\frac{\theta}{2}\|\boldsymbol{x}_T-\hat{\boldsymbol{x}}_T\|_{\boldsymbol{W}_T}^2\right)\mathrm{d}\boldsymbol{x}_T/f\left(\mathcal{I}_T\right)|\ \mathcal{I}_k\right] \tag{5-16}$$

然后使用标准DP理论[123]，后向DP递归规则为

$$\begin{cases}J_T\left(\overline{\boldsymbol{x}}_T\right)=\inf\limits_{\hat{\boldsymbol{x}}_T}\int_{\mathbb{R}^n}\boldsymbol{\alpha}_T\left(\boldsymbol{x}_T\right)\exp\left(\dfrac{\theta}{2}\|\boldsymbol{x}_T-\hat{\boldsymbol{x}}_T\|_{\boldsymbol{W}_T}^2\right)\mathrm{d}\boldsymbol{x}_T/f\left(\mathcal{I}_T\right) & (5\text{-}17)\\ J_k\left(\overline{\boldsymbol{x}}_k\right)=\inf\limits_{\hat{\boldsymbol{x}}_k}\mathbb{E}\left[J_{k+1}\left(\overline{\boldsymbol{x}}_{k+1}\right)|\ \mathcal{I}_k\right] & (5\text{-}18)\end{cases}$$

其中，$k=T-1,\cdots,0$，且期望操作针对$\boldsymbol{y}_{k+1}$和γ_{k+1}。还可以得出最优代价为$J=\mathbb{E}\left[J_0\left(\overline{\boldsymbol{x}}_0\right)\right]$。

在给出最优的 LEQG 估计器之前，需要如下两个引理。

引理 5.1

（文献 [135]，定理 18.1.1）给定一个矩阵$\boldsymbol{R}\in\mathbb{R}^{n\times n}$，一个矩阵$\boldsymbol{S}\in\mathbb{R}^{n\times m}$，一个矩阵$\boldsymbol{T}\in\mathbb{R}^{m\times m}$和一个矩阵$\boldsymbol{U}\in\mathbb{R}^{m\times n}$，如果 $\boldsymbol{R}$ 和 $\boldsymbol{T}$ 为非奇异，那么

$$|\boldsymbol{R}+\boldsymbol{STU}|=|\boldsymbol{R}||\boldsymbol{T}||\boldsymbol{T}^{-1}+\boldsymbol{UR}^{-1}\boldsymbol{S}|$$

引理 5.2

给定$\boldsymbol{x}$、$\boldsymbol{a}\in\mathbb{R}^n,\boldsymbol{b}\in\mathbb{R}^m,\boldsymbol{C}\in\mathbb{R}^{m\times n}$和正定矩阵$\boldsymbol{Y}\in\mathbb{R}^{n\times n},\boldsymbol{Z}\in\mathbb{R}^{m\times m}$，使得$\boldsymbol{Y}+\boldsymbol{C}^{\mathrm{T}}\boldsymbol{ZC}$和$\boldsymbol{Z}^{-1}+\boldsymbol{CY}^{-1}\boldsymbol{C}^{\mathrm{T}}$非奇异，则

$$\begin{aligned}&\|\boldsymbol{x}+\boldsymbol{b}\|_{\boldsymbol{Y}}^2+\|\boldsymbol{Cx}+\boldsymbol{d}\|_{\boldsymbol{Z}}^2\\&=\|\boldsymbol{x}+\left(\boldsymbol{Y}+\boldsymbol{C}^{\mathrm{T}}\boldsymbol{ZC}\right)^{-1}\left(\boldsymbol{Yb}+\boldsymbol{C}^{\mathrm{T}}\boldsymbol{Zd}\right)\|_{\boldsymbol{Y}+\boldsymbol{C}^{\mathrm{T}}\boldsymbol{ZC}}^2+\|\boldsymbol{d}-\boldsymbol{Cb}\|_{\left(\boldsymbol{Z}^{-1}+\boldsymbol{CY}^{-1}\boldsymbol{C}^{\mathrm{T}}\right)^{-1}}^2\end{aligned}$$

证明：

这个引理可以通过完全平方并应用矩阵求逆引理[136]直接证明。■

定理 5.2

对于时域$k\in[0,T]$和数据包到达率为$p\in(0,1)$的 LEQG 估计问题(5-3)，最优估计器为

$$\hat{\boldsymbol{x}}_{k|k}=\overline{\boldsymbol{x}}_k \tag{5-19}$$

且该估计器存在当且仅当对于所有k，满足如下条件：

$$\boldsymbol{\Sigma}_k^{-1} - \theta \boldsymbol{W}_k > 0 \tag{5-20}$$

此外，由式(5-16)定义的值函数$J_k\left(\bar{\boldsymbol{x}}_k\right)$可以明确给定为

$$\begin{cases} J_T\left(\bar{\boldsymbol{x}}_T\right) = (2\pi)^{\frac{n}{2}} \left|\boldsymbol{\Sigma}_T^{-1} - \theta \boldsymbol{W}_T\right|^{-\frac{1}{2}} \boldsymbol{S}_T / f\left(\mathcal{I}_T\right) & (5\text{-}21) \\ J_k\left(\bar{\boldsymbol{x}}_k\right) = (2\pi)^{\frac{n}{2}} \left|\boldsymbol{\Sigma}_k^{-1} - \theta \boldsymbol{W}_k\right|^{-\frac{1}{2}} & \\ \qquad \times \mathbb{E}_\gamma\left[\prod_{i=k+1}^{T} \left|\boldsymbol{I}_n - \theta \boldsymbol{\Sigma}_i \boldsymbol{W}_i\right|^{-\frac{1}{2}} \mid \mathcal{I}_k\right] \boldsymbol{S}_k / f\left(\mathcal{I}_k\right) & (5\text{-}22) \end{cases}$$

其中，$\mathbb{E}_\gamma[\cdot]$表示期望是对于序列$\left\{\gamma_{k+1},\cdots,\gamma_T\right\}$而言。这里，$\boldsymbol{\Sigma}_k$、$\bar{\boldsymbol{x}}_k$和$\boldsymbol{S}_k$的递归由式(5-6)和式(5-8)给出。

证明：

采用数学归纳法证明该定理。首先，证明对于 $k=T$，该结果成立。根据定理 5.1 中的式 (5-5)，值函数 (5-17) 为

$$\begin{aligned} & J_T\left(\bar{\boldsymbol{x}}_T\right) \\ & = \inf_{\hat{\boldsymbol{x}}_T} \int_{\mathbb{R}^n} \boldsymbol{\alpha}_T\left(\boldsymbol{x}_T\right) \exp\left(\frac{\theta}{2} \| \boldsymbol{x}_T - \hat{\boldsymbol{x}}_T \|_{\boldsymbol{W}_T}^2\right) \mathrm{d}\boldsymbol{x}_T / f\left(\mathcal{I}_T\right) \\ & = \inf_{\hat{\boldsymbol{x}}_T} \frac{\boldsymbol{S}_T}{f\left(\mathcal{I}_T\right)} \int_{\mathbb{R}^n} \exp\left(-\frac{1}{2}\left(\| \boldsymbol{x}_T - \bar{\boldsymbol{x}}_T \|_{\boldsymbol{\Sigma}_T^{-1}}^2 - \| \boldsymbol{x}_T - \hat{\boldsymbol{x}}_T \|_{\theta \boldsymbol{W}_T}^2\right)\right) \mathrm{d}\boldsymbol{x}_T \\ & = \inf_{\hat{\boldsymbol{x}}_T} (2\pi)^{\frac{n}{2}} \left|\boldsymbol{\Sigma}_T^{-1} - \theta \boldsymbol{W}_T\right|^{-\frac{1}{2}} \boldsymbol{S}_T \exp\left(\boldsymbol{q}_T / 2\right) / f\left(\mathcal{I}_T\right) \end{aligned}$$

其中，最后一个等式来自完全平方方法：

$$\boldsymbol{q}_T = \| \hat{\boldsymbol{x}}_T \|_{\theta \boldsymbol{W}_T}^2 - \| \bar{\boldsymbol{x}}_T \|_{\boldsymbol{\Sigma}_T^{-1}}^2 + \| \boldsymbol{\Sigma}_T^{-1} \bar{\boldsymbol{x}}_T - \theta \boldsymbol{W}_T \hat{\boldsymbol{x}}_T \|_{\left(\boldsymbol{\Sigma}_T^{-1} - \theta \boldsymbol{W}_T\right)^{-1}}^2 \tag{5-23}$$

以及该积分取值有限当且仅当

$$\boldsymbol{\Sigma}_T^{-1} - \theta \boldsymbol{W}_T > 0 \tag{5-24}$$

由于$\boldsymbol{q}_T$是关于$\bar{\boldsymbol{x}}_T$的指数二次函数，且条件(5-24)保证$\boldsymbol{q}_T$在$\bar{\boldsymbol{x}}_T$上为正，最优估计值$\hat{\boldsymbol{x}}_{T|T}$能够通过求解$\partial \boldsymbol{q}_T / \partial \bar{\boldsymbol{x}}_T = 0$来得到，这使得$\hat{\boldsymbol{x}}_{T|T} = \bar{\boldsymbol{x}}_T$。将最优估计值代回式(5-23)，得$\boldsymbol{q}_T = 0$。因此，可以得到

$$J_T\left(\bar{\boldsymbol{x}}_T\right) = (2\pi)^{\frac{n}{2}} \left|\boldsymbol{\Sigma}_T^{-1} - \theta \boldsymbol{W}_T\right|^{-\frac{1}{2}} \boldsymbol{S}_T / f\left(\mathcal{I}_T\right) \tag{5-25}$$

现在假设 $k+1$ 时该声称为真，也就是

$$J_{k+1}\left(\overline{\boldsymbol{x}}_{k+1}\right)=(2\pi)^{\frac{n}{2}}\left|\boldsymbol{\Sigma}_{k+1}^{-1}-\theta\boldsymbol{W}_{k+1}\right|^{-\frac{1}{2}} \times\mathbb{E}_{\gamma}\left[\prod_{i=k+2}^{T}\left|\boldsymbol{I}_{n}-\theta\boldsymbol{\Sigma}_{i}\boldsymbol{W}_{i}\right|^{-\frac{1}{2}}\mid\mathcal{I}_{k+1}\right]\boldsymbol{S}_{k+1}/f\left(\mathcal{I}_{k+1}\right)$$

为简洁起见，记$\Gamma_{d}^{l}:=\mathbb{E}_{\gamma}\left[\prod_{i=d}^{l}\left|\boldsymbol{I}_{n}-\theta\boldsymbol{\Sigma}_{i}\boldsymbol{W}_{i}\right|^{-\frac{1}{2}}\mid\mathcal{I}_{d-1}\right]$。那么对于$J_{k}\left(\overline{\boldsymbol{x}}_{k}\right)$，有

$$\begin{aligned}J_{k}\left(\overline{\boldsymbol{x}}_{k}\right)&=\inf_{\hat{\boldsymbol{x}}_{k}}\mathbb{E}\left[J_{k+1}\left(\overline{\boldsymbol{x}}_{k+1}\right)\mid\mathcal{I}_{k}\right]\\&=\inf_{\hat{\boldsymbol{x}}_{k}}(2\pi)^{\frac{n}{2}}\underbrace{\mathbb{E}\left[\left|\boldsymbol{\Sigma}_{k+1}^{-1}-\theta\boldsymbol{W}_{k+1}\right|^{-\frac{1}{2}}\Gamma_{k+2}^{T}\boldsymbol{S}_{k+1}/f\left(\mathcal{I}_{k+1}\right)\mid\mathcal{I}_{k}\right]}_{=:\,\Phi}\end{aligned}\tag{5-26}$$

因此，需要计算Φ：

$$\begin{aligned}\Phi&=\mathbb{E}\left[\gamma_{k+1}\left|\boldsymbol{\Sigma}_{k+1,1}^{-1}-\theta\boldsymbol{W}_{k+1}\right|^{-\frac{1}{2}}\Gamma_{k+2,1}^{T}\frac{\boldsymbol{S}_{k+1,1}}{f\left(\boldsymbol{y}_{k+1},\mathcal{I}_{k}\right)}\right.\\&\qquad\left.+\left(1-\gamma_{k+1}\right)\left|\boldsymbol{\Sigma}_{k+1,0}^{-1}-\theta\boldsymbol{W}_{k+1}\right|^{-\frac{1}{2}}\frac{\Gamma_{k+2,0}^{T}\boldsymbol{S}_{k+1,0}}{f\left(\gamma_{k+1}=0,\mathcal{I}_{k}\right)}\mid\mathcal{I}_{k}\right]\\&=\frac{1}{f\left(\mathcal{I}_{k}\right)}\left(p\left|\boldsymbol{\Sigma}_{k+1,1}^{-1}-\theta\boldsymbol{W}_{k+1}\right|^{-\frac{1}{2}}\Gamma_{k+2,1}^{T}\int_{\mathbb{R}^{m}}\boldsymbol{S}_{k+1,1}\mathrm{d}\boldsymbol{y}_{k+1}\right.\\&\qquad\left.+\left|\boldsymbol{\Sigma}_{k+1,0}^{-1}-\theta\boldsymbol{W}_{k+1}\right|^{-\frac{1}{2}}\Gamma_{k+2,0}^{T}\boldsymbol{S}_{k+1,0}\right)\end{aligned}\tag{5-27}$$

其中，$\boldsymbol{\Sigma}_{k+1,1}^{-1}$、$\Gamma_{k+2,1}^{T}$和$\boldsymbol{S}_{k+1,1}$分别代表$\gamma_{k+1}=1$情况下的$\boldsymbol{\Sigma}_{k+1}^{-1}$、$\Gamma_{k+2}^{T}$和$\boldsymbol{S}_{k+1}$。同理，有$\boldsymbol{\Sigma}_{k+1,0}^{-1}$、$\Gamma_{k+2,0}^{T}$和$\boldsymbol{S}_{k+1,0}$。接下来，分别导出$\int_{\mathbb{R}^{m}}\boldsymbol{S}_{k+1,1}\mathrm{d}\boldsymbol{y}_{k+1}$和$\boldsymbol{S}_{k+1,0}$的明确表达式。根据式(5-8)中$\boldsymbol{S}_{k}$的递归，可得

$$\begin{aligned}&\int_{\mathbb{R}^{m}}\boldsymbol{S}_{k+1,1}\mathrm{d}\boldsymbol{y}_{k+1}\\&=\int_{\mathbb{R}^{m}}\boldsymbol{S}_{k}(2\pi)^{-\frac{m}{2}}\left|\boldsymbol{R}_{k+1}\right|^{-\frac{1}{2}}\left|\boldsymbol{Q}_{k}\right|^{-\frac{1}{2}}\left|\overline{\boldsymbol{\Sigma}}_{k+1}\right|^{\frac{1}{2}}\exp\left\{\frac{1}{2}\left[\|\hat{\boldsymbol{x}}_{k}\|_{\theta\boldsymbol{W}_{k}}^{2}-\|\overline{\boldsymbol{x}}_{k}\|_{\boldsymbol{\Sigma}_{k}^{-1}}^{2}-\|\boldsymbol{y}_{k+1}\|_{\boldsymbol{R}_{k+1}^{-1}}^{2}\right.\right.\\&\qquad\left.\left.+\|\boldsymbol{\Sigma}_{k}^{-1}\overline{\boldsymbol{x}}_{k}-\theta\boldsymbol{W}_{k}\hat{\boldsymbol{x}}_{k}\|_{\boldsymbol{\Sigma}_{k+1}}^{2}+\|\overline{\boldsymbol{x}}_{k+1,1}\|_{\boldsymbol{\Sigma}_{k+1,1}^{-1}}^{2}\right]\right\}\mathrm{d}\boldsymbol{y}_{k+1}\\&=\boldsymbol{S}_{k}\left|\boldsymbol{\Sigma}_{k}^{-1}-\theta\boldsymbol{W}_{k}\right|^{-\frac{1}{2}}\left|\boldsymbol{\Sigma}_{k+1,1}\right|^{-\frac{1}{2}}\exp\left(\boldsymbol{q}_{k}/2\right)\end{aligned}\tag{5-28}$$

其中，用到了引理5.1和引理5.2，$\overline{x}_{k+1,1}$代表$\gamma_{k+1}=1$时$\overline{x}_{k+1}$的值，以及

$$q_k=\|\hat{x}_k\|^2_{\theta W_k}-\|\overline{x}_k\|^2_{\Sigma_k^{-1}}+\|\Sigma_k^{-1}\overline{x}_k-\theta W_k\hat{x}_k\|^2_{\left(\Sigma_k^{-1}-\theta W_k\right)^{-1}}$$

另外，通过引理5.1并应用矩阵求逆引理[136]，可得

$$\begin{aligned}S_{k+1,0}&=S_k\left(1-p\right)\left|Q_k\right|^{-\frac{1}{2}}\left|\overline{\Sigma}_{k+1}\right|^{\frac{1}{2}}\exp\left\{\frac{1}{2}\left[\|\hat{x}_k\|^2_{\theta W_k}-\|\overline{x}_k\|^2_{\Sigma_k^{-1}}\right.\right.\\&\quad\left.\left.+\|\Sigma_k^{-1}\overline{x}_k-\theta W_k\hat{x}_k\|^2_{\Sigma_{k+1}}+\|\overline{x}_{k+1,0}\|^2_{\Sigma_{k+1,0}^{-1}}\right]\right\}\\&=S_k\left(1-p\right)\left|\Sigma_k^{-1}-\theta W_k\right|^{-\frac{1}{2}}\left|\Sigma_{k+1,0}\right|^{-\frac{1}{2}}\exp\left(q_k/2\right).\end{aligned}\tag{5-29}$$

将式(5-28)和式(5-29)代入式(5-27)，可得

$$\begin{aligned}\Phi&=\frac{1}{f\left(\mathcal{I}_k\right)}\left(p\left|\Sigma_{k+1,1}^{-1}-\theta W_{k+1}\right|^{-\frac{1}{2}}\left|\Sigma_k^{-1}-\theta W_k\right|^{-\frac{1}{2}}\right.\\&\quad\times\left|\Sigma_{k+1,1}\right|^{-\frac{1}{2}}\Gamma_{k+2,1}^T+\left(1-p\right)\left|\Sigma_{k+1,0}^{-1}-\theta W_{k+1}\right|^{-\frac{1}{2}}\\&\quad\left.\times\left|\Sigma_k^{-1}-\theta W_k\right|^{-\frac{1}{2}}\left|\Sigma_{k+1,0}\right|^{-\frac{1}{2}}\Gamma_{k+2,0}^T\right)\exp\left(q_k/2\right)\\&=\frac{1}{f\left(\mathcal{I}_k\right)}\left|\Sigma_k^{-1}-\theta W_k\right|^{-\frac{1}{2}}\Gamma_{k+1}^T\exp\left(q_k/2\right)\end{aligned}$$

然后，通过求解$\partial q_k/\partial\overline{x}_k=0$，可以获得最优估计$\hat{x}_{k|k}=\overline{x}_k$和值函数$J_k\left(\overline{x}_k\right)=\left(2\pi\right)^{\frac{n}{2}}\left|\Sigma_k^{-1}-\theta W_k\right|^{-\frac{1}{2}}\Gamma_{k+1}^T/f\left(\mathcal{I}_k\right)$。此外，$q_k$在$\hat{x}_k$中为正定当且仅当条件$\Sigma_k^{-1}-\theta W_k>0$被满足，并且该条件保证了不等式(5-7)成立。因此，得到了最优估计器[式(5-19)]的存在性条件(5-20)。证毕。 ■

由于$J=\mathbb{E}\left[J_k\left(\overline{x}_0\right)\right]$，根据上述定理，可以得知最优代价可以进一步写成

$$J=\mathbb{E}_\gamma\left[\prod_{i=0}^T\left|I_n-\theta\Sigma_iW_i\right|^{-\frac{1}{2}}\right]\tag{5-30}$$

其中，期望针对序列$\{\gamma_0,\cdots,\gamma_T\}$。如同文献[126,137]，不能解析地计算出最优代价J，这是因为由式(5-6)知道，Σ_k依赖于$\{\gamma_k\}$序列且是Σ_{k-1}的非线性函数。通过一些进一步的操作，可以用如下类Kalman的结构表示$\hat{x}_{k|k}$和Σ_k的递归：

$$\hat{\boldsymbol{x}}_{k|k} = \hat{\boldsymbol{x}}_{k|k-1} + \gamma_k \boldsymbol{K}_k \left(\boldsymbol{y}_k - \boldsymbol{C}_k \hat{\boldsymbol{x}}_{k|k-1} \right) \tag{5-31}$$

$$\hat{\boldsymbol{x}}_{k+1|k} = \boldsymbol{A}_k \hat{\boldsymbol{x}}_{k|k} \tag{5-32}$$

$$\boldsymbol{P}_{k+1} = \boldsymbol{A}_k \left(\boldsymbol{\Sigma}_k^{-1} - \theta \boldsymbol{W}_k \right)^{-1} \boldsymbol{A}_k^{\mathrm{T}} + \boldsymbol{Q}_k \tag{5-33}$$

$$\boldsymbol{\Sigma}_k^{-1} = \boldsymbol{P}_k^{-1} + \gamma_k \boldsymbol{C}_k^{\mathrm{T}} \boldsymbol{R}_k^{-1} \boldsymbol{C}_k, \boldsymbol{\Sigma}_k^{-1} - \theta \boldsymbol{W}_k > 0 \tag{5-34}$$

$$\boldsymbol{K}_k = \boldsymbol{\Sigma}_k \boldsymbol{C}_k^{\mathrm{T}} \boldsymbol{R}_k^{-1} \tag{5-35}$$

上面递归始于$\hat{\boldsymbol{x}}_{0|-1} = \boldsymbol{\pi}_0$和$\boldsymbol{P}_0 = \boldsymbol{\Pi}_0$。

注释 5.2 在 Kalman 滤波框架中，通过将缺失的测量建模成噪声协方差为无限的测量[126]或使用测量模型$\boldsymbol{y}_k = \gamma_k \boldsymbol{C}_k \boldsymbol{x}_k + \boldsymbol{v}_k$[137]，可以得出具有测量丢包的最优估计器。注意，也可以通过采用与文献 [126,137] 相同的建模方法来获得最优估计器 [式 (5-31) ~ 式 (5-35)]。但是，需要注意的是，由于本节考虑的 LEQG 估计问题是指数形式准则的多阶段决策问题，因此这种等价性不是显而易见的。此外，本节的方法得到了最优代价 [式 (5-30)]。

5.2.2.3 最优估计器的性质

这里简要讨论最优 LEQG 估计器 [式 (5-31) ～式 (5-35)] 与具有随机测量丢包的 Kalman 滤波器[126]、具有随机测量丢包的确定性 H_∞估计器[138]之间的关系。

首先，由式 (5-3) 定义的最优问题可以重新写成如下等价的描述

$$J = \inf_{\hat{x}_{0T}} \frac{1}{\theta} \log \mathbb{E}\left[\exp\left(\theta \boldsymbol{\Psi}_{0,T} \right) \right] \tag{5-36}$$

然后，通过Taylor级数展开和极限$\lim\limits_{z \to 0}(1+z)^{1/z} = e$，可以得知随着$\theta \to 0, \mathbb{E}\left[\boldsymbol{\Psi}_{0,T} \right]$为$\frac{1}{\theta} \log \mathbb{E}\left[\exp\left(\theta \boldsymbol{\Psi}_{0,T} \right) \right]$的极限，这便恢复了MMSE性能标准。随着$\theta \to 0$，最优LEQG估计器[式(5-31)～式(5-35)]降为文献[126]中的Kalman滤波器。

我们还发现，最优 LEQG 估计器与文献 [138] 提出的具有随机测量丢包的 H_∞估计器具有相同的形式，在文献 [138] 中，噪声被视为确定性输入。因此，如果所考虑的系统模型 (5-1) 是定常，则由式 (5-33) 生成 $\{\boldsymbol{P}_k\}$ 的统计收敛属性将是与 H_∞情况中的对应物相同。有关详细结果，请参考文献 [138]。值得一提的是，本节所得的估计器是在高斯随机模型设置中获得的，它与文献 [138] 中的确定性 H_∞描述完全不同。

5.2.3 仿真示例 1

为了说明具有随机测量丢包的最优 LEQG 估计器的鲁棒性，考虑一个具有不确定性的线性 Gauss-Markov 系统

$$\boldsymbol{x}_{k+1}=\left(\boldsymbol{A}_k+\Delta\boldsymbol{A}_k\right)\boldsymbol{x}_k+\boldsymbol{w}_k,\ \boldsymbol{y}_k=\boldsymbol{C}_k\boldsymbol{x}_k+\boldsymbol{v}_k \tag{5-37}$$

且

$$\boldsymbol{A}_k=\begin{bmatrix}0.99 & 0.01 & 0\\ 0 & 0.99 & 0.01\\ 0 & 0 & 0.52\end{bmatrix},\ \Delta\boldsymbol{A}_k=\begin{bmatrix}0 & \delta & 0\\ 0 & 0 & \delta\\ 0 & 0 & 0\end{bmatrix}$$

$$\boldsymbol{C}_k=\begin{bmatrix}1 & -1 & 0\end{bmatrix},\ \boldsymbol{Q}_k=0.1\boldsymbol{I}_3,\ R_k=0.1$$

这里，$\Delta\boldsymbol{A}_k$代表系统不确定性且$|\delta|\leqslant 0.3$，$\boldsymbol{x}_0$是高斯且均值和协方差分别为$\boldsymbol{\pi}_0=\begin{bmatrix}0 & 0 & 0\end{bmatrix}^{\mathrm{T}}$和$\boldsymbol{\Pi}_0=\boldsymbol{I}_3$。选择代价函数中的权重矩阵为$\boldsymbol{W}_k=\boldsymbol{I}_3$。假设测量值的到达率是$p=0.65$，并考虑有限时域$T=150000$。

首先，要找到一个对不确定参数 δ 产生较好鲁棒性的 θ 值，因此固定$\delta=0.3$。然后图 5-2 绘制了所提出的 LEQG 估计器 [式 (5-31) ～式 (5-35)] 和 Kalman 滤波器 [126] 在不同 θ 值下的估计误差，该误差由公式$\dfrac{1}{2T}\sum\limits_{k=1}^{T}\left(\boldsymbol{x}_k-\hat{\boldsymbol{x}}_k\right)^T\left(\boldsymbol{x}_k-\hat{\boldsymbol{x}}_k\right)$计算得出。可以观测到$\theta=4.9\times10^{-4}$是临界值，超过该临界值，估计误差极大。同样，在 T 内，当$\theta\leqslant 4.9\times10^{-4}$时，验证到对于所有 k 条件 (5-20) 满足；当$\theta>4.9\times10^{-4}$时，验证到在某个时刻会违背条件 (5-20)。并且，当$\delta=0.3$时，由于 θ 从较小的值向临界值移动，LEQG 估计器的性能优于 Kalman 滤波器，在$\theta=4.9\times10^{-4}$时，估

计误差最小。然后考虑 δ 取值在 $[-0.3,0.3]$ 范围，并选择 $\theta=4.9\times10^{-4}$。图 5-3 绘制了不同 δ 值下 LEQG 估计器和 Kalman 滤波器的估计误差，并验证了在 T 内满足存在性条件 (5-20)。从图 5-3 中可以明显看出，LEQG 估计器比 Kalman 滤波器具有更好的鲁棒性。

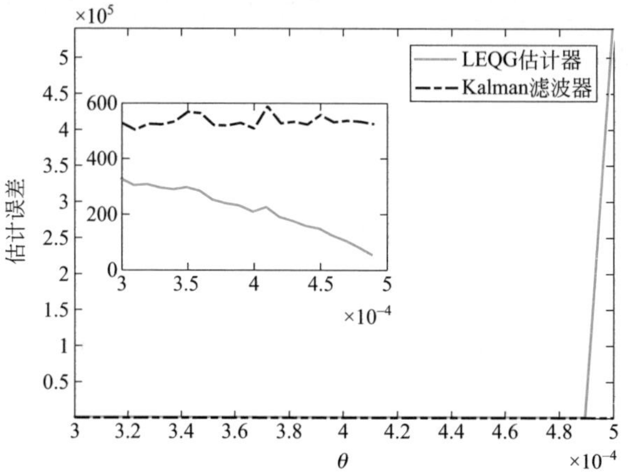

图 5-2　不同 θ 值下 LEQG 估计器估计误差（$\delta=0.3$, p=0.65）

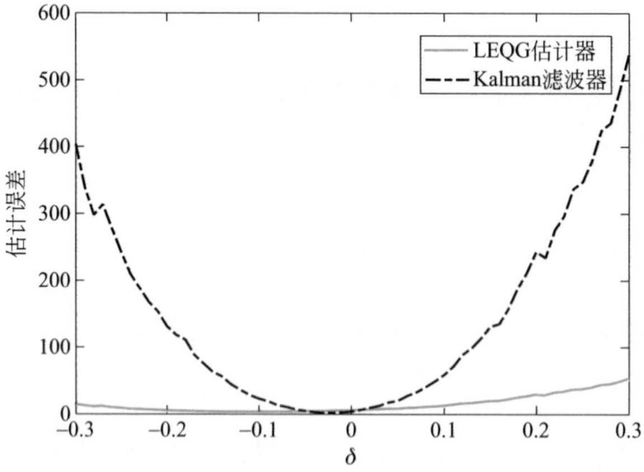

图 5-3　具有系统不确定性时 LEQG 估计器和 Kalman 滤波器之间估计性能比较（p=0.65）

5.3 多个 Markovian 丢包信道下最优稳态状态估计

5.3.1 最优稳态状态估计器

考虑一个离散时间定常系统

$$\boldsymbol{x}(k+1)=\boldsymbol{A}\boldsymbol{x}(k)+\boldsymbol{w}(k),\quad \boldsymbol{x}(0)=\boldsymbol{x}_0 \tag{5-38a}$$

$$\boldsymbol{y}_i(k)=\boldsymbol{C}_i\boldsymbol{x}(k)+\boldsymbol{v}_i(k),\quad 1\leqslant i\leqslant m \tag{5-38b}$$

其中，$\boldsymbol{x}(k)\in\mathbb{R}^n$是系统状态，$\boldsymbol{y}_i(k)\in\mathbb{R}$是第$i$个传感器获得的测量输出。$\boldsymbol{w}(k)$和$\boldsymbol{v}(k)=\mathrm{vec}\{\boldsymbol{v}_1(k),\cdots,\boldsymbol{v}_m(k)\}$是互相独立的零均值白噪声，协方差分别为$\boldsymbol{Q}\geqslant 0$和$\boldsymbol{R}>0$。初始状态$\boldsymbol{x}_0$均值为$\bar{\boldsymbol{x}}_0$，协方差为$\boldsymbol{\Pi}_0$，且独立于$\boldsymbol{w}(k)$和$\boldsymbol{v}(k)$。定义$\boldsymbol{y}(k):=\mathrm{vec}\{y_1(k),\cdots,y_m(k)\}$和$\boldsymbol{C}:=\left[\boldsymbol{C}_1^{\mathrm{T}},\cdots,\boldsymbol{C}_m^{\mathrm{T}}\right]^{\mathrm{T}}$。不失一般性，假设$\boldsymbol{R}$是个对角矩阵。

> **注释 5.3**　请注意，每个传感器的输出测量为一维，不会影响本节的结果一般化到任意维情况。因为本节的关注点是多个 Markovian 数据包丢包过程，所以为了方便考虑，集体测量为$\boldsymbol{y}(k)\in\mathbb{R}^m$。

这里假设集体测量$\boldsymbol{y}(k)$通过遭受丢包的不可靠信道发送，见图 5-4。远程估计器接收到的信号可以描述为

$$\boldsymbol{y}_{\mathrm{r}}(k)=\boldsymbol{\Gamma}(k)\boldsymbol{y}(k) \tag{5-39}$$

其中，$\boldsymbol{\Gamma}(k)\in\mathbb{R}^{m\times m}$代表对角形式的 m 个丢包信道：

$$\boldsymbol{\Gamma}(k)=\mathrm{diag}\{\gamma_1(k),\gamma_2(k),\cdots,\gamma_m(k)\} \tag{5-40}$$

这里，对于$1\leqslant i\leqslant m,\gamma_i(k)\in\{0,1\}$。如果$\gamma_i(k)=1$，则$y_i(k)$到达估计器；否则$\boldsymbol{y}_i(k)$丢失。此外，$\{\gamma_i(k)\}_{i=1}^m$互相独立且每个$\gamma_i(k)$被建模成一个时间齐次两状态的Markov链，其转移概率矩阵（Transition Probability Matrix，TPM）为

$$\boldsymbol{P}_i = \begin{bmatrix} 1-q_i & q_i \\ p_i & 1-p_i \end{bmatrix}, \quad i=1,\cdots,m \tag{5-41}$$

其中，$q_i = \Pr\{\gamma_i(k+1)=1 | \gamma_i(k)=0\}$称为恢复率；$p_i = \Pr\{\gamma_i(k+1)=0 | \gamma_i(k)=1\}$称为失败率。记$\pi_{i,l}(k) := \Pr\{\gamma_i(k)=l-1\}, l \in \{1,2\}$，假设$0 < p_i, q_i < 1$，那么，存在极限概率分布$\boldsymbol{\pi}_i = [\pi_{i,1}\ \pi_{i,2}]$且

$$\pi_{i,1} = \frac{p_i}{p_i+q_i}, \quad \pi_{i,2} = \frac{q_i}{p_i+q_i}$$

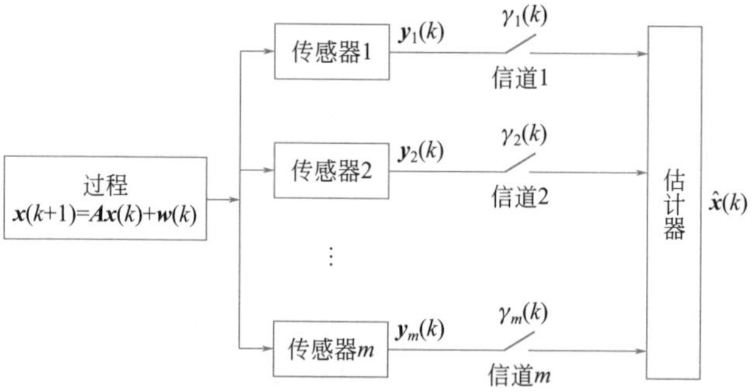

图 5-4　丢包信道下的状态估计问题

不同于现有文献 [127 ～ 132] 研究的 Markovian 丢包下时变 Kalman 滤波器，本节对最优的稳态线性状态估计器感兴趣，该估计器增益可以离线计算，进而减轻估计器实时计算负担。本节将运用 MJLSs 滤波理论[139]来导出这样一个估计器。注意到式 (5-38) 和式 (5-39) 描述的 NCS 可以重写成如下跳变形式：

$$\boldsymbol{x}(k+1) = \boldsymbol{A}\boldsymbol{x}(k) + \boldsymbol{w}(k), \quad \boldsymbol{x}(0) = \boldsymbol{x}_0 \tag{5-42a}$$

$$\boldsymbol{y}_{\mathrm{r}}(k) = \boldsymbol{H}_{\theta(k)}\boldsymbol{x}(k) + \boldsymbol{D}_{\theta(k)}\boldsymbol{v}(k) \tag{5-42b}$$

其中，$\theta(k) \in \mathcal{N} := \{1,\cdots,2^m\}$是Markovian跳变变量，由下式给出

$$\theta(k) = 1 + \sum_{i=1}^{m} 2^{i-1}\gamma_i(k) \tag{5-43}$$

随之可得$\boldsymbol{H}_{\theta(k)} = \boldsymbol{\Gamma}(k)\boldsymbol{C}$和$\boldsymbol{D}_{\theta(k)} = \boldsymbol{\Gamma}(k)$。为了简便，在本节剩余部分记$N = 2^m$。

引理 5.3

给定式 (5-43) 中计算$\theta(k)$的方式，对于 Markov 过程$\{\theta(k)\}$，其 TPM 为

$$\boldsymbol{P} := \boldsymbol{P}_m \otimes \boldsymbol{P}_{m-1} \otimes \cdots \otimes \boldsymbol{P}_1 \tag{5-44}$$

其中，$\boldsymbol{P} = \left[p_{ij} \right], p_{ij} = \Pr\left\{ \theta(k+1) = j | \ \theta(k) = i \right\}, 1 \leqslant i, j \leqslant 2^m$。此外，通过对于每个$i \in \mathcal{N}$，定义$\mu_i(k) := \Pr\left\{ \theta(k) = i \right\}$和$\boldsymbol{\mu}(k) := \left[\mu_1(k) \cdots \mu_N(k) \right]$，则$\theta(k)$具有唯一的稳态分布$\boldsymbol{\mu} = \left[\mu_1 \cdots \quad \mu_N \right]$，也就是对于$i \in \mathcal{N}$，有$\lim\limits_{k \to \infty} \mu_i(k) = \mu_i$，且

$$\boldsymbol{\mu} = \boldsymbol{\pi}_m \otimes \boldsymbol{\pi}_{m-1} \otimes \cdots \otimes \boldsymbol{\pi}_1 \tag{5-45}$$

证明:

下面通过数学归纳法证明，对于任意$m \geqslant 1$，式 (5-44) 成立。明显地，$m = 1$ 时式 (5-44) 成立。假设 $m = l$ 且满足 $l > 1$，可以得到

$$\boldsymbol{P}^{\{l\}} := \left[p_{ij}^{\{l\}} \right] = \boldsymbol{P}_l \otimes \cdots \otimes \boldsymbol{P}_1,\ 1 \leqslant i, j \leqslant 2^l \tag{5-46}$$

此时，记$\vartheta(k)$为Markovian跳变变量，使得$p_{ij}^{\{l\}} = \Pr\left\{ \vartheta(k+1) = j | \ \vartheta(k) = i \right\}$。对于$m = l + 1$，假定与$m = l$情况相比，传感器$l + 1$是新增加的传感器。定义

$$\boldsymbol{P}^{\{l+1\}} := \left[p_{rs}^{\{l+1\}} \right],\ 1 \leqslant r, s \leqslant 2^{l+1} \tag{5-47}$$

为相应的TPM。然后由式(5-43)和$\left\{ \gamma_i(k) \right\}_{i=1}^m$互相独立假设，可得

$$\begin{aligned} & \Pr\left\{ \gamma_{l+1}(k+1), \vartheta(k+1) = j | \ \gamma_{l+1}(k), \vartheta(k) = i \right\} \\ & = p^{\{l+1\}}_{\left(2^l \gamma_{l+1}(k) + i\right)\left(2^l \gamma_{l+1}(k+1) + j\right)} \\ & = \Pr\left\{ \gamma_{l+1}(k+1) | \ \gamma_{l+1}(k) \right\} \Pr\left\{ \vartheta(k+1) = j \,|\, \vartheta(k) = i \right\} \\ & \quad \gamma_{l+1}(k+1), \gamma_{l+1}(k) \in \{0,1\}, 1 \leqslant i, j \leqslant 2^l \end{aligned} \tag{5-48}$$

所以，总结出

$$\boldsymbol{P}^{\{l+1\}} = \boldsymbol{P}_{l+1} \otimes \boldsymbol{P}^{\{l\}} = \boldsymbol{P}_{l+1} \otimes \boldsymbol{P}_l \otimes \cdots \otimes \boldsymbol{P}_1 \tag{5-49}$$

类似地，下式成立

$$\boldsymbol{\mu}(k) = \boldsymbol{\pi}_m(k) \otimes \boldsymbol{\pi}_{m-1}(k) \otimes \cdots \otimes \boldsymbol{\pi}_1(k) \tag{5-50}$$

因此，$\boldsymbol{\mu} = \lim\limits_{k \to \infty} \boldsymbol{\mu}(k) = \boldsymbol{\pi}_m \otimes \boldsymbol{\pi}_{m-1} \otimes \cdots \otimes \boldsymbol{\pi}_1$，唯一性来自极限概率分布$\left\{ \boldsymbol{\pi}_i \right\}_{i=1}^m$。证毕。 ■

令$\hat{\boldsymbol{x}}(k)$为式 (5-42a) 中$\boldsymbol{x}(k)$的状态估计。现在，考虑一个如下描述的

动态 Markovian 跳变线性估计器：

$$\hat{\boldsymbol{x}}(k+1)=\boldsymbol{A}\hat{\boldsymbol{x}}(k)+\boldsymbol{L}_{\theta(k)}\left[\boldsymbol{y}_{\mathrm{r}}(k)-\boldsymbol{H}_{\theta(k)}\hat{\boldsymbol{x}}(k)\right] \tag{5-51}$$

在每个时刻，从一个有限的预先计算值集合中选择估计器增益$\boldsymbol{L}_{\theta(k)}$，即$\boldsymbol{L}_{\theta(k)}\in\left\{\boldsymbol{L}_i\in\mathbb{R}^{n\times m}\right\}_{i\in\mathcal{N}}$，这意味着$\boldsymbol{L}_{\theta(k)}$仅仅依赖于$\theta(k)$（而不是所有过去的模态$\{\theta(0),\cdots,\theta(k)\}$，这对应于时变Kalman滤波器），这是跳变估计器的一个重要特征。目标是寻找一组最优的增益，记为$\left\{\boldsymbol{K}_i\in\mathbb{R}^{n\times m}\right\}_{i\in\mathcal{N}}$，使得在对于每个$i\in\mathcal{N},\boldsymbol{L}_i=\boldsymbol{K}_i$情形下，稳态估计代价

$$J(\infty)=\lim_{k\to\infty}\mathbb{E}\left[\|\boldsymbol{x}(k)-\hat{\boldsymbol{x}}(k)\|^2\right] \tag{5-52}$$

有界且最小。

定理 5.3

对于由式 (5-42) 描述的系统动力学，假设对于如下耦合代数 Riccati 方程（Coupled Algebraic Riccati Equations，CAREs）存在均方镇定解（见 5.3.2 节中定义 5.3）$\boldsymbol{Y}=(\boldsymbol{Y}_1,\cdots,\boldsymbol{Y}_N)$：

$$\boldsymbol{Y}_j=\sum_{i=1}^{N}p_{ij}\left[\boldsymbol{A}\boldsymbol{Y}_i\boldsymbol{A}^{\mathrm{T}}-\boldsymbol{A}\boldsymbol{Y}_i\boldsymbol{H}_i^{\mathrm{T}}\left(\boldsymbol{H}_i\boldsymbol{Y}_i\boldsymbol{H}_i^{\mathrm{T}}+\mu_i\boldsymbol{R}\right)^{-1}\boldsymbol{H}_i\boldsymbol{Y}_i\boldsymbol{A}^{\mathrm{T}}+\mu_i\boldsymbol{Q}\right],\quad j\in\mathcal{N} \tag{5-53}$$

那么，跳变估计器[式(5-51)]的最优稳态增益为

$$\boldsymbol{K}_j=\boldsymbol{K}_j(\boldsymbol{Y}):=\boldsymbol{A}\boldsymbol{Y}_j\boldsymbol{H}_j^{\mathrm{T}}\left(\boldsymbol{H}_j\boldsymbol{Y}_j\boldsymbol{H}_j^{\mathrm{T}}+\mu_j\boldsymbol{R}\right)^{-1},\ j\in\mathcal{N} \tag{5-54}$$

且稳态最优代价为

$$\lim_{k\to\infty}J(k)=\sum_{i=1}^{N}\mathrm{tr}(\boldsymbol{Y}_i)$$

证明：

该结果可以通过使用 MJLSs 滤波理论（文献 [139]，第 5 章）来获得。更清楚些，定义估计误差$\boldsymbol{e}(k):=\boldsymbol{x}(k)-\hat{\boldsymbol{x}}(k)$并令$\boldsymbol{Y}_j(k):=\mathbb{E}\left[\boldsymbol{e}(k)\boldsymbol{e}(k)^{\mathrm{T}}\mathbf{1}_{\theta(k)=j}\right],j\in\mathcal{N}$，使得

$$\mathbb{E}\left[\boldsymbol{e}(k)\boldsymbol{e}(k)^{\mathrm{T}}\right]=\sum_{j=1}^{N}\boldsymbol{Y}_j(k)$$

根据式(5-42)和式(5-51)，有

$$\boldsymbol{e}(k+1)=\left(\boldsymbol{A}-\boldsymbol{L}_{\theta(k)}(k)\boldsymbol{H}_{\theta(k)}\right)\boldsymbol{e}(k)+\boldsymbol{w}(k)-\boldsymbol{L}_{\theta(k)}(k)\boldsymbol{D}_{\theta(k)}\boldsymbol{v}(k) \tag{5-55}$$

这里$\boldsymbol{L}_{\theta(k)}$被替换成$\boldsymbol{L}_{\theta(k)}(k)$。从文献[139]中命题3.35.2可知，

$$\boldsymbol{Y}_j(k+1)=\sum_{i=1}^{N}p_{ij}\Big[\big(\boldsymbol{A}-\boldsymbol{L}_i(k)\boldsymbol{H}_i\big)\boldsymbol{Y}_i(k)\big(\boldsymbol{A}-\boldsymbol{L}_i(k)\boldsymbol{H}_i\big)^{\mathrm{T}} + \mu_i(k)\big(\boldsymbol{Q}+\boldsymbol{L}_i(k)\boldsymbol{D}_i\boldsymbol{R}\boldsymbol{D}_i^{\mathrm{T}}\boldsymbol{L}_i^{\mathrm{T}}(k)\big)\Big],\ j\in\mathcal{N} \tag{5-56}$$

通过对于所有$i\in\mathcal{N}$求解$\partial\mathrm{tr}\big(\boldsymbol{Y}_j(k+1)\big)/\partial\boldsymbol{L}_i(k)=0$，得到$\boldsymbol{L}_i(k)=\boldsymbol{K}_i(k)$且

$$\boldsymbol{K}_i(k)=\boldsymbol{A}\boldsymbol{Y}_i(k)\boldsymbol{H}_i^{\mathrm{T}}\big(\boldsymbol{H}_i\boldsymbol{Y}_i(k)\boldsymbol{H}_i^{\mathrm{T}}+\mu_i(k)\boldsymbol{D}_i\boldsymbol{R}\boldsymbol{D}_i^{\mathrm{T}}\big)^{\dagger} \tag{5-57}$$

上式最小化了$\mathrm{tr}\big(\boldsymbol{Y}_j(k+1)\big)$。这里用到了Moore-Penrose逆，这是因为一般来说矩阵$\boldsymbol{H}_i\boldsymbol{Y}_i\boldsymbol{H}_i^{\mathrm{T}}+\mu_i(k)\boldsymbol{D}_i\boldsymbol{R}\boldsymbol{D}_i^{\mathrm{T}}$是半正定，且对于任意$i\in\mathcal{N}$，关系$\mathcal{R}\{\boldsymbol{H}_i\boldsymbol{Y}_i(k)\boldsymbol{A}^{\mathrm{T}}\}\subseteq\mathcal{R}\{\boldsymbol{H}_i\boldsymbol{Y}_i\boldsymbol{H}_i^{\mathrm{T}}+\boldsymbol{\mu}_i(k)\boldsymbol{D}_i\boldsymbol{R}\boldsymbol{D}_i^{\mathrm{T}}\}$总成立。令$\boldsymbol{L}_i(k)=\boldsymbol{K}_i(k)$，式(5-56)变成

$$\begin{aligned}\boldsymbol{Y}_j(k+1)=\sum_{i=1}^{N}p_{ij}\Big[&\boldsymbol{A}\boldsymbol{Y}_i(k)\boldsymbol{A}^{\mathrm{T}}-\boldsymbol{A}\boldsymbol{Y}_i(k)\boldsymbol{H}_i^{\mathrm{T}}\big(\boldsymbol{H}_i\boldsymbol{Y}_i(k)\boldsymbol{H}_i^{\mathrm{T}}\\&+\mu_i(k)\boldsymbol{D}_i\boldsymbol{R}\boldsymbol{D}_i^{\mathrm{T}}\big)^{\dagger}\boldsymbol{H}_i\boldsymbol{Y}_i(k)\boldsymbol{A}^{\mathrm{T}}+\mu_i(k)\boldsymbol{Q}\Big]\end{aligned} \tag{5-58}$$

随着$k\to\infty$，式(5-58)收敛到CAREs式(5-53)，$\boldsymbol{Y}_j(k+1)\to\boldsymbol{Y}_j$以及$\boldsymbol{K}_i(k)\to\boldsymbol{K}_i$。这里在不改变$\boldsymbol{Y}_j$和$\boldsymbol{K}_i$值的情况下，$\big(\boldsymbol{H}_i\boldsymbol{Y}_i\boldsymbol{H}_i^{\mathrm{T}}+\mu_i\boldsymbol{R}\big)^{-1}$取代了$\big(\boldsymbol{H}_i\boldsymbol{Y}_i\boldsymbol{H}_i^{\mathrm{T}}+\mu_i\boldsymbol{D}_i\boldsymbol{R}\boldsymbol{D}_i^{\mathrm{T}}\big)^{\dagger}$，这是由于$\boldsymbol{\Gamma}(k)$具有特殊形式。另外，稳态最优代价为$\sum_{i=1}^{N}\mathrm{tr}\big(\boldsymbol{Y}_i\big)$。 ■

注释 5.4 从定理 5.3 中可以看出，最优稳态估计器有 $N=2^m$ CAREs 和增益，当传感器数量很大时，这可能会导致实施上的困难。本节将在 5.3.3 小节中通过提出一个局部最优的稳态估计器来解决此问题。

5.3.2 均方镇定解

根据定理 5.3，众所周知，最优的稳态估计器是基于 CAREs 式 (5-53) 的均方镇定解的存在性。为此，本小节将聚焦讨论该存在性问题的充要条件。

5.3.2.1 MJLSs的预备知识

令$\mathbb{H}^{m,n}$代表所有 N 个实矩阵$\boldsymbol{V}=\left(\boldsymbol{V}_1,\cdots,\boldsymbol{V}_N\right),\boldsymbol{V}_i\in\mathbb{R}^{m,n}$序列组成的线性空间。在 $m=n$ 情况下，记$\mathbb{H}^n=\mathbb{H}^{n,n}$，并定义

$$\mathbb{H}^{n*}:=\left\{\boldsymbol{V}=\left(\boldsymbol{V}_1,\cdots,\boldsymbol{V}_N\right)\in\mathbb{H}^n;\boldsymbol{V}_i=\boldsymbol{V}_i^{\mathrm{T}}\forall i\in\mathcal{N}\right\}$$

$$\mathbb{H}_+^n:=\left\{\boldsymbol{V}=\left(\boldsymbol{V}_1,\cdots,\boldsymbol{V}_N\right)\in\mathbb{H}^{n*};\boldsymbol{V}_i\geqslant 0\forall i\in\mathcal{N}\right\}$$

对于$\boldsymbol{V}=\left(\boldsymbol{V}_1,\cdots,\boldsymbol{V}_N\right)\in\mathbb{H}^{n*}$和$\boldsymbol{S}=\left(\boldsymbol{S}_1,\cdots,\boldsymbol{S}_N\right)\in\mathbb{H}^{n*}$，如果$\boldsymbol{V}-\boldsymbol{S}=(\boldsymbol{V}_1-\boldsymbol{S}_1,\cdots,\boldsymbol{V}_N-\boldsymbol{S}_N)\in\mathbb{H}_+^n$（或者$\boldsymbol{V}_i-\boldsymbol{S}_i>0$），写成$\boldsymbol{V}\geqslant\boldsymbol{S}$（或者$\boldsymbol{V}>\boldsymbol{S}$）。已知$\mathbb{H}^{m,n}$可以配备如下内积：

$$\left\langle\boldsymbol{V},\boldsymbol{S}\right\rangle=\sum_{i=1}^{N}\operatorname{tr}\left(\boldsymbol{V}_i^{\mathrm{T}}\boldsymbol{S}_i\right)\tag{5-59}$$

其中，$\boldsymbol{V}=\left(\boldsymbol{V}_1,\cdots,\boldsymbol{V}_N\right)$和$\boldsymbol{S}=\left(\boldsymbol{S}_1,\cdots,\boldsymbol{S}_N\right)$属于$\mathbb{H}^{m,n}$。对于$\boldsymbol{V}=\left(\boldsymbol{V}_1,\cdots,\boldsymbol{V}_N\right)\in\mathbb{H}^n$和$\boldsymbol{L}=\left(\boldsymbol{L}_1,\cdots,\boldsymbol{L}_N\right)\in\mathbb{H}^{n,m}$，分别定义操作算子$\tilde{\boldsymbol{\mathcal{L}}}(\cdot)=\left(\tilde{\boldsymbol{\mathcal{L}}}_1(\cdot),\cdots,\tilde{\boldsymbol{\mathcal{L}}}_N(\cdot)\right)$和$\boldsymbol{\mathcal{L}}(\cdot)=\left(\boldsymbol{\mathcal{L}}_1(\cdot),\cdots,\boldsymbol{\mathcal{L}}_N(\cdot)\right)$为

$$\tilde{\boldsymbol{\mathcal{L}}}_j\left(\boldsymbol{V}\right):=\sum_{i=1}^{N}p_{ij}\left(\boldsymbol{A}+\boldsymbol{L}_i\boldsymbol{H}_i\right)\boldsymbol{V}_i\left(\boldsymbol{A}+\boldsymbol{L}_i\boldsymbol{H}_i\right)^{\mathrm{T}}\tag{5-60}$$

$$\boldsymbol{\mathcal{L}}_j\left(\boldsymbol{V}\right):=\sum_{i=1}^{N}p_{ij}\boldsymbol{A}\boldsymbol{V}_i\boldsymbol{A}^{\mathrm{T}},\ j\in\mathcal{N}\tag{5-61}$$

它们各自的伴随操作算子$\tilde{\boldsymbol{\mathcal{L}}}^*$和$\boldsymbol{\mathcal{L}}^*$为

$$\tilde{\boldsymbol{\mathcal{L}}}_i^*\left(\boldsymbol{V}\right)=\sum_{j=1}^{N}p_{ij}\left(\boldsymbol{A}+\boldsymbol{L}_i\boldsymbol{H}_i\right)^{\mathrm{T}}\boldsymbol{V}_j\left(\boldsymbol{A}+\boldsymbol{L}_i\boldsymbol{H}_i\right)\tag{5-62}$$

$$\boldsymbol{\mathcal{L}}_i^*\left(\boldsymbol{V}\right)=\sum_{j=1}^{N}p_{ij}\boldsymbol{A}^{\mathrm{T}}\boldsymbol{V}_j\boldsymbol{A},\ i\in\mathcal{N}\tag{5-63}$$

且满足如下等式

$$\left\langle\tilde{\boldsymbol{\mathcal{L}}}(\boldsymbol{V}),\boldsymbol{S}\right\rangle=\left\langle\boldsymbol{V},\tilde{\boldsymbol{\mathcal{L}}}^*\left(\boldsymbol{S}\right)\right\rangle\tag{5-64a}$$

$$\left\langle\boldsymbol{\mathcal{L}}(\boldsymbol{V}),\boldsymbol{S}\right\rangle=\left\langle\boldsymbol{V},\boldsymbol{\mathcal{L}}^*\left(\boldsymbol{S}\right)\right\rangle\tag{5-64b}$$

上面等式来自式(5-59)所定义的内积。记

$$\mathbf{A}:=\left(\underbrace{\boldsymbol{A},\cdots,\boldsymbol{A}}_{N}\right),\ \mathbf{Q}:=\left(\underbrace{\boldsymbol{Q},\cdots,\boldsymbol{Q}}_{N}\right)$$

$$\mathbf{H}:=\left(\boldsymbol{H}_1,\cdots,\boldsymbol{H}_N\right),\ \mathbf{p}:=\left\{p_{ij}\right\},i,j\in\mathcal{N}$$

下面介绍一些关于 MJLSs 的概念。

定义 5.2

（文献 [140]，定义 1）如果存在$\boldsymbol{L}=\left(\boldsymbol{L}_1,\cdots,\boldsymbol{L}_N\right)\in\mathbb{H}^{n,m}$使得$\rho\left(\tilde{\mathcal{L}}\right)<1$，那么由式 (5-42) 描述的系统，或者简称$\left(\mathbf{H},\mathbf{A},\mathbf{p}\right)$，被认为是均方可检测。

定义 5.3

（文献 [139]，定义 5.7）如果$\rho\left(\tilde{\mathcal{L}}\right)<1$成立，其中$\boldsymbol{L}_i=-\boldsymbol{K}_i\left(\boldsymbol{Y}\right),i\in\mathcal{N}$，那么$\boldsymbol{Y}=\left(\boldsymbol{Y}_1,\cdots,\boldsymbol{Y}_N\right)\in\mathbb{H}_+^n$被认为是 CAREs 式 (5-53) 的均方镇定解。

遵循文献 [141] 中定理 3 的 MJLSs 可观性定义，引入如下 MJLS 式 (5-42) 的不可控特征值的定义。

定义 5.4

如果存在一个$\mathcal{L}^*$的特征向量$\boldsymbol{V}=\left(\boldsymbol{V}_1,\cdots,\boldsymbol{V}_N\right)\in\mathbb{H}_+^n\setminus\{0\}$，使得

$$\text{(a) } \mathcal{L}^*\left(\boldsymbol{V}\right)=\lambda\boldsymbol{V}\text{，(b) } \boldsymbol{Q}\boldsymbol{V}_i=0\ \ \forall i\in\mathcal{N} \tag{5-65}$$

那么实数$\lambda\geqslant 0$被认为是$(\mathbf{A},\mathbf{Q})$的一个不可控特征值。

5.3.2.2 均方镇定解存在性

下面将从最大解的概念开始研究。对于$\boldsymbol{X}=\left(\boldsymbol{X}_1,\cdots,\boldsymbol{X}_N\right)\in\mathbb{H}^{n*}$，记$\mathcal{P}_j\left(\boldsymbol{X}\right)=\sum_{i=1}^{N}p_{ij}\boldsymbol{X}_j$并定义操作算子$\boldsymbol{\mathcal{X}}(\cdot)=\left(\boldsymbol{\mathcal{X}}_1(\cdot),\cdots,\boldsymbol{\mathcal{X}}_N(\cdot)\right)$和$\boldsymbol{\mathcal{R}}(\cdot)=\left(\boldsymbol{\mathcal{R}}_1(\cdot),\cdots,\boldsymbol{\mathcal{R}}_N(\cdot)\right)$为

$$\boldsymbol{\mathcal{X}}_j\left(\boldsymbol{X}\right)\coloneqq\begin{bmatrix}\boldsymbol{A}\mathbb{P}_j\left(\boldsymbol{X}\right)\boldsymbol{A}^{\mathrm{T}}+\mu_j\boldsymbol{Q}-\boldsymbol{X}_j & \boldsymbol{A}\mathbb{P}_j\left(\boldsymbol{X}\right)\boldsymbol{H}_j^{\mathrm{T}}\\ \boldsymbol{H}_j\mathcal{P}_j\left(\boldsymbol{X}\right)\boldsymbol{A}^{\mathrm{T}} & \mu_j\boldsymbol{R}+\boldsymbol{H}_j\mathcal{P}_j\left(\boldsymbol{X}\right)\boldsymbol{H}_j^{\mathrm{T}}\end{bmatrix} \tag{5-66}$$

$$\boldsymbol{\mathcal{R}}_j\left(\boldsymbol{X}\right)\coloneqq\mu_j\boldsymbol{R}+\boldsymbol{H}_j\mathcal{P}_j\left(\boldsymbol{X}\right)\boldsymbol{H}_j^{\mathrm{T}},\ j\in\mathcal{N} \tag{5-67}$$

然后定义如下集合

$$\varOmega\coloneqq\left\{\boldsymbol{X}\in\mathbb{H}^{n*}\mid\ \boldsymbol{\mathcal{X}}\left(\boldsymbol{X}\right)\geqslant 0,\boldsymbol{\mathcal{R}}\left(\boldsymbol{X}\right)>0\right\}. \tag{5-68}$$

定义 5.5

如果对于任意$\boldsymbol{Y}=\left(\boldsymbol{Y}_1,\cdots,\boldsymbol{Y}_N\right)$其中$\boldsymbol{Y}_j=\mathcal{P}_j\left(\boldsymbol{X}\right),j\in\mathcal{N}$，有$\boldsymbol{X}\in\varOmega,\boldsymbol{Y}^+\geqslant\boldsymbol{Y}$，那么 CAREs 式 (5-53) 解$\boldsymbol{Y}^+=\left(\boldsymbol{Y}_1^+,\cdots,\boldsymbol{Y}_N^+\right)$被称为最大解。

可以通过解决如下凸规划问题来在数值上计算最大解[142]：

$$\max\left\{\operatorname{tr}\left(\sum_{j=1}^{N}\boldsymbol{X}_j\right):\ \boldsymbol{X}=\left(\boldsymbol{X}_1,\cdots,\boldsymbol{X}_N\right)\in\varOmega\right\} \tag{5-69}$$

引理 5.4

（文献 [142]）：回顾定义 5.2 和 5.3。

① 假定 (**H**,**A**,**p**) 均方可检测，那么 CAREs 式 (5-53) 的最大解 $\boldsymbol{Y}^{+}=\left(\boldsymbol{Y}_{1}^{+},\cdots,\boldsymbol{Y}_{N}^{+}\right)\in\mathbb{H}_{+}^{n}$存在。此外，令$\boldsymbol{L}_{i}=-\boldsymbol{K}_{i}\left(\boldsymbol{Y}^{+}\right),1\leqslant i\leqslant N$，则$\rho\left(\tilde{\mathcal{L}}^{*}\right)\leqslant 1$成立。

② CAREs 式 (5-53) 最多存在一个均方镇定解，且该解如果存在便是最大解。

对于凸规划问题 (5-69)，令其解为$\boldsymbol{X}^{+}=\left(\boldsymbol{X}_{1}^{+},\cdots,\boldsymbol{X}_{N}^{+}\right)$，那么最大解为$\boldsymbol{Y}^{+}=\left(\boldsymbol{Y}_{1}^{+},\cdots,\boldsymbol{Y}_{N}^{+}\right)$，其中$\boldsymbol{Y}_{j}^{+}=\mathcal{P}_{j}\left(\boldsymbol{X}^{+}\right)$。根据引理 5.4 ②，$\boldsymbol{Y}^{+}$ 也是 CAREs 式 (5-53) 的均方镇定解（均方镇定解存在情形）。下面先给出一个技术性引理。

引理 5.5

以下两个陈述是正确的：

① 对于任意属于$\mathbb{S}_{+}^{n}$的矩阵 $\boldsymbol{A}$、$\boldsymbol{B}$，有$\operatorname{tr}\left(\boldsymbol{AB}\right)\geqslant 0$，且等式成立当且仅当 $\boldsymbol{AB}=0$。

② 对于任意矩阵$\boldsymbol{A}\in\mathbb{C}^{m\times n},\boldsymbol{AA}^{*}=0$当且仅当 $\boldsymbol{A}=0$。

然后给出下面一个重要结果，可以直接应用于本小节 (5.3.2 小节) 的主要结果。

定理 5.4

CAREs 式 (5-53) 具有均方镇定解，当且仅当以下两条件成立。

① (**H**,**A**,**p**) 均方可检测。

② $\lambda=1$ 不是 (**A**,**Q**) 的一个不可控特征值。

证明：

必要性：很明显，(**H**,**A**,**p**) 的均方可检测性是个必要条件。为了证明条件②也是必要的，相反地，假设条件②不成立，但 CAREs 式 (5-53) 具有均方镇定解 $\boldsymbol{Y}$，意味着在$\boldsymbol{L}_{i}=-\boldsymbol{K}_{i}\left(\boldsymbol{Y}\right)$时$\tilde{\mathcal{L}}^{*}$是个稳定操作算子。由于条件②是错误的，根据定义 5.4，下式成立：

$$\mathcal{L}^{*}\left(\boldsymbol{V}\right)=\boldsymbol{V},\ \boldsymbol{Q}^{1/2}\boldsymbol{V}_{i}=0,\ \forall i\in\mathcal{N} \tag{5-70}$$

其中，$\boldsymbol{V}=\left(\boldsymbol{V}_{1},\cdots,\boldsymbol{V}_{N}\right)\in\mathbb{H}_{+}^{n}\setminus\{0\}$。根据最优状态估计器增益[式(5-54)]和式

(5-61)中$\mathcal{L}_j(\cdot)$的定义，定理5.3中CAREs式(5-53)可以重写为

$$\boldsymbol{Y}_j=\mathcal{L}_j(\boldsymbol{Y})-\sum_{i=1}^{N}p_{ij}\left[\boldsymbol{K}_i(\boldsymbol{Y})\boldsymbol{H}_i\boldsymbol{Y}_i\boldsymbol{A}^{\mathrm{T}}-\mu_i\boldsymbol{Q}\right] \tag{5-71}$$

在上面CAREs的两边同时右乘$\boldsymbol{V}_j$，并运用迹与求和操作，得到

$$\begin{aligned}\sum_{j=1}^{N}\mathrm{tr}\left(\boldsymbol{Y}_j\boldsymbol{V}_j\right)=&\sum_{j=1}^{N}\mathrm{tr}\left\{\mathcal{L}_j(\boldsymbol{Y})\boldsymbol{V}_j\right\}+\sum_{j=1}^{N}\sum_{i=1}^{N}p_{ij}\mu_i\mathrm{tr}\left(\boldsymbol{Q}\boldsymbol{V}_j\right)\\&-\sum_{j=1}^{N}\sum_{i=1}^{N}p_{ij}\mathrm{tr}\left(\boldsymbol{K}_i(\boldsymbol{Y})\boldsymbol{H}_i\boldsymbol{Y}_i\boldsymbol{A}^{\mathrm{T}}\boldsymbol{V}_j\right)\end{aligned} \tag{5-72}$$

根据内积(5-59)和式(5-64b)这一关系，有

$$\sum_{j=1}^{N}\mathrm{tr}\left\{\mathcal{L}_j(\boldsymbol{Y})\boldsymbol{V}_j\right\}=\sum_{j=1}^{N}\mathrm{tr}\left\{\boldsymbol{Y}_j\mathcal{L}_j^{*}(\boldsymbol{V})\right\}$$

然后，结合式(5-70)，可以总结出等式(5-72)等价于

$$\sum_{j=1}^{N}\mathrm{tr}\left(\boldsymbol{Y}_j\boldsymbol{V}_j\right)=\sum_{j=1}^{N}\left[\mathrm{tr}\left(\boldsymbol{Y}_j\boldsymbol{V}_j\right)-\sum_{i=1}^{N}p_{ij}\mathrm{tr}\left\{\boldsymbol{K}_i(\boldsymbol{Y})\boldsymbol{H}_i\boldsymbol{Y}_i\boldsymbol{A}^{\mathrm{T}}\boldsymbol{V}_j\right\}\right]$$

并根据引理5.5①，这意味着$\boldsymbol{K}_i(\boldsymbol{Y})\boldsymbol{H}_i\boldsymbol{Y}_i\boldsymbol{A}^{\mathrm{T}}\boldsymbol{V}_j=0$，$\forall i,j\in\mathcal{N}$。注意到式(5-54)中$\boldsymbol{K}_i(\boldsymbol{Y})$的表达式，进一步根据引理5.5②，可以总结出$\boldsymbol{H}_i\boldsymbol{Y}_i\boldsymbol{A}^{\mathrm{T}}\boldsymbol{V}_j=0$，且因此得到$\boldsymbol{K}_i(\boldsymbol{Y})^{\mathrm{T}}\boldsymbol{V}_j=0$，$\forall i,j\in\mathcal{N}$。所以，通过在操作算子$\tilde{\mathcal{L}}^{*}$中设定$\boldsymbol{L}_i=-\boldsymbol{K}_i(\boldsymbol{Y})$，可以获得

$$\begin{aligned}\tilde{\mathcal{L}}_i^{*}(\boldsymbol{V})&=\sum_{j=1}^{N}p_{ij}\left[\boldsymbol{A}-\boldsymbol{K}_i(\boldsymbol{Y})\boldsymbol{H}_i\right]^{\mathrm{T}}\boldsymbol{V}_j\left[\boldsymbol{A}-\boldsymbol{K}_i(\boldsymbol{Y})\boldsymbol{H}_i\right]\\&=\sum_{j=1}^{N}p_{ij}\boldsymbol{A}^{\mathrm{T}}\boldsymbol{V}_j\boldsymbol{A}=\mathcal{L}_i^{*}(\boldsymbol{V})=\boldsymbol{V}_i\end{aligned} \tag{5-73}$$

这意味着$\tilde{\mathcal{L}}^{*}(\boldsymbol{V})$稳定，其中$\boldsymbol{L}_i=-\boldsymbol{K}_i(\boldsymbol{Y})$，这与均方镇定解的假设矛盾。至此，条件②的必要性证明完毕。

充分性：只需证明在条件①和②下，$\rho\left(\tilde{\mathcal{L}}^{*}\right)<1$。相反地，假设$\rho\left(\tilde{\mathcal{L}}^{*}\right)\geqslant1$。令$\boldsymbol{L}_i=-\boldsymbol{K}_i\left(\boldsymbol{Y}^{+}\right)$，根据引理 5.4 ①，这意味着$\rho\left(\tilde{\mathcal{L}}^{*}\right)=1$。令$(\lambda=1,\boldsymbol{V})$为$\tilde{\mathcal{L}}^{*}$的一个特征值 - 特征向量对，使得$\tilde{\mathcal{L}}^{*}(\boldsymbol{V})=\boldsymbol{V}$。重写 CAREs 式 (5-53) 为

$$\begin{aligned}\boldsymbol{Y}_j^{+}=&\sum_{i=1}^{N}p_{ij}\left\{\left[\boldsymbol{A}-\boldsymbol{K}_i\left(\boldsymbol{Y}^{+}\right)\boldsymbol{H}_i\right]\boldsymbol{Y}_i^{+}\left[\boldsymbol{A}-\boldsymbol{K}_i\left(\boldsymbol{Y}^{+}\right)\boldsymbol{H}_i\right]^{\mathrm{T}}\right.\\&\left.+\mu_i\left[\boldsymbol{K}_i\left(\boldsymbol{Y}^{+}\right)\boldsymbol{R}\boldsymbol{K}_i\left(\boldsymbol{Y}^{+}\right)^{\mathrm{T}}+\boldsymbol{Q}\right]\right\}\end{aligned} \tag{5-74}$$

通过与式(5-72)相同的操作和伴随关系$\sum_{j=1}^{N}\operatorname{tr}\left\{\tilde{\mathcal{L}}_j\left(\boldsymbol{Y}^{+}\right)\boldsymbol{V}_j\right\}=\sum_{j=1}^{N}\operatorname{tr}\left\{\boldsymbol{Y}_j^{+}\tilde{\mathcal{L}}_j^{*}\left(\boldsymbol{V}\right)\right\}$，可以得到

$$\sum_{j=1}^{N}\operatorname{tr}\left(\boldsymbol{Y}_j^{+}\boldsymbol{V}_j\right)=\sum_{j=1}^{N}\operatorname{tr}\left\{\boldsymbol{Y}_j^{+}\tilde{\mathcal{L}}_j^{*}\left(\boldsymbol{V}\right)\right\}+\sum_{j=1}^{N}\sum_{i=1}^{N}p_{ij}\mu_i\operatorname{tr}\left\{\boldsymbol{K}_i\left(\boldsymbol{Y}^{+}\right)\boldsymbol{R}\boldsymbol{K}_i\left(\boldsymbol{Y}^{+}\right)^{\mathrm{T}}\boldsymbol{V}_j+\boldsymbol{Q}\boldsymbol{V}_j\right\}$$

因为$\tilde{\mathcal{L}}^{*}\left(\boldsymbol{V}\right)=\boldsymbol{V}$，上面方程意味着$\boldsymbol{Q}\boldsymbol{V}_j=0$，且根据引理5.5有$\boldsymbol{K}_i\left(\boldsymbol{Y}^{+}\right)^{\mathrm{T}}\boldsymbol{V}_j=0$，进一步得到$\mathcal{L}^{*}\left(\boldsymbol{V}\right)=\boldsymbol{V}$。这与条件②冲突，并完成了充分性证明。证毕。 ■

我们注意到文献 [143] 中推论 14 中给出了与定理 5.4 类似的控制 CAREs 结果。尽管如此，这里对于 CAREs 式 (5-53) 提供了一个独立的证明，其中关于内积的等式 (5-64) 发挥着重要作用。在呈现本小节主要结果之前，需要如下技术性引理。

引理 5.6

令$\lambda>0$，如果$\boldsymbol{A}^{\mathrm{T}}\boldsymbol{X}\boldsymbol{A}=\lambda\boldsymbol{X}$和$\boldsymbol{Q}\boldsymbol{X}=0$有一个解$\boldsymbol{X}\in\mathbb{S}_{+}^{n}\setminus\{0\}$，那么存在一个非零向量$\boldsymbol{x}_0\in\mathbb{C}^n$和$\omega_0\in\mathbb{R}$，使得

$$\boldsymbol{A}^{\mathrm{T}}\boldsymbol{x}_0=\sqrt{\lambda}e^{j\omega_0}\boldsymbol{x}_0,\ \boldsymbol{Q}\boldsymbol{x}_0=0 \tag{5-75}$$

即根据著名的Popov-Belevitch-Hautus(PBH)测试，$\sqrt{\lambda}\mathrm{e}^{j\omega_0}$是$(\boldsymbol{A},\boldsymbol{Q})$的一个不可控特征值。

证明：

$\boldsymbol{X}\in\mathbb{S}_{+}^{n}\setminus\{0\}$这一假设意味着$\boldsymbol{X}=\boldsymbol{G}\boldsymbol{G}^{\mathrm{T}}$，其中$\boldsymbol{G}\in\mathbb{R}^{n\times r},r=\operatorname{rank}\{\boldsymbol{X}\}>0$。随之得到$\lambda\boldsymbol{G}\boldsymbol{G}^{\mathrm{T}}=\boldsymbol{A}^{\mathrm{T}}\boldsymbol{G}\boldsymbol{G}^{\mathrm{T}}\boldsymbol{A}$和$\boldsymbol{Q}\boldsymbol{G}\boldsymbol{G}^{\mathrm{T}}=0$。因此，存在一个正交矩阵$\boldsymbol{U}\in\mathbb{R}^{r\times r}$，使得

$$\sqrt{\lambda}\boldsymbol{G}\boldsymbol{U}=\boldsymbol{A}^{\mathrm{T}}\boldsymbol{G},\ \boldsymbol{Q}\boldsymbol{G}\boldsymbol{U}=0 \tag{5-76}$$

因为一个正交矩阵的所有特征值都在单位圆上，所以对于某个$\omega_0\in\mathbb{R}$，存在一个非零向量$\boldsymbol{u}_0\in\mathbb{C}^r$，使得$\boldsymbol{U}\boldsymbol{u}_0=\mathrm{e}^{j\omega_0}\boldsymbol{u}_0$。在式(5-76)的两个等式中右乘$\boldsymbol{u}_0$，且令$\boldsymbol{x}_0=\boldsymbol{G}\boldsymbol{u}_0$，验证了式(5-75)中特征值-特征向量方程和$\boldsymbol{Q}\boldsymbol{x}_0=0$。证毕。 ■

定理 5.5

CAREs 式 (5-53) 具有均方镇定解当且仅当以下两条件成立：

① **(H,A,p)** 均方可检测。

② $(\boldsymbol{A},\boldsymbol{Q})$ 不存在单位圆上的不可控特征值，也就是

$$\operatorname{rank}\left\{\left[\lambda\boldsymbol{I}_n-\boldsymbol{A}\quad\boldsymbol{Q}\right]\right\}=n,\ \ \forall|\lambda|=1 \tag{5-77}$$

其中，λ是$\boldsymbol{A}$的一个特征值。

证明：

根据定理 5.4 可知，只需证明：1 是 $(\boldsymbol{A},\boldsymbol{Q})$ 的不可控特征值当且仅当 $(\boldsymbol{A},\boldsymbol{Q})$ 存在某个不可控的单位圆上特征值。首先，证明充分性。根据 PBH 测试，存在某个$|\lambda|=1$ 和$0\neq\boldsymbol{v}\in\mathbb{C}^n$，使得

$$\boldsymbol{A}^{\mathrm{T}}v=\lambda\boldsymbol{v},\ \ \boldsymbol{Q}\boldsymbol{v}=0 \tag{5-78}$$

对于所有的$i\in\mathcal{N}$，令$\boldsymbol{V}_i:=\boldsymbol{v}\boldsymbol{v}^*+\overline{\boldsymbol{v}}\overline{\boldsymbol{v}}^*\in\mathbb{S}_+^n$，使得$\boldsymbol{V}=\left(\boldsymbol{V}_1,\cdots,\boldsymbol{V}_N\right)\in\mathbb{H}_+^n$。那么根据式(5-63)有

$$\begin{aligned}\mathcal{L}_i^*\left(\boldsymbol{V}\right)&=\sum_{j=1}^{N}p_{ij}\boldsymbol{A}^{\mathrm{T}}\boldsymbol{V}_i\boldsymbol{A}=\boldsymbol{A}^{\mathrm{T}}\left(\boldsymbol{v}\boldsymbol{v}^*+\overline{\boldsymbol{v}}\overline{\boldsymbol{v}}^*\right)\boldsymbol{A}\\&=\boldsymbol{A}^{\mathrm{T}}\boldsymbol{v}\left(\boldsymbol{A}^{\mathrm{T}}\boldsymbol{v}\right)^*+\boldsymbol{A}^{\mathrm{T}}\overline{\boldsymbol{v}}\left(\boldsymbol{A}^{\mathrm{T}}\overline{\boldsymbol{v}}\right)^*\\&=\boldsymbol{\lambda}\overline{\boldsymbol{\lambda}}\boldsymbol{v}\boldsymbol{v}^*+\boldsymbol{\lambda}\overline{\boldsymbol{\lambda}}\overline{\boldsymbol{v}}\overline{\boldsymbol{v}}^*=V_i,\ \ \forall i\in\mathcal{N}\end{aligned} \tag{5-79}$$

此外，很容易得到$\boldsymbol{Q}\boldsymbol{V}_i=0,\forall i\in\mathcal{N}$。因此，1是$(\boldsymbol{A},\boldsymbol{Q})$的不可控特征值。

为了证明必要性，假设存在$\boldsymbol{V}=\left(\boldsymbol{V}_1,\cdots,\boldsymbol{V}_N\right)\in\mathbb{H}_+^n\setminus\{0\}$，使得$\mathcal{L}_i^*\left(\boldsymbol{V}\right)=\boldsymbol{V}_i$和$\boldsymbol{Q}\boldsymbol{V}_i=0$。回顾式(5-45)中$\boldsymbol{\mu}$的极限概率分布，其中$\boldsymbol{\mu}=\left[\mu_1\quad\mu_2\quad\cdots\quad\mu_N\right]\neq 0$是一个正向量。因为$\boldsymbol{\mu}\boldsymbol{P}=\boldsymbol{\mu}$，下式成立：

$$\mu_j=\sum_{i=1}^{N}\mu_i p_{ij},\ \ \forall j\in\mathcal{N}$$

根据上面等式，$\boldsymbol{V}_i=\mathcal{L}_i^*\left(\boldsymbol{V}\right)$，其中$\mathcal{L}_i^*\left(\boldsymbol{V}\right)$由式(5-63)定义，以及$\boldsymbol{Q}\boldsymbol{V}_i=0,\forall i\in\mathcal{N}$，有

$$\begin{aligned}\boldsymbol{X}:=\sum_{i=1}^{N}\mu_i\boldsymbol{V}_i&=\sum_{i=1}^{N}\mu_i\mathcal{L}_i^*\left(\boldsymbol{V}\right)=\sum_{i=1}^{N}\sum_{j=1}^{N}\mu_i p_{ij}\boldsymbol{A}^{\mathrm{T}}\boldsymbol{V}_j\boldsymbol{A}\\&=\sum_{j=1}^{N}\boldsymbol{A}^{\mathrm{T}}\boldsymbol{V}_j\boldsymbol{A}\left(\sum_{i=1}^{N}\mu_i p_{ij}\right)=\sum_{j=1}^{N}\mu_j\boldsymbol{A}^{\mathrm{T}}\boldsymbol{V}_j\boldsymbol{A}\\&=\boldsymbol{A}^{\mathrm{T}}\left(\sum_{j=1}^{N}\mu_j\boldsymbol{V}_j\right)\boldsymbol{A}=\boldsymbol{A}^{\mathrm{T}}\boldsymbol{X}\boldsymbol{A}\end{aligned}$$

$$\boldsymbol{Q}\boldsymbol{X}=\sum_{i=1}^{N}\mu_i\boldsymbol{Q}\boldsymbol{V}_i=0$$

所以$A^{\mathrm{T}}XA = X$和$QX = 0$成立。运用引理5.6并让$\lambda = 1$，总结出(A,Q)至少有一个单位圆上的不可控特征值。证毕。 ■

定理 5.5 中的条件②表明可控性要求与标准 ARE[144] 的可控性要求相同，这是一个有趣的结果。在定理 5.5 中，矩阵秩条件 (5-77) 可以很容易被检验，而 $(\mathbf{H},\mathbf{A},\mathbf{p})$ 的均方可检测性并不简单，一般地，需要通过解决一个可行性问题来进行数值验证，涉及 2^m 线性矩阵不等式（Linear Matrix Inequalities，LMIs），请参阅下一小节中的式 (5-84)。另一个问题是定理 5.3 中最优稳态估计器的指数复杂度。下一小节将通过探索系统结构，导出均方可检测性的一些充分和必要条件，直接显示系统参数如何影响均方可检测性，并提出一个具有线性复杂度的局部最优稳态估计器。

5.3.3 均方可检测性和局部最优稳态估计器

5.3.3.1 $(\mathbf{H},\mathbf{A},\mathbf{p})$的均方可检测性

文献 [139] 中定理 3.9 描述了 MJLSs 的均方稳定性，基于此也有如下对于系统 (5-42) 的均方可检测性定义，与定义 5.2 完全一致。

定义 5.6

如果存在$\{X_i > 0\}_{i=1}^N$和$L = (L_1,\cdots,L_N) \in \mathbb{H}^{n,m}$，使得下面两个不等式中任意一个成立：

$$X_i > \sum_{j=1}^{N} p_{ij}\left(A + L_j H_j\right)^{\mathrm{T}} X_j \left(A + L_j H_j\right),\ \forall i \in \mathcal{N} \tag{5-80}$$

$$X_j > \sum_{i=1}^{N} p_{ij}\left(A + L_i H_i\right) X_i \left(A + L_i H_i\right)^{\mathrm{T}},\ \forall j \in \mathcal{N} \tag{5-81}$$

那么$(\mathbf{H},\mathbf{A},\mathbf{p})$均方可检测。

在本小节（5.3.3 小节）中，假设 (C,A) 可检测。很明显这比 $(\mathbf{H},\mathbf{A},\mathbf{p})$ 的均方可检测性要弱，其中均方可检测性与信道丢包参数有关。首先，对于均方可检测性提供一个解析的必要条件。

定理 5.6

$(\mathbf{H},\mathbf{A},\mathbf{p})$ 均方可检测，仅当

$$\prod_{i=1}^{N}\left(1-q_i\right)\rho^2\left(\boldsymbol{A}\right)<1. \tag{5-82}$$

证明：

根据式 (5-81)，$(\mathbf{H},\mathbf{A},\mathbf{p})$ 均方可检测意味着存在$\left\{\boldsymbol{X}_i>0\right\}_{i=1}^{N}$，使得

$$\begin{aligned}\boldsymbol{X}_1&>\sum_{i=2}^{N}p_{ij}\left(\boldsymbol{A}+\boldsymbol{L}_i\boldsymbol{H}_i\right)\boldsymbol{X}_i\left(\boldsymbol{A}+\boldsymbol{L}_i\boldsymbol{H}_i\right)^{\mathrm{T}}\\&+\prod_{i=1}^{N}\left(1-q_i\right)\boldsymbol{A}\boldsymbol{X}_1\boldsymbol{A}\geqslant\prod_{i=1}^{N}\left(1-q_i\right)\boldsymbol{A}\boldsymbol{X}_1\boldsymbol{A}^{\mathrm{T}}\end{aligned} \tag{5-83}$$

这意味着$\boldsymbol{A}\sqrt{\prod_{i=1}^{N}\left(1-q_i\right)}$必须是一个Schur稳定矩阵，这给出了不等式(5-82)。证毕。 ■

$(\mathbf{H},\mathbf{A},\mathbf{p})$ 的均方可检测性可以通过求解如下 LMIs 的可行性问题进行数值验证，其中该 LMIs 可以通过使用均方可检测性条件 (5-80) 和重复使用 Schur 补得出。

命题 5.1

记$\boldsymbol{X}_{\mathrm{d}}:=\operatorname{diag}\left\{\boldsymbol{X}_1,\cdots,\boldsymbol{X}_N\right\}$和$\boldsymbol{\Psi}_i\left(\boldsymbol{X},\boldsymbol{\Omega}\right)=\left[\boldsymbol{\Psi}_{i1}\ \ \boldsymbol{\Psi}_{i2}\ \ \cdots\ \ \boldsymbol{\Psi}_{iN}\right]$，其中$\boldsymbol{\Psi}_{ij}=\sqrt{p_{ij}}\left(\boldsymbol{A}^{\mathrm{T}}\boldsymbol{X}_j+\boldsymbol{H}_j^{\mathrm{T}}\boldsymbol{\Omega}_j\right),1\leqslant j\leqslant N$。那么，$(\mathbf{H},\mathbf{A},\mathbf{p})$ 均方可检测，当且仅当存在$\left\{\boldsymbol{X}_i>0\right\}_{i=1}^{N}$和$\left\{\boldsymbol{\Omega}_i\in\mathbb{R}^{n\times m}\right\}_{i=1}^{N}$，使得对于每个$i\in\mathcal{N}$，LMI 式

$$\begin{bmatrix}\boldsymbol{X}_i & \boldsymbol{\Psi}_i\left(\boldsymbol{X},\boldsymbol{\Omega}\right)\\ \boldsymbol{\Psi}_i(\boldsymbol{X},\boldsymbol{\Omega})^{\mathrm{T}} & \boldsymbol{X}_{\mathrm{d}}\end{bmatrix}>0 \tag{5-84}$$

成立。

从命题 5.1 中核对均方可检测性的一个重大挑战在于式 (5-84) 的 $N=2^m$ 个 LMI 造成的复杂性。下面，将聚焦于将复杂度从 $N=2^m$ 降低到 $2m$。这可以通过解耦 $(\mathbf{H},\mathbf{A},\mathbf{p})$ 均方可检测性为 m 个子系统形式来实现。

记状态估计误差为

$$\boldsymbol{e}\left(k\right):=\boldsymbol{x}\left(k\right)-\hat{\boldsymbol{x}}\left(k\right) \tag{5-85}$$

将式(5-38a)和式(5-51)作差得到

$$\boldsymbol{e}\left(k+1\right)=\left(\boldsymbol{A}-\boldsymbol{L}_{\theta(k)}\boldsymbol{H}_{\theta(k)}\right)\boldsymbol{e}\left(k\right) \tag{5-86}$$

其中，噪声项已被移除。那么实现误差动力学[式(5-86)]稳定性的$\left\{\boldsymbol{L}_{\theta(k)}\right\}$的存在性等价于$(\mathbf{H},\mathbf{A},\mathbf{p})$的均方可检测性。不失一般性，对于系统(5-38)，假设$(\boldsymbol{C},\boldsymbol{A})$具有如下Wonham分解形式[145]：

$$
\boldsymbol{A}=\begin{bmatrix}\boldsymbol{A}_1 & 0 & \cdots & 0\\ \boldsymbol{A}_{21} & \boldsymbol{A}_2 & \ddots & \vdots\\ \vdots & \ddots & \ddots & 0\\ \boldsymbol{A}_{m1} & \cdots & \boldsymbol{A}_{m(m-1)} & \boldsymbol{A}_m\end{bmatrix},\ \boldsymbol{C}=\begin{bmatrix}\boldsymbol{c}_1 & 0 & \cdots & 0\\ 0 & \boldsymbol{c}_2 & \ddots & \vdots\\ \vdots & \ddots & \ddots & 0\\ 0 & \cdots & 0 & \boldsymbol{c}_m\end{bmatrix} \tag{5-87}
$$

其中$\boldsymbol{A}_i \in \mathbb{R}^{n_i\times n_i}, \boldsymbol{c}_i \in \mathbb{R}^{1\times n_i}, \sum_{i=1}^m n_i = n$，且在$(\boldsymbol{C},\boldsymbol{A})$可检测下每个$(\boldsymbol{c}_i,\boldsymbol{A}_i)$均可检测。定义

$$
\theta_i(k) := \gamma_i(k)+1,\ \boldsymbol{h}_{i,\theta_i(k)} := \gamma_i(k)\boldsymbol{c}_i
$$

$$
\mathbf{A}_i := (\boldsymbol{A}_i,\boldsymbol{A}_i),\ \mathbf{h}_i := (\boldsymbol{h}_{i,1},\boldsymbol{h}_{i,2}),\ \mathbf{p}_i := \left\{p_{lj}^{(i)}\right\}
$$

其中，$1\leqslant i\leqslant m, p_{lj}^{(i)}$是式 (5-41) 中 TPM $\boldsymbol{P}_i$ 的第 (l, j) 个元素。引入一个具有对角块形式的特定$\overline{\boldsymbol{L}}_{\theta(k)}$：

$$
\overline{\boldsymbol{L}}_{\theta(k)} := \operatorname{diag}\left\{\boldsymbol{\ell}_{\theta_1(k)},\cdots,\boldsymbol{\ell}_{\theta_m(k)}\right\},\ \boldsymbol{\ell}_{\theta_i(k)} \in \mathbb{R}^{n_i} \tag{5-88}
$$

定理 5.7

如果对于所有$1\leqslant i\leqslant m,(\mathbf{h}_i,\mathbf{A}_i,\mathbf{p}_i)$均方可检测，那么 $(\mathbf{H},\mathbf{A},\mathbf{p})$ 均方可检测。

证明：

对于系统(5-42)，考虑一个相似变换$\boldsymbol{x}(k)=\boldsymbol{S}\tilde{\boldsymbol{x}}(k)$，其中

$$
\boldsymbol{S}=\operatorname{diag}\left\{\boldsymbol{I}_{n_1},\epsilon^{-1}\boldsymbol{I}_{n_2},\cdots,\epsilon^{1-m}\boldsymbol{I}_{n_m}\right\},\ \epsilon>0 \tag{5-89}
$$

那么，$(\mathbf{H},\mathbf{A},\mathbf{p})$也就是系统(5-42)的均方可检测性等价于如下系统

$$
\tilde{\boldsymbol{x}}(k+1)=\tilde{\boldsymbol{A}}\tilde{\boldsymbol{x}}(k)+\boldsymbol{S}^{-1}\boldsymbol{w}(k) \tag{5-90a}
$$

$$
\boldsymbol{y}_{\mathrm{r}}(k)=\tilde{\boldsymbol{H}}_{\theta(k)}\tilde{\boldsymbol{x}}(k)+\boldsymbol{D}_{\theta(k)}\boldsymbol{v}(k) \tag{5-90b}
$$

的均方可检测性，其中$\tilde{\boldsymbol{A}}=\boldsymbol{S}^{-1}\boldsymbol{A}\boldsymbol{S}, \tilde{\boldsymbol{H}}_{\theta(k)}=\boldsymbol{H}_{\theta(k)}\boldsymbol{S}=\boldsymbol{\varGamma}_{\theta(k)}\boldsymbol{C}\boldsymbol{S}$是个对角块形式。定义$\tilde{\boldsymbol{L}}_{\theta(k)} := \boldsymbol{S}^{-1}\overline{\boldsymbol{L}}_{\theta(k)}$，其中$\overline{\boldsymbol{L}}_{\theta(k)}$由式(5-88)给出。从式(5-86)得知

$$
\tilde{\boldsymbol{e}}(k+1)=\left(\tilde{\boldsymbol{A}}-\tilde{\boldsymbol{L}}_{\theta(k)}\tilde{\boldsymbol{H}}_{\theta(k)}\right)\tilde{\boldsymbol{e}}(k) \tag{5-91}
$$

其中，$\tilde{\boldsymbol{e}}(k)=\boldsymbol{S}^{-1}\boldsymbol{e}(k)$。具体来说，$\tilde{\boldsymbol{A}}-\tilde{\boldsymbol{L}}_{\theta(k)}\tilde{\boldsymbol{H}}_{\theta(k)}$具有如下下三角形式：

$$
\tilde{\boldsymbol{A}}-\tilde{\boldsymbol{L}}_{\theta(k)}\tilde{\boldsymbol{H}}_{\theta(k)}=\boldsymbol{S}^{-1}\left(\boldsymbol{A}-\overline{\boldsymbol{L}}_{\theta(k)}\boldsymbol{H}_{\theta(k)}\right)\boldsymbol{S}
$$

$$
=\begin{bmatrix} \mathcal{A}_{\theta_1(k)} & 0 & \cdots & 0 \\ \epsilon \boldsymbol{A}_{21} & \mathcal{A}_{\theta_2(k)} & \ddots & \vdots \\ \vdots & \ddots & \ddots & 0 \\ \epsilon^{m-1}\boldsymbol{A}_{m1} & \cdots & \epsilon \boldsymbol{A}_{m(m-1)} & \mathcal{A}_{\theta_m(k)} \end{bmatrix}
$$

其中，$\mathcal{A}_{\theta_i(k)}=\boldsymbol{A}_i-\boldsymbol{\ell}_{\theta_i(k)}\boldsymbol{h}_{i,\theta_i(k)}$。因此，随着$\epsilon\to 0$，估计误差动力学[式(5-91)]趋近于一个对角形式，意味着当有式(5-88)中对角块增益$\overline{\boldsymbol{L}}_{\theta(k)}$时，$\{\mathbf{h}_i,\mathbf{A}_i,\mathbf{p}_i\}_{i=1}^m$的均方可检测性等价于误差动力学[式(5-91)]的均方稳定性。证毕。■

注释 5.5　Wonham 分解 [式 (5-87)] 使得可以从原始可检测系统 $(\boldsymbol{C},\boldsymbol{A})$ 得出 m 个可检测子系统 $(\boldsymbol{c}_i,\boldsymbol{A}_i)$，这在复杂度从 $N=2^m$ 降低到 $2m$ 方面起到一个关键作用。在处理对数量化和 i.i.d. 衰减信道的多输入系统网络稳定性时，这种分解也是一个强大的工具[146-148]。

现在得到了一个解耦均方可检测性的充分条件，但稳态最优估计器依然是指数复杂度。下一子小节将提出一个复杂度也为2m的局部最优稳态估计器。

5.3.3.2　一个局部最优稳态估计器

这里，考虑对角块增益 [式 (5-88)]，而不是估计器 [式 (5-51)] 中的一般性估计器增益$\boldsymbol{L}_{\theta(k)}$。对角块增益将估计器 [式 (5-51)] 降为

$$
\hat{\boldsymbol{x}}(k+1)=\boldsymbol{A}\hat{\boldsymbol{x}}(k)+\overline{\boldsymbol{L}}_{\theta(k)}\left[\boldsymbol{y}_{\mathrm{r}}(k)-\boldsymbol{H}_{\theta(k)}\hat{\boldsymbol{x}}(k)\right] \tag{5-92}
$$

将系统状态$\boldsymbol{x}(k)$写成$\boldsymbol{x}(k)=\left[\boldsymbol{x}_1(\boldsymbol{k})^{\mathrm{T}},\cdots,\boldsymbol{x}_m(k)^{\mathrm{T}}\right]^{\mathrm{T}}$，其中，$\boldsymbol{x}_i(k)\in\mathbb{R}^{n_i}$，将测量$\boldsymbol{y}_{\mathrm{r}}(k)$写成$\boldsymbol{y}_{\mathrm{r}}(k)=\left[\boldsymbol{y}_{\mathrm{r},1}(k),\cdots,\boldsymbol{y}_{\mathrm{r},m}(k)\right]^{\mathrm{T}}$，并令$\hat{\boldsymbol{x}}_i(k)$为$\boldsymbol{x}_i(k)$的状态估计。观测到可以将估计器[式(5-92)]写成如下m个子估计器：

$$
\begin{cases} \hat{\boldsymbol{x}}_1(k+1)=\boldsymbol{A}_1\hat{\boldsymbol{x}}_1(k)+\boldsymbol{\ell}_{\theta_1(k)}\left[\boldsymbol{y}_{\mathrm{r},1}(k)-\boldsymbol{h}_{1,\theta_1(k)}\hat{\boldsymbol{x}}_1(k)\right] \\ \hat{\boldsymbol{x}}_i(k+1)=\boldsymbol{A}_i\hat{\boldsymbol{x}}_i(k)+\boldsymbol{\ell}_{\theta_i(k)}\left[\boldsymbol{y}_{\mathrm{r},i}(k)-\boldsymbol{h}_{i,\theta_i(k)}\hat{\boldsymbol{x}}_i(k)\right] \\ \qquad +\displaystyle\sum_{j=1}^{i-1}\boldsymbol{A}_{ij}\hat{\boldsymbol{x}}_j, 2\leqslant i\leqslant m \end{cases} \tag{5-93}
$$

令$\{\boldsymbol{Q}_i\}_{i=1}^{m}$和$\{\boldsymbol{R}_i\}_{i=1}^{m}$分别是$\boldsymbol{Q}$和$\boldsymbol{R}$的对角上子矩阵。有如下类似于定理5.3的估计结果，该证明被省略。

引理 5.7

假设对于如下 CAREs：

$$\boldsymbol{Z}_{i,r}=\sum_{j=1}^{2}p_{jr}^{(i)}\left\{\boldsymbol{A}_j\boldsymbol{Z}_{i,j}\left[\boldsymbol{I}_{n_i}+\boldsymbol{h}_{i,j}^{\mathrm{T}}\left(\boldsymbol{\pi}_{i,j}\boldsymbol{R}_i\right)^{-1}\boldsymbol{h}_{i,j}\boldsymbol{Z}_{i,j}\right]^{-1}\boldsymbol{A}_j^{\mathrm{T}}\right.$$
$$\left.+\boldsymbol{\pi}_{i,j}\boldsymbol{Q}_j\right\},\ 1\leqslant i\leqslant m,1\leqslant r\leqslant 2, \tag{5-94}$$

存在均方镇定解$\left\{\boldsymbol{Z}_i=\left(\boldsymbol{Z}_{i,1},\boldsymbol{Z}_{i,2}\right)\right\}$。那么对于式(5-93)中每个子估计器$i=1,\cdots,m$，最小化

$$\boldsymbol{J}_i(\infty)=\lim_{k\to\infty}\mathbb{E}\left[\|\boldsymbol{x}_i(k)-\hat{\boldsymbol{x}}_i(k)\|^2\right] \tag{5-95}$$

的最优稳态增益由下式给出：

$$\ell_{\theta_i(k)}=\boldsymbol{K}_i(\boldsymbol{Z}_i):=\boldsymbol{A}_i\boldsymbol{Z}_{i,2}\boldsymbol{c}_i^{\mathrm{T}}\left(\boldsymbol{c}_i\boldsymbol{Z}_{i,2}\boldsymbol{c}_i^{\mathrm{T}}+\boldsymbol{\pi}_{i,2}\boldsymbol{R}_i\right)^{-1} \tag{5-96}$$

上式对$\theta_i(k)=1,2$均成立。

注意到，这里已经特意地让$\theta_i(k)=1$情况的$l_{\theta_i(k)}$取值与$\theta_i(k)=2$情况相同，使得对于每个子估计器仅仅需要一个增益，且这不会对结果造成任何改变。定义$\boldsymbol{\Sigma}(k):=\mathbb{E}\left[\boldsymbol{e}(k)\boldsymbol{e}(k)^{\mathrm{T}}\right]$和$\boldsymbol{\Sigma}_i(k):=\mathbb{E}\left[\boldsymbol{e}(k)\boldsymbol{e}(k)^{\mathrm{T}}\mathbf{1}_{\theta(k)=i}\right]$。注意$\boldsymbol{\Sigma}(k)=\sum_{i=1}^{N}\boldsymbol{\Sigma}_i(k)$。根据引理 5.7，给出如下局部最优估计器。

命题 5.2

对于式 (5-42) 描述的系统动力学，其中 $(\boldsymbol{C},\boldsymbol{A})$ 具有 Wonham 分解形式［式 (5-87)］，一个局部最优稳态估计器由下式给出：

$$\hat{\boldsymbol{x}}(k+1)=\boldsymbol{A}\hat{\boldsymbol{x}}(k)+\overline{\boldsymbol{K}}\left[\boldsymbol{y}_{\mathrm{r}}(k)-\boldsymbol{H}_{\theta(k)}\hat{\boldsymbol{x}}(k)\right] \tag{5-97}$$

其中，$\overline{\boldsymbol{K}}=\operatorname{diag}\left\{\boldsymbol{K}_1(\boldsymbol{Z}_1),\cdots,\boldsymbol{K}_m(\boldsymbol{Z}_m)\right\}$。该局部最优估计器中的子估计器$i$ [式(5-93)]最小化了式 (5-95) 中代价$J_i(\infty)$。此外，相应的稳态误差协方差$\boldsymbol{\Sigma}(\infty)$有如下耦合Lyapunov方程的解给出：

$$\boldsymbol{\Sigma}_j(\infty)=\sum_{i=1}^{N}p_{ij}\left\{\left[\boldsymbol{A}-\overline{\boldsymbol{K}}\boldsymbol{H}_i\right]\boldsymbol{\Sigma}_i(\infty)\left[\boldsymbol{A}-\overline{\boldsymbol{K}}\boldsymbol{H}_i\right]^{\mathrm{T}}+\mu_i\left[\overline{\boldsymbol{K}}\boldsymbol{D}_i\boldsymbol{R}\boldsymbol{D}_i^{\mathrm{T}}\overline{\boldsymbol{K}}^{\mathrm{T}}+\boldsymbol{Q}\right]\right\},j\in\mathcal{N} \tag{5-98}$$

注释 5.6

与最优稳态估计器求解 2^m 个 CARE 的复杂性相比，式 (5-97) 中的估计器仅需要求解 $2m$ 个 CARE 和一个估计器增益 $\overline{K}$。这是通过将估计器增益限制为对角块形式 [式 (5-88)] 并将估计代价限制为局部代价 [式 (5-95)] 来实现的。如同所期望的，与最优估计器相比，式 (5-97) 中的估计器有性能损失，这也将会在仿真中说明。尽管如此，根据定理 5.5 和定理 5.7，如果对于 $1 \leqslant i \leqslant m$，$(\mathbf{h}_i, \mathbf{A}_i, \mathbf{p}_i)$均方可检测，且每个 (A_i, Q_i) 的秩条件 (5-77) 成立，那么式 (5-86) 描述的误差动力学均方稳定，其中$L_{\theta(k)} = \overline{K}$，这意味着局部最优稳态估计器 [式 (5-97)] 收敛到一个有限值$\Sigma(\infty)$。

注释 5.7

如果$m > n$，将不可能得到式 (5-87) 中的 Wonham 分解形式，而是如下形式：

$$A = \begin{bmatrix} A_1 & 0 & \cdots & 0 \\ A_{21} & A_2 & \ddots & \vdots \\ \vdots & \ddots & \ddots & 0 \\ A_{n1} & \cdots & A_{n(n-1)} & A_n \end{bmatrix}, \quad C = \begin{bmatrix} c_1 & 0 & \cdots & 0 \\ 0 & c_2 & \ddots & \vdots \\ \vdots & \ddots & \ddots & 0 \\ 0 & \cdots & 0 & c_n \\ & & C_0 & \end{bmatrix}$$

其中，$C_0 = \left[C_{n+1}^{\mathrm{T}}, \cdots, C_m^{\mathrm{T}} \right]^{\mathrm{T}} T, T$是 Wonham 分解变换矩阵。在该情况下，有可能得不到理想的局部最优估计器 [式 (5-97)]，使得同时拥有低复杂度 $2m$ 和使用所有 m 个传感器测量信息。一个简单方式为仅仅使用 n 个传感器的测量值并丢弃剩下的（$m-n$）个传感器。也可以期望得到一个更好的方法，该方法可以融合 n 个和剩余的（$m-n$）个测量值，这值得将来进一步研究。

因为每个子系统$(\mathbf{h}_i, \mathbf{A}_i, \mathbf{p}_i)$的均方可检测性是 $(\mathbf{H}, \mathbf{A}, \mathbf{p})$ 的均方可检测性充分条件，并且也为局部最优稳态估计器所需要，所以下一子小节将通过提供一些解析的均方可检测性条件来研究$(\mathbf{h}_i, \mathbf{A}_i, \mathbf{p}_i)$的均方可检测性。

5.3.3.3 $(\mathbf{h}_i, \mathbf{A}_i, \mathbf{p}_i)$的均方可检测性

为了简便，下面将省略下标 i，也就是在略有滥用符号情况下记 $\boldsymbol{A}:=\boldsymbol{A}_i$，$\boldsymbol{C}:=\boldsymbol{c}_i$，$\gamma(k):=\gamma_i(k)$，$\theta(k):=1+\gamma(k)$，其中 $i=1,\cdots,m$。另外，现在 Markovian 丢包过程$\{\gamma(k)\}$的 TPM 由下式给出：

$$\boldsymbol{P}=\left[p_{ij}\right]=\begin{bmatrix}1-q & q\\ p & 1-p\end{bmatrix} \tag{5-99}$$

进一步，记$\boldsymbol{C}_{\theta(k)}:=\theta(k)\boldsymbol{C}$和

$$\mathbf{A}:=(\boldsymbol{A},\boldsymbol{A}),\ \mathbf{C}:=(\boldsymbol{C}_1,\boldsymbol{C}_2),\ \mathbf{p}:=\{p_{ij}\},\ i,j\in\{1,2\}$$

为了简便，定义如下操作算子：

$$\phi_1(\boldsymbol{L},\boldsymbol{X}_1,\boldsymbol{X}_2):=(1-q)\boldsymbol{A}^{\mathrm{T}}\boldsymbol{X}_1\boldsymbol{A}+q\boldsymbol{A}_L^{\mathrm{T}}\boldsymbol{X}_2\boldsymbol{A}_L \tag{5-100a}$$

$$\phi_2(\boldsymbol{L},\boldsymbol{X}_1,\boldsymbol{X}_2):=p\boldsymbol{A}^{\mathrm{T}}\boldsymbol{X}_1\boldsymbol{A}+(1-p)\boldsymbol{A}_L^{\mathrm{T}}\boldsymbol{X}_2\boldsymbol{A}_L \tag{5-100b}$$

$$\psi_1(\boldsymbol{L},\boldsymbol{X}_1,\boldsymbol{X}_2):=(1-q)\boldsymbol{A}\boldsymbol{X}_1\boldsymbol{A}^{\mathrm{T}}+p\boldsymbol{A}_L\boldsymbol{X}_2\boldsymbol{A}_L^{\mathrm{T}} \tag{5-100c}$$

$$\psi_2(\boldsymbol{L},\boldsymbol{X}_1,\boldsymbol{X}_2):=q\boldsymbol{A}\boldsymbol{X}_1\boldsymbol{A}^{\mathrm{T}}+(1-p)\boldsymbol{A}_L\boldsymbol{X}_2\boldsymbol{A}_L^{\mathrm{T}} \tag{5-100d}$$

$$\begin{aligned}g_1(\boldsymbol{X}_1,\boldsymbol{X}_2):=&(1-q)\boldsymbol{A}\boldsymbol{X}_1\boldsymbol{A}^{\mathrm{T}}+p\boldsymbol{A}\boldsymbol{X}_2\boldsymbol{A}^{\mathrm{T}}\\&-p\boldsymbol{A}\boldsymbol{X}_2\boldsymbol{C}^{\mathrm{T}}\left(\boldsymbol{C}\boldsymbol{X}_2\boldsymbol{C}^{\mathrm{T}}\right)^{-1}\boldsymbol{C}\boldsymbol{X}_2\boldsymbol{A}^{\mathrm{T}}\end{aligned} \tag{5-100e}$$

$$\begin{aligned}g_2(\boldsymbol{X}_1,\boldsymbol{X}_2):=&q\boldsymbol{A}\boldsymbol{X}_1\boldsymbol{A}^{\mathrm{T}}+(1-p)\boldsymbol{A}\boldsymbol{X}_2\boldsymbol{A}^{\mathrm{T}}\\&-(1-p)\boldsymbol{A}\boldsymbol{X}_2\boldsymbol{C}^{\mathrm{T}}\left(\boldsymbol{C}\boldsymbol{X}_2\boldsymbol{C}^{\mathrm{T}}\right)^{-1}\boldsymbol{C}\boldsymbol{X}_2\boldsymbol{A}^{\mathrm{T}}\end{aligned} \tag{5-100f}$$

其中，$\boldsymbol{A}_L=\boldsymbol{A}+\boldsymbol{L}\boldsymbol{C}$。下面提供一个有用的引理。

引理 5.8

以下表述等价。

a）$(\mathbf{p},\mathbf{C},\mathbf{A})$ 均方可检测。

b）存在$\boldsymbol{X}_1>0,\boldsymbol{X}_2>0$和$\boldsymbol{L}\in\mathbb{R}^{n_i\times m}$，使得对于$i=1,2$有$\boldsymbol{X}_i>\psi_i(\boldsymbol{L},\boldsymbol{X}_1,\boldsymbol{X}_2)$。

c）存在$\boldsymbol{X}_1>0$和$\boldsymbol{X}_2>0$，使得对于$i=1,2$有$\boldsymbol{X}_i>g_i(\boldsymbol{X}_1,\boldsymbol{X}_2)$。

d）存在$\boldsymbol{X}_1>0,\boldsymbol{X}_2>0$和$\boldsymbol{\Omega}\in\mathbb{R}^{n_i\times m}$，使得如下 LMIs 成立：

$$\begin{bmatrix}\boldsymbol{X}_1 & \sqrt{1-q}\boldsymbol{A}^{\mathrm{T}}\boldsymbol{X}_1 & \sqrt{q}\left(\boldsymbol{A}^{\mathrm{T}}\boldsymbol{X}_2+\boldsymbol{C}^{\mathrm{T}}\boldsymbol{\Omega}^{\mathrm{T}}\right)\\ * & \boldsymbol{X}_1 & 0\\ * & 0 & \boldsymbol{X}_2\end{bmatrix}>0$$

$$\begin{bmatrix} X_2 & \sqrt{p}A^{\mathrm{T}}X_1\sqrt{1-p}\left(A^{\mathrm{T}}X_2+C^{\mathrm{T}}\Omega^{\mathrm{T}}\right) \\ * & X_1 & 0 \\ * & 0 & X_2 \end{bmatrix} > 0$$

证明：

a）⇔ b）：根据定义 5.6，这是直截了当的结果。

c）⇒ b）：对于任意$L \in \mathbb{R}^{n_i \times m}$和$X > 0$，有

$$\begin{aligned}(A+LC)X(A+LC)^{\mathrm{T}} &= AXA^{\mathrm{T}} - AXC^{\mathrm{T}}\left(CXC^{\mathrm{T}}\right)^{-1}CXA^{\mathrm{T}} \\ &\quad + \left(L+L_X\right)CXC^{\mathrm{T}}\left(L+L_X\right)^{\mathrm{T}}\end{aligned}$$

其中，$L_X = AXC^{\mathrm{T}}\left(CXC^{\mathrm{T}}\right)^{-1}$。结果，有

$$\psi_1\left(-L_{X_2}, X_1, X_2\right) = g_1\left(X_1, X_2\right)$$

$$\psi_2\left(-L_{X_2}, X_1, X_2\right) = g_2\left(X_1, X_2\right)$$

上面两个等式意味着，对于任意$X_1 > 0, X_2 > 0$和$L \in \mathbb{R}^{n_i \times m}$，有

$$\psi_i\left(L, X_1, X_2\right) \geqslant g_i\left(X_1, X_2\right), \quad i = 1, 2. \tag{5-101}$$

因此，如果对于某个$X_1 > 0$和$X_2 > 0, X_1 > g_1\left(X_1, X_2\right)$和$X_2 > g_2\left(X_1, X_2\right)$成立，那么$X_1 > \psi_1\left(-L_{X_2}, X_1, X_2\right)$及$X_2 > \psi_2\left(-L_{X_2}, X_1, X_2\right)$，这证明了c)⇒b)。

b）⇒ c）：由于$X_1 > \psi_1\left(L, X_1, X_2\right) \geqslant g_1\left(X_1, X_2\right)$和$X_2 > \psi_2\left(L, X_1, X_2\right) \geqslant g_2\left(X_1, X_2\right)$，这明显为真。

a）⇔ d）：这可以由命题 5.1 直接得到。证毕。 ■

定理 5.8

如果

$$\min\{q, 1-p\} > \lambda_{\mathrm{c}} = 1 - \frac{1}{\prod_i \max\left\{\left|\lambda_i(A)\right|^2, 1\right\}} \tag{5-102}$$

其中，$\lambda_i(A)$是A的第i个特征值，那么$(\mathbf{p}, \mathbf{C}, \mathbf{A})$均方可检测。

证明：

回顾（C, A）的可检测性假设。(c_i, A_i) 可检测，其中$\operatorname{rank}\{c_i\} = 1, 1 \leqslant i \leqslant m$。然后根据文献 [137] 中引理 5.4，存在$X > 0$使得

$$\boldsymbol{X} > \boldsymbol{AXA}^{\mathrm{T}} - \lambda \boldsymbol{AXC}^{\mathrm{T}}\left(\boldsymbol{CXC}^{\mathrm{T}}\right)^{-1}\boldsymbol{CXA}^{\mathrm{T}} \tag{5-103}$$

当且仅当$\lambda > \lambda_c$。因此，如果式(5-102)成立，那么对于式(5-103)，存在$\overline{\boldsymbol{X}} > 0$和$\lambda = \min\{q, 1-p\}$，使得

$$\overline{\boldsymbol{X}} > \boldsymbol{A}\overline{\boldsymbol{X}}\boldsymbol{A}^{\mathrm{T}} - q\boldsymbol{A}\overline{\boldsymbol{X}}\boldsymbol{C}^{\mathrm{T}}\left(\boldsymbol{C}\overline{\boldsymbol{X}}\boldsymbol{C}^{\mathrm{T}}\right)^{-1}\boldsymbol{C}\overline{\boldsymbol{X}}\boldsymbol{A}^{\mathrm{T}}$$

$$\overline{\boldsymbol{X}} > \boldsymbol{A}\overline{\boldsymbol{X}}\boldsymbol{A}^{\mathrm{T}} - (1-p)\boldsymbol{A}\overline{\boldsymbol{X}}\boldsymbol{C}^{\mathrm{T}}\left(\boldsymbol{C}\overline{\boldsymbol{X}}\boldsymbol{C}^{\mathrm{T}}\right)^{-1}\boldsymbol{C}\overline{\boldsymbol{X}}\boldsymbol{A}^{\mathrm{T}}$$

令$\overline{\boldsymbol{X}} = \boldsymbol{X}_1 = p\boldsymbol{X}_2 / q$，上面不等式降为

$$\boldsymbol{X}_1 > g_1(\boldsymbol{X}_1, \boldsymbol{X}_2)，\boldsymbol{X}_2 > g_2(\boldsymbol{X}_1, \boldsymbol{X}_2)$$

因此，根据引理5.8中a)和c)的等价性，充分条件(5-102)成立。 ■

当 $m = n$，也就是 $\boldsymbol{A}_i$ 的阶次为1, $1 \leqslant i \leqslant n$，可以得到如下 (**p**,**C**,**A**) 均方可检测性的解析充要条件。

定理 5.9

如果 $m = n$，那么 (**p**,**C**,**A**) 均方可检测，当且仅当

$$q > 1 - \frac{1}{\rho(\boldsymbol{A})^2} \tag{5-104}$$

证明：

必要性论据与定理 5.6 中一样，所以只呈现充分性证明。首先，观察到如果存在一个矩阵 $\boldsymbol{L}$，使得$\boldsymbol{A}_L = \boldsymbol{A} + \boldsymbol{LC} = 0$，那么均方可检测性条件 (5-81) 将变成下面情况：存在$\boldsymbol{X}_1 > 0, \boldsymbol{X}_2 > 0$使得

$$\boldsymbol{X}_1 > (1-q)\boldsymbol{AX}_1\boldsymbol{A}^{\mathrm{T}}，\boldsymbol{X}_2 > p\boldsymbol{AX}_1\boldsymbol{A}^{\mathrm{T}}. \tag{5-105}$$

在这种情况下，如果$q > 1 - \rho(\boldsymbol{A})^{-2}$，则能够找到某个$\boldsymbol{X}_1 > 0$和$\boldsymbol{X}_2 > 0$，使得式(5-105)不等式为真。明显地，当$A$和$C$都是标量时，$L = -A / C$这一选择使得$A_L = 0$。证毕。 ■

注释 5.8　值得提及的是，5.3.2 小节和本子小节的结果能被应用到文献 [149] 研究的对偶控制问题中。文献 [149] 关注的是在镇定性和检测性假设下的收敛性问题。具体来说，考虑由如下描述的线性系统：

$$\boldsymbol{x}(k+1)=\boldsymbol{A}\boldsymbol{x}(k)+\boldsymbol{B}\boldsymbol{u}_{\mathrm{r}}(k),\ \boldsymbol{u}_{\mathrm{r}}(k)=\gamma(k)\boldsymbol{u}(k) \tag{5-106}$$

其中，$\boldsymbol{u}(k)\in\mathbb{R}^m$是控制输入，由远程控制器通过Markovian丢包信道发送到执行器端。根据MJLSs理论[139]，最优控制器为$\boldsymbol{u}(k)=\boldsymbol{F}(\boldsymbol{X})\boldsymbol{x}(k)$，其中$\boldsymbol{F}(\boldsymbol{X})$由下面控制CAREs的镇定解来计算：

$$\boldsymbol{X}_i=\boldsymbol{A}^{\mathrm{T}}\mathcal{E}_i(\boldsymbol{X})\boldsymbol{A}+\boldsymbol{W}+\boldsymbol{A}^{\mathrm{T}}\mathcal{E}_i(\boldsymbol{X})\boldsymbol{B}_i\boldsymbol{F}_i(\boldsymbol{X})$$

$$\boldsymbol{F}(\boldsymbol{X})=-\left(\boldsymbol{U}+\boldsymbol{B}^{\mathrm{T}}\mathcal{E}_i(\boldsymbol{X})\boldsymbol{B}\right)^{-1}\boldsymbol{B}^{\mathrm{T}}\mathcal{E}_i(\boldsymbol{X})\boldsymbol{A},i=1,2 \tag{5-107}$$

这里，$\mathcal{E}_i(\boldsymbol{X})=\sum_{j=1}^{2}p_{ij}\boldsymbol{X}_j,\boldsymbol{B}_1=0,\boldsymbol{B}_2=\boldsymbol{B}$，以及$\boldsymbol{W}\geqslant 0$和$\boldsymbol{U}>0$分别是状态和控制向量的权重矩阵。对于这样一个最优控制问题，根据文献 [140] 中定义 2 的均方可镇定性概念（与均方可检测性对偶），可以得到一些类似的结果。

5.3.4 仿真示例 2

5.3.4.1 均方镇定解存在性

首先，通过一个数值例子说明本节中关于均方镇定解和均方可检测性的理论结果。考虑一个形式如式 (5-38) 的三阶系统，其中$\boldsymbol{C}=\boldsymbol{R}=\boldsymbol{I}_3$，

$$\boldsymbol{A}=\begin{bmatrix}1&0&0\\1&1.2&0\\1&1.5&1.3\end{bmatrix},\ \boldsymbol{Q}=\begin{bmatrix}1&0&0\\0&1&0\\0&0&0\end{bmatrix} \tag{5-108}$$

注意到，（$\boldsymbol{C,A}$）已是Wonham分解形式。明显地，$\lambda=1.3$是（$\boldsymbol{A,Q}$）的一个不可控特征值。尽管如此，根据定理5.5中条件②，仅仅要求$|\lambda|=1$为可控。本例满足该条件。让三个信道的参数为

$$p_1=0.50,\ p_2=0.60,\ p_3=0.70$$

$$q_1=0.20,\ q_2=0.32,\ q_3=0.51$$

所以，$q_1 > 0, q_2 > 1-1.2^{-2} = 0.3056, q_3 > 1-1.3^{-2} = 0.4083$。然后，根据定理5.7和5.9可知，系统均方可检测。因此，根据定理5.5，CAREs式(5-53)的均方镇定解存在。通过使用YALMIP[150]求解凸规划问题(5-69)，可以获得镇定解$\boldsymbol{Y}$。先通过式(5-54)计算出$\left\{\boldsymbol{K}_i\left(\boldsymbol{Y}\right)\right\}_{i=1}^{8}$，然后根据文献[139]中注释3.5，有$\rho\left(\tilde{\mathcal{L}}^*\right) = \rho\left(\mathcal{A}\right) = 0.9297 < 1$，其中

$$\mathcal{A} = \left(\boldsymbol{P}^{\mathrm{T}} \otimes \boldsymbol{I}_9\right) \mathrm{diag}\left\{\left(\boldsymbol{A} - \boldsymbol{K}_i\left(\boldsymbol{Y}\right)\boldsymbol{H}_i\right) \otimes \left(\boldsymbol{A} - \boldsymbol{K}_i\left(\boldsymbol{Y}\right)\boldsymbol{H}_i\right)\right\}_{i \in \mathcal{N}}$$

所以，按照定义5.3，$\boldsymbol{Y}$的确是均方镇定解。

5.3.4.2 最优和局部最优稳态估计器的性能

下面将使用目标跟踪示例[151]分别展现定理5.3和命题5.2中最优和局部最优稳态估计器的估计性能。为简便起见，这两个估计器将分别被缩写为OS估计器和LOS估计器。系统动力学如下描述[151]：

$$\boldsymbol{x}\left(k+1\right) = \begin{bmatrix} 1 & 0 & 0 \\ T & 1 & 0 \\ T^2/2 & T & 1 \end{bmatrix} \boldsymbol{x}\left(k\right) + \boldsymbol{w}\left(k\right) \tag{5-109}$$

其中，$\boldsymbol{T}$是采样周期；$\boldsymbol{w}$(k)是高斯噪声且协方差为

$$\boldsymbol{Q} = 2\alpha\sigma_m^2 \begin{bmatrix} T & T^2/2 & T^3/6 \\ T^2/2 & T^3/3 & T^4/8 \\ T^3/6 & T^4/8 & T^5/20 \end{bmatrix} \tag{5-110}$$

其中，α为机动时间常数的倒数；σ_m^2是目标加速度的方差。首先，$\boldsymbol{x}(k)$的第一、第二和第三个元素分别代表加速度、速度和目标的位置。假定有三个传感器分别测量目标加速度、速度和位置。因此，测量模型由下式给出：

$$\boldsymbol{y}\left(k\right) = \begin{bmatrix} 1 & 0 & 0 \\ 0 & 1 & 0 \\ 0 & 0 & 1 \end{bmatrix} \boldsymbol{x}\left(k\right) + \boldsymbol{v}\left(k\right) \tag{5-111}$$

假设高斯噪声$\boldsymbol{v}\left(k\right)$的协方差为$\boldsymbol{R} = 0.01\boldsymbol{I}_3$，设置其他系统参数为

$T = 1\text{s}, \alpha = 0.01, \sigma_m^2 = 10$。在本例中，文献[151]中的（$\boldsymbol{C},\boldsymbol{A}$）已变换成Wonham分解形式。根据定理5.7和定理5.9，条件$q_1 > 0, q_2 > 0$和$q_3 > 0$足以保证系统的均方可检测性。

设置信道参数为

$$p_1 = 0.20, \quad p_2 = 0.30, \quad p_3 = 0.20$$
$$q_1 = 0.85, \quad q_2 = 0.75, \quad q_3 = 0.80$$

对于OS和LOS估计器，在时域$k \in [1,50]$中均执行了一个50,000次试验的Monte Carlo仿真，来展现估计性能，估计性能由目标位置误差方差来表现。图5-5显示了OS和LOS估计器的实验方差，两者均接近于各自的理论值。不出所料，LOS估计器的性能不如具有更高复杂度的全局最优估计器。另一方面，图5-6显示了目标位置跟踪的数值结果，展示了这两估计器均有相当不错的估计性能。图5-7显示了图5-6中跟踪情况相应的传感器数据丢包序列。

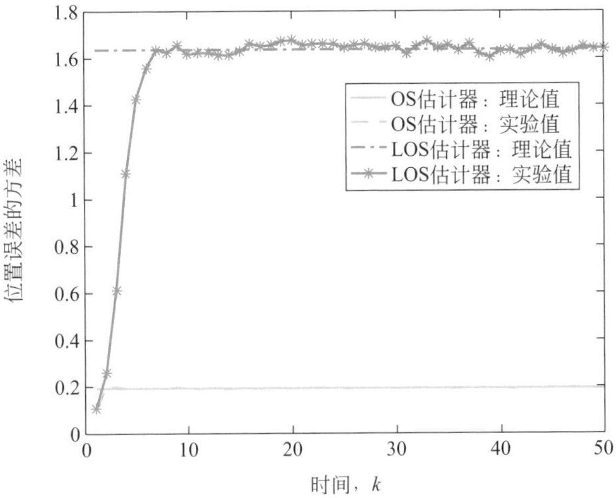

图5-5　OS和LOS估计器的目标位置误差方差：OS估计器的理论值是矩阵$\sum_{i=1}^{N}\boldsymbol{Y}_i$的第(3,3)个元素，LOS估计器的理论值是矩阵$\lim_{k\to\infty}\sum_{i=1}^{N}\boldsymbol{\Sigma}_i(k)$的第(3,3)个元素

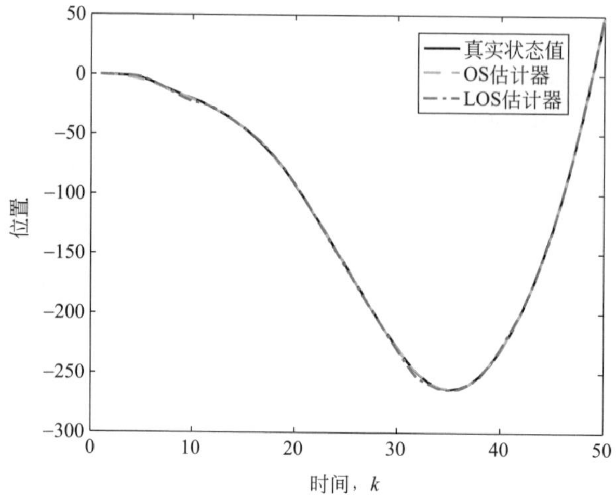

图 5-6　一个 OS 和 LOS 估计器的目标位置估计值的实现情况

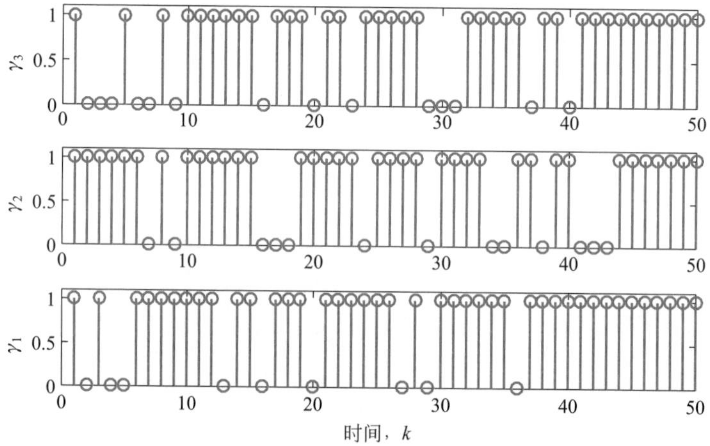

图 5-7　一个丢包情况的采样路径

Digital Wave
Advanced Technology of Industrial Internet

Control of Autonomous Intelligent Systems

自主智能系统控制

第6章

多刚体系统有限时间协同控制

6.1 概述

近年来，对多欧拉 - 拉格朗日系统一致性问题的研究引起了学者们的兴趣。欧拉 - 拉格朗日系统是一类具有刚体性质的特殊多智能体系统。欧拉 - 拉格朗日方程是建模大量机电系统包括机械臂、自主车辆和无人飞行器等的有效方法。由于其广泛的应用背景，有大量关于网络化拉格朗日系统一致性问题的研究文献 [152-153]。本章将进一步从提高收敛性能的角度出发，研究多刚体系统的有限时间一致性问题。由于其在对抗不确定性和干扰方面的良好性质，最近关于有限时间稳定性的研究得到了广泛关注。大多数文献都是基于不连续的控制策略和非光滑分析 [41]。文献 [154] 通过终端滑模的控制方法，研究了网络化的多欧拉 - 拉格朗日系统的有限时间一致性问题。在文献 [155] 中，基于一种非光滑的滑模面技术，对网络化的欧拉 - 拉格朗日系统的有限时间编队控制问题进行了研究。

尽管如此，上述提到的基于滑模控制的多欧拉 - 拉格朗日系统有限时间一致性结果主要的局限性在于不连续的符号函数所造成的抖振现象。为了解决这个问题，连续的螺旋算法是可行的方法之一 [156-157]。基于高阶滑模系统和齐次性的方法，螺旋滑模控制算法可以产生连续的控制信号，还具有包括有限时间收敛和对外部干扰的抑制等优势。然而，大多数螺旋控制算法仅仅针对简单的线性积分器系统，其结果不能直接运用到多智能体系统当中。此外，鉴于欧拉 - 拉格朗日系统的本质非线性，怎样使用螺旋滑模技术实现有限时间一致和抗外界干扰是一个很有挑战的问题。进一步，现有的关于多欧拉 - 拉格朗日系统的有限时间一致性研究都假定智能体之间的通信是连续的 [154]。因此，提出基于事件驱动的有限时间一致性协议，以平衡多智能体网络中的通信资源消耗和快速收敛性能是一个很有价值的研究问题。考虑到欧拉 - 拉格朗日系统的非线性，基于事件驱动的多欧拉 - 拉格朗日系统有限时间一致性并没有被充分研究。

基于欧拉 - 拉格朗日方程的刚体动力学建模方法是在欧几里德空间

下对刚体的位置、速度和加速度进行描述。然而，刚体姿态的完整表述空间是一个非欧几里德流形[78]，姿态的这一属性使得欧拉-拉格朗日方程在姿态的表征上存在固有缺陷，同时也给姿态一致性协议的设计带来了一系列困难。最近，在有限时间姿态一致性方面[154,158-159]以及进阶的固定时间姿态一致性方面[160]都有一些初步的工作，其中固定时间一致性主要是指暂态时间，与系统的初始姿态无关。注意到大多数的工作基于特定的符号或者符号相关的函数设计了有限时间/固定时间姿态一致性协议[154,158-159]。另一方面，在姿态一致性的实际应用中，通信控制成本是一个关键的因素，应当被考虑到一致性协议的设计当中。例如，实时分享和传输信息在空间航天器之间需要非常高的通信功率，这主要是由于很长的空间通信距离[161]。因此，在姿态一致性协议中，考虑基于事件驱动的有限时间/固定时间方法是一个十分有意义的问题。

基于上述讨论，本章致力于研究两类多刚体系统的有限时间一致性问题，即基于事件驱动的多欧拉-拉格朗日系统的有限时间一致性问题与基于事件驱动的固定时间姿态一致性问题。针对第一类问题，首先，基于事件驱动的通信机制设计了有限时间一致性协议，使得一个构造的中间变量在有限时间内达到一致。结合自适应控制律，基于事件驱动的一致性协同不需要利用网络拓扑中的全局信息。此外，运用了推动式的事件驱动策略，避免了事件驱动采样中的连续监测问题。进一步，基于齐次性理论，构造了基于螺旋滑模的控制器，使得网络化的欧拉-拉格朗日系统能在有限时间内收敛到一致。主要贡献概括如下：①为了克服滑模控制的抖振问题，对网络化的欧拉-拉格朗日系统构造了基于螺旋滑模的有限时间控制器。②提出了一种基于事件驱动的有限时间一致性协议，以减少网络中的通信压力，并且提出的协议是全分布式的，也就是协议中不需要利用网络拓扑中的全局信息，如拉普拉斯矩阵的特征值等。

针对第二类问题，利用刚体自身的绝对姿态信息和其邻居的绝对姿态信息，我们设计了基于事件驱动的姿态一致性协议。首先，考虑了在固定拓扑下的事件驱动姿态一致性协议，并证明了姿态一致性集合相对于一个正不变集的固定时间收敛性；进一步，由于多智能体之间的通信链路会发生改变和失效，考虑了时变切换拓扑下的姿态一致性协议。注意到，拓扑切换的发生时刻可能出现在触发间隔中。因此，需要解决当前协议和拓扑之间不同步的问题。主要贡献总结如下：①运用轴角向量的姿态表征方法，提出的一类基于连续可微函数的固定时间姿态一致性协议。值得一提的是，文献 [162] 提出的事件驱动有限时间一致性协议是本章内容的一个特殊情况。②对比现有姿态一致性的工作 [77,163-165]，引入了基于事件驱动的采样机制，大大减轻了网络中的通信负担。由于姿态空间的非欧性，基于欧几里德空间提出的事件驱动协议和条件很难直接推广到姿态一致性当中 [166-167]。此外，文献 [77,164,168] 中的方法同样不能直接应用到事件驱动机制下的姿态一致性。采样误差的存在，造成了证明姿态空间不变性 [77,164,168] 的困难。③对比最近的事件驱动姿态一致性的结果 [166-167]，我们提出了具有固定时间收敛的姿态一致性协议。就我们目前所知，仅有很少的工作关注了基于事件驱动的固定时间姿态一致性。考虑到需要平衡有限时间的收敛性能和事件驱动之间的关系，因此怎样在有限次的协议更新之下确保固定时间的收敛性是一个很有挑战的问题。

本章内容的结构安排如下：第 6.2 节给出了相关引理和问题描述；第 6.3 节、6.4 节介绍了主要理论结果，针对两类多刚体系统有限时间一致性问题，设计基于事件驱动的有限时间一致性协议；第 6.5 节给出了算法的数值仿真验证；第 6.6 节是对本章研究内容的小结。

6.2 相关引理与问题描述

6.2.1 引理

引理 6.1[154]

对于 $y:R\to R$ 和 $\varrho\in\mathbb{R}, \dfrac{\mathrm{d}|y|^{\varrho+1}}{\mathrm{d}t}=(\varrho+1)\mathrm{sig}(y)^{\varrho}\dot{y}$ 且 $\dfrac{\mathrm{d}\left[\mathrm{sig}(y)^{\varrho+1}\right]}{\mathrm{d}t}=(\varrho+1)|y|^{\varrho}\dot{y}$。

引理 6.2[169]

对于任意正数$x_1, x_2, \cdots, x_N$以及$0 < a < 1, b > 1$，有

$$\sum_{i=1}^{N} x_i^a \geqslant \left(\sum_{i=1}^{N} x_i\right)^a \tag{6-1}$$

$$\sum_{i=1}^{N} x_i^b \geqslant N^{1-b}\left(\sum_{i=1}^{N} x_i\right)^b \tag{6-2}$$

引理 6.3[37]

假设一个非线性系统$\dot{\zeta}(t) = f(\zeta(t))$，原点是有限时间稳定的平衡点，如果存在一个正定连续函数$V: D \to \mathbb{R}$满足下面条件

$$\dot{V}(\zeta) + r(V(\zeta))^\gamma \leqslant 0,\ \ \zeta \in U \setminus \{0\} \tag{6-3}$$

其中，$r > 0, \gamma \in (0,1)$且$U \subseteq \mathcal{D}$是原点周围的一个开邻域。除此之外，如果$\mathcal{N} \subseteq \mathcal{D}$是原点周围的一个开邻域，则存在一个有限时间连续函数T满足

$$T(\zeta) \leqslant \frac{1}{r(1-\gamma)} V(\zeta)^{(1-\gamma)},\ \ \zeta \in \mathcal{N} \tag{6-4}$$

进一步，如果$\mathcal{D} = \mathbb{R}^n$，则原点是全局有限时间平衡点，当$\boldsymbol{V}$在$\mathbb{R}^n \setminus \{0\}$中是正定的且$\dot{\boldsymbol{V}}$是负定的。

引理 6.4[170]

假设$F: \mathbb{R}^k \to \mathbb{R}$是一个在集合$S = \{\boldsymbol{x} \in \mathbb{R}^k |\ \boldsymbol{x} \neq 0, \boldsymbol{x}_l \geqslant 0, l = 1, \cdots, k\}$上的正定齐次多项式。那么，可以找到一个整数$n$满足下面形式的多项式

$$P(\boldsymbol{x}) = (x_1 + x_2 + \cdots + x_k)^n F(\boldsymbol{x}) \tag{6-5}$$

是严格正的。

引理 6.5[171]

假设$h(\boldsymbol{x}): \mathbb{R}^n \to \mathbb{R}$相对于$\omega$的齐次度为$\iota < 0$，则当原点是渐近稳定平衡点时，它同时是有限时间平衡点。此外，存在一个驻留时间，满足$T \leqslant -\frac{m}{\iota r} V(\boldsymbol{x}(0))^{-\frac{\iota}{m}}$，其中$V: \mathbb{R}^n \to \mathbb{R}$是一个相对于正常数$m$和$r$的齐次正定连续函数。

证明：

根据定理 7.1 和 7.2[171]，结合引理 6.3，可得结果。证毕 ■

引理 6.6[172]

假设存在一个强连通有向图 $\mathcal{G}$，其对应的拉普拉斯矩阵和邻接矩阵为 $\boldsymbol{\mathcal{L}}$ 和 $\boldsymbol{\mathcal{A}}$，则存在一个对角阵 $\boldsymbol{G} \overset{\text{def}}{=} \text{diag}\{g_i\}$ 满足 $\boldsymbol{G}\boldsymbol{\mathcal{L}} + \boldsymbol{\mathcal{L}}^{\mathrm{T}}\boldsymbol{G} \geqslant 0$，其中，$\boldsymbol{g} = [g_1, \cdots, g_N]^{\mathrm{T}} \in \mathbb{R}^N$是 $\boldsymbol{\mathcal{L}}$ 的左特征值。

6.2.2 欧拉 - 拉格朗日系统有限时间一致性

定义 6.1

给定一个多欧拉 - 拉格朗日系统，如果存在一个有限时间 T 满足下面的条件，$\forall i, j = 1, \cdots, N$

$$\lim_{t \to T} \| \boldsymbol{q}_i(t) - \boldsymbol{q}_j(t) \| = 0 \tag{6-6}$$

其中，$\boldsymbol{q}_i \in \mathbb{R}^n$代表欧拉-拉格朗日系统的状态，称为实现有限时间一致性。

考虑由 N 个智能体组成的多欧拉 - 拉格朗日系统如下式：

$$\boldsymbol{M}_i(\boldsymbol{q}_i)\ddot{\boldsymbol{q}}_i + \boldsymbol{C}_i(\boldsymbol{q}_i, \dot{\boldsymbol{q}}_i)\dot{\boldsymbol{q}}_i + \boldsymbol{g}_i(\boldsymbol{q}_i) = \boldsymbol{\tau}_i + \boldsymbol{\Delta}_i, \quad i = 1, \cdots, N \tag{6-7}$$

其中，$\boldsymbol{q}_i \in \mathbb{R}^n$表示广义坐标向量；$\boldsymbol{M}_i(\boldsymbol{q}_i) \in \mathbb{R}^{n \times n}$和$\boldsymbol{C}_i(\boldsymbol{q}_i, \dot{\boldsymbol{q}}_i) \in \mathbb{R}^n$分别表示对称正定惯性矩阵和科氏力矩阵；$\boldsymbol{g}_i(\boldsymbol{q}_i) \in \mathbb{R}^n$表示重力向量；$\boldsymbol{\tau}_i \in \mathbb{R}^n$表示输入力矩向量；$\boldsymbol{\Delta}_i \in \mathbb{R}^n$表示外部扰动向量满足$\|\boldsymbol{\Delta}_i\| \leqslant \mu_1$和$\|\dot{\boldsymbol{\Delta}}_i\| \leqslant \mu_2$，其中$\mu_1, \mu_2 \in \mathbb{R}$是扰动的上界和扰动的导数上界。

假设 6.1

惯性矩阵是可微的且满足导数有界，即$\dot{\boldsymbol{M}} \leqslant k_{dm}\boldsymbol{I}_n$。

第一类多刚体系统有限时间一致性问题的研究目标是提出一种针对欧拉 - 拉格朗日系统的基于事件驱动通信机制的分布式有限时间算法，使得定义 6.1 中的有限时间一致性实现。为了便于引入事件驱动机制，定义事件驱动时间序列$t_1, t_2, \cdots, t_k$。每个智能体将在触发时刻$t_1^i, t_2^i, \cdots, t_k^i, i = 1, \cdots, N, k \in \mathbb{N}^+$收到来自邻居的信息，且基于一种推动式的触发策略广播自身的状态信息。

6.2.3 多刚体系统固定时间姿态一致性

针对第二类多刚体系统的姿态有限时间一致性问题，每个刚体的轴角姿态运动学方程给出如下，

$$\dot{\boldsymbol{x}}_i = \boldsymbol{J}_{\boldsymbol{x}_i}\boldsymbol{w}_i \tag{6-8}$$

其中，$\boldsymbol{w}_i$是机体坐标系中的角速度向量；$\boldsymbol{J}_{\boldsymbol{x}_i}$是雅可比矩阵，定义为

$$\boldsymbol{J}_{\boldsymbol{x}_i} = \boldsymbol{I}_3 + \frac{\theta_i \boldsymbol{u}_i^{\wedge}}{2} + \left(1 - \frac{\theta_i}{2}\cot\frac{\theta_i}{2}\right)\left(\boldsymbol{u}_i^{\wedge}\right)^2 \tag{6-9}$$

在介绍第二类问题定义之前，我们引入系统 (6-8) 的不变集相关定义。系统 (6-8) 中$\boldsymbol{x}\left(t,\boldsymbol{x}_0,t_0\right)$的初始状态为 $\boldsymbol{x}_0$ 以及初始时刻为 t_0。

定义 6.2[173]

集合$\mathcal{S}$关于系统 (6-8) 是正不变集，如果$\forall t_0 > 0, \boldsymbol{x}_0 \in \mathcal{S}$，解$\boldsymbol{x}\left(t,\boldsymbol{x}_0,t_0\right) \in \mathcal{S}$对所有$t > t_0$成立。

定义 6.3[174]

集合$\mathcal{C}$相对于$\mathcal{S}$是关于系统 (6-8) 有限时间吸引的，如果对于任意的解$\boldsymbol{x}\left(t,\boldsymbol{x}_0,t_0\right), \forall t_0 > 0, \boldsymbol{x}_0 \in \mathcal{S}$在有限时间$T\left(\boldsymbol{x}_0\right) > 0$内到达$\mathcal{C}$并且$\operatorname{dist}\left(\boldsymbol{x}\left(t,\boldsymbol{x}_0,t_0\right),\mathcal{C}\right) \equiv 0$对所有的$t > T\left(\boldsymbol{x}_0\right)$成立。

定义 6.4[174]

集合$\mathcal{C}$相对于$\mathcal{S}$是关于系统 (6-8) 固定时间吸引的，如果它是相对于$\mathcal{S}$有限时间吸引的，且相对于初始值$\boldsymbol{x}_0, T\left(\boldsymbol{x}_0\right)$是一致有界的，即$\exists T_{\max} > 0$满足$T_{\max} \geqslant T\left(\boldsymbol{x}_0\right), \forall \boldsymbol{x}_0 \in \mathcal{S}$。

定义一致性集合为

$$\mathcal{C}_{\mathcal{R}} = \left\{\mathcal{R} = \left\{\mathcal{R}_1, \cdots, \mathcal{R}_N\right\} \in \mathcal{B}_{\pi}\left(\boldsymbol{I}_3\right)^N : \mathcal{R}_1 = \cdots = \mathcal{R}_N\right\}$$

下面给出固定时间姿态一致性的定义。

定义 6.5

对于一个包含N个刚体的多智能体系统，如果集合$\mathcal{C}_{\mathcal{R}}$相对于一个正不变集$\mathcal{S} \subset \mathcal{B}_{\pi}\left(\boldsymbol{I}_3\right)^N$是固定时间吸引的，称其为局部姿态固定

时间一致性。对于轴角表示方法，它等价于$C_x = \left\{ \boldsymbol{x} = \left[\boldsymbol{x}_1^{\mathrm{T}}, \cdots, \boldsymbol{x}_N^{\mathrm{T}} \right]^{\mathrm{T}} \in \mathcal{B}_{\pi}^{'}\left(\mathbf{0}_3\right)^N : \boldsymbol{x}_1 = \cdots = \boldsymbol{x}_N \right\}$是相对于正不变集$\mathcal{S} \subset \mathcal{B}_{\pi}^{'}\left(\mathbf{0}_3\right)^N$固定时间吸引的。

不同于连续时间一致性协议，我们仅在事件驱动的一致性协议当中使用采样的姿态信息。 令时间序列$\left\{t_k^i\right\}, k = 1, 2, \cdots,$表示第 i 个刚体的触发时刻。接下来的主要任务就是提出一个对每个刚体的基于事件驱动的姿态一致性协议和事件驱动条件，以满足定义 6.5 中的固定时间一致性。令$\hat{\boldsymbol{x}}_i(t) = \boldsymbol{x}_i\left(t_k^i\right), i \in \mathcal{V}, t \in \left[t_k^i, t_{k+1}^i\right), k \in \mathbb{N}^+$表示每个刚体的采样姿态信息。令$\hat{\mathcal{L}} = \mathcal{L} \otimes \boldsymbol{I}_3$和$\hat{\mathcal{L}}_i = \mathcal{L}_i \otimes \boldsymbol{I}_3$，其中$\mathcal{L}_i$代表拉普拉斯矩阵$\mathcal{L}$的第 i 列。定义$\boldsymbol{x} = \left[\boldsymbol{x}_1^{\mathrm{T}}, \boldsymbol{x}_2^{\mathrm{T}}, \cdots, \boldsymbol{x}_N^{\mathrm{T}}\right]^{\mathrm{T}}$和$\hat{\boldsymbol{x}} = \left[\hat{\boldsymbol{x}}_1^{\mathrm{T}}, \hat{\boldsymbol{x}}_2^{\mathrm{T}}, \cdots, \hat{\boldsymbol{x}}_N^{\mathrm{T}}\right]^{\mathrm{T}}$。

事件驱动下的固定时间姿态一致性协议可写为下面的形式：

$$\boldsymbol{w}_i = f\left(-\widehat{\mathcal{L}}_i \hat{\boldsymbol{x}}\right) \tag{6-10}$$

其中，函数$\boldsymbol{f}: \mathbb{R}^3 \to \mathbb{R}^3$定义为$f(\boldsymbol{x}) = \dfrac{x\mu\left(\|\boldsymbol{x}\|^2\right)}{\|\boldsymbol{x}\|^2} + \dfrac{x\nu\left(\|\boldsymbol{x}\|^2\right)}{\|\boldsymbol{x}\|^2}$。函数$\mu(\cdot)$是正定的满足$\mu(x_1) \geqslant 0\ \forall x_1 \in \mathbb{R}$和$\mu(x_1) = 0$当且仅当$x_1 = 0$，且满足下面的条件：

① $\forall \boldsymbol{x} \in \mathbb{R}^n, \dfrac{\mu\left(\|\boldsymbol{x}\|^2\right)}{\|\boldsymbol{x}\|^2}$是局部利普希茨连续的。

② $\forall x_1, x_2 \in \mathbb{R}, \mu(x_1 x_2) = \mu(x_1)\mu(x_2)$ 和 $\mu(x_1) + \mu(x_2) > c_\mu \mu(x_1 + x_2)$，其中$c_\mu$是正常数。

③ $\forall s > 0, \int_0^s \dfrac{1}{\mu(V)} \mathrm{d}V$是有限的。

除了性质③，函数$\nu(\cdot)$有和$\mu(\cdot)$相同的性质。

注释 6.1　针对有限时间一致性，我们仅仅需要函数$\boldsymbol{f}$的第一项，即$\boldsymbol{f}(\boldsymbol{x}) = \dfrac{\boldsymbol{x}\mu\left(\|\boldsymbol{x}\|^2\right)}{\|\boldsymbol{x}\|^2}$。对于固定时间一致性，函数$\boldsymbol{f}$的第二项是为了免去暂态时间对初始值的依赖。一个可行的策略是选择$\mu(\cdot)$和$\nu(\cdot)$为幂函数。例如令$\mu\left(\|\boldsymbol{x}\|^2\right) = \|\boldsymbol{x}\|^{1+\alpha}$和$\nu\left(\|\boldsymbol{x}\|^2\right) = \|\boldsymbol{x}\|^{1+\beta}$，其中$0 < \alpha < 1$，$\beta > 1$。

每个刚体的测量误差定义为$\boldsymbol{e}_i(t)=\boldsymbol{x}_i(t_k^i)-\boldsymbol{x}_i(t), t\in[t_k^i,t_{k+1}^i)$。为了确定触发时刻，事件驱动条件给出为

$$t_{k+1}^i=\inf_{t>t_k^i}\left\{t\in\mathbb{R}_{>0}:\left\|\boldsymbol{e}_i(t)\right\|^2\geqslant k_i\left\|\sum_{i=1}^N a_{ij}\left(\boldsymbol{x}_i(t_k^i)-\boldsymbol{x}_j(t_{k'}^j)\right)\right\|^2\right\} \tag{6-11}$$

其中，$k'=\arg\min_{k\in\mathbb{N}^+}\{t\geqslant t_k^i\}$，$k_i$是一个正常数，在后面的证明分析中确定。

注释 6.2

考虑到姿态的非欧性，为了聚焦针对基于事件驱动的姿态一致性的研究，我们在运动学层面考虑了姿态一致性。另一方面，为了使理论结果更容易应用到实际，所设计的事件驱动一致性协议 (6-10) 可被考虑为一个中间输入。接下来，通过推广协议 (6-10)，我们可以进一步在动力学层面设计控制力矩输入。在这一方面，有很多控制方法（如滑模控制和自适应控制）可用于处理姿态的动力学模型[154,175]。

6.3 欧拉 - 拉格朗日系统有限时间一致性

6.3.1 基于事件驱动的有限时间一致性协议

在事件驱动通信机制下的有限时间协议和自适应控制律首先设计如下：

$$\dot{\boldsymbol{\eta}}_i(t)=-\beta_i(t)\operatorname{sig}\left(\sum_{j=1}^N a_{ij}\left(\boldsymbol{\eta}_i(t_k^i)-\boldsymbol{\eta}_j(t_k^i)\right)\right)^\alpha \tag{6-12}$$

$$\dot{\beta}_i(t)=\frac{\sigma_i}{2}\left\|\boldsymbol{P}_i(t_k^i)^\alpha\right\|_2^2 \tag{6-13}$$

其中，$\boldsymbol{\eta}_i(t)$是针对每个拉格朗日系统设计的中间变量，满足$\boldsymbol{\eta}_i(0)=\boldsymbol{q}_i(0)$，$\beta_i(t)$是基于自适应律 (6-13) 更新的自适应分布式增益；$\boldsymbol{P}_i(t_k^i)=\sum_{j=1}^N a_{ij}\left(\boldsymbol{\eta}_i(t_k^i)-\boldsymbol{\eta}_j(t_k^i)\right), i=1,\cdots,N$，$\alpha$是满足$0<\alpha<1$的常数，$\sigma_i$是之后需要确定的正参数。

注释 6.3

在分布式协议 (6-12) 中，$\boldsymbol{\eta}_i$ 的物理含义是一个虚拟的位置变量。每个智能体会根据分布式协议 (6-12)，通过运用自身和周围邻居的信息来更新 $\boldsymbol{\eta}_i$，并且 $\boldsymbol{\eta}_i$ 的初值等于拉格朗日系统的广义坐标的初值。事实上，协议 (6-12) 的目的在于计算每个智能体的有限时间一致性的收敛值。然后，下一步是设计螺旋控制算法，使得每个智能体的广义坐标收敛到该一致性状态值。

此外，对于智能体 i，定义协同测量误差$\boldsymbol{e}_i(t)$和动态变量$\Xi_i(t)$为

$$\boldsymbol{e}_i(t) = \mathrm{sig}\left(\boldsymbol{P}_i\left(t_k^i\right)\right)^{\alpha} - \mathrm{sig}\left(\boldsymbol{P}_i(t)\right)^{\alpha} \tag{6-14}$$

$$\dot{\Xi}_i(t) = -\Lambda_1 \Xi_i^r(t) + \Lambda_2\left(\delta_i \|\boldsymbol{P}_i(t)^{\alpha}\|^2 - \beta_i(t)\|\boldsymbol{e}_i(t)\|^2\right) \tag{6-15}$$

其中，$\Xi_i(0) > 0; r \in (0,1); \Lambda_1$和$\Lambda_2$是在后面证明中确定的正常数。

对于第 i 个智能体，事件驱动时刻$t_0^i, t_1^i, \cdots, t_k^i$由下面的事件驱动条件确定：

$$\|\boldsymbol{e}_i(t)\|^2 \geqslant \frac{1}{\beta_i(t)}\left(\delta_i \|\boldsymbol{P}_i(t)^{\alpha}\|^2 + \Xi_i^r(t)\right) \tag{6-16}$$

其中，δ_i是一个正常数，将会在下面的证明中确定。为了简洁起见，定义变量$\boldsymbol{\xi} \overset{\text{def}}{=} \boldsymbol{P}^{\alpha} \in \mathbb{R}^{Nn}$，且用$\beta_i$、$\boldsymbol{e}_i$、$\Xi_i$、$\boldsymbol{P}_i$表示时间相关的变量。

注释 6.4

协议的增益自适应律 (6-13) 的提出受到了在完全分布式自适应事件驱动一致性协议研究 [176-177] 的启发。值得一提的是大多数关于基于事件驱动一致性的结果 [178-179] 都需要预先知道通信拓扑的全局信息。当通信拓扑的规模变大的时候，这往往很难实现。在本章内容当中，一致性协议和事件驱动条件均没有利用到网络通信拓扑中的全局信息，例如拓扑图拉普拉斯矩阵的特征值等。因此，本章内容的事件驱动协议是完全分布式的，更易推广到节点数量大的网络当中。此外，在文献 [176] 和 [177] 的事件驱动协议设计中，考虑的通信拓扑图是无向图。在本章内容当中，我们考虑了更为一般的有向图。

注释 6.5 动态变量$\boldsymbol{\Xi}_i$的引入是为了更方便证明排除了奇诺现象。有很多事件驱动一致性的结果使用了静态事件驱动条件或者基于状态的条件，并不能完全排除奇诺现象。在文献 [178] 中，当事件驱动条件中$\delta_i = 0$的时候，奇诺现象有可能会发生。在文献 [179] 中，在一致性达到之前排除了奇诺现象，这也意味着当排除奇诺现象时，实现的是有界一致性。为了解决这个问题，文献 [180] 提出了动态事件驱动条件。然而，文献 [180] 考虑的是指数一致性收敛且通信拓扑是无向图，因此所提出的事件驱动协议并不能直接推广到解决有限时间一致性的问题当中。在本章内容中，我们设计了一种新的动态事件驱动条件同时保证了有限时间一致和非奇诺现象。

注意到，在上述设计的事件驱动条件中，每个智能体需要连续接收来自邻居智能体的状态信息去判断是否触发。这显然违背了我们运用事件驱动机制的初衷。考虑到这一点，我们需要设计新的触发策略以避免这个问题。注意到在动态模型 (6-12) 和 (6-13) 中，$\boldsymbol{\eta}_i$ 是分段常值的，可以通过下面的方程预测每个智能体的状态：

$$\begin{aligned}\beta_i(t) &= \beta_i\left(t_k^i\right) + \frac{1}{2}\left(t - t_k^i\right)\|\boldsymbol{P}_i\left(t_k^i\right)\|^2 \\ \boldsymbol{\eta}_i(t) &= \boldsymbol{\eta}_i\left(t_k^i\right) - \left(t - t_k^i\right)\beta_i(t)\operatorname{sig}\left(\boldsymbol{P}_i\left(t_k^i\right)\right)^{\alpha}, \quad t \in \left(t_k^i, t_{k+1}^i\right]\end{aligned} \tag{6-17}$$

因此，通过使用式 (6-17)，当$t > t_k^i$时，智能体 i 的邻居可以预测智能体 i 的状态一直到智能体 i 的下一个触发时刻t_{k+1}^i。基于此，我们有下列触发策略。

① 第 i 个智能体监听来自邻居的新信息。如果智能体 i 的一些邻居在$\left(t_k^i, t_{k+1}^i\right)$中触发，那么智能体 i 将会更新邻居的模型：

$$\begin{aligned}\beta_j(t) &= \beta_j\left(t_k^j\right) + \frac{1}{2}\left(t - t_k^j\right)\|\boldsymbol{P}_j\left(t_k^j\right)\|^2 \\ \boldsymbol{\eta}_j(t) &= \boldsymbol{\eta}_j\left(t_k^j\right) - \left(t - t_k^j\right)\beta_j(t)\operatorname{sig}\left(\boldsymbol{P}_j\left(t_k^j\right)\right)^{\alpha}, \quad j \in \mathcal{N}_i\end{aligned} \tag{6-18}$$

② 如果没有智能体 i 的邻居在$\left(t_k^i, t_{k+1}^i\right)$中触发，那么智能体 i 将会在t_{k+1}^i触发。然后，智能体 i 会更新它的状态$\boldsymbol{\eta}_i\left(t_{k+1}^i\right)$和它的邻居的状态$\boldsymbol{\eta}_j\left(t_{k+1}^j\right)$，并且发送更新后的信息$t_{k+1}^i, \boldsymbol{\eta}_i\left(t_{k+1}^i\right)$和$\boldsymbol{P}_i\left(t_{k+1}^i\right)$给邻居。

通过使用上面的两条规则，可以避免智能体之间的连续通信问题，这也意味着通信网络中的资源消耗会进一步减小。

下面证明，通过运用式(6-12)和式(6-13)，每个智能体会在有限时间内实现一致性，结果陈述如下。

引理 6.7

假设通信拓扑是强连通的，通过使用基于事件驱动有限时间协议式(6-12)和式(6-13)，每个智能体的中间状态会在有限时间内达到一致，即 $\boldsymbol{\eta}_1=\boldsymbol{\eta}_2=\cdots=\boldsymbol{\eta}_N$，并且分布式增益$\beta_i$将会在有限时间内收敛到一个稳定的常数。

证明：

首先，定义李雅普诺夫函数如下

$$
\begin{aligned}
V_P &= V_{P1}+V_{P2}+V_{P3} \\
&= \sum_{\nu=1}^{n}\sum_{i=1}^{N}\frac{g_i}{1+\alpha}\left|\boldsymbol{P}_i^{(\nu)}(t)\right|^{1+\alpha}+\sum_{i=1}^{N}\frac{\lambda\left(\beta_i-c\right)^2}{2k\sigma_i}+\sum_{i=1}^{N}\lambda\varXi_i
\end{aligned}
\tag{6-19}
$$

其中，λ表示$\boldsymbol{G}\boldsymbol{\mathcal{L}}+\boldsymbol{\mathcal{L}}^{\mathrm{T}}\boldsymbol{G},\boldsymbol{g}=\left[g_1,\cdots,g_N\right]^{\mathrm{T}}\in\mathbb{R}^N$的第二小特征值；$c$和$k$是后面证明中确定的常系数。令$\hat{\boldsymbol{\beta}}\overset{\text{def}}{=}\operatorname{diag}\left(\beta_1,\beta_1,\cdots,\beta_N\right),\boldsymbol{G}\overset{\text{def}}{=}\operatorname{diag}\left\{g_i\right\}$，计算$V_P$的导数，并代入式(6-12)和式(6-13)可得：

$$
\begin{aligned}
\dot{V}_P &= \sum_{\nu=1}^{n}\sum_{i=1}^{N}-g_i\operatorname{sig}\left(\boldsymbol{P}_i^{(\nu)}(t)\right)^{\alpha}\cdot\sum_{j=1}^{N}\mathcal{L}_{ij}\left[\beta_j\operatorname{sig}\left(\boldsymbol{P}_j^{(\nu)}(t)\right)^{\alpha}+\beta_j\boldsymbol{e}_j^{(\nu)}\right] \\
&\quad+\sum_{\nu=1}^{n}\sum_{i=1}^{N}\frac{\lambda\left(\beta_i-c\right)}{2k}\boldsymbol{P}_i^{(\nu)}\left(t_k^i\right)^{2\alpha}+\sum_{i=1}^{N}\lambda\dot{\varXi}_i \\
&= \sum_{\nu=1}^{n}\sum_{i=1}^{N}-g_i\boldsymbol{\xi}_i^{(\nu)}\times\sum_{j=1}^{N}\boldsymbol{\mathcal{L}}_{ij}\beta_j\boldsymbol{\xi}_j^{(\nu)}+\sum_{\nu=1}^{n}\sum_{i=1}^{N}-g_i\operatorname{sig}\left(\boldsymbol{P}_i^{(\nu)}(t)\right)^{\alpha} \\
&\quad\times\sum_{j=1}^{N}\boldsymbol{\mathcal{L}}_{ij}\beta_j\boldsymbol{e}_j^{(\nu)}+\sum_{\nu=1}^{n}\sum_{i=1}^{N}\frac{\lambda\left(\beta_i-c\right)}{2k}\boldsymbol{P}_i^{(\nu)}\left(t_k^i\right)^{2\alpha}+\sum_{i=1}^{N}\lambda\dot{\varXi}_i
\end{aligned}
\tag{6-20}
$$

通过将式(6-20)写为向量形式，并使用式(6-16)，我们有

$$
\begin{aligned}
\dot{V}_P &\leqslant -\frac{1}{2}\xi^{\mathrm{T}}\left[\hat{\boldsymbol{\beta}}\left(\boldsymbol{G}\boldsymbol{\mathcal{L}}+\boldsymbol{\mathcal{L}}^{\mathrm{T}}\boldsymbol{G}\right)\otimes\boldsymbol{I}_n\right]\xi \\
&\quad+\frac{1}{2}\boldsymbol{\xi}^{\mathrm{T}}\left[\hat{\boldsymbol{\beta}}\left(\boldsymbol{G}\boldsymbol{\mathcal{L}}+\boldsymbol{\mathcal{L}}^{\mathrm{T}}\boldsymbol{G}\right)\otimes\boldsymbol{I}_n\right]\boldsymbol{e}+\sum_{\nu=1}^{n}\sum_{i=1}^{N}\frac{\lambda}{2k}\left(\beta_i-c\right) \\
&\quad\times\left[\boldsymbol{\xi}_i^{(\nu)}(t)+\boldsymbol{e}_i^{(\nu)}(t)\right]^2+\sum_{i=1}^{N}\varLambda_2\left(\delta_i\|\boldsymbol{\xi}_i\|^2-\beta_i\|\boldsymbol{e}_i\|^2\right)
\end{aligned}
$$

$$-\sum_{i=1}^{N}\Lambda_1\Xi_i^r \tag{6-21}$$

通过使用引理 6.6，基于不等式$\boldsymbol{G}\boldsymbol{\mathcal{L}}+\boldsymbol{\mathcal{L}}^{\mathrm{T}}\boldsymbol{G}\geqslant\lambda\boldsymbol{I}_N$，我们有：

$$\begin{aligned}
&-\frac{1}{2}\boldsymbol{\xi}^{\mathrm{T}}\left[\hat{\boldsymbol{\beta}}\left(\boldsymbol{G}\boldsymbol{\mathcal{L}}+\boldsymbol{\mathcal{L}}^{\mathrm{T}}\boldsymbol{G}\right)\otimes\boldsymbol{I}_n\right]\boldsymbol{\xi}+\frac{1}{2}\boldsymbol{\xi}^{\mathrm{T}}\left[\hat{\boldsymbol{\beta}}\left(\boldsymbol{G}\boldsymbol{\mathcal{L}}+\boldsymbol{\mathcal{L}}^{\mathrm{T}}\boldsymbol{G}\right)\otimes\boldsymbol{I}_n\right]\boldsymbol{e}\\
&\leqslant-\frac{1}{2}\boldsymbol{\xi}^{\mathrm{T}}\left(\lambda\hat{\boldsymbol{\beta}}\otimes\boldsymbol{I}_n\right)\boldsymbol{\xi}+\frac{1}{2}\boldsymbol{\xi}^{\mathrm{T}}\left(\lambda\hat{\boldsymbol{\beta}}\otimes\boldsymbol{I}_n\right)\boldsymbol{e}\\
&\leqslant-\frac{\lambda}{4}\sum_{i=1}^{N}\beta_i\|\boldsymbol{\xi}_i\|^2+\frac{\lambda}{4}\sum_{i=1}^{N}\beta_i\|\boldsymbol{e}_i\|^2
\end{aligned} \tag{6-22}$$

其中，运用柯西不等式得到了第二个不等式。

$$\begin{aligned}
&\sum_{v=1}^{n}\sum_{i=1}^{N}\frac{\lambda\left(\beta_i-c\right)}{2k}\left[\boldsymbol{\xi}_i^{2(v)}(t)+\boldsymbol{e}_i^{2(v)}(t)+2\boldsymbol{\xi}_i^{(v)}(t)\boldsymbol{e}_i^{(v)}(t)\right]\\
&\leqslant\sum_{i=1}^{N}\frac{\lambda\left(\beta_i-c\right)}{2k}\left(\|\boldsymbol{\xi}_i\|^2+\|\boldsymbol{e}_i\|^2\right)+\frac{\lambda}{2k}\sum_{i=1}^{N}\beta_i\left(\|\boldsymbol{\xi}_i\|^2+\|\boldsymbol{e}_i\|^2\right)\\
&\quad+\frac{\lambda c}{k}\sum_{i=1}^{N}\left(\frac{\gamma}{2}\|\boldsymbol{\xi}_i\|^2+\frac{1}{2\gamma}\|\boldsymbol{e}_i\|^2\right)\\
&\leqslant\sum_{i=1}^{N}\left(\frac{\lambda\beta_i}{k}-\frac{c\lambda\epsilon_2}{2k}\right)\|\boldsymbol{\xi}_i\|^2+\sum_{i=1}^{N}\left(\frac{\lambda\beta_i}{k}+\frac{c\lambda\epsilon_1}{2k}\right)\|\boldsymbol{e}_i\|^2
\end{aligned} \tag{6-23}$$

其中，$\epsilon_1=\dfrac{1}{\gamma}-1$；$\epsilon_2=1-\gamma$；$\gamma\in(0,1)$。第一个不等式可从杨氏不等式导出。通过将式 (6-22) 和式 (6-23) 代入式 (6-21)，且通过使用式 (6-16)，可以得到

$$\begin{aligned}
\dot{V}_P\leqslant&-\lambda\sum_{i=1}^{N}\beta_i\left(\frac{1}{4}-\frac{1}{k}-\frac{\Lambda_2\delta_i}{\beta_i}\right)\|\xi_i\|^2\\
&+\frac{c}{2}\sum_{i=1}^{N}\left(\frac{\lambda\epsilon_1\delta_i}{\beta_i k}-\frac{\lambda\epsilon_2}{k}\right)\|\boldsymbol{\xi}_i\|^2+\sum_{i=1}^{N}\left(\frac{\lambda\delta_i}{4}+\frac{\lambda\delta_i}{k}\right)\|\xi_i\|^2\\
&+\sum_{i=1}^{N}\left(\frac{\lambda}{4}+\frac{\lambda}{k}+\frac{c\lambda\epsilon_1}{2k\beta_i}\right)\boldsymbol{\Xi}_i^r-\sum_{i=1}^{N}\Lambda_1\boldsymbol{\Xi}_i^r\\
\leqslant&-\frac{\lambda}{4}\sum_{i=1}^{N}\beta_i\left(\frac{k-4}{k}-\frac{4\Lambda_2\delta_i}{\beta_i}\right)\|\boldsymbol{\xi}_i\|^2\\
&+\sum_{i=1}^{N}\left[\frac{\lambda c}{2k}\left(\epsilon_1\delta_i-\epsilon_2\right)+\frac{\lambda}{4k}\left(\delta_i k+\delta_i\right)\right]\|\boldsymbol{\xi}_i\|^2
\end{aligned}$$

$$
\begin{aligned}
&-\lambda\sum_{i=1}^{N}\left[\Lambda_1-\left(\frac{1}{4}+\frac{1}{k}+\frac{c\epsilon_1}{2k}\right)\right]\varXi_i^r\\
&\leqslant -\frac{\lambda}{4}\sum_{i=1}^{N}\beta_i\left(\frac{k-4}{k}-\frac{4\Lambda_2\delta_i}{\beta_i}\right)\|\xi_i\|^2\\
&+\frac{\lambda}{4k}\sum_{i=1}^{N}\left(-2cm_i+\delta_i k+\delta_i\right)\|\xi_i\|^2
\end{aligned}
\tag{6-24}
$$

其中，$m_i=\delta_i-\dfrac{\delta_i}{\gamma}+1-\gamma$。第一个不等式可由式(6-23)和式(6-15)、式(6-16)得到。令$\Lambda_1>\dfrac{4+k+2c\epsilon_1}{4k}$，可得第三个不等式。通过选择$\gamma=\sqrt{\delta_i}$，易得$m>0$。由于$\beta_i$是单调递增的，当$\Lambda_2<\min\left\{\dfrac{\beta_{\min}(0)(k-4)}{k\delta_{\max}},\dfrac{\beta_{\min}(0)-\delta_{\max}}{4\delta_{\max}}\right\}$，其中$\beta_{\min}(0)\overset{\text{def}}{=}\min\{\beta_i(0)\}$和$\delta_{\max}\overset{\text{def}}{=}\max\{\delta_i\}$时，则能保证$\dfrac{k-4}{k}-\dfrac{4\Lambda_2\delta_i}{\beta_i}>0$。通过选择$k>4$和$c\geqslant\hat{c}+\dfrac{\delta_i(k+1)}{2m}$，其中$\hat{c}$是一个正常数，然后，根据式(6-24)，我们有

$$
\begin{aligned}
\dot{V}_P&\leqslant -\lambda\sum_{i=1}^{N}\left(\epsilon_3\beta_i+\epsilon_4\hat{c}\right)\|\xi_i\|^2\\
&\leqslant -\sum_{i=1}^{N}2\lambda\sqrt{\epsilon_3\epsilon_4\beta_i\hat{c}}\|\xi_i\|^2
\end{aligned}
\tag{6-25}
$$

其中，$\epsilon_3=\dfrac{k-4}{4k}-\dfrac{\Lambda_2\delta_{\max}}{\beta_{\min}(0)}$和$\epsilon_4=\dfrac{\lambda m}{2k}$是两个正常数。定义$V_{P4}\overset{\text{def}}{=}V_{P1}+V_{P3}$，那么下面证明的目标就是找到一个正常数$K$满足$\dot{V}_{P4}\leqslant -KV_{P4}^{\frac{2\alpha}{1+\alpha}}$。根据式(6-22)并结合事件驱动条件(6-16)，我们有，

$$
\begin{aligned}
\dot{V}_{P4}\leqslant &-\frac{\lambda}{4}\sum_{i=1}^{N}\left(\beta_i-\delta_i-4\Lambda_2\delta_i\right)\|\xi_i\|^2\\
&+\frac{\lambda}{4}\sum_{i=1}^{N}\varXi_i^r-\lambda\sum_{i=1}^{N}\Lambda_1\varXi_i^r
\end{aligned}
$$

$$
\begin{aligned}
&\leqslant -\frac{\lambda(\beta_{\min}(0)-\delta_{\max}-4\varLambda_2\delta_{\max})}{4}\sum_{\nu=1}^{n}\sum_{i=1}^{N}\left(\frac{1+\alpha}{g_i}\right)^{\frac{2\alpha}{1+\alpha}} \\
&\quad \times\left(\frac{g_i}{1+\alpha}|\boldsymbol{P}_i^{(\nu)}|^{1+\alpha}\right)^{\frac{2\alpha}{1+\alpha}}-\lambda\sum_{i=1}^{N}\left(\varLambda_1-\frac{1}{4}\right)\boldsymbol{\varXi}_i^r \\
&\leqslant -\frac{\lambda(\beta_{\min}(0)-\delta_{\max}-4\varLambda_2\delta_{\max})}{4}\left(\frac{1+\alpha}{\underline{g}}\right)^{\frac{2\alpha}{1+\alpha}}V_{\mathrm{P4}}^{\frac{2\alpha}{1+\alpha}} \\
&\quad -\lambda\hat{\varLambda}_1\sum_{i=1}^{N}\boldsymbol{\varXi}_i^r \\
&\overset{\text{def}}{=} -K_1V_{P4}^{\frac{2\alpha}{1+\alpha}}-K_2\sum_{i=1}^{N}\boldsymbol{\varXi}_i^r
\end{aligned}
\tag{6-26}
$$

其中，$\underline{g}$是g_i的下界且$\hat{\varLambda}_1$是正常数，基于不等式$\varLambda_1>\dfrac{1}{4}$。

从式(6-25)中可得$\dot{V}_P\leqslant 0$，意味着β_i是有界的。根据式(6-13)，可得β_i是单调递增的。基于此，可证分布式增益β_i会收敛到一个有限值。选择$\delta_i<\beta_i(0)$，因此K_1是一个正常数。定义$\varDelta\overset{\text{def}}{=}\dfrac{-\dot{V}_{P4}}{V_{P4}^{\frac{2\alpha}{1+\alpha}}}$，并且选择$r=\dfrac{2\alpha}{1+\alpha}$，我们可得到$\varDelta\geqslant\dfrac{K_1V_{P4}^{\frac{2\alpha}{1+\alpha}}+K_2\sum_{i=1}^{N}\boldsymbol{\varXi}_i^r}{V_{P4}^{\frac{2\alpha}{1+\alpha}}+\sum_{i=1}^{N}\boldsymbol{\varXi}_i^{\frac{2\alpha}{1+\alpha}}}\geqslant K$，其中，$K=\min\{K_1,K_2\}$。因此，可得$\dot{V}_{P4}\leqslant -KV_{P4}^{\frac{2\alpha}{1+\alpha}}$。基于引理6.3，可得$\|\boldsymbol{P}_i\|$在有限时间$T_1=\dfrac{(1+\alpha)V_{P1}(0)^{\frac{1-\alpha}{1+\alpha}}}{K(1-\alpha)}$内收敛到0。 ■

注释 6.6 引理6.7的证明是基于引理6.6，且假设通信拓扑是强连通的。事实上，上述证明可进一步推广到包含一个有向生成树的有向图当中。证明思路可基于包含有向生成树的拉普拉斯矩阵是一个可退化阵，且可以通过一个合适的排列变换写为Frobenius规范形式[181]。

6.3.2 基于螺旋滑模的一致性算法

进一步，我们需要证明多欧拉 - 拉格朗日系统的状态能在有限时间内趋于一致。为了避免滑模变量 $\boldsymbol{s}_i$ 的抖振现象，设计了下面螺旋滑模的控制算法。在介绍主要结论之前，引入下面的辅助变量和滑模变量：

$$\dot{\boldsymbol{q}}_{ri} = -k_{pi}\text{sig}\left(\boldsymbol{q}_i - \boldsymbol{\eta}_i\right)^{\rho} \tag{6-27}$$

$$\boldsymbol{s}_i = \dot{\boldsymbol{q}}_i - \dot{\boldsymbol{q}}_{ri} \tag{6-28}$$

其中，k_{pi}是一个正增益，ρ是一个常数满足$0<\rho<1$。

通过第 2.3.4 节中的欧拉 - 拉格朗日系统的性质 2，有

$$\boldsymbol{M}_i\left(\boldsymbol{q}_i\right)\ddot{\boldsymbol{q}}_{ri} + \boldsymbol{C}_i\left(\boldsymbol{q}_i,\dot{\boldsymbol{q}}_i\right)\dot{\boldsymbol{q}}_{ri} + \boldsymbol{g}_i\left(\boldsymbol{q}_i\right) = \boldsymbol{Y}_i\left(\ddot{\boldsymbol{q}}_{ri},\dot{\boldsymbol{q}}_{ri},\dot{\boldsymbol{q}}_i,\boldsymbol{q}_i\right)\theta_i,\ i=1,\cdots,N \tag{6-29}$$

其中，θ_i是常数参数。然后，控制力矩可设计如下，当$t\leqslant T_1$时，$\boldsymbol{\tau}_i = -\kappa_d\text{sgn}\left(\dot{\boldsymbol{q}}_i\right) - \kappa_p\boldsymbol{q}_i + \boldsymbol{g}_i\left(\boldsymbol{q}_i\right)$，其中，$\kappa_d>\mu_1$和$\kappa_p$是待确定的正常数。当$t>T_1$时，

$$\boldsymbol{\tau}_i = \boldsymbol{Y}_i\left(\ddot{\boldsymbol{q}}_{ri},\dot{\boldsymbol{q}}_{ri},\dot{\boldsymbol{q}}_i,\boldsymbol{q}_i\right)\theta_i + \boldsymbol{C}_i\left(\boldsymbol{q}_i,\dot{\boldsymbol{q}}_i\right)\boldsymbol{s}_i + \boldsymbol{M}_i\left(\boldsymbol{q}_i\right)\boldsymbol{u}_i \tag{6-30a}$$

$$\boldsymbol{u}_i = -k_1\text{sig}\left(\boldsymbol{s}_i\right)^{\frac{1}{2}} + \boldsymbol{\phi}_i \tag{6-30b}$$

$$\dot{\boldsymbol{\phi}}_i = -k_2\text{sig}\left(\boldsymbol{s}_i\right)^{0} \tag{6-30c}$$

其中，k_1和k_2是待确定的正增益。主要结果陈述如下：

定理 6.1

在强连通有向拓扑下给定多智能体系统 (6-7)。若假设 6.1 成立，通过使用式 (6-30)，可实现控制目标式 (6-6)。

证明：

首先，我们证明当$t\in\left[0,T_1\right)$时，$\dot{\boldsymbol{q}}_i$和 $\boldsymbol{q}_i$ 不会趋于无穷大。通过定义李雅普诺夫函数$V_{T_{1i}} = \frac{1}{2}\dot{\boldsymbol{q}}_i\boldsymbol{M}_i\left(\boldsymbol{q}_i\right)\dot{\boldsymbol{q}}_i + \frac{1}{2}\kappa_p\boldsymbol{q}_i^{\mathrm{T}}\boldsymbol{q}_i, i=1,\cdots,N$，基于第 2.3.4 节中的性质 1，我们有 $\dot{V}_{T_{1i}} = -\dot{\boldsymbol{q}}_i^{\mathrm{T}}\boldsymbol{\varDelta}_i - \kappa_d\dot{\boldsymbol{q}}_i^{\mathrm{T}}\text{sgn}\left(\dot{\boldsymbol{q}}_i\right) \leqslant \|\dot{\boldsymbol{q}}_i\|_1\|\boldsymbol{\varDelta}_i\| - \kappa_d\|\dot{q}_i\|_1 \leqslant 0$，并且可得到$\dot{\boldsymbol{q}}_i$和 $\boldsymbol{q}_i$ 在$t\in\left[0,T_1\right]$是有界的。接下来，我们将证明通过使用式 (6-30)，滑模变量 $\boldsymbol{s}_i$ 在有限时间内收敛到 0。

将式 (6-30) 和式 (6-29) 代入到式 (6-7) 可得

$$\dot{\boldsymbol{s}}_i = -k_1 \operatorname{sig}\left(\boldsymbol{s}_i\right)^{\frac{1}{2}} + \boldsymbol{\phi}_i + \boldsymbol{M}_i^{-1}\boldsymbol{\Delta}_i \tag{6-31}$$

$$\dot{\boldsymbol{\phi}}_i = -k_2 \operatorname{sig}\left(\boldsymbol{s}_i\right)^0 \tag{6-32}$$

为了使得证明更加简洁并且不失一般性，我们选择智能体 1 来证明每个智能体会在有限时间内收敛到一致性的值。令 $\boldsymbol{x}_1 = \boldsymbol{s}_1$ 和 $\boldsymbol{x}_2 = \boldsymbol{\phi}_1 + \boldsymbol{M}_1^{-1}\boldsymbol{\Delta}_1$，系统可被写为下面的形式：

$$\dot{\boldsymbol{x}}_1 = -k_1 \operatorname{sig}\left(\boldsymbol{x}_1\right)^{\frac{1}{2}} + \boldsymbol{x}_2 \tag{6-33a}$$

$$\dot{\boldsymbol{x}}_2 = -k_2 \operatorname{sig}\left(\boldsymbol{x}_1\right)^0 + \boldsymbol{d}_1 \tag{6-33b}$$

其中，$\boldsymbol{d}_1 = \dot{\boldsymbol{M}}_1^{-1}\dot{\boldsymbol{\Delta}}_1$。

注意到，$\dot{\boldsymbol{M}}_1^{-1} = -\boldsymbol{M}_1^{-1}\dot{\boldsymbol{M}}_1\boldsymbol{M}_1$。$\boldsymbol{M}^{-1} \geqslant \bar{k}_M^{-1}\boldsymbol{I}_n, \boldsymbol{M} \leqslant \bar{k}_M \boldsymbol{I}_n$。通过运用第 2.3.4 节中的性质 3，我们有$\boldsymbol{M}^{-1} \geqslant \bar{k}_M^{-1}\boldsymbol{I}_n, M \leqslant \bar{k}_M \boldsymbol{I}_n$。因此，在假设 6.1 下，式 (6-33b) 中的干扰项是有界的$\left|d_{1(v)}\right| \leqslant k_{dm}\mu_2$。

已知系统 (6-33) 是齐次度为$\iota = -1$、权值为$\left(r_1, r_2\right) = \left(2,1\right)$的齐次系统。基于文献 [182] 的结果，我们可设计齐次度为 m=3 的多项式李雅普诺夫函数如下

$$V\left(\boldsymbol{x}\right) = \sum_{i=1}^{n}\left(\epsilon_1\left|x_{1(i)}\right|^{\frac{3}{2}} - \epsilon_2 x_{1(i)}x_{2(i)} + \epsilon_3\left|x_{2(i)}\right|^3\right), \tag{6-34}$$

其中，$\epsilon_1, \epsilon_2, \epsilon_3$是正参数。

计算$V\left(x\right)$的导数，并且令$\dot{V}\left(\boldsymbol{x}\right) = -W\left(\boldsymbol{x}\right)$，我们有

$$\begin{aligned} W\left(\boldsymbol{x}\right) = \sum_{i=1}^{n}\Bigg(& w_1\left|x_{1(i)}\right| - w_2 \operatorname{sig}\left(x_{1(i)}\right)^{\frac{1}{2}} \operatorname{sig}\left(x_{2(i)}\right) \\ & + w_3\left|x_{2(i)}\right|^2 + w_4 \operatorname{sig}\left(x_{1(i)}\right)^0 \operatorname{sig}\left(x_{2(i)}\right)^2 \\ & + w_5 \operatorname{sig}\left(x_{1(i)}\right) d_{1(i)} - w_6 \operatorname{sig}\left(x_{2(i)}\right)^2 d_{1(i)}\Bigg) \end{aligned} \tag{6-35}$$

其中，$w_1 = \dfrac{3}{2}\epsilon_1 k_1 - \epsilon_2 k_2; w_2 = \epsilon_2 k_1 + \dfrac{3}{2}\epsilon_1; w_3 = \epsilon_2; w_4 = 3\epsilon_3 k_2; w_5 = \epsilon_2; w_6 = 3\epsilon_3$。

下面，证明的目标是找到一组正的系数ϵ_1、ϵ_2、ϵ_3和增益 k_1、k_2 去保证 V 和 W 是正的。找到这些参数可分为两步，第一步，为了应用引理 6.4，首先通过象限分析得到正的齐次多项式 V 和 W。第二步，为了确定正系数，使用 SOSTOOLS 用来解平方和问题。一个多变量

多项式$V(x_1,\cdots,x_n)\overset{\text{def}}{=}V(\boldsymbol{x})$可被表示为平方和的形式，如果存在多项式$p_1(\boldsymbol{x}),\cdots,p_m(\boldsymbol{x})$满足$V(\boldsymbol{x})=\sum_{i=1}^{m}p_i^2(\boldsymbol{x})$。显然，若$V(\boldsymbol{x})$是 SOS，则$V(\boldsymbol{x})\geqslant 0$。然而，SOS 仅仅保证了 V 是非负的。根据参考文献 [183]，下面将对新的函数$\bar{\boldsymbol{V}}(\boldsymbol{x})=V(\boldsymbol{x})-\delta\sum_{i=1}^{n}x_i^2$，其中 δ 是一个正标量，找到相应的系数。

□1：通过改变坐标为 z_1 和 z_2，变换 V 和 W 为齐次多项式的形式

$$|\boldsymbol{x}_1|=z_1^2,|\boldsymbol{x}_2|=z_2 \tag{6-36}$$

接下来，分别讨论下面几种情况。

情况I：$\boldsymbol{x}_1>0,\boldsymbol{x}_2>0$，则

$$V_1(\boldsymbol{z})=\sum_{\nu=1}^{n}V_{1(\nu)}=\sum_{\nu=1}^{n}\left(\epsilon_1 z_{1(\nu)}^3-\epsilon_2 z_{1(\nu)}^2 z_{2(\nu)}+\epsilon_3 z_{2(\nu)}^3\right) \tag{6-37}$$

$$\begin{aligned}W_1(\boldsymbol{z})=\sum_{\nu=1}^{n}W_{1(\nu)}=\sum_{\nu=1}^{n}\Big(&w_1 z_{1(\nu)}^2-w_2 z_{1(\nu)}z_{2(\nu)}\\&+(w_3+w_4)z_{2(\nu)}^2+w_5 z_{1(\nu)}^2 d_{1(\nu)}-w_6 z_{2(\nu)}d_{1(\nu)}\Big)\end{aligned} \tag{6-38}$$

由于并不是所有的系数都是正的，根据引理6.4，取$p=2$，可得

$$\begin{aligned}V_1(\boldsymbol{z})&=\sum_{\nu=1}^{n}\left(\left(z_{1(\nu)}+z_{2(\nu)}\right)^2 V_{1(\nu)}\right)\\&=\sum_{\nu=1}^{n}\Big(\epsilon_1 z_{1(\nu)}^5+(2\epsilon_1-\epsilon_2)z_{1(\nu)}^4 z_{2(\nu)}\\&\quad+(\epsilon_3-\epsilon_2)z_{1(\nu)}^2 z_{2(\nu)}^3+(\epsilon_1-2\epsilon_2)z_{1(\nu)}^3 z_{2(\nu)}^2\\&\quad+2\epsilon_3 z_{1(\nu)}z_{2(\nu)}^4+\epsilon_3 z_{2(\nu)}^5\Big)\end{aligned} \tag{6-39}$$

和

$$\begin{aligned}W_1(\boldsymbol{z})&=\sum_{\nu=1}^{n}\left(\left(z_{1(\nu)}+z_{2(i)}\right)^2 W_{1(\nu)}\right)\\&=\sum_{\nu=1}^{n}\Big(w_1 z_{1(\nu)}^4+\left(2w_1-w_2+2w_5 d_{1(\nu)}\right)z_{1(\nu)}^3 z_{2(\nu)}\\&\quad+\big(w_3+w_4-2w_2+w_1+w_5 d_{1(\nu)}\\&\quad-w_6 d_{1(\nu)}\big)z_{1(\nu)}^2 z_{2(\nu)}^2+\left(2w_3+2w_4-w_2-2w_6 d_{1(\nu)}\right)\\&\quad\times z_{1(\nu)}z_{2(\nu)}^3+\left(w_3+w_4-w_6 d_{1(\nu)}\right)z_{1(\nu)}^4\Big)\end{aligned} \tag{6-40}$$

情况II：$\boldsymbol{x}_1 > 0, \boldsymbol{x}_2 < 0$，则

$$V_2(\boldsymbol{z}) = \sum_{\nu=1}^{n} V_{1(\nu)} = \sum_{\nu=1}^{n} \left(\epsilon_1 z_{1(\nu)}^3 + \epsilon_2 z_{1(\nu)}^2 z_{2(\nu)} + \epsilon_3 z_{2(\nu)}^3 \right) \tag{6-41}$$

$$\begin{aligned} W_2(\boldsymbol{z}) = \sum_{\nu=1}^{n} W_{2(\nu)} = \sum_{\nu=1}^{n} \Big(& \left(w_1 + w_5 d_{1(\nu)} \right) z_{1(\nu)}^2 + w_2 z_{1(\nu)} z_{2(\nu)} \\ & + \left(w_3 - w_4 + w_6 d_{1(\nu)} \right) z_{2(\nu)}^2 \Big) \end{aligned} \tag{6-42}$$

另外两种情况与上面两种情况类似，因此省略。

□2：针对多项式 W_i，考虑两种不同的极端情况$\boldsymbol{d}_1 > 0$和$\boldsymbol{d}_1 < 0$。通过代入$z_{i(\nu)} = y_{i(\nu)}^2, i = 1,2$ 和 $\nu = 1, \cdots, N$，可定义 SOS 问题如下

$$\overline{\boldsymbol{V}}_i(\boldsymbol{y}) = V_i(\boldsymbol{y}) - \delta \sum_{j=1}^{2} y_j^{2l}，\overline{\boldsymbol{W}}_i(\boldsymbol{y}) = W_i(\boldsymbol{y}) - \delta \sum_{j=1}^{2} y_j^{2l} \tag{6-43}$$

其中，l是$z_j, j = 1,2$的最大指数。然后，固定$k_{dm}\mu_2 = 0.5, k_1 = 2$和$k_2 = 1$，通过使用SOSTOOLS软件工具，可以找到表6-1中第一列的系数ϵ，以使$\left\{ \overline{V}_i, \overline{W}_i \right\}$中的所有形式都可以表示为SOS的形式。选择S1中的系统，$V(\boldsymbol{x})$和$W(\boldsymbol{x})$的正定性分别如图6-1和图6-2所示。

表6-1　增益和系数举例

i	ϵ_1	ϵ_2	ϵ_3	k_1	k_2	μ
S1	115.3	110.91	23.66	2	1	0.5
S2	40.76	27.73	2.98	2.83	2	1
S3	14.41	6.93	0.37	4	4	2

因此，经过上面的步骤，我们已经找到了式 (6-33) 的一个李雅普诺夫函数 (6-34)。通过使用引理 6.5，可以得到存在一个常数 r，使得式 (6-33) 的原点收敛到 0，当$t > \frac{3}{r} V(0)^{\frac{1}{3}} \stackrel{\text{def}}{=} T_2$。

接下来，根据式 (6-28)，可以得到$\dot{\boldsymbol{q}}_i = -k_{p_i} \operatorname{sig}(\boldsymbol{q}_i - \boldsymbol{\eta}_i)^{\rho}$（当$t > T_2$）。通过使用引理 6.7，可以证明当$t > T_1$时，$\boldsymbol{\eta}_1 = \boldsymbol{\eta}_2 = \cdots = \boldsymbol{\eta}_N = \boldsymbol{\eta}^*$。选择$T_3 \stackrel{\text{def}}{=} \max\{T_1, T_2\}$以及表示$\tilde{\boldsymbol{q}}_i = \boldsymbol{q}_i - \boldsymbol{\eta}^*$，我们可以得到$\dot{\tilde{\boldsymbol{q}}}_i = -k_{p_i} \operatorname{sig}(\tilde{\boldsymbol{q}}_i)^{\rho}$（当$t > T_3$）。通过考虑李雅普诺夫函数$V_{q_i} = \frac{1}{2} \tilde{\boldsymbol{q}}_i^{\mathrm{T}} \tilde{\boldsymbol{q}}_i$，以及它的导数$\dot{V}_{q_i} = -k_{q_i} \sum_{\nu=1}^{n} \left| \tilde{\boldsymbol{q}}_i^{(\nu)} \right|^{\rho+1} \leqslant$

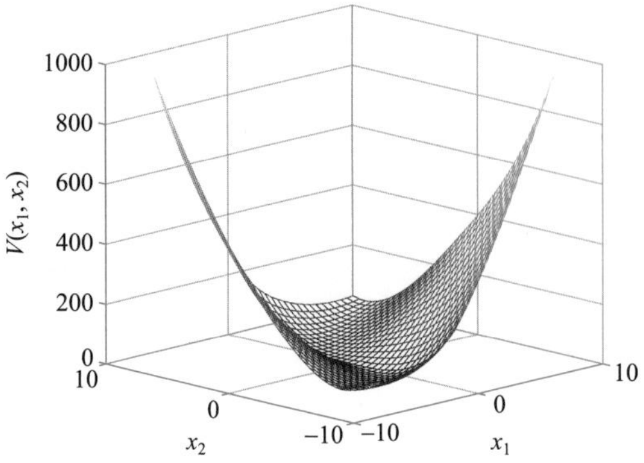

图 6-1　李雅普诺夫函数 [式 (6-34)]

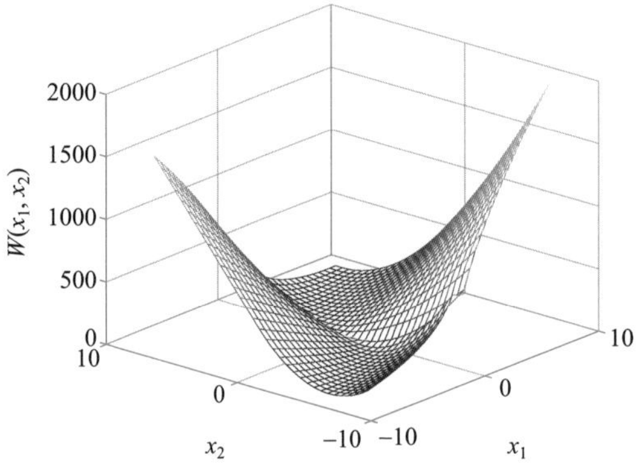

图 6-2　李雅普诺夫导数 [式 (6-35)]

$-k_{\boldsymbol{q}_i}\left(\sum_{\nu=1}^{n}\left|\tilde{\boldsymbol{q}}_i^{(\nu)}\right|^2\right)^{\frac{\rho+1}{2}} \leqslant -k_{\boldsymbol{q}_i}V_{\boldsymbol{q}_i}^{\frac{\rho+1}{2}}$。根据引理 6.3，可以得到在一个有限时间 $T_4 = \dfrac{2}{k_{\boldsymbol{q}_i(1-\rho)}}V_{\boldsymbol{q}_i}(0)^{\frac{1-\rho}{2}}$内，$\tilde{\boldsymbol{q}}_i \to 0$。因此，通过定义$T = T_4 + T_3$，可得当$t > T$时，我们有$\boldsymbol{q}_1 = \boldsymbol{q}_2 = \cdots = \boldsymbol{q}_N = \boldsymbol{\eta}^*$。证毕。 ■

注释 6.7　从上面的证明中可以看出，总的驻留时间 T 取决于两个部分 T_1 和$T_2 + T_4$。T_1表征的是 $\boldsymbol{\eta}_i$ 的有限收敛时间，取决于通信拓扑图的性质、$\boldsymbol{\eta}_i$ 的初值和控制参数 α。第二部分$T_2 + T_4$是欧拉 - 拉格朗日系统状态的收敛时间，取决于一致性的值 $\boldsymbol{\eta}^*$，控制增益 k_{pi}、k_1、k_2 和参数 ρ。

通过使用 SOSTOOLS，可以得到一组针对干扰的系数满足$\left|d_{1(\nu)}\right| \leqslant 0.5$。为了补偿更一般的干扰，可以成倍地增大或者减小调整多项式李雅普诺夫函数的系数和增益，以适应更大或者更小的扰动。

定义新的变量，$\boldsymbol{z}_1 = L\boldsymbol{x}_1, \boldsymbol{z}_2 = L\boldsymbol{x}_2$，其中$L>0$是一个正的增益。然后，有

$$\dot{\boldsymbol{z}}_1 = -k_1^{\mathrm{T}}\operatorname{sig}\left(\boldsymbol{z}_1\right)^{\frac{1}{2}} + \boldsymbol{z}_2$$

$$\dot{\boldsymbol{z}}_2 = -k_2^{\mathrm{T}}\operatorname{sig}\left(\boldsymbol{z}_1\right)^{0} + \overline{\boldsymbol{d}}_1$$

其中，$k_1^{\mathrm{T}} = k_1 L^{\frac{1}{2}}; k_2^{\mathrm{T}} = k_2 L, \overline{\boldsymbol{d}}_1 = L\boldsymbol{d}_1$。

为了获得多项式李雅普诺夫函数的系数，通过将式 (6-34) 中的变量代换为 $\boldsymbol{z}_1$、$\boldsymbol{z}_2$，可得$\epsilon_1^{\mathrm{T}} = \epsilon_1 L^{-\frac{3}{2}}, \epsilon_2^{\mathrm{T}} = \epsilon_2 L^{-2}, \epsilon_3^{\mathrm{T}} = \epsilon_3 L^{-3}$。对应的增益和系数在表 6-1 中列出。

6.3.3 非奇诺现象分析

本节中，我们将证明奇诺现象不会出现。通过运用不等式$\|\boldsymbol{v}-\boldsymbol{\omega}\|^2 \leqslant 2\|\boldsymbol{v}\|^2 + 2\|\boldsymbol{\omega}\|^2, \forall \boldsymbol{v}, \boldsymbol{\omega} \in \mathbb{R}^n$，可以得到$\beta_i\|\boldsymbol{e}_i(t)\|^2 \leqslant 2\beta_i\left(\|\boldsymbol{\xi}_i\left(t_k^i\right)\|^2 + \|\xi_i(t)\|^2\right)$。下面将证明$\boldsymbol{\varXi}_i^r$在有限时间内收敛到 0。从式 (6-15) 和式 (6-16) 可以得出，$\dot{\varXi}_i \geqslant -\left(\varLambda_1 + \varLambda_2\right)\boldsymbol{\varXi}_i^r$，并且$x(t)$是微分方程$\dot{x} = -\left(\varLambda_1 + \varLambda_2\right)x^r$的一个解，等价于微分方程$\dfrac{\mathrm{d}\left[x^{1-r}\right]}{\mathrm{d}t} = -\left(\varLambda_1 + \varLambda_2\right)(1-r)$。解这个不等式，有$x^{1-r} = -\left(\varLambda_1 + \varLambda_2\right)(1-r)t + x_0^{1-r}$。通过定义新变量$y \overset{\text{def}}{=} x^r$，则$y^{\frac{1}{r}-1} = -\left(\varLambda_1 + \varLambda_2\right)(1-r)t + x_0^{1-r}$。通过运用比较引理和指数函数的性质，可以得到$\boldsymbol{\varXi}_i^r \geqslant \left(-h_1 t + h_2\right)^{\frac{r}{1-r}}$，其中$h_1 = \left(\varLambda_1 + \varLambda_2\right)(1-r), h_2 = x_0^{1-r}$。基于此，可得事件驱动条件 (6-16) 的一个充分条件

$$2\beta_i\|\boldsymbol{\xi}_i\left(t_k^i\right)\|^2 + \left(2\beta_i - \delta_i\right)\|\xi_i(t)\|^2 \leqslant \left(-h_1 t + h_2\right)^{\frac{r}{1-r}} \tag{6-44}$$

由于$\|\boldsymbol{\xi}_i\|$是有界的，则在第k个触发时刻，存在一个有界常数$M_1\left(t_k^i\right) = \left(4\beta_i - \delta_i\right)\|\boldsymbol{\xi}_i\left(t_k^i\right)\| \geqslant \left(-h_1 t_k^i + h_2\right)^{\frac{r}{1-r}}$。注意到$\|\boldsymbol{\xi}_i\|$是递减的，则有另一个有界常数$M_2\left(t_{k+1}^i\right) < M_1\left(t_k^i\right)$。显然，下一次触发时刻必定在条件

$M_2\left(t_{k+1}^i\right) \geqslant \left(-h_1 t_{k+1}^i + h_2\right)^{\frac{r}{1-r}}$满足，且式 (6-44) 的右边是可微的，那么存在一个常数$\varrho\left(t_k^i\right)$满足

$$\left(t_{k+1}^i - t_k^i\right)\varrho\left(t_k^i\right) > M_1\left(t_k^i\right) - M_2\left(t_{k+1}^i\right) \tag{6-45}$$

其中，ϱ是式 (6-44) 右边的导数。因此，$t_{k+1}^i - t_k^i > \dfrac{M_1\left(t_k^i\right) - M_2\left(t_{k+1}^i\right)}{\varrho\left(t_k^i\right)}$。基于此，可以推断出所设计的协议不会引起奇诺现象。

注释 6.8 与基于事件驱动的拉格朗日系统的一致性问题 [184] 相比，这里主要研究了有限时间一致性，保证了更优的收敛性能。从上面的证明中可发现，$\boldsymbol{x}_1$ 收敛到 0 保证了有限时间一致性得以实现，$\boldsymbol{x}_2$ 收敛到 0 意味着$\boldsymbol{M}_i\boldsymbol{\phi}_i$会在有限时间内对干扰予以补偿。此外，与有限时间的欧拉 - 拉格朗日系统的一致性结果相比较，我们考虑了基于事件驱动的通信机制，大大减小了通信负担。另外，以上理论结果的通信拓扑图是比无向图 [154,184-185] 更为一般的有向图。不同于连续时间螺旋滑模控制器的工作 [183]，仅仅考虑了二阶单个系统的有限时间稳定，我们针对多欧拉 - 拉格朗日系统研究了有限时间一致性算法。

6.4 多刚体系统固定时间姿态一致性

在上一节当中，基于欧拉 - 拉格朗日方程，我们研究了多刚体系统的有限时间一致性问题。然而，由于欧拉 - 拉格朗日方程的广义坐标向量本质上属于欧几里德空间，而刚体姿态的完整表征空间是一个不微分同胚于欧几里德空间的三维正交群。因此，欧拉 - 拉格朗日方程在刚体的姿态运动表征方面存在固有缺陷。另一方面，注意到在上一节的有限

时间一致性结果当中，暂态收敛时间与系统状态的初值相关，这意味着当系统初始状态偏离一致性状态较大时，暂态收敛时间也会随之增加。因此，需要进一步考虑暂态收敛时间与系统初始状态无关的有限时间一致性，也即固定时间一致性。

本节中，针对多刚体系统的固定时间姿态一致性问题，首先我们考虑了固定无向图的情况。可以证明当使用设计的一致性协议时，一致性集合相对于一个在姿态空间上的正不变集是固定时间吸引的。接下来，我们将结果推广到了联合连通图。为了使陈述变得简洁，忽略了时间依赖函数中的时间变量 t，但仍然保留解中的时间变量如 $\boldsymbol{x}\left(t,\boldsymbol{x}_0,t_0\right)$。

6.4.1 固定拓扑的情况

在陈述主要定理之前，首先引入下面的正不变集合。

引理 6.8

考虑到一个在无向图下的，包含 N 个刚体的多智能体系统。如果选择 $\bar{k}<\dfrac{1}{\lambda_{\max}\left(\mathcal{L}\right)^2}$，其中 $\bar{k}=\max_i\left\{k_i\right\}$，并且使用事件驱动协议 (6-10) 和事件驱动条件 (6-16)，$\mathcal{S}=\{\boldsymbol{x}\in\mathbb{R}^{3N}$：$\sum_{i=1}^{N}\|\boldsymbol{x}_i\left(t_0\right)\|^2<\pi^2\}$ 是一个正不变集，对于解 $\boldsymbol{x}\left(t,\boldsymbol{x}_0,t_0\right)$。

证明：

考虑李雅普诺夫函数 $V=\dfrac{1}{2}\sum_{i=1}^{N}\|\boldsymbol{x}_i\|^2$，取李雅普诺夫函数 V 在时间 $\left(t_k^i,t_{k+1}^i\right)$ 的导数和 V 在触发时刻 $\left\{t_k^i\right\},k\in\mathbb{N}^+$ 的右侧导数，可得

$$\dot{V}=-\sum_{i=1}^{N}\frac{\boldsymbol{x}_i^{\mathrm{T}}\left(\hat{\mathcal{L}}_i\hat{\boldsymbol{x}}\right)\mu\left(\|\hat{\mathcal{L}}_i\hat{\boldsymbol{x}}\|^2\right)}{\|\hat{\mathcal{L}}_i\hat{\boldsymbol{x}}\|^2}-\sum_{i=1}^{N}\frac{\boldsymbol{x}_i^{\mathrm{T}}\left(\hat{\mathcal{L}}_i\hat{\boldsymbol{x}}\right)\nu\left(\|\hat{\mathcal{L}}_i\hat{\boldsymbol{x}}\|^2\right)}{\|\hat{\mathcal{L}}_i\hat{\boldsymbol{x}}\|^2}$$

其中用到了 $\boldsymbol{x}_i^{\mathrm{T}}\boldsymbol{J}_{\boldsymbol{x}_i}=\boldsymbol{x}_i^{\mathrm{T}}$ 获得等式。令 $\dfrac{\mu\left(\|\hat{\mathcal{L}}_i\hat{\boldsymbol{x}}\|^2\right)}{\|\hat{\mathcal{L}}_i\hat{\boldsymbol{x}}\|^2}=\varGamma_i$ 和 $\dfrac{\nu\left(\|\hat{\mathcal{L}}_i\hat{\boldsymbol{x}}\|^2\right)}{\|\hat{\mathcal{L}}_i\hat{x}\|^2}=\varLambda_i$，

可得

$$
\begin{aligned}
\dot{V} \leqslant & -\sum_{i=1}^{N}\hat{\boldsymbol{x}}_i^{\mathrm{T}}\Gamma_i\left(\hat{\mathcal{L}}_i\hat{\boldsymbol{x}}\right)+\sum_{i=1}^{N}\boldsymbol{e}_i^{\mathrm{T}}\Gamma_i\left(\hat{\mathcal{L}}_i\hat{\boldsymbol{x}}\right) \\
& -\sum_{i=1}^{N}\hat{\boldsymbol{x}}_i^{\mathrm{T}}\Lambda_i\left(\hat{\mathcal{L}}_i\hat{\boldsymbol{x}}\right)+\sum_{i=1}^{N}\boldsymbol{e}_i^{\mathrm{T}}\Lambda_i\left(\hat{\mathcal{L}}_i\hat{\boldsymbol{x}}\right) \\
\leqslant & -\hat{\boldsymbol{x}}^{\mathrm{T}}\left(\hat{\Gamma}\otimes\boldsymbol{I}_3\hat{\mathcal{L}}\right)\hat{\boldsymbol{x}}+\sum_{i=1}^{N}\Gamma_i\left(\frac{\gamma}{2}\|\boldsymbol{e}_i\|^2+\frac{1}{2\gamma}\hat{\boldsymbol{x}}\hat{\mathcal{L}}_i^{\mathrm{T}}\hat{\mathcal{L}}_i\hat{\boldsymbol{x}}\right) \\
& -\hat{\boldsymbol{x}}^{\mathrm{T}}\left(\hat{\Lambda}\otimes\boldsymbol{I}_3\hat{\mathcal{L}}\right)\hat{\boldsymbol{x}}+\sum_{i=1}^{N}\Lambda_i\left(\frac{\gamma}{2}\|\boldsymbol{e}_i\|^2+\frac{1}{2\gamma}\hat{\boldsymbol{x}}\hat{\mathcal{L}}_i^{\mathrm{T}}\hat{\mathcal{L}}_i\hat{\boldsymbol{x}}\right)
\end{aligned}
$$

其中，用到杨氏不等式$\forall a,b\in\mathbb{R},\gamma>0,ab\leqslant\frac{a^2}{2\gamma}+\frac{\gamma b^2}{2}$得到最后一个等式。$\hat{\Gamma}=\mathrm{diag}\{\Gamma_i\}$和$\hat{\Lambda}=\mathrm{diag}\{\Lambda_i\}$为对角矩阵，相应的对角元素为$\{\Gamma_i\}$和$\{\Lambda_i\},i=1,\cdots,N$。在事件驱动条件 (6-16) 下，可得

$$
\begin{aligned}
\dot{V} \leqslant & -\hat{\boldsymbol{x}}^{\mathrm{T}}\left(\hat{\Gamma}\otimes\boldsymbol{I}_3\hat{\mathcal{L}}\right)\hat{\boldsymbol{x}}+\frac{\lambda_{\max}(\mathcal{L})}{2\gamma}\hat{\boldsymbol{x}}^{\mathrm{T}}\left(\hat{\Gamma}\otimes\boldsymbol{I}_3\right)\hat{\mathcal{L}}\hat{\boldsymbol{x}} \\
& +\frac{\lambda_{\max}(\mathcal{L})\gamma\hat{k}}{2}\hat{\boldsymbol{x}}^{\mathrm{T}}\left(\hat{\Gamma}\otimes\boldsymbol{I}_3\right)\hat{\mathcal{L}}\hat{\boldsymbol{x}}-\hat{\boldsymbol{x}}^{\mathrm{T}}\left(\hat{\Lambda}\otimes\boldsymbol{I}_3\hat{\mathcal{L}}\right)\hat{\boldsymbol{x}} \\
& +\frac{\lambda_{\max}(\mathcal{L})}{2\gamma}\hat{\boldsymbol{x}}^{\mathrm{T}}\left(\hat{\Lambda}\otimes\boldsymbol{I}_3\right)\hat{\mathcal{L}}\hat{\boldsymbol{x}}+\frac{\lambda_{\max}(\mathcal{L})\gamma\bar{k}}{2}\hat{\boldsymbol{x}}^{\mathrm{T}}\left(\hat{\Lambda}\otimes\boldsymbol{I}_3\right)\hat{\mathcal{L}}\hat{\boldsymbol{x}} \\
\leqslant & -\left(\frac{1}{2}-\frac{\lambda_{\max}(\mathcal{L})^2\bar{k}}{2}\right)\hat{\boldsymbol{x}}^{\mathrm{T}}\left(\hat{\Gamma}\otimes\boldsymbol{I}_3\hat{\mathcal{L}}\right)\hat{\boldsymbol{x}}-\left(\frac{1}{2}-\frac{\lambda_{\max}(\mathcal{L})^2\bar{k}}{2}\right)\hat{\boldsymbol{x}}^{\mathrm{T}}\left(\hat{\Lambda}\otimes\boldsymbol{I}_3\hat{\mathcal{L}}\right)\hat{\boldsymbol{x}}
\end{aligned}
$$

其中，$\bar{k}=\max_i\{k_i\}$，选择$\gamma=\lambda_{\max}(\mathcal{L})$可得最后一个不等式。根据上述不等式，令$\bar{k}<\frac{1}{\lambda_{\max}(\mathcal{L})^2}$，则有$\dot{V}\leqslant 0$，这意味着$V$是随着时间非增的。由于$\boldsymbol{x}_i(t_0)\in\mathcal{B}_{\pi}^{'}(\boldsymbol{0}_3)$，对所有$t>t_0$，解$\boldsymbol{x}(t,\boldsymbol{x}_0,t_0)$将永远在集合$\mathcal{S}=\left\{\boldsymbol{x}\in\mathbb{R}^{3N}:\sum_{i=1}^{N}\|\boldsymbol{x}_i(t_0)\|^2<\pi^2\right\}$中。因此，$\mathcal{S}=\left\{\boldsymbol{x}\in\mathbb{R}^{3N}:\sum_{i=1}^{N}\|\boldsymbol{x}_i(t_0)\|^2<\pi^2\right\}$是正不变集。证毕。 ■

下面介绍主要结果。

定理 6.2

考虑在连通的无向图下，一个包含有 N 个刚体的多智能系统，当使用协议 (6-10) 和事件驱动条件 (6-16)，若$k_i < \min\left\{\dfrac{1}{\lambda_{\max}\left(\mathcal{L}\right)^2}, \kappa_i\right\}$满足，则$\mathcal{C}_R$是固定时间吸引的，相对于$\mathcal{S} = \left\{\mathcal{R} \in \mathbb{SO}(3)^N : \sum_{i=1}^{N} d^2\left(\mathcal{R}_i\left(t_0\right), \boldsymbol{I}_3\right) < \pi^2\right\}$，且暂态时间为

$$T_1 = \frac{1}{\underline{c} c_\mu k_\mu}\int_0^1 \frac{1}{\mu(V)}\mathrm{d}V + \frac{1}{\underline{c} c_\nu k_\nu}\int_1^\infty \frac{1}{\nu(V)}\mathrm{d}V \tag{6-46}$$

其中，$\underline{c} = \min_i\left(\lambda_{\min}\left(\boldsymbol{J}_{x_i}\right) - \dfrac{\gamma k_i \lambda_{\max}\left(\mathcal{L}^{\mathrm{T}}\mathcal{L}\right)}{2} - \dfrac{\lambda\left(\boldsymbol{J}_{x_i}^{\mathrm{T}}\boldsymbol{J}_{x_i}\right)}{2\gamma}\right); \gamma = \dfrac{\lambda_{\max}\left(\boldsymbol{J}_{x_i}^{\mathrm{T}}\boldsymbol{J}_{x_i}\right)}{2\left(\lambda_{\min}\left(\boldsymbol{J}_{x_i}\right) - a\right)}$，

$\kappa_i = \dfrac{\left(\lambda_{\min}\left(\boldsymbol{J}_{x_i}\right) - a\right)a}{\lambda_{\max}\left(\mathcal{L}^{\mathrm{T}}\mathcal{L}\right)\lambda_{\max}\left(\boldsymbol{J}_{x_i}^{\mathrm{T}}\boldsymbol{J}_{x_i}\right)}$，其中$0 < a < \lambda_{\min}\left(\boldsymbol{J}_{x_i}\right)$。

证明：

根据引理 6.8，我们知道$\mathcal{S}$是正不变集，这意味着每个刚体的姿态将会待在集合$\mathcal{S}$中，对所有$t > t_0$。考虑李雅普诺夫函数$V = \dfrac{1}{2}\boldsymbol{x}^{\mathrm{T}}\hat{\mathcal{L}}\boldsymbol{x}$。为了让证明更加简洁，定义$\tilde{\boldsymbol{x}}_i = \hat{\mathcal{L}}_i\hat{\boldsymbol{x}}$，$\tilde{\boldsymbol{e}}_i = \hat{\mathcal{L}}_i\boldsymbol{e}, \tilde{\boldsymbol{x}} = \left[\tilde{\boldsymbol{x}}_1^{\mathrm{T}}, \tilde{\boldsymbol{x}}_2^{\mathrm{T}}, \cdots, \tilde{\boldsymbol{x}}_N^{\mathrm{T}}\right]^{\mathrm{T}}$和$\tilde{\boldsymbol{e}} = \left[\tilde{\boldsymbol{e}}_1^{\mathrm{T}}, \tilde{\boldsymbol{e}}_2^{\mathrm{T}}, \cdots, \tilde{\boldsymbol{e}}_N^{\mathrm{T}}\right]^{\mathrm{T}}$。对李雅普诺夫函数在式 (6-8) 上求导，可得

$$\begin{aligned}\dot{V} &= -\sum_{i=1}^{N}\boldsymbol{x}_i^{\mathrm{T}}\sum_{j=1}^{N}\mathcal{L}_{ij}\boldsymbol{J}_{x_j}\tilde{\boldsymbol{x}}_j\frac{\mu\left(\|\tilde{\boldsymbol{x}}_j\|^2\right)}{\|\tilde{\boldsymbol{x}}_j\|^2} \\ &\quad -\sum_{i=1}^{N}\boldsymbol{x}_i^{\mathrm{T}}\sum_{j=1}^{N}\mathcal{L}_{ij}\boldsymbol{J}_{x_j}\tilde{\boldsymbol{x}}_j\frac{\nu\left(\|\tilde{\boldsymbol{x}}_j\|^2\right)}{\|\tilde{\boldsymbol{x}}_j\|^2} \\ &= \overline{V}_1 + \overline{V}_2,\end{aligned} \tag{6-47}$$

其中，$\overline{\boldsymbol{V}}_1 = -\sum_{i=1}^{N}\boldsymbol{x}_i^{\mathrm{T}}\sum_{j=1}^{N}\mathcal{L}_{ij}\boldsymbol{J}_{x_j}\tilde{\boldsymbol{x}}_j\dfrac{\mu\left(\|\tilde{\boldsymbol{x}}_j\|^2\right)}{\|\tilde{\boldsymbol{x}}_j\|^2}$；$\overline{\boldsymbol{V}}_2 = -\sum_{i=1}^{N}\boldsymbol{x}_i^{\mathrm{T}}\sum_{j=1}^{N}\mathcal{L}_{ij}\boldsymbol{J}_{x_j}\tilde{\boldsymbol{x}}_j\dfrac{\nu\left(\|\tilde{\boldsymbol{x}}_j\|^2\right)}{\|\tilde{\boldsymbol{x}}_j\|^2}$。那么，

$$
\begin{aligned}
\overline{\boldsymbol{V}}_1 &= -\sum_{i=1}^{N}\tilde{\boldsymbol{x}}_i^{\mathrm{T}}\boldsymbol{J}_{x_i}\tilde{\boldsymbol{x}}_i\frac{\mu\left(\|\tilde{\boldsymbol{x}}_i\|^2\right)}{\|\tilde{\boldsymbol{x}}_i\|^2}+\sum_{i=1}^{N}\tilde{\boldsymbol{e}}_i^{\mathrm{T}}\boldsymbol{J}_{x_i}\tilde{\boldsymbol{x}}_i^{\mathrm{T}}\frac{\mu\left(\|\tilde{\boldsymbol{x}}_i\|^2\right)}{\|\tilde{\boldsymbol{x}}_i\|^2} \\
&\leqslant -\sum_{i=1}^{N}\lambda_{\min}\left(\boldsymbol{J}_{x_i}\right)\mu\left(\|\tilde{\boldsymbol{x}}_i\|^2\right)+\sum_{i=1}^{N}\left(\frac{\gamma}{2}\|\tilde{\boldsymbol{e}}_i\|^2+\frac{1}{2\gamma}\tilde{\boldsymbol{x}}_i^{\mathrm{T}}\boldsymbol{J}_{x_i}^{\mathrm{T}}\boldsymbol{J}_{x_i}\tilde{\boldsymbol{x}}_i\right)\frac{\mu\left(\|\tilde{\boldsymbol{x}}_i\|^2\right)}{\|\tilde{\boldsymbol{x}}_i\|^2} \\
&\leqslant -\sum_{i=1}^{N}\lambda_{\min}\left(\boldsymbol{J}_{x_i}\right)\mu\left(\|\tilde{\boldsymbol{x}}_i\|^2\right)+\sum_{i=1}^{N}\left(\frac{\gamma\lambda_{\max}\left(\mathcal{L}^{\mathrm{T}}\mathcal{L}\right)}{2}\|\boldsymbol{e}_i\|^2+\frac{\lambda_{\max}\left(\boldsymbol{J}_{x_i}^{\mathrm{T}}\boldsymbol{J}_{x_i}\right)}{2\gamma}\|\tilde{\boldsymbol{x}}_i\|^2\right)\frac{\mu\left(\|\tilde{\boldsymbol{x}}_i\|^2\right)}{\|\tilde{\boldsymbol{x}}_i\|^2}
\end{aligned}
$$

其中，γ是一个正常数。根据事件驱动条件 (6-16)，可得

$$
\overline{\boldsymbol{V}}_1\leqslant -\sum_{i=1}^{N}\left(\lambda_{\min}\left(\boldsymbol{J}_{x_i}\right)-\frac{\gamma k_i\lambda_{\max}\left(\mathcal{L}^{\mathrm{T}}\mathcal{L}\right)}{2}-\frac{\lambda_{\max}\left(\boldsymbol{J}_{x_i}^{\mathrm{T}}\boldsymbol{J}_{x_i}\right)}{2\gamma}\right)\mu\left(\|\tilde{\boldsymbol{x}}_i\|^2\right)=-\sum_{i=1}^{N}c_i\mu\left(\|\tilde{\boldsymbol{x}}_i\|^2\right) \tag{6-48}
$$

与上述过程相似，可得 $\overline{V}_2\leqslant -\sum_{i=1}^{N}c_i\nu\left(\|\tilde{\boldsymbol{x}}_i\|^2\right)$。不失一般性，选择 $\gamma=\dfrac{\lambda_{\max}\left(\boldsymbol{J}_{x_i}^{\mathrm{T}}\boldsymbol{J}_{x_i}\right)}{2(\lambda_{\min}\left(\boldsymbol{J}_{x_i}\right)-a)}$，其中 $0<a<\lambda_{\min}\left(\boldsymbol{J}_{x_i}\right)$，令 $k_i<\dfrac{\left(\lambda_{\min}\left(\boldsymbol{J}_{x_i}\right)-a\right)a}{\lambda_{\max}\left(\mathcal{L}^{\mathrm{T}}\mathcal{L}\right)\lambda_{\max}\left(\boldsymbol{J}_{x_i}^{\mathrm{T}}\boldsymbol{J}_{x_i}\right)}$ $=\kappa_i$，可保证$c_i>0$。

结合$\overline{\boldsymbol{V}}_1\leqslant -\sum_{i=1}^{N}c_i\mu\left(\|\tilde{\boldsymbol{x}}_i\|^2\right)$和$\overline{\boldsymbol{V}}_2\leqslant -\sum_{i=1}^{N}c_i\nu\left(\|\tilde{\boldsymbol{x}}_i\|^2\right)$，可总结到

$$
\dot{V}\leqslant -\underline{c}c_{\mu}\mu\left(\|\tilde{\boldsymbol{x}}\|^2\right)-\underline{c}c_{\nu}\nu\left(\|\tilde{\boldsymbol{x}}\|^2\right) \tag{6-49}
$$

其中，$\underline{c}=\min_i\{c_i\}$是一个正常数。根据

$$
\begin{aligned}
\boldsymbol{x}^{\mathrm{T}}\hat{\mathcal{L}}\boldsymbol{x} &= \hat{\boldsymbol{x}}^{\mathrm{T}}\hat{\mathcal{L}}\hat{\boldsymbol{x}}-2\boldsymbol{e}^{\mathrm{T}}\hat{\mathcal{L}}\hat{\boldsymbol{x}}+\boldsymbol{e}^{\mathrm{T}}\hat{\mathcal{L}}\boldsymbol{e} \\
&\leqslant \hat{\boldsymbol{x}}^{\mathrm{T}}\hat{\mathcal{L}}\hat{\boldsymbol{x}}+2\boldsymbol{e}^{\mathrm{T}}\hat{\mathcal{L}}\boldsymbol{e}+2\hat{\boldsymbol{x}}^{\mathrm{T}}\hat{\mathcal{L}}\hat{\boldsymbol{x}}+\boldsymbol{e}^{\mathrm{T}}\hat{\mathcal{L}}\boldsymbol{e} \\
&\leqslant 3\hat{\boldsymbol{x}}^{\mathrm{T}}\hat{\mathcal{L}}\hat{\boldsymbol{x}}+3\lambda_{\max}\left(\mathcal{L}\right)\sum_{i=1}^{N}\|\boldsymbol{e}_i\|^2 \\
&\leqslant 3\left(1+\lambda_{\max}\left(\mathcal{L}\right)^2\overline{k}\right)\hat{\boldsymbol{x}}^{\mathrm{T}}\hat{\mathcal{L}}\hat{\boldsymbol{x}}
\end{aligned} \tag{6-50}
$$

其中，$\overline{k}=\max_i\{k_i\}$是正常数。第一个不等式可由$\left|\boldsymbol{a}^{\mathrm{T}}\boldsymbol{\Phi}\boldsymbol{b}\right|\leqslant\left|\boldsymbol{a}^{\mathrm{T}}\boldsymbol{\Phi}\boldsymbol{a}\right|+\left|\boldsymbol{b}^{\mathrm{T}}\boldsymbol{\Phi}\boldsymbol{b}\right|$对所有的$\boldsymbol{a},\boldsymbol{b}\in\mathbb{R}^N$和对称矩阵$\boldsymbol{\Phi}\in\mathbb{R}^{N\times N}$得到。

注意到，$\|\tilde{\boldsymbol{x}}\|^2=\hat{\boldsymbol{x}}^{\mathrm{T}}\hat{\mathcal{L}}^{\mathrm{T}}\hat{\mathcal{L}}\hat{\boldsymbol{x}}\geqslant\lambda_2\left(\mathcal{L}\right)\hat{\boldsymbol{x}}^{\mathrm{T}}\hat{\mathcal{L}}\boldsymbol{x}$，其中$\lambda_2\left(\mathcal{L}\right)$是拉普拉斯矩阵的第二小特征值。根据式 (6-49) 和式 (6-50)，可得

$$\begin{aligned}\dot{V} &\leqslant -\underline{c}c_{\mu}\mu\left(\lambda_2(\mathcal{L})\hat{\boldsymbol{x}}^{\mathrm{T}}\hat{\mathcal{L}}\hat{\boldsymbol{x}}\right)-\underline{c}c_{\nu}\nu\left(\lambda_2(\mathcal{L})\hat{\boldsymbol{x}}^{\mathrm{T}}\hat{\mathcal{L}}\hat{\boldsymbol{x}}\right)\\ &\leqslant -\underline{c}c_{\mu}\mu\left(\sigma\lambda_2(\mathcal{L})\boldsymbol{x}^{\mathrm{T}}\hat{\mathcal{L}}\boldsymbol{x}\right)-\underline{c}c_{\nu}\nu\left(\sigma\lambda_2(\mathcal{L})\boldsymbol{x}^{\mathrm{T}}\hat{\mathcal{L}}\boldsymbol{x}\right)\\ &\leqslant -\underline{c}c_{\mu}k_{\mu}\mu\left(\frac{1}{2}\boldsymbol{x}^{\mathrm{T}}\hat{\mathcal{L}}\boldsymbol{x}\right)-\underline{c}c_{\nu}k_{\nu}\nu\left(\frac{1}{2}\boldsymbol{x}^{\mathrm{T}}\hat{\mathcal{L}}\boldsymbol{x}\right)\end{aligned}\tag{6-51}$$

其中，$\sigma=3\left(1+\lambda_{\max}(\mathcal{L})^2\overline{k}\right)$，利用第6.2.3节中函数$\mu\left(\|\boldsymbol{x}\|^2\right)$和$\nu\left(\|\boldsymbol{x}\|^2\right)$的性质2，可得到最后一个不等式。$k_{\mu}$和$k_{\nu}$都是正参数依赖于$\sigma$和$\lambda_2(\mathcal{L})$。令$c_1=\underline{c}c_{\mu}k_{\mu}$和$c_2=\underline{c}c_{\nu}k_{\nu}$。根据式(6-51)，可知

$$\dot{V}\leqslant -c_1\mu(V)-c_2\nu(V)\tag{6-52}$$

考虑V的反函数，即$V^{-1}(t)=t(V)$，我们有$\dot{t}(V)\geqslant\dfrac{1}{-c_1\mu(V)-c_2\nu(V)}$。由于$V$是递减的且有一个下界，通过两边积分反函数，可得

$$\begin{aligned}T_1-t_0 &\leqslant \int_{V(T_1)}^{V(t_0)}\frac{1}{c_1\mu(V)+c_2\nu(V)}\mathrm{d}V\\ &\leqslant \int_0^{\infty}\frac{1}{c_1\mu(V)+c_2\nu(V)}\mathrm{d}V\\ &\leqslant \frac{1}{c_1}\int_0^1\frac{1}{\mu(V)}\mathrm{d}V+\frac{1}{c_2}\int_1^{\infty}\frac{1}{\nu(V)}\mathrm{d}V\\ &=t^*\end{aligned}\tag{6-53}$$

其中，$T_1=t_0+t^*$是暂态时间，满足对于解$\boldsymbol{x}(t,\boldsymbol{x}_0,t_0),V\left(\boldsymbol{x}(t,\boldsymbol{x}_0,t_0)\right)=0,t\geqslant T_1$。同样注意到$\forall t>t_0,\left\{\boldsymbol{x}\in\mathbb{R}^{3N}:V\left(\boldsymbol{x}(t,\boldsymbol{x}_0,t_0)\right)=0\right\}=\{\boldsymbol{x}\in\mathbb{R}^{3N}:\boldsymbol{x}_1=\boldsymbol{x}_2=\cdots=\boldsymbol{x}_N\}$。因此，可得$\mathcal{C}_{\boldsymbol{x}}$是相对于$\mathcal{S}$固定时间收敛的，同时意味着$\mathcal{C}_{\boldsymbol{R}}$是相对于$\mathcal{S}$固定时间收敛的。因此，可得固定时间姿态一致性。证毕。 ■

定理 6.3

给定系统 (6-8) 和控制协议 (6-10)，当使用事件驱动条件 (6-16) 时，奇诺现象不会发生在时间$\left[t_0,T_1^-\right)\cup\left(T_1^+,+\infty\right)$。

证明：

首先考虑$\|\boldsymbol{e}_i\|^2$的演化过程：

$$\begin{aligned}\frac{\mathrm{d}\left(\boldsymbol{e}_i^{\mathrm{T}}\boldsymbol{e}_i\right)}{\mathrm{d}t}&=-2\boldsymbol{e}_i^{\mathrm{T}}\boldsymbol{J}_{\boldsymbol{x}_i}\left[\left(\hat{\mathcal{L}}_i\boldsymbol{x}\right)\frac{\mu\left(\|\tilde{\boldsymbol{x}}_i\|^2\right)}{\|\tilde{\boldsymbol{x}}_i\|^2}+\left(\hat{\mathcal{L}}_i\boldsymbol{x}\right)\frac{\nu\left(\|\tilde{\boldsymbol{x}}_i\|^2\right)}{\|\tilde{\boldsymbol{x}}_i\|^2}\right]\\&\leqslant\left(\|\boldsymbol{e}_i\|^2+\tilde{\boldsymbol{x}}_i^{\mathrm{T}}\boldsymbol{J}_{x_i}^{\mathrm{T}}\boldsymbol{J}_{x_i}\tilde{\boldsymbol{x}}_i\right)\frac{\mu\left(\|\tilde{\boldsymbol{x}}_i\|^2\right)}{\|\tilde{\boldsymbol{x}}_i\|^2}+\left(\|\boldsymbol{e}_i\|^2+\tilde{\boldsymbol{x}}_i^{\mathrm{T}}\boldsymbol{J}_{x_i}^{\mathrm{T}}\boldsymbol{J}_{x_i}\tilde{\boldsymbol{x}}_i\right)\frac{\nu\left(\|\tilde{\boldsymbol{x}}_i\|^2\right)}{\|\tilde{\boldsymbol{x}}_i\|^2}\\&\leqslant\left(k_i+\lambda_{\max}\left(\boldsymbol{J}_{x_i}^{\mathrm{T}}\boldsymbol{J}_{x_i}\right)\right)\mu\left(\|\tilde{\boldsymbol{x}}_i\|^2\right)+\left(k_i+\lambda_{\max}\left(\boldsymbol{J}_{x_i}^{\mathrm{T}}\boldsymbol{J}_{x_i}\right)\right)\nu\left(\|\tilde{\boldsymbol{x}}_i\|^2\right)\end{aligned}\tag{6-54}$$

其中，第二个不等式可通过事件驱动条件 (6-16) 得到。根据式(6-54)，可得

$$\frac{\mathrm{d}\left(\boldsymbol{e}_i^{\mathrm{T}}\boldsymbol{e}_i\right)}{\mathrm{d}t}\leqslant\omega_i\mu\left(\|\tilde{\boldsymbol{x}}_i\|^2\right)+\omega_i\nu\left(\|\tilde{\boldsymbol{x}}_i\|^2\right)\tag{6-55}$$

其中，$\omega_i=\left(k_i+\lambda_{\max}\left(\boldsymbol{J}_{x_i}^{\mathrm{T}}\boldsymbol{J}_{x_i}\right)\right)$。因此，如果没有智能体$i$的邻居智能体在$t\in\left[t_k^i,t_{k+1}^i\right)$内触发，基于$\boldsymbol{e}_i\left(t_k^i\right)=0$，可得$\|\boldsymbol{e}_i\|^2\leqslant\left(\omega_i\mu\left(\|\tilde{\boldsymbol{x}}_i\|^2\right)+\omega_i\nu\left(\|\tilde{\boldsymbol{x}}_i\|^2\right)\right)\left(t-t_k^i\right)$。如果存在邻居智能体$j\in\mathcal{N}_i$在时刻$t_{k_l}^j\in\left[t_k^i,t_{k+1}^i\right)$、$l\in\mathbb{N}^+$内触发，可知在这些有限个小的时间区间内$I_l=\left[t_{k_l}^j,t_{k_{l+1}}^j\right),l\in\mathbb{N}^+$，$\|\tilde{\boldsymbol{x}}_i\|^2$是常数。令$I=\left\{\left[t_k^i,t_{k_1}^j\right),\left[t_{k_1}^j,t_{k_2}^j\right),\cdots,\left[t_{k_l}^j,t_{k_{l+1}}^j\right),\left[t_{k_{l+1}}^j,t_{k+1}^i\right)\right\}$，且$\max_{t\in I}\tilde{\boldsymbol{x}}_i(t)=\|\tilde{\boldsymbol{x}}_{i,\max}\|^2$，可获得上界

$$\|\boldsymbol{e}_i\|^2\leqslant\left(\omega_i\mu\left(\|\tilde{\boldsymbol{x}}_{i,\max}\|^2\right)+\omega_i\nu\left(\|\tilde{\boldsymbol{x}}_{i,\max}\|^2\right)\right)\left(t-t_k^i\right)\tag{6-56}$$

根据定理 6.2，我们知道对所有的$t\geqslant T_1$，存在一个暂态时间 T_1 满足$\boldsymbol{x}\in\mathcal{C}_{\boldsymbol{x}}$。这意味着对所有的$t\geqslant T_1$，有$\dot{\boldsymbol{x}}_1=\cdots=\dot{\boldsymbol{x}}_N$。根据控制协议 (6-10)，可得$\hat{\boldsymbol{x}}^{\mathrm{T}}\hat{\mathcal{L}}\hat{\boldsymbol{x}}\equiv0$（对所有的$t\geqslant T_1$）。那么我们知道$\|\tilde{\boldsymbol{x}}_i\|\equiv0,i\in\mathcal{V}$（对所有的$t\geqslant T_1$）。

下面，首先考虑当$t\in\left(T_1^+,+\infty\right)$时，根据式(6-56)，则有$\|\boldsymbol{e}_i(t)\|=0$。在这个例子当中，当$t\in\left(T_1^+,+\infty\right)$时，根据事件驱动条件 (6-16)，智能体 i 将不会触发。对于第二种情况，当$t\in\left[t_0,T_1^-\right)$时，根据式 (6-56) 和事件驱动条件，可得触发间隔必定大于一个界$t_{k+1}^i-t_k^i>\dfrac{k_{\min}\|\tilde{\boldsymbol{x}}_{i,\max}\|^2}{\omega_i\mu\left(\|\tilde{\boldsymbol{x}}_{i,\max}\|^2\right)+\omega_i\nu\left(\|\tilde{\boldsymbol{x}}_{i,\max}\|^2\right)}=\bar{\tau}_i$。注意到，随着$t\to T_1^-$，当$\tilde{\boldsymbol{x}}_i$收敛到 0，$\bar{\tau}_i$可能收敛到 0。这意味着 T_1 之前不会出现奇诺现象。因此，我们仅保证了奇诺现象在时间段

$\left(0,T_1^-\right)\cup\left(T_1^+,+\infty\right)$内不会出现。证毕。 ■

注释 6.9 从上面的奇诺现象可以看出，事件间的触发间隔依赖于函数$\mu\left(\|\tilde{\boldsymbol{x}}\|^2\right)$和$\nu\left(\|\tilde{\boldsymbol{x}}\|^2\right)$，并且对实现固定时间一致性至关重要。一个可行的选择是令$\mu\left(\|\tilde{\boldsymbol{x}}\|^2\right)=\left(\|\tilde{\boldsymbol{x}}\|^2\right)^{\frac{\alpha+1}{2}}$和$\nu\left(\|\tilde{\boldsymbol{x}}\|^2\right)=\left(\|\tilde{\boldsymbol{x}}\|^2\right)^{\frac{\beta+1}{2}}$，其中$0<\alpha<1$和$\beta>1$。根据上述分析，$\bar{\tau}_i$随着我们选择更大的$\alpha$和$\beta$将会变得更大 。进一步，如果我们选择$\alpha>1$和$\beta>1$，$\bar{\tau}_i$将趋于发散，这意味着触发次数将会随着时间增加显著减少。然而，在这种情况下，根据式(6-51)，我们只能保证渐近一致性的实现。否则，如果我们选择$0<\alpha<1$和$\beta>1$，收敛率会提高，同时也会增加触发次数。这也说明了我们需要权衡收敛律和触发次数之间的关系。另一方面，相比于连续的通信协议，尽管奇诺现象发生在暂态时间T_1，但我们注意到通信频率和控制器更新都会显著地减少。最近，文献[186]研究了基于有限时间事件驱动的一致性问题，提出了一个动态事件驱动条件。主要实现有限时间一致性的思想是引入的动态变量具有有限时间收敛的性质，然而，必须指出的是，如果动态变量早于一致性完成，那么仍然会发生奇诺现象。因此，如何完全避免奇诺现象在基于事件驱动的有限时间一致性中，仍没有得到完全的解决。

注释 6.10 在文献[79]中，提出了一个有限时间姿态一致性协议。然而，协议是不连续的，这可能会导致抖振现象的发生。本章提出了更为一般的基于事件驱动的固定时间姿态一致性协议。如果选择$\mu\left(\|\tilde{\boldsymbol{x}}\|^2\right)=\left(\|\tilde{\boldsymbol{x}}\|^2\right)^{\frac{\alpha+1}{2}}$（对于任意$0<\alpha<1$）。可以看出，文献[162]中的基于事件驱动的有限时间一致性是定理6.2中的一个特殊情况。

注释 6.11 注意到定理 6.2 中的固定时间一致性结果是有一些保守的，这是由于引理 6.8 中的不变集过小，原因是由于事件驱动条件集合$\mathcal{B}_{\pi}(\mathbf{0}_3)$的正不变性难以得到保证。采样误差的引入使得证明变得很困难。在文献 [77,168] 中，集合$\mathcal{B}_{\pi}(\mathbf{0}_3)$的不变性可通过证明在一个多智能体中最大的姿态范数是非增的来实现。然而，由于存在采样误差，这种方法难以应用于事件驱动的姿态一致性协议中。此外，通过单个状态的李雅普诺夫函数导出事件驱动条件同样困难。尽管如此，值得一提的是集合 $\mathcal{S}$ 可以进一步扩大成为在$\mathbb{SO}(3)$上的最大凸集。通过使用球心投影的方法，$\mathcal{B}_{\frac{\pi}{2}}(\boldsymbol{I}_3)$上的凸包可以被投影为 $\mathbb{R}^3$ 上的凸多面体。因此，基于李雅普诺夫函数$V=\frac{1}{2}\boldsymbol{y}^{\mathrm{T}}\hat{\mathcal{L}}\boldsymbol{y}$，其中 $\boldsymbol{y}$ 是一个欧氏平面上的投影状态，与定理 6.2 的证明类似，可得在事件驱动条件 (6-16) 下，$\mathcal{B}_{\frac{\pi}{2}}(\boldsymbol{I}_3)$是一个正不变集。通过推广定理 6.2 的证明，不难得到这个结果。

6.4.2 切换拓扑的情况

本小节我们考虑智能体之间的拓扑图是时变的情况。定义一个切换信号$\sigma(t):\mathbb{R}_{>0}\to\mathcal{P}$，其中$\mathcal{P}=\{1,2,\cdots,M\}$。切换子图表示为$\hat{\mathcal{G}}_1,\hat{\mathcal{G}}_2,\cdots,\hat{\mathcal{G}}_M,\hat{\mathcal{G}}_i=\left\{\hat{\mathcal{N}}_i,\hat{\mathcal{E}}_i\right\},i=1,\cdots,M$，其中$\hat{\mathcal{N}}_i$和$\hat{\mathcal{E}}_i$是对应的节点集和边集。令$\{\tau_k\},k\in\mathbb{N}^{+}$表示切换时刻，假设在任意两个切换时刻之间存在一个正驻留时间，即$\tau_{k+1}-\tau_k>\tau_D$。在时间区间$[t,t+\tau]$的联合连通图定义为$\mathcal{G}_{[t,t+\tau]}=\cup\mathcal{G}_{\sigma(t)},t\in[t,t+\tau]$。如果存在一个正的时间常数 ΔT，使得联合图$\mathcal{G}_{[t,t+\Delta T]}$是连通的，则称这一并图是联合连通的。

假设 6.2[164]

存在一个常数 ΔT 满足无向图$\mathcal{G}_{[t,t+\Delta T]},\forall t>0$是联合连通的。

令 $\mathcal{L}_{\sigma(t)}^{i}$ 表示在切换信号下的 $\mathcal{L}_{\sigma(t)}$ 矩阵的第 i 行，定义

$\lambda_{\max M}=\max\left\{\lambda_{\max}\left(\mathcal{L}_p\right),p\in\mathcal{P}\right\}$和$\hat{\mathcal{L}}_{\sigma(t)}^i=\mathcal{L}_{\sigma(t)}^i\otimes I_3$，在联合连通图的条件下设计姿态一致性协议如下

$$\boldsymbol{w}_i=\boldsymbol{f}\left(-\hat{\mathcal{L}}_{\sigma(t)}^i\hat{\boldsymbol{x}}\right) \tag{6-57}$$

其中，$\boldsymbol{f}:\mathbb{R}^3\to\mathbb{R}^3$具有定理6.2中相同的性质。触发时刻由下面的事件驱动条件确定：

$$t_{k+1}^i=\inf\left\{t>t_k^i:\|\boldsymbol{e}_i\|^2>k_i\|\tilde{\boldsymbol{x}}_i\|^2\ \text{ or } \mathcal{N}_i\left(t_k^i\right)\neq\mathcal{N}_i\left(t\right)\right\} \tag{6-58}$$

其中，$\tilde{\boldsymbol{x}}_i=\sum_{i=1}^N a_{ij}\left(t\right)\left(\hat{\boldsymbol{x}}_i-\hat{\boldsymbol{x}}_j\right)$，$k_i$是在下面的证明中确定的正常数。

注释 6.12　根据事件驱动条件，可以看出每个智能体将会在两种情况下触发。一种情况是测量误差$\boldsymbol{e}_i$超过一定的门限，即$\|\boldsymbol{e}_i\|^2>k_i\|\tilde{\boldsymbol{x}}_i\|^2$。另一种情况是当拓扑切换的时刻。需要注意的是这些额外的触发是必要的，因为在切换时刻的触发保证了每个智能体的协议与当前的通信拓扑是一致的。否则，可能存在在一些特别的切换情况下，两个节点之间没有一条连通的有向路径，这就会导致多智能体最终不会趋于一致。

下面的定理说明了协议 (6-57) 可用来实现在切换拓扑下的固定时间姿态一致性。

定理 6.4

考虑一个包含了N个刚体的多智能体系统，假设通信拓扑满足假设 6.2。当使用姿态一致性协议 (6-57) 和事件驱动条件 (6-58) 时，如果 $k_i<\min\left\{\dfrac{1}{\lambda_{\max M}^2},\kappa_i\right\}$，其中 $\kappa_i=\dfrac{\left(\lambda_{\min}\left(\boldsymbol{J}_{\boldsymbol{x}_i}\right)-a\right)a}{\lambda_{\max M}^2\lambda_{\max}\left(\boldsymbol{J}_{\boldsymbol{x}_i}^{\mathrm{T}}\boldsymbol{J}_{\boldsymbol{x}_i}\right)},0<a<\lambda_{\min}\left(\boldsymbol{J}_{\boldsymbol{x}_i}\right)$，则$\mathcal{C}_{\mathcal{R}}$相对于$\mathcal{S}=\left\{\mathcal{R}\in\mathbb{SO}(3)^N:\sum_{i=1}^N d^2\left(\mathcal{R}_i\left(t_0\right),\boldsymbol{I}_3\right)<\pi^2\right\}$是固定时间收敛的。

证明：

根据引理 6.8，我们首先考虑李雅普诺夫函数$V=\dfrac{1}{2}\sum_{i=1}^N\|\boldsymbol{x}_i\|^2$。相似地，可以证明每个刚体的姿态将会待在集合$\mathcal{S}$内，意味着

$\mathcal{S}=\left\{\mathcal{R}\in\mathbb{SO}(3)^N:\sum_{i=1}^N d^2\left(\mathcal{R}_i\left(t_0\right),\boldsymbol{I}_3\right)<\pi^2\right\}$是一个正不变集。考虑李雅普诺夫函数

$$W=\frac{1}{2}\sum_{i,j=1}^N\|\boldsymbol{x}_i-\boldsymbol{x}_j\|^2 \tag{6-59}$$

再一次，基于与定理6.2相似的过程和事件驱动条件 (6-58)，可得W是一个非增函数：

$$\dot{W}\leqslant -c_3c_\mu^{\sigma(t)}k_\mu^{\sigma(t)}\mu\left(W_{\sigma(t)}\right)-c_3c_\nu^{\sigma(t)}k_\nu^{\sigma(t)}\nu\left(W_{\sigma(t)}\right) \tag{6-60}$$

其中，$c_3=\min_i\left(\lambda_{\min}\left(\boldsymbol{J}_{\boldsymbol{x}_i}\right)-\frac{\gamma k_i\lambda_{\max M}^2}{2}-\frac{\lambda_{\max}\left(\boldsymbol{J}_{\boldsymbol{x}_i}^{\mathrm{T}}\boldsymbol{J}_{\boldsymbol{x}_i}\right)}{2\gamma}\right)$；$W_{\sigma(t)}=\frac{1}{2}\boldsymbol{x}^{\mathrm{T}}\hat{\mathcal{L}}_{\sigma(t)}\boldsymbol{x}$表示对应每个切换子图$\hat{\mathcal{G}}_{\mathcal{P}}$的李雅普诺夫函数；$c_\mu^{\sigma(t)}$和$c_\nu^{\sigma(t)}$是取决于切换信号$\sigma(t)$的正常数。选择$\gamma=\frac{\lambda_{\max}\left(\boldsymbol{J}_{\boldsymbol{x}_i}^{\mathrm{T}}\boldsymbol{J}_{\boldsymbol{x}_i}\right)}{2\left(\lambda_{\min}\left(\boldsymbol{J}_{\boldsymbol{x}_i}\right)-a\right)}$，其中$0<a<\lambda_{\min}\left(\boldsymbol{J}_{\boldsymbol{x}_i}\right)$，且令$k_i<\frac{\left(\lambda_{\min}\left(\boldsymbol{J}_{\boldsymbol{x}_i}\right)-a\right)a}{\lambda_{\max M}^2\lambda_{\max}\left(\boldsymbol{J}_{\boldsymbol{x}_i}^{\mathrm{T}}\boldsymbol{J}_{\boldsymbol{x}_i}\right)}=\kappa_i$，可保证$c_3>0$。

下面的证明过程将分为两步。第一步，证明在联合连通的拓扑条件下，姿态的渐近一致性。再进一步给出暂态时间的估计。注意到，如果初始状态$\boldsymbol{x}(t_0)\in\mathcal{C}_x=\{\boldsymbol{x}=\left[\boldsymbol{x}_1^{\mathrm{T}},\cdots,\boldsymbol{x}_N^{\mathrm{T}}\right]^{\mathrm{T}}\in\mathcal{B}_\pi^{'}\left(\boldsymbol{0}_3\right)^N:\boldsymbol{x}_1=\cdots=\boldsymbol{x}_N\}$，解$\boldsymbol{x}\left(t,\boldsymbol{x}(t_0),t_0\right),t>t_0$将不会逃离这个集合，因为$\mathcal{C}_x$是一个不变集，同时也说明实现了姿态一致性。接下来，考虑情况$\boldsymbol{x}(t_0)\notin\mathcal{C}_x$，首先证明下面的结论 1。

结论 1. 假定初始状态$\boldsymbol{x}(t_0)\in\boldsymbol{C}_{\mathcal{B}_\pi^{'}(\boldsymbol{0}_3)^N}\mathcal{C}_x$，那么存在一个正的增函数$\rho(t)$和一个正常数 ΔT 满足

$$W\left(\boldsymbol{x}\left(t_0+\tilde{\tau}\right)\right)-W\left(\boldsymbol{x}\left(t_0\right)\right)<-\rho\left(\tilde{\tau},\boldsymbol{x}\left(t_0\right)\right) \tag{6-61}$$

其中，$\tilde{\tau}>\Delta T$是一个正常数。

不失一般性，假定 t_0 是一个切换时刻，由于任意两次切换之间存在一个驻留时间 τ_D，则有一个$\varepsilon>0$满足$\sigma(t)$是固定的，且$W\left(\boldsymbol{x}(t)\right)$在时间$t\in\left(\tau_k,\tau_k+\varepsilon\right)$是可微的，其中$\hat{\tau}_k$是一个任意切换常数。定义一个集合$\mathcal{C}_{\mathcal{P}}=\left\{\boldsymbol{x}_{\mathcal{P}}=\{\boldsymbol{x}_i\}\in\mathbb{R}^{3\left|\hat{\mathcal{N}}_{\mathcal{P}}\right|}:\boldsymbol{x}_i=\boldsymbol{x}_j,i,j\in\hat{\mathcal{N}}_{\mathcal{P}}\right\}$，根据式 (6-60)，可得如果

$\boldsymbol{x}(\tau_k) \in \boldsymbol{C}_{\mathcal{B}_\pi^{'}(\boldsymbol{0}_3)^N} \boldsymbol{C}_{\mathcal{P}}$，可得

$$\lim_{t \to \tau_k^+} \dot{W}(\boldsymbol{x}(t)) < 0 \tag{6-62}$$

因此，存在一个$\tilde{\tau} > 0$满足

$$W(\boldsymbol{x}(\tau_k + \tilde{\tau})) - W(\boldsymbol{x}(\tau_k)) < 0 \tag{6-63}$$

由于并图$\mathcal{G}_{[t,t+\Delta T]}$是联合连通的，可得$\bigcap \boldsymbol{C}_{\mathcal{B}_\pi^{'}(\boldsymbol{0}_3)^N} \boldsymbol{C}_{\sigma(\tau_k,\tau_k+\Delta T)} = \boldsymbol{C}_{\mathcal{B}_\pi^{'}(\boldsymbol{0}_3)^N} \left(\bigcup \boldsymbol{C}_{\sigma(\tau_k,\tau_k+\Delta T)}\right) = \boldsymbol{C}_{\mathcal{B}_\pi^{'}(0_3)^N} \boldsymbol{C}_{\boldsymbol{x}}$，则必定存在一个$\bar{t} > \Delta T$，满足当$t \in (\tau_k, \tau_k + \bar{t})$时$W(\boldsymbol{x}(t))$是递减的。因此，我们可总结到

$$W(\boldsymbol{x}(\tau_k + t)) - W(\boldsymbol{x}(\tau_k)) < -\rho(t, \boldsymbol{x}(\tau_k)) \tag{6-64}$$

其中，$\rho(t, \boldsymbol{x}(\tau_k))$是一个正函数且当$t > \Delta T$时随着时间递增。由于我们考虑的是任意的切换时刻，可得

$$W(\boldsymbol{x}(t_0 + \tilde{\tau})) - W(\boldsymbol{x}(t_0)) < -\rho(\tilde{\tau}, \boldsymbol{x}(t_0)) \tag{6-65}$$

基于式 (6-61)，可得$\boldsymbol{x}(t)$渐近收敛到$\mathcal{C}_{\boldsymbol{x}}$。

首先，根据式 (6-60)，我们得到$W(\boldsymbol{x}(t))$是非增的且$W(\boldsymbol{x}(t))$有一个下界满足$\lim_{t \to \infty} W(\boldsymbol{x}(t)) = a$。根据文献 [173] 中的引理 4.1，存在一个非空、紧不变集$\mathcal{D}^+$满足$\boldsymbol{x}(t) \to \mathcal{D}^+$。

注意到如果不变集仅仅包含 $\boldsymbol{C}_{\boldsymbol{x}}$，即 $a = 0$，证毕。

下面我们关注当不变集中存在点$\bar{\boldsymbol{x}} \in \mathcal{D}^+$满足$\bar{\boldsymbol{x}} \notin C_{\boldsymbol{x}}$且$a > 0$的情况。我们将说明这种情况下会与$\lim_{t \to \infty} W(\boldsymbol{x}(t)) = a$的结果导致矛盾。假设$\bar{\boldsymbol{x}} \in \mathcal{D}^+ \bigcap \boldsymbol{C}_{\mathcal{B}_\pi^{'}(0_3)^N} \boldsymbol{C}_{\boldsymbol{x}}$，我们有$W(\bar{\boldsymbol{x}}) = a$且$a > 0$。考虑初始状态为$\bar{\boldsymbol{x}}$，初始时间为$\bar{t}_0$的解$\boldsymbol{x}(t, \bar{\boldsymbol{x}}, \bar{t}_0)$。基于式 (6-61)，有

$$W(\boldsymbol{x}(\bar{t}_0 + \Delta t, \bar{\boldsymbol{x}}, \bar{t}_0)) < a - \rho(\Delta t, \bar{\boldsymbol{x}}) \tag{6-66}$$

其中，$\Delta t = t - \bar{t}_0$。由于$\bar{\boldsymbol{x}} \in \mathcal{D}^+$，可得针对任意的$\epsilon > 0$，存在一个$\delta(\epsilon, \tilde{t}) > 0$和一个时间序列$\boldsymbol{x}'(\bar{t}_0, \boldsymbol{x}_0, t_0)$满足$\| \boldsymbol{x}'(\bar{t}_0, \boldsymbol{x}_0, t_0) - \bar{\boldsymbol{x}} \| \leqslant \delta(\epsilon, \tilde{t})$。现在，不失一般性，我们令$\bar{t}_0 = t_0 + \tilde{t}$。因为$W(\boldsymbol{x}(t, \boldsymbol{x}_0, t_0))$对于状态$\boldsymbol{x}$是利普希茨连续的且对于时间$t$是分段时间连续的，通过运用初值依赖定理，有

$$\| W(\boldsymbol{x}(t, \boldsymbol{x}', \bar{t}_0) - W(\boldsymbol{x}(t, \bar{\boldsymbol{x}}, \bar{t}_0)) \| < \epsilon \tag{6-67}$$

不失一般性，选择$\epsilon = \frac{\rho\left(\Delta t, \bar{x}\right)}{2}$，结合式(7-35)和式(6-67)，有

$$W\left(\boldsymbol{x}\left(t, \boldsymbol{x}_0, t_0\right)\right) < a - \frac{\rho\left(\Delta t, \bar{\boldsymbol{x}}\right)}{2} \tag{6-68}$$

由于$\rho > 0$，如果$a > 0$，则显然$W\left(\boldsymbol{x}\left(t, \boldsymbol{x}_0, t_0\right)\right)$的下界不成立，与原假设矛盾。因此，我们可得$\lim\limits_{t\to\infty} W\left(\boldsymbol{x}\left(t\right)\right) = 0$，这意味着$\boldsymbol{x}\left(t\right)$将会渐近收敛到$\mathcal{C}_x$。

接下来，我们将证明在一个暂态时间内$\boldsymbol{x}\left(t\right)$收敛到$\mathcal{C}_x$，通过定理6.2中证明相似的方法，假定在时间$\left[t_0, +\infty\right)$内存在有限$p$+1个切换时刻$\tau_0, \tau_1, \cdots, \tau_p$。不失一般性，假定$\tau_0 = t_0$，则对于每个切换时间间隔$\left[\tau_k, \tau_{k+1}\right), k \in \mathbb{N}^+$，根据式(6-60)，可得

$$\dot{W} = \begin{cases} \dot{W}_1 \leqslant -a_1 \mu\left(W_1\right) - b_1 \nu\left(W_1\right), & t \in \left[\tau_0, \tau_1\right) \\ \dot{W}_2 \leqslant -a_2 \mu\left(W_2\right) - b_2 \nu\left(W_2\right), & t \in \left[\tau_1, \tau_2\right) \\ \qquad\vdots \\ \dot{W}_p \leqslant -a_p \mu\left(W_p\right) - b_p \nu\left(W_p\right), & t \in \left[\tau_p, +\infty\right) \end{cases} \tag{6-69}$$

其中，$a_{\sigma(t)}$和$b_{\sigma(t)}$是两个与切换信号相关的正参数。在每个切换时间间隔内$\left[\tau_k, \tau_{k+1}\right)$，定义

$$\bar{W}_{\sigma(t)} = \frac{1}{2} \sum_{(i,j) \in \mathcal{B}_{\mathcal{V}\times\mathcal{V}} \mathcal{E}_{\sigma(t)}} \left(\boldsymbol{x}_i - \boldsymbol{x}_j\right)^2, \quad t \in \left[\tau_k, \tau_{k+1}\right) \tag{6-70}$$

由于集合$\boldsymbol{C}_{\mathcal{V}\times\mathcal{V}} \mathcal{E}_{\sigma(t)}$的边在子图$\hat{\mathcal{G}}_p$是不连通的，$\overline{W}_{\sigma(t)}$在一个时间区间$\left[\tau_k, \tau_{k+1}\right)$内将保持为常数。考虑到$W$是一个反函数，即$W^{-1}\left(t\right) = t\left(W\right)$，有

$$\dot{t}\left(W\right) \geqslant \frac{1}{-a_{\sigma(t)} \mu\left(W - \overline{W}_{\sigma(t)}\right) - b_{\sigma(t)} \nu\left(W - \overline{W}_{\sigma(t)}\right)} = g\left(W, \sigma\left(t\right)\right)$$

对于时间$t \in \left[\tau_k, \tau_{k+1}\right)$，可得

$$\begin{aligned} \int_{\tau_k}^{\tau_{k+1}} \mathrm{d}t &\leqslant \int_{W\left(\tau_{k+1}\right)}^{W\left(\tau_k\right)} g\left(W, \sigma\left(\tau_k\right)\right) \mathrm{d}\left(W - \overline{W}_{\sigma\left(\tau_k\right)}\right) \\ &\leqslant \frac{2a_{\sigma\left(\tau_k\right)}}{1-\alpha} \left(W\left(\tau_k\right) - \overline{W}_{\sigma\left(\tau_k\right)}\right)^{\frac{1-\alpha}{2}} + \frac{2b_{\sigma\left(\tau_k\right)}}{1-\beta} \left(W\left(\tau_k\right) - \overline{W}_{\sigma\left(\tau_k\right)}\right)^{\frac{1-\beta}{2}} \\ &\quad - \frac{2a_{\sigma\left(\tau_{k+1}\right)}}{1-\alpha} \left(W\left(\tau_{k+1}\right) - \overline{W}_{\sigma\left(\tau_{k+1}\right)}\right)^{\frac{1-\alpha}{2}} + \frac{2b_{\sigma\left(\tau_{k+1}\right)}}{1-\beta} \left(W\left(\tau_{k+1}\right) - \overline{W}_{\sigma\left(\tau_{k+1}\right)}\right)^{\frac{1-\beta}{2}} \end{aligned} \tag{6-71}$$

进而，

$$T_2-t_0\leqslant\int_{W(\infty)}^{W(\tau_p)}g\left(W,\sigma\left(\tau_p\right)\right)\mathrm{d}\left(W-\overline{W}_{\sigma(\tau_p)}\right)+\cdots+\int_{W(\tau_{k+1})}^{W(\tau_k)}g\left(W,\sigma\left(\tau_k\right)\right)\mathrm{d}\left(W-\overline{W}_{\sigma(\tau_k)}\right)$$
$$+\cdots+\int_{W(\tau_1)}^{W(\tau_0)}g\left(W,\sigma\left(\tau_0\right)\right)\mathrm{d}\left(W-\overline{W}_{\sigma(\tau_0)}\right) \tag{6-72}$$

因此，可得固定时间的估计：

$$\begin{aligned}T_2-t_0\leqslant&\frac{1}{a_{\sigma(\tau_p)}}\int_0^{W(\tau_p)}\frac{1}{\mu\left(W-\overline{W}_{\sigma(\tau_p)}\right)}\mathrm{d}\left(W-\overline{W}_{\sigma(\tau_p)}\right)\\&+\cdots\\&+\frac{1}{a_{\sigma(\tau_k)}}\int_{W(\tau_{k+1})}^{W(\tau_k)}\frac{1}{\mu\left(W-\overline{W}_{\sigma(\tau_k)}\right)}\mathrm{d}\left(W-\overline{W}_{\sigma(\tau_k)}\right)\\&+\frac{1}{b_{\sigma(\tau_k)}}\int_{W(\tau_{k+1})}^{W(\tau_k)}\frac{1}{\nu\left(W-\overline{W}_{\sigma(\tau_k)}\right)}\mathrm{d}\left(W-\overline{W}_{\sigma(\tau_k)}\right)\\&+\cdots\\&+\frac{1}{b_{\sigma(\tau_0)}}\int_{W(\tau_1)}^{\infty}\frac{1}{\nu\left(W-\overline{W}_{\sigma(\tau_0)}\right)}\mathrm{d}\left(W-\overline{W}_{\sigma(\tau_0)}\right)\\=&t^*\end{aligned} \tag{6-73}$$

其中，上述不等式可根据$\overline{W}_{\sigma(\tau_p)}=0$得到。基于此，根据第6.2.3节中函数$\mu(\cdot)$和$\nu(\cdot)$的性质3，可得$t^*$是一个正常数。因此，固定时间姿态一致性可在暂态时间$T_2=t_0+t^*$之前实现。

基于事件驱动条件 (6-58)，可得触发将在测量误差超过门限或者拓扑发生切换的时候发生。首先考虑$t\in\left[t_0,T_2^-\right)$的情况。如果两次触发时刻之间没有切换，则根据定理 6.3 的证明，已知一定存在一个触发间隔的下界$t_{k+1}^i-t_k^i>\overline{\tau}_i$。如果$t_{k+1}^i$是触发时刻非切换时刻，$t_k^i$是切换时刻$\tau_k$，我们可以发现这个情况与上面的情况类似。如果$t_k^i$是触发时刻非切换时刻，$t_{k+1}^i$是切换时刻$\tau_{k+1}$，那么触发间隔的下界为$\min\left\{\overline{\tau}_i,\tau_{k+1}-t_k^i\right\}$。如果两次触发时刻都是切换时刻，则触发间隔的下界即为驻留时间间隔τ_D。因此，可以总结，当$t\in\left[t_0,T_2^-\right)$时，奇诺现象将不会发生。对于$t\in\left(T_2^+,+\infty\right)$

的情况，根据定理 6.3 的证明，可以推断出触发将仅仅发生在拓扑是切换的时候，这也意味着奇诺现象不会发生。最后，可得奇诺现象不会在 $\left[t_0,T_2^-\right)\cup\left(T_2^+,+\infty\right)$发生。证毕。 ■

注释 6.13 根据上面的分析，暂态时间 T_2 取决于图的代数连通性和切换时刻。一致性的暂时时间是固定的，原因是其并不依赖于刚体的初始姿态。文献 [164,168] 研究了在切换拓扑下的连续时间姿态一致性。不同于渐近稳定的结果 [164,168]，我们证明了集合$\mathcal{C}_{\mathcal{R}}$相对于一个正不变集是固定时间吸引的。

6.5 仿真示例

首先，给出一个数值仿真例子来验证基于欧拉 - 拉格朗日系统的有限时间一致性算法的有效性。考虑一个包含 7 个 2 自由度平面机器人的网络化欧拉 - 拉格朗日系统，2 自由度平面机器人如图 6-3 所示。通信网络的拓扑是强连通图，与之相关联的拉普拉斯矩阵为：

$$
\mathcal{H}=\begin{bmatrix}1&0&-1&0&0&0&0\\-1&3&-1&0&0&0&-1\\0&0&1&-1&0&0&0\\-1&-1&0&2&0&0&0\\0&-1&0&0&2&-1&0\\0&0&-1&0&0&1&0\\-1&0&0&0&-1&0&2\end{bmatrix}
$$

机器人的模型如下：

$$
\boldsymbol{M}_i(\boldsymbol{q}_i)\ddot{\boldsymbol{q}}_i+\boldsymbol{C}_i(\boldsymbol{q}_i,\dot{\boldsymbol{q}}_i)\dot{\boldsymbol{q}}_i+\boldsymbol{g}_i(\boldsymbol{q}_i)=\boldsymbol{\tau}_i+\boldsymbol{\varDelta}_i,\quad i=1,\cdots,7
$$

其中，关节位置$\boldsymbol{q}_i=\left[q_{1i},q_{2i}\right]^{\mathrm{T}}$，惯性矩阵$\boldsymbol{M}_i(\boldsymbol{q}_i)$和科氏力矩阵$\boldsymbol{C}_i(\dot{\boldsymbol{q}}_i,\boldsymbol{q}_i)$可写为

$$
\boldsymbol{M}_i(\boldsymbol{q}_i)=\begin{bmatrix}\theta_1+2\theta_2\cos(q_{2i}) & \theta_3+\theta_2\cos(q_{2i})\\ \theta_3+\theta_2\cos(q_{2i}) & \theta_3\end{bmatrix}
$$

和

$$\boldsymbol{C}_i\left(\boldsymbol{q}_i,\dot{\boldsymbol{q}}_i\right)=\begin{bmatrix}-\theta_2\sin\left(q_{2i}\right)\dot{\boldsymbol{q}}_{2i} & -\theta_2\sin\left(q_{2i}\right)\left(\dot{\boldsymbol{q}}_{1i}+\dot{\boldsymbol{q}}_{2i}\right)\\ \theta_2\sin\left(q_{2i}\right)\dot{\boldsymbol{q}}_{1i} & 0\end{bmatrix}$$

物理参数$\theta^{\mathrm{T}}=[\theta_1,\theta_2,\theta_3]$为

$$\begin{bmatrix}\theta_1\\ \theta_2\\ \theta_3\end{bmatrix}=\begin{bmatrix}I_1+m_1l_{c,1}^2+m_2l_1^2+I_2+m_2l_{c,2}^2\\ m_2l_1l_{c,2}\\ I_2+m_2l_{c,2}^2\end{bmatrix}$$

其中，机器人手臂的物理参数如表6-2所示。

表6-2　机械臂的物理参数

描述	变量	数值	单位
关节 1 的质量	m_1	6.5225	kg
关节 2 的质量	m_2	2.0458	kg
关节 1 的长度	l_1	0.26	m
关节 2 的长度	l_2	0.26	m
关节 1 质心距离	l_{c1}	0.0983	m
关节 2 质心距离	l_{c2}	0.0229	m
关节 1 质心惯性矩	I_1	0.1213	kg • m^2
关节 2 质心惯性矩	I_2	0.0116	kg • m^2
重力加速度	g	9.81	m/s^2

在仿真中，控制器的参数被选为$k_1=2,k_2=1,k_{p_i}=2,k_{\xi_i}=1,\rho=0.6$，有限时间一致性协议的参数被选为$\alpha=0.5,\beta_i\left(0\right)=1,\sigma_i=0.05,i=1,\cdots,N$。事件驱动条件参数$\delta_i$被选为0.5，动态参数由$\Lambda_1=0.5$、$\Lambda_2=0.125$给出，动态变量的初值$\Xi_i\left(0\right)$为0.5。外界扰动$\boldsymbol{\Delta}_i$定义为$\left[0.5\sin t,0.5\cos t\right]^{\mathrm{T}}$。每个智能体的初值为$\boldsymbol{q}_i\left(0\right)=\boldsymbol{\eta}_i\left(0\right)=\left[0.5i,i\right]^{\mathrm{T}},i=1,\cdots,N$。

数值仿真结果由图6-3～图6-5给出。事件驱动有限时间一致性的收敛轨迹如图6-3所示。从图中可知，每个智能体在很短的有限时间内收敛到了一致点，表明了有限时间一致性协议的有效性。每个智能体的自适应耦合变量β_i轨迹如图6-4(a)所示，显示了增益趋于一个稳定的有限值。在图6-4(b)中显示了每个智能体的触发时刻。令$\hat{\boldsymbol{\Delta}}_i=\boldsymbol{M}_i\boldsymbol{\phi}_i$，$\hat{\boldsymbol{\Delta}}_i$两个分量的演化轨迹如图6-5所示，可以看出$\hat{\boldsymbol{\Delta}}_i$对干扰的补偿。

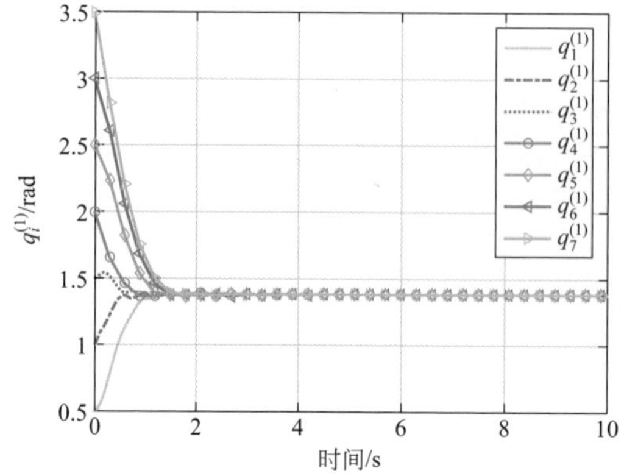

(a) 机器人第一个分量的位置轨迹

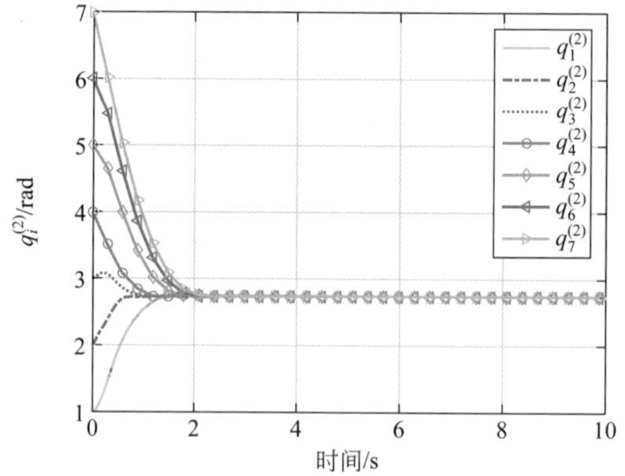

(b) 机器人第二个分量的位置轨迹

图 6-3 事件驱动有限时间一致性的收敛轨迹

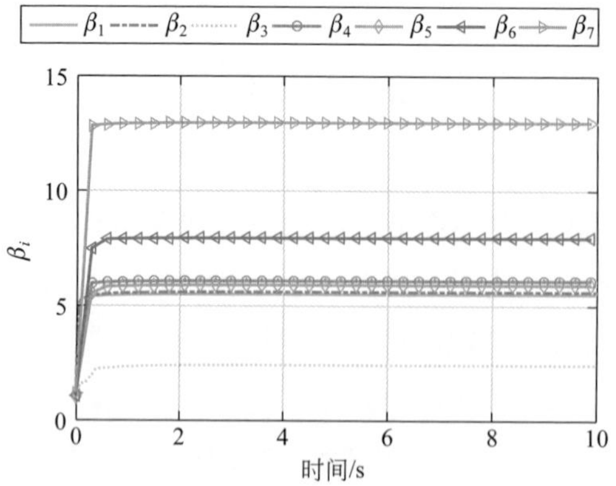

(a) 每个智能体的自适应耦合β_i

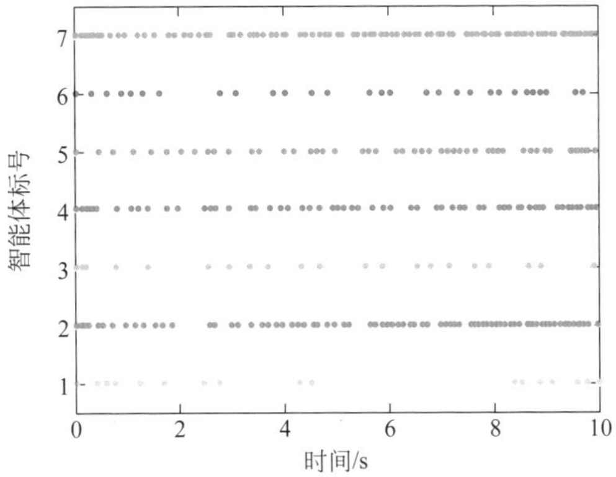

(b) 每个智能体的触发时刻

图 6-4　每个智能体的自适应耦合 β_i 和触发时刻

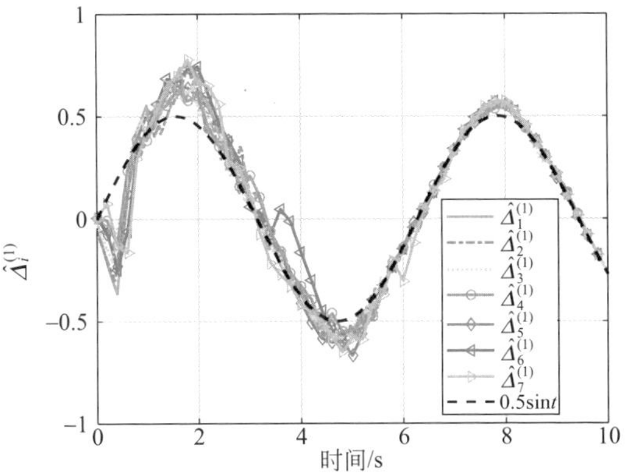

(a) 每个机械臂第一个坐标的干扰补偿（实线）和干扰（虚线）

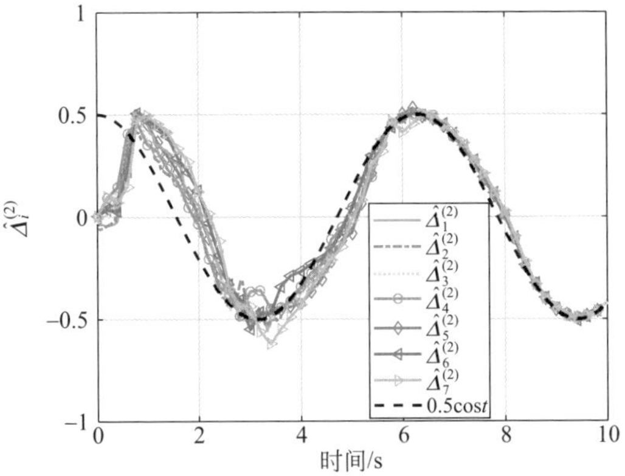

(b) 每个机械臂第二个坐标的干扰补偿（实线）和干扰（虚线）

图 6-5　每个机械臂第一和第二个坐标的干扰补偿（实线）和干扰（虚线）

针对多刚体系统姿态一致性问题，考虑由四个刚体组成的多智能体系统。我们首先考虑固定拓扑的情况。固定拓扑的边集合为

$$\varepsilon=\{(1,2),(2,1),(2,3),(3,2),(3,4),(4,3),(1,4),(4,1)\} \tag{6-74}$$

每个刚体的姿态由$\mathcal{B}_{\pi}^{'}(\mathbf{0}_3)$中的轴角向量来表征。我们设定姿态初值为

$$\boldsymbol{x}_1(0)=[-0.03\ 0.11-0.19]^{\mathrm{T}} \tag{6-75}$$

$$\boldsymbol{x}_2(0)=[0.51-0.08-0.01]^{\mathrm{T}} \tag{6-76}$$

$$\boldsymbol{x}_3(0)=[0.35\ 0.35\ 0.03]^{\mathrm{T}} \tag{6-77}$$

$$\boldsymbol{x}_4(0)=[0.33-0.17-0.01]^{\mathrm{T}} \tag{6-78}$$

且都包含在$S=\left\{\boldsymbol{x}\in\mathbb{R}^{3N}:\sum_{i=1}^{N}\|\boldsymbol{x}_i(t_0)\|^2<\pi^2\right\}$。在这个仿真中，选择$\mu(\|\tilde{\boldsymbol{x}}\|^2)=(\|\tilde{\boldsymbol{x}}\|^2)^{\frac{\alpha+1}{2}}$和$\nu(\|\tilde{\boldsymbol{x}}\|^2)=(\|\tilde{\boldsymbol{x}}\|^2)^{\frac{\beta+1}{2}}$，其中$\alpha=0.8,\beta=4$。为了设定事件驱动条件的参数，计算初始雅可比矩阵的最小特征值$\lambda_{\min}(\boldsymbol{J}_{\boldsymbol{x}_i}(0))$，以及最大特征值$\lambda_{\max}(\boldsymbol{J}_{\boldsymbol{x}_i}^{\mathrm{T}}(0)\boldsymbol{J}_{\boldsymbol{x}_i}(0))$和$\lambda_{\max}(\mathcal{L}^{\mathrm{T}}\mathcal{L})$。根据理论结果，$k$的最小值应该小于0.025，则需要定义

$$k=[0.022\ 0.02\ 0.02\ 0.015]^{\mathrm{T}} \tag{6-79}$$

在这个仿真中，我们使用欧拉法得到微分方程的数值解。采样步长选择为0.001，仿真时间为10 s。考虑到计算的精度问题，我们假定当测量误差小于10^{-9}的时候为0。

固定拓扑情况下的仿真结果如图 6-6 所示。每个刚体的姿态轨迹如图 6-6(a) 所示，可以看出刚体的姿态在有限时间内实现一致。图 6-6(b) 显示了每个刚体的触发时间。我们可以看到在暂态时间之前，每个刚体的触发时间都比周期采样时间大得多，因此排除了奇诺现象。下面，将考虑切换拓扑的情况。将切换信号定义为周期性的：

$$
\sigma(t)=\begin{cases}1, & \dfrac{4nT}{m}<t\leqslant\dfrac{(4n+1)T}{m}\\ 2, & \dfrac{4nT+1}{m}<t\leqslant\dfrac{(4n+2)T}{m}\\ 3, & \dfrac{4nT+2}{m}<t\leqslant\dfrac{(4n+3)T}{m}\\ 4, & \dfrac{4nT+3}{m}<t\leqslant\dfrac{(4n+4)T}{m}\end{cases} \tag{6-80}
$$

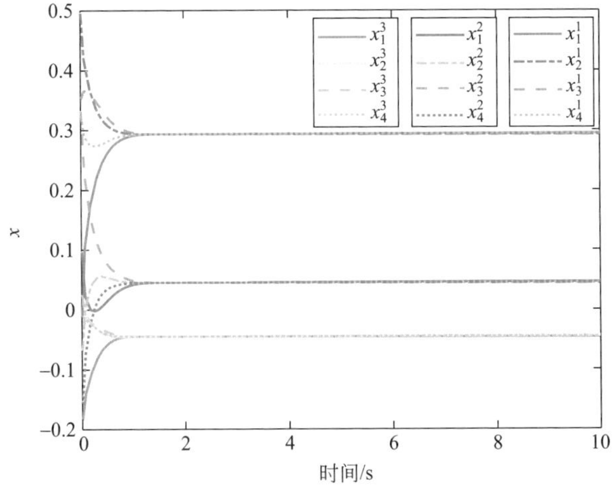

(a) 固定拓扑下每个刚体的轴角姿态

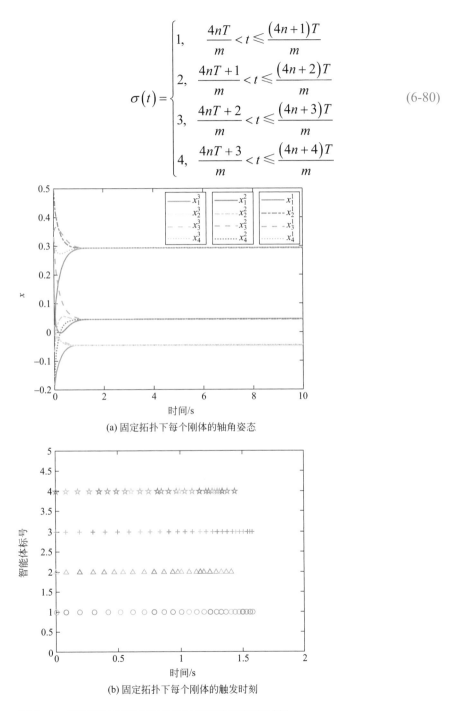

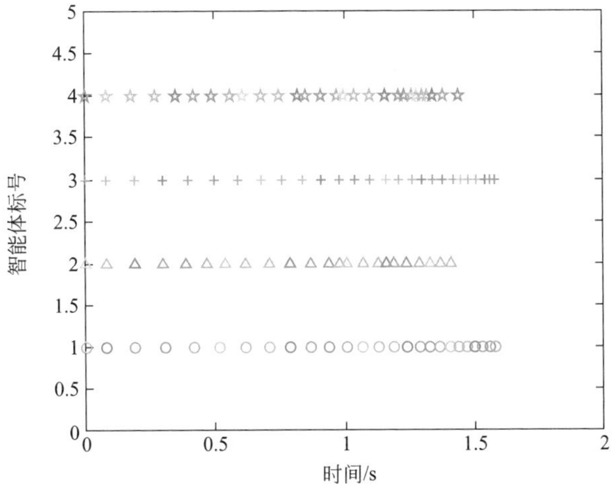

(b) 固定拓扑下每个刚体的触发时刻

图 6-6　固定拓扑下每个刚体的轴角姿态和触发时刻

其中，$n=1,2,\cdots$，仿真中令$m=100, T=10$。设定切换拓扑的边集合为

$$\mathcal{E}_1=\{(1,2),(2,1)\}，\mathcal{E}_2=\{(1,3),(3,1)\}，\mathcal{E}_3=\{(3,4),(4,3)\}，\mathcal{E}_4=\{(2,4),(4,2)\} \tag{6-81}$$

如图 6-7 所示，对每个切换拓扑，$\lambda_{\max}\left(\mathcal{L}_{\sigma(t)}^{\mathrm{T}}\mathcal{L}_{\sigma(t)}\right)=4$，因此可设 $\boldsymbol{k}=[0.09\ 0.08\ 0.095\ 0.09]^{\mathrm{T}}$，可以看出满足定理 6.4 的条件。

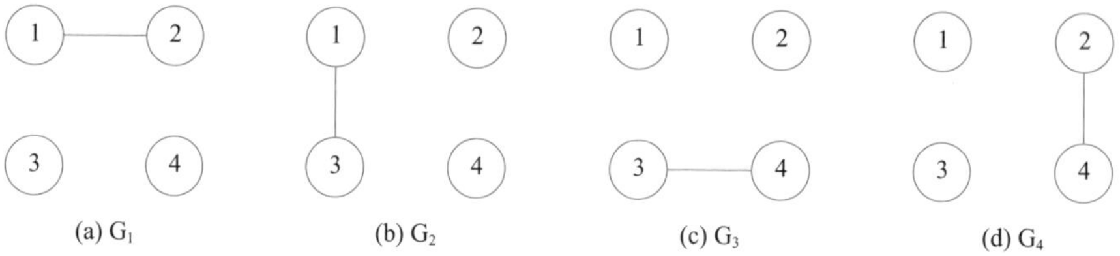

图 6-7 切换拓扑

仿真结果见图 6-8(a)、(b)。图 6-8(a) 证明了在切换拓扑下姿态的轨迹可以达到固定时间一致。图 6-8(b) 中给出了每个刚体的触发时刻。可以看出，触发间隔会大于固定拓扑的情况。这是因为对于每个切换子图，特征值$\lambda_{\max}\left(\mathcal{L}_{\sigma(t)}^{\mathrm{T}}\mathcal{L}_{\sigma(t)}\right)$较小。因此，我们选择的事件驱动条件参数 k_i 要大于固定拓扑的情况。

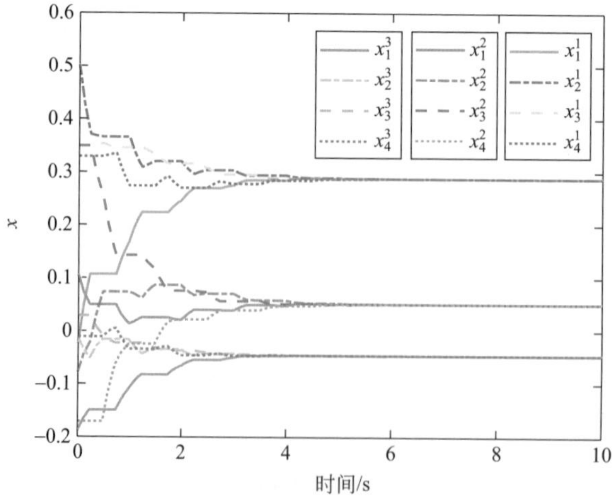

(a) 切换拓扑下每个刚体的轴角姿态

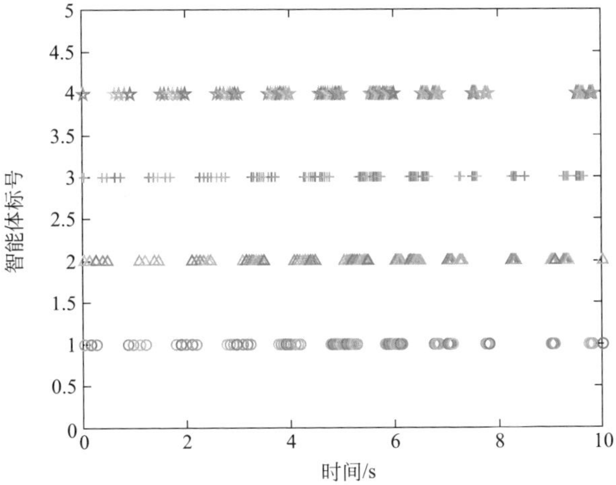

(b) 切换拓扑下每个刚体的触发时刻

图 6-8　切换拓扑下每个刚体的轴角姿态和触发时刻

Digital Wave
Advanced Technology of Industrial Internet

Control of Autonomous Intelligent Systems

自主智能系统控制

第7章

多刚体系统姿态同步控制

7.1 概述

在上一章的内容当中，我们对具有固定时间收敛性能的姿态一致性协议进行了初步的研究。可以发现，在姿态的收敛范围方面仍存在不足。此外，针对姿态一致性的某些应用，如深空探测、相机或光学导航传感器只能测量相对姿态信息，无法精确获得绝对姿态信息 [187]，因此本章将分别对绝对和相对姿态测量下的姿态一致性问题进行进一步的研究。

此外，在实际中由于传感器成本和失效的问题，使得对角速度的直接测量并不总是有效的和可靠的 [188]。在过去的几年中，有不少文献研究了关于无角速度的姿态一致性 [188-191]。在这些文章中，主要有两种方法：基于观测器的方法 [188,190] 和基于虚拟系统的方法 [189]。对于第一种方法，可以设计观测器从一个参考向量或者姿态中得到对角速度的估计，从而避免了在控制协议中的角速度测量。需要注意的是，一些研究提出了一个有限时间 / 固定时间的反馈结合一个有限时间 / 固定时间的观测器来保证满足分离原理 [192-193]。第二种方法设计了一个虚拟系统来对角速度和相对角速度给出了一个间接的估计，它实际上是给姿态控制协议提供了一个阻尼项 [194-195]。这种方法的优点在于不依赖分离原理，每个虚拟系统都是与闭环系统一起进行设计和分析。文献 [195] 研究了基于四元数的多刚体系统姿态一致性问题。通过引入一个基于四元数运动学模型的虚拟系统，避免了所设计的协同控制协议中对角速度的测量。然而，大多无角速度测量的姿态一致性协议是基于局部的姿态表征或是四元数姿态表征。对于局部的姿态表征会出现姿态表征奇异的问题 [77]；四元数虽然是全局的姿态表征，但也会出现表征不唯一的问题 [196]。最近，一些文献基于旋转矩阵研究了无角速度测量的领航者姿态跟踪问题 [197]。尽管如此，由于旋转矩阵属于一个非欧空间，传统欧氏空间的一致性协议会存在不理想的平衡点。因此，使用这类协议可能会陷入这些不理想平衡点，从而无法实现一致性 [198]。

另外，由于姿态的非欧几里德性质，大多数无速度姿态一致性结果都集中在固定拓扑图[194-195,197]。应该注意的是，由于不可靠的通信约束，通信链接可能会在深空任务中发生变化或者失效[199]。在文献 [77] 中通过使用局部姿态表征的方法研究了基于切换拓扑的姿态一致性问题。但是，需要有角速度测量才能实现同步，并且它需要任意两个连续切换之间的驻留时间有一个固定一致下界。

目前，在航天航空领域中，姿态一致性的应用范围逐渐增多，如小卫星网络和分级航天器。在这些场景当中，每个单独的系统都与无线通信网络连接。数据采样和控制动作信号全部通过无线信道传输，这意味着当多个子系统共享公共通道时，拥塞可能会急剧增加。而且，由于功率限制和远距离通信，航天器之间的无线通信带宽非常有限[200]。因此，在设计姿态一致性协议时考虑通信和控制效率有实际必要性。在事件驱动采样机制下，事件驱动控制协议的更新仅仅在一个预设计的事件驱动条件被违背时的某些离散时刻发生，因而，使得计算和通信资源的利用率大大提升。另外注意到，每个智能体可以通过使用相机获得相对姿态信息且不需要进行通信，这意味着可以进一步降低通信成本。因此，这促使我们设计仅仅依靠相对姿态信息的事件驱动一致性协议。近几年，一些文献研究了基于事件驱动的领航者姿态跟踪问题[201-202]。文献 [202]，提出一类事件驱动的自适应控制器，用于保证系统容错性，同时降低通信负担。文献 [200] 针对多航天器系统，考虑了带有模型不确定性和外界扰动的事件驱动姿态同步问题。然而，由于姿态空间的非线性，很少有文献考虑事件驱动的几乎全局姿态一致性问题。另一方面，在最近的一些有关几乎全局姿态一致性的研究成果中，几乎全局姿态一致性问题可分别通过使用旋转矩阵[203-204]和局部的姿态表征[17,164]来解决。在文献 [203-204] 中，通过引入带特定约束条件的候选成本函数，保证了所有一致集中的理想平衡点是唯一的渐近平衡点，而其他非理想平衡点都是不稳定的。但是，引入的约束条件可能会非常保守，因此在事件驱动的采样机制下很难满足。另一方面的研究中，用轴角向量来几乎全局表征姿态。为了实现几乎全局一致性，通过使用一类局部的李雅普诺夫函数，局部姿态表征的集合被证明是正不变集。然而，这种方法不能用来解决基于事件驱动的几乎全局姿态一致性问题。其难点在于分析由姿态运动学与事件驱动采样带来的混杂非线性问题。具体地，局部姿态表征集合的正不变性很难在事件驱动采样机制下保证。进一步，现有的事件驱动姿态控制问题都依赖一个全局坐标系[167,205]。实际上，在姿态一致性问题当中，同时在一致性协议和事件驱动条件中避免使用绝对姿态信息是一个难点问题。此外，最近分布式事件驱动控制[176,206]有了新的发展。从可拓展性的角度来看，主要关注的问题之一是提出一种完全分布式的事件驱动机制。该机制的重点在于避免协议和 ETC 中[176]中使用全局信息。注意到，由于几何拓扑的限制，在姿态一致性协议的设计中需要一些全局信息，例如图的最大度数[197-198]。因此，如何在姿态一致性问题中不使用全局信息的情况下设计事件驱动协议和 ETC 是一个具有挑战性的问题。基于上述讨论，本章内容分别考虑了在绝对和相对姿态测量下基于事件驱动的姿态一致性以及在缺少角速度测量下的姿态一致性问题。针对绝对和相对姿态测量下的事件驱动姿态一致性问题，第一种方法的优势在于可以实现几乎全局姿态一致性，相比之下，第二种方法姿态的初值需要限制在 $\mathbb{SO}(3)$ 上的一个凸集内。第二种协议的好处在于可应用于只有相对姿态测量的场合。进一步，我们将第二种事件驱动协议推广到了自触发的版本，避免了在事件驱动机制下的连续监测问题。主要贡献如下：①通过运用球心投影的方法，$\mathbb{SO}(3)$ 上的姿态表征可以被几乎全局地投影到欧几里德超平面。基于此，提出了一种基于投影的事件驱动姿态一致性协议，实现了几乎全局的姿态一致性。②基于黎曼梯度下降法，提出了一种仅使用相对姿态测量的事件驱动姿态一致性协议。此外，基于 $\mathbb{SO}(3)$ 的几何属性，我们将结果扩展到自触发情况，以避免事件监测中的连续测量问题。③两种事件驱动姿态一致性协议都是完全分布式的，意味着在事件驱动协议和条件中不需要使用全局拓扑信息。

针对无角速度测量下的事件驱动姿态一致性问题，首先，在固定图的情况下研究事件驱动的一致性协议，然后将其扩展到时变拓扑图。主要的贡献如下：①提出了一种无需角速度测量的事件驱动姿态一致性协议。与现有无速度策略的姿态一致性研究[190,194-195]不同，本章内容的收敛结果基于一个在 $\mathbb{SO}(3)$ 上的正不变开集，避免了局部姿态表示的奇异性问题和单位四元数的 unwinding 现象。此外，由于 $\mathbb{SO}(3)$ 的非欧几里德性质，如何在没有角速度反馈的情况下保证姿态空间是正不变的，也带来了一定技术上的困难。②为了降低通信成本，刚体之间的通信是基于时钟的触发机制控制的。设计的事件驱动协议和 ETC 仅取决于每个刚体的本地信息，没有利用任何全局信息。与文献 [207] 中事件驱动的姿态一致性研究相比，该一致性协议设计为运动学的角速度输入，在本章内容中，我们考虑了姿态动力学模型和没有角速度反馈的一致性协议。此外，基于时钟的触发机制，我们给出一个统一的最小事件间隔，该间隔仅与每个刚体的局部信息有关。③将提出的免角速度测量一致性协议进一步推广到了切换拓扑的情况。与现有基于切换拓扑的姿态一致性问题相比 [77,164]，基于一个平均驻留时间的概念，我们进一步放宽了两次连续切换之间需要一个固定驻留时间下界的假设，且在切换拓扑的姿态一致性协议中，没有用到角速度测量信息。

7.2 相关引理和问题描述

7.2.1 引理

引理 7.1[208]

给定一个包含 N 节点的强连通图 $\mathcal{G}$ 和相关的拉普拉斯矩阵 $\mathcal{L}\in\mathbb{R}^{N\times N}$，存在一个正向量 $\boldsymbol{p}=\left[p_1\ p_2\ \cdots\ p_N\right]^{\mathrm{T}}$ 满足 $\boldsymbol{p}^{\mathrm{T}}\mathcal{L}=0$。进一步，$\hat{\boldsymbol{p}}\mathcal{L}+\mathcal{L}^{\mathrm{T}}\hat{\boldsymbol{p}}$ 是一个对称非负矩阵，其中 $\hat{\boldsymbol{p}}=\operatorname{diag}\left\{p_i\right\},i=1,\cdots,N$。

引理 7.2[77]

给定两个旋转矩阵 $\mathcal{R}_i$ 和 $\mathcal{R}_j,i,j\in\mathcal{V}$，位于开球 $\mathcal{B}_{\pi}\left(\boldsymbol{I}_3\right)$ 中，运动学 $\boldsymbol{x}_{ij}=\log\left(\mathcal{R}_i^{\mathrm{T}}\mathcal{R}_j\right)^{\vee}$ 由下式给出

$$\dot{\boldsymbol{x}}_{ij}=\left[\boldsymbol{I}_3+\frac{1}{2}\theta_{ij}\boldsymbol{u}_{ij}^{\wedge}+\left(1-\frac{\theta_{ij}}{2}\cot\frac{\theta_{ij}}{2}\right)\left(\boldsymbol{u}_{ij}^{\wedge}\right)^2\right]\boldsymbol{w}_{ij} \tag{7-1}$$

其中，$\boldsymbol{w}_{ij}=\boldsymbol{w}_j-\mathcal{R}_i^{\mathrm{T}}\mathcal{R}_j\boldsymbol{w}_i$。

证明：

定义相对旋转向量为$\boldsymbol{x}_{ij}=\theta_{ij}\boldsymbol{u}_{ij}$，其中$\boldsymbol{u}_{ij}=\dfrac{1}{2\sin\theta_{ij}}\left(\mathcal{R}_{ij}-\mathcal{R}_{ij}^{\mathrm{T}}\right)^{\wedge}$和$\theta_{ij}=\arccos\dfrac{\mathrm{tr}\left(\mathcal{R}_{ij}\right)-1}{2}$。我们首先考虑$\theta_{ij}$和$\mathcal{R}_{ij}$的导数，

$$\begin{aligned}\dot{\mathcal{R}}_{ij}&=\left(-\mathcal{R}_i^{\mathrm{T}}\dot{\mathcal{R}}_i\mathcal{R}_i^{\mathrm{T}}\right)\mathcal{R}_j+\mathcal{R}_i^{\mathrm{T}}\dot{\mathcal{R}}_j=-\boldsymbol{w}_i^{\wedge}\mathcal{R}_i^{\mathrm{T}}\mathcal{R}_j+\mathcal{R}_i^{\mathrm{T}}\mathcal{R}_j\boldsymbol{u}_i^{\mathrm{T}}\\&=-\mathcal{R}_i^{\mathrm{T}}\mathcal{R}_j\left(\mathcal{R}_j^{\mathrm{T}}\mathcal{R}_i\boldsymbol{w}_i\right)^{\wedge}+\mathcal{R}_i^{\mathrm{T}}\mathcal{R}_j\boldsymbol{w}_j^{\wedge}=-\boldsymbol{w}_i^{\wedge}\mathcal{R}_i^{\mathrm{T}}\mathcal{R}_j+\mathcal{R}_i^{\mathrm{T}}\mathcal{R}_j\boldsymbol{w}_j^{\wedge}\\&=\mathcal{R}_i^{\mathrm{T}}\mathcal{R}_j\left[\boldsymbol{w}_j-\mathcal{R}_j^{\mathrm{T}}\mathcal{R}_i\boldsymbol{w}_i\right]^{\wedge}\end{aligned}\tag{7-2}$$

和

$$\begin{aligned}\dot{\theta}_{ij}&=\frac{-\mathrm{tr}\left(\dot{\mathcal{R}}_{ij}\right)}{2\sqrt{1-\left(\dfrac{\mathrm{tr}\left(\mathcal{R}_{ij}\right)-1}{2}\right)^2}}=\frac{-\mathrm{tr}\left(\dot{\mathcal{R}}_{ij}\right)}{2\sqrt{1-\cos^2\theta_{ij}}}\\&=\frac{-\mathrm{tr}\left(\boldsymbol{w}_{ij}^{\wedge}+\sin\theta_{ij}\boldsymbol{u}_{ij}^{\wedge}\boldsymbol{w}_{ij}^{\wedge}+\left(1-\cos\theta_{ij}\right)\left(\boldsymbol{u}_{ij}^{\wedge}\right)^2\boldsymbol{w}_{ij}^{\wedge}\right)}{2\sin\theta_{ij}}=\frac{2\boldsymbol{u}_{ij}^{\mathrm{T}}\boldsymbol{w}_{ij}\sin\theta_{ij}}{2\sin\theta_{ij}}=\boldsymbol{u}_{ij}^{\mathrm{T}}\boldsymbol{w}_{ij}\end{aligned}\tag{7-3}$$

这里，代入了罗德里格斯公式(2-36)去获得第三个等式并且使用性质$\mathrm{tr}\left(\boldsymbol{u}^{\wedge}\boldsymbol{w}^{\wedge}\right)=-2\boldsymbol{u}^{\mathrm{T}}\boldsymbol{w}$和$\mathrm{tr}\left(\left(\boldsymbol{u}^{\wedge}\right)^2\boldsymbol{w}^{\wedge}\right)=0$去获得最后一个等式。

那么，取$\boldsymbol{u}_{ij}$的导数，有

$$\begin{aligned}\dot{\boldsymbol{u}}_{ij}&=-\frac{\cos\theta_{ij}}{2\sin^2\theta_{ij}}\dot{\theta}_{ij}\left(\mathcal{R}_{ij}-\mathcal{R}_{ij}^{\mathrm{T}}\right)^{\wedge}+\frac{1}{2\sin\theta_{ij}}\left(\mathcal{R}_{ij}-\mathcal{R}_{ij}^{\mathrm{T}}\right)^{\wedge}\\&=\frac{-\cos\theta_{ij}}{2\sin^2\theta_{ij}}\boldsymbol{u}_{ij}^{\mathrm{T}}\boldsymbol{w}_{ij}2\sin\theta_{ij}\boldsymbol{u}_{ij}+\frac{1}{2\sin\theta_{ij}}\\&\quad\times\left[2+\sin\theta_{ij}\boldsymbol{u}_{ij}\boldsymbol{u}_{ij}^{\wedge}-\left(1-\cos\theta_{ij}\right)\left(1+\boldsymbol{u}_{ij}\boldsymbol{u}_{ij}^{\mathrm{T}}\right)\right]\boldsymbol{w}_{ij}\\&=\frac{-\cos\theta_{ij}}{\sin\theta_{ij}}\boldsymbol{u}_{ij}\boldsymbol{u}_{ij}^{\mathrm{T}}\boldsymbol{w}_{ij}+\frac{1}{2\sin\theta_{ij}}\\&\quad\times\left[2\cos\theta_{ij}\boldsymbol{I}_3+\sin\theta_{ij}\boldsymbol{u}_{ij}^{\wedge}-\left(1-\cos\theta_{ij}\right)\boldsymbol{u}_{ij}^2\right]\boldsymbol{w}_{ij}\end{aligned}$$

$$
\begin{aligned}
&=\left(-\frac{1}{2}\boldsymbol{u}_{ij}^{\wedge}-\frac{1+\cos\theta_{ij}}{2\sin\theta_{ij}}\boldsymbol{u}_{ij}^{\wedge 2}\right)\boldsymbol{w}_{ij}\\
&=\left(-\frac{1}{2}\boldsymbol{u}_{ij}^{\wedge}-\frac{1}{2}\cot\frac{\theta_{ij}}{2}\boldsymbol{u}_{ij}^{\wedge 2}\right)\boldsymbol{w}_{ij}
\end{aligned}
\tag{7-4}
$$

因此，有

$$
\begin{aligned}
\dot{\boldsymbol{x}}_{ij}&=\dot{\theta}_{ij}\boldsymbol{x}_{ij}+\theta_{ij}\dot{\boldsymbol{x}}_{ij}\\
&=\boldsymbol{u}_{ij}^{\mathrm{T}}\boldsymbol{u}_{ij}\boldsymbol{w}_{ij}+\theta_{ij}\left(\frac{1}{2}\boldsymbol{u}_{ij}^{\wedge}-\frac{1}{2}\cot\frac{\theta_{ij}}{2}\boldsymbol{u}_{ij}\right)\boldsymbol{w}_{ij}\\
&=\left[\boldsymbol{I}_3+\frac{1}{2}\theta_{ij}\boldsymbol{u}_{ij}^{\wedge}+\left(1-\frac{\theta_{ij}}{2}\cot\frac{\theta_{ij}}{2}\right)\left(\boldsymbol{u}_{ij}^{\wedge}\right)^2\right]\boldsymbol{w}_{ij}
\end{aligned}
\tag{7-5}
$$

证毕。 ■

引理 7.3[209]

给定一个函数$f:[0,+\infty)\to\mathbb{R}$在$[t_k,t_{k+1}),k\in\mathbb{N}^+$中是连续的，且在$(t_k,t_{k+1}),k\in\mathbb{N}^+$连续可微，如果$f(t)$在$[0,+\infty)$是黎曼可积的，且在$(t_k,t_{k+1}),k\in\mathbb{N}^+$一致有界，则随着$t\to+\infty$，$f(t)\to 0$。

7.2.2 问题描述 1

定义一致性集合$\mathcal{A}\subseteq\mathbb{SO}(3)^N$即

$$
\mathcal{A}=\left\{\mathcal{R}=\left\{\mathcal{R}_1,\cdots,\mathcal{R}_N\right\}:\mathcal{R}_i=\mathcal{R}_j,\ \forall i,j=1,\cdots,N\right\}
\tag{7-6}
$$

姿态一致性指的是在一个多刚体系统中，所有的刚体姿态集合$\mathcal{R}\in\mathbb{SO}(3)^N$趋于集合$\mathcal{A}$。下面给出几乎全局姿态一致性和局部姿态一致性的定义。

定义 7.1[77]

考虑一个包含N个刚体的多智能体系统。假定每个刚体的初始姿态$\mathcal{R}_i(0)$位于一个正不变集$\mathcal{B}_\pi(\boldsymbol{I}_3)$，如果$\mathcal{R}(t)\to\mathcal{A}$，随着$t\to\infty$，对于$i=1,\cdots,N,t\geqslant 0$，则称实现了几乎全局姿态一致性。

定义 7.2[77]

考虑一个包含N个刚体的多智能体系统。假定每个刚体的初始姿态$\mathcal{R}_i(0)$位于一个正不变集$\mathcal{B}_{\frac{\pi}{2}}(\bar{\mathcal{R}})$，如果$\mathcal{R}(t)\to\mathcal{A}$，随着$t\to\infty$，对于$i=1,\cdots,N,t\geqslant 0$，则称实现了局部姿态一致性。

对于局部的姿态表征方法$\boldsymbol{y}_i=f(\mathcal{R}_i)$，如果姿态$\mathcal{R}_i$位于开球$\mathcal{B}_r(\boldsymbol{I}_3)$中，其中$r\leqslant\pi$，则在$\mathbb{SO}(3)$上基于黎曼测度的姿态一致性等价于在$\mathbb{R}^3$上基于欧几里德测度的姿态一致性。更加清楚地，定义一个集合$\mathcal{A}_y\in\mathbb{R}^{3N}$，即

$$\mathcal{A}_y=\left\{\boldsymbol{y}=\{\boldsymbol{y}_1,\cdots,\boldsymbol{y}_N\}:\boldsymbol{y}_i=\boldsymbol{y}_j,\ i,j=1,\cdots,N\right\} \tag{7-7}$$

那么，如果$\boldsymbol{y}(t)\in\mathcal{B}_{r'}^{'}(\boldsymbol{0}_3)^N$，随着$t\to\infty$，我们有$\boldsymbol{y}(t)\to\mathcal{A}_y$，等价于随着$t\to\infty$，$\mathcal{R}(t)\to\mathcal{A}$。

文献 [77] 中的姿态一致性问题是对每个智能体找到一个连续反馈控制，使得随着$t\to\infty$，$\boldsymbol{y}(t)\to\mathcal{A}_y$。然而，每个智能体需要连续获得邻居的状态信息会造成通信资源的浪费。为了解决这个问题，在本章内容中，我们引入事件驱动通信机制，即对每个智能体设计一个分布式协议$\boldsymbol{w}_i(t)$和事件驱动条件，协议中仅仅使用到邻居智能体的采样信息，使得满足上述定义中的姿态一致性得以完成。在绝对姿态和相对测量下的反馈控制协议的形式如下：

$$\boldsymbol{w}_i\left(t,\boldsymbol{\psi}_i\left(t,\mathcal{R}(t)\right)\right)=\sum_{j\in\mathcal{N}_i}\boldsymbol{\psi}_i\left(\mathcal{R}_i\left(t_k^i\right),\mathcal{R}_j\left(t_{k^{\mathrm{T}}}^j\right)\right) \tag{7-8}$$

和

$$\boldsymbol{w}_i\left(t,\boldsymbol{\psi}_i\left(t,\mathcal{R}(t)\right)\right)=\sum_{j\in\mathcal{N}_i}\boldsymbol{\psi}_i\left(\mathcal{R}_i^{\mathrm{T}}\left(t_k^i\right)\mathcal{R}_j\left(t_k^i\right)\right) \tag{7-9}$$

其中，$\boldsymbol{\psi}_i\left(t,\mathcal{R}(t)\right):\mathbb{SO}(3)^N\to\mathbb{R}^3$表示测量函数。

在控制协议 (7-8) 和 (7-9) 中，分别考虑了基于绝对姿态测量和相对姿态测量的事件驱动姿态一致性问题。为了确定智能体的触发时刻t_k^i，我们需要设计事件驱动条件：

$$t_{k+1}^i=\min_{t\geqslant r_k^i}\left\{t\in\mathbb{R}:\boldsymbol{H}\left(\boldsymbol{e}_i(t),\boldsymbol{\psi}_i\left(t,\mathcal{R}(t)\right)\right)>0\right\} \tag{7-10}$$

其中，$\boldsymbol{H}\left(\boldsymbol{e}_i(t),\boldsymbol{\psi}_i\left(t,\mathcal{R}(t)\right)\right)$是门限函数且$\boldsymbol{e}_i(t)$是所需要设计的测量误差。基于事件驱动的姿态一致性框图如图7-1所示。

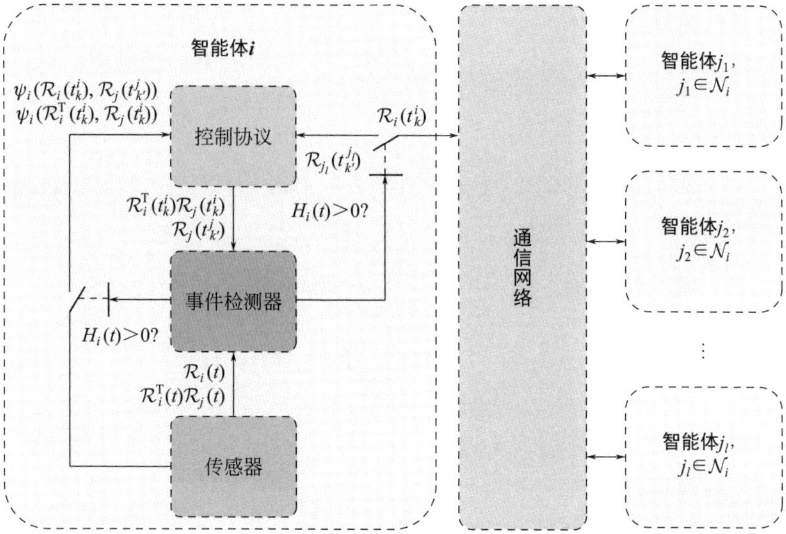

图 7-1　基于事件驱动的姿态一致性

7.2.3　问题描述 2

本章内容考虑了多刚体系统 (2-37) 的姿态一致性问题，其定义在下面给出。

定义 7.3

给定一个多刚体系统 (2-37)，姿态一致性能实现，当且仅当随着 $t \to +\infty$，$\mathcal{R}_i = \mathcal{R}_j, i, j \in \mathcal{V}$，等价于$\boldsymbol{x}_i = \boldsymbol{x}_j$，且$\boldsymbol{w}_i = 0$。

7.3
绝对和相对姿态测量下的姿态一致性

7.3.1　基于绝对姿态的事件驱动姿态一致性

在绝对姿态测量的情况中，我们运用球心映射的方法来解决在事件驱动条件下的几乎全局姿态一致性问题。在介绍主定理之前，我们首先在下面的引理中引入从$\mathbb{SO}(3)$到欧几里德投影平面的映射。

引理 7.4[77,210]

球心投影可定义为下面的微分同胚映射$f:\mathcal{B}_r\left(\boldsymbol{I}_3\right)\rightarrow\mathcal{B}_{r'}^{'}\left(\mathbf{0}_3\right)\in\mathbb{R}^3$其中$r=\pi$和$r'=+\infty$。

进一步，基于此映射的姿态运动学方程为

$$\dot{\boldsymbol{y}}_i\left(t\right)=\frac{1}{2}\boldsymbol{h}_i\left(\boldsymbol{y}_i\left(t\right)\right)\boldsymbol{w}_i\left(t\right) \tag{7-11}$$

其中，$\boldsymbol{h}_i\left(\boldsymbol{y}_i\left(t\right)\right)=\left(\boldsymbol{I}_3+\boldsymbol{y}_i^{\wedge}\left(t\right)+\boldsymbol{y}_i\left(t\right)\boldsymbol{y}_i^{\mathrm{T}}\left(t\right)\right)$；$\boldsymbol{w}_i\left(t\right)$是在机体坐标系下的角速度向量。

证明：

该结果可基于文献[77,210]得到。首先，运用四元数，姿态可表示为

$$\boldsymbol{q}_i=\left[\cos\frac{\Theta_i}{2},\sin\frac{\Theta_i}{2}\boldsymbol{u}_i^{\mathrm{T}}\right]^{\mathrm{T}}\in\mathbb{S}^3 \tag{7-12}$$

之后，基于球心投影将$\mathbb{S}^3$上的点投影到位于球$\boldsymbol{p}=\begin{bmatrix}-1 & 0 & 0 & 0\end{bmatrix}^{\mathrm{T}}\in\mathbb{R}^4$上南极点的高维切平面上。基于此，可得投影为$\boldsymbol{y}_i=\tan\frac{\Theta_i}{2}\boldsymbol{u}_i\in\mathbb{R}^3$。注意到当$\Theta_i=\pi$，投影$\boldsymbol{y}_i$将会趋于无穷远，这意味着球心投影仅仅覆盖在开球$\mathcal{B}_\pi\left(\boldsymbol{I}_3\right)$上。由于我们已知变换$\mathcal{R}_i\rightarrow\boldsymbol{y}_i$是一个测地映射，这意味着$f:\mathcal{B}_r\left(\boldsymbol{I}_3\right)\rightarrow\mathcal{B}_{r'}\left(\mathbf{0}\right)$，其中$r=\pi$和$r'=+\infty$是一个微分同胚映射[77]。最后，姿态的运动学方程(7-11)可从四元数的运动学方程[210]中推导得出。证毕。 ■

在后文中，使用微分同胚映射$\mathcal{R}_i\rightarrow\boldsymbol{y}_i$，我们将认同姿态$\mathcal{R}_i\in\mathcal{B}_r\left(\boldsymbol{I}_3\right)$和投影$\boldsymbol{y}_i=\tan\frac{\Theta_i}{2}\boldsymbol{u}_i$。利用采样的绝对姿态信息所设计的一致性控制协议(7-8) 如下：

$$\boldsymbol{w}_i\left(t\right)=\sum_{j=1}^{N}a_{ij}\left(\boldsymbol{y}_j\left(t_{k'}^j\right)-\boldsymbol{y}_i\left(t_k^i\right)\right),t\in\left[t_k^i,t_{k+1}^i\right) \tag{7-13}$$

其中，$\boldsymbol{y}_i\left(t_k^i\right)$和$\boldsymbol{y}_j\left(t_{k'}^j\right)$分别是刚体$i$和$j$在各自触发时刻的采样姿态信息。这里需要指出的是，绝对姿态信息可由惯性传感器获得。考虑到刚体i的

测量误差，定义为$\boldsymbol{e}_i(t)=\boldsymbol{y}_i(t_k^i)-\boldsymbol{y}_i(t), t\in[t_k^i, t_{k+1}^i)$。事件驱动时刻可由下面的公式确定：

$$t_{k+1}^i=\min_{t\geqslant t_k^i}\left\{t\in\mathbb{R}:\|\boldsymbol{e}_i(t)\|>\alpha_i\|\boldsymbol{z}_i(t)\|, i=1,\cdots,N\right\} \tag{7-14}$$

其中，$\boldsymbol{z}_i(t)=\sum_{j=1}^N a_{ij}\left(\boldsymbol{y}_j(t_{k'}^j)-\boldsymbol{y}_i(t_k^i)\right); t\in[t_k^i, t_{k+1}^i); \alpha_i$是需要在后面证明中确定的正常数。

注释 7.1 事件驱动条件 (7-14) 的优势有两个方面，其中一个是当判断事件驱动条件 (7-14) 时，第 i 个刚体不需要连续测量邻居的姿态信息$\boldsymbol{y}_j(t), j\in\mathcal{N}_i$，这意味着在触发判断中智能体 i 不需要连续地与邻居进行通信。

下面，我们给出基于事件驱动姿态一致性的第一个主要结果。

定理 7.1

考虑一组 N 个刚体组成的多智能体系统，其初始的姿态位于一个不变开球$\mathcal{B}_\pi(\boldsymbol{I}_3)^N$中。当使用基于事件驱动的姿态一致性协议 (7-13) 和事件驱动条件 (7-14) 满足$\alpha_i<\dfrac{1}{2\bar{a}|\mathcal{N}_i|}$，则所有的刚体姿态会收敛到一致性集合 $\mathcal{A}$，当通信拓扑是强连通时，奇诺现象不会发生。

证明：

考虑李雅普诺夫函数

$$V(t)=\sum_{i=1}^N p_i\|\boldsymbol{y}_i(t)\|^2 \tag{7-15}$$

通过对其求导，代入式(7-11)，可得

$$\dot{V}(\boldsymbol{y}(t))=\sum_{i=1}^N p_i\left(\|\boldsymbol{y}_i(t)\|^2+1\right)\left(\boldsymbol{y}_i^{\mathrm{T}}(t)\boldsymbol{z}_i(t)\right) \tag{7-16}$$

其中，我们使用$\boldsymbol{y}_i^{\mathrm{T}}(t)\boldsymbol{y}_i^{\wedge}(t)=0$得到上述等式。根据$\boldsymbol{e}_i(t)=\boldsymbol{y}_i(t_k^i)-\boldsymbol{y}_i(t)$和$\boldsymbol{p}^{\mathrm{T}}\mathcal{L}=0$，可得

$$\begin{aligned}\dot{V}\left(\boldsymbol{y}(t)\right)&=\sum_{i=1}^{N}p_i\left(\|\boldsymbol{y}_i(t)\|^2+1\right)\left(\boldsymbol{y}_i^{\mathrm{T}}\left(t_k^i\right)\boldsymbol{z}_i(t)\right)\\&\quad-\sum_{i=1}^{N}p_i\left(\|\boldsymbol{y}_i(t)\|^2+1\right)\left(\boldsymbol{e}_i^{\mathrm{T}}(t)\boldsymbol{z}_i(t)\right)\\&\leqslant-\frac{1}{2}\sum_{i=1}^{N}p_i\hat{\boldsymbol{\varphi}}_i(t)-\sum_{i=1}^{N}p_i\boldsymbol{e}_i^{\mathrm{T}}(t)\boldsymbol{z}_i(t)\\&\leqslant-\frac{1}{2}\sum_{i=1}^{N}p_i\sum_{j=1}^{N}a_{ij}\|\boldsymbol{y}_j\left(t_{k'}^j\right)-\boldsymbol{y}_i\left(t_k^i\right)\|^2\\&\quad+\bar{a}\sum_{i=1}^{N}p_i\alpha_i\left|\mathcal{N}_i\right|\sum_{j=1}^{N}a_{ij}\|\boldsymbol{y}_j\left(t_{k'}^j\right)-\boldsymbol{y}_i\left(t_k^i\right)\|^2\\&\leqslant-\min\left\{c_i\right\}\sum_{i=1}^{N}p_i\sum_{j=1}^{N}\|\boldsymbol{y}_j\left(t_{k'}^j\right)-\boldsymbol{y}_i\left(t_k^i\right)\|^2\end{aligned}\tag{7-17}$$

其中，$\hat{\boldsymbol{\varphi}}_i(t)=\sum_{j=1}^{N}a_{ij}\|\boldsymbol{y}_j\left(t_{k'}^j\right)-\boldsymbol{y}_i\left(t_k^i\right)\|^2;t\in\left[t_k^i,t_{k+1}^i\right)$和$c_i=\dfrac{1}{2}-\bar{a}\alpha_i\left|\mathcal{N}_i\right|$。其中我们使用不等式

$$\|\boldsymbol{z}_i(t)\|^2\leqslant\bar{a}\left|\mathcal{N}_i\right|\sum_{j=1}^{N}a_{ij}\|\boldsymbol{y}_j\left(t_{k^{\mathrm{T}}}^j\right)-\boldsymbol{y}_i\left(t_k^i\right)\|^2\tag{7-18}$$

和事件驱动条件(7-14)获得第二个不等式。

令$\boldsymbol{\varphi}_i(t)=\sum_{j=1}^{N}a_{ij}\|\boldsymbol{y}_j(t)-\boldsymbol{y}_i(t)\|^2$和$\hat{\boldsymbol{y}}(t)=\left[\boldsymbol{y}_1^{\mathrm{T}}\left(t_{k_1}^1\right)\boldsymbol{y}_2^{\mathrm{T}}\left(t_{k_2}^2\right)\cdots\boldsymbol{y}_N^{\mathrm{T}}\left(t_{k_N}^N\right)\right]^{\mathrm{T}}$，$k_i\in\mathbb{N}^+$，其中$t_{k_i}^i$为刚体 i 的触发时刻，则有

$$\begin{aligned}\sum_{i=1}^{N}p_i\boldsymbol{\varphi}_i(t)&=2\boldsymbol{y}^{\mathrm{T}}(t)\left(\hat{\boldsymbol{p}}\mathcal{L}\otimes\boldsymbol{I}_3\right)\boldsymbol{y}(t)\\&=2\left(\hat{\boldsymbol{y}}(t)-\boldsymbol{e}(t)\right)^{\mathrm{T}}\left(\hat{\boldsymbol{p}}\mathcal{L}\otimes\boldsymbol{I}_3\right)\left(\hat{\boldsymbol{y}}(t)-\boldsymbol{e}(t)\right)\\&=2\hat{\boldsymbol{y}}^{\mathrm{T}}(t)\left(\hat{\boldsymbol{p}}\mathcal{L}\otimes\boldsymbol{I}_3\right)\hat{\boldsymbol{y}}(t)+2\boldsymbol{e}^{\mathrm{T}}(t)\left(\hat{\boldsymbol{p}}\mathcal{L}\otimes\boldsymbol{I}_3\right)\boldsymbol{e}(t)\\&\quad-2\boldsymbol{e}^{\mathrm{T}}(t)\left(\boldsymbol{\Phi}\otimes\boldsymbol{I}_3\right)\hat{\boldsymbol{y}}(t)\\&\leqslant2\hat{\boldsymbol{y}}^{\mathrm{T}}(t)\left(\hat{\boldsymbol{p}}\mathcal{L}\otimes\boldsymbol{I}_3\right)\hat{\boldsymbol{y}}(t)+2\boldsymbol{e}^{\mathrm{T}}(t)\left(\hat{\boldsymbol{p}}\mathcal{L}\otimes\boldsymbol{I}_3\right)\boldsymbol{e}(t)\\&\quad+\left|\boldsymbol{e}^{\mathrm{T}}(t)\left(\boldsymbol{\Phi}\otimes\boldsymbol{I}_3\right)\boldsymbol{e}(t)\right|+\left|\hat{\boldsymbol{y}}^{\mathrm{T}}(t)\left(\boldsymbol{\Phi}\otimes\boldsymbol{I}_3\right)\hat{\boldsymbol{y}}(t)\right|\\&=4\hat{\boldsymbol{y}}^{\mathrm{T}}(t)\left(\hat{\boldsymbol{p}}\mathcal{L}\otimes\boldsymbol{I}_3\right)\hat{\boldsymbol{y}}(t)+4\boldsymbol{e}^{\mathrm{T}}(t)\left(\hat{\boldsymbol{p}}\mathcal{L}\otimes\boldsymbol{I}_3\right)\boldsymbol{e}(t)\\&\leqslant2\sum_{i=1}^{N}p_i\hat{\boldsymbol{\varphi}}_i(t)+\frac{4\hat{\lambda}}{\min\left\{p_i\right\}}\sum_{i=1}^{N}\alpha_i^2p_i\|\boldsymbol{z}_i(t)\|^2\end{aligned}$$

$$\leqslant \left(2+\frac{4\hat{\lambda}\bar{a}N\max\{\alpha_i\}^2}{\min\{p_i\}}\right)\sum_{i=1}^{N}p_i\hat{\boldsymbol{\varphi}}_i(t) \tag{7-19}$$

其中，$\Phi=\hat{\boldsymbol{p}}\mathcal{L}+\mathcal{L}^{\mathrm{T}}\hat{\boldsymbol{p}}$是对称负定矩阵，$\hat{\lambda}$是矩阵$\hat{\boldsymbol{p}}\mathcal{L}$的最大特征值。第一个不等式由$\left|\boldsymbol{a}^{\mathrm{T}}\Phi\boldsymbol{b}\right|\leqslant\left|\boldsymbol{a}^{\mathrm{T}}\Phi\boldsymbol{a}\right|+\left|\boldsymbol{b}^{\mathrm{T}}\Phi\boldsymbol{b}\right|$可得，其中$\boldsymbol{a},\boldsymbol{b}\in\mathbb{R}^n$。

根据式 (7-17) 和式 (7-19)，有

$$\dot{V}(\boldsymbol{y}(t))\leqslant-\frac{\min\{c_i\}}{2+\dfrac{4\hat{\lambda}\bar{a}N\max\{\alpha_i\}^2}{\min\{p_i\}}}\sum_{i=1}^{N}p_i\boldsymbol{\varphi}_i(t) \tag{7-20}$$

进一步，基于不变集原理，我们证明$\boldsymbol{y}(t)$将会收敛到$\mathcal{A}$。由于$\dot{V}(\boldsymbol{y}(t))\leqslant 0$，可得$V(\boldsymbol{y}(t))\leqslant V(\boldsymbol{y}(0))$，对$t>0$。因此，$\boldsymbol{y}(t)$是有界的且$\boldsymbol{y}(t)\in\mathcal{B}_{r'}'(\boldsymbol{0}_3)^N$，对$t>0$。由此可得任何初值在集合$\mathcal{B}_{\pi}(\boldsymbol{I}_3)^N$的解将会永远待在这个集合内，即$\mathcal{B}_{\pi}(\boldsymbol{I}_3)^N$是不变集。

在强连通拓扑的条件下，我们有$\left\{\boldsymbol{y}(t)\in\mathbb{R}^{3N}:\dot{V}(\boldsymbol{y}(t))=0,\forall t>0\right\}=\left\{\boldsymbol{y}\in\mathbb{R}^{3N}:\boldsymbol{y}_1=\boldsymbol{y}_2=\cdots=\boldsymbol{y}_N\right\}$。因此，根据不变集原理，我们知道随着$t\to\infty$，$\boldsymbol{y}(t)$会收敛到集合$\mathcal{A}_y=\left\{\boldsymbol{y}\in\mathbb{R}^{3N}:\boldsymbol{y}_1=\boldsymbol{y}_2=\cdots=\boldsymbol{y}_N\right\}$。

接下来，我们将证明奇诺现象不会发生。考虑$\dfrac{\|\boldsymbol{e}_i(t)\|}{\|z_i(t)\|}$的导数在$t\in\left[t_k^i,t_{k+1}^i\right)$，有

$$\frac{\mathrm{d}}{\mathrm{d}t}\frac{\|\boldsymbol{e}_i(t)\|}{\|\boldsymbol{z}_i(t)\|}=\frac{\boldsymbol{e}_i^{\mathrm{T}}(t)\dot{\boldsymbol{e}}_i(t)}{\|\boldsymbol{e}_i(t)\|\|\boldsymbol{z}_i(t)\|}-\frac{\boldsymbol{z}_i^{\mathrm{T}}(t)\dot{\boldsymbol{z}}_i(t)}{\|\boldsymbol{z}_i(t)\|^3} \tag{7-21}$$

首先，我们考虑没有刚体 i 的邻居在$t\in\left[t_k^i,t_{k+1}^i\right)$触发的情况。在这种情况下，$\dot{\boldsymbol{z}}_i(t)=0$，有

$$\begin{aligned}\frac{\mathrm{d}}{\mathrm{d}t}\frac{\|\boldsymbol{e}_i(t)\|}{\|\boldsymbol{z}_i(t)\|}&\leqslant\frac{\|\dot{\boldsymbol{y}}_i(t)\|}{\|\boldsymbol{z}_i(t)\|}\\&\leqslant\frac{\|\boldsymbol{h}(\boldsymbol{y}_i)\|\|\boldsymbol{z}_i(t)\|}{2\|\boldsymbol{z}_i(t)\|}\\&=\frac{\|\boldsymbol{h}(\boldsymbol{y}_i)\|}{2}\end{aligned} \tag{7-22}$$

由于已经知道$\boldsymbol{y}_i(t)\in\mathcal{B}_{r'}(\boldsymbol{0})$对于$t>0$，则必然存在一个上界满足$\|\boldsymbol{h}(\boldsymbol{y}_i(t))\|\leqslant M_i$，其中$M_i$是一个正常数。根据上述不等式，可得出上界：

$$\frac{\|\boldsymbol{e}_i(t)\|}{\|\boldsymbol{z}_i(t)\|} \leqslant \int_{t_k^i}^{t_{k+1}^i} \frac{M_i}{2}\mathrm{d}t = \frac{M_i}{2}\left(t_{k+1}^i - t_k^i\right) \tag{7-23}$$

由于在不等式$\frac{\|\boldsymbol{e}_i(t)\|}{\|\boldsymbol{z}_i(t)\|} > \alpha_i$被满足之前，事件不会触发，则触发事件的间隔上界为$\tau = t_{k+1}^i - t_k^i \geqslant \frac{2\alpha_i}{M_i}$。对于存在一个刚体$j, j \in \mathcal{N}_i$在$t_{k'}^j \in \left[t_k^i, t_{k+1}^i\right)$时刻触发的情况，我们有$\dot{\boldsymbol{z}}_i(t) = 0$，$t \in \left[t_k^i, t_{k'}^j\right)$。因此，存在一个下界满足$\tau^* = t_{k+1}^i - t_{k'}^j \geqslant \frac{2\alpha_i}{M_i}$。结合上面两个情况，可求得触发时间间隔的下界为$\tau \geqslant \tau^* = \frac{2\alpha_i}{M_i}$。基于此，奇诺现象不会发生。

注释 7.2

为了实现几乎全局姿态一致性，运用了一种球心投影的方法将姿态几乎全局投影到欧氏空间。事实上，投影的姿态表征又称罗德里格斯参数，是一种局部姿态表征方法。但是，局部姿态表征和$\mathbb{SO}(3)$的关系即微分同胚映射并没有在现有的事件驱动姿态一致性中涉及 [167]，因此现有工作并没有解决几乎全局姿态一致性问题。

注释 7.3

球心投影的方法有一个重要的几何性质是球面上的曲线对应投影面上的直线。基于这个性质，我们可在欧几里德平面上讨论姿态一致性协议而不需要考虑$\mathbb{SO}(3)$上的曲线特征。注意到球心投影的方法已经在一些文献被应用。在文献 [211-212] 中，球心投影分别被用来研究绝对信息下的球面一致性问题 [212] 和相对信息下的姿态一致性问题 [211]。然而，文献 [211-212] 考虑的是连续时间的一致性协议。此外，在文献 [211] 中，一种升维的方法被提出，用来将球面上的运动学转换到欧几里德空间。这类方法适用于在协议是连续时间的情况。事件驱动采样机制本质上给姿态运动学带来了混杂特征，因此上述的方法难以用于解决事件驱动机制下几乎全局姿态一致性问题。另一方面，为了实现一致性，文献 [212] 需要知道通信拓扑的全局信息，使得一致性协议不是完全分布式的。

注释 7.4　值得一提的是大多数基于事件驱动的一致性算法是基于正定李雅普诺夫函数设计的，如$V=\frac{1}{2}\boldsymbol{x}^{\mathrm{T}}\mathcal{L}\boldsymbol{x}$或者$V=\frac{1}{2}\sum_{i=1}^{N}(\boldsymbol{x}-\bar{\boldsymbol{x}})^2$[208]。然而，基于上述李雅普诺夫函数导出的事件驱动条件需要利用全局的通信拓扑图信息，如拉普拉斯矩阵的特征值。因此，受到文献 [172] 的李雅普诺夫函数设计的启发，以及拉塞尔不变集原理，我们设计了与拓扑图信息无关的事件驱动条件。

7.3.2　基于相对姿态的事件驱动姿态一致性

到现在，定理 7.1 给出了运用绝对姿态信息的事件驱动姿态一致性协议和事件驱动条件。然而，考虑到$\boldsymbol{y}_{ij}\neq\boldsymbol{y}_j-\boldsymbol{y}_i$一般并不成立，控制协议 (7-13) 不能被用到相对姿态的测量情况上来。因此，在这个小节当中，我们想提出一种基于相对姿态信息的事件驱动姿态一致性协议。为了使得叙述更方便，我们首先引入一些记号。在$\mathbb{SO}(3)$上的测地距离的梯度向量定义为

$$\frac{1}{2}\mathrm{grad}_{\mathcal{R}_i}d^2\left(\mathcal{R}_i,\mathcal{R}_j\right)=-\mathcal{R}_i\log\left(\mathcal{R}_i^{\mathrm{T}}\mathcal{R}_j\right),\quad\forall i,j=1,\cdots,N\tag{7-24}$$

其中，测地距离$d\left(\mathcal{R}_i,\mathcal{R}_j\right)$满足$d\left(\mathcal{R}_i,\mathcal{R}_j\right)=d\left(\mathcal{R}_j,\mathcal{R}_i\right)$且$\log\left(\mathcal{R}_i^{\mathrm{T}}\mathcal{R}_j\right)$: $\mathbb{SO}(3)\to\mathfrak{so}(3)$是对数映射。一个多刚体系统的一致性误差函数定义如下

$$\phi=\sum_{i=1}^{N}\phi_i=\frac{1}{2}\sum_{i=1}^{N}\sum_{j=1}^{N}a_{ij}d^2\left(\mathcal{R}_i,\mathcal{R}_j\right)\tag{7-25}$$

这个函数的梯度向量为

$$\mathrm{grad}_{\mathcal{R}_i}\phi=\sum_{i=1}^{N}\mathcal{R}_i\nabla_{\mathcal{R}_i}\phi_i \tag{7-26}$$

其中，$\nabla_{\mathcal{R}_i}\phi_i:=-\sum_{j=1}^{N}a_{ij}\log\left(\mathcal{R}_i^{\mathrm{T}}\mathcal{R}_j\right)$。

下面，设计如下的事件驱动一致性协议：

$$\begin{aligned}\boldsymbol{w}_i^{\wedge}(t)&=-\nabla_{\mathcal{R}_i}\phi_i\left(t_k^i\right)\\&=\sum_{j=1}^{N}a_{ij}\log\left(\mathcal{R}_i^{\mathrm{T}}\mathcal{R}_j\right)\left(t_k^i\right),\ t\in\left[t_k^i,t_{k+1}^i\right)\end{aligned} \tag{7-27}$$

接下来的问题就是仅仅使用相对姿态信息要怎么设计事件驱动条件。一种可行的策略是在$\mathbb{SO}(3)$的切空间上考虑事件驱动采样误差：

$$\begin{aligned}\boldsymbol{E}_i(t)&=\nabla_{\mathcal{R}_i}\phi_i(t)-\nabla_{\mathcal{R}_i}\phi_i\left(t_k^i\right)\\&=\sum_{j=1}^{N}a_{ij}\log\left(\mathcal{R}_i^{\mathrm{T}}\mathcal{R}_j\right)\left(t_k^i\right)-\sum_{j=1}^{N}a_{ij}\log\left(\mathcal{R}_i^{\mathrm{T}}\mathcal{R}_j\right)(t),\ t\in\left[t_k^i,t_{k+1}^i\right)\end{aligned} \tag{7-28}$$

注意到式(7-28)实际表示的是梯度向量的采样误差，并且仅仅依赖于相对姿态信息。可以发现，测量误差$\boldsymbol{E}_i(t)$可以视为偏离一致性误差函数的梯度方向的度量。对式(7-28)的设计实际上源于一个事实，即梯度向量$\nabla_{\mathcal{R}_i}\phi_i(t)$永远指向邻居智能体形成的凸包几何的内侧[213]。如果这种偏离过大，那么事件驱动一致性协议(7-27)的梯度向量就需要更新。

此外，为了排除奇诺现象，我们在事件驱动的条件中引入了动态事件驱动的条件[180]。定义一个内部的动态变量η_i：

$$\dot{\eta}_i(t)=-\lambda_i\eta_i(t)+\beta_i\left(\alpha_i\|\nabla_{\mathcal{R}_i}^{\vee}\phi_i(t)\|^2-\|\boldsymbol{E}_i^{\vee}(t)\|^2\right) \tag{7-29}$$

其中，$\eta_i(0)>0,\lambda_i>0,\beta_i\in\left[0,\dfrac{1}{2}\right]$且$\alpha_i\in[0,1]$是非负参数。则事件驱动时刻可由下面的动态事件驱动条件确定：

$$t_{k+1}^i=\min_{t\geqslant t_k^i}\left\{t\in\mathbb{R}:\eta_i(t)+\theta_i\left(\alpha_i\|\nabla_{\mathcal{R}_i}^{\vee}\phi_i(t)\|^2-\|\boldsymbol{E}_i^{\vee}(t)\|^2\right)\leqslant 0,t\in\left[t_k^i,t_{k+1}^i\right)\right\} \tag{7-30}$$

其中，$\theta_i\in\left[\dfrac{1-\beta_i}{\lambda_i},+\infty\right)$。

引理 7.5

考虑一个内部动态变量$\eta_i(t)$由式 (7-29) 给出。可得在事件驱动条件 (7-30) 下，有$\eta_i(t) \geqslant 0$。

证明：

根据式 (7-30)，可以保证对$t > 0$，

$$\left(\alpha_i \| \nabla^{\vee}_{\mathcal{R}_i} \phi_i(t) \|^2 - \| \boldsymbol{E}_i^{\vee}(t) \|^2\right) \geqslant -\frac{\beta_i}{\theta_i}\eta_i(t) \tag{7-31}$$

由式 (7-29)，可得

$$\dot{\eta}_i(t) \geqslant -\lambda_i \eta_i(t) - \frac{\beta_i}{\theta_i}\eta_i(t), \quad \eta_i(0) > 0 \tag{7-32}$$

令$\chi_i(t)$表示微分方程$\dot{\chi}_i(t) = -\lambda_i \chi_i(t) - \frac{\beta_i}{\theta_i}\chi_i(t), \chi_i(0) = \eta_i(0)$的解。基于比较引理[173]，我们有

$$\eta_i(t) \geqslant \chi_i(t) = \eta_i(0) \mathrm{e}^{-\left(\lambda_i + \frac{\beta_i}{\theta_i}\right)t} \tag{7-33}$$

因此，可得对$t > 0$，有$\eta_i(t) \geqslant 0$。证毕。 ■

基于协议 (7-27) 和事件驱动条件 (7-30)，有如下定理。

定理 7.2

考虑一组 N 个刚体组成的多智能体系统，其初始的姿态位于一个以$\bar{\mathcal{R}}$为球心的不变开球集合$\mathcal{B}_{\frac{\pi}{2}}(\bar{\mathcal{R}})$中。当使用基于事件驱动的姿态一致性协议 (7-27) 和事件驱动条件 (7-30) 时，$\mathcal{R}(t)$会收敛到一致性集合 $\mathcal{A}$ 且当通信拓扑是无向连通的，奇诺现象不会发生。

证明：

下面的证明主要分为两个部分。首先，需要证明$\mathcal{B}_{\frac{\pi}{2}}(\bar{\mathcal{R}})$是一个不变集，即$\mathcal{R}_i(t) \in \mathcal{B}_{\frac{\pi}{2}}(\bar{\mathcal{R}})$，$t > 0$。我们考虑李雅普诺夫函数

$$W(t) = \sum_{i=1}^{N} \phi_i(t) + \sum_{i=1}^{N} \eta_i(t) \tag{7-34}$$

求得其导数为

$$
\begin{aligned}
\dot{W}(t) &= 2\sum_{i=1}^{N}\left\langle \mathcal{R}_i(t)\nabla_{\mathcal{R}_i}\phi_i(t), \mathcal{R}_i(t)\boldsymbol{w}_i(t)\right\rangle + \sum_{i=1}^{N}\dot{\eta}_i(t) \\
&= 2\sum_{i=1}^{N}\Big[\left\langle \mathcal{R}_i(t)\nabla_{\mathcal{R}_i}\phi_i(t), -\mathcal{R}_i(t)\nabla_{\mathcal{R}_i}\phi_i(t)\right\rangle \\
&\quad + \left\langle \mathcal{R}_i(t)\nabla_{\mathcal{R}_i}\phi_i(t), -\mathcal{R}_i(t)E_i(t)\right\rangle\Big] + \sum_{i=1}^{N}\dot{\eta}_i(t) \\
&= 2\sum_{i=1}^{N}\left[\nabla_{\mathcal{R}_i}^{\vee}\phi_i(t)\right]^{\mathrm{T}}\left[-\nabla_{\mathcal{R}_i}^{\vee}\phi_i(t) - \boldsymbol{E}_i^{\vee}(t)\right] + \sum_{i=1}^{N}\dot{\eta}_i(t)
\end{aligned}
$$

其中，最后一个不等式根据式(2-28)得出。使用杨氏不等式，可得

$$
\begin{aligned}
\dot{W}(t) &\leqslant -\sum_{i=1}^{N}\|\nabla_{\mathcal{R}_i}^{\vee}\phi_i(t)\|^2 + \sum_{i=1}^{N}\|\boldsymbol{E}_i^{\vee}(t)\|^2 \\
&\quad -\sum_{i=1}^{N}\lambda_i\eta_i(t) + \sum_{i=1}^{N}\beta_i\left(\alpha_i\|\nabla_{\mathcal{R}_i}^{\vee}\phi_i(t)\|^2 - \|\boldsymbol{E}_i^{\vee}(t)\|^2\right) \\
&\leqslant -\sum_{i=1}^{N}(1-\alpha_i)\|\nabla_{\mathcal{R}_i}^{\vee}\phi_i(t)\|^2 - \sum_{i=1}^{N}\lambda_i\eta_i(t) \\
&\quad + \sum_{i=1}^{N}(\beta_i - 1)\left(\alpha_i\|\nabla_{\mathcal{R}_i}^{\vee}\phi_i(t)\|^2 - \|\boldsymbol{E}_i^{\vee}(t)\|^2\right) \\
&\leqslant -\sum_{i=1}^{N}(1-\alpha_i)\|\nabla_{\mathcal{R}_i}^{\vee}\phi_i(t)\|^2 - \eta_i(t)\left(\lambda_i + \frac{\beta_i - 1}{\theta_i}\right)
\end{aligned}
\tag{7-35}
$$

根据式(7-35)，有$\dfrac{\mathrm{d}W_i}{\mathrm{d}t}\leqslant 0$，意味着$W_i(t)$是非增的。因此，我们有

$$
W(t)\leqslant W(0)，\ \forall t>0 \tag{7-36}
$$

这也意味着$\sum_{i=1}^{N}\phi_i(t)<W(t)\leqslant W(0)$，则一致性误差$\frac{1}{2}\sum_{i=1}^{N}\sum_{j=1}^{N}a_{ij}d^2\left(\mathcal{R}_i(t),\mathcal{R}_j(t)\right)$的上界为$W(0)$。选择

$$
W(0) = \frac{1}{2}\sum_{i=1}^{N}\sum_{j=1}^{N}a_{ij}d^2\left(\mathcal{R}_i(0),\mathcal{R}_j(0)\right) + \sum_{i=1}^{N}\eta_i(0) \tag{7-37}
$$

满足$\mathcal{R}_i(0)\in\mathcal{B}_{\frac{\pi}{2}}(\bar{\mathcal{R}})$，那么可得$\mathcal{R}_i(t)$将永远属于$\mathcal{B}_{\frac{\pi}{2}}(\bar{\mathcal{R}}),\forall t>0$，这意味着$\mathcal{B}_{\frac{\pi}{2}}(\bar{\mathcal{R}})$是一个正不变集。接下来，我们证明$\mathcal{R}_i(t)$将收敛到一致集 $\mathcal{A}$。基于$\mathcal{B}_{\frac{\pi}{2}}(\bar{\mathcal{R}})$是一个$\mathbb{SO}(3)$上的凸集，因此$\phi(t)$必然存在一个全局

最小值[213]。根据$\phi(t)$的非负性，可知$\phi(t)$唯一的全局极小值就是一致性点，满足$\mathcal{R}_i=\mathcal{R}_j,\forall i,j=1,\cdots,N$。因此，$W(t)$在$\mathcal{B}_{\frac{\pi}{2}}(\bar{\mathcal{R}})$上是正定的。根据式(7-35)，可得随着$t\to\infty$，$\phi(t)$将会收敛到0，这意味着随着$t\to\infty$，$\eta_i(t)$和$W(t)$收敛到0。因此，$\mathcal{R}_i(t)$将会收敛到一致性集$\mathcal{A}$。下面我们考虑动态事件驱动的奇诺现象。通过使用式(7-33)和$\alpha_i>0$，我们可以导出触发条件(7-30)的充分条件。

$$\|\boldsymbol{E}_i^{\vee}(t)\|\leqslant\sqrt{\frac{\eta_i(0)}{\theta_i}}e^{-\frac{1}{2}\left(\lambda_i+\frac{1}{\theta_i}\right)t}\tag{7-38}$$

令$\log\left(\mathcal{R}_i^{\mathrm{T}}\mathcal{R}_j\right)=\Theta_{ij}\boldsymbol{u}_{ij}^{\wedge}$，可以计算在$t\in\left[t_k^i,t_{k+1}^i\right),\boldsymbol{E}_i^{\vee}(t)$的导数如下：

$$\frac{\mathrm{d}\left[\boldsymbol{E}_i^{\vee}(t)\right]}{\mathrm{d}t}=\sum_{j=1}^{N}a_{ij}\left[\frac{\mathrm{d}\Theta_{ij}(t)}{\mathrm{d}t}\boldsymbol{u}_{ij}(t)+\Theta_{ij}(t)\frac{\mathrm{d}\boldsymbol{u}_{ij}(t)}{\mathrm{d}t}\right]\tag{7-39}$$

为了得到式(7-39)，首先计算

$$\begin{aligned}\frac{\mathrm{d}\Theta_{ij}(t)}{\mathrm{d}t}&=\left\langle-\boldsymbol{u}_{ij}^{\wedge}(t),\nabla_{\mathcal{R}_i}\phi_i\left(t_k^i\right)\right\rangle+\left\langle\boldsymbol{u}_{ij}^{\wedge}(t),\nabla_{\mathcal{R}_j}\phi_j\left(t_{k'}^j\right)\right\rangle\\&=\boldsymbol{u}_{ij}^{\mathrm{T}}(t)\left(\nabla_{\mathcal{R}_j}^{\vee}\phi_j\left(t_{k'}^j\right)-\nabla_{\mathcal{R}_i}^{\vee}\phi_i\left(t_k^i\right)\right)\end{aligned}\tag{7-40}$$

其中，用到了$\nabla_{\mathcal{R}_i}\Theta_{ij}(t)=-\boldsymbol{u}_{ij}$和$\nabla_{\mathcal{R}_i}\Theta_{ij}(t)=-\nabla_{\mathcal{R}_j}\Theta_{ij}(t)$两个等式。且

$$\begin{aligned}\frac{\mathrm{d}u_{ij}(t)}{\mathrm{d}t}&=\left[\boldsymbol{D}_{ij}(t)\frac{\mathrm{d}\left(\mathcal{R}_i^{\mathrm{T}}(t)\mathcal{R}_j(t)\right)}{\mathrm{d}t}\right]^{\vee}\\&=\left[\boldsymbol{D}_{ij}(t)\left(-\nabla_{\mathcal{R}_i}^{\mathrm{T}}\phi_i\left(t_k^i\right)\mathcal{R}_i^{\mathrm{T}}(t)\mathcal{R}_j(t)-\mathcal{R}_i^{\mathrm{T}}(t)\mathcal{R}_j(t)\nabla_{\mathcal{R}_j}\phi_j\left(t_{k'}^j\right)\right)\right]^{\vee}\end{aligned}\tag{7-41}$$

其中，$\boldsymbol{D}_{ij}(t)$是$\boldsymbol{u}_{ij}^{\wedge}(t)$关于$\mathcal{R}_i^{\mathrm{T}}(t)\mathcal{R}_j(t)$的导数。这个矩阵$\boldsymbol{D}_{ij}(t)$可通过罗德里格斯公式以及对$\log\left(\mathcal{R}_i^{\mathrm{T}}\mathcal{R}_j\right)(t)$求导来确定。由于下面我们仅需要知道它的谱范数，所以这里省略了求导过程。将式(7-40)和式(7-41)代入到式(7-39)，可得

$$\begin{aligned}\left\|\frac{\mathrm{d}\left[\boldsymbol{E}_i^{\vee}(t)\right]}{\mathrm{d}t}\right\|\leqslant&\sum_{j=1}^{N}a_{ij}\left[\left(\|\nabla_{\mathcal{R}_i}^{\vee}\phi_i\left(t_k^i\right)\|+\|\nabla_{\mathcal{R}_j}^{\vee}\phi_j\left(t_{k^{\mathrm{T}}}^j\right)\|\right)\right.\\&\left.+\Theta_{ij}(t)\|\boldsymbol{D}_{ij}(t)\|\left(\|\nabla_{\mathcal{R}_i}^{\vee}\phi_i\left(t_k^i\right)\|+\|\nabla_{\mathcal{R}_j}^{\vee}\phi_j\left(t_{k^{\mathrm{T}}}^j\right)\|\right)\right]\end{aligned}\tag{7-42}$$

其中，$\|\boldsymbol{D}_{ij}(t)\|$可通过谱分解获得。根据文献 [203] 的命题1,$\boldsymbol{D}_{ij}(t)$的特征值集合为$\left\{0,\frac{\Theta_{ij}}{2}\cot\left(\frac{\Theta_{ij}}{2}\right)\pm\frac{\Theta_{ij}}{2}\boldsymbol{j}\right\}$，则当$\Theta_{ij}\in[0,\pi)$时，存在$\|\boldsymbol{D}_{ij}(t)\|$的上界，即

$$\|\boldsymbol{D}_{ij}(t)\|=\sqrt{\frac{\Theta_{ij}^2(t)}{4}\cot^2\frac{\Theta_{ij}(t)}{2}+\frac{\Theta_{ij}^2(t)}{4}}\leqslant\overline{D}=1+\frac{\pi}{2} \tag{7-43}$$

则我们可获得当$t\in\left[t_k^i,t_{k+1}^i\right)$时，$\|\boldsymbol{E}_i^{\vee}(t)\|$的上界如下，

$$\begin{aligned}\|\boldsymbol{E}_i^{\vee}(t)\|&\leqslant\int_{t_k^i}^{t}\left\|\frac{\mathrm{d}\left[\boldsymbol{E}_i^{\vee}(t)\right]}{\mathrm{d}t}\right\|\mathrm{d}t\\&=\int_{t_k^i}^{t}\sum_{j=1}^{N}a_{ij}\mu\left(\|\nabla_{\mathcal{R}_i}^{\vee}\phi_i\left(t_k^i\right)\|+\|\nabla_{\mathcal{R}_j}^{\vee}\phi_j\left(t_{k'}^j\right)\|\right)\mathrm{d}t\end{aligned} \tag{7-44}$$

其中，$\mu=1+\pi\overline{D}$是一个正常数。根据式(7-44)，可导出事件驱动条件的一个充分条件：

情况 1. 对于刚体 i，如果没有邻居智能体在$\left[t_k^i,t_{k+1}^i\right)$中触发，则有

$$\sum_{j=1}^{N}a_{ij}\mu\left(\|\nabla_{\mathcal{R}_i}^{\vee}\phi_i\left(t_k^i\right)\|+\|\nabla_{\mathcal{R}_j}^{\vee}\phi_j\left(t_{k'}^j\right)\|\right)\times\left(t_{k+1}^i-t_k^i\right)\leqslant\sqrt{\frac{\eta_i(0)}{\theta_i}}\mathrm{e}^{-\frac{1}{2}\left(\lambda_i+\frac{1}{\theta_i}\right)t_k^i} \tag{7-45}$$

情况 2. 对于刚体 i，如果有邻居智能体在$\left[t_k^i,t_{k+1}^i\right)$中触发，则令 ρ_j 表示智能体 j 的触发次数，$t_0^j,t_1^j,\cdots,t_{\rho_j}^j$表示触发时刻，其中$t_0^j=t_k^i$。之后，有

$$\begin{aligned}&\sum_{j=1}^{N}a_{ij}\mu\left(\|\nabla_{\mathcal{R}_i}^{\vee}\phi_i\left(t_k^i\right)\|\left(t_{k+1}^i-t_k^i\right)+\sum_{r=0}^{\rho_j-1}\|\nabla_{\mathcal{R}_j}^{\vee}\phi_j\left(t_r^j\right)\|\left(t_{r+1}^j-t_r^i\right)+\right.\\&\left.\|\nabla_{\mathcal{R}_j}^{\vee}\phi_j\left(t_{\rho_j}^j\right)\|\left(t_{k+1}^i-t_{\rho_j}^j\right)\right)\leqslant\sqrt{\frac{\eta_i(0)}{\theta_i}}\mathrm{e}^{-\frac{1}{2}\left(\lambda_i+\frac{1}{\theta_i}\right)t_k^i}\end{aligned} \tag{7-46}$$

通过结合上面的两种情况，可得事件驱动条件的一个充分条件为

$$\sum_{j=1}^{N} a_{ij}\mu\left(\| \nabla_{\mathcal{R}_i}^{\vee}\phi_i\left(t_k^i\right)\| + \max_{r=0,\cdots,\rho_j}\{\| \nabla_{\mathcal{R}_j}^{\vee}\phi_j\left(t_r^j\right)\|\}\right)\left(t_{k+1}^i - t_k^i\right) \leqslant \sqrt{\frac{\eta_i(0)}{\theta_i}}\mathrm{e}^{-\frac{1}{2}\left(\lambda_i+\frac{1}{\theta_i}\right)t_k^i} \tag{7-47}$$

用$\tilde{t}_{k+1}^i$表示由式(7-46)确定的下一个触发时刻，则

$$\begin{aligned} t_{k+1}^i - t_k^i &\geqslant \tilde{t}_{k+1}^i - t_k^i \\ &= \sqrt{\frac{\eta_i(0)}{\theta_i \mu^2 L_i^2}}\mathrm{e}^{-\frac{1}{2}\left(\lambda_i+\frac{\beta_i}{\theta_i}\right)\tilde{t}_{k+1}} \end{aligned} \tag{7-48}$$

其中，

$$L_i = \sum_{j=1}^{N} a_{ij}\left(\| \nabla_{\mathcal{R}_i}^{\vee}\phi_i\left(t_k^i\right)\| + \max_{r=0,\cdots,\rho_j}\left\{\| \nabla_{\mathcal{R}_j}^{\vee}\phi_j\left(t_r^j\right)\|\right\}\right) \tag{7-49}$$

由于在式(7-48)的右边，L_i、$\eta_i(0)$、θ_i都是有界正数，因此，可知奇诺现象不会发生。■

注释 7.5 定理 7.2 基于黎曼梯度下降的方法研究了$\mathbb{SO}(3)$上的基于事件驱动的姿态一致性问题。与运用绝对姿态测量的事件驱动姿态一致性工作[167,205]相比，本节提出的事件驱动姿态一致性协议仅使用了相对姿态信息。因此，当使用协议时，不需要建立一个全局坐标系。另一方面，旋转矩阵的非欧性给分析奇诺现象带来了困难。针对这个情况，我们引入了动态事件驱动机制，排除了奇诺现象。进一步，动态事件驱动条件 (7-30) 并不依赖任何全局信息，因此事件驱动协议 (7-27) 是完全分布式的。与基于相对姿态测量的姿态一致性算法[203]相比，为了确定一致性算法的步长，通信拓扑中的全局信息如节点的最大度数需要已知。

7.3.3 自触发姿态一致性

在上述的事件驱动条件 (7-30) 中，注意到包含了连续相对姿态信息。这意味着每个刚体需要连续测量与邻居之间的相对姿态，以用来判断事件驱动条件 (7-30)。为了规避这个问题，一个有效的方法是使用式 (7-44) 去估计连续的梯度$\nabla_{\mathcal{R}_i}\phi_i(t)$信息。为了达到这个目的，我们首先需要一个新的事

件驱动条件 (7-30) 的充分条件。令$\beta_i=0$和$\alpha_i=0$，可得下面的触发条件

$$t_{k+1}^i=\min_{t\geqslant t_k^i}\left\{t\in\mathbb{R}\,\|\,\boldsymbol{E}_i^{\vee}(t)\,\|^2\geqslant\frac{\eta_i(0)}{\theta}\boldsymbol{e}^{-\lambda t},\ \ t\in\left[t_k^i,t_{k+1}^i\right)\right\} \tag{7-50}$$

其中，λ和$\eta_i(0)$都是正常数。

基于上述触发条件，给出下面推论：

推论 7.1

考虑一个由 N 个刚体组成的多智能体系统，其初始姿态$\mathcal{R}_i(0)$包含在以$\bar{\mathcal{R}}$为球心的开球$\mathcal{B}_{\frac{\pi}{2}}(\bar{\mathcal{R}})$中。当使用事件驱动协议 (7-27) 和事件驱动体条件 (7-50) 时，如果通信拓扑是无向连通图，$\mathcal{R}(t)$会收敛到一致性集 $\mathcal{A}$ 中。

相应的自触发策略总结如算法 7.1 所示。

算法 7.1

自触发姿态一致性算法。

```
初始化：
设定触发时刻 t_1^i = 0, i = 1,…,N，每个刚体获得其邻居的个数 |N_i|;
for 刚体 i = 1 ～ N do
    接收来自邻居的信息；
    if 一个新的信息 ∇_{R_j} φ_j(t*), j ∈ N_i 在 t* 时刻被收到 then
        赋值 k' = k' + 1, t_{k'}^j = t*；
        用 ∇_{R_j} φ_j(t_{k'}^j) 更新式 (7-44) 中的 ∇_{R_j} φ_j(t*)；
    end
    通过式 (7-44) 计算 ‖E_i^∨(t)‖ 的上界；
    if 式 (7-50) 中的条件在 t* 时刻被违背 then
        赋值 k = k + 1, t_k^i = t*；
        采样与邻居的相对姿态信息 (R_i^T R_j)(t_k^i)；
        用 ∇_{R_i} φ_i(t*) 更新控制协议 w_i(t) = −∇_{R_i} φ_i(t_k^i), t ∈ [t_k^i, t_{k+1}^i)；
        广播信息 ∇_{R_i} φ_i(t_k^i) 给自己所有的邻居 j ∈ N_i；
    end
end
```

7.4 无角速度测量下的姿态一致性

我们首先考虑固定拓扑的情况，并为每个刚体设计两个辅助系统。基于辅助系统的输出，提出了基于动力学的事件驱动一致性协议。接下来，基于不变性原理，将结果扩展到联合连通切换拓扑的情况。

7.4.1 固定拓扑下的事件驱动姿态一致性

首先引入一个辅助系统，其中状态为$\boldsymbol{q}_i = \log \boldsymbol{Q}_i^{\vee}$。则有

$$\dot{\boldsymbol{q}}_i = \boldsymbol{L}_{\boldsymbol{q}_i} \boldsymbol{\beta}_i \tag{7-51}$$

其中，$\boldsymbol{\beta}_i$是待确定的控制输入。

令辅助系统的相对状态为$\boldsymbol{q}_{ij} = \log\left(\boldsymbol{Q}_i^{\mathrm{T}} \boldsymbol{Q}_j\right)^{\vee}$。则根据引理 7.2，有

$$\dot{\boldsymbol{q}}_{ij} = \boldsymbol{L}_{\boldsymbol{q}_{ij}} \left(\boldsymbol{\beta}_j - \boldsymbol{Q}_j^{\mathrm{T}} \boldsymbol{Q}_i \boldsymbol{\beta}_i\right), (i, j) \in \mathcal{E} \tag{7-52}$$

对于辅助系统 i，定义下列时间序列：

① $\left\{t_{k_i}\right\} \in \mathbb{R}, k_i \in \mathbb{N}^+, i \in \mathcal{V}$表示每个辅助系统 i 输入的更新时刻。

② $\left\{t_{k_{ij}}\right\} \in \mathbb{R}, k_{ij} \in \mathbb{N}^+, (i, j) \in \mathcal{E}$表示辅助系统 i 和 j 的相对状态的更新时刻。

根据上面的定义，我们知道$\left\{t_{k_{ij}}\right\}$是$\left\{t_{k_i}\right\}$的子列。令时刻$t_{l_{ij}} = \min_{t \geqslant k_{k_{ij}}, k_{ij} \in \mathbb{N}^+} t_{k_{ij}}$表示当前从邻居$j \in \mathcal{N}_i$获得最新信息的时刻，且$\hat{\boldsymbol{q}}_{ij} = \boldsymbol{q}_{ij}\left(t_{l_{ij}}\right)$表示从邻居获得的最新信息。对于$t \in \left[t_{k_i}, t_{k_i+1}\right)$，控制输入$\boldsymbol{\beta}_i(t)$设计为

$$\boldsymbol{\beta}_i(t) = k_i \sum_{i=1}^{N} a_{ij} \hat{\boldsymbol{q}}_{ij} \tag{7-53}$$

其中，k_i是正增益。因此，$\boldsymbol{\beta}_i(t)$是一个分段常值输入。

为了确定每个辅助系统的触发时刻，我们给每一对刚体 i 和 j 都分配一个时钟变量。触发条件表示为

$$t_{k_i+1} = \min_{t \geqslant t_{k_i}} \left\{t \in \mathbb{R} : \mathit{\Gamma}_{ij} \leqslant 0, j \in \mathcal{N}_i\right\} \tag{7-54}$$

$\mathit{\Gamma}_{ij}$的动态定义为$\dot{\mathit{\Gamma}}_{ij} = \gamma_{ij}$，且

$$\gamma_{ij} = -k_i \left|\mathcal{N}_i\right| \overline{\boldsymbol{\lambda}} \left(\mathit{\Gamma}_{ij}^2 + a_i \mathit{\Gamma}_{ij}^2 + 1\right) - \epsilon_{ij} \tag{7-55}$$

其中，$\bar{\lambda}=\max_{i,j}\bar{\lambda}_{ij}$；$a_i$是一个正常数，满足$a_i>\dfrac{|\mathcal{N}_i|(\bar{\lambda}+1)}{2}$；$\epsilon_{ij}$和$\bar{\lambda}_{ij}$是在下面证明分析过程中待确定的正常数。

触发条件背后的想法是当事件被触发时，时钟变量Γ_{ij}将被设置为上界参数$\bar{\Gamma}_{ij}$，然后单调下降直到为 0。触发条件 (7-54) 的优点在于，对于任何触发间隔，它都自然保证了一个统一的正下界。事件驱动算法如算法 7.2 所示。

算法 7.2

事件驱动算法。

初始化：

设定参数$t_{k_i}=0, i\in\mathcal{V}, t_{k_{ij}}=0, (i,j)\in\varepsilon$；

$\hat{\boldsymbol{q}}_{ij}(0)=\log(\boldsymbol{Q}_i^{\mathrm{T}}\boldsymbol{Q}_j)(0)$；

$\Gamma_{ij}=\bar{\Gamma}_{ij}$对所有的$(i,j)\in\varepsilon$；

for 辅助系统 $i=1\sim \boldsymbol{N}$ **do**

　接收来自邻居的相对信息；

　if 一个新的相对信息$\hat{\boldsymbol{q}}_{ij}(t^*)$，$j\in\mathcal{N}_i$在 t^* 时刻收到 **then**

　　令$k_i=k_i+1$且$t_{k_i}=t^*$；

　　令$k_{ij}=k_{ij}+1$且$t_{k_{ij}}=t^*$；

　　在式 (7-53) 中，根据$\hat{\boldsymbol{q}}_{ij}(t^*)$更新 $\boldsymbol{\beta}_i$；

　end

　用式 (7-55) 更新Γ_{ij}；

　if 对于一对刚体$(i,j)\in\varepsilon$，条件 (7-54) 在时刻 t^* 被违反 **then**

　　令$k_i=k_i+1$和$t_{k_i}=t^*$；

　　令$k_{ij}=k_{ij}+1$和$t_{k_{ij}}=t^*$；

　　采样相对姿态信息$\hat{\boldsymbol{q}}_{ij}(t^*)$；

　　更新控制输入 $\boldsymbol{\beta}_i$ 为$\hat{\boldsymbol{q}}_{ij}(t^*)$；

　　更新Γ_{ij}为$\bar{\Gamma}_{ij}$；

　　广播$\hat{\boldsymbol{q}}_{ij}(t^*)$给邻居智能体$j\in\mathcal{N}_i$；

　end

end

接下来的任务是根据上述辅助系统的输出，在动力学的层面上设计姿态一致性协议$\boldsymbol{\tau}_i$。每个刚体和它的辅助系统的偏差可以定义为$\tilde{\boldsymbol{q}}_i = \log \tilde{\mathcal{R}}_i^{\vee}$，其中$\tilde{\mathcal{R}}_i = \boldsymbol{Q}_i^{\mathrm{T}} \mathcal{R}_i$。根据引理 7.2，有

$$\dot{\tilde{\boldsymbol{q}}}_i = \boldsymbol{L}_{\tilde{\boldsymbol{q}}_i} \left(\boldsymbol{w}_i - \tilde{\mathcal{R}}_i^{\mathrm{T}} \boldsymbol{\beta}_i \right) = \boldsymbol{L}_{\tilde{\boldsymbol{q}}_i} \tilde{\boldsymbol{w}}_i \tag{7-56}$$

其中，$\tilde{\boldsymbol{w}}_i = \boldsymbol{w}_i - \tilde{\mathcal{R}}_i^{\mathrm{T}} \boldsymbol{\beta}_i$。定义$\overline{\boldsymbol{\beta}}_i = \tilde{\mathcal{R}}_i^{\mathrm{T}} \boldsymbol{\beta}_i$并且取$\boldsymbol{\beta}_i$在时刻$\left\{ t_{k_i} \right\}, k_i \in \mathbb{N}^{+}$的右边导数，有

$$\begin{aligned} \boldsymbol{J}_i \dot{\tilde{\boldsymbol{w}}} &= \boldsymbol{J}_i \left(\dot{\boldsymbol{w}}_i + \tilde{\mathcal{R}}_i^{\mathrm{T}} S\left(\tilde{\mathcal{R}}_i \boldsymbol{w}_i \right) \boldsymbol{\beta}_i \right) \\ &= \boldsymbol{J}_i \left(\dot{\boldsymbol{w}}_i + S\left(\tilde{\boldsymbol{w}}_i \right) \overline{\boldsymbol{\beta}}_i \right) \end{aligned}$$

其中我们利用了$S(\boldsymbol{x})\boldsymbol{x} = 0$和$S(\boldsymbol{\mathcal{R}}\boldsymbol{x}) = \mathcal{R} S(\boldsymbol{x}) \mathcal{R}^{\mathrm{T}}$获得等式。根据式(2-37)，有

$$\begin{aligned} \tilde{\boldsymbol{w}}_i^{\mathrm{T}} \boldsymbol{J}_i \dot{\tilde{\boldsymbol{w}}} &= \tilde{\boldsymbol{w}}_i^{\mathrm{T}} \left[\boldsymbol{\tau}_i - S\left(\overline{\boldsymbol{\beta}}_i \right) \boldsymbol{J}_i \tilde{\boldsymbol{w}}_i + \boldsymbol{J}_i \boldsymbol{S}\left(\tilde{\boldsymbol{w}}_i \right) \overline{\boldsymbol{\beta}}_i - S\left(\overline{\boldsymbol{\beta}}_i \right) \boldsymbol{J}_i \overline{\boldsymbol{\beta}}_i \right] \\ &= \tilde{\boldsymbol{w}}_i^{\mathrm{T}} \left[\boldsymbol{\tau}_i - S\left(\overline{\boldsymbol{\beta}}_i \right) \boldsymbol{J}_i \overline{\boldsymbol{\beta}}_i \right] \end{aligned} \tag{7-57}$$

其中，$S(\boldsymbol{x})\boldsymbol{y} = -S(\boldsymbol{y})\boldsymbol{x}$用来获得最后一个不等式。

考虑一个新的辅助系统如下，

$$\dot{\boldsymbol{p}}_i = \boldsymbol{L}_{\boldsymbol{p}_i} \boldsymbol{r}_i \tag{7-58}$$

其中，$\boldsymbol{p}_i = \log \boldsymbol{P}_i^{\vee}$和$\boldsymbol{r}_i$是待确定的控制输入。定义$\boldsymbol{P}_i$和$\tilde{\mathcal{R}}_i$的偏差$\tilde{\boldsymbol{p}}_i = \log\left(\boldsymbol{P}_i^{\mathrm{T}} \tilde{\mathcal{R}}_i \right)^{\vee}$，有

$$\dot{\tilde{\boldsymbol{p}}}_i = \boldsymbol{L}_{\tilde{\boldsymbol{p}}_i} \left(\tilde{\boldsymbol{w}}_i - \tilde{\mathcal{R}}_i^{\mathrm{T}} \boldsymbol{P}_i \boldsymbol{r}_i \right) \tag{7-59}$$

则控制输入$\boldsymbol{r}_i$可以定义为

$$\boldsymbol{r}_i = \tilde{\boldsymbol{p}}_i \tag{7-60}$$

基于上述分析，我们可以设计多刚体系统(2-37)的姿态一致性协议$\boldsymbol{\tau}_i$如下：

$$\tau_i = -\kappa_1 \tilde{\boldsymbol{q}}_i - \kappa_2 \tilde{\boldsymbol{p}}_i + S\left(\overline{\boldsymbol{\beta}}_i \right) \boldsymbol{J}_i \overline{\boldsymbol{\beta}}_i, i \in \mathcal{V} \tag{7-61}$$

注释 7.6 基于上面提出的两个辅助系统的状态来设计控制转矩[式(7-61)]。第一项是驱动每个刚体的姿态收敛到辅助状态$\boldsymbol{Q}_i$，从而在控制输入[式(7-53)]下趋于姿态一致性。在控制输入

[式(7-60)]下，第二个辅助系统的辅助状态 $\boldsymbol{P}_i$ 趋向于$\tilde{\mathcal{R}}_i$。因此，根据式(7-58)，控制转矩[式(7-61)]右侧的第二项可用于提供阻尼项，从而避免了控制转矩设计中的角速度反馈。

现在我们可以阐述定理如下。

定理 7.3

考虑多刚体系统(2-37)在控制输入[式(7-61)]和ETC[式(7-54)]下，如果初始姿态$\mathcal{R}_i(0)\in\mathcal{B}_{\frac{\pi}{2}}(\boldsymbol{Q}_i),i\in\mathcal{V}$和辅助系统的初始状态$\boldsymbol{Q}_i(0)\in\mathcal{B}_{\frac{\pi}{2}}(\boldsymbol{I}_3),i\in\mathcal{V}$，那么有

$$\left\{\mathcal{R}_i(t),\boldsymbol{Q}_i(t)\right\}\in\mathcal{B}_{\pi}(\boldsymbol{I}_3)\times\mathcal{B}_{\frac{\pi}{2}}(\boldsymbol{I}_3),\ \ i\in\mathcal{V} \tag{7-62}$$

则对任意$t>0$，姿态一致性可以实现，即对于$t\to+\infty$，有$\mathcal{R}_i\to\mathcal{R}_j$和$\boldsymbol{w}_i\to\boldsymbol{0}$。此外，触发间隔的一致下界为

$$\delta_{\min}=\min_{i,j}\frac{1}{\sqrt{n_{ij}m_i}}\arctan\frac{\overline{\Gamma}_{ij}\sqrt{m_i}}{\sqrt{n_{ij}}} \tag{7-63}$$

其中，$m_i=k_i\left|\mathcal{N}_i\right|\overline{\lambda}(1+a_i)$和$n_{ij}=k_i\left|\mathcal{N}_i\right|\overline{\lambda}+\epsilon_{ij}$。

证明：

令$\boldsymbol{e}_{ij}=\hat{\boldsymbol{q}}_{ij}-\boldsymbol{q}_{ij},(i,j)\in\mathcal{E},t\in\left[t_{k_{ij}},t_{k_{ij}+1}\right)$表示测量误差。考虑李雅普诺夫函数，

$$V=V_1+V_2 \tag{7-64}$$

其中

$$V_1=\frac{1}{2}\sum_{i=1}^{N}\sum_{j=1}^{N}a_{ij}\left(\boldsymbol{q}_{ij}^{\mathrm{T}}\boldsymbol{q}_{ij}+\Gamma_{ij}\boldsymbol{e}_{ij}^{\mathrm{T}}\boldsymbol{e}_{ij}\right) \tag{7-65}$$

和

$$V_2=\frac{1}{2}\sum_{i=1}^{N}\tilde{\boldsymbol{w}}_i^{\mathrm{T}}\boldsymbol{J}_i\tilde{\boldsymbol{w}}_i+\frac{1}{2}\sum_{i=1}^{N}\kappa_1\tilde{\boldsymbol{q}}_i^{\mathrm{T}}\tilde{\boldsymbol{q}}_i+\frac{1}{2}\sum_{i=1}^{N}\kappa_2\tilde{\boldsymbol{p}}_i^{\mathrm{T}}\tilde{\boldsymbol{p}}_i \tag{7-66}$$

注意到V在时间间隔$\left[t_{k_i},t_{k_i+1}\right),k_i\in\mathbb{N}^{+}$上是分段连续可微的。我们考虑$V_1$在$\left[t_{k_i},t_{k_i+1}\right)$上的一般导数和$V_1$在$t_{k_i+1}$上的右边导数。则

$$
\begin{aligned}
\dot{V}_1 &= \sum_{i=1}^{N}\sum_{j=1}^{N} a_{ij}\boldsymbol{q}_{ij}^{\mathrm{T}}\left(\boldsymbol{\beta}_j - \boldsymbol{Q}_j^{\mathrm{T}}\boldsymbol{Q}_i\boldsymbol{\beta}_i\right) + \frac{1}{2}\sum_{i=1}^{N}\sum_{j=1}^{N} a_{ij}\boldsymbol{e}_{ij}^{\mathrm{T}}\dot{\varGamma}_{ij}\boldsymbol{e}_{ij} \\
&\quad + \sum_{i=1}^{N}\sum_{j=1}^{N} a_{ij}\varGamma_{ij}\boldsymbol{e}_{ij}^{\mathrm{T}}\left(\boldsymbol{L}_{\boldsymbol{q}_{ji}} + \boldsymbol{L}_{\boldsymbol{q}_{ij}}\right)\boldsymbol{\beta}_i \\
&= -\sum_{i=1}^{N}\sum_{j=1}^{N} a_{ij}\boldsymbol{q}_{ij}^{\mathrm{T}}\boldsymbol{\beta}_i + \frac{1}{2}\sum_{i=1}^{N}\sum_{j=1}^{N} a_{ij}\gamma_{ij}\boldsymbol{e}_{ij}^{\mathrm{T}}\boldsymbol{e}_{ij} + \sum_{i=1}^{N}\sum_{j=1}^{N} a_{ij}\varGamma_{ij}\boldsymbol{e}_{ij}^{\mathrm{T}}\left(\boldsymbol{L}_{\boldsymbol{q}_{ji}} + \boldsymbol{L}_{\boldsymbol{q}_{ij}}\right)\boldsymbol{\beta}_i
\end{aligned}
$$

其中，我们利用了$\boldsymbol{Q}_i^{\mathrm{T}}\boldsymbol{Q}_j\boldsymbol{q}_{ij} = \boldsymbol{q}_{ij}$去获得最后一个等式。为了使得后面的证明更加清晰，定义向量

$$
\bar{\boldsymbol{q}}_i = \left[\boldsymbol{q}_{ij_1}^{\mathrm{T}}\ \boldsymbol{q}_{ij_2}^{\mathrm{T}}\ \ldots \boldsymbol{q}_{ij_{|\mathcal{N}_i|}}^{\mathrm{T}}\right]^{\mathrm{T}} \in \mathbb{R}^{3|\mathcal{N}_i|},\ \ \bar{\boldsymbol{e}}_i = \left[\boldsymbol{e}_{ij_1}^{\mathrm{T}}\ \boldsymbol{e}_{ij_2}^{\mathrm{T}}\ \ldots \boldsymbol{e}_{ij_{|\mathcal{N}_i|}}^{\mathrm{T}}\right]^{\mathrm{T}} \in \mathbb{R}^{3|\mathcal{N}_i|} \tag{7-67}
$$

和矩阵$\boldsymbol{\varPi}_{\mathcal{N}_i} = \mathbf{1}_{\mathcal{N}_i}\mathbf{1}_{\mathcal{N}_i}^{\mathrm{T}}, \boldsymbol{L}_{\boldsymbol{q}_{ij}}^{\mathrm{T}} = \boldsymbol{L}_{\boldsymbol{q}_{ij}} + \boldsymbol{L}_{\boldsymbol{q}_{ji}}$。根据式(7-1)，我们知道$\boldsymbol{L}_{\boldsymbol{q}_{ij}}^{\mathrm{T}}$是对称正定的。

则我们有

$$
\begin{aligned}
\dot{V}_1 &= -\sum_{i=1}^{N} k_i\bar{\boldsymbol{q}}_i^{\mathrm{T}}\boldsymbol{\varPi}_{\mathcal{N}_i}\bar{\boldsymbol{q}}_i - \sum_{i=1}^{N} k_i\bar{\boldsymbol{e}}_i^{\mathrm{T}}\boldsymbol{\varPi}_{\mathcal{N}_i}\bar{\boldsymbol{q}}_i \\
&\quad + \frac{1}{2}\sum_{i=1}^{N}\sum_{j=1}^{N} a_{ij}\gamma_{ij}\boldsymbol{e}_{ij}^{\mathrm{T}}\boldsymbol{e}_{ij} + \sum_{i=1}^{N} k_i\sum_{j=1}^{N} a_{ij}\varGamma_{ij}\boldsymbol{e}_{ij}^{\mathrm{T}}\boldsymbol{L}_{\boldsymbol{q}_{ij}}^{\mathrm{T}}\mathbf{1}_{\mathcal{N}_i}^{\mathrm{T}}\bar{\boldsymbol{e}}_i \\
&\quad + \sum_{i=1}^{N} k_i\sum_{j=1}^{N} a_{ij}\varGamma_{ij}\boldsymbol{e}_{ij}^{\mathrm{T}}\boldsymbol{L}_{\boldsymbol{q}_{ij}}^{\mathrm{T}}\mathbf{1}_{\mathcal{N}_i}^{\mathrm{T}}\bar{\boldsymbol{q}}_i \\
&\leqslant -\sum_{i=1}^{N} k_i\bar{\boldsymbol{q}}_i^{\mathrm{T}}\boldsymbol{\varPi}_{\mathcal{N}_i}\bar{\boldsymbol{q}}_i + \sum_{i=1}^{N} k_i\left(\frac{a}{2}\bar{\boldsymbol{e}}_i^{\mathrm{T}}\bar{\boldsymbol{e}}_i + \frac{|\mathcal{N}_i|}{2a}\bar{\boldsymbol{q}}_i^{\mathrm{T}}\boldsymbol{\varPi}_{\mathcal{N}_i}\bar{\boldsymbol{q}}_i\right) \\
&\quad + \sum_{i=1}^{N} k_i\frac{|\mathcal{N}_i|}{2}\sum_{j=1}^{N} a_{ij}\left(\varGamma_{ij}^2\bar{\lambda}_{ij} + a_i\varGamma_{ij}^2\bar{\lambda}_{ij} + \bar{\lambda}_{ij}\right)\|\boldsymbol{e}_{ij}\|^2 \\
&\quad + \sum_{i=1}^{N} k_i\frac{\bar{\lambda}_{ij}|\mathcal{N}_i|}{2a}\bar{\boldsymbol{q}}_i^{\mathrm{T}}\boldsymbol{\varPi}_{\mathcal{N}_i}\bar{\boldsymbol{q}}_i + \frac{1}{2}\sum_{i=1}^{N}\sum_{j=1}^{N} a_{ij}\gamma_{ij}\boldsymbol{e}_{ij}^{\mathrm{T}}\boldsymbol{e}_{ij}
\end{aligned}
$$

其中，$\bar{\lambda}_{ij}$是矩阵$\boldsymbol{L}_{\boldsymbol{q}_{ij}}^{\mathrm{T}}$的最大特征值，则选择$a_i > \dfrac{|\mathcal{N}_i|\left(\bar{\lambda}+1\right)}{2}$且令

$c = \min_{i,j}\left(a_i - \dfrac{|\mathcal{N}_i|\left(\bar{\lambda}_{ij}+1\right)}{2}\right)$，有

$$\dot{V}_1 \leqslant -\sum_{i=1}^{N} k_i c \bar{\boldsymbol{q}}_i^{\mathrm{T}} \boldsymbol{\Pi}_{\mathcal{N}_i} \bar{\boldsymbol{q}}_i + \frac{1}{2}\sum_{i=1}^{N}\sum_{j=1}^{N} a_{ij}\gamma_{ij}\boldsymbol{e}_{ij}^{\mathrm{T}}\boldsymbol{e}_{ij}$$
$$+\sum_{i=1}^{N} k_i \sum_{j=1}^{N} a_{ij} \frac{|\mathcal{N}_i|}{2}\left(\Gamma_{ij}^2 \bar{\lambda}_{ij} + a_i \Gamma_{ij}^2 \bar{\lambda}_{ij} + \bar{\lambda}_{ij}\right)\|\boldsymbol{e}_{ij}\|^2$$

为了使$\dot{V}_1 < 0$，令

$$\gamma_{ij} = -k_i |\mathcal{N}_i| \bar{\lambda}\left(\Gamma_{ij}^2 + a_i \Gamma_{ij}^2 + 1\right) - \epsilon_{ij} \tag{7-68}$$

其中，$\bar{\lambda} = \max_{i,j} \bar{\lambda}_{ij}$和$\epsilon_{ij}$是一个正常数。因此，根据式(7-54)，有

$$\dot{V}_1 \leqslant -\sum_{i=1}^{N} k_i c \bar{\boldsymbol{q}}_i^{\mathrm{T}} \boldsymbol{\Pi}_{\mathcal{N}_i} \bar{\boldsymbol{q}}_i - \frac{1}{2}\sum_{i=1}^{N}\sum_{j=1}^{N} a_{ij}\epsilon_{ij}\boldsymbol{e}_{ij}^{\mathrm{T}}\boldsymbol{e}_{ij} \tag{7-69}$$

从式(7-69)我们知道$\dot{V}_1 \leqslant 0$且V_1在任意时间间隔$\left[t_{k_i}, t_{k_i+1}\right)$是非增的。当触发发生，有跳变的状态$\boldsymbol{q}_{ij}\left(t_{k_i+1}^+\right) = \boldsymbol{q}_{ij}\left(t_{k_i+1}^-\right), \boldsymbol{e}_{ij}\left(t_{k_i+1}^+\right) = 0$且$\Gamma_{ij}\left(t_{k_i+1}^+\right) = \bar{\Gamma}_{ij}$，这意味着$V_1\left(t_{k_i+1}^+\right) - V_1\left(t_{k_i+1}^-\right) \leqslant 0$。然后，我们可得$V_1(t)$对$t>0$是非增的。因此，

$$\begin{aligned} V_1(t) \leqslant V_1(0) &= \frac{1}{2}\sum_{i=1}^{N}\sum_{j=1}^{N} a_{ij}\boldsymbol{q}_{ij}(0)^{\mathrm{T}}\boldsymbol{q}_{ij}(0) \\ &= \frac{1}{2}\sum_{i=1}^{N}\sum_{j=1}^{N} a_{ij} d^2\left(\boldsymbol{Q}_i(0), \boldsymbol{Q}_j(0)\right) \end{aligned} \tag{7-70}$$

定义凸包$\mathrm{Cov}\{\boldsymbol{Q}_1, \cdots, \boldsymbol{Q}_N\}$。根据式(7-70)，我们知道对所有的$t>0$，这个凸包将会一直被包含在由初始状态组成的凸包内，即

$$\mathrm{Cov}\{\boldsymbol{Q}_1, \cdots, \boldsymbol{Q}_N\} \subseteq \mathrm{Cov}\{\boldsymbol{Q}_1(0), \cdots, \boldsymbol{Q}_N(0)\} \subseteq \mathcal{B}_{\frac{\pi}{2}}(\boldsymbol{I}_3)$$

因此，对所有的$t>0$，若$\boldsymbol{Q}_i(0) \in \mathcal{B}_{\frac{\pi}{2}}(\boldsymbol{I}_3)$，我们可得$\boldsymbol{Q}_i, i \in \mathcal{V}$总是在开球$\mathcal{B}_{\frac{\pi}{2}}(\boldsymbol{I}_3)$内。

下面，我们将证明收敛结果。考虑式(7-69)的第一项，如果$\bar{\boldsymbol{q}}_i^{\mathrm{T}} \boldsymbol{\Pi}_{\mathcal{N}_i} \bar{\boldsymbol{q}}_i = 0$，有

$$\sum_{i=1}^{N}\sum_{j=1}^{N} a_{ij}\boldsymbol{q}_{ij}^{\mathrm{T}}\boldsymbol{q}_{ij} = \sum_{i=1}^{N}\sum_{j=1}^{N} a_{ij} d^2\left(\boldsymbol{Q}_i, \boldsymbol{Q}_j\right) = 0$$

注意到，由于$\mathbb{SO}(3)$上的凸性[213]，V_1的全局最小值在$\mathcal{B}_{\frac{\pi}{2}}(\boldsymbol{I}_3)$上是唯一

的。由于$d(\ ,\ )$总是正的，对于$j\in\mathcal{N}_i$，唯一的全局最小值是$\boldsymbol{q}_{ij}=\boldsymbol{0}$，因此，对于$\dot{V}_1=0$的最大不变集是$\{(\boldsymbol{q},\boldsymbol{e})\in\mathbb{R}^{6(N+1)}$：$\boldsymbol{q}_1=\boldsymbol{q}_2=\cdots=\boldsymbol{q}_N,\boldsymbol{e}=0\}$，这意味着$\boldsymbol{q}_i\to\boldsymbol{q}_j,i,j\in\mathcal{V}$，对于$t\to+\infty$。

下面，我们给出触发间隔的下界。根据$\dot{\varGamma}_{ij}=\gamma_{ij}$和式 (7-55)，有

$$\dot{\varGamma}_{ij}=-\varGamma_{ij}^2k_i\left|\mathcal{N}_i\right|\bar{\lambda}\left(1+a_i\right)-k_i\left|\mathcal{N}_i\right|\bar{\lambda}-\epsilon_{ij}$$

通过从$\varGamma_{ij}(0)$到 0 对等式两边积分，有

$$t_{k_{ij}+1}-t_{k_{ij}}=\frac{1}{\sqrt{n_{ij}m_i}}\arctan\frac{\bar{\varGamma}_{ij}\sqrt{m_i}}{\sqrt{n_{ij}}}$$

其中，$m_i=k_i\left|\mathcal{N}_i\right|\bar{\lambda}\left(1+a_i\right)$；$n_{ij}=k_i\left|\mathcal{N}_i\right|\bar{\lambda}+\epsilon_{ij}$。因此，触发间隔的下界为$\delta_{\min}=\min_{i,j}\dfrac{1}{\sqrt{n_{ij}m_i}}\arctan\dfrac{\bar{\varGamma}_{ij}\sqrt{m_i}}{\sqrt{n_{ij}}}$。

在第二步中，通过计算 V_2 的导数以及使用式 (7-69)，有

$$\begin{aligned}\dot{V}\leqslant&\sum_{i=1}^{N}\tilde{\boldsymbol{w}}_i^{\mathrm{T}}\boldsymbol{J}_i\dot{\tilde{\boldsymbol{w}}}+\sum_{i=1}^{N}\kappa_1\tilde{\boldsymbol{q}}_i^{\mathrm{T}}\tilde{\boldsymbol{w}}_i+\sum_{i=1}^{N}\kappa_2\tilde{\boldsymbol{p}}_i^{\mathrm{T}}\left(\tilde{\boldsymbol{w}}_i-\tilde{\mathcal{R}}_i^{\mathrm{T}}\boldsymbol{P}_i\boldsymbol{r}_i\right)\\&-\sum_{i=1}^{N}k_ic\bar{\boldsymbol{q}}_i^{\mathrm{T}}\boldsymbol{\Pi}_{\mathcal{N}_i}\bar{\boldsymbol{q}}_i-\frac{1}{2}\sum_{i=1}^{N}\sum_{j=1}^{N}a_{ij}\epsilon_{ij}\boldsymbol{e}_{ij}^{\mathrm{T}}\boldsymbol{e}_{ij}\end{aligned}$$

代入式(7-57)和式(7-61)，有

$$\dot{V}\leqslant-\sum_{i=1}^{N}\kappa_2\tilde{\boldsymbol{p}}_i^{\mathrm{T}}\tilde{\boldsymbol{p}}_i-\sum_{i=1}^{N}k_ic\bar{\boldsymbol{q}}_i^{\mathrm{T}}\boldsymbol{\Pi}_{\mathcal{N}_i}\bar{\boldsymbol{q}}_i-\frac{1}{2}\sum_{i=1}^{N}\sum_{j=1}^{N}a_{ij}\epsilon_{ij}\boldsymbol{e}_{ij}^{\mathrm{T}}\boldsymbol{e}_{ij}\tag{7-71}$$

下面我们将证明对所有的$t>0$，如果$\boldsymbol{x}_i(0)\in\mathcal{B}_{\frac{\pi}{2}}(\boldsymbol{0})$，每个刚体的姿态$\boldsymbol{x}_i$总是在集合$\mathcal{B}_{\pi}(\boldsymbol{0})$内。由于$\dot{V}\leqslant0$，这意味着$V(t)\leqslant V(0)$，对所有的$t>0$，可得$\tilde{\boldsymbol{q}}_i(t)\leqslant\tilde{\boldsymbol{q}}_i(0)$，对所有的$t>0$且$\tilde{\boldsymbol{q}}_i$是单调下降的。根据$\tilde{\boldsymbol{q}}_i=\log\tilde{\mathcal{R}}_i^{\vee}$，其中$\tilde{\mathcal{R}}_i=\boldsymbol{Q}_i^{\mathrm{T}}\mathcal{R}_i$，如果$\tilde{\mathcal{R}}_i(0)\in\mathcal{B}_{\frac{\pi}{2}}(\boldsymbol{Q}_i)$，可得对每个刚体，$\mathcal{R}_i$将一直在集合内$\mathcal{B}_{\frac{\pi}{2}}(\boldsymbol{Q}_i),i\in\mathcal{V}$。因此，对于每个刚体的姿态$\mathcal{R}_i$，可得$\mathcal{B}_{\frac{\pi}{2}}(\boldsymbol{Q}_i)$是正不变的。

现在，引入矩阵$\tilde{\mathcal{R}}_i$的对于四元数$\boldsymbol{\rho}_i^1=\left[\cos\dfrac{\tilde{\theta}^i}{2},\boldsymbol{\mu}_1^{\mathrm{T}}\sin\dfrac{\tilde{\theta}^i}{2}\right]^{\mathrm{T}}$，其中

$\tilde{\theta}_i \in \left[0, \frac{\pi}{2}\right)$，和矩阵 $\boldsymbol{Q}_i$ 的四元数$\boldsymbol{\rho}_i^2 = \left[\cos\frac{\eta_i}{2}, \boldsymbol{\mu}_2^{\mathrm{T}} \sin\frac{\eta_i}{2}\right]^{\mathrm{T}}$，其中$\eta_i \in \left[0, \frac{\pi}{2}\right)$。根据四元数的组合法则，有

$$\boldsymbol{\rho}_i^1 \odot \boldsymbol{\rho}_i^2 = \begin{bmatrix} \cos\frac{\tilde{\theta}_i}{2}\cos\frac{\eta_i}{2} - \sin\frac{\tilde{\theta}_i}{2}\sin\frac{\eta_i}{2}\boldsymbol{\mu}_1^{\mathrm{T}}\boldsymbol{\mu}_2 \\ \cos\frac{\tilde{\theta}_i}{2}\boldsymbol{\mu}_2 + \cos\frac{\eta_i}{2}\boldsymbol{\mu}_1 + S(\boldsymbol{\mu}_1)\boldsymbol{\mu}_2 \end{bmatrix}$$

由于$\boldsymbol{Q}_i\tilde{\mathcal{R}}_i = \mathcal{R}_i$，有

$$\begin{aligned} \cos\frac{\theta_i}{2} &= \cos\frac{\tilde{\theta}_i}{2}\cos\frac{\eta_i}{2} - \sin\frac{\tilde{\theta}_i}{2}\sin\frac{\eta_i}{2}\boldsymbol{\mu}_1^{\mathrm{T}}\boldsymbol{\mu}_2 \\ &\leqslant \cos\frac{1}{2}\left(\tilde{\theta}_i + \eta_i\right) \end{aligned}$$

因此，$0 < \cos\frac{\theta_i}{2} \leqslant 1$。由于$\theta_i \geqslant 0$，那么我们将得到$\theta_i \in [0, \pi)$总是成立的，这意味着对所有的$t > 0$，$\boldsymbol{x}_i \in \mathcal{B}_{\pi}(\boldsymbol{0})$。

为了证明收敛的结果，首先证明对$t \to +\infty$，有$\dot{V} \to 0$。根据式(7-71)，有$V \geqslant 0$且$\dot{V} \leqslant 0$，那么存在一个正的极限满足$V \to V^*$对$t \to +\infty$。进一步，可得V是一致有界的，这同样意味着$\boldsymbol{q}_{ij}$、$\boldsymbol{e}_{ij}$、$\tilde{\boldsymbol{p}}_i$、$\tilde{\boldsymbol{q}}_i$和$\tilde{\boldsymbol{w}}_i$都是一致有界的。那么对于触发间隔$\left[t_{k_i}, t_{k_i+1}\right)$，根据式(7-59)和式(7-52)，可得$\dot{\tilde{\boldsymbol{p}}}_i$和$\dot{\tilde{\boldsymbol{q}}}_i$都是一致有界的，这意味着$\ddot{V}$在触发间隔$\left[t_{k_i}, t_{k_i+1}\right)$上是一致有界的。因此，通过运用引理7.3，对于$t \to +\infty$，有$\dot{V} \to 0$，这意味着$\boldsymbol{q}_{ij} \to \boldsymbol{0}, \tilde{\boldsymbol{p}}_i \to \boldsymbol{0}$。因此，我们可得对所有的$t \to +\infty$，$\boldsymbol{r}_i \to \boldsymbol{0}$且$\boldsymbol{\beta}_i \to \boldsymbol{0}$。

接下来，可得$\boldsymbol{\tau}_i$、$\tilde{\boldsymbol{w}}_i$是一致有界的，根据式(7-56)、式(7-57)和式(7-59)，$\dot{\tilde{\boldsymbol{w}}}_i$和$\dot{\tilde{\boldsymbol{q}}}_i$都是一致有界的。那么，根据式(7-59)，得$\ddot{\tilde{\boldsymbol{p}}}_i$在$\left[t_{k_i}, t_{k_i+1}\right)$上是一致有界的。基于$\boldsymbol{\beta}_i$在$\left[t_{k_i}, t_{k_i+1}\right)$上是分段常值的且$\dot{\tau}_i$是一致有界的，我们可得$\ddot{\tilde{\boldsymbol{w}}}_i$在$\left[t_{k_i}, t_{k_i+1}\right)$上是一致有界的。再一次，通过运用引理7.3，可得对于$t \to +\infty$，有$\dot{\tilde{\boldsymbol{p}}}_i \to 0$。根据式(7-59)和矩阵$\boldsymbol{L}_{\tilde{p}_i}$是正定的，对于$t \to +\infty$，有$\tilde{\boldsymbol{w}}_i \to 0$。进一步，基于$\ddot{\tilde{\boldsymbol{w}}}_i$是一致有界的对于$t \to +\infty$，有$\dot{\tilde{\boldsymbol{w}}}_i \to 0$。那么根据式(7-57)，我们对于$t \to +\infty$有$\boldsymbol{\tau}_i \to 0$，这意味着对于$t \to +\infty$有$\tilde{\boldsymbol{q}}_i \to 0$。因此，对于$t \to +\infty$有$\tilde{\mathcal{R}}_i \to \boldsymbol{I}_3$和$\boldsymbol{w}_i \to 0$，这意味着对于$t \to +\infty$ $\boldsymbol{q}_i \to \boldsymbol{x}_i$。则，由于$\boldsymbol{q}_i \to \boldsymbol{x}_i, i \in \mathcal{V}$，可得对于$t \to +\infty$，$\boldsymbol{x}_i \to \boldsymbol{x}_j, i, j \in \mathcal{V}$。证毕。 ■

注释 7.7 时钟变量 Γ_{ij} 和动态 [式 (7-55)] 的设计是基于下面的考虑。注意到采样误差 $\boldsymbol{e}_{ij}(t)$ 包含连续相对旋转 $\boldsymbol{q}_{ij}(t)$。因此，为了免除事件检测中的连续监测问题，我们引入一个时钟变量 Γ_{ij} 来控制触发，并且它的动态的设计不依赖于采样误差 $\boldsymbol{e}_{ij}(t)$ 和连续的相对旋转 $\boldsymbol{q}_{ij}(t)$。此外，根据式 (7-55)，保证了触发间隔的一个一致的下界，且对于每个智能体的触发间隔都可以由参数$\overline{\Gamma}_{ij}$和ϵ_{ij}来调整。值得一提的是，这种设计的本质原因是旋转 $\boldsymbol{Q}_i$ 的非欧几里德性质，且测地距离用来描述多刚体姿态之间的不一致。因此，另一种可行的方法是利用球心投影的方法，将测地曲线和欧几里德空间的直线对应起来。控制协议的姿态和事件驱动条件可通过推广文献 [207] 中的结果获得。同样注意到 Γ_{ij} 的动态设计仅仅使用了局部的信息，保证了触发策略的完全分布式。

7.4.2 切换拓扑下的事件驱动姿态一致性

在这个部分中，我们将姿态一致性的协议推广至切换拓扑的情况。令$\sigma:[0,+\infty)\to\mathcal{P}$代表分段常值信号，其中$\mathcal{P}=\{1,2,\cdots\}$。一个切换信号被称为有一个平均驻留时间 τ_{ADT}，如果对于任意的时间间隔(t_1,t_2)，切换次数$N_\sigma(t_2,t_1)$满足

$$N_\sigma(t_2,t_1)\leqslant N_0+\frac{t_2-t_1}{\tau_{\mathrm{ADT}}} \tag{7-72}$$

其中，N_0代表抖振的界。

在时间区间$[t,t+\tau]$内的并图定义为$\mathcal{G}_{[t,t+\tau]}=\cup\mathcal{G}_{\sigma(t)}$，$t\in[t,t+\tau]$，其中$\tau>0$是一个正常数。如果存在一个正的常数$\Delta T>0$满足对所有$t>0$，并图$\mathcal{G}_{[t,t+\Delta T]}$是连通的，则连通图被称为联合连通的。

假设 7.1

并图是联合连通的，即在平均驻留时间条件 (7-72) 中，任意的平均驻留时间$\tau_{\mathrm{ADT}}>0$和任意的抖振界$N_0>0$。

同样地，在这个小节中，考虑辅助系统 (7-51) 和控制输入

$$\boldsymbol{\beta}_i(t)=k_i\sum_{j\in\mathcal{N}_i^{\sigma(t)}}\hat{\boldsymbol{q}}_{ij},t\in\left[t_{k_i},t_{k_i+1}\right) \tag{7-73}$$

其中，k_i是一个正增益。第二个辅助系统设计如式(7-58)和式(7-60)所示，则切换拓扑下的ETC设计为

$$t_{k_i+1}=\min_{t\geqslant t_{k_i}}\left\{t\in\mathbb{R}:\Gamma_{ij}\leqslant 0\text{ or }\mathcal{N}_i\left(t_{k_i}\right)\neq\mathcal{N}_i(t),j\in\mathcal{N}_i^{\sigma}\right\} \tag{7-74}$$

其中，$\dot{\Gamma}_{ij}=\gamma_{ij}$的动态为

$$\gamma_{ij}=-k_i\left|\mathcal{N}_i^{\sigma}\right|\bar{\lambda}\left(\Gamma_{ij}^2+a_i\Gamma_{ij}^2+1\right)-\epsilon_{ij} \tag{7-75}$$

注释 7.8 根据 ETC[式 (7-74)]，我们可以知道，每个刚体将会在两种情况下触发。一种是当时钟变量满足条件$\Gamma_{ij}\leqslant 0$的时候。另一种情况是当拓扑切换的时候。在第二种触发情况下，我们要求每个刚体去采样与它当前邻居的相对旋转$\boldsymbol{q}_{ij},j\in\mathcal{N}_i$。注意，这些额外的触发在算法 7.1 的同步实现中是必要的。否则，对于一个节点，在某些特别的切换下可能不存在一个与其他节点的联合连通路径，这可能导致姿态一致性的失败。

定理 7.4

在假设 7.1 下，考虑多刚体系统 (2-37) 以及控制输入 [式 (7-61)、式 (7-73) 和式 (7-60)]。如果初始姿态$\mathcal{R}_i(0)\in\mathcal{B}_{\frac{\pi}{2}}(\boldsymbol{Q}_i),i\in\mathcal{V}$和初始状态$\boldsymbol{Q}_i(0)\in\mathcal{B}_{\frac{\pi}{2}}(\boldsymbol{I}_3),i\in\mathcal{V}$，那么满足$\left\{\mathcal{R}_i(t),\boldsymbol{Q}_i(t)\right\}\in\mathcal{B}_{\pi}(\boldsymbol{I}_3)\times\mathcal{B}_{\frac{\pi}{2}}(\boldsymbol{I}_3),i\in\mathcal{V}$，对所有的$t>0$且姿态一致性能实现，即$\mathcal{R}_i\to\mathcal{R}_j$且$\boldsymbol{w}_i\to\mathbf{0}$，对$t\to+\infty$。此外，奇诺现象在 ETC[式 (7-74)] 下不会发生。

证明：

考虑多李雅普诺夫函数

$$W_1^{\sigma}=\frac{1}{2}\sum_{i\in\mathcal{V}^{\sigma}}\sum_{j\in\mathcal{N}_i^{\sigma}}\left(\boldsymbol{q}_{ij}^{\mathrm{T}}\boldsymbol{q}_{ij}+\Gamma_{ij}\boldsymbol{e}_{ij}^{\mathrm{T}}\boldsymbol{e}_{ij}\right) \tag{7-76}$$

考虑W_1^{σ}在$\left[t_{k_i},t_{k_i+1}\right)$上的导数。

在这之前，我们首先证明存在无限多个的时间间隔$\left[t_{k_i},t_{k_i+1}\right)$满足

$\sigma \in \mathcal{P}$是一个常数。根据 ETC[式 (7-74)]，在序列$\left\{t_{k_i}\right\}, k_i \in \mathbb{N}^+, i \in \mathcal{V}$中的时刻要么是违背了条件$\Gamma_{ij} \leqslant 0$且没有切换发生，要么就是在切换时刻。对于没有发生切换的时刻t_{k_i}或者t_{k_i+1}，很明显，σ在$\left[t_{k_i}, t_{k_i+1}\right)$是个常数。如果$t_{k_i}$和$t_{k_i+1}$都是切换时刻，根据平均驻留时间条件，总有$t_{k_i}+\tau_k \geqslant t_{k_i+1}$，其中，$\tau_k \in \left(0, \tau_{\mathrm{ADT}}\right)$满足$\sigma$在$\left[t_{k_i}, t_{k_i+1}\right)$上是常数。

那么在 ETC[式 (7-74)] 下，有

$$\dot{W}_1^\sigma \leqslant -\sum_{i=1}^{N} k_i c \left(\bar{\boldsymbol{q}}_i^\sigma\right)^{\mathrm{T}} \boldsymbol{\Pi}_{\mathcal{N}_i^\sigma} \bar{\boldsymbol{q}}_i^\sigma - \frac{1}{2}\sum_{i=1}^{N} \epsilon_{ij} \left(\bar{\boldsymbol{e}}_i^\sigma\right)^{\mathrm{T}} \bar{\boldsymbol{e}}_i^\sigma \tag{7-77}$$

其中，$\bar{q}_i^\sigma = \mathrm{col}\left\{\boldsymbol{q}_{ij}\right\}$；$j \in \mathcal{N}_i^\sigma$；$\bar{\boldsymbol{e}}_i^\sigma = \mathrm{col}\left\{\boldsymbol{e}_{ij}\right\}, j \in \mathcal{N}_i^\sigma$。由于对每个$\sigma \in \mathcal{P}$，有$\dot{W}_1^\sigma \leqslant 0$。由于图是联合连通的，与定理7.3的证明相似，对于解$\boldsymbol{Q}_i, \forall i \in \mathcal{V}$我们可以得到集合$\mathcal{B}_{\frac{\pi}{2}}\left(\boldsymbol{I}_3\right)$的不变性。

下面，随着$t \to +\infty$，有$\boldsymbol{q}_i \to \boldsymbol{q}_j$。让我们选择无限的时间区间$\left[t_{k_i}, t_{k_i+1}\right)$满足$\sigma$在这些区间上的值是一样的，并且定义这些区间的并集为$\mathcal{M}^\sigma = \bigcup_{k_i \in \mathbb{N}^+} \left[t_{k_i}, t_{k_i+1}\right)$。因此，有

$$\begin{aligned} \int_0^{+\infty} \dot{W}_1^\sigma(t)\mathrm{d}t &\leqslant \int_{\mathcal{M}^\sigma} \dot{W}_1^\sigma(t)\mathrm{d}t \\ &= W_1^\sigma\left(t_{k_i}\right) - W_1^\sigma\left(t_{k_{\max}}\right) < +\infty \end{aligned} \tag{7-78}$$

根据式 (7-78)，$\bar{\boldsymbol{q}}_i^\sigma$在$\left[t_{k_i}, t_{k_i+1}\right)$上是有界的，意味着$\dot{\boldsymbol{q}}_i^\sigma$在$\left[t_{k_i}, t_{k_i+1}\right)$上是有界的。因此，$\ddot{W}_1(t)$在$\left[t_{k_i}, t_{k_i+1}\right)$上是有界的。则根据引理 7.3，随着$t \to +\infty$，有$\dot{W}_1^\sigma \to 0$，这意味着随着$t \to +\infty$，$\boldsymbol{q}_{ij} \to 0, (i, j) \in \mathcal{E}^\sigma$。

下面的证明与定理 1 相似，考虑

$$W^\sigma = W_1^\sigma + W_2^\sigma \tag{7-79}$$

对于$\sigma \in \mathcal{P}$，其中

$$W_2^\sigma = \frac{1}{2}\sum_{i \in \mathcal{V}^\sigma} \tilde{\boldsymbol{w}}_i^{\mathrm{T}} \boldsymbol{J}_i \tilde{\boldsymbol{w}}_i + \frac{1}{2}\sum_{i \in \mathcal{V}^\sigma} \kappa_1 \tilde{\boldsymbol{q}}_i^{\mathrm{T}} \tilde{\boldsymbol{q}}_i + \frac{1}{2}\sum_{i \in \mathcal{V}^\sigma} \kappa_2 \tilde{\boldsymbol{p}}_i^{\mathrm{T}} \tilde{\boldsymbol{p}}_i \tag{7-80}$$

然后，对W_2^σ在$\left(t_{k_i}, t_{k_i+1}\right)$上求导，对$W_2^\sigma$在$t_{k_i+1}$上求右边导数，有$\dot{W}^\sigma \leqslant 0$。由于图是联合连通的，可以得到每个刚体的姿态$\boldsymbol{x}_i$将永远待在集合$\mathcal{B}_\pi(\boldsymbol{0})$中。然后，通过运用引理7.3和结果$\boldsymbol{q}_{ij} \to 0, (i, j) \in \mathcal{E}^\sigma$，对$t \to +\infty$，有$\boldsymbol{x}_i \to \boldsymbol{x}_j, (i, j) \in \mathcal{E}^\sigma$。最后，由于通信拓扑是联合连通的，必然存在

$\sigma_s \in \mathcal{P}$满足$\bigcup_s \mathcal{E}^{\sigma_s} = \mathcal{E}$。因此，通过重复上述的过程，我们可得结果$\boldsymbol{x}_i \to \boldsymbol{x}_j, (i,j) \in \mathcal{E}^{\sigma}$，这意味着随着$t \to +\infty$，$\boldsymbol{x}_i \to \boldsymbol{x}_j, i, j \in \mathcal{V}$。

奇诺现象的排除基于下面两个事实：

① 如果$t_{k_{ij}}$和$t_{k_{ij}+1}$都是没有发生切换的触发时刻。根据定理 7.1 的证明，我们可以保证触发间隔是一致有界的，即

$$\delta_{\min} = \min_{i,j} \frac{1}{\sqrt{n_{ij} m_i}} \arctan \frac{\overline{\Gamma}_{ij} \sqrt{m_i}}{\sqrt{n_{ij}}} \tag{7-81}$$

其中，$m_i = k_i \max_{\sigma \in \mathcal{P}} \left|\mathcal{N}_i^{\sigma}\right| \overline{\lambda}(1 + a_i), n_{ij} = k_i \max_{\sigma \in \mathcal{P}} \left|\mathcal{N}_i^{\sigma}\right| \overline{\lambda} + \epsilon_{ij}$ 和 $a_i > \dfrac{\max_{\sigma \in \mathcal{P}} \left|\mathcal{N}_i^{\sigma}\right| \left(\overline{\lambda} + 1\right)}{2}$。

② 如果$t_{k_{ij}}$和$t_{k_{ij}+1}$是切换时刻，由于切换信号存在一个平均驻留时间，则存在一个下界$\tau_k \in (0, \tau_{\text{ADT}})$满足$\tau_k \geqslant t_{k_{ij}+1} - t_{k_{ij}}$。结合上述的事实，我们可以总结出存在一个触发间隔的下界，证毕。 ■

注释 7.9　文献 [197] 通过旋转矩阵研究了免速度测量的领航者姿态跟踪问题。但是，该方法无法解决多刚体系统的通信拓扑图为环图时的问题。其原因是当拓扑为环图时，在协议中仅仅包含相对姿态的项会存在不属于一致集的不理想平衡点。为了解决这个问题，我们证明了通过使用协同控制输入 [式 (7-73)]，辅助系统的所有旋转$\boldsymbol{Q}_i(t), i \in \mathcal{V}$都将会待在凸集$\mathcal{B}_{\frac{\pi}{2}}(\boldsymbol{I}_3)$内（对所有的$t > 0$），因此辅助系统的状态最终会收敛到唯一的平衡点，满足$\boldsymbol{Q}_i = \boldsymbol{Q}_j, i, j \in \mathcal{V}$。另一方面，与切换拓扑下的姿态一致性问题相比 [77,164]，本章内容研究了基于事件驱动的一致性协议。注意到切换有可能发生在触发间隔内，这就会造成控制协议与当前的通信拓扑不一致的情况。进一步，基于切换系统的不变集原理，固定驻留时间的假设被放宽到了平均驻留时间的条件。

7.5
仿真示例

7.5.1 仿真示例 1

我们首先考虑在强连通拓扑下包含五个刚体的多智能体系统。邻接矩阵表示为

$$\boldsymbol{A}_1=\begin{bmatrix}0&0.5&0&0&0\\0&0&0.8&0&0\\0&0&0&0.6&0\\0&0&0&0&0.3\\1&0&0&0&0\end{bmatrix}\tag{7-82}$$

在这个例子中，姿态由局部表征方式$\boldsymbol{y}_i=\tan\dfrac{\Theta_i}{2}\boldsymbol{u}_i$来描述。每个刚体的旋转单位向量随机给出，选取的初始旋转角度满足$\Theta_i\in(0,\pi)$。给定事件驱动的参数为$\alpha=[0.3\ 0.3\ 0.4\ 0.2\ 0.4]^{\mathrm{T}}$。可以看出，满足定理 7.1 中的条件$\alpha_i<\dfrac{1}{2\bar{a}|\mathcal{N}_i|}$。通过使用事件驱动姿态协议 (7-13) 和事件驱动条件 (7-14)，仿真结果如图 7-2 所示。刚体的姿态轨迹如图 7-2(a) 所示。从图中可以看出所有刚体姿态向量的三个分量趋于同样的值，表明实现了姿态一致性。令每个刚体的触发次数为 $\varDelta_i$，最小触发间隔为$\min_k\{\tau_k^i\},k\in\mathbb{N}^+$。在图 7-2(b) 中，可以看出，每个刚体的触发次数都远远小于仿真中的迭代次数。由此可见，明显降低了通信和控制协议的更新次数。每个刚体的最小触发间隔$\min_k\{\tau_k^i\}$都大于仿真中的离散步长 10^{-2}，表明没有发生奇诺现象。

为了验证定理 7.2，考虑一个无向图。邻接矩阵为

$$\boldsymbol{A}_2=\begin{bmatrix}0&0.5&0&0&1\\0.5&0&0.8&0&0\\0&0.8&0&0.6&0\\0&0&0.6&0&0.3\\1&0&0&0.3&0\end{bmatrix}\tag{7-83}$$

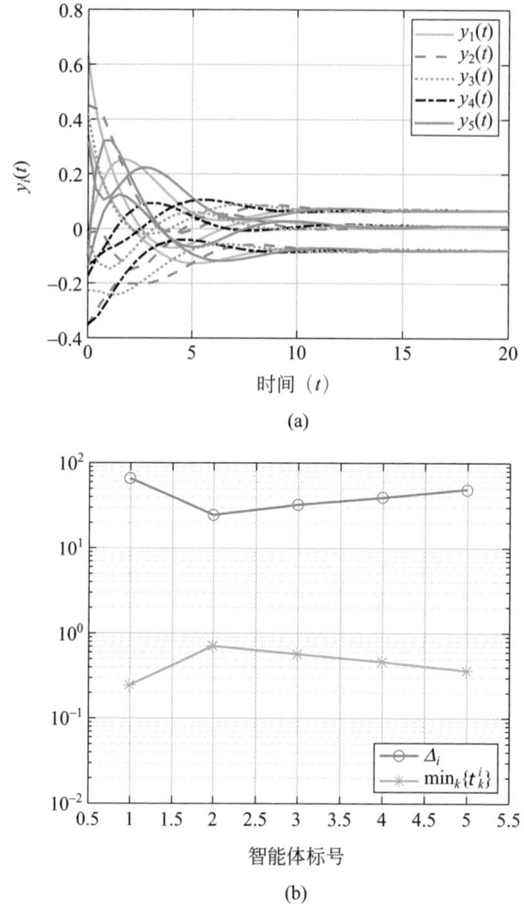

图 7-2 绝对姿态测量下的仿真结果

在这个例子中，每个刚体的姿态由一个旋转矩阵$\mathcal{R}_i, i=1,\cdots,N$来描述，其中初始的姿态包含在集合$\mathcal{B}_{\frac{\pi}{2}}\left(\boldsymbol{I}_3\right)$中。不失一般性，控制参数设定为$\alpha_i=1, \beta_i=0.2, \lambda=0.3, \theta_i=5$。$\eta_i$的初始值设定为$\eta_i\left(0\right)=2$。协议 (7-27) 的仿真结果如图 7-3(a) 所示。可以看出，$\mathcal{R}_1$与每个旋转矩阵的误差谱范数都趋于 0，表明实现了姿态一致性。与第一种情况类似，可看出在事件驱动的机制下，与连续时间一致性协议相比，多智能体系统在网络中的通信次数和协议更新次数都大大减少，且奇诺现象不会发生。

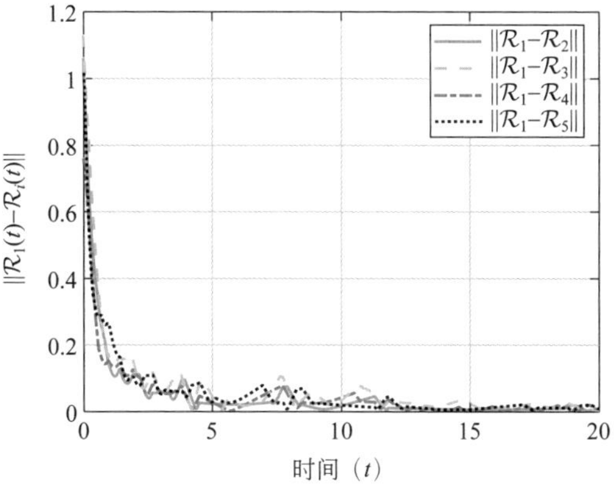

(a)

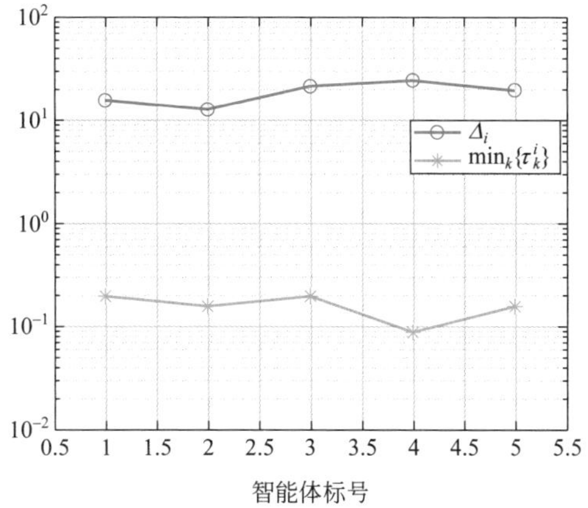

(b)

图 7-3　相对姿态测量下的仿真结果

7.5.2　仿真示例 2

本部分，我们将给出无角速度测量下的姿态一致性协议的仿真验证。考虑一个多刚体系统，通信拓扑图是一个连通图，其中边的集合为

$$\mathcal{E}=\{(1,2),(2,1),(2,3),(3,2),(3,4),(4,3),(1,4),(4,1)\} \tag{7-84}$$

每个刚体的惯性矩阵为$\boldsymbol{J}_i=\operatorname{diag}\{5,5,5\}, i\in\mathcal{V}$。每个辅助系统的初始旋转向量为

$$\boldsymbol{q}_1=[0.21\ 0.64\ 0.01]^{\mathrm{T}} \tag{7-85}$$

$$\boldsymbol{q}_2=[-0.02\ 0.86-0.26]^{\mathrm{T}} \tag{7-86}$$

$$\boldsymbol{q}_3=[-0.01\ 1.11\ 0.18]^{\mathrm{T}} \tag{7-87}$$

$$\boldsymbol{q}_4=[1.07\ 0.79\ 0.22]^{\mathrm{T}} \tag{7-88}$$

包含在开集$\mathcal{B}_{\frac{\pi}{2}}(\mathbf{0})$中。取$\boldsymbol{p}_i=[0\ 0\ 0]^{\mathrm{T}},i\in\mathcal{V}$。取初始旋转向量为

$$\boldsymbol{x}_1=[0.32\ 0.93\ 0.02]^{\mathrm{T}} \tag{7-89}$$

$$\boldsymbol{x}_2=[-0.03\ 1.16-0.35]^{\mathrm{T}} \tag{7-90}$$

$$\boldsymbol{x}_3=[-0.01\ 1.42\ 0.23]^{\mathrm{T}} \tag{7-91}$$

$$\boldsymbol{x}_4=[1.32\ 0.97\ 0.27]^{\mathrm{T}} \tag{7-92}$$

初始的角速度为$\boldsymbol{w}_i=[0.1\ 0.1\ 0.1]^{\mathrm{T}},i\in\mathcal{V}$。为了确定参数$\overline{\lambda}_{ij}$，我们通过设定$\theta_{ij}$为π并计算矩阵$\boldsymbol{L}'_{\boldsymbol{q}_{ij}}$的最大特征值，选取一致性协议的参数为$\boldsymbol{k}_i=[0.3\ 0.4\ 0.1\ 0.2]^{\mathrm{T}},\kappa_1=150,\kappa_2=150$，$a=2,\overline{\lambda}=2$。不失一般性，我们对每对刚体考虑相同的参数，取$\epsilon_{ij}=0.1,\overline{\varGamma}_{ij}=5,(i,j)\in\mathcal{E}$。

对于固定拓扑的情况，仿真结果如图 7-4 所示。图 7-4(a) 显示了所有刚体的旋转向量都趋于相同的值。触发时间和最小触发间隔见图 7-4(b)，其中每个刚体的总触发次数用圆圈线表示，最小触发间隔用星号线表示。可以看出，所有智能体的最小触发事件之间的间隔为 0.02，比定理 7.1 中的一致下界要大。

对于切换拓扑的情况，我们给定一个周期的切换信号

$$\sigma(t)=\begin{cases}1, & \dfrac{4nT}{m}<t\leqslant\dfrac{(4n+1)T}{m}\\ 2, & \dfrac{4nT+1}{m}<t\leqslant\dfrac{(4n+2)T}{m}\\ 3, & \dfrac{4nT+2}{m}<t\leqslant\dfrac{(4n+3)T}{m}\\ 4, & \dfrac{4nT+3}{m}<t\leqslant\dfrac{(4n+4)T}{m}\end{cases} \tag{7-93}$$

其中，$n=0,1,2,\cdots$，并且在仿真中选择$m=10,T=10$。切换拓扑的边集给

定为

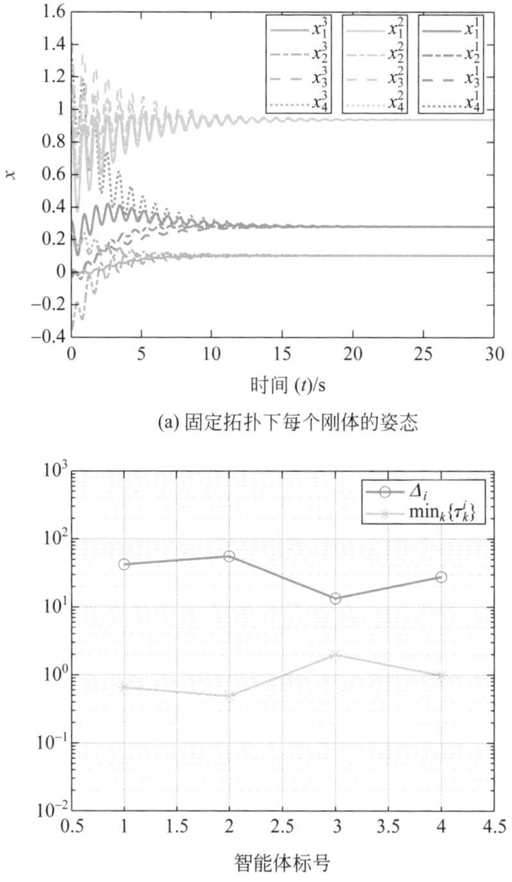

(a) 固定拓扑下每个刚体的姿态

(b) 固定拓扑下每个刚体的触发时刻

图 7-4　固定拓扑下的仿真结果

$$
\begin{aligned}
\mathcal{E}_1 &= \{(1,2),(2,1),(3,4),(4,3)\} \\
\mathcal{E}_2 &= \{(2,3),(3,2),(3,4),(4,3)\} \\
\mathcal{E}_3 &= \{(2,3),(3,2)\} \\
\mathcal{E}_4 &= \{(1,4),(4,1),(2,3),(3,2)\}
\end{aligned}
\tag{7-94}
$$

仿真结果由图7-5(a)、(b)给出。从图7-5(a)可看出，所有旋转向量都会趋于一致。图7-5(b)显示了每个刚体的触发次数和最小触发间隔。注意到，最小触发间隔比采样周期0.01大很多，这意味着不会发生奇诺现象。

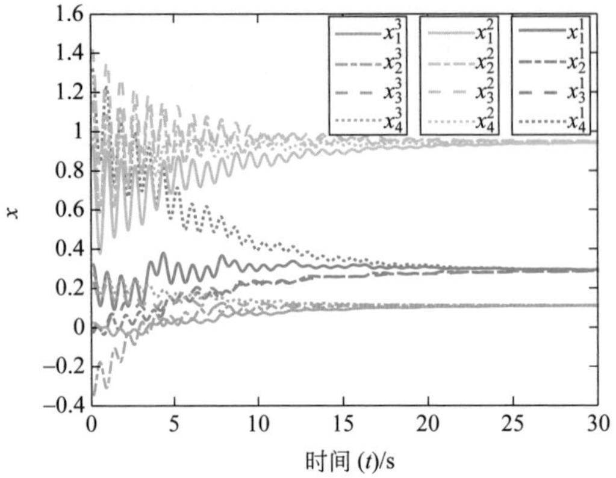

(a) 切换拓扑下每个刚体的姿态

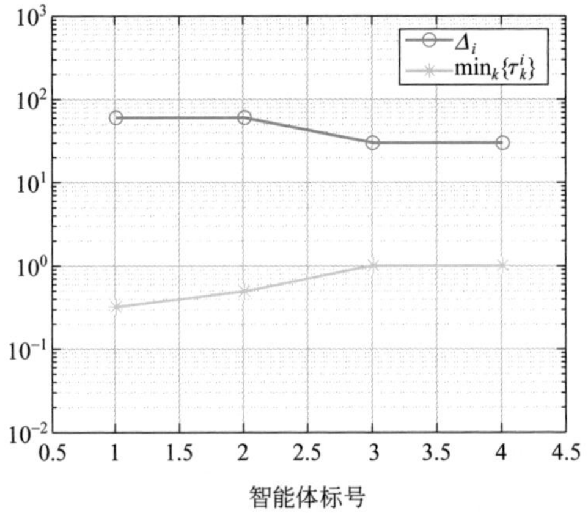

(b) 切换拓扑下每个刚体的触发时刻

图 7-5　切换拓扑下的仿真结果

Digital Wave
Advanced Technology of
Industrial Internet

Control of
Autonomous Intelligent Systems

自主智能系统控制

第8章 基于事件触发强化学习的多刚体系统最优一致性

8.1 概述

随着数字传感器技术的飞速发展与计算机运算能力的大幅度提升，携带着系统信息的大规模数据可以被很好地收集与处理，从而使得机器学习在控制领域发挥着越来越重要的作用。作为机器学习的三大分支之一，强化学习（Reinforcement Learning，RL）能够使智能体通过观察环境来选择自己的行动，从而追求最大化的长期回报[214]。与传统预先设计的控制策略不同，强化学习使得智能体在与环境进行信息交互的同时能够不断地在线学习，根据环境对不同的控制信号所产生的反馈信号来改进自己的控制策略，进而得到最优的控制策略。强化学习的优势在于可以解决大规模繁琐复杂的多级决策问题，尤其是当系统已知的模型信息较少或者是未知模型信息的时候。因此，本章针对多刚体系统，基于强化学习的方法对其最优一致性控制问题展开研究。

8.2 基于增广系统的无模型事件触发方法

8.2.1 问题描述

对于一个由 N 个节点组成的多刚体系统，每一个刚体可以借助下面的欧拉－拉格朗日方程进行建模：

$$\boldsymbol{M}_i\left(\boldsymbol{\eta}_i(t)\right)\ddot{\boldsymbol{\eta}}_i(t)+\boldsymbol{C}_i\left(\boldsymbol{\eta}_i(t),\dot{\boldsymbol{\eta}}_i(t)\right)\dot{\boldsymbol{\eta}}_i(t)+\boldsymbol{G}_i\left(\boldsymbol{\eta}_i(t)\right)=\boldsymbol{\tau}_i(t),\ i\in\mathcal{V} \tag{8-1}$$

其中，$\boldsymbol{\eta}_i(t)\in\mathbb{R}^n$表示广义坐标，即在广义坐标系下第 i 个刚体的状态向量；$\dot{\boldsymbol{\eta}}_i(t)$和$\ddot{\boldsymbol{\eta}}_i(t)$分别表示$\boldsymbol{\eta}_i(t)$关于时间 t 的一阶导数和二阶导数，为了表达的简洁性，我们在后文中将省略掉非必要的 t；$\boldsymbol{\tau}_i\in\mathbb{R}^n$表示第 i 个刚体的控制输入向量；$\boldsymbol{M}_i(\boldsymbol{\eta}_i):\mathbb{R}^n\rightarrow\mathbb{R}^{n\times n}$表示惯性矩阵；$\boldsymbol{C}_i(\boldsymbol{\eta}_i,\dot{\boldsymbol{\eta}}_i):\mathbb{R}^n\rightarrow\mathbb{R}^{n\times n}$表示科里奥利和离心转矩矩阵；$\boldsymbol{G}_i(\boldsymbol{\eta}_i):\mathbb{R}^n\rightarrow\mathbb{R}^n$表

示重力转矩向量。

对于上述的多刚体系统，有下面三条假设：

假设 8.1[215]

对于每个刚体，其位置信息$\boldsymbol{\eta}_i$和速度信息$\dot{\boldsymbol{\eta}}_i$是可测量的，加速度信息$\ddot{\boldsymbol{\eta}}_i$是可用的，即能通过传感器测量或者观测器观测得到$\ddot{\boldsymbol{\eta}}_i$。

假设 8.2[215]

惯性矩阵$\boldsymbol{M}_i\left(\boldsymbol{\eta}_i\right)$是对称的、正定的，并且对任意的$\boldsymbol{\xi}\in\mathbb{R}^n$，$\underline{m}_i\|\boldsymbol{\xi}\|^2\leqslant\boldsymbol{\xi}^{\mathrm{T}}\boldsymbol{M}_i\left(\boldsymbol{\eta}_i\right)\boldsymbol{\xi}\leqslant\overline{m}_i\|\boldsymbol{\xi}\|^2$均成立。其中，$\underline{m}_i\in\mathbb{R}_{>0}$是一个已知的正实数，$\overline{m}_i\in\mathbb{R}_{>0}$是一个已知的正有界函数。矩阵$\boldsymbol{M}_i^{-1}\left(\boldsymbol{\eta}_i\right)$存在并且是可以确定的，即$\|\boldsymbol{M}_i^{-1}\left(\boldsymbol{\eta}_i\right)\|\leqslant\overline{\boldsymbol{M}}$。其中，$\overline{\boldsymbol{M}}\in\mathbb{R}_{>0}$是一个可确定的正常数。

假设 8.3[215]

矩阵$\dot{\boldsymbol{M}}_i\left(\boldsymbol{\eta}_i\right)-2\boldsymbol{C}_i\left(\boldsymbol{\eta}_i,\dot{\boldsymbol{\eta}}_i\right)$满足斜对称的性质，即对任意的$\boldsymbol{\xi}\in\mathbb{R}^n$，都有$\boldsymbol{\xi}^{\mathrm{T}}\left(\dot{\boldsymbol{M}}_i\left(\boldsymbol{\eta}_i\right)-2\boldsymbol{C}_i\left(\boldsymbol{\eta}_i,\dot{\boldsymbol{\eta}}_i\right)\right)\boldsymbol{\xi}=0$。

为了后文分析的便利，现将上述欧拉－拉格朗日方程描述的多刚体系统的动力学模型写成紧凑的形式，如下所示：

$$\boldsymbol{M}\left(\boldsymbol{\eta}\right)\ddot{\boldsymbol{\eta}}+\boldsymbol{C}\left(\boldsymbol{\eta},\dot{\boldsymbol{\eta}}\right)\dot{\boldsymbol{\eta}}+\boldsymbol{G}\left(\boldsymbol{\eta}\right)=\boldsymbol{\tau} \tag{8-2}$$

其中：

$$\boldsymbol{\eta}=\left[\boldsymbol{\eta}_1^{\mathrm{T}}\ \boldsymbol{\eta}_2^{\mathrm{T}}\ \ldots\boldsymbol{\eta}_N^{\mathrm{T}}\right]^{\mathrm{T}}\in\mathbb{R}^{Nn}$$

$$\boldsymbol{\tau}=\left[\boldsymbol{\tau}_1^{\mathrm{T}}\ \boldsymbol{\tau}_2^{\mathrm{T}}\ \ldots\boldsymbol{\tau}_N^{\mathrm{T}}\right]^{\mathrm{T}}\in\mathbb{R}^{Nn}$$

$$\boldsymbol{M}\left(\boldsymbol{\eta}\right)=\mathrm{diag}\left\{\boldsymbol{M}_1\left(\boldsymbol{\eta}_1\right),\boldsymbol{M}_2\left(\boldsymbol{\eta}_2\right),\cdots,\boldsymbol{M}_N\left(\boldsymbol{\eta}_N\right)\right\}\in\mathbb{R}^{Nn\times Nn}$$

$$\boldsymbol{C}\left(\boldsymbol{\eta},\dot{\boldsymbol{\eta}}\right)=\mathrm{diag}\left\{\boldsymbol{C}_1\left(\boldsymbol{\eta}_1,\dot{\boldsymbol{\eta}}_1\right),\boldsymbol{C}_2\left(\boldsymbol{\eta}_2,\dot{\boldsymbol{\eta}}_2\right),\ldots,\boldsymbol{C}_N\left(\boldsymbol{\eta}_N,\dot{\boldsymbol{\eta}}_N\right)\right\}\in\mathbb{R}^{Nn\times Nn}$$

$$\boldsymbol{G}=\left[\boldsymbol{G}_1^{\mathrm{T}}\left(\boldsymbol{\eta}_1\right)\boldsymbol{G}_2^{\mathrm{T}}\left(\boldsymbol{\eta}_2\right)\ldots\boldsymbol{G}_N^{\mathrm{T}}\left(\boldsymbol{\eta}_N\right)\right]^{\mathrm{T}}\in\mathbb{R}^{Nn}$$

多刚体系统一致性控制的目标是驱动所有节点的状态趋于一致。因此，为了便于分析，可以引入局部一致性误差$v_{1,i}\in\mathbb{R}^n$，它被定义为下面的形式：

$$\boldsymbol{v}_{1,i}=\sum_{j\in\mathcal{N}_i}e_{ij}\left(\boldsymbol{\eta}_i-\boldsymbol{\eta}_j\right) \tag{8-3}$$

由于式 (8-1) 描述的刚体系统是一个二阶系统，考虑到将位置信息和速度信息融合，定义以下辅助状态向量 [216]：

$$\boldsymbol{\zeta}_i = c\boldsymbol{\eta}_i + \dot{\boldsymbol{\eta}}_i \tag{8-4}$$

其中，$c \in \mathbb{N}_{>0}$是一个正常数。

因此，参照式 (8-3)，可以定义以下形式的辅助一致性误差：

$$\begin{aligned}\boldsymbol{v}_{2,i} &= \sum_{j\in\mathcal{N}_i} e_{ij}\left(\boldsymbol{\zeta}_i - \boldsymbol{\zeta}_j\right) = \sum_{j\in\mathcal{N}_i} e_{ij}\left(c\boldsymbol{\eta}_i + \dot{\boldsymbol{\eta}}_i - c\boldsymbol{\eta}_j - \dot{\boldsymbol{\eta}}_j\right)\\ &= c\sum_{j\in\mathcal{N}_i} e_{ij}\left(\boldsymbol{\eta}_i - \boldsymbol{\eta}_j\right) + \sum_{j\in\mathcal{N}_i} e_{ij}\left(\dot{\boldsymbol{\eta}}_i - \dot{\boldsymbol{\eta}}_j\right)\\ &= c\boldsymbol{v}_{1,i} + \dot{\boldsymbol{v}}_{1,i}\end{aligned} \tag{8-5}$$

辅助一致性误差$\boldsymbol{v}_{2,i}$可以用来衡量多刚体系统是否实现了状态一致性。当$\lim\limits_{t\to\infty}\boldsymbol{v}_{2,i} = 0$成立时，可以推导出$\lim\limits_{t\to\infty}\boldsymbol{v}_{1,i} = 0$与$\lim\limits_{t\to\infty}\dot{\boldsymbol{v}}_{1,i} = 0$。也就是说，在多刚体系统的通信拓扑是强连通的情况下，所有刚体的状态达成了一致。

为了更简洁地表示多刚体系统的一致性控制问题，式 (8-3) 和式 (8-5) 可以被重写为下面紧凑的形式：

$$\boldsymbol{v}_1 = \boldsymbol{\Sigma}\boldsymbol{\eta}, \tag{8-6}$$

$$\boldsymbol{v}_2 = \dot{\boldsymbol{v}}_1 + \bar{\boldsymbol{c}}\boldsymbol{v}_1 \tag{8-7}$$

其中：

$$\begin{aligned}\boldsymbol{v}_1 &= \left[\boldsymbol{v}_{1,1}^{\mathrm{T}}\ \boldsymbol{v}_{1,2}^{\mathrm{T}} \dots \boldsymbol{v}_{1,N}^{\mathrm{T}}\right] \in \mathbb{R}^{Nn}\\ \boldsymbol{v}_2 &= \left[\boldsymbol{v}_{2,1}^{\mathrm{T}}\ \boldsymbol{v}_{2,2}^{\mathrm{T}} \dots \boldsymbol{v}_{2,N}^{\mathrm{T}}\right] \in \mathbb{R}^{Nn}\\ \boldsymbol{\Sigma} &= \boldsymbol{\mathcal{L}} \otimes \boldsymbol{I}_n \in \mathbb{R}^{Nn\times Nn}\\ \tilde{\boldsymbol{c}} &= c\boldsymbol{I}_N \in \mathbb{R}^{N\times N}\\ \bar{\boldsymbol{c}} &= \tilde{\boldsymbol{c}} \otimes \boldsymbol{I}_n \in \mathbb{R}^{Nn\times Nn}\end{aligned}$$

将式 (8-5) 的两边分别对时间 t 求导，可以得到辅助一致性误差 $\boldsymbol{v}_{2,i}$ 的动态特性，如下所示：

$$\dot{\boldsymbol{v}}_{2,i} = \boldsymbol{\Sigma}_i\ddot{\boldsymbol{\eta}} + \bar{\boldsymbol{c}}_i\boldsymbol{\Sigma}\dot{\boldsymbol{\eta}} = \boldsymbol{\Psi}(t) + \boldsymbol{\Phi}(t)\boldsymbol{\tau} \tag{8-8}$$

其中，$\boldsymbol{\Psi}(t) = -\boldsymbol{\Sigma}_i\boldsymbol{M}^{-1}\left(\boldsymbol{C}\dot{\boldsymbol{\eta}} + \boldsymbol{G}\right) + \bar{\boldsymbol{c}}_i\boldsymbol{\Sigma}\dot{\boldsymbol{\eta}}$；$\boldsymbol{\Phi}(t) = \boldsymbol{\Sigma}_i\boldsymbol{M}^{-1}$；$\boldsymbol{\Sigma}_i = \boldsymbol{\mathcal{L}}_i \otimes \boldsymbol{I}_n \in \mathbb{R}^{n\times Nn}$；$\bar{\boldsymbol{c}}_i = \tilde{\boldsymbol{c}}_i \otimes \boldsymbol{I}_n \in \mathbb{R}^{n\times Nn}$；$\boldsymbol{\mathcal{L}}_i$和$\tilde{\boldsymbol{c}}_i$均表示行向量；$\boldsymbol{\mathcal{L}}_i$表示拉普拉斯矩阵$\boldsymbol{\mathcal{L}}$的第$i$行，$\widetilde{\boldsymbol{c}_i}$表示矩阵$\tilde{\boldsymbol{c}}$的第$i$行。

因此，我们可以将多刚体系统的一致性控制问题概括成为每个刚体设计合适的控制策略 $\boldsymbol{\tau}_i$，从而使得辅助一致性误差 $\boldsymbol{v}_{2,i}$ 趋于 0。在本章中，我们的目标是设计一种不依赖于系统模型矩阵的控制策略，在实现多刚体系统的一致性的同时，还能够将性能成本降至最低。进一步地，我们还考虑了融入事件触发控制，从而降低计算资源的消耗。

8.2.2 增广系统的设计

为了克服最优控制器对模型信息矩阵的依赖，本小节构造了一个增广系统，它包含了辅助一致性误差的动态特性 [式 (8-8)] 和一个预先设计的补偿器 [式 (8-9)]。首先，对于刚体 i，参照文献 [217] 与文献 [218]，设计如下所示的预补偿器：

$$\dot{\boldsymbol{\tau}}_i = \boldsymbol{\gamma}_i\left(\boldsymbol{\tau}_i\right) + \boldsymbol{\pi}_i\left(\boldsymbol{\tau}_i\right)\boldsymbol{\omega}_i \tag{8-9}$$

它是一个非线性仿射微分方程。其中，状态$\boldsymbol{\tau}_i$为刚体i原本的控制输入，控制输入$\boldsymbol{\omega}_i$则是一个新定义的变量。向量$\boldsymbol{\tau}_i$与向量$\boldsymbol{\omega}_i$具有相同的维度，即有$\boldsymbol{\omega}_i \in \mathbb{R}^n$。函数$\boldsymbol{\gamma}_i\left(\boldsymbol{\tau}_i\right)$与函数$\boldsymbol{\pi}_i\left(\boldsymbol{\tau}_i\right)$的设计需要保证补偿器[式(8-9)]是可控的。在本章中，$\boldsymbol{\gamma}_i\left(\boldsymbol{\tau}_i\right)$与$\boldsymbol{\pi}_i\left(\boldsymbol{\tau}_i\right)$分别被设计为：$\boldsymbol{\gamma}_i\left(\boldsymbol{\tau}_i\right) = -2\boldsymbol{\tau}_i$，$\boldsymbol{\pi}_i(\boldsymbol{\tau}_i) = \operatorname{diag}\{\cos^2(\boldsymbol{\tau}_i^{(1)}), \cos^2(\boldsymbol{\tau}_i^{(2)}), \cdots, \cos^2(\boldsymbol{\tau}_i^{(n)})\}$。

通过结合式 (8-8) 与式 (8-9)，可以构造出以下形式的增广系统：

$$\dot{\bar{\boldsymbol{v}}}_{2,i} = \boldsymbol{\Gamma}_i\left(\bar{\boldsymbol{v}}_{2,i}\right) + \boldsymbol{\Pi}_i\left(\bar{\boldsymbol{v}}_{2,i}\right)\boldsymbol{\omega}_i \tag{8-10}$$

其中，增广辅助误差向量$\bar{\boldsymbol{v}}_{2,i} = \left[\boldsymbol{v}_{2,i}^{\mathrm{T}}\ \boldsymbol{\tau}_i^{\mathrm{T}}\right]^{\mathrm{T}} \in \mathbb{R}^{2n}$。对于增广系统 (8-10)，预补偿器[式(8-9)]的控制输入$\boldsymbol{\omega}_i$即为其控制输入。$\boldsymbol{\Gamma}_i\left(\bar{\boldsymbol{v}}_{2,i}\right)$与$\boldsymbol{\Pi}_i\left(\bar{\boldsymbol{v}}_{2,i}\right)$的形式如下：

$$\boldsymbol{\Gamma}_i\left(\bar{\boldsymbol{v}}_{2,i}\right) = \begin{bmatrix} \boldsymbol{\Psi}(t) + \boldsymbol{\Phi}(t)\boldsymbol{\tau} \\ \boldsymbol{\gamma}_i\left(\boldsymbol{\tau}_i\right) \end{bmatrix} \in \mathbb{R}^{2n} \boldsymbol{\Pi}_i\left(\bar{\boldsymbol{v}}_{2,i}\right) = \begin{bmatrix} \boldsymbol{O}_{n\times n} \\ \boldsymbol{\pi}_i\left(\boldsymbol{\tau}_i\right) \end{bmatrix} \in \mathbb{R}^{2n\times n}$$

根据预先设计的$\boldsymbol{\pi}_i\left(\boldsymbol{\tau}_i\right)$，我们可以得到$\boldsymbol{\Pi}_i\left(\bar{\boldsymbol{v}}_{2,i}\right)$是有界的。也就是说，存在常数$\Pi_M \in \mathbb{R}_{>0}$，使得$\|\boldsymbol{\Pi}_i\left(\bar{\boldsymbol{v}}_{2,i}\right)\| \leqslant \Pi_M$成立。

假设 8.4

存在常数$\boldsymbol{\Gamma}_M \in \mathbb{R}_{>0}$，使得增广系统 (8-10) 中的函数$\boldsymbol{\Gamma}_i\left(\bar{\boldsymbol{v}}_{2,i}\right)$满足$\boldsymbol{\Gamma}_i\left(\bar{\boldsymbol{v}}_{2,i}\right) \leqslant \boldsymbol{\Gamma}_M \|\bar{\boldsymbol{v}}_{2,i}\|$。

定义如下所示的性能成本函数，它包括增广辅助误差的二次型以及增广系统的控制输入向量的二次型：

$$K_i\left(\bar{\boldsymbol{v}}_{2,i}(0),\boldsymbol{\omega}_i,\boldsymbol{\omega}_{-i}\right)=\int_0^{\infty}\left(\bar{\boldsymbol{v}}_{2,i}^{\mathrm{T}}\boldsymbol{X}_i\bar{\boldsymbol{v}}_{2,i}+\boldsymbol{\omega}_i^{\mathrm{T}}\boldsymbol{Z}_i\boldsymbol{\omega}_i\right)\mathrm{d}t \tag{8-11}$$

其中，$\boldsymbol{X}_i\in\mathbb{R}^{2n\times 2n}$；$\boldsymbol{Z}_i\in\mathbb{R}^{n\times n}$；$\boldsymbol{X}_i\geqslant 0$；$\boldsymbol{Z}_i>0$；$\boldsymbol{\omega}_{-i}=\left\{\boldsymbol{\omega}_j \mid j\in\mathcal{N}_i\right\}$。从式 (8-9) 和式 (8-10) 中可以看出，$\bar{\boldsymbol{v}}_{2,i}$是受 $\boldsymbol{\omega}_i$ 与 $\boldsymbol{\omega}_{-i}$ 的驱动的。所以，等式 (8-11) 左边的自变量同时包含 $\boldsymbol{\omega}_i$ 与 $\boldsymbol{\omega}_{-i}$。

综上所述，多刚体系统的最优一致性控制问题可以被描述成为所有刚体设计一组控制协议$\boldsymbol{\omega}_1^*,\boldsymbol{\omega}_2^*,\cdots,\boldsymbol{\omega}_N^*$，它们在驱动增广辅助误差$\bar{\boldsymbol{v}}_{2,i}$趋于 0 的同时，还能够保证性能成本函数 (8-11) 达到最小值。

8.2.3 无模型最优控制器的设计

根据性能成本函数 (8-11)，可以定义以下形式的价值函数：

$$\boldsymbol{F}_i\left(\bar{\boldsymbol{v}}_{2,i}(t)\right)=\int_t^{\infty}\left(\bar{\boldsymbol{v}}_{2,i}^{\mathrm{T}}(s)\boldsymbol{X}_i\bar{\boldsymbol{v}}_{2,i}(s)+\boldsymbol{\omega}_i^{\mathrm{T}}(s)\boldsymbol{Z}_i\boldsymbol{\omega}_i(s)\right)\mathrm{d}s \tag{8-12}$$

其中，s表示积分变量。

通过对价值函数 (8-12) 求导可以得到相应的哈密顿函数，如下所示：

$$\begin{aligned}&\mathcal{H}_i\left(\bar{\boldsymbol{v}}_{2,i},\nabla F_i,\boldsymbol{\omega}_i,\boldsymbol{\omega}_{-i}\right)\\&=\bar{\boldsymbol{v}}_{2,i}^{\mathrm{T}}\boldsymbol{X}_i\bar{\boldsymbol{v}}_{2,i}+\boldsymbol{\omega}_i^{\mathrm{T}}\boldsymbol{Z}_i\boldsymbol{\omega}_i+\left(\nabla F_i\right)^{\mathrm{T}}\left(\boldsymbol{\Gamma}_i\left(\bar{\boldsymbol{v}}_{2,i}\right)+\boldsymbol{\Pi}_i\left(\bar{\boldsymbol{v}}_{2,i}\right)\boldsymbol{\omega}_i\right)\\&=0\end{aligned} \tag{8-13}$$

其中，$\nabla F_i=\dfrac{\partial \boldsymbol{F}_i}{\partial \bar{\boldsymbol{v}}_{2,i}}$。

通过使用$\dfrac{\partial \mathcal{H}_i}{\partial \boldsymbol{\omega}_i}=0$，可以推导出最优控制器的隐式解为：

$$\boldsymbol{\omega}_i^*=-\frac{1}{2}\boldsymbol{Z}_i^{-1}\boldsymbol{\Pi}_i^{\mathrm{T}}\left(\bar{\boldsymbol{v}}_{2,i}\right)\nabla F_i^* \tag{8-14}$$

其中，$\nabla F_i^*=\dfrac{\partial \boldsymbol{F}_i^*}{\partial \bar{\boldsymbol{v}}_{2,i}}$，$\boldsymbol{F}_i^*$表示最优价值函数。

当刚体 i 以及它的邻居的控制器都采用式 (8-14) 给出的最优控制器时，可以得到如下所示的耦合 HJB 方程：

$$
\begin{aligned}
&\mathcal{H}_i\left(\bar{\boldsymbol{v}}_{2,i}, \nabla F_i^*, \boldsymbol{\omega}_i^*, \boldsymbol{\omega}_{-i}^*\right) \\
&= \bar{\boldsymbol{v}}_{2,i}^{\mathrm{T}} \boldsymbol{X}_i \bar{\boldsymbol{v}}_{2,i} + \left(\boldsymbol{\omega}_i^*\right)^{\mathrm{T}} \boldsymbol{Z}_i \boldsymbol{\omega}_i^* + \left(\nabla F_i^*\right)^{\mathrm{T}} \left(\boldsymbol{\Gamma}_i\left(\bar{\boldsymbol{v}}_{2,i}\right) + \boldsymbol{\Pi}_i\left(\bar{\boldsymbol{v}}_{2,i}\right)\boldsymbol{\omega}_i^*\right) \\
&= \bar{\boldsymbol{v}}_{2,i}^{\mathrm{T}} \boldsymbol{X}_i \bar{\boldsymbol{v}}_{2,i} - \frac{1}{4}\left(\nabla \boldsymbol{F}_i^*\right)^{\mathrm{T}} \boldsymbol{\Pi}_i\left(\bar{\boldsymbol{v}}_{2,i}\right) \boldsymbol{Z}_i^{-1} \boldsymbol{\Pi}_i^{\mathrm{T}}\left(\bar{\boldsymbol{v}}_{2,i}\right) \nabla F_i^* + \left(\nabla F_i^*\right)^{\mathrm{T}} \boldsymbol{\Gamma}_i\left(\bar{\boldsymbol{v}}_{2,i}\right) = 0
\end{aligned}
\tag{8-15}
$$

从式 (8-14) 中可以看出，最优控制器的隐式解包含$\boldsymbol{\Pi}_i\left(\bar{\boldsymbol{v}}_{2,i}\right)$与最优价值函数$\boldsymbol{F}_i^*$。相比于第 3 章，本章的最优控制器避免了使用模型信息矩阵$\boldsymbol{M}_i\left(\boldsymbol{\eta}_i\right)$，转而使用了增广系统的矩阵$\boldsymbol{\Pi}_i\left(\bar{\boldsymbol{v}}_{2,i}\right)$。因为矩阵$\boldsymbol{\Pi}_i\left(\bar{\boldsymbol{v}}_{2,i}\right)$是预先构造的，所以使用增广系统克服了对模型信息的依赖。在 8.3 节中，我们将给出一个无模型事件触发的强化学习算法在线获取$\boldsymbol{F}_i^*$，该算法只需要使用事件触发时刻的测量数据。因此，本章的最优控制器的实现不需要预先知道刚体的模型信息矩阵$\boldsymbol{M}_i\left(\boldsymbol{\eta}_i\right)$。

8.2.4 事件触发机制的引入

不同于按照固定周期采样的时间触发控制，事件触发控制只在一些离散的时刻采样。本小节中，我们考虑在静态事件触发机制下实现多刚体系统的最优一致性控制，以减少计算资源的消耗。

在静态事件触发机制下，我们只需要在一些满足要求的离散时刻$\left\{t_i^q\right\}_{q\in\mathbb{N}}$更新控制器，其中，对任意的$q\in\mathbb{N}_{>0}$，都有$t_i^{q+1} > t_i^q > 0$，初始的事件触发时刻被设置为$t_i^0 = 0$，在任意的事件触发间隔期间，即$t\in\left(t_i^q, t_i^{q+1}\right)$，控制器保持不变。

定义增广系统状态的测量误差如下：

$$
\boldsymbol{\epsilon}_i(t) = \hat{\bar{\boldsymbol{v}}}_{2,i} - \bar{\boldsymbol{v}}_{2,i}(t),\quad \forall t\in\left[t_i^q, t_i^{q+1}\right) \tag{8-16}
$$

其中，$\bar{\boldsymbol{v}}_{2,i}(t)$表示实时的增广辅助误差，$\hat{\bar{\boldsymbol{v}}}_{2,i}$表示在事件触发时刻$t_i^q$时的增广辅助误差$\bar{\boldsymbol{v}}_{2,i}\left(t_i^q\right)$。在每一个事件触发时刻，$\boldsymbol{\epsilon}_i(t)$被重新设置为0。

因此，在静态事件触发机制下，增广系统 (8-10) 可以转化为：

$$
\dot{\bar{\boldsymbol{v}}}_{2,i} = \boldsymbol{\Gamma}_i\left(\bar{\boldsymbol{v}}_{2,i}\right) + \boldsymbol{\Pi}_i\left(\bar{\boldsymbol{v}}_{2,i}\right)\hat{\boldsymbol{\omega}}_i \tag{8-17}
$$

其中，$\hat{\boldsymbol{\omega}}_i = \boldsymbol{\omega}_i\left(t_i^q\right)$表示刚体$i$在事件触发时刻$t_i^q$时的控制输入向量。

定义 8.1

对于增广辅助误差$\bar{\boldsymbol{v}}_{2,i}$的动态特性 (8-17)，如果控制输入$\boldsymbol{\omega}_i$是分段连续的，且$\boldsymbol{\omega}_i(0)=0$，当$\boldsymbol{\omega}_i$能够使得下面两个条件被满足时，则称控制输入$\boldsymbol{\omega}_i$为事件触发容许控制。

① 系统 (8-17) 是渐近稳定的，即$\lim\limits_{t\to\infty}\bar{\boldsymbol{v}}_{2,i}=0$。

② 性能成本函数 (8-11) 是有限的，即$K_i<+\infty$。

当增广系统状态的测量误差违背下面的不等式时：

$$\|\boldsymbol{\epsilon}_i(t)\|\leqslant\epsilon_i^H(t) \tag{8-18}$$

就代表一个新的事件触发时刻的到来。$\epsilon_i^H(t)$表示静态事件触发阈值，它的设计会直接影响到增广系统的稳定性以及奇诺现象的发生与否。我们将在8.3.3小节给出合适的静态事件触发阈值。

根据最优控制器 (8-14)，我们可以获取静态事件触发机制下的最优控制器，如下所示：

$$\hat{\boldsymbol{\omega}}_i^*=-\frac{1}{2}\boldsymbol{Z}_i^{-1}\boldsymbol{\varPi}_i^{\mathrm{T}}\left(\hat{\bar{\boldsymbol{v}}}_{2,i}\right)\nabla\hat{F}_i^* \tag{8-19}$$

假设 8.5

在事件触发间隔$t\in\left(t_i^q,t_i^{q+1}\right)$期间，连续时间容许控制$\boldsymbol{\omega}_i$是利普希茨连续的，$\boldsymbol{\omega}_i$依赖于$\bar{\boldsymbol{v}}_{2,i}$，$\hat{\boldsymbol{\omega}}_i$依赖于$\hat{\bar{\boldsymbol{v}}}_{2,i}$，即$\boldsymbol{\omega}_i=\boldsymbol{\omega}_i\left(\bar{\boldsymbol{v}}_{2,i}\right),\hat{\boldsymbol{\omega}}_i=\boldsymbol{\omega}_i\left(\hat{\bar{\boldsymbol{v}}}_{2,i}\right)$，有：

$$\begin{aligned}\|\boldsymbol{\omega}_i-\hat{\boldsymbol{\omega}}_i\|&=\|\boldsymbol{\omega}_i\left(\bar{\boldsymbol{v}}_{2,i}\right)-\boldsymbol{\omega}_i\left(\hat{\bar{\boldsymbol{v}}}_{2,i}\right)\|\\&=\|\boldsymbol{\omega}_i\left(\bar{\boldsymbol{v}}_{2,i}\right)-\boldsymbol{\omega}_i\left(\bar{\boldsymbol{v}}_{2,i}+\boldsymbol{\epsilon}_i\right)\|\\&\leqslant Y\|\boldsymbol{\epsilon}_i\|\end{aligned} \tag{8-20}$$

其中，Y表示利普希茨常数，它的选取需要不小于$\left\|\dfrac{\partial\boldsymbol{\omega}_i}{\partial\bar{\boldsymbol{v}}_{2,i}^{\mathrm{T}}}\right\|$的最大值[219]。

在静态事件触发机制下，可以得到新的耦合 HJB 方程，如下所示：

$$\begin{aligned}&\mathcal{H}_i\left(\bar{\boldsymbol{v}}_{2,i},\nabla\hat{\boldsymbol{F}}_i^*,\hat{\boldsymbol{\omega}}_i^*,\hat{\boldsymbol{\omega}}_{-i}^*\right)\\&=\bar{\boldsymbol{v}}_{2,i}^{\mathrm{T}}\boldsymbol{X}_i\bar{\boldsymbol{v}}_{2,i}+\left(\hat{\boldsymbol{\omega}}_i^*\right)^{\mathrm{T}}\boldsymbol{Z}_i\hat{\boldsymbol{\omega}}_i^*+\left(\nabla\hat{F}_i^*\right)^{\mathrm{T}}\left(\boldsymbol{\varGamma}_i\left(\bar{\boldsymbol{v}}_{2,i}\right)+\boldsymbol{\varPi}_i\left(\bar{\boldsymbol{v}}_{2,i}\right)\hat{\boldsymbol{\omega}}_i^*\right)=0\end{aligned} \tag{8-21}$$

在静态事件触发机制下，对于所有的事件触发容许控制，HJB 方程 (8-21) 的解$\hat{F}_i^*$能够使得控制器 [式 (8-19)] 是最优的。对比时间触发控制，事件触发控制使得系统计算资源的消耗减少了，但是性能成本却增加了。也就是说，该方法以牺牲少量的性能成本为代价大幅度地减少了系统计算资源的消耗。当事件触发阈值$\epsilon_i^H(t)$趋于 0 时，方程 (8-21) 的解将与方程 (8-15) 的解趋于一致 [220]。

8.3 基于事件触发强化学习的算法实现

8.3.1 无模型事件触发的强化学习算法

为了求解静态事件触发机制下的 HJB 方程 (8-21)，本节提出了一个无模型事件触发的强化学习算法，如算法 8.1 所示。在每个事件触发时刻，重复地执行策略评估与策略改进，直到算法收敛。每个刚体通过在事件触发时刻执行算法 8.1，从而能够实现多刚体系统的最优一致性控制。下面，我们给出定理 8.1 来保证算法 8.1 的收敛性。

算法 8.1

无模型事件触发的强化学习算法。

① 初始化增广系统的控制输入向量$\hat{\boldsymbol{\omega}}_i^{(0)}$，$i\in\mathcal{V}$，并设置迭代索引的初始值为$q=0$。

② 在每个事件触发时刻重复执行以下步骤：

a．策略评估：对于刚体$i,i\in\mathcal{V}$，通过使用给定的$\hat{\boldsymbol{\omega}}_i^{(q)}$求解下述事件触发耦合 HJB 方程 (8-21) 的解$\hat{\boldsymbol{F}}_i^{(q+1)}$：

$$\mathcal{H}_i\left(\bar{\boldsymbol{v}}_{2,i},\nabla\hat{F}_i^{(q+1)},\hat{\boldsymbol{\omega}}_i^{(q)},\hat{\boldsymbol{\omega}}_{-i}^{(q)}\right)$$
$$=\bar{\boldsymbol{v}}_{2,i}^{\mathrm{T}}\boldsymbol{X}_i\bar{\boldsymbol{v}}_{2,i}+\left(\hat{\boldsymbol{\omega}}_i^{(q)}\right)^{\mathrm{T}}\boldsymbol{Z}_i\hat{\boldsymbol{\omega}}_i^{(q)}+\left(\nabla\hat{F}_i^{(q+1)}\right)^{\mathrm{T}}\left(\boldsymbol{\Gamma}_i\left(\bar{\boldsymbol{v}}_{2,i}\right)+\boldsymbol{\Pi}_i\left(\bar{\boldsymbol{v}}_{2,i}\right)\hat{\boldsymbol{\omega}}_i^{(q)}\right)=0 \tag{8-22}$$

b．策略改进：

$$\hat{\boldsymbol{\omega}}_i^{(q+1)}=-\frac{1}{2}\boldsymbol{Z}_i^{-1}\boldsymbol{\Pi}_i^{\mathrm{T}}\left(\hat{\bar{\boldsymbol{v}}}_{2,i}\right)\nabla\hat{F}_i^{(q+1)} \tag{8-23}$$

c．当$\|\hat{\boldsymbol{F}}_i^{(q)}-\hat{\boldsymbol{F}}_i^{(q+1)}\|\leqslant\delta$时，退出循环，其中$\delta\in\mathbb{R}_{>0}$表示一个较小的正常数。否则，令$q=q+1$，继续循环。

定理 8.1

当N个节点构成的多刚体系统对应的通信拓扑是强连通时，假设每个刚体都执行算法 8.1，则有下面的两条结论成立：

① 方程 (8-22) 的解逼近于最优价值函数，即$\lim\limits_{q\to\infty}\hat{\boldsymbol{F}}_i^{(q)}=\hat{\boldsymbol{F}}_i^{*}$；

② 控制策略 (8-23) 逼近于最优控制策略，即$\lim\limits_{q\to\infty}\hat{\boldsymbol{\omega}}_i^{(q)}=\hat{\boldsymbol{\omega}}_i^{*}$。

证明：

对于刚体i，我们可以得到：

$$\begin{aligned}&\mathcal{H}_i\left(\bar{\boldsymbol{v}}_{2,i},\nabla\hat{F}_i^{(q)},\hat{\boldsymbol{\omega}}_i^{(q-1)},\hat{\boldsymbol{\omega}}_{-i}^{(q-1)}\right)\\&=\bar{\boldsymbol{v}}_{2,i}^{\mathrm{T}}\boldsymbol{X}_i\bar{\boldsymbol{v}}_{2,i}+\left(\hat{\boldsymbol{\omega}}_i^{(q-1)}\right)^{\mathrm{T}}\boldsymbol{Z}_i\hat{\boldsymbol{\omega}}_i^{(q-1)}+\left(\nabla\hat{F}_i^{(q)}\right)^{\mathrm{T}}\left(\boldsymbol{\Gamma}_i\left(\bar{\boldsymbol{v}}_{2,i}\right)+\boldsymbol{\Pi}_i\left(\bar{\boldsymbol{v}}_{2,i}\right)\hat{\boldsymbol{\omega}}_i^{(q-1)}\right)\\&=0\end{aligned}\tag{8-24}$$

以及

$$\begin{aligned}&\mathcal{H}_i\left(\bar{\boldsymbol{v}}_{2,i},\nabla\hat{F}_i^{(q+1)},\hat{\boldsymbol{\omega}}_i^{(q)},\hat{\boldsymbol{\omega}}_{-i}^{(q)}\right)\\&=\bar{\boldsymbol{v}}_{2,i}^{\mathrm{T}}\boldsymbol{X}_i\bar{\boldsymbol{v}}_{2,i}+\left(\hat{\boldsymbol{\omega}}_i^{(q)}\right)^{\mathrm{T}}\boldsymbol{Z}_i\hat{\boldsymbol{\omega}}_i^{(q)}+\left(\nabla\hat{F}_i^{(q+1)}\right)^{\mathrm{T}}\left(\boldsymbol{\Gamma}_i\left(\bar{\boldsymbol{v}}_{2,i}\right)+\boldsymbol{\Pi}_i\left(\bar{\boldsymbol{v}}_{2,i}\right)\hat{\boldsymbol{\omega}}_i^{(q)}\right)\\&=0\end{aligned}\tag{8-25}$$

当增广辅助误差$\bar{\boldsymbol{v}}_{2,i}$是由$\hat{\boldsymbol{\omega}}_i^{(q)}$驱动时，即$\dot{\bar{\boldsymbol{v}}}_{2,i}=\boldsymbol{\Gamma}_i\left(\bar{\boldsymbol{v}}_{2,i}\right)+\boldsymbol{\Pi}_i\left(\bar{\boldsymbol{v}}_{2,i}\right)\hat{\boldsymbol{\omega}}_i^{(q)}$，结合等式 (8-24) 与等式 (8-25)，有下式成立：

$$\begin{aligned}&\dot{\hat{\boldsymbol{F}}}_{i(q)}-\dot{\hat{\boldsymbol{F}}}_{i(q+1)}\\&=\left(\nabla\hat{F}_i^{(q)}\right)^{\mathrm{T}}\dot{\bar{\boldsymbol{v}}}_{2,i}-\left(\nabla\hat{\boldsymbol{F}}_i^{(q+1)}\right)^{\mathrm{T}}\dot{\bar{\boldsymbol{v}}}_{2,i}\\&=\left(\nabla\hat{F}_i^{(q)}\right)^{\mathrm{T}}\left(\boldsymbol{\Gamma}_i\left(\bar{\boldsymbol{v}}_{2,i}\right)+\boldsymbol{\Pi}_i\left(\bar{\boldsymbol{v}}_{2,i}\right)\hat{\boldsymbol{\omega}}_i^{(q)}\right)-\left(\nabla\hat{F}_i^{(q+1)}\right)^{\mathrm{T}}\left(\boldsymbol{\Gamma}_i\left(\bar{\boldsymbol{v}}_{2,i}\right)+\boldsymbol{\Pi}_i\left(\bar{\boldsymbol{v}}_{2,i}\right)\hat{\boldsymbol{\omega}}_i^{(q)}\right)\\&=\left(\hat{\boldsymbol{\omega}}_i^{(q)}\right)^{\mathrm{T}}\boldsymbol{Z}_i\hat{\boldsymbol{\omega}}_i^{(q)}-\left(\hat{\boldsymbol{\omega}}_i^{(q-1)}\right)^{\mathrm{T}}\boldsymbol{Z}_i\hat{\boldsymbol{\omega}}_i^{(q-1)}+\left(\nabla\hat{F}_i^{(q)}\right)^{\mathrm{T}}\boldsymbol{\Pi}_i\left(\bar{\boldsymbol{v}}_{2,i}\right)\left(\hat{\boldsymbol{\omega}}_i^{(q)}-\hat{\boldsymbol{\omega}}_i^{(q-1)}\right)\end{aligned}\tag{8-26}$$

根据等式 (8-23)，我们可以得到：

$$\left(\nabla \hat{F}_i^{(q)}\right)^{\mathrm{T}} \boldsymbol{\Pi}_i\left(\overline{\boldsymbol{v}}_{2,i}\right) = -2\left(\hat{\boldsymbol{\omega}}_i^{(q)}\right)^{\mathrm{T}} \boldsymbol{Z}_i \tag{8-27}$$

因此，等式 (8-26) 可以转换为：

$$\begin{aligned}\dot{\hat{\boldsymbol{F}}}_i^{(q)} - \dot{\hat{\boldsymbol{F}}}_i^{(q+1)} &= \left(\hat{\boldsymbol{\omega}}_i^{(q)}\right)^{\mathrm{T}} \boldsymbol{Z}_i \hat{\boldsymbol{\omega}}_i^{(q)} - \left(\hat{\boldsymbol{\omega}}_i^{(q-1)}\right)^{\mathrm{T}} \boldsymbol{Z}_i \hat{\boldsymbol{\omega}}_i^{(q-1)} - 2\left(\hat{\boldsymbol{\omega}}_i^{(q)}\right)^{\mathrm{T}} \boldsymbol{Z}_i \left(\hat{\boldsymbol{\omega}}_i^{(q)} - \hat{\boldsymbol{\omega}}_i^{(q-1)}\right) \\ &= -\left(\hat{\boldsymbol{\omega}}_i^{(q)} - \hat{\boldsymbol{\omega}}_i^{(q-1)}\right)^{\mathrm{T}} \boldsymbol{Z}_i \left(\hat{\boldsymbol{\omega}}_i^{(q)} - \hat{\boldsymbol{\omega}}_i^{(q-1)}\right)\end{aligned} \tag{8-28}$$

因为矩阵$\boldsymbol{Z}_i$满足$\boldsymbol{Z}_i > 0$，所以有不等式$\dot{\hat{\boldsymbol{F}}}_i^{(q)} \leqslant \dot{\hat{\boldsymbol{F}}}_i^{(q+1)}$成立。

在区间$[t,+\infty)$上对不等式$\dot{\hat{\boldsymbol{F}}}_i^{(q)} \leqslant \dot{\hat{\boldsymbol{F}}}_i^{(q+1)}$积分，我们可以得到：

$$\hat{\boldsymbol{F}}_i^{(q+1)} \leqslant \hat{\boldsymbol{F}}_i^{(q)} \tag{8-29}$$

因此，$\hat{\boldsymbol{F}}_i^{(q)}$是一个非递增函数，即有$\lim\limits_{q\to\infty} \hat{\boldsymbol{F}}_i^{(q)} = \hat{\boldsymbol{F}}_i^{\infty} \leqslant \hat{\boldsymbol{F}}_i^{*}$成立。

因为$\hat{\boldsymbol{F}}_i^{*}$为静态事件触发机制下的最优价值函数，根据最优性准则，有下面的不等式成立：

$$\hat{\boldsymbol{F}}_i^{(q)} \geqslant \int_0^{\infty}\left(\overline{\boldsymbol{v}}_{2,i}^{\mathrm{T}} \boldsymbol{X}_i \overline{\boldsymbol{v}}_{2,i} + \left(\hat{\boldsymbol{\omega}}_i^{*}\right)^{\mathrm{T}} \boldsymbol{Z}_i \hat{\boldsymbol{\omega}}_i^{*}\right)\mathrm{d}t = \hat{\boldsymbol{F}}_i^{*} \tag{8-30}$$

根据式 (8-30)，可以很容易地得到$\hat{\boldsymbol{F}}_i^{\infty} \geqslant \hat{\boldsymbol{F}}_i^{*}$。所以，结合$\hat{\boldsymbol{F}}_i^{\infty} \leqslant \hat{\boldsymbol{F}}_i^{*}$与$\hat{\boldsymbol{F}}_i^{\infty} \geqslant \hat{\boldsymbol{F}}_i^{*}$，我们可以得到$\hat{\boldsymbol{F}}_i^{\infty} = \hat{\boldsymbol{F}}_i^{*}$。因此，$\hat{\boldsymbol{F}}_i^{(q)}$最终逼近于$\hat{\boldsymbol{F}}_i^{*}$。同时，易证$\hat{\boldsymbol{\omega}}_i^{(q)}$逼近于$\hat{\boldsymbol{\omega}}_i^{*}$。证毕。 ■

8.3.2 基于神经网络的在线算法实现

在目前已经存在的方法中，需要借助评估神经网络近似价值函数，然后直接使用已知的模型信息矩阵$\boldsymbol{M}_i(\boldsymbol{\eta}_i)$或者借助额外的神经网络来实现最优控制器。在本章中，增广系统的引入使得最优控制器的隐式解中不再包含刚体系统的模型信息，所以我们不再需要借助额外的神经网络就能实现无模型的效果。因此，本章只需要使用一个评估神经网络来近似价值函数$\boldsymbol{F}_i\left(\overline{\boldsymbol{v}}_{2,i}\right)$，评估神经网络只在一些满足要求的离散时刻$t_i^q$进行更新，从而大幅地减少计算资源的消耗。

对于刚体 i，我们使用下面的评估神经网络来描述价值函数$\boldsymbol{F}_i\left(\overline{\boldsymbol{v}}_{2,i}\right)$：

$$\boldsymbol{F}_i\left(\overline{\boldsymbol{v}}_{2,i}\right) = \boldsymbol{W}_{F,i}^{\mathrm{T}} \boldsymbol{\phi}_{F,i}\left(\boldsymbol{w}_{F,i}^{\mathrm{T}} \overline{\boldsymbol{v}}_{2,i}\right) + \boldsymbol{\varepsilon}_{F,i}\left(\overline{\boldsymbol{v}}_{2,i}\right) \tag{8-31}$$

其中，$\boldsymbol{W}_{F,i} \in \mathbb{R}^{h_{F,i}}$为评估网络的隐藏层与输出层之间的目标权重，$h_{F,i}$为隐藏层的维度；$\boldsymbol{w}_{F,i}$为评估网络的输入层与隐藏层之间的目标权重，为了后文分析的方便，同样将$\boldsymbol{w}_{F,i}$设为单位矩阵，即$\boldsymbol{w}_{F,i} = \boldsymbol{I}_{2n}$；$\boldsymbol{\phi}_{F,i}(\cdot): \mathbb{R}^{2n} \to \mathbb{R}^{h_{F,i}}$表示激活函数；$\boldsymbol{\varepsilon}_{F,i}(\cdot)$表示重构误差。

在每个事件触发时刻，$\boldsymbol{F}_i\left(\bar{\boldsymbol{v}}_{2,i}\right)$的估计值可以表示为：

$$\hat{\boldsymbol{F}}_i\left(\bar{\boldsymbol{v}}_{2,i}\right) = \hat{\boldsymbol{W}}_{F,i}^{\mathrm{T}} \boldsymbol{\phi}_{F,i}\left(\bar{\boldsymbol{v}}_{2,i}\right) \tag{8-32}$$

其中，$\hat{\boldsymbol{W}}_{F,i}$表示评估网络目标权重的估计值。

定义评估网络的残差信号为：

$$\boldsymbol{\sigma}_{F,i} = \bar{\boldsymbol{v}}_{2,i}^{\mathrm{T}} \boldsymbol{X}_i \bar{\boldsymbol{v}}_{2,i} + \boldsymbol{\omega}_i^{\mathrm{T}} \boldsymbol{Z}_i \boldsymbol{\omega}_i + \boldsymbol{W}_{F,i}^{\mathrm{T}} \nabla \boldsymbol{\phi}_{F,i} \dot{\bar{\boldsymbol{v}}}_{2,i} \tag{8-33}$$

其中，$\nabla \boldsymbol{\phi}_{F,i} = \dfrac{\partial \boldsymbol{\phi}_{F,i}}{\partial \boldsymbol{v}_{2,i}^{\mathrm{T}}}$。在每个事件触发时刻，残差的估计值为$\hat{\boldsymbol{\sigma}}_{F,i} = \bar{\boldsymbol{v}}_{2,i}^{\mathrm{T}} \boldsymbol{X}_i \bar{\boldsymbol{v}}_{2,i} + \boldsymbol{\omega}_i^{\mathrm{T}} \boldsymbol{Z}_i \boldsymbol{\omega}_i + \hat{\boldsymbol{W}}_{F,i}^{\mathrm{T}} \nabla \boldsymbol{\phi}_{F,i} \dot{\bar{\boldsymbol{v}}}_{2,i}$。

在静态事件触发机制下，$\hat{\boldsymbol{W}}_{F,i}$只在一些违背条件 (8-18) 的离散时刻$t_i^q$进行更新，在满足条件 (8-18) 的$\left(t_i^q, t_i^{q+1}\right)$期间保持不变。采用梯度下降法最小化目标函数$\boldsymbol{E}_{F,i} = \dfrac{1}{2} \hat{\boldsymbol{\sigma}}_{F,i}^{\mathrm{T}} \hat{\boldsymbol{\sigma}}_{F,i}$，我们可以得到$\hat{\boldsymbol{W}}_{F,i}$的更新律，如下所示：

$$\dot{\hat{\boldsymbol{W}}}_{F,i} = 0, t \in \left(t_i^q, t_i^{q+1}\right) \tag{8-34a}$$

$$\hat{\boldsymbol{W}}_{F,i}^{+} = \hat{\boldsymbol{W}}_{F,i} - l_{F,i} \boldsymbol{\rho}_{F,i}\left(\bar{\boldsymbol{v}}_{2,i}^{\mathrm{T}} \boldsymbol{X}_i \bar{\boldsymbol{v}}_{2,i} + \boldsymbol{\omega}_i^{\mathrm{T}} \boldsymbol{Z}_i \boldsymbol{\omega}_i + \boldsymbol{\rho}_{F,i}^{\mathrm{T}} \hat{\boldsymbol{W}}_{F,i}\right), \quad t = t_i^q \tag{8-34b}$$

其中，$\hat{\boldsymbol{W}}_{F,i}^{+}$表示在第$q$个事件触发时刻后评估网络目标权重的估计值，常数$l_{F,i} \in \mathbb{R}_{>0}$表示评估网络的学习率，$\boldsymbol{\rho}_{F,i} = \nabla \boldsymbol{\phi}_{F,i} \dot{\bar{\boldsymbol{v}}}_{2,i}$。

然后，定义$\tilde{\boldsymbol{W}}_{F,i} = \boldsymbol{W}_{F,i} - \hat{\boldsymbol{W}}_{F,i}$表示评估网络的权重估计误差。根据式 (8-34)，可以推导出$\tilde{\boldsymbol{W}}_{F,i}$的更新律如下：

$$\dot{\tilde{\boldsymbol{W}}}_{F,i} = 0, t \in \left(t_i^q, t_i^{q+1}\right) \tag{8-35a}$$

$$\tilde{\boldsymbol{W}}_{F,i}^{+} = \tilde{\boldsymbol{W}}_{F,i} + l_{F,i} \boldsymbol{\rho}_{F,i}\left(\boldsymbol{\sigma}_{F,i} - \boldsymbol{\rho}_{F,i}^{\mathrm{T}} \tilde{\boldsymbol{W}}_{F,i}\right), \quad t = t_i^q \tag{8-35b}$$

根据式 (8-23) 与式 (8-32)，我们可以得到以下形式的事件触发控制器：

$$\hat{\boldsymbol{\omega}}_i = -\frac{1}{2} \boldsymbol{Z}_i^{-1} \boldsymbol{\Pi}_i\left(\hat{\bar{\boldsymbol{v}}}_{2,i}\right)\left(\nabla \boldsymbol{\phi}_{F,i}\right)^{\mathrm{T}} \hat{\boldsymbol{W}}_{F,i} \tag{8-36}$$

因为增广系统中的矩阵$\boldsymbol{\Pi}_i\left(\hat{\bar{\boldsymbol{v}}}_{2,i}\right)$是预先设计的，所以我们可以直接

获得控制器，而不再需要使用额外的神经网络。下面，我们将给出基于神经网络的在线训练算法 8.2 来实现算法 8.1。根据算法 8.2，我们可以只使用测量数据就能够在线获得事件触发最优控制器。

算法 8.2

基于神经网络的在线训练算法。

① 初始化评估网络目标权重的估计值$\hat{\boldsymbol{W}}_{F,i}^{(0)}$，$i \in \mathcal{V}$，并设置迭代索引的初始值为$q=0$。

② 在每个事件触发时刻重复执行以下步骤：

a．计算增广辅助误差$\bar{\boldsymbol{v}}_{2,i}^{(q)}$。

b．使用式 (8-36) 更新控制器$\hat{\boldsymbol{\omega}}_i^{(q)}$。

c．分别使用式 (8-37) 与式 (8-38) 更新$\hat{\boldsymbol{W}}_{F,i}^{(q+1)}$以及$\hat{\boldsymbol{F}}_i^{(q+1)}$：

$$\hat{\boldsymbol{W}}_{F,i}^{(q+1)} = \hat{\boldsymbol{W}}_{F,i}^{(q)} - l_{F,i}\boldsymbol{\rho}_{F,i}\left(\left(\bar{\boldsymbol{v}}_{2,i}^{(q)}\right)^{\mathrm{T}}\boldsymbol{X}_i\bar{\boldsymbol{v}}_{2,i}^{(q)} + \left(\hat{\boldsymbol{\omega}}_i^{(q)}\right)^{\mathrm{T}}\boldsymbol{Z}_i\hat{\boldsymbol{\omega}}_i^{(q)} + \boldsymbol{\rho}_{F,i}^{\mathrm{T}}\hat{\boldsymbol{W}}_{F,i}^{(q)}\right) \tag{8-37}$$

$$\hat{\boldsymbol{F}}_i^{(q+1)} = \left(\hat{\boldsymbol{W}}_{F,i}^{(q+1)}\right)^{\mathrm{T}}\boldsymbol{\phi}_{F,i}\left(\bar{\boldsymbol{v}}_{2,i}^{(q)}\right) \tag{8-38}$$

d．当$\|\hat{\boldsymbol{F}}_i^{(q)} - \hat{\boldsymbol{F}}_i^{(q+1)}\| \leqslant \delta$时，退出循环，其中$\delta \in \mathbb{R}_{>0}$表示一个较小的正常数。否则，令$q = q+1$，继续循环。

与评估网络目标权重的估计值$\hat{\boldsymbol{W}}_{F,i}$相同，控制器$\hat{\boldsymbol{\omega}}_i$也只在事件触发时刻更新。在两个相邻的事件触发时刻期间，$\hat{\boldsymbol{\omega}}_i$通过使用零阶保持器（Zero-Order Hold, ZOH）来保持不变。到此处为止，我们已经可以通过算法 8.2 在线求取最优控制器。但是，我们尚未提供静态事件触发条件的具体形式，这个问题将在下一小节中得以解决。

为了后文的分析，给出以下必要的假设。

假设 8.6

对于上述的评估神经网络，以下所有变量都是有界的：

① 目标权重$\boldsymbol{W}_{F,i}$，即$\|\boldsymbol{W}_{F,i}\| \leqslant W_{FM}$；

② 激活函数$\boldsymbol{\phi}_{F,i}$，即$\|\boldsymbol{\phi}_{F,i}\| \leqslant \phi_{FM}$；

③ 重构误差$\boldsymbol{\varepsilon}_{F,i}$，即$\|\boldsymbol{\varepsilon}_{F,i}\| \leqslant \varepsilon_{FM}$；

④ 激活函数的梯度$\nabla \boldsymbol{\phi}_{F,i}$，即$\| \nabla \boldsymbol{\phi}_{F,i} \| \leqslant \nabla \phi_{FM}$；

⑤ 重构误差的梯度$\nabla \boldsymbol{\varepsilon}_{F,i}$，即$\| \nabla \boldsymbol{\varepsilon}_{F,i} \| \leqslant \nabla \varepsilon_{FM}$；

⑥ 残差$\boldsymbol{\sigma}_{F,i}$，即$\| \boldsymbol{\sigma}_{F,i} \| \leqslant \sigma_{FM}$。

其中，常数$W_{FM} \in \mathbb{R}_{>0}$，$\phi_{FM} \in \mathbb{R}_{>0}$，$\varepsilon_{FM} \in \mathbb{R}_{>0}$，$\nabla \phi_{FM} \in \mathbb{R}_{>0}$，$\nabla \varepsilon_{FM} \in \mathbb{R}_{>0}$，$\sigma_{FM} \in \mathbb{R}_{>0}$。

假设 8.7

根据持续激励的条件[221]，存在常数$\rho_{FM} \in \mathbb{R}_{>0}$使得$\| \boldsymbol{\rho}_{F,i} \| \leqslant \rho_{FM}$恒成立。

8.3.3 事件触发条件的设计

在本节中，我们将会给出静态事件触发条件的具体形式，它只依赖于增广系统的实时状态。同时，该静态事件触发条件保证了增广辅助误差$\bar{\boldsymbol{v}}_{2,i}$以及评估网络的权重估计误差$\tilde{\boldsymbol{W}}_{F,i}$都满足 UUB。

定理 8.2

考虑增广系统 (8-17)，以及性能成本函数 (8-11)，并且假设 8.1 ～假设 8.7 成立。在每一个事件触发时刻，如果使用式 (8-34) 对评估网络目标权重的估计值$\hat{\boldsymbol{W}}_{F,i}$进行更新，使用式 (8-36) 对控制器$\hat{\boldsymbol{\omega}}_i$进行更新，然后使用下面的静态事件触发条件来确定每一个事件触发时刻，

$$\| \boldsymbol{\epsilon}_i (t) \| \leqslant \sqrt{\frac{2\lambda_{\min}(\boldsymbol{X}_i)\alpha_i}{\Pi_M^2 Y^2}} \| \bar{\boldsymbol{v}}_{2,i} \| = \epsilon_i^H (t) \tag{8-39}$$

$\bar{\boldsymbol{v}}_{2,i}$与$\tilde{\boldsymbol{W}}_{F,i}$都满足 UUB。其中，正常数$\alpha_i$的取值范围为区间$(0,1)$。

证明：

我们将通过使用稳定性理论来证明$\bar{\boldsymbol{v}}_{2,i}$与$\tilde{\boldsymbol{W}}_{F,i}$都满足 UUB。在静态事件触发机制下，需要分别考虑条件 (8-39) 被满足与被违背两种不同的情况。因此，完整的证明可以分为两个部分。

第一部分：当静态事件触发条件 (8-39) 被满足时，也就是说，在两个相邻的事件触发时刻的间隔期间，有$t \in \left(t_i^q, t_i^{q+1}\right)$。

令$\boldsymbol{L}_i$表示李雅普诺夫候选函数，它被设计为下面的形式：

$$\boldsymbol{L}_i = \boldsymbol{L}_{F,i} + \boldsymbol{L}_{W,i} \tag{8-40}$$

其中，$\boldsymbol{L}_{F,i}=\boldsymbol{F}_i\left(\bar{\boldsymbol{v}}_{2,i}\right)$表示价值函数，$\boldsymbol{L}_{W,i}=\dfrac{1}{l_{F,i}}\operatorname{tr}\left(\tilde{\boldsymbol{W}}_{F,i}^{\mathrm{T}}\tilde{\boldsymbol{W}}_{F,i}\right)$。

通过对等式(8-40)求导，可以得到：

$$\dot{\boldsymbol{L}}_i=\dot{\boldsymbol{L}}_{F,i}+\dot{\boldsymbol{L}}_{W,i}=\dot{\boldsymbol{F}}_i\left(\bar{\boldsymbol{v}}_{2,i}\right)+\frac{2}{l_{F,i}}\operatorname{tr}\left(\tilde{\boldsymbol{W}}_{F,i}^{\mathrm{T}}\dot{\tilde{\boldsymbol{W}}}_{F,i}\right) \tag{8-41}$$

根据式(8-35)，我们可以得到$\dot{\tilde{\boldsymbol{W}}}_{F,i}=0$。因此，上式可以变换为：

$$\begin{aligned}\dot{\boldsymbol{L}}_i&=\dot{\boldsymbol{F}}_i\left(\bar{\boldsymbol{v}}_{2,i}\right)=\left(\nabla\boldsymbol{F}_i\right)^{\mathrm{T}}\left(\boldsymbol{\Gamma}_i\left(\bar{\boldsymbol{v}}_{2,i}\right)+\boldsymbol{\Pi}_i\left(\bar{\boldsymbol{v}}_{2,i}\right)\hat{\boldsymbol{\omega}}_i\right)\\&=\left(\nabla\boldsymbol{F}_i\right)^{\mathrm{T}}\left(\boldsymbol{\Gamma}_i\left(\bar{\boldsymbol{v}}_{2,i}\right)+\boldsymbol{\Pi}_i\left(\bar{\boldsymbol{v}}_{2,i}\right)\boldsymbol{\omega}_i\right)-\left(\nabla\boldsymbol{F}_i\right)^{\mathrm{T}}\boldsymbol{\Pi}_i\left(\bar{\boldsymbol{v}}_{2,i}\right)\left(\boldsymbol{\omega}_i-\hat{\boldsymbol{\omega}}_i\right)\end{aligned} \tag{8-42}$$

根据式(8-13) 与式(8-31)，我们可以分别得到：

$$\left(\nabla\boldsymbol{F}_i\right)^{\mathrm{T}}\left(\boldsymbol{\Gamma}_i\left(\bar{\boldsymbol{v}}_{2,i}\right)+\boldsymbol{\Pi}_i\left(\bar{\boldsymbol{v}}_{2,i}\right)\boldsymbol{\omega}_i\right)=-\bar{\boldsymbol{v}}_{2,i}^{\mathrm{T}}\boldsymbol{X}_i\bar{\boldsymbol{v}}_{2,i}-\boldsymbol{\omega}_i^{\mathrm{T}}\boldsymbol{Z}_i\boldsymbol{\omega}_i \tag{8-43}$$

与

$$\nabla\boldsymbol{F}_i=\left(\boldsymbol{W}_{F,i}^{\mathrm{T}}\nabla\boldsymbol{\phi}_{F,i}+\nabla\boldsymbol{\varepsilon}_{F,i}\right)^{\mathrm{T}} \tag{8-44}$$

其中，$\nabla\boldsymbol{\phi}_{F,i}=\dfrac{\partial\boldsymbol{\phi}_{F,i}}{\partial\bar{\boldsymbol{v}}_{2,i}^{\mathrm{T}}}$，$\nabla\boldsymbol{\varepsilon}_{F,i}=\dfrac{\partial\boldsymbol{\varepsilon}_{F,i}}{\partial\bar{\boldsymbol{v}}_{2,i}^{\mathrm{T}}}$。

将等式 (8-43) 与等式 (8-44) 代入到等式 (8-42) 中，有下式成立：

$$\dot{\boldsymbol{L}}_i=-\bar{\boldsymbol{v}}_{2,i}^{\mathrm{T}}\boldsymbol{X}_i\bar{\boldsymbol{v}}_{2,i}-\boldsymbol{\omega}_i^{\mathrm{T}}\boldsymbol{Z}_i\boldsymbol{\omega}_i-\left(\boldsymbol{W}_{F,i}^{\mathrm{T}}\nabla\boldsymbol{\phi}_{F,i}+\nabla\boldsymbol{\varepsilon}_{F,i}\right)\boldsymbol{\Pi}_i\left(\bar{\boldsymbol{v}}_{2,i}\right)\left(\boldsymbol{\omega}_i-\hat{\boldsymbol{\omega}}_i\right) \tag{8-45}$$

因为$\bar{\boldsymbol{v}}_{2,i}^{\mathrm{T}}\boldsymbol{X}_i\bar{\boldsymbol{v}}_{2,i}\geqslant\boldsymbol{\lambda}_{\min}\left(\boldsymbol{X}_i\right)\|\bar{\boldsymbol{v}}_{2,i}\|^2$，$\boldsymbol{\omega}_i^{\mathrm{T}}\boldsymbol{Z}_i\boldsymbol{\omega}_i>0$，并且根据假设8.5有$\|\boldsymbol{\omega}_i-\hat{\boldsymbol{\omega}}_i\|\leqslant Y\|\boldsymbol{\epsilon}_i\left(t\right)\|$成立，所以$\dot{\boldsymbol{L}}_i$ 可以转化为：

$$\begin{aligned}\dot{\boldsymbol{L}}_i&\leqslant-\lambda_{\min}\left(\boldsymbol{X}_i\right)\|\bar{\boldsymbol{v}}_{2,i}\|^2+0.5\|\boldsymbol{W}_{F,i}^{\mathrm{T}}\nabla\boldsymbol{\phi}_{F,i}+\nabla\boldsymbol{\varepsilon}_{F,i}\|^2+0.5\|\boldsymbol{\Pi}_i\left(\bar{\boldsymbol{v}}_{2,i}\right)\left(\boldsymbol{\omega}_i-\hat{\boldsymbol{\omega}}_i\right)\|^2\\&\leqslant-\lambda_{\min}\left(\boldsymbol{X}_i\right)\|\bar{\boldsymbol{v}}_{2,i}\|^2+W_{FM}^2\left(\nabla\phi_{FM}\right)^2+\left(\nabla\varepsilon_{FM}\right)^2+0.5\boldsymbol{\Pi}_M^2Y^2\|\boldsymbol{\epsilon}_i\left(t\right)\|^2.\end{aligned} \tag{8-46}$$

将静态事件触发条件 (8-39) 代入到上式中，可以推导出$\dot{\boldsymbol{L}}_i$ 满足下面的不等式：

$$\dot{\boldsymbol{L}}_i\leqslant-\left(1-\alpha_i\right)\lambda_{\min}\left(\boldsymbol{X}_i\right)\|\bar{\boldsymbol{v}}_{2,i}\|^2+\Omega_{FM} \tag{8-47}$$

其中，$\Omega_{FM}=W_{FM}^2\left(\nabla\phi_{FM}\right)^2+\left(\nabla\varepsilon_{FM}\right)^2$。

为了保证$\dot{\boldsymbol{L}}_i<0$成立，需要有下式成立：

$$\|\bar{\boldsymbol{v}}_{2,i}\|>\sqrt{\frac{\Omega_{FM}}{\left(1-\alpha_i\right)\lambda_{\min}\left(\boldsymbol{X}_i\right)}}. \tag{8-48}$$

因此，在事件触发间隔期间，$\bar{\boldsymbol{v}}_{2,i}$满足 UUB。因为评估网络目标权重的估计值$\hat{\boldsymbol{W}}_{F,i}$在事件触发间隔期间保持不变，所以$\tilde{\boldsymbol{W}}_{F,i}$也保持不变。因此，在事件触发间隔期间，$\tilde{\boldsymbol{W}}_{F,i}$也满足 UUB。综上所述，当静态事件触发条件 (8-39) 被满足时，$\bar{\boldsymbol{v}}_{2,i}$与$\tilde{\boldsymbol{W}}_{F,i}$均满足 UUB。

第二部分：当条件 (8-39) 被违背时，也就是说，在每一个事件触发时刻到来时，有$t = t_i^q$。

在这一部分中，李雅普诺夫候选函数$\boldsymbol{L}_i$的设计与第一部分相同。在事件触发时刻，等式 (8-40) 可以转化为：

$$\begin{aligned}\Delta \boldsymbol{L}_i &= \Delta \boldsymbol{L}_{F,i} + \Delta \boldsymbol{L}_{W,i} \\ &= \boldsymbol{F}_i\left(\bar{\boldsymbol{v}}_{2,i}^{+}\right) - \boldsymbol{F}_i\left(\bar{\boldsymbol{v}}_{2,i}\right) + \frac{1}{l_{F,i}}\operatorname{tr}\left(\left(\tilde{\boldsymbol{W}}_{F,i}^{+}\right)^{\mathrm{T}}\tilde{\boldsymbol{W}}_{F,i}^{+}\right) - \frac{1}{l_{F,i}}\operatorname{tr}\left(\tilde{\boldsymbol{W}}_{F,i}^{\mathrm{T}}\tilde{\boldsymbol{W}}_{F,i}\right).\end{aligned} \tag{8-49}$$

因为增广系统 (8-17) 的状态$\bar{\boldsymbol{v}}_{2,i}$是连续的，即有$\bar{\boldsymbol{v}}_{2,i}^{+} = \bar{\boldsymbol{v}}_{2,i}$，所以$\boldsymbol{F}_i\left(\bar{\boldsymbol{v}}_{2,i}^{+}\right) - \boldsymbol{F}_i\left(\bar{\boldsymbol{v}}_{2,i}\right) = 0$。因此，式 (8-49) 可以转化为：

$$\Delta \boldsymbol{L}_i = \frac{1}{l_{F,i}}\operatorname{tr}\left(\left(\tilde{\boldsymbol{W}}_{F,i}^{+}\right)^{\mathrm{T}}\tilde{\boldsymbol{W}}_{F,i}^{+}\right) - \frac{1}{l_{F,i}}\operatorname{tr}\left(\tilde{\boldsymbol{W}}_{F,i}^{\mathrm{T}}\tilde{\boldsymbol{W}}_{F,i}\right) \tag{8-50}$$

将式(8-35) 代入到式 (8-50) 中，可以得到：

$$\begin{aligned}\Delta \boldsymbol{L}_i &= \frac{1}{l_{F,i}}\operatorname{tr}\left[\left(\tilde{\boldsymbol{W}}_{F,i} + l_{F,i}\boldsymbol{\rho}_{F,i}\left(\boldsymbol{\sigma}_{F,i} - \boldsymbol{\rho}_{F,i}^{\mathrm{T}}\tilde{\boldsymbol{W}}_{F,i}\right)\right)^{\mathrm{T}}\left(\tilde{\boldsymbol{W}}_{F,i} + l_{F,i}\boldsymbol{\rho}_{F,i}(\boldsymbol{\sigma}_{F,i} - \boldsymbol{\rho}_{F,i}^{\mathrm{T}}\tilde{\boldsymbol{W}}_{F,i})\right)\right] \\ &\quad - \frac{1}{l_{F,i}}\operatorname{tr}\left(\tilde{\boldsymbol{W}}_{F,i}^{\mathrm{T}}\tilde{\boldsymbol{W}}_{F,i}\right) \\ &= 2\operatorname{tr}\left(\tilde{\boldsymbol{W}}_{F,i}^{\mathrm{T}}\boldsymbol{\rho}_{F,i}\boldsymbol{\sigma}_{F,i} - \tilde{\boldsymbol{W}}_{F,i}^{\mathrm{T}}\boldsymbol{\rho}_{F,i}\boldsymbol{\rho}_{F,i}^{\mathrm{T}}\tilde{\boldsymbol{W}}_{F,i}\right) \\ &\quad + l_{F,i}\operatorname{tr}\left(\left(\boldsymbol{\sigma}_{F,i} - \boldsymbol{\rho}_{F,i}^{\mathrm{T}}\tilde{\boldsymbol{W}}_{F,i}\right)^{\mathrm{T}}\boldsymbol{\rho}_{F,i}^{\mathrm{T}}\boldsymbol{\rho}_{F,i}\left(\boldsymbol{\sigma}_{F,i} - \boldsymbol{\rho}_{F,i}^{\mathrm{T}}\tilde{\boldsymbol{W}}_{F,i}\right)\right) \\ &= 2\|\tilde{\boldsymbol{W}}_{F,i}^{\mathrm{T}}\boldsymbol{\rho}_{F,i}\boldsymbol{\sigma}_{F,i}\| - 2\|\tilde{\boldsymbol{W}}_{F,i}^{\mathrm{T}}\boldsymbol{\rho}_{F,i}\|^2 + l_{F,i}\|\boldsymbol{\rho}_{F,i}\boldsymbol{\sigma}_{F,i} - \boldsymbol{\rho}_{F,i}\boldsymbol{\rho}_{F,i}^{\mathrm{T}}\tilde{\boldsymbol{W}}_{F,i}\|^2\end{aligned} \tag{8-51}$$

因为$\|\tilde{\boldsymbol{W}}_{F,i}^{\mathrm{T}}\boldsymbol{\rho}_{F,i}\|^2 > 0$，所以存在一个常数$\rho_{Fm} \in \mathbb{R}_{>0}$使得$\|\tilde{\boldsymbol{W}}_{F,i}^{\mathrm{T}}\boldsymbol{\rho}_{F,i}\|^2 \geqslant \rho_{Fm}\|\tilde{\boldsymbol{W}}_{F,i}\|^2$成立。然后，我们可以推导出：

$$\begin{aligned}\Delta \boldsymbol{L}_i &\leqslant \sigma_{FM}\left(\|\tilde{\boldsymbol{W}}_{F,i}\|^2 + \rho_{FM}^2\right) - 2\rho_{Fm}\|\tilde{\boldsymbol{W}}_{F,i}\|^2 + 2l_{F,i}\sigma_{FM}\rho_{FM}^2 + 2l_{F,i}\rho_{FM}^4\|\tilde{\boldsymbol{W}}_{F,i}\|^2 \\ &= -\left(2\rho_{Fm} - \sigma_{FM} - 2l_{F,i}\rho_{FM}^4\right)\|\tilde{\boldsymbol{W}}_{F,i}\|^2 + \Omega_{WM}\end{aligned} \tag{8-52}$$

其中，$\Omega_{WM} = \sigma_{FM}\rho_{FM}^2 + 2l_{F,i}\sigma_{FM}\rho_{FM}^2$，并且学习率$l_{F,i}$需要满足下式：

$$l_{F,i} < \frac{2\rho_{Fm} - \sigma_{FM}}{2\rho_{FM}^4} \tag{8-53}$$

因此，当下面不等式被满足时，可以推导出$\Delta \boldsymbol{L}_i < 0$成立：

$$\|\tilde{\boldsymbol{W}}_{F,i}\| > \sqrt{\frac{\Omega_{WM}}{2\rho_{Fm} - \sigma_{FM} - 2l_{F,i}\rho_{FM}^4}} \tag{8-54}$$

所以，当静态事件触发条件(8-39)被违背时，$\bar{\boldsymbol{v}}_{2,i}$与$\tilde{\boldsymbol{W}}_{F,i}$也满足UUB。

综合第一部分和第二部分两种情况，可以总结得到$\bar{\boldsymbol{v}}_{2,i}$与$\tilde{\boldsymbol{W}}_{F,i}$始终满足 UUB。也就是说，增广系统 (8-17) 是稳定的，并且评估网络目标权重的估计值是收敛的。最终，所有的刚体将达成状态一致性。 ■

下面，我们将给出定理 8.3，该定理表明了静态事件触发条件 (8-39) 能够确保奇诺现象不会发生。

定理 8.3

当每一个事件触发时刻都是通过静态事件触发条件 (8-39) 来确定时，事件触发间隔是有下界的，即对于任意的$q \in \mathbb{N}$以及任意的$i \in \mathcal{V}$，均有$t_i^{q+1} - t_i^q > T_m$。其中，$T_m \in \mathbb{R}_{>0}$。

证明：

根据式 (8-16) 与式 (8-17)，有：

$$\begin{aligned} \|\dot{\boldsymbol{\epsilon}}_i(t)\| &= \|\dot{\hat{\boldsymbol{v}}}_{2,i} - \dot{\bar{\boldsymbol{v}}}_{2,i}\| \\ &= \|\dot{\bar{\boldsymbol{v}}}_{2,i}\| \\ &= \|\boldsymbol{\Gamma}_i(\bar{\boldsymbol{v}}_{2,i}) + \boldsymbol{\Pi}_i(\bar{\boldsymbol{v}}_{2,i})\hat{\boldsymbol{\omega}}_i\| \\ &\leqslant \|\boldsymbol{\Gamma}_i(\bar{\boldsymbol{v}}_{2,i})\| + \|\boldsymbol{\Pi}_i(\bar{\boldsymbol{v}}_{2,i})\hat{\boldsymbol{\omega}}_i\|, t \in [t_i^q, t_i^{q+1}). \end{aligned} \tag{8-55}$$

根据假设 8.4 以及等式 (8-36)，可以将上式转化为：

$$\begin{aligned} \|\dot{\boldsymbol{\epsilon}}_i(t)\| &\leqslant \Gamma_M \|\bar{\boldsymbol{v}}_{2,i}\| + \Pi_M \|\frac{1}{2}\boldsymbol{Z}_i^{-1}\boldsymbol{\Pi}_i(\hat{\bar{\boldsymbol{v}}}_{2,i})(\nabla\boldsymbol{\phi}_{F,i})^{\mathrm{T}}\hat{\boldsymbol{W}}_{F,i}\| \\ &\leqslant \Gamma_M \|\boldsymbol{\epsilon}_i(t)\| + \Omega_{\epsilon,i} \end{aligned} \tag{8-56}$$

其中，$\Omega_{\epsilon,i} = \Gamma_M \|\hat{\bar{\boldsymbol{v}}}_{2,i}\| + \Pi_M \|\frac{1}{2}\boldsymbol{Z}_i^{-1}\boldsymbol{\Pi}_i(\hat{\bar{\boldsymbol{v}}}_{2,i})(\nabla\boldsymbol{\phi}_{F,i})^{\mathrm{T}}\hat{\boldsymbol{W}}_{F,i}\|$。根据假设8.6以及定理8.2，可以得到$\Omega_{\epsilon,i}$是有界的。

根据比较引理 [222]，我们可以得到：

$$\|\boldsymbol{\epsilon}_i(t)\| \leqslant \exp\{\Gamma_M(t - t_i^q)\}\boldsymbol{\epsilon}_i(t_i^q) + \frac{1}{2}\int_{t_i^q}^{t}\exp\{\Gamma_M(t-s)\}\Omega_{\epsilon,i}\mathrm{d}s. \tag{8-57}$$

在每个事件触发时刻t_i^q，$\epsilon_i(t)$被设置为 0，即$\epsilon_i(t_i^q)=0$。所以，不等式 (8-57) 等价为：

$$\|\epsilon_i(t)\| \leqslant \frac{\Omega_{\epsilon,i}}{2\Gamma_M}\left(\exp\left\{\Gamma_M\left(t-t_i^q\right)\right\}-1\right), t\in\left[t_i^q,t_i^{q+1}\right) \tag{8-58}$$

根据静态事件触发条件 (8-39)，有：

$$\|\epsilon_i\left(t_i^{q+1}\right)\| \leqslant \frac{\sqrt{2\lambda_{\min}\left(\boldsymbol{X}_i\right)\alpha_i}}{\Pi_M Y}\|\bar{\boldsymbol{v}}_{2,i}\| \tag{8-59}$$

在事件触发时刻t_i^{q+1}，结合式 (8-58) 与式(8-59)，可以得到：

$$\frac{\sqrt{2\lambda_{\min}\left(\boldsymbol{X}_i\right)\alpha_i}}{\Pi_M Y}\|\bar{\boldsymbol{v}}_{2,i}\| \leqslant \frac{\Omega_{\epsilon,i}}{2\Gamma_M}\left(\exp\left\{\Gamma_M\left(t_i^{q+1}-t_i^q\right)\right\}-1\right) \tag{8-60}$$

因此，对于任意的$q\in\mathbb{N}$以及任意的$i\in\mathcal{V}$，事件触发间隔$t_i^{q+1}-t_i^q$满足下式：

$$t_i^{q+1}-t_i^q \geqslant \frac{1}{\Gamma_M}\log\left(1+\frac{2\Gamma_M\sqrt{2\lambda_{\min}\left(\boldsymbol{X}_i\right)\alpha_i}}{\Pi_M Y\Omega_{\epsilon,i}}\|\bar{\boldsymbol{v}}_{2,i}\|\right)=T_m. \tag{8-61}$$

从上式可知，$T_m>0$。因此，任意两个事件触发时刻之间的间隔$t_i^{q+1}-t_i^q$均有一个正的下界T_m。也就是说，当在无模型事件触发的强化学习算法 8.1 中使用静态事件触发条件 (8-39) 时，能够将奇诺现象排除在外。 ■

8.4 算法验证与分析

为了验证上述方法的可行性，本节给出了一个数值仿真案例。考虑一个由六个节点组成的多刚体系统，每个节点均为一个 2 自由度的机械臂系统，如图 8-1 所示。图 8-2 是一个强连通图，它描述了任意两个机械臂之间的信息交互关系。然后，我们给出了图 8-2 对应的拉普拉斯矩阵，如下所示：

$$\mathcal{L}=\begin{bmatrix} 2 & -2 & 0 & 0 & 0 & 0 \\ -1 & 2 & 0 & -1 & 0 & 0 \\ 0 & -2 & 2 & 0 & 0 & 0 \\ 0 & 0 & -1 & 2 & 0 & -1 \\ 0 & 0 & -2 & 0 & 2 & 0 \\ 0 & 0 & 0 & -1 & -1 & 2 \end{bmatrix}$$

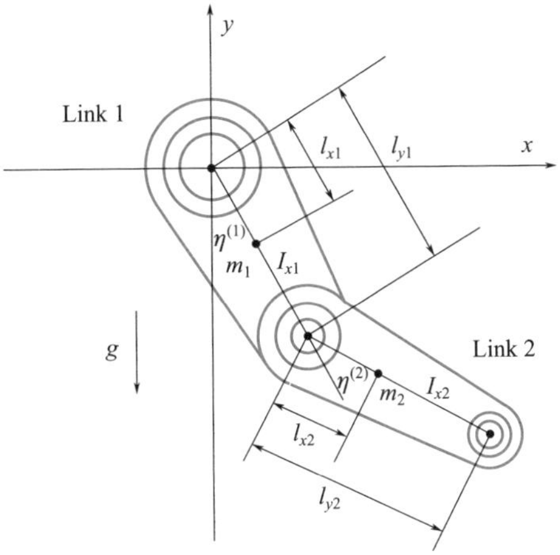

图 8-1　2 自由度机械臂

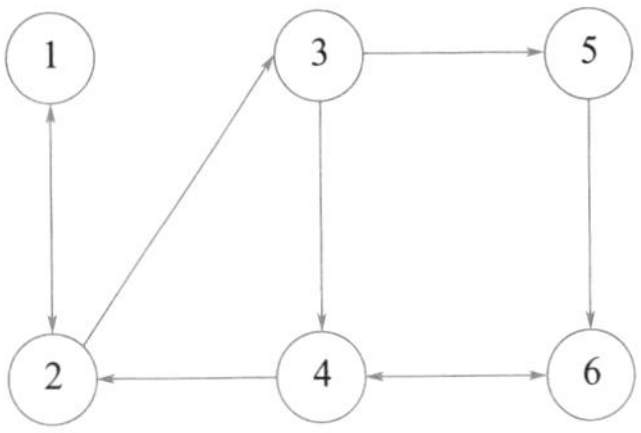

图 8-2　多刚体系统的通信拓扑

对于每一个机械臂系统，可以借助下面的欧拉 - 拉格朗日方程进行建模：

$$\boldsymbol{M}_i(\boldsymbol{\eta}_i)\ddot{\boldsymbol{\eta}}_i + \boldsymbol{C}_i(\boldsymbol{\eta}_i, \dot{\boldsymbol{\eta}}_i)\dot{\boldsymbol{\eta}}_i + \boldsymbol{G}_i(\boldsymbol{\eta}_i) = \boldsymbol{\tau}_i,\quad i = 1,2,3,4,5,6$$

其中，$\boldsymbol{\eta}_i = \left[\eta_i^{(1)}\ \eta_i^{(2)}\right]^{\mathrm{T}}$表示状态向量。参照文献[223]，给出上述模型中使用的矩阵$\boldsymbol{M}_i(\boldsymbol{\eta}_i)$和$\boldsymbol{C}_i(\boldsymbol{\eta}_i, \dot{\boldsymbol{\eta}}_i)$，如下：

$$\boldsymbol{M}_i(\boldsymbol{\eta}_i) = \begin{bmatrix} 2\chi_2\cos\eta_i^{(2)} + \chi_1 & \chi_2\cos\eta_i^{(2)} + \chi_3 \\ \chi_2\cos\eta_i^{(2)} + \chi_3 & \chi_3 \end{bmatrix}$$

$$\boldsymbol{C}_i(\boldsymbol{\eta}_i, \dot{\boldsymbol{\eta}}_i) = \begin{bmatrix} -\chi_2\dot{\eta}_i^{(2)}\sin\eta_i^{(2)} & -\chi_2\left(\dot{\eta}_i^{(1)} + \dot{\eta}_i^{(2)}\right)\sin\eta_i^{(2)} \\ \chi_2\dot{\eta}_i^{(1)}\sin\eta_i^{(2)} & 0 \end{bmatrix}$$

其中，$\chi_1 = m_1 l_{x1}^2 + m_2 l_{y1}^2 + m_2 l_{x2}^2 + I_{x1} + I_{x2}$，$\chi_2 = m_2 l_{y1} l_{x2}$，$\chi_3 = m_2 l_{x2}^2 + I_{x2}$。

图 8-1 表示的 2 自由度机械臂的物理参数可以从表 8-1 中查看，并且将重力加速度设为 9.8m/s^2。

表8-1　2自由度机械臂的物理参数

物理意义	符号	数值	单位
第一个关节的质量	m_1	1.2	kg
第一个关节的长度	l_{y1}	0.75	m
第一个关节的质心的位置	l_{x1}	0.26	m
第一个关节的转动惯量	I_{x1}	0.125	kg • m^2
第二个关节的质量	m_2	1	kg
第二个关节的长度	l_{y2}	0.75	m
第二个关节的质心的位置	l_{x2}	0.5	m
第二个关节的转动惯量	I_{x2}	0.188	kg • m^2

在本节的数值仿真过程中，总时长设置为 50s，采样周期设置为 0.01s。性能成本函数中的矩阵$\boldsymbol{X}_i = \boldsymbol{I}_4, i = 1,2,3,4,5,6$，矩阵$\boldsymbol{Z}_i$分别设置为$\boldsymbol{Z}_1 = \boldsymbol{Z}_3 = \boldsymbol{Z}_5 = \boldsymbol{I}_2$，$\boldsymbol{Z}_2 = 0.5\boldsymbol{I}_2$，$\boldsymbol{Z}_4 = 2\boldsymbol{I}_2$，$\boldsymbol{Z}_6 = 3\boldsymbol{I}_2$。一些必要的参数选取为：$c = 0.3$，$l_{F,i} = 0.5$，$\alpha_i = 0.8$，$\Pi_M = \sqrt{2}$，$Y = 1.25$。增广系统的状态向量$\bar{\boldsymbol{v}}_{2,i}$可以表示为$\bar{\boldsymbol{v}}_{2,i} = \left[v_{2,i}^{(1)}\ v_{2,i}^{(2)}\ \tau_i^{(1)}\ \tau_i^{(2)} \right]^{\mathrm{T}} \in \mathbb{R}^4$，激活函数$\boldsymbol{\phi}_{F,i}\left(\bar{\boldsymbol{v}}_{2,i}\right)$的选取如下所示：

$$\boldsymbol{\phi}_{F,i}\left(\bar{\boldsymbol{v}}_{2,i}\right) = \Big[\left(v_{2,i}^{(1)}\right)^2\ v_{2,i}^{(1)}v_{2,i}^{(2)}\ v_{2,i}^{(1)}\tau_i^{(1)}\ v_{2,i}^{(1)}\tau_i^{(2)}\ \left(v_{2,i}^{(2)}\right)^2$$
$$v_{2,i}^{(2)}\tau_i^{(1)}\ v_{2,i}^{(2)}\tau_i^{(2)}\ \left(\tau_i^{(1)}\right)^2\ \tau_i^{(1)}\tau_i^{(2)}\ \left(\tau_i^{(2)}\right)^2 \Big]^{\mathrm{T}} \in \mathbb{R}^{10}.$$

表8-2　事件触发控制与时间触发控制的对比

物理意义	事件触发次数	性能成本（事件触发控制）	性能成本（时间触发控制）
机械臂 1	53	28.5717	16.1431
机械臂 2	82	9.6510	7.7079
机械臂 3	61	28.1983	22.0636
机械臂 4	38	14.3522	4.7351
机械臂 5	74	18.6576	13.9130
机械臂 6	51	11.4330	3.3038
平均值	59.83	18.4773	11.3111

每个机械臂的初始状态被设置为：

$$\boldsymbol{\eta}_1=\begin{bmatrix}1\\-0.5\end{bmatrix},\ \boldsymbol{\eta}_2=\begin{bmatrix}0.5\\-1\end{bmatrix},\ \boldsymbol{\eta}_3=\begin{bmatrix}-1\\0.5\end{bmatrix},\ \boldsymbol{\eta}_4=\begin{bmatrix}-0.5\\1\end{bmatrix},$$

$$\boldsymbol{\eta}_5=\begin{bmatrix}0.2\\-0.2\end{bmatrix},\ \boldsymbol{\eta}_6=\begin{bmatrix}-0.2\\0.2\end{bmatrix},\ \dot{\boldsymbol{\eta}}_i=\ddot{\boldsymbol{\eta}}_i=\begin{bmatrix}0\\0\end{bmatrix},\ i=1,2,3,4,5,6.$$

将第 8.3 节中介绍的无模型事件触发的强化学习算法用到上述的多刚体系统中，可以得到如图 8-3 ～图 8-7 所示的仿真结果。图 8-3(a) 显示了每个机械臂的事件触发时刻，图 8-3(b) 描述了每个机械臂的事件触发总数以及最小事件触发间隔。在式 (8-39) 给出的静态事件触发条件下，每个机械臂在 50s 的仿真时间内平均有 59.83 次事件触发时刻。对比时间触发控制固定频率的采样与更新，事件触发控制大幅度地减少了神经网络以及控制器的更新次数，从而节省了大量的计算资源。同时，六个机械臂的最小事件触发间隔分别为：0.05s, 0.04s, 0.04s, 0.21s, 0.06s, 0.02s。很显然，在式 (8-39) 给出的静态事件触发条件下，奇诺现象不会出现。从表 8-2 中我们可以看出，事件触发控制的性能成本是比时间触发控制的性能成本要高的，这符合我们在假设 8.5 中的猜测。也就是说，对于所有的事件触发容许控制，本章得到的控制器是它们中的最优解。但对比于时间触发容许控制中的最优解，本章给出的最优控制器则会带来性能成本的增加。因为在静态事件触发机制下，我们只在很少量的事件触发时刻对控制器进行更新，从而不可避免地导致了性能成本的增加。

图 8-4(a) 与图 8-4(b) 分别描述了增广系统的控制输入以及每个机械臂原本的控制输入。从图 8-4(a) 中可以看出，增广系统的控制输入$\boldsymbol{\omega}_i$只在图 8-3(a) 中对应的事件触发时刻更新。从图 8-4(b) 中可以看出，作为增广辅助误差$\bar{\boldsymbol{v}}_{2,i}$的一部分，机械臂原本的控制输入$\boldsymbol{\tau}_i$最终是收敛到零的，满足 UUB。

图 8-5(a) 与图 8-5(b) 分别描述了每个机械臂的位置向量的曲线以及角速度向量的曲线。从中可以看出，在接近 15s 的时候，所有机械臂的状态达成了一致，即实现了状态一致性。图 8-6 描述了每个机械臂的辅助一致性误差的轨迹，从中我们可以得到相同的结论。图 8-7 描述了评估神经网络目标权重的估计值，它也只在图 8-3(a) 中对应的事件触发时

刻更新，并最终收敛。根据图 8-6 与图 8-7，我们可以验证辅助一致性误差$\boldsymbol{v}_{2,i}$与评估网络的权重估计误差$\tilde{\boldsymbol{W}}_{F,i}$都是满足 UUB 的。因为机械臂系统原本的控制输入$\boldsymbol{\tau}_i$也满足 UUB，所以定理 8.2 得以验证。因此，本章提出的无模型事件触发的强化学习算法能够有效地实现模型信息未知的多刚体系统的最优一致性控制。

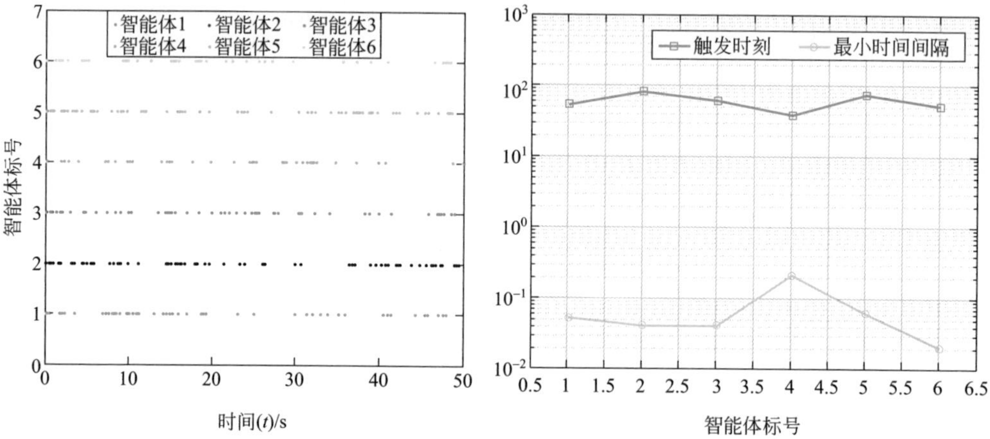

(a) 每个机械臂的事件触发时刻　　(b) 事件触发总数与最小事件触发间隔

图 8-3　每个机械臂的事件触发控制

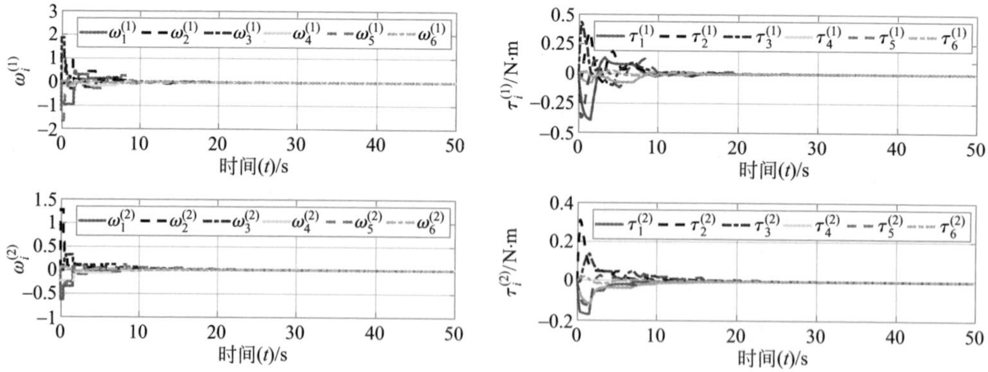

(a) 每个机械臂对应增广系统的控制输入　　(b) 每个机械臂原本的控制输入

图 8-4　每个机械臂的控制输入

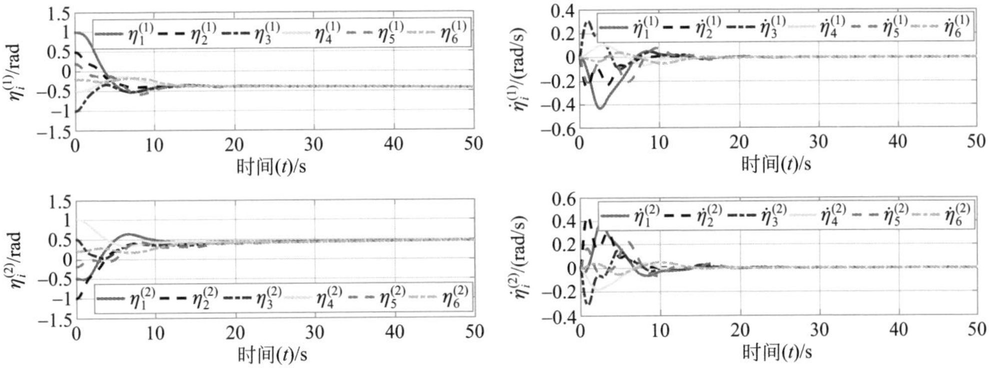

(a) 每个机械臂的位置向量

(b) 每个机械臂的角速度向量

图 8-5　每个机械臂的状态向量

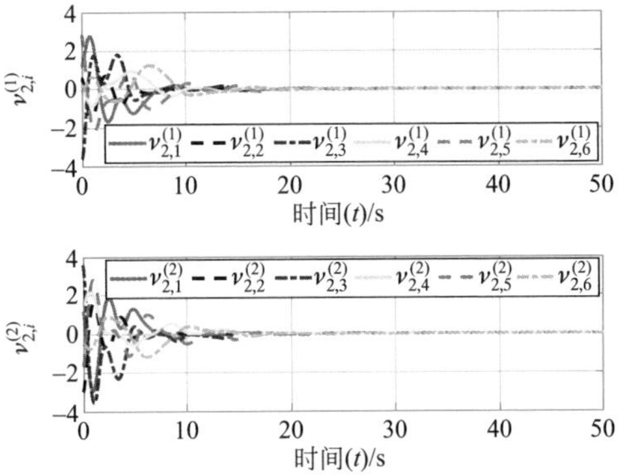

图 8-6　每个机械臂的辅助一致性误差向量

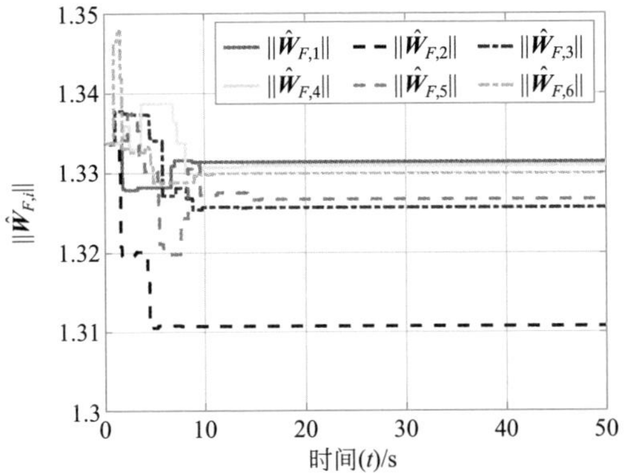

图 8-7　神经网络目标权重的估计值

Digital Wave
Advanced Technology of Industrial Internet

Control of Autonomous Intelligent Systems

自主智能系统控制

第9章

单个刚体系统的安全性能

9.1 概述

在众多信息物理系统问题中，欠驱动刚体的姿态控制以及非完整机器人系统的控制受到极大的关注[224-225]。此外，信息物理系统中的通信过程是通过数字平台以离散的方式进行的，即系统数据可以通过共享的网络通道在某些离散时刻进行信息传输，而这种信息传输方式有很多优点，比如同时节省了系统的通信和计算资源[226]。因此，为保证网络通信的可靠性、鲁棒性和高效性，研究利用（可能遭受攻击的）网络信道进行传输数据的安全性就显得尤为重要[227-228]。实际上，由于计算和通信基础设施的紧密集成，网络通道极易受到拒绝服务攻击[229]，即信道被切断使得采样数据不能及时传输。如果实时采样信息的传播被中断，使得控制器得不到更新，那么整个系统可能因得不到及时控制而崩溃，因此研究网络系统的安全性是非常重要的。然而由于这类问题具有相当大的复杂性和挑战性，目前对拒绝服务攻击下的研究还很不完善，如欠驱动刚体的姿态安全控制和非完整机器人系统的安全控制，这激发了本章节致力于研究此类安全问题的动力。

本章节考虑了单个刚体系统的安全控制问题，即欠驱动刚体的姿态系统和非完整机器人系统的安全控制问题。针对欠驱动垂直起降飞行器的姿态控制问题，通过设计基于事件驱动的采样策略，9.2 节建立了拒绝服务攻击下的弹性时变姿态跟踪混杂框架，其中混杂系统理论为建立连接离散和连续动力学的数学模型提供了技术支撑[230]。作为一个活跃的研究领域，混杂系统理论可以用于各种系统的建模和仿真，包括飞机控制和机器人，特别是针对飞机姿态的控制系统。因为基于事件触发的采样机制和基于二进制逻辑变量的混杂控制策略能更加凸显混杂控制系统的优势。9.2 节的贡献可以概括如下：①在理论方面，9.2 节设计了事件触发策略，使姿态系统的时变姿态跟踪具有抵抗拒绝服务攻击的能力。在该方案中，通过全测量姿态数据来评估触发条件，在触发时刻传输数据进而更新控制输入，其中在没有拒绝服务攻击的情况下数据会传输成功，但在受到拒绝服务攻击时会传输失败。②结合基于逻辑变量的混杂控制策略、事件触发策略和拒绝服务攻击特征的数学描述，本章节为姿态追踪系统建立了一般的混杂模型。其实如何在混杂模型中恰当地衔接这几种互有关联的因素并不是一件容易的工作。在实际应用方面，9.2 节用一个小型四旋翼原型的跟踪性能验证了理论结果的有效性。

针对非完整机器人系统的追踪控制问题，与仅在运动学水平上进行控制的设计相比，尽管考虑动态水平在理论上更具挑战性，但在实际操作中却是必要的，因为实际控制输入是用于驱动机器人车轮的电机、电流和电压[225,231-233]。实际上，从机器人的速度到电压的静态映射可能会严重恶化控制性能[234]。然而，在设计的控制器中，电机的动力学特性往往被忽略，因为电机的动力学响应通常比机器人的动力学响应快得多，所以 9.3 节中将应用于机器人车轮上的力矩作为动态水平的控制输入是合理的。除此之外，9.3 节将设计适当的通信协议，使得即使在通信网络受到拒绝服务攻击（而不是物理攻击）的情况下，移动机器人也可以即时跟踪参考轨迹，即移动机器人可以从任何初始位置以目标速度渐近地追踪时变的目标轨迹。换句话说，9.3 节设计的传输协议使控制律对拒绝服务攻击具有弹性，也就是即使在有攻击存在的情况下，机器人的跟踪控制依然不会被破坏。9.3 节的主要贡献如下：①建立一个混杂系统模型来研究受到拒绝服务攻击的移动机器人的跟踪控制问题，该模型考虑了拒绝服务攻击的显式描述。在拒绝服务攻击下，本章节结合混杂控制理论设计了一种事件触发策略，其主要结果具有低保守性，这是因为与状态相关的参数可以在没有拒绝服务攻击的情况下灵活地扩大事件触发间隔。②根据所设计的策略，9.3 节在拒绝服务攻击存在的情况下进行实验，实验结果表明，在拒绝服务攻击的约束下，通过设计适当的采样频率可以实现弹性跟踪特性。

9.2
基于拒绝服务攻击的姿态系统弹性跟踪控制算法

9.2.1 姿态系统在攻击下的动力学模型

基于 Goebel 等人[230]研究中的理论，混杂系统模型可以表示为

$$\mathcal{H}:\begin{cases}\dot{\xi}\in\mathcal{F}(\xi),\ \ \xi\in C\\ \xi^{+}\in\mathcal{G}(\xi),\ \ \xi\in D\end{cases} \tag{9-1}$$

其中，$\boldsymbol{\xi}\in\mathbb{R}^n$是系统状态；流动集$C$和跳变集$D$是$\mathbb{R}^n$中的闭子集；流动集值映射$\mathcal{F}$在闭集$C$上是非空连续的并且是凸的，跳变集值映射$\mathcal{G}$在集合$D$上是非空连续的，因此混杂系统$\mathcal{H}$是适定的。如果$E=\bigcup_{j=0}^{J-1}\left(\left[t_j,t_{j+1}\right],j\right)$，其中有限时间序列满足$0=t_0\leqslant t_1\leqslant\cdots\leqslant t_J$，则集合$E\subset\mathbb{R}_{\geqslant0}\times\mathbb{Z}_{\geqslant0}$是紧的混杂时间域。如果对任意$(T,J)\in E,E\bigcap\left(\left[0,T\right]\times\left\{0,\cdots,J\right\}\right)$是紧的混杂时间域，则集合$E$是混杂时间域。如果$E$是混杂时间域并且$\boldsymbol{\xi}(\cdot,j)$是局部绝对连续的，其中$j\in\{1,2,\cdots,J-1\}$，则函数$\boldsymbol{\xi}:E\to\mathbb{R}^n$是一个混杂弧。当满足下列三个条件时，混杂弧$\xi$：$\mathrm{dom}\boldsymbol{\xi}\to\mathbb{R}^n$是式(9-1)的解。条件：①$\boldsymbol{\xi}(0,0)\in C\bigcup D$；②对任意$j\in\mathbb{Z}_{\geqslant0}$以及几乎所有$t$使得$(t,j)\in\mathrm{dom}\,\boldsymbol{\xi},\boldsymbol{\xi}(t,j)\in C$以及$\dot{\boldsymbol{\xi}}(t,j)\in\mathcal{F}\left(\boldsymbol{\xi}(t,j)\right)$；③对任意$(t,j)\in\mathrm{dom}\boldsymbol{\xi}$满足$(t,j+1)\in\mathrm{dom}\,\boldsymbol{\xi}$，有$\boldsymbol{\xi}(t,j)\in D$和$\boldsymbol{\xi}(t,j+1)=\mathcal{G}\left(\boldsymbol{\xi}(t,j)\right)$。如果该解不能被扩展，则混杂系统$\mathcal{H}$的解$\boldsymbol{\xi}$是最大的；如果$\mathrm{dom}\,\xi$是无界的，则该解是完备的。

飞行器姿态$\boldsymbol{R}\in\mathbb{R}^{3\times3}$是机体$\mathcal{F}_b$相对于惯性参照系$\mathcal{F}_o$的旋转，并且属于李群$\mathbb{SO}(3)$，即

$$\mathbb{SO}(3)=\left\{\boldsymbol{R}\in\mathbb{R}^{3\times3}:\boldsymbol{R}^{\mathrm{T}}\boldsymbol{R}=\boldsymbol{I},\det\boldsymbol{R}=1\right\} \tag{9-2}$$

其中，$\boldsymbol{I}$为该李群的单位元素，即单位矩阵，因此姿态矩阵$\boldsymbol{R}\in\mathbb{SO}(3)$都是行列式为1的正交矩阵。姿态系统的运动学和动力学模型为：

$$\dot{\boldsymbol{R}}=\boldsymbol{R}\boldsymbol{S}(\omega),\ \ \boldsymbol{J}\dot{\boldsymbol{\omega}}=\boldsymbol{S}(\boldsymbol{J}\boldsymbol{\omega})\boldsymbol{\omega}+\boldsymbol{u}_\tau \tag{9-3}$$

其中，$\boldsymbol{R}\in\mathbb{SO}(3)$和$\boldsymbol{\omega}\in\mathbb{R}^3$分别表示姿态矩阵和角速度，惯性矩阵$\boldsymbol{J}\in\mathbb{R}^{3\times3}$满足$\boldsymbol{J}=\boldsymbol{J}^{\mathrm{T}}>0$，力矩控制输入为$\boldsymbol{u}_\tau\in\mathbb{R}^3$。任意两个向量$\boldsymbol{a}=(\boldsymbol{a}_1\ \boldsymbol{a}_2\ \boldsymbol{a}_3)^{\mathrm{T}}\in\mathbb{R}^3$和$\boldsymbol{b}=(\boldsymbol{b}_1\ \boldsymbol{b}_2\ \boldsymbol{b}_3)^{\mathrm{T}}\in\mathbb{R}^3$的外积为$\boldsymbol{a}\times\boldsymbol{b}=\boldsymbol{S}(\boldsymbol{a})\boldsymbol{b}$，其中

$$\boldsymbol{S}(\boldsymbol{a})=\begin{bmatrix}0 & -a_3 & a_2\\ a_3 & 0 & -a_1\\ -a_2 & a_1 & 0\end{bmatrix}\tag{9-4}$$

令$S^n=\{\boldsymbol{x}\in\mathbb{R}^{n+1}:\boldsymbol{x}^{\mathrm{T}}\boldsymbol{x}=1\}$以及$\boldsymbol{i}=(1\,0\,0\,0)\in S^3$为$S^3$的单位元素。根据罗德里格斯公式$\mathcal{R}(\boldsymbol{q})=\boldsymbol{I}_3+2\boldsymbol{\eta}\boldsymbol{S}(\boldsymbol{\epsilon})+2\boldsymbol{S}^2(\boldsymbol{\epsilon})$和单位四元数$\boldsymbol{q}=(\boldsymbol{\eta}\ \boldsymbol{\epsilon}^{\mathrm{T}})^{\mathrm{T}}\in S^3$，可知$\mathbb{SO}(3)$的每个元素都可以被两个对跖单位四元数参数化，即$\mathcal{R}(-\boldsymbol{q})=\mathcal{R}(\boldsymbol{q})$，这是因为$\mathcal{R}(\cdot)$是个双覆盖映射而且$\mathcal{R}$是局部微分同胚映射。当姿态矩阵$\boldsymbol{R}\in\mathbb{SO}(3)$被单位四元数$\boldsymbol{q}\in S^3$参数化的时候，基于覆盖空间的基本性质，可以采用路径提升机制[235]，即式(9-3)的运动学方程需要提升到S^3上来克服双覆盖映射带来的不便。本章节中用$\boldsymbol{q}=(\boldsymbol{\eta},\boldsymbol{\epsilon})$表示单位四元数，因而任意两个单位四元数之间的乘积表达式为

$$\boldsymbol{q}_1\otimes\boldsymbol{q}_2=\begin{bmatrix}\boldsymbol{\eta}_1\boldsymbol{\eta}_2-\boldsymbol{\epsilon}_1^{\mathrm{T}}\boldsymbol{\epsilon}_2\\ \boldsymbol{\eta}_1\boldsymbol{\epsilon}_2+\boldsymbol{\eta}_2\boldsymbol{\epsilon}_1+\boldsymbol{S}(\boldsymbol{\epsilon}_1)\boldsymbol{\epsilon}_2\end{bmatrix}\tag{9-5}$$

因此式(9-3)的运动学方程可以表示为

$$\dot{\boldsymbol{q}}=\begin{pmatrix}\dot{\boldsymbol{\eta}}\\ \dot{\boldsymbol{\epsilon}}\end{pmatrix}=\frac{1}{2}\boldsymbol{q}\otimes\boldsymbol{v}(\boldsymbol{\omega})=\frac{1}{2}\begin{pmatrix}-\boldsymbol{\epsilon}^{\mathrm{T}}\\ \boldsymbol{\eta}I+\boldsymbol{S}(\boldsymbol{\epsilon})\end{pmatrix}\boldsymbol{\omega}\tag{9-6}$$

其中，$\boldsymbol{v}(\boldsymbol{\omega})=(0,\boldsymbol{\omega}^{\mathrm{T}})^{\mathrm{T}},\boldsymbol{\omega}\in\mathbb{R}^3$。

本章节的主要目标是通过设计反馈控制策略$\boldsymbol{u}_\tau$，对于任意$M>0$和紧集$\Omega\subset\mathbb{R}^3$，使得姿态$\boldsymbol{R}$和角速度$\boldsymbol{\omega}$能够分别渐近追踪期望姿态$\boldsymbol{R}_d\in\mathbb{SO}(3)$和期望角速度$\boldsymbol{\omega}_d\in\Omega$，即$(\boldsymbol{R},\boldsymbol{\omega})\to(\boldsymbol{R}_d,\boldsymbol{\omega}_d)$，并且期望角速度满足$\dot{\boldsymbol{\omega}}_d\in M\mathbb{B}$，其中$\mathbb{B}$是个单位球。

令$\boldsymbol{q}_d=(\boldsymbol{\eta}_d,\boldsymbol{\epsilon}_d)\in S^3$表示对应于期望姿态$\boldsymbol{R}_d$的参数化单位四元数，即$\mathcal{R}(\boldsymbol{q}_d)=\boldsymbol{R}_d$。基于式(9-6)，有$\dot{\boldsymbol{q}}_d=\dfrac{1}{2}\boldsymbol{q}_d\otimes\boldsymbol{v}(\boldsymbol{\omega}_d)$。定义追踪误差为

$$\tilde{\boldsymbol{q}}=\boldsymbol{q}_d^{-1}\otimes\boldsymbol{q},\quad\tilde{\boldsymbol{\omega}}=\boldsymbol{\omega}-\overline{\boldsymbol{\omega}}_d\tag{9-7}$$

其中，$\overline{\boldsymbol{\omega}}_d=\tilde{\boldsymbol{R}}^{\mathrm{T}}\boldsymbol{\omega}_d,\mathcal{R}(\tilde{\boldsymbol{q}})=\tilde{\boldsymbol{R}}$和$\tilde{\boldsymbol{q}}=(\tilde{\boldsymbol{\eta}},\tilde{\boldsymbol{\epsilon}})$。因而有

$$\dot{\tilde{q}} = \frac{1}{2}\tilde{q} \otimes v(\tilde{\omega}) \tag{9-8}$$

其中，$\dot{\tilde{\eta}} = -\frac{1}{2}\tilde{\epsilon}^{\mathrm{T}}\tilde{\omega}$以及$\dot{\tilde{\epsilon}} = \frac{1}{2}\left(\tilde{\eta}I + S(\tilde{\epsilon})\right)\tilde{\omega}$，进而可以得到$J\dot{\tilde{\omega}} = \Sigma\left(\tilde{\omega},\overline{\omega}_d\right)\tilde{\omega} - S\left(\overline{\omega}_d\right)J\overline{\omega}_d - J\tilde{R}^{\mathrm{T}}\dot{\omega}_d + u_\tau$，其中$\Sigma\left(\tilde{\omega},\overline{\omega}_d\right) = S(J\tilde{\omega}) + S\left(J\overline{\omega}_d\right) - S\left(\overline{\omega}_d\right)J - JS\left(\overline{\omega}_d\right)$是个反对称矩阵。

本章节考虑的是完整状态测量输出，控制输入u_τ在离散的更新时刻$\{t_r\}_{r\in\mathbb{Z}_{\geqslant 0}}$进行更新，其中$0 \leqslant t_0 < t_1 < \cdots < t_r < t_{r+1}$并且$0 < \varDelta \leqslant t_{r+1} - t_r$。本章节忽略了数据采样、传输和更新过程中的时滞。定义$y \overset{\text{def}}{=} (q,\omega)$，在每一个采样时刻$t_r$，采样值$\hat{y}(t_r)$经过网络通道到达控制器端进而更新控制器$u_\tau$，与此同时期望姿态$y_d \overset{\text{def}}{=} (q_d,\omega_d)$也会伴随着$y$一起通过网络通道同步传输到控制器端进而用于更新控制器u_τ。在$t \in [t_r,t_{r+1}]$期间，零阶保持器使得采样值$(\hat{y}(t_r),\hat{y}_d(t_r))$保持不变，即

$$\dot{\hat{y}}(t) = 0,\ \dot{\hat{y}}_d(t) = 0$$

基于此，定义$t \in [t_r,t_{r+1}]$期间的采样误差为

$$e(t) = \left(\hat{\tilde{q}}(t_r) - \tilde{q}(t),\hat{\tilde{\omega}}(t_r) - \tilde{\omega}(t)\right) \tag{9-9}$$

其中，$\hat{\tilde{q}} = \hat{q}_d^{-1} \otimes \hat{q}$和$\hat{\tilde{\omega}} = \hat{\omega} - \mathcal{R}\left(\hat{\tilde{q}}\right)^{\mathrm{T}}\hat{\omega}_d$分别表示$\tilde{q}$和$\tilde{\omega}$的采样值。在每一个更新时刻$t_r$，$(\hat{y}(t_r),\hat{y}_d(t_r))$被重置为$(y(t_r),y_d(t_r))$，即

$$\left(\hat{y}(t_r^+),\hat{y}_d(t_r^+)\right) = \left(y(t_r),y_d(t_r)\right)$$

因此

$$\begin{cases} \dot{e}(t) = g(e(t)), t \in [t_r,t_{r+1}] \\ e(t_{r+1}^+) = 0 \end{cases} \tag{9-10}$$

其中，$g(e) = \left(-\dot{\tilde{q}},-\dot{\tilde{\omega}}\right)$。定义$x_p = (q,\omega,q_d,\omega_d) \in \mathbb{X}_0$和$\tilde{x}_p = (\tilde{q},\tilde{\omega},q_d,\omega_d) \in \mathbb{X}_0$，其中，$\mathbb{X}_0 = \mathcal{S}^3 \times \mathbb{R}^3 \times \mathcal{S}^3 \times \varOmega$以及$e \in \mathbb{R}^e$，因此本章节的姿态追踪目标是通过设计控制策略和通信时刻来保证紧集$\mathcal{A}_0 = \{x_p \in \mathbb{X}_0 \times \mathbb{R}^e |\ \tilde{q} = \pm i, \tilde{\omega} = 0, e = 0\}$的渐近稳定性。

拒绝服务攻击阻塞了网络通道，导致任何采样值$(y(t_r),y_d(t_r)) \overset{\text{def}}{=} \mathfrak{J}(t_r)$都不能成功传输。定义$\{\mathcal{L}_n\}_{n\in\mathbb{Z}_{\geqslant 0}}$为拒绝服务攻击发生的一系列时间

区间，即 $\mathcal{L}_n \overset{\text{def}}{=} \{\iota_n\} \cup [\iota_n, \iota_n + \varepsilon_n)$，其中 $\iota_n \leqslant \iota_n + \varepsilon_n < \iota_{n+1}$，即 $\varepsilon_n \in \mathbb{R}_{\geqslant 0}$ 表示每次拒绝服务攻击持续的时间长度[85]，这些参数的含义可以参见图 9-1。在图 9-1 中，拒绝服务攻击的开始符号为↑，而↓则表示攻击结束符号。在 $t \in [0,10]$ 时间段内：$\varepsilon_0 = 2$，$\varepsilon_1 = 1.5$，$\varepsilon_2 = 1.5$，$\iota_0 = 1$，$\iota_1 = 4.3$，$\iota_2 = 7.6$，$n(0,0.8) = 0$，$n(1,5) = 2$，$n(5,9) = 1$，$\Xi(0,0,8) = \varnothing$，$\Xi(1,5) = [1,3) \cup [4.3,5)$，$\Xi(5,9) = [5,5.8) \cup [7.6,9)$。对任意给定的时间区间$[T_1, T_2]$，其中$0 \leqslant T_1 \leqslant T_2$，定义攻击发生的总时间为$\mathcal{T} \overset{\text{def}}{=} \bigcup_{n \in \mathbb{Z}_{\geqslant 0}} \mathcal{L}_n$，那么$\Xi(T_1, T_2) \overset{\text{def}}{=} \left(\bigcup_{n \in \mathbb{Z}_0} \mathcal{L}_n \right) \cap [T_1, T_2]$和$\Theta(T_1, T_2) \overset{\text{def}}{=} [T_1, T_2] \setminus \Xi(T_1, T_2)$分别表示没有攻击时间和有攻击的时间。因此$\mathfrak{F}(t_r)$的更新可以表示为

$$\hat{\mathfrak{J}}(t_k^+) = \begin{cases} \mathfrak{J}(t_r), & t_r \notin \mathcal{T} \\ \hat{\mathfrak{J}}(t_r), & t_r \in \mathcal{T} \end{cases}$$

其中，$r \in \mathbb{Z}_{\geqslant 0}$。在不引起混淆的情况下，$\Theta(T_1, T_2)$和$\Xi(T_1, T_2)$可以分别表示为$\Theta$和$\Xi$。在任意区间$(T_1, T_2)$内，我们需要限制拒绝服务攻击发生的频率和总时间长度，否则拒绝服务攻击有可能发生在所有的传输时刻上，使得任何控制策略在任何通信时刻都不能对该拒绝服务攻击有韧性，正如Dolk等人[83]、De和Tesi[85]以及Feng等人[236]研究中分析的那样。定义$n(T_1, T_2)$为拒绝服务攻击在(T_1, T_2)发生的次数，而$|\Xi(T_1, T_2)|$为拒绝服务攻击发生的总时间，基于此，接下来我们给出拒绝服务攻击的限制条件。

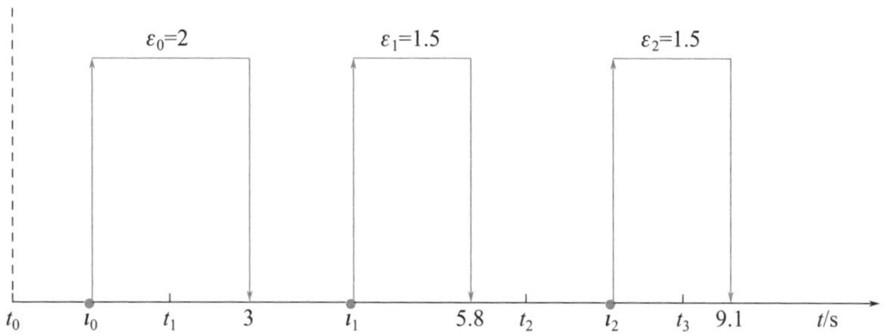

图 9-1　拒绝服务攻击的例子

假设 9.1

存在非负常数ν、$\varsigma \in \mathbb{R}_{\geqslant 0}, \tau_f \in \mathbb{R}_{>\Delta}$和$\tau_d \in \mathbb{R}_{>1}$，满足：①拒绝服务攻击

频率$n\left(T_1,T_2\right)\leqslant v+\frac{T_2-T_1}{\tau_f}$；②拒绝服务攻击时长$\left|\Xi\left(T_1,T_2\right)\right|\leqslant\varsigma+\frac{T_2-T_1}{\tau_d}$，其中$T_1,T_2\in\mathbb{R}_{\geqslant0}$满足$T_1\leqslant T_2$。

注释 9.1 由上述假设可知，拒绝服务攻击的频率和持续时间特征受到约束，可以通过适当选择上述涉及的参数来满足这个约束，其中，v表示振动界限；τ_f表示连续两次拒绝服务攻击的均滞留时间。比如$n(T_1,T_2)=1$时（即拒绝服务攻击发生在$T_1=T_2$时刻上），那么v就是必须的。假设 9.1 中②表示，如果$\left|\Xi\left(l_n,t_n+\varepsilon_n\right)\right|=\varepsilon_n$［即$n\left(T_1,T_2\right)=1$时］，那么$\varepsilon_n\geqslant\frac{\varepsilon_n}{\tau_d}$，因为$\tau_d>1$。因此参数$\varsigma$保证了假设 9.1 中②的可行性。

本章节中采样值$\mathfrak{J}$被连续传输给事件的驱动策略来产生一系列离散的传输时刻，即一旦事件驱动策略被触发，采样值会经过网络通道传输给控制器：如果有拒绝服务攻击存在，那么采样数据将不能经过网络传输给控制器；如果没有拒绝服务攻击，那么传输将会成功。基于事件驱动的采样通信机制为

$$t_{r+1}=\inf_{r\in\mathbb{Z}_{\geqslant0}}\left\{t:t\geqslant t_r+\tau_{\text{miet}}^m\mid\chi\left(t\right)\leqslant0,r\in\mathbb{Z}_{\geqslant0},t_0=0\right\}\tag{9-11}$$

其中，函数$\chi\left(t\right)$表示事件触发函数，布尔变量$m\in\left\{0,1\right\}$被定义为：①$m=0$表示传输尝试会成功，其中，传输尝试时刻t_{r+1}会在自上一次传输时刻t_r再经过时间τ_{miet}^0之后且$\chi\left(t\right)\leqslant0$的时候到来；②$m=1$表示传输尝试时刻$t_{r+1}$会在自上一次传输时刻$t_r$再经过$\tau_{\text{miet}}^1$之后且$\chi\left(t\right)\leqslant0$的时候到来，但是数据不会被成功传输给控制器，因为此时网络通道因拒绝服务攻击而不可用。值得注意的是，τ_{miet}^m表示最小允许传输间隔并满足$0<\tau_{\text{miet}}^1\leqslant\tau_{\text{miet}}^0$，这可以帮助排除奇诺现象。触发时刻方程(9-11)表示事件触发机制只有可能自上次传输时刻再经过τ_{miet}^m之后被触发，而且参数τ_{miet}^m使得在攻击发生期间传输更频繁，使得事件驱动机制尽早知道攻击何时结束。本章节在控制器和事件驱动策略之间采用确认机制，使事件触发机制知道控制器是否接收到采样值，为此我们遵循基于事件触发策略(9-11)的

设计机理，这确保了控制系统能够对不可预测的变化（由拒绝服务攻击引起）做出响应，以保持跟踪性能（弹性控制），而正是数字信息和通信技术使我们能够实现这一目标，进而使飞行器姿态控制系统更加智能化。事件驱动函数$\chi(t)$可以设计为

$$\begin{cases} \mathrm{d}\chi(t)=\Psi\left(m(t),\ \boldsymbol{o}(t),\ \chi(t)\right)\mathrm{d}t, & t\in[t_r,t_{r+1}) \\ \chi^+(t_r)=\begin{cases}\tilde{\chi}(t_r), & t_r\in\Theta_Z \\ \chi(t_r), & t_r\in\Xi_Z\end{cases} \end{cases} \tag{9-12}$$

其中，$\boldsymbol{o}=(\boldsymbol{e},\tau,\phi)\in\mathbb{R}^e\times\mathbb{R}_{\geqslant 0}\times\left[\lambda,\lambda^{-1}\right]$表示部分变量，其中$\lambda\in(0,1)$。时间变量$\tau$表示自上次传输尝试之后经过的时间，函数$\phi$将用于推导最小允许传输间隔（将在后面的章节中分析），而函数Ψ和$\tilde{\chi}$将会在下文中给出具体表达式。

9.2.2 姿态系统的追踪控制分析

为实现全局姿态控制，本节设计了一种基于二进制逻辑变量的四元数混杂控制策略，该策略基于逻辑变量的切换进而实现反馈控制律之间基于迟滞宽度的切换。

基于单位四元数的混杂控制策略可以设计为

$$u_\tau=S\left(\overline{\boldsymbol{\omega}}_d\right)J\overline{\boldsymbol{\omega}}_d+J\tilde{R}^{\mathrm{T}}\dot{\boldsymbol{\omega}}_d-\kappa_b h\tilde{\boldsymbol{\epsilon}}-K\tilde{\boldsymbol{\omega}} \tag{9-13}$$

其中，$\kappa_b\in\mathbb{R}_{>0}$和$\boldsymbol{K}\in\mathbb{R}^{3\times 3}$表示控制增益，并且$\boldsymbol{K}^{\mathrm{T}}=\boldsymbol{K}>0$，逻辑变量$h$的动态变化为

$$\begin{cases}\dot{h}=0, & x\in C_h \\ h^+\in\overline{\mathrm{sgn}}(\tilde{\eta}), & x\in D_h\end{cases} \tag{9-14}$$

其中，集值映射$\overline{\mathrm{sgn}}:\mathbb{R}\rightrightarrows\{-1,1\}$为

$$\overline{\mathrm{sgn}}(s)=\begin{cases}\mathrm{sgn}(s), & s\neq 0 \\ \{-1,1\}, & s=0\end{cases} \tag{9-15}$$

其中，$\mathrm{sgn}(s)=1$，当$s>0;\mathrm{sgn}(s)=-1$，当$s<0$。闭集合C_h和D_h分别为

$$C_h=\left\{\boldsymbol{x}\in\mathbb{X}:h\tilde{\eta}\geqslant-\delta\right\},D_h=\left\{\boldsymbol{x}\in\mathbb{X}:h\tilde{\boldsymbol{\eta}}\leqslant-\delta\right\} \tag{9-16}$$

其中，$\boldsymbol{x}\overset{\mathrm{def}}{=}\left(\boldsymbol{x}_p,h\right)\in\mathbb{X},h\in\{-1,1\},\mathbb{X}=\mathbb{X}_0\times\{-1,1\}$，而$\delta\in(0,1)$表示用户定

义的滞后宽度。

注释 9.2 由式 (9-14) 可知，基于四元数和逻辑变量的混杂控制法则 (9-13) 会根据条件 (9-16) 进行切换。用户定义的滞后宽度 δ 可视为一个缓冲间隔，起决定当前控制律何时切换的作用，而且滞后宽度 δ 可以增加混杂控制法则 (9-13) 对外界干扰和系统误差的鲁棒性。正如 Mayhew 等人 [237] 的研究中，切换条件 (9-16) 表明滞后宽度 $\delta \in (0,1)$ 越大，混杂控制法则 (9-13) 的鲁棒性越好，但会降低控制效率，而当 $\delta = 0$ 时会带来由四元数和旋转矩阵之间的双覆盖映射引起的退绕现象。因此，滞后宽度 δ 权衡着基于噪声的鲁棒性和少量的滞后导致的低效率。

在采样区间 $[t_r, t_{r+1}]$ 内，基于 9.2.1 节中的分析可知控制器 u_τ [式 (9-13)] 可以表示为：$u_\tau = S\left(\hat{\overline{\boldsymbol{\omega}}}_d\right) J \hat{\overline{\boldsymbol{\omega}}}_d - \kappa_b h \hat{\tilde{\boldsymbol{\epsilon}}} - K \hat{\tilde{\boldsymbol{\omega}}}$。接下来我们分析这三项（$\hat{\overline{\omega}}_d, \hat{\tilde{\epsilon}}$ 和 $\hat{\tilde{\omega}}$）：定义 $\boldsymbol{e} = \left(\hat{\tilde{\boldsymbol{q}}} - \tilde{\boldsymbol{q}}, \hat{\tilde{\boldsymbol{\omega}}} - \tilde{\boldsymbol{\omega}}\right) = \left(\hat{\tilde{\boldsymbol{q}}} - \tilde{\boldsymbol{q}}, (\hat{\boldsymbol{\omega}} - \boldsymbol{\omega}) - \left(\hat{\overline{\boldsymbol{\omega}}}_d - \overline{\boldsymbol{\omega}}_d\right)\right) \stackrel{\text{def}}{=} \left(\boldsymbol{e}_{\tilde{q}}, \boldsymbol{e}_{\tilde{\omega}}\right)$，其中 $\hat{\tilde{\boldsymbol{q}}} - \tilde{\boldsymbol{q}} \stackrel{\text{def}}{=} \boldsymbol{e}_{\tilde{q}} = \left(\boldsymbol{e}_{\tilde{q}}^{\tilde{\eta}}, \boldsymbol{e}_{\tilde{q}}^{\tilde{\epsilon}}\right)$，以及 $(\hat{\boldsymbol{\omega}} - \boldsymbol{\omega}) - \left(\hat{\overline{\boldsymbol{\omega}}}_d - \overline{\boldsymbol{\omega}}_d\right) \stackrel{\text{def}}{=} \boldsymbol{e}_\omega - \boldsymbol{e}_{\overline{\omega}_d} \stackrel{\text{def}}{=} \boldsymbol{e}_{\tilde{\omega}}$ 满足 $\hat{\boldsymbol{\omega}} - \boldsymbol{\omega} \stackrel{\text{def}}{=} \boldsymbol{e}_\omega$ 和 $\hat{\overline{\boldsymbol{\omega}}}_d - \overline{\boldsymbol{\omega}}_d = \boldsymbol{e}_{\overline{\omega}_d}$，进而得到 $\hat{\overline{\boldsymbol{\omega}}}_d = \overline{\boldsymbol{\omega}}_d + \boldsymbol{e}_{\overline{\omega}_d}, \hat{\tilde{\boldsymbol{\omega}}} = \tilde{\boldsymbol{\omega}} + \boldsymbol{e}_{\tilde{\omega}}$，以及 $\hat{\tilde{\boldsymbol{\epsilon}}} = \tilde{\boldsymbol{\epsilon}} + \boldsymbol{e}_{\tilde{q}}^{\tilde{\epsilon}}$，根据 $\hat{\tilde{\boldsymbol{q}}} - \tilde{\boldsymbol{q}} = \left(\hat{\tilde{\boldsymbol{\eta}}} - \tilde{\boldsymbol{\eta}}, \hat{\tilde{\boldsymbol{\epsilon}}} - \tilde{\boldsymbol{\epsilon}}\right)$。因此在采样区间 $[t_r, t_{r+1}]$ 内，控制器 u_τ [式 (9-13)] 可以表示为：

$$
\begin{aligned}
& u_\tau\left(\tilde{\boldsymbol{q}} + \boldsymbol{e}_{\tilde{q}}, \tilde{\boldsymbol{\omega}} + \boldsymbol{e}_{\tilde{\omega}}\right) \\
& = S\left(\overline{\boldsymbol{\omega}}_d + \boldsymbol{e}_{\overline{\omega}_d}\right) J\left(\overline{\boldsymbol{\omega}}_d + \boldsymbol{e}_{\overline{\omega}_d}\right) - \kappa_b h\left(\tilde{\epsilon} + \boldsymbol{e}_{\tilde{q}}^{\tilde{\epsilon}}\right) - \boldsymbol{K}\left(\tilde{\boldsymbol{\omega}} + \boldsymbol{e}_{\tilde{\omega}}\right) \\
& = S\left(\overline{\boldsymbol{\omega}}_d\right) \boldsymbol{J} \overline{\boldsymbol{\omega}}_d + S\left(\overline{\boldsymbol{\omega}}_d\right) \boldsymbol{J} \boldsymbol{e}_{\overline{\omega}_d} + S\left(\boldsymbol{e}_{\overline{\omega}_d}\right) J\left(\overline{\boldsymbol{\omega}}_d + \boldsymbol{e}_{\overline{\omega}_d}\right) \\
& -\kappa_b h \tilde{\boldsymbol{\epsilon}} - \kappa_b h \boldsymbol{e}_{\tilde{q}}^{\tilde{\epsilon}} - \boldsymbol{K} \tilde{\boldsymbol{\omega}} - \boldsymbol{K} \boldsymbol{e}_{\tilde{\omega}}
\end{aligned} \tag{9-17}
$$

在每一个更新时刻 t_{r+1}，控制输入 u_τ 会基于采样值 $\hat{\overline{\boldsymbol{\omega}}}_d(t_{r+1})$、$\hat{\tilde{\epsilon}}(t_{r+1})$ 以及 $\hat{\tilde{\boldsymbol{\omega}}}(t_{r+1})$ 进行更新。以上分析表明本章节中存在两种跳跃，一种是基于逻辑变量 h[式(9-14)]的混杂控制器 u_τ 的切换，另一种是基于采样时刻

$\{t_r\}_{r\in\mathbb{Z}_{\geqslant 0}}$的采样更新。

基于上述分析，接下来我们建立一个带有反馈控制策略 (9-17) 的闭环系统的混杂形式，进而推导出姿态跟踪的弹性事件触发条件。令 $\boldsymbol{\xi}=(\tilde{\boldsymbol{x}},\boldsymbol{e},\chi,\phi,\tau,s,m)\in\tilde{\mathbb{X}}$，其中 $\tilde{\boldsymbol{x}}=(\tilde{\boldsymbol{x}}_p,h)\in\mathbb{X}$，$\tilde{\mathbb{X}}=\mathbb{X}\times\mathbb{R}^e\times\mathbb{R}_{\geqslant 0}\times[\lambda,\lambda^{-1}]\times\mathbb{R}_{\geqslant 0}\times\mathbb{R}_{\geqslant 0}\times\{0,1\}$，而变量 s 表示整体时间。基于 Goebel 等人 [230] 研究中的混杂模型，本章节所考虑的姿态追踪控制可以建模成混杂形式：

$$
\mathcal{H}:\begin{cases}\dot{\boldsymbol{\xi}}\in\mathcal{F}(\boldsymbol{\xi}), & \boldsymbol{\xi}\in C\\ \boldsymbol{\xi}^+\in\mathcal{G}(\boldsymbol{\xi}), & \boldsymbol{\xi}\in D\end{cases}\tag{9-18}
$$

其中，

$$
\begin{aligned}
&C=C_h\cap C_t, D=D_1\cup D_2\cup D_3\cup D_4\cup D_5,\\
&D_1=D_h\cap D_{t0}, D_2=D_h\cap D_{t1}, D_3=D_h\cap C_t, D_4=C_h\cap D_{t0}, D_5=C_h\cap D_{t1},\\
&C_t=\left\{\boldsymbol{\xi}\in\tilde{\mathbb{X}}|\ \tau\leqslant\tau_{\text{miet}}^m \text{ or } \chi\geqslant 0\right\}, D_t=\left\{\boldsymbol{\xi}\in\tilde{\mathbb{X}}|\ \tau\geqslant\tau_{\text{miet}}^m,\chi\leqslant 0\right\},\\
&D_{t0}=D_t\cap\left\{\boldsymbol{\xi}\in\tilde{\mathbb{X}}|\ s\notin\mathcal{T}\right\}, D_{t1}=D_t\cap\left\{\boldsymbol{\xi}\in\tilde{\mathbb{X}}|\ s\in\mathcal{T}\right\},
\end{aligned}\tag{9-19}
$$

流动映射 $\mathcal{F}(\boldsymbol{\xi})=(\boldsymbol{f}(\tilde{\boldsymbol{x}},\boldsymbol{e}),g(\boldsymbol{e}),\varPsi,\varPhi,1,1,0)$，其中 $\dot{\tilde{\boldsymbol{x}}}\overset{\text{def}}{=}f(\tilde{\boldsymbol{x}},\boldsymbol{e})=\left(\boldsymbol{f}(\tilde{\boldsymbol{x}}_p,\boldsymbol{e}),\tilde{h}\right)=\left(\frac{1}{2}\tilde{\boldsymbol{q}}\otimes\boldsymbol{v}(\tilde{\boldsymbol{\omega}}),\boldsymbol{f}\left(\tilde{\boldsymbol{\omega}},u_\tau\left(\tilde{\boldsymbol{q}}+\boldsymbol{e}_{\tilde{q}},\tilde{\boldsymbol{\omega}}+\boldsymbol{e}_{\tilde{\omega}}\right)\right),\frac{1}{2}\boldsymbol{q}_d\otimes\boldsymbol{v}(\boldsymbol{\omega}_d),M\mathbb{B},\tilde{h}\right)$，$\boldsymbol{f}\left(\tilde{\boldsymbol{\omega}},u_\tau\left(\tilde{\boldsymbol{q}}+\boldsymbol{e}_{\tilde{q}},\tilde{\boldsymbol{\omega}}+\boldsymbol{e}_{\tilde{\omega}}\right)\right)=\Sigma\left(\tilde{\boldsymbol{\omega}},\overline{\boldsymbol{\omega}}_d\right)\tilde{\boldsymbol{\omega}}-\boldsymbol{J}\tilde{\boldsymbol{R}}^{\mathrm{T}}\dot{\boldsymbol{\omega}}_d+S\left(\overline{\boldsymbol{\omega}}_d\right)\boldsymbol{J}\boldsymbol{e}_{\bar{\omega}_d}+S\left(\boldsymbol{e}_{\bar{\omega}_d}\right)\boldsymbol{J}\left(\overline{\boldsymbol{\omega}}_d+\boldsymbol{e}_{\bar{\omega}_d}\right)-\kappa_b h\tilde{\boldsymbol{\varepsilon}}-\kappa_b h\boldsymbol{e}_{\tilde{q}}^{\tilde{\varepsilon}}-\boldsymbol{K}\tilde{\boldsymbol{\omega}}-\boldsymbol{K}\boldsymbol{e}_{\tilde{\omega}}$，其中 $\varPhi$ 会在下文中详细给出，而跳变映射 $\mathcal{G}(\boldsymbol{\xi})=\left\{\mathcal{G}_{D_1}(\boldsymbol{\xi}),\mathcal{G}_{D_2}(\boldsymbol{\xi}),\mathcal{G}_{D_3}(\boldsymbol{\xi}),\mathcal{G}_{D_4}(\boldsymbol{\xi}),\mathcal{G}_{D_5}(\boldsymbol{\xi})\right\}$可以表达为

$$
\begin{cases}
\mathcal{G}_{D_1}(\boldsymbol{\xi})=\left(\tilde{\boldsymbol{x}}_p,\overline{\operatorname{sgn}}(\tilde{\eta}),0,\tilde{\chi},\lambda^{-1},0,s,0\right), & \boldsymbol{\xi}\in D_1,\\
\mathcal{G}_{D_2}(\boldsymbol{\xi})=\left(\tilde{\boldsymbol{x}}_p,\overline{\operatorname{sgn}}(\tilde{\eta}),\boldsymbol{e},\chi,\phi,\tau,s,1\right), & \boldsymbol{\xi}\in D_2,\\
\mathcal{G}_{D_3}(\boldsymbol{\xi})=\left(\tilde{\boldsymbol{x}}_p,\overline{\operatorname{sgn}}(\tilde{\eta}),\boldsymbol{e},\chi,\phi,\tau,s,m\right), & \boldsymbol{\xi}\in D_3,\\
\mathcal{G}_{D_4}(\boldsymbol{\xi})=\left(\tilde{\boldsymbol{x}}_p,h,0,\tilde{\chi},\lambda^{-1},0,s,0\right), & \boldsymbol{\xi}\in D_4,\\
\mathcal{G}_{D_5}(\boldsymbol{\xi})=\left(\tilde{\boldsymbol{x}}_p,h,e,\chi,\phi,\tau,s,1\right), & \boldsymbol{\xi}\in D_5.
\end{cases}\tag{9-20}
$$

所以本章节的主要目标是保证紧集$\boldsymbol{\mathcal{A}}=\left\{\boldsymbol{\xi}\in\tilde{\mathbb{X}}|\ \tilde{\boldsymbol{q}}=\operatorname{sgn}(h)\boldsymbol{i},\tilde{\boldsymbol{\omega}}=0,\boldsymbol{e}=0\right\}$的全局渐近稳定性，即实现姿态系统 (9-3) 基于弹性事件驱动策略的追踪控制。为此我们首先给出以下定义。

定义 9.1[230]

混杂系统$\mathcal{H}$ [式 (9-18)] 相对于紧集$\boldsymbol{\mathcal{A}}=\{\xi\in\tilde{\mathbb{X}}|\ \tilde{\boldsymbol{q}}=\operatorname{sgn}(h)\boldsymbol{i},\tilde{\boldsymbol{\omega}}=0,\boldsymbol{e}=0\}$是一致全局稳定的，如果所有最大解$\xi$是完备的并满足$\boldsymbol{\xi}(0,0)\in\tilde{\mathbb{X}}_0$，并且存在函数$\beta\in\mathcal{K}_\infty$，使得对任意初始值$\boldsymbol{\xi}(0,0)\in\tilde{\mathbb{X}}_0$，解$\xi$满足

$$\left|\boldsymbol{\xi}(t,k)\right|_{\mathcal{A}}\leqslant\beta\left(\left|\boldsymbol{\xi}(0,0)\right|_{\mathcal{A}},t,k\right),\ (t,k)\in\operatorname{dom}\boldsymbol{\xi}$$

其中，$\tilde{\mathbb{X}}_0=\left\{\boldsymbol{\xi}\in\tilde{\mathbb{X}}|\ \tau\geqslant\tau_{\text{miet}}^0,\chi=0,\phi=\phi_{\text{miet}},s=0\right\}$。

接下来我们给出关于网络通信系统的基本限制条件。给定$S\subset\mathbb{R}^n$，令$V\in\mathcal{D}(\mathcal{H}\cap S)$，如果满足$V:\mathbb{R}^n\to\mathbb{R}_{\geqslant0}$在闭集合$D\cap S$上连续且在开集$C\cap S$上处处可微；如果$S=\mathbb{R}^n$，则$V\in\mathcal{D}(\mathcal{H})$。

假设 9.2

存在函数$W\in D(\mathcal{H})$，其中$W:\mathbb{R}^e\to\mathbb{R}_{\geqslant0}$，连续有界函数$H:\mathbb{X}_0\to\mathbb{R}_{\geqslant0}$，函数$\underline{c}_W,\bar{c}_W\in\mathcal{K}_\infty$和常数$L>0$，使得

① 对任意$e\in\mathbb{R}^e$，有$\underline{c}_W(|\boldsymbol{e}|)\leqslant W(\boldsymbol{e})\leqslant\bar{c}_W(|\boldsymbol{e}|)$；

② 对任意$\tilde{x}_p\in\mathbb{X}_0$和几乎所有$e\in\mathbb{R}^e$，有$\langle\nabla W(\boldsymbol{e}),g(\boldsymbol{e})\rangle\leqslant LW(e)+H(\tilde{\boldsymbol{x}}_p)$。

假设 9.3

存在函数$V\in\mathcal{D}(\mathcal{H})$，其中$V:\mathbb{X}\to\mathbb{R}_{\geqslant0}$，函数$\underline{\alpha}$、$\bar{\alpha}\in\mathcal{K}_\infty$和$\mathcal{P}:\mathbb{R}^{2n_y}\to\mathbb{R}_{\geqslant0}$，正常数$\rho_V$、$\gamma$和$\rho_W$满足$\rho_W\leqslant\gamma^2$，使得

① 对任意$\tilde{x}\in\mathbb{X}$，有$\underline{\alpha}(|\tilde{\boldsymbol{x}}|)\leqslant V(\tilde{\boldsymbol{x}})\leqslant\bar{\alpha}(|\tilde{\boldsymbol{x}}|)$；

② 对几乎所有$\tilde{x}\in\mathbb{X}$和所有$e\in\mathbb{R}^e$,$V\left(\tilde{\boldsymbol{x}}_p,\overline{\operatorname{sgn}}(\tilde{\eta})\right)\leqslant V(\tilde{\boldsymbol{x}}_p,h)$；

③ 对几乎所有$\tilde{x}\in\mathbb{X}$和所有$e\in\mathbb{R}^e$，有$\langle\nabla V(\tilde{x}),\boldsymbol{f}(\tilde{\boldsymbol{x}},\boldsymbol{e})\rangle\leqslant-\rho_V V(\tilde{\boldsymbol{x}})-\mathcal{P}(|\mathfrak{J}|)-H^2(\tilde{x}_p)+(\gamma^2-\rho_W)W^2(e)$。

实际上，假设 9.2 可以用来保证关于网络诱导误差子系统 e 的稳定性，而假设 9.3 则是用来保证子系统 x 是 W 到 H 的 $\mathcal{L}_2$ 稳定性。考虑到式 (9-11) 中的事件驱动策略在建立驱动函数$\chi(t)$的具体表达式之前需要首先确定最小传输间隔τ_{miet}^m，为此我们首先给出$\Phi:\mathbb{R}_{\geqslant0}\to\mathbb{R}_{\geqslant0}$，其中$\dot{\phi}=\Phi$参见混杂系统$\mathcal{H}$ [式 (9-18)]，即

$$\frac{\mathrm{d}\phi}{\mathrm{d}\tau}=\begin{cases}-2L\phi-\gamma\left(\left(m\epsilon+1\right)\phi^2+\dfrac{m\left(\tilde{\gamma}-\gamma\right)+\gamma}{\gamma}\right), & \tau\in\left[0,\tau_{\text{miet}}^m\right]\\ 0, & \tau>\tau_{\text{miet}}^m\end{cases} \tag{9-21}$$

其中，$\epsilon\geqslant 0$和$\tilde{\gamma}\geqslant\gamma$，常数$L$和$\gamma$来自假设9.2、假设9.3，并满足$\lim\limits_{t\to\tau_{\text{miet}}}\phi\left(t\right)=\phi\left(\tau_{\text{miet}}\right)$使得$\phi$在$\tau_{\text{miet}}$处连续。于是，式(9-21)中的$\tau_{\text{miet}}^m$可以取值为

$$\tau_{\text{mati}}^m=\begin{cases}\dfrac{1}{Lr}\arctan\left(\varGamma\right), & \varLambda>L\\ \tilde{\varGamma}, & \varLambda=L\\ \dfrac{1}{Lr}\operatorname{arctanh}\left(\varGamma\right), & \varLambda<L\end{cases} \tag{9-22}$$

其中，$\lambda\in\left(0,1\right)$来自式(9-12)，$r=\sqrt{\left|\dfrac{\varLambda^2}{L^2}-1\right|}$，$\varLambda=\sqrt{\gamma\left(1+m\epsilon\right)\left[m\tilde{\gamma}+\gamma\left(1-m\right)\right]}$，$\tilde{\varGamma}=\dfrac{1}{\lambda\gamma\left(1+m\epsilon\right)+L}-\dfrac{\lambda}{\lambda L+\gamma\left(1+m\epsilon\right)}$，$\varGamma=\dfrac{r\left(1-\lambda\right)}{\dfrac{\lambda}{1+\lambda}\left[\dfrac{\gamma(1+m(\epsilon)+m\tilde{\gamma}+\gamma\left(1-m\right)}{L}-2\right]+1+\lambda}$。

接下来给出关于式(9-22)的引理。

引理 9.1[83]

假设τ_{mati}^m满足式(9-22)，那么方程

$$\dot{\tilde{\phi}}=-2L\tilde{\phi}-\gamma\left[\left(m\epsilon+1\right)\tilde{\phi}^2+m\frac{\tilde{\gamma}}{\gamma}+1-m\right] \tag{9-23}$$

对应初始值$\tilde{\phi}\left(0\right)=\lambda^{-1}$的解满足$\tilde{\phi}\left(t\right)\in\left[\lambda,\lambda^{-1}\right]$，其中$t\in\left[0,\tau_{\text{mati}}^m\right]$，$\tilde{\phi}\left(\tau_{\text{mati}}^m\right)=\lambda$。

基于式(9-22)和引理9.1，τ_{miet}^1和τ_{miet}^0可以取值为：$\tau_{\text{miet}}^1\leqslant\tau_{\text{mati}}^1\leqslant\tau_{\text{mati}}^0$以及$\tau_{\text{miet}}^1\leqslant\tau_{\text{miet}}^0\leqslant\tau_{\text{mati}}^0$。为方便进一步的分析，我们定义$\phi_{\text{miet}}=\tilde{\phi}(\tau_{\text{miet}}^0)$，其中$\tilde{\phi}$来自式(9-23)，并满足$\tilde{\phi}\left(0\right)=\lambda^{-1}$和$\phi_{\text{miet}}\geqslant\lambda$。

注释 9.3

跟Postoyan等人[84]的研究相比，式(9-23)额外考虑两个参数ϵ和$\tilde{\gamma}$，目的是尽早告知事件驱动策略拒绝服务攻击何时结束[83]，因为这两个参数会加快在拒绝服务攻击发生时的传输尝试频率，所以可以根据需要选择这两个参数ϵ和$\tilde{\gamma}$。ϵ或者$\tilde{\gamma}$的值越大会使得$m=1$时的传输间隔更小，即拒绝服务攻

击发生时的传输尝试频率更高，进而更有利于确认机制，帮助事件驱动策略察觉攻击何时结束。

基于以上分析，我们可以得到紧集$\mathcal{A}$的基于事件驱动策略的弹性控制结果。

定理 9.1

考虑混杂系统$\mathcal{H}$ [式 (9-18)] 满足假设 9.1 ～假设 9.3。假设：

① 对于满足假设 9.1 的拒绝服务攻击，拒绝服务攻击的频率和时间长度的参数τ_f和τ_d满足

$$\varpi^0 > \left(\tau_{\text{miet}}^1 / \tau_f + 1 / \tau_d\right)\left(\varpi^0 + \varpi^1\right) \tag{9-24}$$

其中，$\varpi^0 = \min\left\{\rho_V, \frac{\lambda\rho_W}{\gamma}\right\}, \varpi^1 = \frac{\bar{\gamma} - \rho_W}{\gamma\lambda}$以及$\bar{\gamma} = \gamma\left[2\lambda L + \gamma\left(1 + \lambda^2\right)\right]$。

② 对于参数满足假设 9.2、假设 9.3 中约束的事件驱动策略：存在函数$\delta_\chi \in \mathcal{K}_\infty$满足$\delta_\chi(s) \geqslant \varpi^0 s$，而且有

$$\Psi = \begin{cases} \rho\left(|\Im|\right) - \delta_\chi(\chi), & m = 0, 0 \leqslant \tau \leqslant \tau_{\text{miet}}^0 \\ \mathcal{P}\left(|\Im|\right) - \bar{\gamma}W^2(\boldsymbol{e}) - \delta_\chi(\chi), & m = 0, \tau > \tau_{\text{miet}}^0 \\ 0, & m = 1, 0 \leqslant \tau \leqslant \tau_{\text{miet}}^1 \\ -1, & m = 1, \tau > \tau_{\text{miet}}^1 \end{cases} \tag{9-25}$$

其中，$\bar{\gamma}$参考①；③函数$\tilde{\chi}$满足$\tilde{\chi} = \gamma\lambda W^2(\boldsymbol{e})$。如果满足以上假设，那么紧集$\boldsymbol{\mathcal{A}} = \left\{\boldsymbol{\xi} \in \tilde{\mathbb{X}} \mid \tilde{\boldsymbol{q}} = \text{sgn}(h)\boldsymbol{i}, \tilde{\boldsymbol{\omega}} = 0, \boldsymbol{e} = 0\right\}$是一致全局渐近稳定的，即实现了姿态系统(9-3)的基于事件驱动策略的弹性追踪控制。

证明：

首先为一些变量提供对应的速记符号，如$V \overset{\text{def}}{=} V(\tilde{x}), \mathcal{P} \overset{\text{def}}{=} \mathcal{P}\left(|\Im|\right), W \overset{\text{def}}{=} W(\boldsymbol{e}), H \overset{\text{def}}{=} H(\tilde{\boldsymbol{x}}), \Psi \overset{\text{def}}{=} \Psi\left(m(t), o(t), \chi(t)\right)$。类似文献 [230] 中定义 6-15，令$\mathcal{R}(\mathbb{X}_0)$表示系统状态的可达集，其中$\xi(0,0) \in \tilde{\mathbb{X}}_0$，而$\tilde{\mathbb{X}}_0$参见定义 9.1。进而可得到[83]：对任意$\boldsymbol{\xi} \in \mathcal{R}\left(\tilde{\mathbb{X}}_0\right)$，基于引理 9.1 和事件驱动策略 [式 (9-11)、式 (9-12)]，有 ① $\tau \geqslant \tau_{\text{miet}}^m \Leftrightarrow \phi = \phi_{\text{miet}}$； ② $\lambda^{-1} \geqslant \phi \geqslant \phi_{\text{miet}} \geqslant \lambda$；

③$\chi \geqslant 0$。

结合假设9.2、假设9.3，对任意$\xi \in C \cup D$，定义$\mathcal{V}(\boldsymbol{\xi})=V+\gamma\phi W^2+\chi$，进而可知，存在函数$\underline{\beta}_{\mathcal{V}},\overline{\beta}_{\mathcal{V}} \in \mathcal{K}_\infty$，使得对任意$\boldsymbol{\xi} \in \mathcal{R}\left(\tilde{\mathbb{X}}_0\right)$，$\underline{\beta}_{\mathcal{V}}\left(\left|\boldsymbol{\xi}(t,k)\right|_{\mathcal{A}}\right) \leqslant \mathcal{V}\left(\boldsymbol{\xi}(t,k)\right) \leqslant \overline{\beta}_{\upsilon}\left(\left|\boldsymbol{\xi}(t,k)\right|_{\mathcal{A}}\right)$成立，其中$\underline{\beta}_{\upsilon}: s \to \min\left\{\underline{\alpha}\left(\frac{s}{2}\right), \gamma\lambda \underline{c}_W^2\left(\frac{s}{2}\right)\right\}, \overline{\beta}_{\mathcal{V}}: s \to \max\left\{2\overline{\alpha}(s), 2\frac{\gamma}{\lambda}\overline{c}_W^{2}(s)\right\}$。

由混杂系统$\mathcal{H}$[式(9-18)]可知，如果$\xi \in D_h$，有$h^+ \in \overline{\operatorname{sgn}}(\tilde{\eta})$和$\tilde{\boldsymbol{x}}_p^+ = \tilde{\boldsymbol{x}}_p$，而根据假设9.3可知$V^+ = V\left(\tilde{\boldsymbol{x}}^+\right) = V\left(\tilde{\boldsymbol{x}}_p^+, h^+\right) = V\left(\tilde{\boldsymbol{x}}_p, \overline{\operatorname{sgn}}(\tilde{\eta})\right) \leqslant V\left(\tilde{\boldsymbol{x}}_p, h\right) = V\left(\tilde{\boldsymbol{x}}\right) = V$，即$V^+ \leqslant V$；如果$\xi \in D_{t0}$，有$\left(\boldsymbol{e}^+, \chi^+, \phi^+,\right) = \left(0, \tilde{\chi}, \lambda^{-1}\right)$，其中$\tilde{\chi} = \gamma\lambda W^2$；如果$\boldsymbol{\xi} \in D_{t1}$，有$\left(\boldsymbol{e}^+, \chi^+, \phi^+\right) = \left(\boldsymbol{e}, \chi, \phi\right)$；接下来，考虑$D = D_1 \cup D_2 \cup D_3 \cup D_4 \cup D_5$。结合式(9-20)以及给定的初始条件$\tilde{\mathbb{X}}_0 = \left\{\boldsymbol{\xi} \in \tilde{\mathbb{X}} \mid \tau \geqslant \tau_{\text{miet}}^0, \chi = 0, \phi = \phi_{\text{miet}}, s = 0\right\}$，可以得到

(J_1)如果$\boldsymbol{\xi} \in D_1 \cap \mathcal{R}\left(\tilde{\mathbb{X}}_0\right), D_1 = D_h \cap D_{t0}$，有$\mathcal{V}\left(\boldsymbol{\xi}^+\right) - \mathcal{V}(\boldsymbol{\xi}) = V^+ + \gamma\phi^+\left(W^+\right)^2 + \chi^+ - (V + \gamma\phi W^2 + \chi) = V^+ - V + 0 - \gamma\phi W^2 + \gamma\lambda W^2 - 0 \leqslant -\gamma\left(\phi_{\text{miet}} - \lambda\right) W^2 \leqslant 0$，因为$V^+ \leqslant V$以及$\phi \geqslant \phi_{\text{miet}} \geqslant \lambda$，即，$\mathcal{V}\left(\boldsymbol{\xi}^+\right) \leqslant \mathcal{V}(\boldsymbol{\xi})$。

(J_2)如果$\boldsymbol{\xi} \in D_2 \cap \mathcal{R}\left(\tilde{\mathbb{X}}_0\right), D_2 = D_h \cap D_{t1}$，有$\mathcal{V}\left(\boldsymbol{\xi}^+\right) - \mathcal{V}(\boldsymbol{\xi}) = V^+ + \gamma\phi^+\left(W^+\right)^2 + \chi^+ - (V + \gamma\phi W^2 + \chi) = V^+ - V + \gamma\phi W^2 - \gamma\phi W^2 + \chi - \chi = V^+ - V \leqslant 0$，因为$V^+ \leqslant V$，即，$\mathcal{V}\left(\boldsymbol{\xi}^+\right) \leqslant \mathcal{V}(\boldsymbol{\xi})$。

(J_3)如果$\boldsymbol{\xi} \in D_3 \cap \mathcal{R}\left(\tilde{\mathbb{X}}_0\right), D_3 = D_h \cap C_t$，有$\mathcal{V}\left(\boldsymbol{\xi}^+\right) - \mathcal{V}(\boldsymbol{\xi}) = V^+ + \gamma\phi^+\left(W^+\right)^2 + \chi^+ - \left(V + \gamma\phi W^2 + \chi\right) = V^+ - V + \gamma\phi W^2 - \gamma\phi W^2 + \chi - \chi \leqslant V^+ - V \leqslant 0$，因为$V^+ \leqslant V$，即，$\mathcal{V}\left(\boldsymbol{\xi}^+\right) \leqslant \mathcal{V}(\boldsymbol{\xi})$。

(J_4)如果$\boldsymbol{\xi} \in D_4 \cap \mathcal{R}\left(\tilde{\mathbb{X}}_0\right), D_4 = C_h \cap D_{t0}$，有$\mathcal{V}\left(\boldsymbol{\xi}^+\right) - \mathcal{V}(\boldsymbol{\xi}) = V^+ + \gamma\phi^+\left(W^+\right)^2 + \chi^+ - \left(V + \gamma\phi W^2 + \chi\right) = V - V + \gamma\lambda W^2 - \gamma\phi W^2 + \gamma\lambda W^2 - 0 = -\gamma\left(\phi_{\text{miet}} - \lambda\right) W^2 \leqslant 0$，因为$\phi \geqslant \phi_{\text{miet}} \geqslant \lambda$，即，$\mathcal{V}\left(\boldsymbol{\xi}^+\right) = \mathcal{V}(\boldsymbol{\xi})$。

(J_5)如果$\boldsymbol{\xi} \in D_5 \cap \mathcal{R}\left(\tilde{\mathbb{X}}_0\right)$，其中$D_5 = C_h \cap D_{t1}$，有$\mathcal{V}\left(\boldsymbol{\xi}^+\right) - \mathcal{V}(\boldsymbol{\xi}) = V^+ + \gamma\phi^+\left(W^+\right)^2 + \chi^+ - \left(V + \gamma\phi W^2 + \chi\right) = V - V + \gamma\phi W^2 - \gamma\phi W^2 + \chi - \chi = 0$，即$\mathcal{V}\left(\boldsymbol{\xi}^+\right) = \mathcal{V}(\boldsymbol{\xi})$。

因此，当$\boldsymbol{\xi} \in D$时，有$\mathcal{V}\left(\boldsymbol{\xi}^+\right) \leqslant \mathcal{V}(\boldsymbol{\xi})$成立。

基于混杂系统$\mathcal{H}$ [式 (9-18) ～式 (9-20)]，接下来考虑函数$\mathcal{V}(\boldsymbol{\xi})$在$\boldsymbol{\xi}\in C$上的动态变化，其中$C=C_h\cap C_t$。接下来考虑$\mathcal{V}(\boldsymbol{\xi})$在流动集上的情况，根据传输尝试成功与否（没有拒绝服务攻击，即$m=0$或者有拒绝服务攻击，即$m=1$）。

$(F_0)m=0$：基于式 (9-25) 中事件驱动策略 (9-11) 和 (9-12) 的动态变化，有以下分析：

$\left(F_0^a\right)$如果$\tau\in\left[0,\tau_{\text{miet}}^0\right]$，式 (9-21) 就变为$\dot{\phi}=-2L\phi-\gamma\left(\phi^2+1\right)$。结合$\boldsymbol{\Psi}=\varrho-\delta_\chi$，假设 9.2、假设 9.3 和不等式$\Lambda:2\gamma\phi WH\leqslant\gamma^2\phi^2W^2+H^2$，有$\left\langle\nabla\mathcal{V}(\boldsymbol{\xi}),\mathcal{F}(\boldsymbol{\xi})\right\rangle\leqslant-\rho_V V-\varrho-H^2+\left(\gamma^2-\rho_W\right)W^2+\gamma\left(-2L\phi-\gamma\left(\phi^2+1\right)\right)W^2+2\gamma\phi W\left(LW+H\right)+\Psi\leqslant-\varpi^0\mathcal{V}(\boldsymbol{\xi})$，因为$\phi\geqslant\phi_{\text{miet}}\geqslant\lambda$以及$\delta_\chi(\chi)\geqslant\varpi^0\chi$，其中$\varpi^0=\min\left\{\rho_V,\dfrac{\lambda\rho_W}{\gamma}\right\}$。

$\left(F_0^b\right)$如果$\tau>\tau_{\text{miet}}^0$，式 (9-21) 变成$\dot{\phi}=0$。结合$\Psi=\varrho-\bar{\gamma}W^2-\delta_\chi$和$\bar{\gamma}=\gamma\left(2\lambda L+\gamma\left(1+\lambda^2\right)\right)$，假设 9.2、假设 9.3 以及上述不等式$\Lambda$，有$\left\langle\nabla\mathcal{V}(\boldsymbol{\xi}),\mathcal{F}(\boldsymbol{\xi})\right\rangle\leqslant-\rho_V V-\varrho-H^2+\left(\gamma^2-\rho_W\right)W^2+2\gamma\phi W\left(LW+H\right)+\Psi\leqslant-\varpi^0\mathcal{V}(\boldsymbol{\xi})$，因为$\phi\geqslant\phi_{\text{miet}}\geqslant\lambda$和$\delta_\chi(\chi)\geqslant\varpi^0\chi$，其中$\varpi^0=\min\left\{\rho_V,\dfrac{\lambda\rho_W}{\gamma}\right\}$，所以，对于几乎所有的$\boldsymbol{\xi}\in C\cap\mathcal{R}\left(\tilde{\mathbb{X}}_0\right)$，其中$m=0$，有$\left\langle\mathcal{V}(\boldsymbol{\xi}),\mathcal{F}(\boldsymbol{\xi})\right\rangle\leqslant-\varpi^0\mathcal{V}(\boldsymbol{\xi})$，其中$\varpi^0=\min\left\{\rho_V,\dfrac{\lambda\rho_W}{\gamma}\right\}$。

$(F_1)\ m=1$：不同于(F_0)，这里的传输尝试不能成功。当$m=1$的时候，基于不同的事件驱动策略 [式 (9-11)、式 (9-12)]，式 (9-25) 中的$\dot{\chi}<0$表明，一旦$\tau>\tau_{\text{miet}}^0$，设计的事件驱动 [式 (9-11)、式 (9-12)] 会被触发，因此有以下分析：

$\left(F_1^a\right)$如果$\tau>\tau_{\text{miet}}^0$，那么式 (9-21) 表明$\dot{\phi}=0$，所以式 (9-25) 表明$\dot{\chi}<0$。结合$\boldsymbol{\Psi}=-1$和$\bar{\gamma}=\gamma\left[2\lambda L+\gamma\left(1+\lambda^2\right)\right]$以及假设 9.2、假设 9.3，有$\left\langle\nabla\mathcal{V}(\boldsymbol{\xi}),\mathcal{F}(\boldsymbol{\xi})\right\rangle\leqslant-\rho_V V-\varrho+\left(\bar{\gamma}-\rho_W\right)W^2\leqslant\varpi^1\mathcal{V}(\boldsymbol{\xi})$，因为$\phi\geqslant\phi_{\text{miet}}\geqslant\lambda$，其中$\varpi^1=\dfrac{\bar{\gamma}-\rho_W}{\gamma\lambda}>0$，而$\rho_W\leqslant\gamma^2$满足假设 9.3。

$\left(F_1^b\right)$如果$\tau\in\left[0,\tau_{\text{miet}}^1\right]$满足$\dot{\chi}=0$ [式 (9-25)]，那么式 (9-21) 变

成 $\dot{\phi}=-2L\phi-\gamma\phi^2-\gamma\epsilon\phi^2-\tilde{\gamma}$。结合 $\Psi=0$ 以及假设 9.2、假设 9.3，有 $\langle\nabla\mathcal{V}(\boldsymbol{\xi}),\mathcal{F}(\boldsymbol{\xi})\rangle\leqslant-\rho_V V-\varrho+(\gamma^2-\rho_W)W^2\leqslant\varpi^1\mathcal{V}(\boldsymbol{\xi})$，其中 $\phi\geqslant\phi_{\text{miet}}\geqslant\lambda$ 以及 $\varpi^1=\dfrac{\bar{\gamma}-\rho_W}{\gamma\lambda}$。因此，对于几乎所有的 $\boldsymbol{\xi}\in C\cap\mathcal{R}(\tilde{\mathbb{X}}_0)$，当 $m=1$，有 $\langle\nabla\mathcal{V}(\boldsymbol{\xi}),\mathcal{F}(\boldsymbol{\xi})\rangle\leqslant\varpi^1\mathcal{V}(\boldsymbol{\xi})$，其中 $\varpi^1=\dfrac{\bar{\gamma}-\rho_W}{\gamma\lambda}$ 和 $\bar{\gamma}=\gamma\left[2\lambda L+\gamma(1+\lambda^2)\right]$。

混杂系统 $\mathcal{H}$ 不会出现奇诺现象，一方面因为事件驱动策略中严格正的最小驱动间隔 τ_{miet}^m，另一方面是以上关于流动集和跳变集的分析表明，系统没有有限逃逸时间，即 $\tilde{\boldsymbol{x}}$、$\boldsymbol{e}$、χ、ϕ、τ、s 以及 m。类似于文献 [83] 中定理 2，可知混杂系统 $\mathcal{H}$ 的最大解 ξ 是完备的，其中 $\boldsymbol{\xi}(0,0)\in\tilde{\mathbb{X}}_0$。对于任意最大解 ξ，定义：$\bar{\Theta}_\xi(T_1,T_2)=\{\tilde{t}\in(T_1,T_2)|\ \tilde{t}\notin\mathcal{T},\forall k\in\mathbb{Z}_{\geqslant0},(\tilde{t},k)\in\text{dom}\,\xi,m(\tilde{t})=0\}$。因此，混杂系统 $\mathcal{H}$ 在 $t\in\bar{\Theta}_\xi$ 中是稳定的，但在 $\bar{\Xi}_q(T_1,T_2)=[T_1,T_2]/\bar{\Theta}_q(T_1,T_2)$ 中并不一定稳定。也就是说，对于任意 $T_1,T_2\in\mathbb{R}_{\geqslant0}$ 满足 $T_1\leqslant T_2$，有 $\bar{\Theta}_\xi(T_1,T_2)\cup\bar{\Xi}_\xi(T_1,T_2)=[T_1,T_2]$，因此 $\bar{\Xi}_q(T_1,T_2)$ 和 $\bar{\Theta}_q(T_1,T_2)$ 可以表示为 $\bar{\Xi}_\xi(T_1,T_2)=\bigcup_{p_k\in\mathbb{Z}_{\geqslant0}}\tilde{Z}_{p_k}\cap[T_1,T_2]$ 和 $\bar{\Theta}_\xi(T_1,T_2)=\bigcup_{p_k\in\mathbb{Z}_{\geqslant0}}\tilde{W}_{p_k-1}\cap[T_1,T_2]$，对任意 $p_k\in\mathbb{Z}_{\geqslant0}$，其中 $\tilde{Z}_{p_k}$ 和 $\tilde{W}_{p_k}$ 满足

$$\tilde{Z}_{p_k}\overset{\text{def}}{=}\begin{cases}[\zeta_{p_k},\zeta_{p_k}+v_{p_k}), & v_{p_k}>0,\\ \{\zeta_{p_k}\}, & v_{p_k}=0,\end{cases}$$

$$\tilde{W}_{p_k}\overset{\text{def}}{=}\begin{cases}[\zeta_{p_k}+v_{p_k},\zeta_{p_k+1}), & v_{p_k}>0,\\ (\zeta_{p_k},\zeta_{p_k+1}), & v_{p_k}=0,\end{cases}$$

其中，$v_{p_k}\geqslant0$ 是时间变量，而且 $\zeta_0=h_0$ 和 $\tilde{W}_{-1}=[0,\zeta_0)$，$h_0>0$，否则 $\tilde{W}_{-1}=\varnothing$。

另外假设 9.1 表明 $\left|\bar{\Xi}_\xi(T_1,T_2)\right|\leqslant\left|\Xi(T_1,T_2)\right|+n(T_1,T_2)\tau_{\text{miet}}^1$，正如 Feng 和 Tesi[236] 的分析。基于假设 9.2，上边这个不等式可以写成 $\left|\bar{\Xi}_\xi(T_1,T_2)\right|\leqslant\varsigma_*+\dfrac{T_2-T_1}{T_*}$，其中 $\varsigma_*=\varsigma+v\tau_{\text{miet}}^1,T_*=\dfrac{\tau_f\tau_d}{\tau_f+\tau_d\tau_{\text{miet}}^1}$。接下来分析系统 (9-18) 的稳定性。对任意 $(t,k)\in(\text{dom}\,\boldsymbol{\xi})\cap(\tilde{W}_{p_k}\times\mathbb{Z}_{\geqslant0})$ 满足 $p_k\in\mathbb{Z}_{\geqslant0}\cup\{-1\}$，以及任意 $(t,k)\in(\text{dom}\,\boldsymbol{\xi})\cap(\tilde{Z}_{p_k}\times\mathbb{Z}_{\geqslant0})$ 满足 $p_k\in\mathbb{Z}_{\geqslant0}$，可以得到如果 $m=0$，则 $\mathcal{V}(\boldsymbol{\xi}(t,k))\leqslant\exp(-\varpi^0(t-\zeta_{p_k}-v_{p_k}))\mathcal{V}(\boldsymbol{\xi}(\zeta_{p_k}+v_{p_k},k))$，

以及如果 $m=1$，则 $\mathcal{V}(\boldsymbol{\xi}(t,k)) \leqslant \exp\left(\varpi^1\left(t-\zeta_{p_k^{'}}\right)\right)\mathcal{V}\left(\boldsymbol{\xi}\left(\zeta_{p_k}+v_{p_k},k\right)\right)$。上述不等式右侧给出的$\mathcal{V}$的两种上界用于导出混杂系统 (9-18) 的镇定性。基于文献 [85] 引理 3 和文献 [83] 引理 4，我们做一下分析，将以上两种模式联系起来：对任意$(t,k)\in \mathrm{dom}\,\xi$，根据上面$m=0$和$m=1$时候的$\mathcal{V}(\xi(t,k))$的不等式，有$\mathcal{V}(\boldsymbol{\xi}(t,k)) \leqslant Y(0,t)\mathcal{V}(\boldsymbol{\xi}(0,0))$，其中$Y(t_1,t_2)=\mathrm{e}^{-\varpi^0\left|\Theta_\xi(t_1,t_2)\right|}\mathrm{e}^{\varpi^1\left|\Xi_\xi(t_1,t_2)\right|}$以及$t_1\leqslant t_2$。因此可以得到$Y(T_1,T_2)\leqslant \mathrm{Im}\mathrm{e}^{-\hat{\beta}(T_2-T_1)}$，其中，$\mathrm{Im}=\mathrm{e}^{\varsigma_*\left(\varpi^0+\varpi^1\right)},\hat{\beta}=\varpi^0-\dfrac{\varpi^0+\varpi^1}{T_*}>0$。因此，混杂系统 (9-18) 的稳定性可以用$\mathcal{V}$在两种模式下的上界进行评估，即

$$\mathcal{V}(\boldsymbol{\xi}(t,k)) \leqslant \mathrm{Im}\mathrm{e}^{-\hat{\beta}t}\mathcal{V}(\boldsymbol{\xi}(0,0)) \tag{9-26}$$

其中，$\mathrm{Im}=\mathrm{e}^{\varsigma_*(\varpi_0+\varpi_1)},\varsigma_*=\varsigma+v,\hat{\beta}=\varpi_0-\dfrac{\varpi_0+\varpi_1}{T_*},T_*=\dfrac{\tau_d\tau_f}{\Delta\tau_d+\tau_f}$。因此$\left|\boldsymbol{\xi}(t,k)\right|_{\mathcal{A}}\leqslant \underline{\beta}_{\mathcal{V}}^{-1}\left(\mathrm{Im}\mathrm{e}^{-\hat{\beta}t}\overline{\beta}_{\mathcal{V}}\left(\left|\boldsymbol{\xi}(0,0)\right|_{\mathcal{A}}\right)\right)$，或者令$\underline{\beta}_{\mathcal{V}}^{-1}\left(\mathrm{Im}\mathrm{e}^{-\hat{\beta}t}\overline{\beta}_{\mathcal{V}}(\cdot)\right)\overset{\text{def}}{=}\beta(\cdot,t,k)$，其中$\beta(\cdot,t,k)\in\mathcal{KLL}$，即根据定义9.1可知$\mathcal{A}$是一致全局渐近稳定的。证毕。 ■

注释 9.4 结合假设 9.1 和不等式 (9–24) 可知，最小允许传输间隔τ_{miet}^m负相关于拒绝服务攻击的次数和持续时间，其中这一负相关关系体现在参数$\dfrac{1}{\tau_f}$和$\dfrac{1}{\tau_d}$。不等式 (9–26) 表明参数$\hat{\beta}>0$是追踪控制的收敛率，并与$\dfrac{1}{\tau_f}$和$\dfrac{1}{\tau_d}$负相关。也就是说，拒绝服务攻击的次数 / 持续时间越大，最小传输间隔τ_{miet}^m越小（即需要更多的通信次数），越不易实现追踪控制收敛。这与 Wang 等人在文献 [238] 备注 1、文献 [239] 备注 3 以及文献 [240] 备注 1 中分析的采样间隔、收敛速度和通信计算资源之间的关系是一致的。

在本章节中，二元逻辑变量 h 和基于事件触发的采样策略意味着本章节同时考虑了式 (9-19) 中的两种跳跃 / 流动闭集，这将带来更多的复

杂性，使理论分析比以往关于姿态控制问题的工作更具有挑战性。本章节的跳变集 D[式 (9-18)] 包含最全的 5 种情况，这表明总的通信 / 更新时刻不仅仅取决于事件驱动而且取决于 h 的切换，即通信策略应该为

$$t_{r+1}=\inf_{r\in\mathbb{Z}_{\geqslant 0}}\left\{t:\left(t\geqslant t_r+\tau_{\text{miet}}^{m}\mid \chi\left(t\right)\leqslant 0\right)\vee\left(t\geqslant t_r:x\in D_h\right),t_0=0\right\},$$

其中，$\vee$代表逻辑词“或者”。另一方面，假设9.3表明事件驱动函数会受到h切换的影响，所以本章节中当h切换的时候，事件驱动函数未必更新，因此本章节考虑的跳变集D[式(9-18)]的5种情况也是合理的。但为了确保降低分析复杂性的同时保证基于h切换的全局姿态控制效果，我们令事件驱动函数在每一个通信时刻都更新，即事件驱动函数$\chi\left(t\right)$不仅在$\chi\left(t\right)\leqslant 0$的时候被重置，而且也会在$h$切换的时候被重置，即服从式(9-18)中所考虑的5种跳变情况，除了D_3。

本章节的通信策略也可以避免奇诺现象，这是因为基于事件驱动策略式 (9-11) 的通信时刻有最小触发间隔，而滞后宽度保证了 h 不会连续切换。当事件驱动和 h 的切换同时发生的时候，数据也只被通信一次，但这并不会导致奇诺现象。首先，与 Mayhew 等人 [237] 在姿态控制问题上的研究相比较，本章节设计了一种动态事件触发策略来实现依赖状态的非周期采样，使采样更加灵活，这样有利于节约通信资源 [240]。其次，二进制逻辑变量 h 使得控制律是根据滞后宽度切换的，这是为了克服 $\mathbb{SO}(3)$ 的拓扑约束从而实现全局追踪控制，这不同于 Du 和 Li[241-242]，以及 Abdessameud 等人 [243] 研究的连续控制律（将旋转角度限制在一定范围内，如$\left[-\pi,+\pi\right]$，控制结果是局部的）。综上所述，同时考虑两种跳跃 / 流动闭集，本章节解决了 $\mathbb{SO}(3)$ 的拓扑约束和通信资源限制问题。为实现全局稳定，Berkane 和 Tayebi 等人 [244-245] 还设计了具有协同势函数的混杂控制律。从整体追踪效果的角度来看，不能绝对地说基于逻辑变量的混杂方法是否优于另一种基于协同势函数的混杂方法，但是直观上看，基于逻辑变量的方法似乎更简单，这两种方法的潜在应用和优势需要在未来的工作中进一步探索。

9.2.3 算法验证与分析 1

本部分的数值仿真利用小型四旋翼原型来验证主要结果的有效性，其中惯性矩阵为$\boldsymbol{J}=\mathrm{diag}\{0.13,0.13,0.04\}$（参见 Abdessameud 和 Tayebi[246] 的研究），控制增益为$\kappa_b=3$以及$K=0.013I$，控制器切换的滞后宽度为$\delta=0.45$。

接下来首先验证假设 9.2、假设 9.3：对于任意$\xi\in C\cap D$，选择

$$W(\boldsymbol{e})=|\boldsymbol{e}|\text{和}V(\tilde{\boldsymbol{x}})=2(1-h\tilde{\eta})+\frac{1}{2}\tilde{\boldsymbol{\omega}}^{\mathrm{T}}\boldsymbol{J}\tilde{\boldsymbol{\omega}}$$

其中，$\tilde{\boldsymbol{x}}=(\tilde{\boldsymbol{x}}_p,h);\tilde{\boldsymbol{x}}_p=(\tilde{\boldsymbol{q}},\tilde{\boldsymbol{\omega}},\boldsymbol{q}_d,\boldsymbol{\omega}_d)$。首先我们考虑$W(\boldsymbol{e})=|\boldsymbol{e}|$:
基于式 (9-10)、式 (9-17)、式 (9-18) 和假设 9.2，可知

$$\begin{aligned}
&\langle\nabla W(\boldsymbol{e}),\boldsymbol{g}(\boldsymbol{e})\rangle\\
&=\left\langle\sqrt{\boldsymbol{e}^{\mathrm{T}}\boldsymbol{e}},\boldsymbol{g}(\boldsymbol{e})\right\rangle\\
&=\frac{1}{2}\times\frac{2\boldsymbol{e}^{\mathrm{T}}\cdot\boldsymbol{g}(\boldsymbol{e})}{\sqrt{\boldsymbol{e}^{\mathrm{T}}\boldsymbol{e}}}\\
&\leqslant|\boldsymbol{g}(\boldsymbol{e})|\\
&\leqslant\left\{\frac{1}{2}\left|\tilde{\epsilon}^{\mathrm{T}}\tilde{\boldsymbol{\omega}}\right|+\frac{1}{2}\left|\left(\tilde{\eta}\boldsymbol{I}+S(\tilde{\epsilon})\right)\tilde{\boldsymbol{\omega}}\right|+\left|\Sigma\left(\tilde{\boldsymbol{\omega}},\overline{\boldsymbol{\omega}}_d\right)\tilde{\boldsymbol{\omega}}-\boldsymbol{J}\tilde{\boldsymbol{R}}^{\mathrm{T}}\dot{\boldsymbol{\omega}}_d-\kappa_b h\tilde{\epsilon}-\boldsymbol{K}\tilde{\boldsymbol{\omega}}\right|\right\}\\
&+\left\{\left|\boldsymbol{S}\left(\overline{\boldsymbol{\omega}}_d\right)\boldsymbol{J}\boldsymbol{e}_{\tilde{\omega}_d}+S\left(\boldsymbol{e}_{\tilde{\boldsymbol{\omega}}_d}\right)\boldsymbol{J}\left(\overline{\boldsymbol{\omega}}_d+\boldsymbol{e}_{\tilde{\omega}_d}\right)-\kappa_b h\boldsymbol{e}_{\tilde{q}}^{\tilde{\epsilon}}-K\boldsymbol{e}_{\tilde{\omega}}\right|\right\}\\
&\leqslant LW(\boldsymbol{e})+H(\tilde{\boldsymbol{x}}_p),
\end{aligned}$$

其中根据式(9-6)和式(9-8)，有$L=\max\{\rho(\boldsymbol{J}),|\kappa_b|,|\boldsymbol{K}|\}=3$和$|\tilde{\boldsymbol{\omega}}|+\left|\boldsymbol{J}\tilde{\boldsymbol{R}}^{\mathrm{T}}\dot{\boldsymbol{\omega}}_d\right|+|\kappa_b h\tilde{\epsilon}|+|\boldsymbol{K}\tilde{\boldsymbol{\omega}}|\leqslant H(\tilde{\boldsymbol{x}}_p)$。现在考虑$V(\tilde{x})$:

基于式 (9-17)、式 (9-18) 和假设 9.2、假设 9.3，有

$$\begin{aligned}
&\langle V(\tilde{\boldsymbol{x}}),\boldsymbol{f}(\tilde{\boldsymbol{x}},\boldsymbol{e})\rangle\\
&=-2h\dot{\tilde{\eta}}+\tilde{\boldsymbol{\omega}}^{\mathrm{T}}\boldsymbol{J}\dot{\tilde{\boldsymbol{\omega}}}\\
&=\widetilde{\omega}^{\mathrm{T}}\left(h\tilde{\epsilon}-\kappa_b h\tilde{\epsilon}-\boldsymbol{K}\tilde{\boldsymbol{\omega}}-\boldsymbol{J}\tilde{\boldsymbol{R}}^{\mathrm{T}}\dot{\boldsymbol{\omega}}_d\right)\\
&+\tilde{\omega}^{\mathrm{T}}\left(S\left(\overline{\boldsymbol{\omega}}_d\right)\boldsymbol{J}\boldsymbol{e}_{\bar{\omega}_d}+S\left(\boldsymbol{e}_{\bar{\omega}_d}\right)J\left(\overline{\boldsymbol{\omega}}_d+\boldsymbol{e}_{\bar{\omega}_d}\right)-\kappa_b h\boldsymbol{e}_{\tilde{q}}^{\tilde{\epsilon}}-K\boldsymbol{e}_{\bar{\omega}}\right)\\
&\leqslant-\rho_V V(\tilde{\boldsymbol{x}})-\varrho(|\Im|)-H^2(\tilde{\boldsymbol{x}}_{\boldsymbol{p}})+(\gamma^2-\rho_W)W^2(\boldsymbol{e})
\end{aligned}$$

其中

$$\tilde{\boldsymbol{\omega}}^{\mathrm{T}}\left(h\tilde{\epsilon}-\kappa_b h\tilde{\epsilon}-\boldsymbol{K}\tilde{\boldsymbol{\omega}}-\boldsymbol{J}\tilde{\boldsymbol{R}}^{\mathrm{T}}\dot{\boldsymbol{\omega}}_d\right)\leqslant-\rho_V V\left(\tilde{\boldsymbol{x}}\right)-\varrho\left(\left|\Im\right|\right)-H^2\left(\tilde{x}_p\right)$$

$$\tilde{\boldsymbol{\omega}}^{\mathrm{T}}\left(S\left(\overline{\omega}_d\right)\boldsymbol{J}\boldsymbol{e}_{\overline{\omega}_d}+S\left(\boldsymbol{e}_{\overline{\omega}_d}\right)J\left(\overline{\omega}_d+\boldsymbol{e}_{\overline{\omega}_d}\right)-\kappa_b h\boldsymbol{e}_{\tilde{q}}^{\tilde{\varepsilon}}-\boldsymbol{K}\boldsymbol{e}_{\tilde{\omega}}\right)\leqslant\left(\gamma^2-\rho_W\right)W^2\left(\boldsymbol{e}\right)$$

因此可以选择$\rho_V\leqslant 2\dfrac{|\boldsymbol{K}|}{\rho(J)}\leqslant 0.2, \varrho=0$和$\gamma^2-\rho_W=\max\left\{\rho\left(\boldsymbol{J}\right),\left|\boldsymbol{\kappa}_b\right|,\left|\boldsymbol{K}\right|\right\}=3$，即$\rho_W=1$。

基于以上分析，选择 $\lambda=0.2, \epsilon=0.125, \tilde{\gamma}=4$ 和 $\gamma=2$，其中 $\overline{\gamma}=\gamma\left[2\lambda L+\gamma\left(1+\lambda^2\right)\right]=4.4$，也就是如果 $m=1, \Lambda=L$；如果 $m=0$，$\Lambda<L$，由式 (9-22) 可以得到 $\tau_{\text{miet}}^0=0.274881$ 和 $\tau_{\text{miet}}^1=0.219680$，以及 $\varpi^0=\min\left\{\rho_V, \dfrac{\lambda\rho_W}{\gamma}\right\}=0.2, \varpi^1=\dfrac{\overline{\gamma}-\rho_W}{\gamma\lambda}=8.4$。为了尽可能获得较好的跟踪性能，$\epsilon$和$\tilde{\gamma}$可以适当地选择较大的数值，这样$\tau_{\text{mati}}^m$会更小。现在，考虑拒绝服务攻击对所考虑的姿态系统跟踪性能的影响，其中$T_1=5$ 和 $T_2=25$。取 $v=8, \varsigma=8$ 以及 $\hat{\beta}=\dfrac{1}{2}\varpi^0>0$，于是有 $\dfrac{1}{T_*}=\dfrac{1}{2}\times\dfrac{\varpi^0}{\varpi^0+\varpi^1}$，因为 $\hat{\beta}=\varpi^0-\dfrac{\varpi^0+\varpi^1}{T_*}$。因此验证了定理 9.1 中的式 (9-24)，即 $\dfrac{1}{T_*}=\tau_{\text{miet}}^1/\tau_f+1/\tau_d=\dfrac{1}{2}\times\dfrac{\varpi^0}{\varpi^0+\varpi^1}\leqslant\dfrac{\varpi^0}{\varpi^0+\varpi^1}$。基于假设 9.1 中的$\tau_f\in\mathbb{R}_{>\Delta}$，选择 $\tau_f=5\tau_{\text{miet}}^0=1.374405$，也即 $n\left(T_1, T_2\right)\leqslant v+\dfrac{T_2-T_1}{\tau_f}=22.551751$ 和 $\left|\varXi\left(T_1, T_2\right)\right|\leqslant 8+\dfrac{T_2-T_1}{\tau_d}$。

定理 9.1 中的事件驱动策略 (9-25) 对不可预知的拒绝服务攻击具有智能的弹性：对于不存在拒绝服务攻击的情况，控制器可以根据接收到的实时数据，以事件驱动的方式进行更新 [式 (9-25) 的前两个方程]；如果拒绝服务攻击不可预测地发生了，那么控制器就不能及时更新，因为新采样数据不能成功通过通信网络传输。为避免奇诺现象，我们设置一个时间变量参数τ_{miet}^i，那么下一次传输尝试只能在上一次传输的τ_{miet}^1时刻之后：当$m=1$的时候（不考虑 h 切换的情况下），如果$0\leqslant\tau\leqslant\tau_{\text{miet}}^1$，有$\varPsi=0$；如果$\tau>\tau_{\text{miet}}^1$，$\varPsi=-1$。在 h 切换存在的情况下，即使两次连续驱动之间的间隔可以无限小，这种无限小也不会连续发生，这是因为 h 的切换是基于滞后宽度的，而以上分析的事件驱动是可以避免奇诺现象的，所以即使将 h 的切换考虑进去，依旧可以排除奇诺现象。

定义参照轨迹$\boldsymbol{q}_d(0)=(1\,0\,0\,0)^{\mathrm{T}}$和$\boldsymbol{\omega}_d=2\sin(0.1\pi t)(1\,1\,1)^{\mathrm{T}}$。这里展示了 20s 的数值仿真，其中$\boldsymbol{q}(0)=(0\,0\,1\,0)^{\mathrm{T}}$和$\boldsymbol{\omega}=(0\,0\,0)^{\mathrm{T}}$。图 9-2(a) ～ (d) 中那些垂直的灰色条纹表示拒绝服务攻击发生的时间。图 9-2(a)、(b) 表示式 (9-7) 中的追踪误差$\tilde{q}$和$\tilde{\omega}$的收敛情况，其中$\tilde{\boldsymbol{q}}=(\tilde{\eta},\tilde{\boldsymbol{\epsilon}})$，$\tilde{\boldsymbol{\epsilon}}=(\tilde{\epsilon}_1\;\tilde{\epsilon}_2\;\tilde{\epsilon}_3)^{\mathrm{T}}$和$\tilde{\boldsymbol{\omega}}=(\tilde{\omega}_1\;\tilde{\omega}_2\;\tilde{\omega}_3)^{\mathrm{T}}$实现了姿态追踪的弹性收敛。值得注意的是，图 9-2(a)、(b) 分别代表的是姿态系统的运动学收敛（即姿态轨迹）和动力学收敛（即角速度轨迹），而且所设计的力矩控制器直接作用在角速度的收敛上，所以最终角速度的收敛效果会更好。图 9-2(c) 展示了输入控制u_τ和采样误差e的状态变化，也就是说随着网络诱导的采样

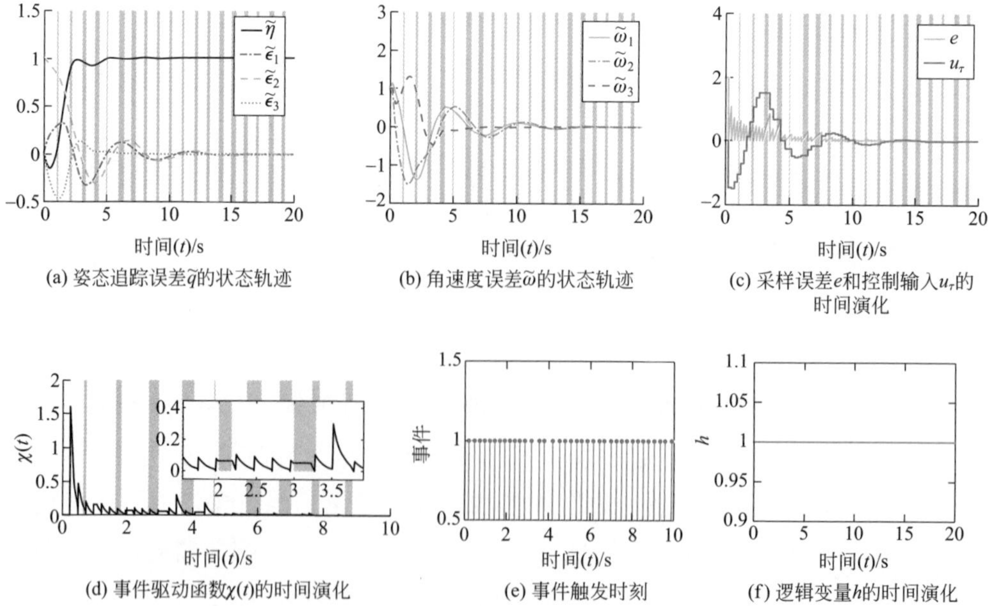

(a) 姿态追踪误差$\tilde{q}$的状态轨迹
(b) 角速度误差$\tilde{\omega}$的状态轨迹
(c) 采样误差e和控制输入u_τ的时间演化
(d) 事件驱动函数$\chi(t)$的时间演化
(e) 事件触发时刻
(f) 逻辑变量h的时间演化

图 9-2　小型四旋翼机状态轨迹的追踪收敛图

误差趋于零（如注释 9.3 所分析），闭环混合系统 (9-18) 的状态趋于稳定，即在保证采样子系统e稳定性的前提下，实现了姿态跟踪。相应地，事件触发函数$\chi(t)$的动态变化参见图 9-2(d)，而图 9-2(e) 则表示事件驱动触发时刻（同时也是通信时刻，因为在此仿真中h并没有切换），从图 9-2(d) 可知，在拒绝服务攻击发生的时间段内，函数$\chi(t)$保持不变 / 递减，这正好符合定理 9.1 中式 (9-25) 的第 3 个和第 4 个方程，即$m=1$的时候，$\dot{\chi}(t)=\Psi=0$（如果$0\leqslant\tau\leqslant\tau_{\text{miet}}^{1}$）；$\dot{\chi}(t)=\Psi=-1$（$\tau>\tau_{\text{miet}}^{1}$）。此外，图 9-2(d) 也表示在没有攻击的时候，函数$\chi(t)$有两个下降阶段，即第一个阶段是因为用于排除奇诺现象的最小触发间隔τ_{miet}^{0}，对应于定理 9.1 中的式 (9-25)，即$\dot{\chi}(t)=\Psi=\varrho(|\Im|)-\delta_{\chi}(\chi)$（当$m=0$和$0\leqslant\tau\leqslant\tau_{\text{miet}}^{0}$）；而第二个阶段对应于定理 9.1 中式 (9-25) 的第二个等式，即$\dot{\chi}(t)=\Psi=\varrho(|\Im|)-\overline{\gamma}W^{2}(\boldsymbol{e})-\delta_{\chi}(\chi)$（当$m=0$和$\tau>\tau_{\text{miet}}^{0}$）。很显然，$\varrho(|\Im|)-\overline{\gamma}W^{2}(\boldsymbol{e})-\delta_{\chi}(\chi)\leqslant\varrho(|\Im|)-\delta_{\chi}(\chi)$，这表明事件触发函数$\chi(t)$的轨迹在这两个阶段有不同的下降趋势，所以这种排除奇诺现象的方法使本章节的事件驱动策略有别于其他的方法，如Li等人 [247] 的研究。图 9-2(a)、(b) 表明，通过基于事件驱动的控制律 (9-17)，事件控制系统对网络通信通道上发起的拒绝服务攻击具有弹性，其中图 9-2(f) 展示了逻辑变量 h 的时间演化。

9.3 基于拒绝服务攻击的非饱和移动机器人系统跟踪控制

9.3.1 移动机器人系统在攻击下的动力学模型

移动机器人的运动学和动力学模型可以表示为 [232,248]

$$\begin{aligned}&\dot{x}=v\cos\theta,\ \dot{y}=v\sin\theta,\ \dot{\theta}=\omega\\&\overline{M}\dot{\overline{\omega}}=-\overline{C}(\omega)\overline{\omega}-\overline{D}\overline{\omega}+\overline{B}(\tau+\tau_{E})\end{aligned}\tag{9-27}$$

其中，位置变量$(x,y)^{\mathrm{T}}\in\mathbb{R}^2$表示机器人左右驱动轮重心$O_b$的坐标；$\theta$表示机器人在地球固定参考系$O_E X_E Y_E$中的航向角；$v$表示机器人的线性速度；$\omega=\mathrm{col}(\omega_1,\omega_2)$中的$\omega_1$和$\omega_2$是机器人驱动轮的角速度；$\tau=\mathrm{col}(\tau_1,\tau_2)$中的$\tau_1$和$\tau_2$是用于驱动机器人轮子的控制力矩；$\tau_E=\mathrm{col}(\tau_{E1},\tau_{E2})$是扰动力矩载荷（如摩擦力）。更详细的系统信息为：

$$\overline{\boldsymbol{\omega}}=\mathrm{col}(v,\omega)$$

$$\overline{\boldsymbol{M}}=\mathrm{diag}(m_{11}+m_{12},m_{11}-m_{12})$$

$$\overline{\boldsymbol{D}}=\begin{pmatrix}\frac{1}{2}(d_{11}+d_{22}) & \frac{b}{2}(d_{11}-d_{22})\\ \frac{1}{2b}(d_{11}-d_{22}) & \frac{1}{2}(d_{11}+d_{22})\end{pmatrix}$$

$$\overline{\boldsymbol{C}}(\boldsymbol{\omega})=\begin{pmatrix}0 & -bc\omega\\ \frac{c}{b}\omega & 0\end{pmatrix},\overline{B}=\frac{r}{2}\begin{pmatrix}1 & 1\\ \frac{1}{b} & -\frac{1}{b}\end{pmatrix}$$

更详细地，

$$m_{11}=\frac{r^2}{4b^2}(mb^2+I)+I_\omega,\ \ m_{12}=\frac{r^2}{4b^2}(mb^2-I)$$

$$c=\frac{r^2}{2b}m_c a,m=m_c+2m_\omega,\ \ I=m_c a^2+2m_\omega b^2+I_c+2I_m$$

其中，m_c和m_ω分别表示带有发动机的车身和车轮的质量；I_c、I_ω和I_m分别表示物体通过质心P_c绕垂直轴的转动惯量、带电机转子的轮子绕轮轴的转动惯量，以及带电机转子的轮子绕轮径的转动惯量；参数r、a和b参见图9-3；正常数d_{11}和d_{22}是阻尼系数。考虑参考轨迹的运动学和动力学方程为

$$\dot{x}_r=v_r\cos\theta_r,\ \ \dot{y}_r=v_r\sin\theta_r,\ \ \dot{\theta}_r=\omega_r,\ \ \dot{v}_r=\chi_{1r},\ \ \dot{\omega}_r=\chi_{2r},$$

其中，v_r和ω_r分别为机器人线速度和角速度的时变参考信号，有界函数χ_{1r}和χ_{2r}是连续的。

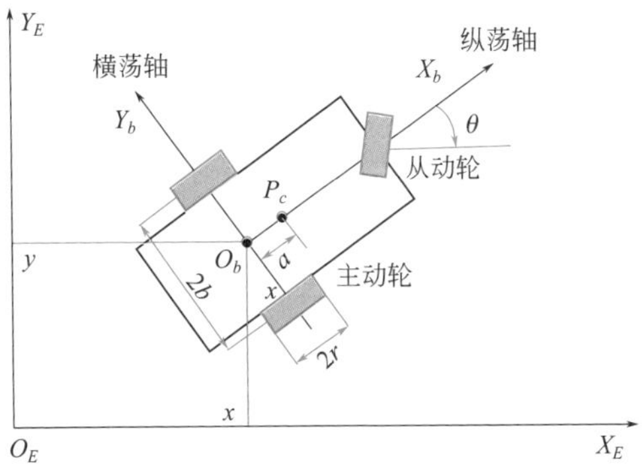

图 9-3　移动机器人坐标参数

基于 Postoyan 等人[249]的研究，相应的误差系统可以表示为

$$\begin{pmatrix} x_e \\ y_e \\ \theta_e \end{pmatrix} = \begin{pmatrix} \cos\theta & \sin\theta & 0 \\ -\sin\theta & \cos\theta & 0 \\ 0 & 0 & 1 \end{pmatrix} \begin{pmatrix} x_r - x \\ y_r - y \\ \theta_r - \theta \end{pmatrix} \tag{9-28}$$

于是有$\dot{x}_e = \omega y_e - v + v_r \cos\theta_e, \dot{y}_e = -\omega x_e + v_r \sin\theta_e$，和$\dot{\theta}_e = \omega_r - \omega$。

接下来给出参考速度信号的限制条件。

假设 9.4

① 存在正常数 α、β_1、β_2、β_3 和 β_4，使得 $0 < \alpha \leqslant |v_r(t)| \leqslant \beta_1$，$|\omega_r(t)| \leqslant \beta_2, |\dot{v}_r(t)| \leqslant \beta_3$，$|\dot{\omega}_r(t)| \leqslant \beta_4$，其中$t \in \mathbb{R}_{\geqslant 0}$。②载荷向量$\tau_E$是有界常数或慢时变函数，其对时间的导数可以忽略不计。

注释 9.5

假设 9.4 的假设是合理的，比如可以采用饱和函数 arctan 或者 tanh[225] 来限制参照系统的参数，条件$v_r(t) \geqslant \alpha > 0$是为了排除机器人被镇定在一个固定点的情形，而假设 9.4 中②约束了机器人扰动的类型，通常认为是摩擦力$k_g g$，其中k_g是一个常数，g 是重力加速度。因此以上关于τ_E的假设条件能够实现机器人系统全局渐近跟踪参考轨迹[232]。

令$\xi = (x_e, y_e, \theta_e)$，转化形式［式 (9-28)］表明，如果$\xi$收敛到原点的一个小邻域，那么$(x, y, \theta) \to (x_r, y_r, \theta_r)$。也就是说，我们的目标是为$v$和$\omega$设计一个适当的控制器使$\xi = 0$渐近稳定，进而实现移动机器人的跟踪控制。假设所设计的控制器通过网络通道与机器人［式 (9-27)］以及参考轨迹通信，为保证$\boldsymbol{\xi}$的全局收敛，本章节采用 Postoyan 等人 [249] 的研究中设计的控制器，即

$$
\begin{aligned}
&v = v_1 + c_1 \bar{x}_e, \bar{x}_e = x_e - c_3 \omega y_e \\
&\omega = \omega_r + \gamma y_e v_r \sin\theta_e + c_2 \gamma \theta_e \\
&v_1 = v_r \cos\theta_e - c_3 \dot{\omega} y_e + c_3 \omega \left(\omega x_e - v_r \sin\theta_e \right)
\end{aligned}
\tag{9-29}
$$

其中，c_1, c_2, c_3和γ为正参数。正如Jiang和Nijmeijer[231]所分析的那样，假设9.4用来保证闭环系统[式(9-28)、式(9-29)]中$\boldsymbol{\xi} = 0$的稳定性。

本章节采用网络化数字平台传输参考轨迹和传感器测量数据，而且传输只发生在一系列离散的传输时刻$\{t_k\}_{k\in\mathbb{Z}_{\geqslant 0}}$，而零阶保持器使得在任意两个连续传输期间，$(x, y, \theta, x_r, y_r, \theta_r, v_r, \omega_r)$的采样值可以保持常数（采样值通常用“$\hat{}$”表示），即令$\boldsymbol{Re} \overset{\text{def}}{=} (x, y, \theta, x_r, y_r, \theta_r, v_r, \omega_r)$，那么$\widehat{\mathrm{Re}}(t) = \mathrm{Re}(t_k)$，其中$t \in [t_k, t_{k+1}]$。对传输时刻$\{t_k\}_{k\in\mathbb{Z}_{\geqslant 0}}$，假设$0 < \Delta_0 \leqslant t_{k+1} - t_k \leqslant T$和$0 \leqslant t_0 < \cdots < t_k < \cdots$ [250]，那么，对任意$t \in [t_k, t_{k+1}]$,式 (9-28) 中的误差的动态为

$$
\begin{aligned}
&\dot{x}_e = \omega \left(y_e + e_{y_e} \right) - v + \left(v_r + e_{v_r} \right) \cos\left(\theta_e + e_{\theta_e} \right) \\
&\dot{y}_e = -\omega \left(x_e + e_{x_e} \right) + \left(v_r + e_{v_r} \right) \sin\left(\theta_e + e_{\theta_e} \right) \\
&\dot{\theta}_e = \left(\omega_r + e_{\omega_r} \right) - \omega
\end{aligned}
$$

其中，

$$
\begin{aligned}
&e_{x_e} = \hat{x}_e - x_e, e_{y_e} = \hat{y}_e - y_e, e_{\theta_e} = \hat{\theta}_e - \theta_e \\
&e_{v_r} = \hat{v}_r - v_r, e_{\omega_r} = \hat{\omega}_r - \omega_r
\end{aligned}
$$

根据式(9-28)，采样值$\hat{x}_e$、$\hat{y}_e$和$\hat{\theta}_e$的表达式为

$$
\begin{aligned}
&\hat{x}_e = (\hat{x}_r - x)\cos\theta + (\hat{y}_r - y)\sin\theta \\
&\hat{y}_e = -(\hat{x}_r - x)\sin\theta + (\hat{y}_r - y)\cos\theta \\
&\hat{\theta}_e = \hat{\theta}_r - \theta
\end{aligned}
$$

为了使研究更具一般性，本章节考虑在每一个传输时刻，节点根据传输协议访问通信网络，即

$$
\widehat{\boldsymbol{Re}}\left(t_k^+\right) = \boldsymbol{Re}\left(t_k\right) + h\left(\kappa, \boldsymbol{e}\left(t_k\right)\right) \tag{9-30}
$$

其中，$\boldsymbol{e}=\left(\boldsymbol{e}_q,\boldsymbol{e}_d,\boldsymbol{e}_{ff}\right)\in\mathbb{R}^{n_e}$可以表示为

$$\begin{aligned}\boldsymbol{e}_q&=\left(e_x,e_y,e_\theta\right)=\left(\hat{x}-x,\hat{y}-y,\hat{\theta}-\theta\right)\\ \boldsymbol{e}_d&=\left(e_{x_r},e_{y_r},e_{\theta_r}\right)=\left(\hat{x}_r-x,\hat{y}_r-y,\hat{\theta}_r-\theta\right)\\ \boldsymbol{e}_{ff}&=\left(e_{v_r},e_{\omega_r}\right)=\left(\hat{v}_r-v_r,\hat{\omega}_r-\omega_r\right)\end{aligned}$$

而 $h\left(\kappa,\boldsymbol{e}\left(t_k\right)\right)=\left(h_x\left(\kappa,\boldsymbol{e}\left(t_k\right)\right),\ h_y\left(\kappa,\boldsymbol{e}\left(t_k\right)\right),\ h_\theta\left(\kappa,\boldsymbol{e}\left(t_k\right)\right),\ h_{x_r}\left(\kappa,\boldsymbol{e}\left(t_k\right)\right),\ h_{y_r}\left(\kappa,\boldsymbol{e}\left(t_k\right)\right),\ h_{\theta_r}\left(\kappa,\boldsymbol{e}\left(t_k\right)\right),\ h_{v_r}\left(\kappa,\boldsymbol{e}\left(t_k\right)\right),\ h_{\omega_r}\left(\kappa,\boldsymbol{e}\left(t_k\right)\right)\right)$表示传输协议[251]。因此根据式(9-30)，网络传输协议可以表示为$\boldsymbol{e}\left(t_k^+\right)=h\left(\kappa,\boldsymbol{e}\left(t_k\right)\right)$。

注释 9.6　由于本章节的重点是为受拒绝服务攻击影响的跟踪控制设计合适的事件触发策略，所以我们忽略可能存在的网络延迟。在网络化控制系统中，通信确实经常受到网络延迟的影响，这可能会恶化协调性能[80-81]，所以关于时滞的解决方法可以按照 Heemels 等人[82]的思路进行分析。本章节中，网络分配给每个节点的传输优先级由传输协议$h\left(\kappa,\boldsymbol{e}\left(t_k\right)\right)$决定，而传输时刻是由中央协调器根据动态系统的当前状态通过实时检测事件触发条件来决定的，所以接下来我们会设计合适的事件触发策略。

拒绝服务攻击表示网络通信在攻击期间被阻塞，因此采样值$\mathrm{Re}\left(t_k\right)$不能经过网络成功传输。定义一系列时间区间$\left\{H_n\right\}_{n\in\mathbb{Z}_{\geqslant0}}$表示拒绝服务攻击发生的时间段，即$H_n=\left\{h_n\right\}\cup\left[h_n,h_n+\tau_n\right)$，其中$\tau_n\in\mathbb{R}_{\geqslant0}$表示一次拒绝服务攻击的持续时间长度，且满足$h_n\leqslant h_n+\tau_n<h_{n+1}$，令$\mathcal{T}=\bigcup_{n\in\mathbb{Z}_{\geqslant0}}H_n$表示拒绝服务攻击的总时间，那么拒绝服务攻击的集合可以表示为$\varXi\left(T_1,T_2\right)=\left(\bigcup_{n\in\mathbb{Z}_0}H_n\right)\cap\left[T_1,T_2\right]$，其中任意给定的区间$\left[T_1,T_2\right]$满足$0\leqslant T_1\leqslant T_2$，相应地，$\Theta\left(T_1,T_2\right)=\left[T_1,T_2\right]\backslash\varXi\left(T_1,T_2\right)$表示$\left[T_1,T_2\right]$中没有攻击的时间段。根据上面的描述，在每次传输尝试时刻t_k，$k\in\mathbb{Z}_{\geqslant0}$，$\mathbf{Re}$的状态更新可以进一步表示为

$$\widehat{\mathbf{Re}}\left(t_k^+\right)=\begin{cases}\boldsymbol{Re}\left(t_k\right)+h\left(\kappa,\boldsymbol{e}\left(t_k\right)\right), & t_k\notin\mathcal{T}\\ \widehat{\mathbf{Re}}\left(t_k\right), & t_k\in\mathcal{T}\end{cases}$$

在不引起歧义的情况下，$\Theta\left(T_1,T_2\right)$和$\Xi\left(T_1,T_2\right)$可以分别简写为$\Theta$和$\Xi$。

相应地，传输误差可以表示为

$$e\left(t_k^+\right)=\begin{cases}h\left(\kappa,e\left(t_k\right)\right), & t_k\notin\mathcal{T}\\ e\left(t_k\right), & t_k\in\mathcal{T}\end{cases}$$

其中，$k\in\mathbb{Z}_{\geqslant 0}$。

正如 Dolk 等人[83] 以及 De 和 Tesi[85] 所述，要对拒绝服务攻击的发生频率和持续时间做限制，来阻止拒绝服务攻击恰好发生在所有的传输时刻 $t_k,k\in\mathbb{Z}_{\geqslant 0}$。令$n\left(T_1,T_2\right)$表示拒绝服务攻击在区间$\left(T_1,T_2\right)$发生的次数，$\left|\Xi\left(T_1,T_2\right)\right|$表示拒绝服务攻击发生的总时间长度，接下来对这两个参数进行如下限制。

假设 9.5

存在常数v、$\varsigma\in\mathbb{R}_{\geqslant 0},\tau_f\in\mathbb{R}_{>\Delta}$和$\tau_d\in\mathbb{R}_{>1}$，使得对任意$T_1$、$T_2\in\mathbb{R}_{\geqslant 0}$满足$T_1\leqslant T_2$，满足以下条件：①攻击发生的频率$n\left(T_1,T_2\right)\leqslant v+\dfrac{T_2-T_1}{\tau_f}$；②攻击发生的总时间$\left|\Xi\left(T_1,T_2\right)\right|\leqslant\varsigma+\dfrac{T_2-T_1}{\tau_d}$。

注释 9.7

基于平均驻留时间的定义 [252]，假设 9.5 考虑了一定时间内，拒绝服务攻击发生的时长和频率。事实上，假设 9.5 ①中的参数τ_f可以被看作是连续两次攻击之间的平均驻留时间，而且，如果$n\left(T_1,T_2\right)=1$且$T_1=T_2$（即攻击发生在$T_1=T_2$时刻），那么参数v是保证①中的不等式有意义所必需的参数（或称为浮动界限）；类似地，ς保证了假设 9.5 ②中不等式的可行性，比如如果$\left|\Xi\left(h_n,h_n+\tau_n\right)\right|=\tau_n$，那么$\tau_n\geqslant\dfrac{\tau_n}{\tau_d}$，因为$\tau_d>1$，所以这时候需要参数$\varsigma$来保证不等式的可行性。

根据上面的分析，对攻击信号的频率和持续时间进行约束之后，可以设计一个算法以确保系统的正常运行。也就是说，基于控制法则 (9-29)，设计合适的拒绝服务攻击下的通信协议，移动机器人可以从任意初始值移动到期望轨迹，并以期望的速度v_r和ω_r沿目标轨迹移动。接下来，我们将设计基于事件触发的通信协议，保证机器人控制系统对拒绝服务攻击的容忍度 / 弹性。下面分别介绍有或没有拒绝服务攻击两种情况下的传输尝试，即

$$t_{k+1}=\begin{cases}\inf\left\{t:t\geqslant t_k+\underline{\Delta}_0 \mid m=0,\eta(t)=a_\lambda\lambda^2+b_\lambda,k\in\mathbb{Z}_{\geqslant0}\right\}\\ \inf\left\{t:t\geqslant t_k+\underline{\Delta}_1 \mid m=1,k\in\mathbb{Z}_{\geqslant0}\right\}\end{cases} \tag{9-31}$$

其中，$t_0=0$；函数$\eta(t)$表示事件驱动函数，正参数$b_\lambda\in\left(0,a_\lambda\left(1-\lambda^2\right)\right)$；$a_\lambda\in(\lambda,+\infty)(\lambda\in[0,1))$；$\max\left\{\underline{\Delta}_0,\underline{\Delta}_1\right\}\leqslant t_{k+1}-t_k\leqslant T$；$0\leqslant\underline{\Delta}_1\leqslant\underline{\Delta}_0\leqslant T$[83,85]。参数$\lambda\in[0,1)$会在下文中给出具体分析。本章节使用确认机制，即事件触发机制可以知道控制器端是否成功接收数据包[83]。对于$m\in\{0,1\}$：①$m=0$表示没有拒绝服务攻击的情况，即采样数据可以成功传输，因而有$\boldsymbol{e}\left(t_k^+\right)=h\left(\kappa,\boldsymbol{e}\left(t_k\right)\right)$；②$m=1$表示存在拒绝服务攻击，采样数据不能成功传输，因而有$\boldsymbol{e}\left(t_k^+\right)=\boldsymbol{e}\left(t_k\right)$。另一方面，驱动函数$\eta(t)$满足

$$\begin{cases}\dot{\eta}(t)=\Psi\left(m,\boldsymbol{o}(t),\eta(t)\right), & t\in\left[t_k,t_{k+1}\right)\\ \eta^+\left(t_k\right)=\begin{cases}a_\lambda, & t_k\notin\mathcal{T}\\ \eta\left(t_k\right), & t_k\in\mathcal{T}\end{cases}\end{cases}$$

其中，$\boldsymbol{o}(t)$代表部分信息变量。函数$\Psi\left(m,\boldsymbol{o}(t),\eta(t)\right)$会在下文中给出具体表达式。如上所示，事件触发策略依赖时钟变量η，即当函数η下降到$a_\lambda\lambda^2+b_\lambda$的时候，事件驱动策略会被触发进而传输采样数据。也就是说，事件触发间隔对应着η从a_λ下降到$a_\lambda\lambda^2+b_\lambda$所经过的时间。

9.3.2 移动机器人系统的跟踪控制分析

根据式 (9-28) 和$\boldsymbol{e}=\left(\boldsymbol{e}_q,\boldsymbol{e}_d,\boldsymbol{e}_{ff}\right)$，令$\boldsymbol{\xi}=\left(x_e,y_e,\theta_e\right)$。根据式 (9-27) ～式 (9-30)，有

$$\begin{cases}\dot{\boldsymbol{\xi}}=\boldsymbol{f}_\xi(\boldsymbol{\xi},\boldsymbol{e})=\left(\omega\left(y_e+e_{y_e}\right)-v+\left(v_r+e_{v_r}\right)\cos\left(\theta_e+e_{\theta_e}\right)\right.\\ \qquad -\omega\left(x_e+e_{x_e}\right)+\left(v_r+e_{v_r}\right)\sin\left(\theta_e+e_{\theta_e}\right)\\ \qquad \left.\left(\omega_r+e_{\omega_r}\right)-\omega,\chi_{1r}-\chi_1,\chi_{2r}-\chi_2\right)\\ \dot{\boldsymbol{e}}=\boldsymbol{g}_e(\boldsymbol{\xi},\boldsymbol{e})=-\left(v\cos\left(\theta+e_\theta\right),v\sin\left(\theta+e_\theta\right),\omega,\left(v_r+e_{v_r}\right)\cos\left(\theta_r\right.\right.\\ \qquad \left.\left.+e_{\theta_r}\right),\left(v_r+e_{v_r}\right)\sin\left(\theta_r+e_{\theta_r}\right),\omega_r+e_{\omega_r},\chi_{1r},\chi_{2r}\right)\end{cases} \tag{9-32}$$

其中，$\dot{\boldsymbol{\omega}}\overset{\text{def}}{=}\chi=\operatorname{col}\left(\chi_1,\chi_2\right)$；$\overline{\boldsymbol{M}}\chi=-\overline{\boldsymbol{C}}(\omega)\overline{\boldsymbol{\omega}}-\overline{\boldsymbol{D}}\overline{\boldsymbol{\omega}}+\overline{\boldsymbol{B}}\left(\tau+\tau_E\right)$。

接下来我们给出部分参数的约束条件，以便进行后续分析。

假设 9.6

存在局部利普希茨函数$W:\mathbb{Z}_{\geqslant 0}\times\mathbb{R}^{n_e}\to\mathbb{R}_{\geqslant 0},V:\mathbb{R}^{n_\xi}\to\mathbb{R}_{\geqslant 0}$，以及连续函数$H:\mathbb{R}^{n_\xi}\to\mathbb{R}_{\geqslant 0}$，连续有界函数$L,G:\mathbb{R}^{n_\xi}\times\mathbb{R}^{n_e}\to\mathbb{R}_{\geqslant 0}$，函数$\underline{\alpha}_W$、$\overline{\alpha}_W$、$\underline{\alpha}_V$、$\overline{\alpha}_V\in\mathcal{K}_\infty$和非负常数$\rho_V\in\mathbb{R}_{\geqslant 0},\lambda\in[0,1)$，使得以下条件满足

① 对于任意$(\boldsymbol{\xi},\boldsymbol{\kappa})\in\mathbb{R}^{n_\xi}\times\mathbb{Z}_{\geqslant 0}$和几乎所有的$e\in\mathbb{R}^{n_e}$，有

$$\begin{cases}\underline{\alpha}_W(|\boldsymbol{e}|)\leqslant W(\kappa,\boldsymbol{e})\leqslant\overline{\alpha}_W(|\boldsymbol{e}|)\\ W(\kappa+1,h(\kappa,\boldsymbol{e}))\leqslant\lambda W(\kappa,\boldsymbol{e})\\ \left\langle\dfrac{\partial W(k,\boldsymbol{e})}{\partial e},\boldsymbol{g}_e(\boldsymbol{\xi},\boldsymbol{e})\right\rangle\leqslant L(\boldsymbol{\xi},\boldsymbol{e})W(\kappa,\boldsymbol{e})+H(\boldsymbol{\xi})\end{cases}\tag{9-33}$$

② 对于任意$(\boldsymbol{e},\boldsymbol{\kappa})\in\mathbb{R}^{n_e}\times\mathbb{Z}_{\geqslant 0}$和几乎所有的$\xi\in\mathbb{R}^{n_\xi}$，有

$$\begin{cases}\underline{\alpha}_V(|\boldsymbol{\xi}|)\leqslant V(\boldsymbol{\xi})\leqslant\overline{\alpha}_V(|\boldsymbol{\xi}|)\\ \langle\nabla V(\boldsymbol{\xi}),\boldsymbol{f}_\xi(\boldsymbol{\xi},\boldsymbol{e})\rangle\leqslant-\rho_V V(\boldsymbol{\xi})-H^2(\boldsymbol{\xi})+G(\boldsymbol{\xi},\boldsymbol{e})W^2(\kappa,\boldsymbol{e})\end{cases}\tag{9-34}$$

注释 9.8

假设 9.6 ①保证了传输协议$h(\kappa,\boldsymbol{e}(t_k))$的一致全局渐近稳定性 [251]。如果$\lambda=0$，则表示没有节点通过传输网络（即$h(\kappa,\boldsymbol{e})=0$），所以假设 9.6 ①的第二个不等式不是必需的，但是其他两个不等式却是保证采样误差系统稳定性的前提 [250]。正如 Postoyan 等人 [253] 的分析，假设 9.6 ②表明，状态依赖函数L和G可以保证子系统ξ从W到H的L_2稳定性，即通过定义李雅普诺夫函数$V(\boldsymbol{\xi})$证明子系统ξ是稳定的，类似于输入到状态的稳定，其中采样误差可以看作干扰（参见文献 [254] 第 10 章了解更多细节）。

注释 9.9

对于上述假设，由于机器人的运动学和动力学 [式 (9-27)] 的复杂性，我们需要通过复杂的计算一步一步来验证这些条件：首先，我们选择合适的函数$V(\boldsymbol{\xi})$和$W(\boldsymbol{e})$（更多细节见

Nesic 等人[251]的研究），然后根据假设 9.6 中的$\dot{\boldsymbol{\xi}}=\boldsymbol{f}_{\xi}(\boldsymbol{\xi},\boldsymbol{e})$和$\dot{\boldsymbol{e}}=\boldsymbol{g}_{e}(\boldsymbol{\xi},\boldsymbol{e})$的表达式以及控制器[式(9-29)]，进一步得到$\left\langle\frac{\partial W(k,\boldsymbol{e})}{\partial \boldsymbol{e}},g_{e}(\boldsymbol{\xi},\boldsymbol{e})\right\rangle$和$\langle\nabla V(\boldsymbol{\xi}),\boldsymbol{f}_{\xi}(\boldsymbol{\xi},\boldsymbol{e})\rangle$，最后我们根据假设 9.4 和方程(9-27)的状态信息，估计出这些参数的上界（参见 Postoyan 等人[253]的分析）。我们将在实验部分给出具体的计算过程。

因此机器人系统跟踪控制问题的脉冲模型可以建模为：

$$\begin{cases}\dot{\boldsymbol{\xi}}=\boldsymbol{f}_{\xi}(\boldsymbol{\xi},\boldsymbol{e}), & t\in[t_k,t_{k+1}]\\ \dot{\boldsymbol{e}}=\boldsymbol{g}_{e}(\boldsymbol{\xi},\boldsymbol{e}), & t\in[t_k,t_{k+1}]\\ \dot{\eta}=\boldsymbol{\varPsi}, & t\in[t_k,t_{k+1}]\end{cases}$$

$$\begin{cases}\boldsymbol{\xi}^{+}(t_k)=\boldsymbol{\xi}(t_k), & t_k\in\varTheta\setminus\varXi\\ \boldsymbol{e}^{+}(t_k)=h(k,\boldsymbol{e}(t_k)), & t_k\in\varTheta\setminus\varXi\\ \eta^{+}(t_k)=a_{\lambda}, & t_k\in\varTheta\setminus\varXi\end{cases}\quad\begin{cases}\boldsymbol{\xi}^{+}(t_k)=\boldsymbol{\xi}(t_k), & t_k\in\varXi\\ \boldsymbol{e}^{+}(t_k)=\boldsymbol{e}(t_k), & t_k\in\varXi\\ \eta^{+}(t_k)=\eta(t_k), & t_k\in\varXi\end{cases}$$

其中，函数$\boldsymbol{f}_{\xi}(\boldsymbol{\xi},\boldsymbol{e})$和$\boldsymbol{g}_{\xi}(\boldsymbol{\xi},\boldsymbol{e})$可以由式(9-32)得到。$\Psi_i$会在下文中给出分析。此脉冲模型表明：①如果$t_k\in\Theta\setminus\Xi$（即$m=0$），那么通过实时数据，可以对事件触发策略进行连续正常评估，进而正常更新控制器数据；②如果$t_k\in\Xi$（即$m=1$），那么确认机制会通知事件驱动策略进而停止评估，但是会按照式(9-31)中第二个等式产生周期性的通信时刻，直到攻击消失然后数据成功传输给控制器，然后事件驱动被告知控制器端收到新数据，进而转回正常的驱动机制，即式(9-31)中第一个等式根据系统状态继续产生非周期性的通信时刻。

根据 Goebel 等人[230]研究中的混杂模型和以上关于机器人系统的脉冲模型，接下来我们建立一个在拒绝服务攻击下机器人追踪系统的混杂模型。定义$\boldsymbol{q}=(\boldsymbol{\xi},\boldsymbol{e},\eta,\kappa,m,\tau_1,\tau_2)$，变量$\kappa\in\mathbb{Z}_{\geqslant 0}$表示传输尝试的次数，$\tau_1\in\mathbb{R}_{\geqslant 0}$表示自最近一次传输尝试之后所经过的时间，$\tau_2\in\mathbb{R}_{\geqslant 0}$表示整体连续时间。基于 Goebel 等人[230]和 Postoyan 等人[84]的研究，拒绝服务攻击下移动机器人跟踪控制问题可以建模为混杂模型，即

$$\mathcal{H}_{\text{dos}}:\begin{cases}\dot{\boldsymbol{q}}=\mathcal{F}(\boldsymbol{q}),\ \boldsymbol{q}\in C\\ \boldsymbol{q}^{+}=\mathcal{G}(\boldsymbol{q}),\ \boldsymbol{q}\in D\end{cases} \tag{9-35}$$

其中

$$\begin{cases}C=\left\{\boldsymbol{q}:\eta\in\left[a_\lambda\lambda^2+b_\lambda,a_\lambda\right]\right\}\in\mathbb{R}^{n_\xi}\times\mathbb{R}^{n_e}\times\left[a_\lambda\lambda^2+b_\lambda\right.\\ \qquad \left.a_\lambda\right]\times\mathbb{Z}_{\geqslant 0}\times\{0,1\}\times\mathbb{R}_{\geqslant 0}\times\mathbb{R}_{\geqslant 0}\\ D=D_0\cup D_1\\ D_0=\left\{\boldsymbol{q}:\left(\eta=a_\lambda\lambda^2+b_\lambda\right)\wedge\left(\tau_1\geqslant\varDelta_0\right),m=0\right\}\\ D_1=\left\{\boldsymbol{q}:\left(\eta\in\left[a_\lambda\lambda^2+b_\lambda,a_\lambda\right]\right)\wedge\left(\tau_1=\varDelta_0\right),m=1\right\}\end{cases} \tag{9-36}$$

另外，映射$\mathcal{F}(\boldsymbol{q})$和$\mathcal{G}(\boldsymbol{q})$为：

$$\mathcal{F}(\boldsymbol{q})=\left(\boldsymbol{f}_\xi(\boldsymbol{\xi},\boldsymbol{e}),\boldsymbol{g}_e(\boldsymbol{\xi},\boldsymbol{e}),(m-1)\left(2\eta L(\boldsymbol{\xi},\boldsymbol{e})+\eta^2+G(\boldsymbol{\xi},\boldsymbol{e})+\varepsilon\right),0,0,1,1\right)$$

$$\mathcal{G}(\boldsymbol{q})=\begin{cases}\left(\boldsymbol{\xi},h(\kappa,\boldsymbol{e}),a_\lambda,\kappa+1,1,0,\tau_2\right),\ q\in D_0\\ \left(\boldsymbol{\xi},\boldsymbol{e},\eta,\kappa,0,\tau_1,\tau_2\right),\qquad q\in D_1\end{cases} \tag{9-37}$$

其中，$\varepsilon>0$，函数$\boldsymbol{f}_\xi(\boldsymbol{\xi},\boldsymbol{e})$和$\boldsymbol{g}_e(\boldsymbol{\xi},\boldsymbol{e})$满足假设9.6。

注释 9.10　为了使理论研究更具一般性，本章节假设数据是经过无线网的双通道传输的，即从传感器到控制器（事件驱动策略）以及从控制器（事件驱动策略）到执行器，但这在实际中不易实现。如果拒绝服务攻击只发生在控制器到执行器这个通道，那么设计的事件驱动策略将不会对攻击有任何弹性作用，所以实际上本章节考虑的是传感器到控制器的网络通道（即采样误差$\boldsymbol{e}=(e_q,e_d)$），而且当攻击存在的时候，确认机制会告诉事件驱动使其不再被检测是否要触发，从而使事件驱动函数保持不变（即$\dot{\eta}=0$，这是为了节省计算资源），但是数据会以周期$\varDelta_1$的形式进行传输，直到攻击消失。在实际应用中我们也只考虑传感器到控制器的无线通道（比如后边的实验），即传感器数据连续测量事件驱动策略来决定数据何时经过无线通道传输进而更新控制器。

实际上，可以用其他建模方法来设计拒绝服务攻击下的事件触发策略，如 De 和 Tesi[85] 的研究，所以当在拒绝服务攻击下建立系统模型时，可以选择 De 和 Tesi[85] 中的一般形式，或者 Dolk 等人 [83] 研究中（或本章节）的标准混杂形式，其实它们之间没有本质的区别，都是依赖传输误差决定是否传输数据，我们不能评估哪种形式能使事件驱动建模策略更好，但是标准混杂形式在事件触发调度方面表现出很强的系统性，通过适当选择参数a_λ和b_λ可以避免奇诺现象，而且这些参数也使标准混杂形式下的事件触发机制更加灵活。

对于混杂系统$\mathcal{H}_{\text{dos}}$ [式 (9-35)]，以下定义将用于得到基于混杂理论的机器人系统式 (9-27) 的跟踪控制结果。

定义 9.2[230]

混杂系统$\mathcal{H}_{\text{dos}}$ [式 (9-35)] 相对于紧集$\boldsymbol{\mathcal{A}}=\{\boldsymbol{q}:\boldsymbol{\xi}=0,\boldsymbol{e}=0\}$是一致全局渐近稳定的，如果式 (9-35) 的最大解都是完备的，而且存在函数$\beta\in\mathcal{KLL}$使得对任意初始条件$\left|\boldsymbol{q}(0,0)\right|_{\mathcal{A}}<r(r>0)$，有

$$\left|\boldsymbol{q}(t,k)\right|_{\boldsymbol{\mathcal{A}}}\leqslant\beta\left(\left|\boldsymbol{q}(0,0)\right|_{\boldsymbol{\mathcal{A}}},t,k\right),\ (t,k)\in\operatorname{dom}\boldsymbol{q}\tag{9-38}$$

基于事件驱动传输策略 (9-31)，接下来为受到攻击的机器人系统 (9-27) 建立跟踪控制结果。

定理 9.2

考虑系统 (9-35)、(9-36) 满足假设 9.4 ～假设 9.6 和$T\geqslant\underline{\Delta}_0\geqslant\underline{\Delta}_1$，以及以下条件

$$\frac{\varpi_0}{\varpi_0+\varpi_1}>\frac{\underline{\Delta}_1}{\tau_f}+\frac{1}{\tau_d}\tag{9-39}$$

其中，τ_f和τ_d满足假设 9.5，$\varpi_0=\min\left\{\rho_V,\dfrac{\varepsilon}{a_\lambda}\right\}$，$\varpi_1=\sup\{2L(\xi,e)+a_\lambda+\dfrac{G(\boldsymbol{\xi},\boldsymbol{e})}{a_\lambda\lambda^2+b_\lambda}\}$，那么紧集$\mathcal{A}=\left\{\boldsymbol{q}:(\boldsymbol{\xi},\boldsymbol{e})=0,\eta\in\left[a_\lambda\lambda^2+b_\lambda,a_\lambda\right]\right\}$是一直全局渐近稳定的，而且式(9-35)的每一个解ϕ都有一致正传输间隔，即实现了移动机器人[式(9-27)]在遭受拒绝服务攻击时基于事件的跟踪控制。

证明：基于假设9.4～假设9.6以及式(9-36)，对任意$\boldsymbol{q}\in C\cup D$，定义

$\mathcal{R}(\boldsymbol{q})=V(\boldsymbol{\xi})+\eta W^2(\kappa,\boldsymbol{e})$。假设9.6和连续映射$\mathcal{F}$以及$\mathcal{G}$表明，$\mathcal{R}$是连续且局部利普希茨的。令$\boldsymbol{q}\in C\cup D$，那么式(9-36)和假设9.5、假设9.6表明

$$\underline{\beta}\left(|\boldsymbol{q}|_{\boldsymbol{\mathcal{A}}}\right)\leqslant\mathcal{R}(\boldsymbol{q})\leqslant\overline{\beta}\left(|\boldsymbol{q}|_{\boldsymbol{\mathcal{A}}}\right)\tag{9-40}$$

其中，$\boldsymbol{\mathcal{A}}=\{\boldsymbol{q}:\boldsymbol{\xi}=0,\boldsymbol{e}=0\}$；$\underline{\beta}:s\to\min\left\{\underline{\alpha}_V\left(\dfrac{s}{2}\right),\dfrac{a_\lambda\lambda^2+b_\lambda}{2}\underline{\alpha}_W^2 s\right\}$以及$\overline{\beta}:s\to\max\left\{2\overline{\alpha}_V(s),2a_\lambda\overline{\alpha}_W^2 s\right\}$，因此$\underline{\beta},\overline{\beta}\in\mathcal{K}_\infty$。

为了研究混杂系统的稳定性 [式 (9-35)]，下面分析函数 $\mathcal{R}$ 是如何随着跳跃集和流动集而演化的（有或没有拒绝服务攻击）。

① 跳变集 D　如果$m(t)=0$且$\boldsymbol{q}\in D_0$，那么式 (9-37)（其中$\eta^+=a_\lambda$）和假设 9.6 表明$\mathcal{R}^+(\boldsymbol{q})=V(\boldsymbol{\xi})+a_\lambda\left(\lambda W(k,\boldsymbol{e})\right)^2\leqslant\mathcal{R}(\boldsymbol{q})$；如果$m(t)=1$且$q\in D_1$，那么$\boldsymbol{\xi}^+=\boldsymbol{\xi},\boldsymbol{e}^+=\boldsymbol{e}$以及$\eta^+=\eta$ [式 (9-37)] 表明$\mathcal{R}^+(\boldsymbol{q})=V(\boldsymbol{\xi})+\eta W^2(k,\boldsymbol{e})=\mathcal{R}(\boldsymbol{q})$。因此有

$$\mathcal{R}^+(\boldsymbol{q})\leqslant\mathcal{R}(\boldsymbol{q})\tag{9-41}$$

② 流动集 C　如果$m(t)=0$即$\dot{\eta}=-\left(2\eta L(\boldsymbol{\xi},\boldsymbol{e})+\eta^2+G(\boldsymbol{\xi},\boldsymbol{e})+\varepsilon\right),\eta\in\left[a_\lambda\lambda^2+b_\lambda,a_\lambda\right]$，假设9.6和不等式$\Lambda:2\eta W(\boldsymbol{k},\boldsymbol{e})H(\boldsymbol{\xi})\leqslant H^2(\boldsymbol{\xi})+\eta^2W^2(k,\boldsymbol{e})$表明

$$\begin{aligned}\mathcal{R}^\circ(\boldsymbol{q})&=-\rho_V V(\boldsymbol{\xi})-H^2(\boldsymbol{\xi})+G(\boldsymbol{\xi},e)W^2(k,\boldsymbol{e})\\&-\left(2\eta L(\boldsymbol{\xi},\boldsymbol{e})+\eta^2+G(\boldsymbol{\xi},\boldsymbol{e})+\varepsilon\right)W^2(k,\boldsymbol{e})\\&+2\eta W(k,e)\left(L(\xi,e)W(k,e)+H(\xi)\right)\\&\underbrace{\leqslant}_{\Lambda}-\rho_V V(\boldsymbol{\xi})-\varepsilon W^2(k,\boldsymbol{e})\\&\leqslant-\varpi_0\mathcal{R}(\boldsymbol{q})\end{aligned}\tag{9-42}$$

其中，$\varpi_0=\min\left\{\rho_V,\dfrac{\varepsilon}{a_\lambda}\right\}$。如果$m(t)=1$即$\dot{\eta}=0$，结合$\eta\in\left[a_\lambda\lambda^2+b_\lambda,a_\lambda\right]$和假设9.6可知

$$\begin{aligned}\mathcal{R}^\circ(\boldsymbol{q})&=-\rho_V V(\boldsymbol{\xi})-H^2(\boldsymbol{\xi})+G(\boldsymbol{\xi},\boldsymbol{e})W^2(k,\boldsymbol{e})\\&\quad+2\eta W(k,\boldsymbol{e})\left(L(\boldsymbol{\xi},\boldsymbol{e})W(k,\boldsymbol{e})+H(\boldsymbol{\xi})\right)\\&\underbrace{\leqslant}_{\Lambda}-\rho_V V(\boldsymbol{\xi})+G(\boldsymbol{\xi},\boldsymbol{e})W^2(k,\boldsymbol{e})+2\eta L(\boldsymbol{\xi},\boldsymbol{e})\\&\quad\times W^2(k,\boldsymbol{e})+\eta^2W^2(k,\boldsymbol{e})\end{aligned}\tag{9-43}$$

$$\leqslant -\rho_V V(\boldsymbol{\xi}) + \eta W^2(k,\boldsymbol{e})(2L(\boldsymbol{\xi},\boldsymbol{e}) + a_\lambda + \frac{G(\boldsymbol{\xi},\boldsymbol{e})}{a_\lambda \lambda^2 + b_\lambda})$$
$$\leqslant \varpi_1 \mathcal{R}$$

其中，$\varpi_1 = \sup\left\{2L(\boldsymbol{\xi},\boldsymbol{e}) + a_\lambda + \frac{G(\boldsymbol{\xi},\boldsymbol{e})}{a_\lambda \lambda^2 + b_\lambda}\right\}$。

实际上，混杂系统$\mathcal{H}_{\text{dos}}$的解不会出现奇诺现象，因为严格正的最小传输间隔满足$T \geqslant \underline{\Delta}_0 \geqslant \underline{\Delta}_1 > 0$；另一方面，式 (9-40) ～式 (9-43) 表明混杂系统也不会出现有限逃逸时间，这是因为$\boldsymbol{\xi}$和$\boldsymbol{e}$有界且η、τ_1、τ_2以及m无有限逃逸时间，即$\sup_t E \to \infty$。基于 De 和 Postoyan[226]，以及 Goebel 等人[230]的研究，以上分析也表明每个解$\phi \in S_{\mathcal{H}_{\text{dos}}}$是完备的。确实，在区间$[t_k, t_{k+1})$其中$m(t)=0$（即没有攻击），根据$\eta = -(2\eta L(\boldsymbol{\xi},\boldsymbol{e}) + \eta^2 + G(\boldsymbol{\xi},\boldsymbol{e}) + \varepsilon)$，对任意$t \in [t_k, t_{k+1})$，式 (9-37) 表明事件驱动函数会被连续监测；而且式 (9-37) 表明事件驱动被触发，一旦η减小到$a_\lambda \lambda^2 + b_\lambda$，进而$\eta$被重置为$a_\lambda$。从而可知$t_{k+1} - t_k > 0$，因为$\eta$是严格连续下降的，从$\lim\limits_{t \to t_k^+} \eta(t) = a_\lambda$到$\lim\limits_{t \to t_{k+1}^-} \eta(t) = a_\lambda \lambda^2 + b_\lambda$，其中$a_\lambda \lambda^2 + b_\lambda < a_\lambda$，所以总有$\underline{\Delta}_0 > 0$，即没有攻击时候的传输间隔是严格正的，即可以排除奇诺现象。另一方面，式 (9-37) 表明传输尝试将会失败，因为事件驱动函数在攻击时保持不变，而数据会进行周期传输（周期为$\underline{\Delta}_1$），这可以当成是拒绝服务攻击而引起的驱动延迟[85]，而且传输周期$\underline{\Delta}_1$满足$\underline{\Delta}_1 \leqslant \underline{\Delta}_0$，这是为了尽早根据确认机制觉察攻击在何时结束，进而尽早进行事件驱动策略的正常评估，因此式 (9-35) 的解ϕ具有严格正的一致传输间隔。

根据式 (9-40) ～式 (9-43)，接下来考虑混杂系统$\mathcal{H}_{\text{dos}}$ [式 (9-35)] 的最大解给定最大解$\boldsymbol{q}$和区间$[T_1, T_2]$，没有拒绝服务攻击的总时间定义为：

$$\overline{\Theta}_q(T_1,T_2) = \left\{\tilde{t} \in (T_1,T_2) \mid \tilde{t} \notin \mathcal{T},\ \forall k \in \mathbb{Z}_{\geqslant 0}\ \ (\tilde{t},k) \in \operatorname{dom} \boldsymbol{q}, m(\tilde{t}) = 0\right\} \tag{9-44}$$

即如果$t \in \overline{\Theta}_q$，混杂系统$\mathcal{H}_{\text{dos}}$处在稳定模式，且满足式(9-42)。根据式(9-44)，我们定义相应的攻击总时间，在此时间内混杂系统$\mathcal{H}_{\text{dos}}$处在不稳定模式下，满足式(9-43)：

$$\overline{\Xi}_q(T_1,T_2) = [T_1,T_2] \setminus \overline{\Theta}_q(T_1,T_2) \tag{9-45}$$

那么对任意$T_1, T_2 \in \mathbb{R}_{\geqslant 0}$，其中$T_1 \leqslant T_2$，有$\overline{\Theta}_q(T_1, T_2) \cup \overline{\Xi}_q(T_1, T_2) = [T_1, T_2]$，因此$\overline{\Xi}_q(T_1, T_2)$和$\overline{\Theta}_q(T_1, T_2)$可以表示为

$$\begin{cases} \overline{\Xi}_q(T_1, T_2) = \bigcup\limits_{p_k \in \mathbb{Z}_{\geqslant 0}} \tilde{Z}_{p_k} \cap [T_1, T_2] \\ \overline{\Theta}_q(T_1, T_2) = \bigcup\limits_{p_k \in \mathbb{Z}_{\geqslant 0}} \tilde{W}_{p_k - 1} \cap [T_1, T_2] \end{cases} \tag{9-46}$$

对任意$p_k \in \mathbb{Z}_{\geqslant 0}$，其中$\tilde{Z}_{p_k}$和$\tilde{W}_{p_k}$满足

$$\begin{gathered} \tilde{Z}_{p_k} = \begin{cases} [\zeta_{p_k}, \zeta_{p_k} + v_{p_k}), & v_{p_k} > 0 \\ \{\zeta_{p_k}\}, & v_{p_k} = 0 \end{cases} \\ \tilde{W}_{p_k} = \begin{cases} [\zeta_{p_k} + v_{p_k}, \zeta_{p_k + 1}), & v_{p_k} > 0 \\ (\zeta_{p_k}, \zeta_{p_k + 1}), & v_{p_k} = 0 \end{cases} \end{gathered} \tag{9-47}$$

其中，$v_{p_k} \geqslant 0$表示自ζ_{p_k}到下一次成功传输所经过的时间。如果$v_{p_k} = 0$，那么$\tilde{Z}_{p_k} = \{\zeta_{p_k}\}$和$\tilde{W}_{p_k} = [\zeta_{p_k}, \zeta_{p_k+1})$，即此时攻击是脉冲形式。另外，如果$h_0 > 0$，有$\zeta_0 = h_0$和$\tilde{W}_{-1} = [0, \zeta_0)$；否则$\tilde{W}_{-1} = \varnothing$。

结合假设 9.5，可知

$$\left|\overline{\Xi}_q(T_1, T_2)\right| \leqslant \left|\Xi(T_1, T_2)\right| + n(T_1, T_2)\Delta_1 \tag{9-48}$$

这意味着$n(T_1, T_2)\Delta_1$延长了攻击的总时间。基于假设9.5，可以将式(9-48)表示为

$$\begin{aligned} \left|\overline{\Xi}_q(T_1, T_2)\right| &\leqslant \varsigma + \frac{T_2 - T_1}{\tau_d} + \left(v + \frac{T_2 - T_1}{\tau_f}\right)\Delta_1 \\ &\leqslant \varsigma_* + \frac{T_2 - T_1}{T_*} \end{aligned} \tag{9-49}$$

其中，$\varsigma_* = \varsigma + v\Delta_1, T_* = \dfrac{\tau_f \tau_d}{\tau_f + \tau_d \Delta_1}$。

基于以上分析，接下来分析混杂系统 (9-35) 的稳定性。对任意$(t, k) \in (\operatorname{dom} \boldsymbol{q}) \cap (\tilde{W}_{p_k} \times \mathbb{Z}_{\geqslant 0})$（没有攻击），其中$p_k \in \mathbb{Z}_{\geqslant 0} \cup \{-1\}$，和任意$(t, k) \in (\operatorname{dom} \boldsymbol{q}) \cap (\tilde{Z}_{p_k} \times \mathbb{Z}_{\geqslant 0})$（有攻击），其中$p_k \in \mathbb{Z}_{\geqslant 0}$，有

$$\mathcal{R}\left(\boldsymbol{q}(t,k)\right)\leqslant\begin{cases}\exp\left(-\varpi^{0}\left(t-\zeta_{p_k}-v_{p_k}\right)\right)\mathcal{R}\left(\boldsymbol{q}\left(\zeta_{p_k}+v_{p_k},k\right)\right) & \text{无攻击}\\ \exp\left(\varpi^{1}\left(t-\zeta_{p_k'}\right)\right)\mathcal{R}\left(\boldsymbol{q}\left(\zeta_{p_k}+v_{p_k},k\right)\right) & \text{有攻击}\end{cases}$$

(9-50)

不等式 (9-50) 的右边分别反映了函数 $\mathcal{R}$ 在稳定模式（没有攻击）和不稳定模式（有攻击）的上界。下面通过结合$(t,k)\in\text{dom}\, q$的两种模式，用 $\mathcal{R}$ 的上界来评估混杂系统 (9-35) 的稳定性。基于此，我们给出连接两种模式的桥梁，具体证明可以参见 De Persis 和 Tesi[85]，以及 Dolk 等人[83] 的研究。

引理 9.2

对任意$(t,k)\in\text{dom}\, q$，有

$$\mathcal{R}\left(q(t,k)\right)\leqslant\mathcal{Y}(0,t)\mathcal{R}\left(q(0,0)\right) \tag{9-51}$$

其中，$\mathcal{Y}(s,t)=\mathrm{e}^{-\varpi^{0}\left|\bar{\Theta}_q(s,t)\right|}\mathrm{e}^{\varpi^{1}\left|\Xi_q(s,t)\right|}$。

因此，根据式 (9-44) ～式 (9-47)，有

$$\begin{aligned}\mathcal{Y}(\ T_1,T_2)&=\mathrm{e}^{-\bar{\omega}_0\left|\bar{\Theta}_q(T_1,T_2)\right|}\mathrm{e}^{\bar{\omega}_1\left|\Xi_q(T_1,T_2)\right|}\\&\leqslant\mathrm{e}^{-\varpi_0(T_2-T_1)}\mathrm{e}^{(\omega_0+\omega_1)\left(\varsigma_*+\frac{T_2-T_1}{T_*}\right)}\\&\leqslant\mathrm{Ime}^{-\hat{\beta}(T_2-T_1)}\end{aligned} \tag{9-52}$$

其中，$\mathrm{Im}=e^{\varsigma_*\left(\varpi^{0}+\varpi^{1}\right)},\hat{\beta}=\varpi^{0}-\dfrac{\varpi^{0}+\varpi^{1}}{T_*}>0$。

因此结合假设 9.5，不等式 (9-40) ～式 (9-43) 以及式 (9-51)、式 (9-52)，可以将式 (9-50) 写成

$$\mathcal{R}\left(\boldsymbol{q}(t,k)\right)\leqslant\mathrm{Ime}^{-\hat{\beta}t}\mathcal{R}\left(\boldsymbol{q}(0,0)\right) \tag{9-53}$$

其中，$\mathrm{Im}=\mathrm{e}^{\varsigma_*(\varpi_0+\varpi_1)}$；$\varsigma_*=\varsigma+v$；$\hat{\beta}=\varpi_0-\dfrac{\varpi_0+\varpi_1}{T_*}$；$T_*=\dfrac{\tau_d\tau_f}{\Delta_1\tau_d+\tau_f}$。基于式(9-40)和定理9.2的条件，有$\left|\boldsymbol{q}(t,k)\right|_{\mathcal{A}}\leqslant\underline{\beta}^{-1}\left(\mathrm{Ime}^{-\hat{\beta}t}\overline{\beta}\left(\left|\boldsymbol{q}(0,0)\right|_{\mathcal{A}}\right)\right)$。令$\underline{\beta}^{-1}\left(\mathrm{Ime}^{-\hat{\beta}t}\overline{\beta}(\cdot)\right)\overset{\text{def}}{=}\beta(\cdot,t,k)$，其中$\beta(\cdot,t,k)\in\mathcal{KLL}$，所以根据定义9.2可知$\boldsymbol{\mathcal{A}}$是一致全局渐近稳定的。证毕。 ■

注释 9.11　事实上，已有文献对跟踪控制问题进行研究并给出很好的研究结果[84,249]，而且 Dolk 等人[83]，以及 De 和 Tesi[85] 也对拒绝服务攻击进行了很彻底的研究。但在本章节，我们考虑如何在拒绝服务攻击存在的情况下依靠事件驱动策略决定传输时刻，使得机器人系统的追踪控制对攻击具有弹性作用，而且将理论成果付诸实践。与 Postoyan 等人在文献 [249] 和文献 [84] 的研究相比，本章节为使机器人系统具有抵抗拒绝服务攻击的能力设计了一种事件触发策略，以获得离散的采样时刻来节省通信和计算时刻。与 Dolk 等人[83] 的研究，以及 De 和 Tesi[85] 的研究相比，本章节展示了拒绝服务攻击对移动机器人跟踪性能的影响，这对解决现实中机器人系统的服务器故障问题有一定的指导意义。

9.3.3 算法验证与分析 2

接下来，我们基于所提出的事件触发策略，考虑 Amigobot 机器人的跟踪控制问题。在本实验中，我们在一个基准测试上实现所提出的策略，其中控制器安装在 RaspberryPi（树莓派，简写为 RP_i）上，通过有线链路直接控制 Amigobot 机器人。机器人的状态和参考轨迹的数据由 Vicon 三维运动捕捉系统测量，通过有线连接并传输到服务器，然后传到电脑（PC）并进一步通过 XBee 广播到 RaspberryPi。测量到的数据在电脑端被用来评估事件触发条件进而决定何时进行数据的传输，最后在 RaspberryPi 上更新控制器直接控制 Amigobot 机器人，在无线传输过程中，XBee 电台之间的无线信道可能会受到拒绝服务攻击（DoS 攻击），导致数据不能够成功传输，正如图 9-4 中所描绘的硬件体系结构和图 9-5 描述的软件实现过程。值得注意的是，在理论研究中，我们假设控制器是通过无线网控制机器人，这是为了使研究更具一般性，所以在本实验中，数据经过无线网传给控制器（由事件驱动策略决定何时传输），而控制器有线直接控制机器人也是合理的。

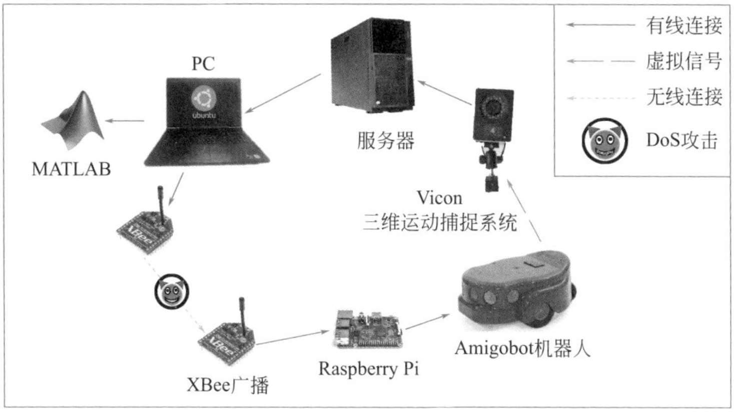

图 9-4　实验台和硬件架构

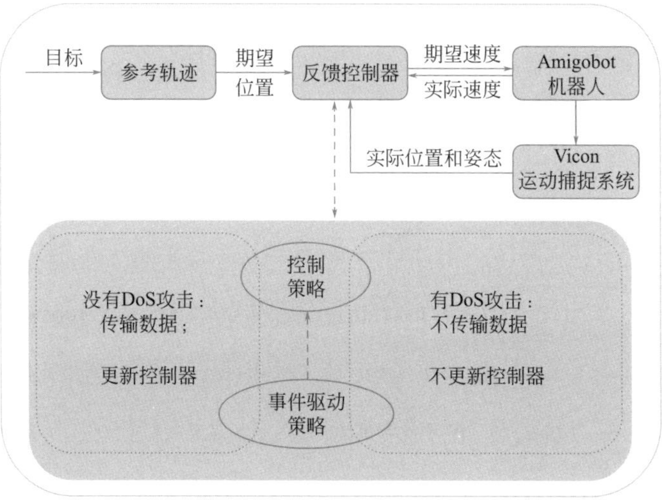

图 9-5　软件实现过程

当$m(t)=0$（没有拒绝服务攻击）时，数据通过传输来评估事件触发条件，一旦违背事件触发条件，控制器将更新。当$m(t)=1$（有拒绝服务攻击）时，数据会以周期方式进行传输，但却不能成功，所以控制器将不会更新。这里我们考虑系统模型 [式 (9-27)、式 (9-28)] 和控制法则 [式 (9-29)]，参照系

统状态为 $x_r = 5\cos(0.08t), y_r = 5\sin(0.08t), v_r = 0.4\mathrm{m/s}$ 和 $\omega_r = 0.08\mathrm{rad/s}$，其中 $\beta_1 = 1$，$\beta_2 = 0.56, \beta_3 = 0, \beta_4 = 0, c_1 = 2.0, c_2 = 0.6, c_3 = 6.0, \gamma = 1.5$，也就是机器人会做匀速圆周运动，其中 $\theta = 0.08t$。机器人系统的物理参数如下：$b = 0.14\mathrm{m}, a = 0.1\mathrm{m}, r = 0.05\mathrm{m}, m_c = 3.6\mathrm{kg}$，$m_\omega = 0.3\mathrm{kg}, I_c = 5.2080\mathrm{kg}\cdot\mathrm{m}^2$，$I_\omega = 0.0167\mathrm{kg}\cdot\mathrm{m}^2, I_m = 0.0008\mathrm{kg}\cdot\mathrm{m}^2, d_{11} = d_{22} = 1.6667\mathrm{kg/s}$，$\tau_E = k_g g$，其中，$k_g = 0.01$ 和 $g = 9.8\ \mathrm{m/s^2}$。基于假设 9.6，令 $W(e) = |e|$，其中 $\lambda = 0, b_\lambda = 5$，$a_\lambda = 10$ 和 $\varepsilon = 10$ [251]。令 $V(\xi) = \frac{1}{2}\bar{x}_e^2 + \frac{1}{2}y_e^2 + \frac{1}{2}\theta_e^2$，其中 $\bar{x}_e$ 和 γ 在式 (9-29) 中有定义。基于 Postoyan 等人 [84] 的研究，有 $\left|\frac{\partial W(k,\boldsymbol{e})}{\partial \boldsymbol{e}}\right| \leqslant \sqrt{2}$。结合式 (9-33) 和式 (9-34)，选择 $L(\xi,\boldsymbol{e}) = 2\sqrt{2}, \rho_V = 1$，$H(\xi) = 2|v| + 2\beta_1 + |\omega| + \beta_2 + \beta_3 + \beta_4$ 以及 $G(\xi,\boldsymbol{e}) = (x_e - c_3\omega y_e)(\omega + \cos\theta_e + c_3\omega(\omega + \sin\theta_e)) + y_e(\omega + \sin\theta_e) + \theta_e$。

基于假设 9.6，选择 $W(\boldsymbol{e}) = |\boldsymbol{e}|$，其中，$\lambda = 0, b_\lambda = 5, a_\lambda = 10$ 和 $\varepsilon = 10$ [251]，以及 $V(\xi) = \frac{1}{2}\bar{x}_e^2 + \frac{1}{2}y_e^2 + \frac{1}{2}\theta_e^2$。由此，我们得到 $\left|\frac{\partial W(k,\boldsymbol{e})}{\partial \boldsymbol{e}}\right| \leqslant \sqrt{2}$ [84] 和

$$
\begin{aligned}
|\dot{\boldsymbol{e}}| &= |\boldsymbol{g}_e(\xi,\boldsymbol{e})| \\
&\leqslant |(v\cos(\theta + e_\theta), v\sin(\theta + e_\theta), \omega, (v_r + e_{v_r}) \\
&\quad \times \cos(\theta_r + e_{\theta_r}), (v_r + e_{v_r})\sin(\theta_r + e_{\theta_r}), \omega_r + e_{\omega_r}, \chi_{1r}, \chi_{2r})| \\
&\leqslant |(v\cos(\theta + e_\theta)| + |v\sin(\theta + e_\theta)| + |\omega| + |(v_r + e_{v_r})\cos(\theta_r \\
&\quad + e_{\theta_r})| + |(v_r + e_{v_r})\sin(\theta_r + e_{\theta_r})| + |\omega_r + e_{\omega_r}| + |\chi_{1r}| \\
&\quad + |\chi_{2r}| \leqslant 2|v| + 2\beta_1 + |\omega| + \beta_2 + |e_\omega| + 2|e_{v_r}| + |e_{\omega_r}| + \beta_3 + \beta_4,
\end{aligned}
$$

其中，$v = v_r\cos\theta_e - c_3\dot{\omega}y_e + c_3\omega(\omega x_e - v_r\sin\theta_e) + c_1(x_e - c_3\omega y_e), \omega = \omega_r + \gamma y_e v_r\sin\theta_e + c_2\gamma\theta_e$，$x_e = (x_r - x)\cos\theta + (y_r - y)\sin\theta, y_e = -(x_r - x)\sin\theta + (y_r - y)\cos\theta, \theta_e = \theta_r - \theta, e_{v_r} = \hat{v}_r - v_r$，$e_{\omega_r} = \hat{\omega}_r - \omega_r, e_\theta = \hat{\theta} - \theta$ 以及 $e_{\theta_r} = \hat{\theta}_r - \theta_r$。基于 Postoyan 等人 [84] 的研究，式 (9-33) 表明 $L(\xi,\boldsymbol{e}) \leqslant 2\sqrt{2}, \rho_V = 1, H(\xi) = 2|v| + 2\beta_1 + |\omega| + \beta_2 + \beta_3 + \beta_4$。接下来我们检验假设 9.6 中的不等式 (9-34)。基于参数 $H(\xi)$ 和 $V(\xi) = \frac{1}{2}\bar{x}_e^2 + \frac{1}{2}y_e^2 + \frac{1}{2}\theta_e^2$，有

$$
\begin{aligned}
\dot{V}(\boldsymbol{\xi}) &= \bar{x}_e\dot{\bar{x}}_e + y_e\dot{y}_e + \theta_e\dot{\theta}_e \\
&= (x_e - c_3\omega y_e)(\dot{x}_e - c_3\dot{\omega}y_e - c_3\omega y_e) + y_e\dot{y}_e + \theta_e\dot{\theta}_e \\
&= (x_e - c_3\omega y_e)\Big\{\omega\left(y_e + e_{y_e}\right) - v + \left(v_r + e_{v_r}\right)\cos\left(\theta_e + e_{\theta_e}\right) - c_3\dot{\omega}y_e \\
&\quad - c_3\omega\Big[-\omega\left(x_e + e_{x_e}\right) + \left(v_r + e_{v_r}\right)\sin\left(\theta_e + e_{\theta_e}\right)\Big]\Big\} + y_e\Big[-\omega\left(x_e + e_{x_e}\right) \\
&\quad + \left(v_r + e_{v_r}\right)\sin\left(\theta_e + e_{\theta_r}\right)\Big] + \theta_e\Big[\left(\omega_r + e_{\omega_r}\right) - \omega\Big] \\
&\leqslant (x_e - c_3\omega y_e)\times\Big\{\omega y_e - v + v_r\cos\left(\theta_e + e_{\theta_e}\right) - c_3\dot{\omega}y_e - c_3\omega\big[-\omega x_e \\
&\quad + v_r\sin\left(\theta_e + e_{\theta_e}\right)\big]\Big\} + y_e\Big[-\omega x_e + v_r\sin\left(\theta_e + e_{\theta_e}\right)\Big] + \theta_e(\omega_r - \omega) \\
&\quad + (x_e - c_3\omega y_e)\Big\{\omega e_{y_e} + e_{v_r}\cos\left(\theta_e + e_{\theta_e}\right) - c_3\omega\Big[-\omega e_{x_e} + e_{v_r}\sin\left(\theta_e + e_{\theta_e}\right)\Big]\Big\} \\
&\quad + y_e\Big[-\omega e_{x_e} + e_{v_r}\sin\left(\theta_e + e_{\theta_e}\right)\Big] + \theta_e e_{\omega_r} \\
&\leqslant -\left[\frac{1}{2}(x_e - c_3\omega y_e)^2 + \frac{1}{2}y_e^2 + \frac{1}{2}\theta_e^2\right] - \left(2|v| + 2\beta_1 + |\omega| + \beta_2 + \beta_3 + \beta_4\right)^2 \\
&\quad + \Big\{(x_e - c_3\omega y_e)\big[\omega + \cos\theta_e + c_3\omega(\omega + \sin\theta_e)\big] + y_e(\omega + \sin\theta_e) + \theta_e\Big\} \\
&\quad \times\Big[(\hat{x} - x)^2 + (\hat{y} - y)^2 + \left(\hat{\theta} - \theta\right)^2 + (\hat{x}_r - x)^2 \\
&\quad + (\hat{y}_r - y)^2 + \left(\hat{\theta}_r - \theta\right)^2 + (\hat{v}_r - v_r)^2 + (\hat{\omega}_r - \omega_r)^2\Big],
\end{aligned}
$$

其中，$v = v_r\cos\theta_e - c_3\dot{\omega}y_e + c_3\omega(\omega x_e - v_r\sin\theta_e) + c_1(x_e - c_3\omega y_e), \omega = \omega_r + \gamma y_e v_r\sin\theta_e + c_2\gamma\theta_e$，$x_e = (x_r - x)\cos\theta + (y_r - y)\sin\theta$，$y_e = -(x_r - x)\sin\theta + (y_r - y)\cos\theta, \theta_e = \theta_r - \theta, e_{x_e} = \hat{x}_e - x_e, e_{y_e} = \hat{y}_e - y_e, e_{\theta_e} = \hat{\theta}_e - \theta_e, \boldsymbol{e} = \left(\boldsymbol{e}_q, \boldsymbol{e}_d, \boldsymbol{e}_{ff}\right)$，$\boldsymbol{e}_q = \left(e_x, e_y, e_\theta\right) = \left(\hat{x} - x, \hat{y} - y, \hat{\theta} - \theta\right), \boldsymbol{e}_d = \left(e_{x_r}, e_{y_r}, e_{\theta_r}\right) = \left(\hat{x}_r - x, \hat{y}_r - y, \hat{\theta}_r - \theta\right)$ 以及 $\boldsymbol{e}_{ff} = \left(e_{v_r}, e_{\omega_r}\right) = (\hat{v}_r - v_r, \hat{\omega}_r - \omega_r)$。因此根据式 (9-33) 和式 (9-34)，有 $L(\boldsymbol{\xi}, \boldsymbol{e}) = 2\sqrt{2}, \rho_V = 1, H(\boldsymbol{\xi}) = 2|v| + 2\beta_1 + |\omega| + \beta_2 + \beta_3 + \beta_4$ 以及 $G(\boldsymbol{\xi}, \boldsymbol{e}) = (x_e - c_3\omega y_e)\big[\omega + \cos\theta_e + c_3\omega(\omega + \sin\theta_e)\big] + y_e(\omega + \sin\theta_e) + \theta_e$。

根据定理 9.2，可以得到 $\varpi_0 = 1, \varpi_1 = 16.1041$。基于式 (9-37)，可以得到 $\varDelta_0 = 0.0467$（没有拒绝服务攻击），进而可以选择 $\varDelta_1 = \frac{1}{2}\varDelta_0 = 0.0467 \times 0.5 = 0.0234$。基于式 (9-53)，选择 $\hat{\beta} = \frac{1}{4}\varpi_0$，即有 $\frac{1}{T_*} \leqslant \frac{3}{4} \times \frac{\varpi_0}{\varpi_0 + \varpi_1} = 0.0439$。因此有 $\frac{\varDelta_1}{\tau_f} + \frac{1}{\tau_d} \leqslant 0.0439$。根据假设 9.5 的

$T_1 = 0$和$T_2 = 150$，选择$\varsigma = 20, v = 0, \frac{1}{\tau_f} = \frac{1}{8}$以及$\frac{1}{\tau_d} = 0.0408$，并且满足$\frac{\Delta_1}{\tau_f} + \frac{1}{\tau_d} = 0.0437 \leqslant 0.0439$，因而验证了式(9-39)，那么有$n(0,150) \leqslant v + \frac{150}{\tau_f} = 18.75$和$\left|\Xi(0,150)\right| \leqslant \varsigma + \frac{150}{\tau_d} = 20 + 6.1200 = 26.1200$。

因此，基于上述分析，在拒绝服务攻击下的跟踪控制性能如图 9-6 所示，其中灰色区域表示拒绝服务攻击发生的区间，其中$n(0,150) = 15$和$\left|\Xi(0,150)\right| = 25.1510$。在图 9-6 中，机器人系统的初始条件为$(x(0), y(0), v(0), \omega(0)) = (0\text{m}, 0\text{m}, 0\text{m/s}, 0\text{rad/s})$，进而表明机器人在所

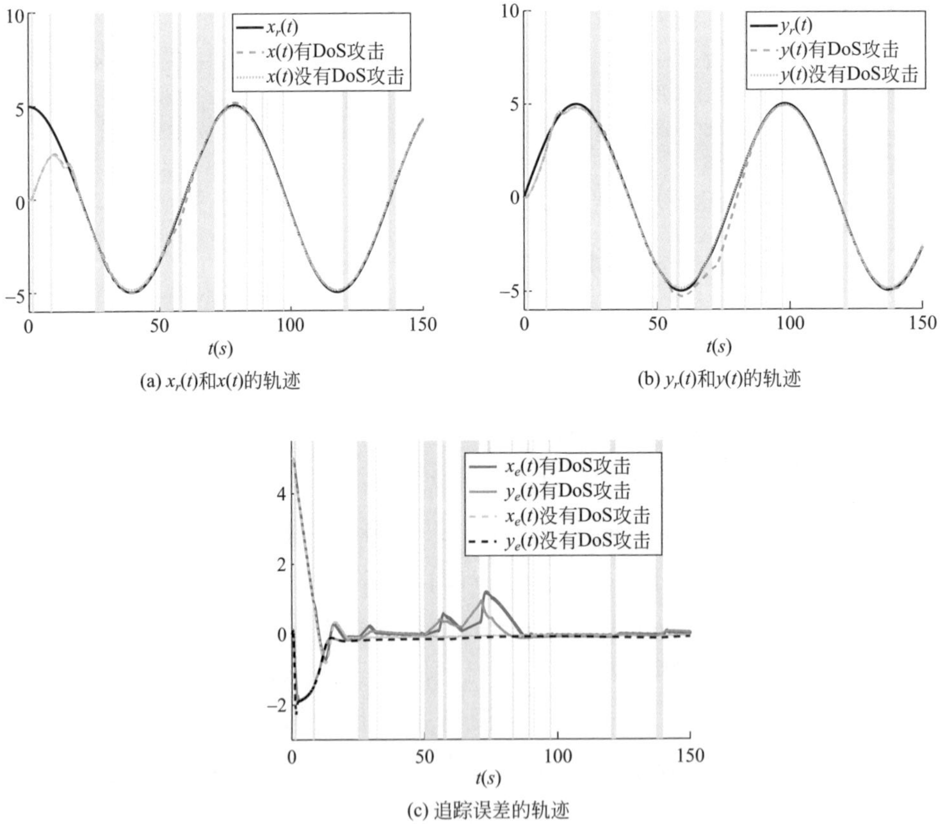

(a) $x_r(t)$和$x(t)$的轨迹

(b) $y_r(t)$和$y(t)$的轨迹

(c) 追踪误差的轨迹

图 9-6　机器人系统状态轨迹的追踪收敛图

设计的事件驱动控制器下能够跟踪参照轨迹。实际上，为了获得更好的跟踪性能，可以选择更大的λ，这样就可以选择更小的$\underline{\Delta}_0$和$\underline{\Delta}_1$，以便尽早检测到拒绝服务攻击何时结束，从而新采样数据可以尽早传输到控制器进行更新。

Digital Wave
Advanced Technology of Industrial Internet

Control of Autonomous Intelligent Systems

自主智能系统控制

第 10 章

多个智能体系统的安全协同性能

10.1 概述

网络通信渠道使信息传输更加有利[255-256]，但同时也更容易受到网络攻击，比如数据很容易被对手篡改[257]，不同于拒绝服务攻击[83,85,236]，这种数据被篡改的欺诈攻击会对系统造成更严重的破坏，也因此值得深入地探索和研究。如果控制器始终接收敌方注入的虚假信息，则控制信号将与当前物理特性不相符[258-259]，毫无疑问，这很不利于网络多智能体的编队跟踪控制。当前的大部分工作主要集中在设计各种检测或估计算法来检测和识别欺诈攻击[260-261]，然而，狡猾的攻击者总是可以通过精心设计攻击策略来欺骗检测器和鲁棒估计器，使得任何算法都可能不起作用。基于此，在不需检测网络攻击（包括欺诈攻击和拒绝服务攻击）的情况下，本章节设计了弹性控制策略来实现多智能体的编队跟踪。此外，在网络多智能体中，控制器通过通信网络与传感器 / 执行器在离散时刻通信[262-265]，其中这些离散时刻可以通过建立基于系统状态的事件驱动策略来产生，该策略比时间触发控制更有效[226,266]。一般地，在基于事件的采样策略中会增加一个正常数来补偿奇诺现象，使得只有在特定的时间期限之后（即最小触发间隔，小于或等于最大允许传输间隔[267]），触发条件才可能被触发[83]，这就使本章节与 Ding 等人[268]研究中的事件触发方式不同。

本章节旨在为定向信息传播拓扑的时变多智能体的编队（同时跟踪各领导者的凸组合）设计合适的控制策略，使其对网络攻击行为（包括拒绝服务攻击和欺诈攻击）具有弹性作用。作为一个活跃的研究领域，混合系统理论为连接离散和连续动力学的数学模型提供了技术支撑[230]。基于此，本章节建立了具有网络攻击的多智能体系统通用混杂模型，主要贡献可以概括如下。

① 10.2 节建立了一种通用的混杂模型，用于研究欺诈攻击下网络群体的编队跟踪控制。首先给出欺诈攻击的显式描述，包括它们的频率、持续时间和大小，接着设计基于边的没有奇诺现象的分布式弹性事件驱

动策略，使得在不需要检测欺诈攻击的情况下也可以实现编队跟踪的弹性控制。

② 利用每条边的内邻集和外邻集，考虑边之间的有向信息流。根据攻击模式，10.2 节分别考虑三种情况，即异步欺诈攻击、基于邻近边的同步欺诈攻击、基于所有边的同步欺诈攻击。

③ 10.3 节建立了非线性多智能体系统，为非线性多智能体系统的编队控制建立了在拒绝服务攻击下基于分布式事件触发的混杂模型。通过考虑智能体之间的有向通信，拒绝服务攻击可以攻击某个多智能体的接收端或者发送端。

④ 基于事件触发策略中的附加辅助参数，10.3 节给出拒绝服务攻击下数据传输间隔的计算方法。如果存在拒绝服务攻击，此计算方法有利于更频繁地安排传输尝试，以便更早地确定拒绝服务攻击何时结束，进而继续进行基于事件驱动的数据传输。相比之下，在早期关于拒绝服务攻击的研究中，De 和 Tesi[85]，以及后来的 Dolk 等人[83]、Senejohnny 等人[229]和 Feng 等人[236]的研究中都给出相对完善的理论结果。标准混杂理论首次在 De 等人[226]的研究中用于处理无拒绝服务攻击的多智能体协调问题，而 Senejohnny 等人[229]研究了拒绝服务攻击下的多智能体协调问题。为了将 Dolk 等人[83]的研究扩展到非线性多智能体的情形，我们需要通过考虑内邻域和外邻域之间的局部信息交流来重新设计分布式事件触发条件，这就给基于事件的混杂模型的建立和分析带来困难。此外，为了将没有拒绝服务攻击的 Wang 等人[264]的研究拓展到有拒绝服务攻击的情形，我们必须分析多智能体局部的广播信息，解决有向拓扑下拒绝服务攻击的影响，设计基于辅助函数的没有奇诺现象的触发条件，进而建立基于事件驱动的满足混杂基本条件的混杂模型。10.3 节不需要邻居之间连续地检测，因为 10.3 节采用的是基于点的分布式事件驱动策略而非 De 等人[85]的研究中基于边的分布式事件驱动策略。10.3 节中，当智能体 j 将其数据广播给邻居 i 时，该值将被 i 保持为常数，然后在智能体 i 的下一个更新时刻到来时使用。

10.2 欺诈攻击下网络化多智能体系统的编队控制

令 $G=\left(\mathcal{N},\mathcal{E}_{\mathrm{cm}},\tilde{\boldsymbol{A}}\right)$ 为加权图，其中节点集为 $\mathcal{N}=\{1,2,\cdots,N\}$，边集为 $\mathcal{E}_{\mathrm{cm}}\subseteq\mathcal{N}\times\mathcal{N}$，加权邻接矩阵为 $\tilde{\boldsymbol{A}}=\left(a_{ij}\right)_{\mathcal{N}\times\mathcal{N}}$（矩阵元素非负）。令 (i,j) 表示 $\varepsilon_{\mathrm{cm}}$ 中的有向边，即节点 j 可以收到节点 i 广播过来的信息。有向路径是一个有向边序列，其形式为 $(i,j),(j,k),\cdots$，其中 $\{i,j,k,\cdots\}\subseteq\mathcal{N}$。$l_{ij}\overset{\mathrm{def}}{=}(i,j)$ 表示有向加权图 G 的边，并令 $l_{ij}=l_{i,j}$。本节中假设前 M 个节点为跟随者，其余为领导者。令 $F=\{1,2,\cdots,M\}$ 和 $E=\{M+1,M+2,\cdots,N\}$ 分别表示跟随者集和领导者集，即 $\mathcal{N}=E\cup F$。加权邻接矩阵 $\tilde{\boldsymbol{A}}$ 满足：① $a_{ij}>0\Leftrightarrow(j,i)\in\mathcal{E}_{\mathrm{cm}}$。如果 $j\in E$，那么 $a_{ij}\overset{\mathrm{def}}{=}b_j>0$，即节点 i 可以收到领导者 j 的信息。② $a_{ij}=0\Leftrightarrow(j,i)\notin\mathcal{E}_{\mathrm{cm}}$ 或者 i=j。拉普拉斯矩阵 $\boldsymbol{L}=\left(\ell_{ij}\right)_{N\times N}$，定义为：① $\ell_{ij}=-a_{ij},i\neq j$；② $\ell_{ii}=-\sum_{j=1,j\neq i}^{N}\ell_{ij}$，$i,j\in\mathcal{N}$，即 $\sum_{j=1}^{N}\ell_{ij}=0$。于是有 $\boldsymbol{L}=\begin{bmatrix}\boldsymbol{L}_1 & \boldsymbol{L}_2\\ 0 & 0\end{bmatrix}_{N\times N}$，其中，$\boldsymbol{L}_1\in\mathbb{R}^{M\times M}$ 表示跟随者之间的通信；$\boldsymbol{L}_2\in\mathbb{R}^{M\times(N-M)}$ 表示信息从领导者到跟随者的传输。为了符号简便，定义：① $Z_{l_{ij}}$ 为边 l_{ij} 的内邻居，即节点 i 是个信息中转站。集合 $Z_{l_{ij}}$ 中边的信息流方向跟边 l_{ij} 的信息流方向一致，其中 $Z_{l_{ij}}=\{(k,i)|(k,i)\in\mathcal{E}_{\mathrm{cm}},k\in\mathcal{N}\}$ 以及 $\overline{Z}_{l_{ij}}=Z_{l_{ij}}\cup l_{ij}$。② $U_{l_{ij}}$ 为 l_{ij} 的外邻居，即节点 j 是个信息中转站。集合 $Z_{l_{ij}}$ 中边的信息流方向跟边 l_{ij} 的信息流方向一致，其中 $U_{l_{ij}}=\{(j,k)|(j,k)\in\mathcal{E}_{\mathrm{cm}},k\in\mathcal{N}\}$ 以及 $\overline{U}_{l_{ij}}=U_{l_{ij}}\cup l_{ij}$。③ $Z_j=\{i\in\mathcal{N}|(i,j)\in\mathcal{E}_{\mathrm{cm}}\}$ 是节点 j 的内邻居，即节点 j 可以收到 Z_j 中的节点的信息传播，其中 $\overline{Z}_j=\{j\}\cup Z_j$。类似地，定义 U_j 为节点 j 的外邻居。定义 $|\mathcal{E}_{\mathrm{cm}}|$、$|U_{l_{ij}}|$、$|Z_j|$、$|U_j|$ 和 $|\overline{Z}_j|$ 分别为集合 $\mathcal{E}_{\mathrm{cm}}$、$U_{l_{ij}}$、$Z_j$、$U_j$ 和 $\overline{Z}_j$ 的基数。

10.2.1 多智能体系统在攻击下的动力学模型

本节研究了具有 N 个多智能体的时变编队包围控制，其中跟随者组成的编队随着领导者组成的凸多边形进行移动。假设在采

样、传输、更新和广播之间没有延迟。令 $\left\{t_r^i\right\}_{r\in\mathbb{Z}_{\geqslant 0}}$ 表示智能体 i 的一系列采样时刻，而 $\left\{t_r^{ij}\right\}_{r\in\mathbb{Z}_{\geqslant 0}}$ 表示边 (i,j) 的一系列通信时刻，满足 $0\leqslant t_0^i<\cdots<t_r^i<\cdots$ 和 $0\leqslant t_0^{ij}<\cdots<t_r^{ij}<\cdots$，其中 $(i,j)\in\mathcal{E}_{\mathrm{cm}}$。更具体地，对任意 i，$j\in U_i$，智能体 i 的采样时刻 t_r^i 是由一系列时刻 t_r^{ij} 组成的，等价地，$t_r^i\in\left\{t_r^{i1},t_r^{i2},\cdots,t_r^{i|U_i|}\right\}$（参见图 10-1）。在图 10-1 中，$U_i=\left\{j,j'\right\},t_r^i\in\left\{t_r^{i1},t_r^{i2},\cdots,t_r^{i|U_i|}\right\},\left|U_i\right|=2$。欺诈攻击的开始标识为↑，而↓为结束标识。在 $t\in[0,10]$ 期间：$\tau_0^{ij}=2,\tau_1^{ij}=1.5,\tau_2^{ij}=1.5,\tau_0^{ij'}=2$ 和 $\tau_1^{ij'}=1.3$。在智能体 i 的每一个传输时刻 t_r^{ji} [即边（i,j）被触发，智能体 i 将广播数据给外邻居 $j\in U_i$]，①智能体 i 需要采样它自己的信息，并基于此采样值 $\hat{x}_i$ 和最新传播值 $\hat{x}_{\bar{Z}_i}$ 来更新控制器 u_i，同时将信息广播给外邻居 j；②外邻居 j 获取 i 在时刻 t_r^{ij} 广播过来的数据，立刻保存该数据并在 j 的更新时刻到来时使用该数据 (j 的更新时刻为 $t_r^{ji'},i'\in U_j$)。在这个更新和数据传输过程中，可以看作是在 t_r^{ij} 和 $t_r^{ji'}$ 之间有个时间缓冲，即从时刻 t_r^{ij}（i 在此刻广播当前采样数据给 j）到智能体 j 紧接着的下一个更新时刻 $t_r^{ji'}$，其中 $j\in U_i$ 和 $i'\in U_j$。时刻 t_r^{ij} 是智能体 i 的采样时刻，此时智能体 i 广播数据给外邻居 $j\in U_i$；而 $t_r^{ji'}$ 是智能体 j 的更新时刻，智能体 j 在此刻广播数据给外邻居 $i'\in U_j$。如果 $i\notin U_j$，那么不存在更新时刻 t_r^{ji}，因此也就不存

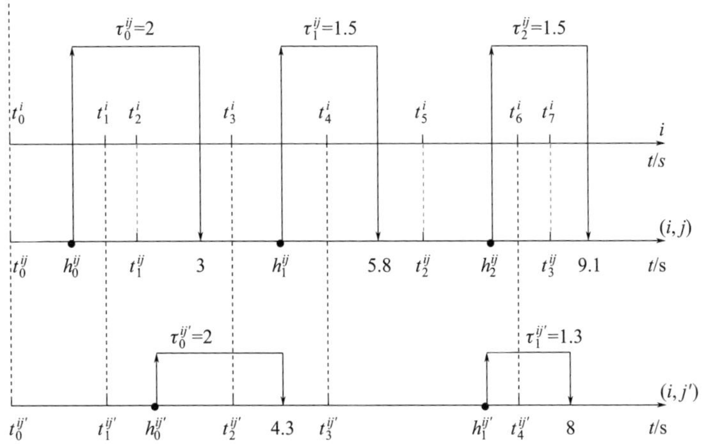

图 10-1　在欺诈攻击下，智能体 i 和边 (i,j) 以及 (i,j') 的更新时刻的例子

在t_r^{ji}与t_r^{ji}相等与否。因此，智能体i（包括领导者和追随者）的状态方程可以建模为

$$\begin{aligned}&\dot{\boldsymbol{x}}_d(t)=\tilde{\boldsymbol{f}}_d\left(\boldsymbol{x}_d(t),\boldsymbol{\omega}_d(t)\right), \qquad d\in E\\&\dot{\boldsymbol{x}}_i(t)=\boldsymbol{f}_i\left(\boldsymbol{x}_i(t),\boldsymbol{u}_i(t),\boldsymbol{\omega}_i(t)\right),\ i\in F,t\in\left[t_r^i,t_{r+1}^i\right]\\&\boldsymbol{u}_i(t)=\boldsymbol{g}_i\left(\hat{\boldsymbol{x}}_{\bar{Z}_i}(t)\right),\boldsymbol{x}_i(t_0)=\boldsymbol{x}_{i0},i\in F,t\in\left[t_r^i,t_{r+1}^i\right]\end{aligned} \tag{10-1}$$

其中，$\boldsymbol{x}_d(t)\in\mathbb{R}^n$和$\boldsymbol{x}_i(t)\in\mathbb{R}^n$分别代表领导者$d$和跟随者$i$的状态；$\boldsymbol{u}_i(t)\in\mathbb{R}^m$表示跟随者$i$的控制输入；$\boldsymbol{\omega}_d(t)$、$\boldsymbol{\omega}_i(t)\in\mathbb{R}^{m'}$表示$\mathcal{L}_\infty$空间中的干扰输入。局部利普希茨函数$\tilde{\boldsymbol{f}}_d,\boldsymbol{f}_i\in\mathbb{R}^n$满足$\tilde{\boldsymbol{f}}_d(0,0,0)=0$以及$\boldsymbol{f}_i(0,0,0)=0$，函数$\boldsymbol{g}_i\in\mathbb{R}^m$满足$\boldsymbol{g}_i(0)=0$。另外，$\hat{\boldsymbol{x}}_{\bar{Z}_i}(t)=\{\hat{\boldsymbol{x}}_i(t)\}\cup\hat{\boldsymbol{x}}_{Z_i}$，其中$\hat{\boldsymbol{x}}_{Z_i}=\left\{\hat{x}_{i_1},\cdots,\hat{x}_{i_{|Z_i|}}\right\}$表示$Z_i$中的智能体的最新传播值。如果智能体$i'$到跟随者$i$存在有向路（即$i$可以收到$i'$的信息），那么假设$\boldsymbol{u}_i$与$\hat{\boldsymbol{x}}_{\bar{Z}_i}$相关，其中$i'\in\bar{Z}_i$。因此，对任意智能体$i\in\mathcal{N},\boldsymbol{u}_i(t)$跟$\hat{\boldsymbol{x}}_i(t)$相关，其中$\hat{\boldsymbol{x}}_i(t)$在传输期间$t\in\left[t_r^i,t_{r+1}^i\right]$内靠零阶保持器保持常数，即$\hat{\boldsymbol{x}}_i(t)=\boldsymbol{x}_i\left(t_r^i\right)$。在智能体$i$的下一个更新时刻$t_{r+1}^i,\hat{\boldsymbol{x}}_i\left(t_r^i\right)$将被重置为$\boldsymbol{x}_i\left(t_{r+1}^i\right)$，即$\hat{\boldsymbol{x}}_i^+\left(t_{r+1}^i\right)=\boldsymbol{x}_i\left(t_{r+1}^i\right)$，与此同时，当智能体$j$的更新时刻到来时，智能体$j\in U_i\setminus E$会根据$i$的广播值更新控制器$\boldsymbol{u}_j$。总的来说，就是在每个更新时刻$t_{r+1}^i$，控制器$\boldsymbol{u}_i(t)$会根据当前采样值$\boldsymbol{x}_i\left(t_{r+1}^i\right)$和最新传输值$\hat{\boldsymbol{x}}_{Z_i}$进行更新，而且保持$\boldsymbol{u}_i\left(t_{r+1}^i\right)$直到下一个更新时刻的到来。在智能体$i$和$j\left(j\in U_i\right)$的信息传输过程中，真实的传输值有可能被恶意攻击者用错误的数据替代，即欺诈攻击，由此图10-2表明：当欺诈攻击存在的时候（开关s_{i1}被拨动），那么这些信息就会被截获和篡改，即

$$\hat{\boldsymbol{x}}_i^+\left(t_r^{ij}\right)=\boldsymbol{x}_i\left(t_r^{ij}\right)+\mathcal{Q}_{i_j}\left(\boldsymbol{x}_i\left(t_r^{ij}\right)\right) \tag{10-2}$$

其中，非零欺诈函数$\mathcal{Q}_{i_j}(\cdot)$会在后文中给出进一步解释；当没有攻击的时候（开关s_{i2}被拨动），那么数据可以经过网络安全传输，即

$$\hat{\boldsymbol{x}}_i^+\left(t_r^{ij}\right)=\boldsymbol{x}_i\left(t_r^{ij}\right) \tag{10-3}$$

接下来我们首先给出关于非线性多智能体 [式 (10-1)] 时变编队跟踪的定义。

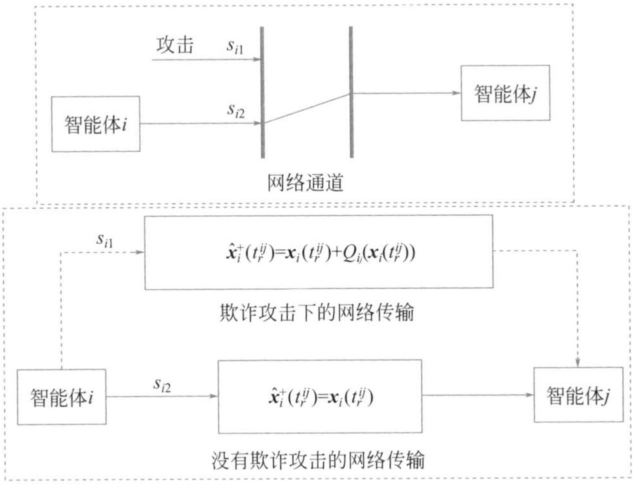

图 10-2　相邻智能体之间的通信，其中两个切换开关 s_{i1} 和 s_{i2} 分别表示有或没有欺诈攻击的情况

定义 10.1

非线性多智能体 [式 (10-1)] 实现了时变编队跟踪，如果对任意给定的初始状态，存在正常数 α_d 满足 $\sum\limits_{d\in E}\alpha_d=1$，使得

$$\lim_{t\to\infty}\|\boldsymbol{x}_i(t)-\boldsymbol{r}_i(t)-\sum_{d\in E}\alpha_d\boldsymbol{x}_d\|=0,\quad i\in\mathcal{N} \tag{10-4}$$

其中，分段连续函数$\boldsymbol{r}_i(t)$表示智能体i的当前状态和目标状态的相对距离。

在进一步分析之前，对有向图 G 的拓扑结构做以下限制[269]。

条件 10.2.1：至少有一个领导者有一条通向每一个追随者的有向路径。

对任意$i\in\mathcal{N}$，定义$\boldsymbol{\theta}_i(t)=\boldsymbol{x}_i(t)-\boldsymbol{r}_i(t)$和$\boldsymbol{\mathcal{E}}_f=\boldsymbol{\theta}_f+\tilde{\boldsymbol{L}}\tilde{\boldsymbol{x}},\boldsymbol{\mathcal{E}}_f\in\mathbb{R}^{Nn\times1}$，其中$\boldsymbol{\theta}_f=(\theta_1,\cdots,\theta_N)\in\mathbb{R}^{Nn\times1},\tilde{\boldsymbol{x}}=(\overline{x},\overline{x})\in\mathbb{R}^{2(N-M)n\times1},\overline{\boldsymbol{x}}=(\boldsymbol{x}_{M+1},\boldsymbol{x}_{M+2},\cdots,\boldsymbol{x}_N)\in\mathbb{R}^{Mn\times1}$，以及

$$\tilde{\boldsymbol{L}}=\begin{bmatrix}\left(\boldsymbol{L}_1^{-1}\boldsymbol{L}_2\otimes\boldsymbol{I}_n\right)_{(N-M)n\times Mn} & 0_{(N-M)n\times Mn}\\ 0_{Mn\times Mn} & -\boldsymbol{I}_{Mn\times Mn}\end{bmatrix}$$

由此可知，$\boldsymbol{L}_1$是可逆的（当且仅当条件10.2.1成立[269]）。为实现系统 (10-1) 的时变编队，由定义10.1可知，我们只需保证$\boldsymbol{\mathcal{E}}_f\to0$，其中$-\boldsymbol{L}_1^{-1}\boldsymbol{L}_2\otimes\boldsymbol{I}_n$的每一行和为1。基于$\boldsymbol{\varepsilon}_f=\boldsymbol{\theta}_f+\tilde{\boldsymbol{L}}\tilde{\boldsymbol{x}}$，可以将$\boldsymbol{\mathcal{E}}_f$表示为$\boldsymbol{\mathcal{E}}_f=(\boldsymbol{\xi}_1,\cdots,\boldsymbol{\xi}_i,\cdots,\boldsymbol{\xi}_N)$，其中$\boldsymbol{\xi}_i=\boldsymbol{x}_i-\boldsymbol{r}_i-\sum\limits_{d\in E}\alpha_d\boldsymbol{x}_d,i\in\mathcal{N}$，而$-\sum\limits_{d\in E}\alpha_d$是矩阵

$\tilde{\boldsymbol{L}}$的每一行的和。由此可知，对任意$i\in F$，有$\boldsymbol{\xi}_i=\boldsymbol{x}_i-\boldsymbol{r}_i-\sum\limits_{d\in E}\alpha_d\boldsymbol{x}_d$；而对任意$i\in E$，有$\boldsymbol{\xi}_i=\boldsymbol{x}_i-\boldsymbol{r}_i-\boldsymbol{x}_i=0$，其中$\boldsymbol{r}_i$=0。定义$\boldsymbol{\xi}_{ij}=\left(\boldsymbol{\xi}_i,\boldsymbol{\xi}_j\right)\in\mathbb{R}^{2n}$，其中，$(i,j)\in\mathcal{E}_{\text{cm}}$，以及$\boldsymbol{e}_{ij}(t)=\hat{\boldsymbol{\xi}}_{ij}\left(t_r^{ij}\right)-\boldsymbol{\xi}_{ij}(t)$，其中，$t\in\left[t_r^{ij},t_{r+1}^{ij}\right]$，及其相对输出为$z_{ij}\left(\boldsymbol{\xi}_{ij}\right)=\mathcal{O}_{ij}\left(\boldsymbol{\xi}_{ij}\right)$。基于系统(10-1)下边的分析可知，$\hat{\boldsymbol{\xi}}_{ij}^+\left(t_r^{ij}\right)=\left(\hat{\boldsymbol{\xi}}_i^+\left(t_r^{ij}\right),\hat{\boldsymbol{\xi}}_j^+\left(t_r^{ij}\right)\right)$，其中$\hat{\boldsymbol{\xi}}_i^+\left(t_r^{ij}\right)=\boldsymbol{\xi}_i\left(t_r^{ij}\right)$。值得注意的是，$\hat{\boldsymbol{\xi}}_j(t)$也许不会在$t\in\left[t_r^{ij},t_{r+1}^{ij}\right)$内保持不变，因为$j$的更新时刻$t_r^{ji}$也许会发生在期间$\left[t_r^{ij},t_{r+1}^{ij}\right)$内，即$\hat{\boldsymbol{\xi}}_j^+(t)=\hat{\boldsymbol{\xi}}_j(t)$（前提是智能体$j$在$\left[t_r^{ij},t_{r+1}^{ij}\right)$内没更新）或者$\hat{\boldsymbol{\xi}}_j^+(t)\neq\hat{\boldsymbol{\xi}}_j(t)$（前提是智能体$j$也在时刻$t_r^{ij}$更新)。本章节中$\boldsymbol{\xi}_i$和$\boldsymbol{\xi}_j$都在包含在$e_{ij}$中，这不会影响以下分析（参见图10-1）。图10-2表示相邻智能体之间的通信，其中两个切换开关s_{i1}和s_{i2}分别表示有或没有欺诈攻击的情况。基于图10-2，以及式(10-2)和式(10-3)可知：在开关s_{i2}处，有

$$\begin{aligned}\boldsymbol{e}_{ij}^+\left(t_r^{ij}\right)&=\left(\hat{\boldsymbol{\xi}}_{ij}^+\left(t_r^{ij}\right)-\boldsymbol{\xi}_{ij}\left(t_r^{ij}\right)\right)\\&=\left(\hat{\boldsymbol{\xi}}_i^+\left(t_r^{ij}\right)-\boldsymbol{\xi}_i\left(t_r^{ij}\right),\hat{\boldsymbol{\xi}}_j^+\left(t_r^{ij}\right)-\boldsymbol{\xi}_j\left(t_r^{ij}\right)\right)\\&=\left(0,\hat{\boldsymbol{\xi}}_j^+\left(t_r^{ij}\right)-\boldsymbol{\xi}_j\left(t_r^{ij}\right)\right)\overset{\text{def}}{=}\left(0,0_j\right)\end{aligned}\tag{10-5}$$

其中，如果智能体j在t_r^{ij}时刻也更新，则$\left(0,0_j\right)=(0,0)$；否则$\left(0,0_j\right)\neq(0,0)$。在开关$s_{i1}$处，有$\hat{\boldsymbol{\xi}}_{ij}^+=\boldsymbol{\xi}_{ij}+\mathcal{Q}_{i_{j0}}\left(\boldsymbol{\xi}_{ij}\right)$，其中$\mathcal{Q}_{i_{j0}}(\cdot)=\left(\mathcal{Q}_{i_j}(\cdot),0\right)$。因此，有

$$\begin{aligned}\boldsymbol{e}_{ij}^+\left(t_r^{ij}\right)&=\left(\hat{\boldsymbol{\xi}}_{ij}^+\left(t_r^{ij}\right)-\boldsymbol{\xi}_{ij}\left(t_r^{ij}\right)\right)\\&=\left(\hat{\boldsymbol{\xi}}_i^+\left(t_r^{ij}\right)-\boldsymbol{\xi}_i\left(t_r^{ij}\right),\hat{\boldsymbol{\xi}}_j^+\left(t_r^{ij}\right)-\boldsymbol{\xi}_j\left(t_r^{ij}\right)\right)\\&=\left(\mathcal{Q}_{i_j}\left(\boldsymbol{\xi}_i\left(t_r^{ij}\right)\right),0_j\right)\overset{\text{def}}{=}\tilde{\mathcal{Q}}_{ij}\left(\boldsymbol{e}_{ij}\left(t_r^{ij}\right)\right)\neq\left(0,0_j\right)\end{aligned}\tag{10-6}$$

基于式(10-5)和式(10-6)，有

$$\boldsymbol{e}_{ij}^+=\begin{cases}\boldsymbol{0}_{i_j}, & \text{没有攻击}\\\tilde{\mathcal{Q}}_{ij}\left(\boldsymbol{e}_{ij}\right), & \text{有攻击}\end{cases}\tag{10-7}$$

其中，$\boldsymbol{0}_{i_j}\overset{\text{def}}{=}\left(0_i,0_j\right)$和$0_i=0,i_j\in\mathcal{E}_{\text{cm}}$，非零欺诈函数$\tilde{\mathcal{Q}}_{ij}(\cdot)$会在后文中分析。

对于有向边$(i,j)\in\mathcal{E}_{\text{cm}}$，定义一系列时间间隔$\left\{H_n^{ij}\right\}_{n\in\mathbb{Z}_{\geqslant0}}$表示欺诈攻击发生的区间。考虑到这些攻击也可能是一个单脉冲序列，我们定义

$H_n^{ij}=\left\{h_n^{ij}\right\}\bigcup\left[h_n^{ij},h_n^{ij}+\tau_n^{ij}\right)$，其中$\tau_n^{ij}\in\mathbb{R}_{\geqslant 0}$，$n\in\mathbb{Z}_{\geqslant 0}$，也就是说边$(i,j)$在区间$\left[h_n^{ij},h_n^{ij}+\tau_n^{ij}\right)$内被持续攻击且攻击长度为$\tau_n^{ij}$，始于时刻$h_n^{ij}$，结束于时刻$h_n^{ij}+\tau_n^{ij}$（参考图 10-1）。假设$h_n^{ij}\leqslant h_n^{ij}+\tau_n^{ij}<h_{n+1}^{ij}$，即任意两个连续的欺诈攻击不会重叠。对于任意区间$[T_1,T_2]$，其中$0\leqslant T_1\leqslant T_2$，定义$\Xi^{ij}(T_1,T_2)$为边$(i,j)$受到的攻击事件的集合，即$\Xi^{ij}(T_1,T_2)=\left(\bigcup_{n\in\mathbb{Z}_{\geqslant 0}}H_n^{ij}\right)\bigcap[T_1,T_2]$；相应地，$\Theta^{ij}(T_1,T_2)$表示没有攻击的时间集，即$\Theta^{ij}(T_1,T_2)=[T_1,T_2]\backslash\Xi^{ij}(T_1,T_2)$。令$n^{ij}(T_1,T_2)$表示边 (i,j) 在区间 (T_1,T_2) 内发生的欺诈攻击的次数，参见图 10-1，关于$\Xi^{ij}(T_1,T_2)$和$n^{ij}(T_1,T_2)$的例子。

假设 10.1

对任意$(i,j)\in\mathcal{E}_{cm}$和所有的$T_1,T_2\in\mathbb{R}_{\geqslant 0}$（$T_1\geqslant T_2$）。

(A1) 攻击频率 / 时长：存在常数$v^{ij}\in\mathbb{R}_{\geqslant 0},\tau_f^{ij}\in\mathbb{R}_{>\varDelta_{ij}},\varsigma^{ij}\in\mathbb{R}_{\geqslant 0}$以及$\tau_d^{ij}\in\mathbb{R}_{>1}$满足

$$n^{ij}(T_1,T_2)\leqslant v^{ij}+\frac{T_2-T_1}{\tau_f^{ij}},\left|\Xi^{ij}(T_1,T_2)\right|\leqslant\varsigma^{ij}+\frac{T_2-T_1}{\tau_d^{ij}} \tag{10-8}$$

(A2) 攻击幅度：存在常数$\chi_{ij}\in\mathbb{R}_{\geqslant 0}$，使得对任意$e_{ij},e_{ij}^{\mathrm{T}}\in\mathbb{R}$，有

$$\left|\tilde{\mathcal{Q}}_{ij}(e_{ij})-\tilde{\mathcal{Q}}_{ij}(e_{ij}^{\mathrm{T}})\right|\leqslant\chi_{ij}\left|e_{ij}-e_{ij}^{\mathrm{T}}\right|. \tag{10-9}$$

注释 10.1

类似 Feng 和 Tesi[236] 的研究，假设 10.1(A1) 中的τ_f^{ij}可以看作是两次连续欺诈攻击的平均驻留时间，而v^{ij}表示振动界，具体地说，$n^{ij}(T_1,T_2)=1$且$T_1=T_2$，那么参数v^{ij}保证了假设 10.1(A1) 的第一个不等式的合理性和可行性；为了保证假设 10.1(A1) 中第二个不等式的可行性，参数ς^{ij}在调节欺诈攻击的时间长度方面起着类似的作用，比如，如果$\left|\Xi^{ij}\left(h_n^{ij},h_n^{ij}+\tau_n^{ij}\right)\right|=\tau_n^{ij}$，那么$\tau_n^{ij}>\tau_n^{ij}/\tau_d^{ij}$，因为$\tau_d^{ij}>1$。对于假设 10.1(A2)，如果欺诈攻击是一系列脉冲点，那么式 (10-9) 退化为$\left|\tilde{\mathcal{Q}}_{ij}(\boldsymbol{e}_{ij})\right|\leqslant\chi_{ij}$，而如果不存在欺诈攻击，则$\chi_{ij}=0$。

下面介绍每条边$(i,j)\in\mathcal{E}_{cm}$的事件触发通信机制，即

$$t_{r+1}^{ij}=\inf\left\{t:t\geqslant t_r^{ij}+\tau_{\text{miet}}^{ij}\mid k\in\mathbb{Z}_{\geqslant 0},\eta_{ij}(t)<0\right\} \tag{10-10}$$

其中，$\underline{\Delta}_{ij} \geqslant \tau_{\text{miet}}^{ij}$，参数$\underline{\Delta}_{ij}$满足假设 10.1，$\eta_{ij}(t)$为有向边$(i,j)$的事件驱动函数，$\tau_{\text{miet}}^{ij}$为 (i,j) 的最小传输间隔，后文会给出具体分析。事件驱动机制 (10-10) 表明，一旦$\eta_{ij}(t)<0$成立，i 到 j 进行有向信息传输。有向边(i,j)的事件驱动函数 $\eta_{ij}(t)$ 的动态满足为

$$\begin{cases}\dot{\eta}_{ij}(t)=\varPsi_{ij}\left(\boldsymbol{o}_{ij}(t),\eta_{ij}(t)\right),\ t\in\left[t_r^{ij},t_{r+1}^{ij}\right)\\ \eta_{ij}^{+}\left(t_r^{ij}\right)=\tilde{\eta}_{ij}\left(t_r^{ij}\right)\end{cases} \tag{10-11}$$

其中，$\boldsymbol{o}_{ij}=\left(\boldsymbol{e}_{ij},\tau_{ij},\phi_{ij}\right)\in\mathbb{R}^{2n}\times\mathbb{R}_{\geqslant 0}\times\left[\delta_{ij},\delta_{ij}^{-1}\right]$表示部分信息变量，其中$\delta_{ij}\in(0,1)$。辅助时间变量$\tau_{ij}\in\mathbb{R}_{\geqslant 0}$和辅助变量$\phi_{ij}\in\left[\delta_{ij},\delta_{ij}^{-1}\right]$以及函数$\varPsi_{ij}$和$\tilde{\eta}_{ij}$会在下文中逐步给出具体分析。

注释 10.2　如上所示，每条边都配备一个事件，即所谓的分布式事件触发，式 (10-10) 意味着事件只会在经过最近一次传输之后的特定时间（即最小触发间隔）之后，当 η_{ij}=0 时才触发。本节的正参数最小触发间隔与 Ding 等人 [268]、Ge 和 Han[270] 研究中的采样区间，以及 Ge 等人 [271] 研究中的补偿发挥了类似的作用。此处基于边的事件触发机制不同于 Ding 等人 [268,272] 的研究，以及 Wang 和 Lemmon[264] 的研究中基于点的触发机制，本节的触发函数η_{ij}可以不依赖某个特定的形式。对于基于节点的事件驱动，Ding 等人 [272] 做了总结分析，其中最好的一种情形是每一个节点的触发时刻不依赖于其他节点的触发时刻 [273]，因为这样有利于减少整体的触发次数，而在 Ding 等人 [268] 的研究中，触发策略只在某些非连续采样时刻使用邻居的数据，而不是在所有连续采样时刻使用邻居的数据，这样就能在保证系统稳定的情况下减少邻居之间的通信，以节省通信资源。Hu 等人 [273] 和 Ding 等人 [268] 研究中使用的策略，都可以用于解决在 Ding 等人 [272] 研究中所讨论的关于智能体有较多邻居而导致触发频率高的问题。本章节基于边的触发策略也可以解决类似的触发频率较高的问题，也就是说，章节 10.2 中描述的节点更新和通信过程也具有 Hu 等

人[273]研究中的优势。此外，式 (10-10) 中的最小触发间隔使得本章节中基于边的触发策略也具有在 Ding 等人[268]研究中的优势。正如 Ding 等人在文献 [272] 的 Table I 中分析的那样，在某些基于节点的事件触发策略中添加一个正常数来排除奇诺现象（不包含 Wang 和 Lemmon[264] 的研究）；在没有类似的正常数的情况下（如 Wang 和 Lemmon[264] 的研究），系统的收敛性能更好，因为采样频率比正常数存在的情况下更高；而本节中边事件驱动策略中的正常数也可以达到类似的效果，即可以更有效地排除奇诺现象。如 Ge 等人[271]研究中分析的那样，在节省通信资源（即低采样频率）和尽快达到预期收敛性能之间有一个权衡。综上所述，基于节点的事件驱动和基于边的事件驱动策略各有优点，在今后的研究中，我们也会设计基于节点的事件触发策略，来处理具有切换拓扑[274]或者非完整链式动态[275]的多智能体编队[276]，非完整移动机器人的集体行为[270,275,277]，随机系统的$\mathcal{H}_\infty$滤波问题[262]，以及文献 [265,271-272,278-279] 中所描述的其他问题。

接下来首先建立基于边的混杂事件驱动多智能体 [式 (10-1)] 在欺诈攻击下的脉冲模型，即

$$
\begin{cases}
\dot{\boldsymbol{\xi}}_{ij} = F_{ij}\left(\boldsymbol{\xi}_{\bar{z}_{l_{ij}}}, \boldsymbol{e}_{\bar{z}_{l_{ij}}}, \boldsymbol{\omega}_{ij}\right), & t \in \left[t_r^{ij}, t_{r+1}^{ij}\right] \\
\dot{\boldsymbol{e}}_{ij} = G_{ij}\left(\boldsymbol{\xi}_{\bar{z}_{l_{ij}}}, \boldsymbol{e}_{\bar{z}_{l_{ij}}}, \boldsymbol{\omega}_{ij}\right), & t \in \left[t_r^{ij}, t_{r+1}^{ij}\right] \\
\dot{\eta}_{ij} = \Psi_{ij}, & t \in \left[t_r^{ij}, t_{r+1}^{ij}\right] \\
\dot{\phi}_{ij} = \Phi_{ij}, & t \in \left[t_r^{ij}, t_{r+1}^{ij}\right]
\end{cases}
$$

$$
\begin{cases}
\boldsymbol{\xi}_{ij}^{+} = \boldsymbol{\xi}_{ij}, & t_r^{ij} \in \Theta^{ij} \\
\boldsymbol{e}_{ij}^{+} = 0_{ij}, & t_r^{ij} \in \Theta^{ij} \\
\eta_{ij}^{+} = \tilde{\eta}_{ij}, & t_r^{ij} \in \Theta^{ij} \\
\phi_{ij}^{+} = \delta_{ij}^{-1}, & t_r^{ij} \in \Theta^{ij}
\end{cases}
\quad
\begin{cases}
\boldsymbol{\xi}_{ij}^{+} = \boldsymbol{\xi}_{ij}, & t_r^{ij} \in \Xi^{ij} \\
\boldsymbol{e}_{ij}^{+} = \tilde{\mathcal{Q}}_{ij}\left(\boldsymbol{e}_{ij}\right), & t_r^{ij} \in \Xi^{ij} \\
\eta_{ij}^{+} = \tilde{\eta}_{ij}, & t_r^{ij} \in \Xi^{ij} \\
\phi_{ij}^{+} = \delta_{ij}^{-1}, & t_r^{ij} \in \Xi^{ij}
\end{cases}
\tag{10-12}
$$

其中，$\boldsymbol{\omega}_{ij} = \left(\omega_i, \omega_j\right) \in \mathbb{R}^{2m'}$，$\tilde{\mathcal{Q}}_{ij}\left(e_{ij}\right)$表示欺诈函数。连续函数$F_{ij}$和$G_{ij}$可

以由式(10-1) 得到，变量Ψ_{ij},Φ_{ij}，$\tilde{\eta}_{ij}$以及δ_{ij}^{-1}会在10.2.2节中给出详细介绍。

接下来，令τ_{ij}表示$l_{ij}\in\mathcal{E}_{\text{cm}}$的自上一次最新传输尝试所经过的时间。类似地，对于边$l_{ij}\in\mathcal{E}_{\text{cm}}$，时间变量$s_{ij}$表示整体时间。定义$\boldsymbol{q}=\left(\boldsymbol{q}_{ij}\right)_{(i,j)\in\mathcal{E}_{\text{cm}}}=\left(\boldsymbol{\xi},\boldsymbol{e},\eta,\phi,\tau,s\right)\in\mathbb{X}$，其中$\boldsymbol{q}_{ij}=\left(\boldsymbol{\xi}_{ij},\boldsymbol{e}_{ij},\eta_{ij},\phi_{ij},\tau_{ij},s_{ij}\right),\mathbb{X}=\left(\mathbb{R}^{2n}\right)^{|\mathcal{E}_{\text{cm}}|}\times\left(\mathbb{R}^{2n}\right)^{|\mathcal{E}_{\text{cm}}|}\times\mathbb{R}_{\geqslant 0}^{|\mathcal{E}_{\text{cm}}|}\times\left[\delta_{ij},\delta_{ij}^{-1}\right]^{|\mathcal{E}_{\text{cm}}|}\times\mathbb{R}_{\geqslant 0}^{|\mathcal{E}_{\text{cm}}|}\times\mathbb{R}_{\geqslant 0}^{|\mathcal{E}_{\text{cm}}|}$，$\boldsymbol{\xi}=\left(\boldsymbol{\xi}_{ij}\right)_{(i,j)\in\mathcal{E}_{\text{cm}}},\boldsymbol{e}=\left(\boldsymbol{e}_{ij}\right)_{(i,j)\in\mathcal{E}_{\text{cm}}},\boldsymbol{\eta}=\left(\eta_{ij}\right)_{(i,j)\in\mathcal{E}_{\text{cm}}}$，$\boldsymbol{\phi}=\left(\phi_{ij}\right)_{(i,j)\in\mathcal{E}_{\text{cm}}},\boldsymbol{\tau}=\left(\tau_{ij}\right)_{(i,j)\in\mathcal{E}_{\text{cm}}},\boldsymbol{s}=\left(s_{ij}\right)_{(i,j)\in\mathcal{E}_{\text{cm}}}$。利用 De 和 Postoyan [226]，以及 Goebel 等人 [230] 研究中的混杂建模技术，可以建立脉冲系统 (10-12) 在欺诈攻击下基于事件的混杂形式，即

$$\mathcal{H}_{\text{dept}}:\begin{cases}\dot{\boldsymbol{q}}\in\boldsymbol{\mathcal{F}}\left(\boldsymbol{q},\boldsymbol{\omega}\right), & q\in C\\ \boldsymbol{q}^{+}\in\mathcal{G}_{\text{dept}}\left(\boldsymbol{\Omega}\right), & q\in D\end{cases}\tag{10-13}$$

其中，$\boldsymbol{\Omega}=\left(\boldsymbol{\xi},\boldsymbol{e},\boldsymbol{\eta},\boldsymbol{\phi},\boldsymbol{\tau},\boldsymbol{s}\right),C=\left\{\boldsymbol{q}\in\mathbb{X}|\ \tau_{ij}\leqslant\tau_{\text{miet}}^{ij}\text{或}\eta_{ij}\geqslant 0,\forall(i,j)\in\mathcal{E}_{\text{cm}}\right\}$，$D=\left\{\boldsymbol{q}\in\mathbb{X}|\ \tau_{ij}\geqslant\tau_{\text{miet}}^{ij}\ \text{且}\ \eta_{ij}\leqslant 0,\exists(i,j)\in\mathcal{E}_{\text{cm}}\right\}$，集值映射$\mathcal{F}\left(\boldsymbol{q},\boldsymbol{\omega}\right)=\left(\mathcal{F}_{ij}\left(q_{ij},\omega_{ij}\right)\right)_{(i,j)\in\mathcal{E}_{\text{cm}}}$和$\mathcal{G}_{\text{dept}}\left(\boldsymbol{\Omega}\right)$为

$$\begin{aligned}\mathcal{F}\left(\boldsymbol{q},\boldsymbol{\omega}\right)=&\Big(\left(F_{ij}\right)_{(i,j)\in\mathcal{E}_{\text{cm}}},\left(G_{ij}\right)_{(i,j)\in\mathcal{E}_{\text{cm}}},\left(\Psi_{ij}\right)_{(i,j)\in\mathcal{E}_{\text{cm}}}\\&\left(\Phi_{ij}\right)_{(i,j)\in\mathcal{E}_{\text{cm}}},\left(1_{ij}\right)_{(i,j)\in\mathcal{E}_{\text{cm}}},\left(1_{ij}\right)_{(i,j)\in\mathcal{E}_{\text{cm}}}\Big)\end{aligned}$$

$$\mathcal{G}_{\text{dept}}\left(\boldsymbol{\Omega}\right)=\left\{\mathcal{G}_{l_{ij}}\mid l_{ij}\in\mathcal{E}_{\text{cm}},\eta_{l_{ij}}<0\right\}$$

$$\begin{aligned}\mathcal{G}_{l_{ij}}=&\left(\left(\boldsymbol{\xi}_{l_{ij}}\right)_{l_{ij}\in\mathcal{E}_{\text{cm}}},0_1,\cdots,\frac{\chi_{ij}}{\chi_{ij}}\tilde{\mathcal{Q}}_{l_{ij}}\left(\boldsymbol{e}_{l_{ij}}\right)+\left(1-\frac{\chi_{ij}}{\chi_{ij}}\right)\boldsymbol{0}_{i_j},\cdots,\boldsymbol{0}_{\Lambda_{l_{ij}}},\eta_1,\cdots\right.\\&\eta_{l_{ij}-1},\tilde{\eta}_{l_{ij}},\eta_{l_{ij}+1},\cdots,\eta_{\Lambda_{l_{ij}}},\phi_1,\cdots,\phi_{l_{ij}-1},\frac{1}{\delta_{l_{ij}}},\phi_{l_{ij}+1},\cdots,\phi_{\Lambda_{l_{ij}}}\\&\left.\tau_1,\cdots,\tau_{l_{ij}-1},0,\tau_{l_{ij}+1},\cdots,\tau_{\Lambda_{l_{ij}}},s_1,\cdots,s_{\Lambda_{l_{ij}}}\right)\end{aligned}$$

其中，$\Lambda_{l_{ij}}$表示边集$\mathcal{E}_{\text{cm}}$的基数；$\tilde{\mathcal{Q}}_{ij}\left(\boldsymbol{e}_{ij}\right)$表示满足假设10.1(A2)的欺诈函数。

定义$\frac{\chi_{ij}}{\chi_{ij}}=1$，则$\chi_{ij}\neq 0$（有欺诈攻击）；以及$\frac{\chi_{ij}}{\chi_{ij}}=0$（没有欺诈攻击）。

注释 10.3 以上建立的混杂系统$\mathcal{H}_{\text{dept}}$ [式 (10-13)] 的模型是合理的并且具有一般性：作为$\mathcal{G}_{\text{dept}}(\Omega)$中的元素，$\mathcal{G}_{l_{ij}}$同时包含了在有或者没有欺诈攻击下基于事件的跳跃情况，反映在$\eta_{l_{ij}}<0$和$\boldsymbol{e}_{l_{ij}}^{+}=\frac{\chi_{ij}}{\chi_{ij}}\tilde{\mathcal{Q}}_{l_{ij}}\left(\boldsymbol{e}_{l_{ij}}\right)$。如果$\eta_{l_{ij}}<0$，$\phi_{l_{ij}}^{+}=\frac{1}{\delta_{l_{ij}}}$表明边 l_{ij} 的驱动条件被触发进而通信。同时，$e_{l_{ij}}^{+}=\frac{\chi_{ij}}{\chi_{ij}}\tilde{\mathcal{Q}}_{l_{ij}}\left(e_{l_{ij}}\right)$表明边 l_{ij} 被 / 不被攻击，即不需要检测何时被攻击都能使建立的事件驱动策略对攻击有弹性作用，进而实现编队追踪的弹性控制。

为了研究混杂模型$\mathcal{H}_{\text{dept}}$ [式 (10-13)] 的稳定性，对任意$(i,j)\in\mathcal{E}_{\text{cm}}$，我们定义初始条件为$\mathbb{X}_0=\{q\in\mathbb{X}|\tau_{ij}\geqslant\tau_{\text{miet}}^{ij}$，$\eta_{ij}=0$，$\phi_{ij}=\phi_{ij\text{miet}}$，$\varsigma_{ij}=0\}$，其中 $\phi_{ij,\text{miet}}$ 会在后文中给出分析。定义 $\|\boldsymbol{\omega}\|_{\mathcal{L}_\infty}=\sup_{j\in\mathbb{Z}_{\geqslant 0}}\left(\text{ess}\sup_{t\in\mathbb{R}_{\geqslant 0}|(t,j)\in\text{dom}_{\omega}}\|\boldsymbol{\omega}(t,j)\|\right)$。基于 Dolk 等人[83]的研究，我们给出以下定义来继续研究混杂系统 (10-13) 的稳定性。

定义 10.2

混杂系统$\mathcal{H}_{\text{dept}}$ [式 (10-13)] 相对于初始状态集$\mathbb{X}_0$是持续流动的，如果对任意满足$\boldsymbol{q}(0,0)\in\mathbb{X}_0$的最大解 $\boldsymbol{q}$ 在 t 方向上有无界的时间域，即 $\sup_t \text{dom}\,\boldsymbol{q}=\infty$。

定义 10.3

对于混杂系统$\mathcal{H}_{\text{dept}}$ [式 (10-13)]，紧集$\boldsymbol{\mathcal{A}}=\{\boldsymbol{q}\in\mathbb{X}|\ \boldsymbol{\xi}=0,\boldsymbol{e}=0\}$是一致全局渐近稳定的，关于初始状态$\mathbb{X}_0$，如果系统$\mathcal{H}_{\text{dept}}$相对于初始状态集$\mathbb{X}_0$是持续流动的，而且存在函数$\beta\in\mathcal{KLL}$使得对任意初始条件$\boldsymbol{q}(0,0)\in\mathbb{X}_0,\mathcal{H}_{\text{dept}}$的所有解 $\boldsymbol{q}$ 满足$|\boldsymbol{q}(t,r)|_{\boldsymbol{\mathcal{A}}}\leqslant\beta\left(|\boldsymbol{q}(0,0)|_{\boldsymbol{\mathcal{A}}},t,r\right)$，$(t,r)\in\text{dom}\,q$。

定义 10.4

对混杂系统$\mathcal{H}_{\text{dept}}$ [式 (10-13)]，紧集$\boldsymbol{\mathcal{A}}=\{\boldsymbol{q}\in\mathbb{X}|\ \boldsymbol{\xi}=0,\boldsymbol{e}=0\}$从输入干扰$\boldsymbol{\omega}$到输出 $\boldsymbol{z}$ 是$\mathcal{L}_\infty$稳定的，其中增益L_∞小于等于$\hat{\theta}$，如果系统$\mathcal{H}_{\text{dept}}$相对于初始状态集$\mathbb{X}_0$是持续流动的而且存在函数$\beta\in\mathcal{K}_\infty$使得对任意初始条件$\boldsymbol{q}(0,0)\in\mathbb{X}_0$和外部干扰输入$\boldsymbol{\omega}\in\mathcal{L}_\infty$，则系统$\mathcal{H}_{\text{dept}}$的所有解 q 满足

$$\| z(t,r) \|_{\mathcal{L}_\infty} \leqslant \beta\left(\left|q(0,0)\right|_{\mathcal{A}}\right) + \hat{\theta} \| \omega \|_{\mathcal{L}\infty}, (t,r) \in \operatorname{dom} q。$$

10.2.2 三种欺诈攻击下的分布式事件驱动策略

接下来我们介绍三种不同基于边的欺诈攻击情况：

情形Ⅰ. 任何边都可能以异步方式被攻击，即对给定的$n \in \mathbb{Z}_{\geqslant 0}, \left\{H_n^{ij}\right\}_{(i,j)\in\mathcal{E}_{\text{cm}}}$因$(i,j)$不同而不同。

情形Ⅱ. 只有边l_{ij}的内邻集中的边同步受到欺诈攻击，对任意$(i,j) \in \mathcal{E}_{\text{cm}}$，即$\left\{H_n^{ki}\right\}_{(k,i)\in Z_{l_{ij}}}$对于$(k,i) \in Z_{l_{ij}}$是相同的，其中$\chi_{ki} \neq 0$。

情形Ⅲ. 所有的边都可能同时受到欺诈攻击，即对给定的$n \in \mathbb{Z}_{\geqslant 0}, \left\{H_n^{ij}\right\}_{(i,j)\in\mathcal{E}_{\text{cm}}}$对任意$(i,j) \in \mathcal{E}_{\text{cm}}$都是相同的，其中$\chi_{ij} \neq 0$。

情形Ⅰ： 异步传输攻击，即对给定的$n \in \mathbb{Z}_{\geqslant 0}, \left\{H_n^{ij}\right\}_{(i,j)\in\mathcal{E}_{\text{cm}}}$因$(i,j)$不同而不同。

接下来我们将介绍式 (10-13) 中子系统 e 的稳定性和指数增长条件所需的基本条件。

假设 10.2

对任意$(i,j) \in \mathcal{E}_{\text{cm}}$，存在连续可微函数$W_{ij}: \mathbb{R}^{n_{eij}} \to \mathbb{R}_{\geqslant 0}$，连续有界函数$H_{ij}: \mathbb{R}^n \times \mathbb{R}^{m+n} \times \mathbb{R}^{m'} \to \mathbb{R}$，常数$c_{ij}, L_{ij} \in [0,1)$，函数$\underline{c}_{W_{ij}}, \bar{c}_{W_{ij}} \in \mathcal{K}_\infty$，使得

i）对任意$\boldsymbol{e}_{ij} \in \mathbb{R}^{2n}$，根据式 (10-7)，有

$$\begin{aligned} &\underline{c}_{W_{ij}}\left(\left|\boldsymbol{e}_{ij}\right|\right) \leqslant W_{ij}\left(\boldsymbol{e}_{ij}\right) \leqslant \bar{c}_{W_{ij}}\left(\left|\boldsymbol{e}_{ij}\right|\right) \\ &W_{ij}\left(0_{i_j}\right) \leqslant c_{ij} W_{ij}\left(\boldsymbol{e}_{ij}\right),\ W_{ij}\left(\tilde{\mathcal{Q}}_{ij}\left(\boldsymbol{e}_{ij}\right)\right) \leqslant \tilde{L}_{ij} W_{ij}\left(\boldsymbol{e}_{ij}\right) \end{aligned} \tag{10-14}$$

ii）对于任意$\xi_i \in \mathbb{R}^n$和几乎所有的$e_{ij} \in \mathbb{R}^{2n}$，有

$$\left\langle \nabla W_{ij}\left(\boldsymbol{e}_{ij}\right), G_{ij}\left(\boldsymbol{\xi}_{ij}, \boldsymbol{e}_{ij}, \boldsymbol{\omega}_{ij}\right)\right\rangle \leqslant L_{ij} W_{ij}\left(\boldsymbol{e}_{ij}\right) + H_{ij}\left(\boldsymbol{\xi}_{ij}, \boldsymbol{e}_{ij}, \boldsymbol{\omega}_{ij}\right) + \tilde{L}_{ij} W_{ij}\left(\boldsymbol{e}_{ij}\right) \tag{10-15}$$

其中，$\tilde{L}_{ij} = \dfrac{\chi_{ij}}{\chi_{ij}+1}$和$\chi_{ij}$满足假设10.1。

假设 10.3

对任意$(i,j) \in \mathcal{E}_{\text{cm}}$，存在连续可微函数$V_{ij}: \mathbb{R}^{2n} \to \mathbb{R}_{>0}$，函数$\underline{\alpha}^{ij}, \overline{\alpha}^{ij}$，$\varrho_{ij} \in \mathcal{K}_\infty$，正常数$\rho_{ij}, \gamma_{ij}$和$\tilde{\rho}_{ij}$满足

i）对任意$\boldsymbol{\xi}_{ij} \in \mathbb{R}^{2n}$和输出$z_{ij}\left(\boldsymbol{\xi}_{ij}\right) = \mathcal{O}_{ij}\left(\boldsymbol{\xi}_{ij}\right)$，有

$$c_{z_{ij}}\left|\mathcal{O}_{ij}\left(\boldsymbol{\xi}_{ij}\right)\right|^2 \leqslant V_{ij}\left(\boldsymbol{\xi}_{ij}\right), \underline{\alpha}^{ij}\left(\left|\boldsymbol{\xi}_{ij}\right|\right) \leqslant V_{ij}\left(\boldsymbol{\xi}_{ij}\right) \leqslant \overline{\alpha}^{ij}\left(\left|\boldsymbol{\xi}_{ij}\right|\right) \quad (10\text{-}16)$$

ii）对任意$e_{ij} \in \mathbb{R}^{2n}$和几乎所有的$\xi_{ij} \in \mathbb{R}^{2n}$，有

$$\begin{aligned}
&\left\langle \nabla V_{ij}\left(\boldsymbol{\xi}_{ij}\right), F_{ij}\left(\boldsymbol{\xi}_{\overline{Z}_{l_{ij}}}, \boldsymbol{e}_{\overline{Z}_{l_{ij}}}, \boldsymbol{\omega}_{ij}\right)\right\rangle \\
&\leqslant -\rho_{ij} V_{ij}\left(\boldsymbol{\xi}_{ij}\right) - \varrho_{ij}\left(\left|\boldsymbol{\xi}_{ij}\right|\right) - H_{ij}^2\left(\boldsymbol{\xi}_{ij}, \boldsymbol{e}_{ij}, \boldsymbol{\omega}_{ij}\right) + \theta_{ij}^2\left|\boldsymbol{\omega}_{ij}\right|^2 + \left(\gamma_{ij}^2 - \tilde{\rho}_{ij}\right) W_{ij}^2\left(\boldsymbol{e}_{ij}\right) \\
&\quad - \sum_{(i',i) \in Z_{l_{ij}}} \left[\varrho_{i'i}\left(\left|\boldsymbol{\xi}_{ij}\right|\right) + \theta_{i'i}^2\left|\boldsymbol{\omega}_{i'i}\right|^2 + \left(\gamma_{i'i}^2 - \tilde{\rho}_{i'i}\right) W_{i'i}^2\left(\boldsymbol{e}_{i'i}\right) - H_{i'i}^2\left(\boldsymbol{\xi}_{i'i}, \boldsymbol{e}_{i'i}, \boldsymbol{\omega}_{i'i}\right)\right]
\end{aligned} \quad (10\text{-}17)$$

并且在$\tilde{L}_{ij} \neq 0$的情况下，有$\tilde{\rho}_{ij} < \gamma_{ij}\tilde{L}_{ij}$。

注释 10.4 $\tilde{L}_{ij} \neq 0$表示有欺诈攻击，$\tilde{L}_{ij} = 0$表示没有欺诈攻击，以上分析主要用于推导事件触发条件而非检测欺诈攻击的存在。假设 10.3 是为了保证子系统$\dot{\boldsymbol{\xi}}_{ij} = F_{ij}\left(\boldsymbol{\xi}_{ij}, \boldsymbol{e}_{ij}, \boldsymbol{\omega}_{ij}\right)$关于采样误差的输入状态稳定[254]。事实上，非线性系统很难对上述两个假设做出严格的验证，这在后边的数值仿真中会给出相关的验证过程，但对于某些非线性系统（如姿态控制系统），通常可以在转矩控制输入的详细表达式中找到两个函数W_{ij}和V_{ij}，这在未来关于姿态控制系统的工作中会进行进一步的研究。

正如上面提到的，对任意$(i,j) \in \mathcal{E}_{\text{cm}}$，应该选择合适的$\tau_{\text{miet}}^{ij}$使得基于边事件的混杂系统 (10-13) 对欺诈攻击有弹性作用。为此，式 (10-13) 中的函数$\Phi_{ij}: \mathbb{R}_{\geqslant 0} \to \mathbb{R}_{\geqslant 0}$满足

$$\frac{\mathrm{d}\phi_{ij}}{\mathrm{d}\tau_{ij}} = \begin{cases} -2L_{ij}\phi_{ij} - \gamma_{ij}\left(\lambda_{ij}\phi_{ij}^2 + 1\right), \tau_{ij} \in \left[0, \tau_{\text{miet}}^{ij}\right] \\ 0, \qquad \tau_{ij} > \tau_{\text{miet}}^{ij} \end{cases} \quad (10\text{-}18)$$

其中，函数L_{ij}和γ_{ij}满足式 (10-15) 和式 (10-17)；$\lim_{t \to \tau_{\text{miet}}^{ij}} \phi_{ij}(t) = \phi_{ij}\left(\tau_{\text{miet}}^{ij}\right)$。另外$\lambda_{ij} = \| \nabla \psi_{ij}\left(\left|\boldsymbol{\xi}_{ij}\right|\right) \| = \lambda_{ji} > 0$，其中$\psi_{ij}$是连续可微的增函数。因此$\tau_{\text{miet}}^{ij}$可以根据最大允许传输间隔$\tau_{\text{mati}}^{ij}$进行取值，即$\tau_{\text{miet}}^{ij} \leqslant \tau_{\text{mati}}^{ij}$。基于 Carnevale 等人[267]的研究，最大允许传输间隔τ_{mati}^{ij}为

$$\tau_{\text{mati}}^{ij}=\begin{cases}\dfrac{1}{L_{ij}r_{ij}}\arctan\left(\Gamma_{ij}\right), & \gamma_{ij}\sqrt{\lambda_{ij}}>L_{ij}\\ \tilde{\Gamma}_{ij}, & \gamma_{ij}\sqrt{\lambda_{ij}}=L_{ij}\\ \dfrac{1}{L_{ij}r_{ij}}\operatorname{arctanh}\left(\Gamma_{ij}\right), & \gamma_{ij}\sqrt{\lambda_{ij}}<L_{ij}\end{cases} \tag{10-19}$$

其中，$\lambda_{ij}>0, 1>\delta_{ij}>\tilde{L}_{ij}\geqslant 0$，以及

$$r_{ij}=\sqrt{\left|\frac{\gamma_{ij}^2\lambda_{ij}}{L_{ij}^2}-1\right|}, \Gamma_{ij}=\frac{r_{ij}\left(1-\delta_{ij}\right)}{\dfrac{\delta_{ij}}{1+\delta_{ij}}\left(\dfrac{\gamma_{ij}+\gamma_{ij}\lambda_{ij}}{L_{ij}}-2\right)+1+\delta_{ij}},$$

$$\tilde{\Gamma}_{ij}=\frac{1}{\delta_{ij}\gamma_{ij}\lambda_{ij}+L_{ij}}-\frac{\delta_{ij}}{\delta_{ij}L_{ij}+\gamma_{ij}\lambda_{ij}}。$$

注释 10.5　由此可见，式 (10-19) 中的最小触发间隔依赖于假设 10.2、假设 10.3 中的参数，参数$\lambda_{ij}>0$相关于$\psi_{ij}\left(\left|\xi_{ij}\right|\right)$，其中函数$\psi_{ij}$正相关于参数$\left|\xi_{ij}\right|$。数值$\left|\xi_{ij}\right|$的增长表明智能体偏离预期目标，而为了弱化这种偏离，需要智能体之间进行频繁的通信，即需要更小的τ_{mati}^{ij}，这可以通过增大参数λ_{ij}的值实现。另外，$1>\delta_{ij}>\tilde{L}_{ij}\geqslant 0$是为了保证能设计出合适的对欺诈攻击有弹性的事件驱动策略，正如 De 和 Tesi[85] 所分析的那样。

基于 Carnevale 等人 [267] 的研究，下面的引理给出了相关方程 (10-18) 来推导出τ_{miet}^{ij}和τ_{mati}^{ij}，对于每条有向边$(i,j)\in\mathcal{E}_{\text{cm}}$。

引理 10.1

对任意$(i,j)\in\mathcal{E}_{\text{cm}}$，假设$\tau_{\text{mati}}^{ij}$满足式 (10-19)，那么方程

$$\dot{\tilde{\phi}}_{ij}=-2L_{ij}\tilde{\phi}_{ij}-\gamma_{ij}\left(\lambda_{ij}\tilde{\phi}_{ij}^2+1\right),\ \tilde{\phi}_{ij}(0)=\delta_{ij}^{-1} \tag{10-20}$$

的解满足$\tilde{\phi}_{ij}(t)\in\left[\delta_{ij},\delta_{ij}^{-1}\right]$，对任意$t\in\left[0,\tau_{\text{mati}}^{ij}\right]$。

对任意$(i,j)\in\mathcal{E}_{\text{cm}}$，定义

$$\phi_{ij,\text{miet}}=\tilde{\phi}_{ij}\left(\tau_{\text{miet}}^{ij}\right), \tag{10-21}$$

其中，$\tilde{\phi}_{ij}$满足式 (10-20) 和$\tilde{\phi}_{ij}(0)=\delta_{ij}^{-1}$，因此由$\tau_{\text{miet}}^{ij}\leqslant\tau_{\text{mati}}^{ij}$可知

$\phi_{ij,\text{miet}} \geqslant \delta_{ij}$。

接下来，我们推导混杂系统 (10-13) 在欺骗攻击下基于边事件的弹性性能保证。

定理 10.1

混杂系统 (10-13) 满足条件 10.2.1 和假设 10.1 ～假设 10.3。对任意 $(i,j)\in\mathcal{E}_{\text{cm}}$，假设以下条件成立：

(1a) 对于满足假设 10.1 的欺诈攻击：存在两个正常数系列 $\left\{\kappa_{ij}^{0}\right\}_{(i,j)\in\mathcal{E}_{\text{cm}}}$ 和 $\left\{\kappa_{ij}^{1}\right\}_{(i,j)\in\mathcal{E}_{\text{cm}}}$ 满足

$$\kappa_{ij}^{0}\varpi_{ij}^{0} > \left(\frac{\tau_{\text{miet}}^{ij}}{\tau_{f}^{ij}} + \frac{1}{\tau_{d}^{ij}}\right)\left(\kappa_{ij}^{0}\varpi_{ij}^{0} + \kappa_{ij}^{1}\varpi_{ij}^{1}\right) \tag{10-22}$$

其中，τ_{f}^{ij} 和 τ_{d}^{ij} 满足假设 10.1(A1), $\varpi_{ij}^{0} = \min\left\{\rho_{ij}, \frac{\delta_{ij}\tilde{\rho}_{ij}}{\gamma_{ij}}\right\}$，以及 $\varpi_{ij}^{1} = \frac{2\gamma_{ij}\frac{1}{\delta_{ij}}\tilde{L}_{ij} - \tilde{\rho}_{ij}}{\gamma_{ij}\delta_{ij}}$，其中 $\tilde{L}_{ij} = \frac{\chi_{ij}}{\chi_{ij}+1}$ 满足假设10.1（A2）而 $\delta_{ij} \leqslant \phi_{ij,\text{miet}}$ 满足式 (10-21)。

(1b) 式 (10-11) 中设计的参数满足假设 10.2、假设 10.3 的事件驱动策略：基于 $\sigma_{ij}\in\mathcal{K}_{\infty}$，其中 $\sigma_{ij}\left(\eta_{ij}\right) \geqslant \varpi_{ij}^{0}\eta_{ij}$，有

$$\Psi_{ij} = \begin{cases} \hat{\Psi}_{ij}, 0 \leqslant \tau_{ij} \leqslant \tau_{\text{miet}}^{ij} \\ \tilde{\Psi}_{ij}, \tau_{ij} \geqslant \tau_{\text{miet}}^{ij} \end{cases} \tag{10-23}$$

其中，$\hat{\Psi}_{ij} = \left(\left|U_{l_{ij}}\right|+1\right)\left[\varrho_{ij}\left(\left|\boldsymbol{\xi}_{ij}\right|\right) + \left(\lambda_{ij}-1\right)\gamma_{ij}^{2}W_{ij}^{2}\phi_{ij}^{2} - \sigma_{ij}\left(\eta_{ij}\right)\right]$ 以及 $\tilde{\Psi}_{ij} = \left(\left|U_{l_{ij}}\right|+1\right)\left[\varrho_{ij}\left(\left|\boldsymbol{\xi}_{ij}\right|\right) - \left(\gamma_{ij}^{2} + 2\gamma_{ij}\phi_{ij}L_{ij} + \gamma_{ij}^{2}\phi_{ij}^{2}\right)W_{ij}^{2} - \sigma_{ij}\left(\eta_{ij}\right)\right]$。

(1c) $\tilde{\eta}_{ij} = \left(\left|U_{l_{ij}}\right|+1\right)\gamma_{ij}W_{ij}^{2}\left(\delta_{ij} - \frac{c_{ij}^{2}}{\delta_{ij}}\right)$，其中 $\delta_{ij} > c_{ij} \geqslant \tilde{L}_{ij}$，而 γ_{ij}、W_{ij} 和 $\tilde{L}_{ij}$ 满足假设 10.2，γ_{ij} 满足式 (10-18)，$\delta_{ij} \leqslant \phi_{ij,\text{miet}}$ 满足式 (10-21)。

因此根据定义 10.3 可知，当 $\boldsymbol{\omega}=0$ 时，紧集 $\mathcal{A} = \left\{q\in\mathbb{X} \mid \xi=0, e=0\right\}$ 是一致全局渐近稳定的。紧集 $\boldsymbol{\mathcal{A}} = \left\{\boldsymbol{q}\in\mathbb{X} \mid \boldsymbol{\xi}=0, \boldsymbol{e}=0\right\}$ 关于非零输入干扰 $\boldsymbol{\omega}$ 和输

出 $\boldsymbol{z}$ 是$\mathcal{L}_\infty$稳定的，而且有限增益$\mathcal{L}_\infty$小于等于$\max_{(i,j)\in\mathcal{E}_{\text{cm}}}\left\{\theta_{ij}\sqrt{\frac{\left(\left|U_{l_{ij}}\right|+1\right)\text{Im}_{ij}}{\hat{\beta}_{ij}c_{z_{ij}}}}\right\}$，

其中，$\text{Im}_{ij}=\text{e}^{\varsigma_*^{ij}\left(\kappa_{ij}^0\varpi_{ij}^0+\kappa_{ij}^1\varpi_{ij}^1\right)}$，$\hat{\beta}_{ij}=\kappa_{ij}^0\varpi_{ij}^0-\frac{\kappa_{ij}^0\varpi_{ij}^0+\kappa_{ij}^1\varpi_{ij}^1}{T_*^{ij}}$，$\varsigma_*^{ij}=\varsigma^{ij}+\nu^{ij}\tau_{\text{miet}}^{ij}$ 以及

$T_*^{ij}=\frac{\tau_f^{ij}\tau_d^{ij}}{\tau_f^{ij}+\tau_d^{ij}\tau_{\text{miet}}^{ij}}$。

证明：

接下来的证明主要考虑基于弹性事件触发的混杂系统$\mathcal{H}_{\text{dept}}$［式(10-13)］，来推导出紧集$\boldsymbol{\mathcal{A}}=\{\boldsymbol{q}\in\mathbb{X}|\ \boldsymbol{\xi}=0,\boldsymbol{e}=0\}$的一致全局渐近稳定性和$\mathcal{L}_\infty$稳定性。令$\mathcal{R}(\mathbb{X}_0)$表示系统$\mathcal{H}_{\text{dept}}$的可达状态，其中$\boldsymbol{q}(0,0)\in\mathbb{X}_0$，那么对任意$\boldsymbol{q}\in\mathcal{R}(\mathbb{X}_0)$，基于引理 10.1，有(P1)$\left\{\tau_{ij}\geqslant\tau_{\text{miet}}^{ij}\right\}\Leftrightarrow\phi_{ij}=\phi_{ij,\text{miet}}$；(P2) $\delta_{ij}^{-1}\geqslant\phi_{ij}\geqslant\phi_{ij,\text{miet}}$；(P3)$\eta_{ij}\geqslant0$。

基于假设 10.1 ～假设 10.3，对任意$q\in C\cup D$，考虑李雅普诺夫函数

$$\mathcal{V}(\boldsymbol{q})=\sum_{(i,j)\in\mathcal{E}_{\text{cm}}}\mathcal{V}_{ij}(\boldsymbol{q}_{ij})\tag{10-24}$$

其中，$\mathcal{V}_{ij}(\boldsymbol{q}_{ij})=\mathcal{V}_{ij}(\boldsymbol{\xi}_{ij})+\left(\left|U_{l_{ij}}\right|+1\right)\gamma_{ij}\phi_{ij}W_{ij}^2(\boldsymbol{e}_{ij})+\eta_{ij}$。结合(P1)～(P3)和假设10.2、假设10.3，可知存在函数$\underline{\beta}_{\mathcal{V}_{ij}},\overline{\beta}_{\mathcal{V}_{ij}},\underline{\beta}_{\mathcal{V}},\overline{\beta}_{\mathcal{V}}\in\mathcal{K}_\infty$，使得对任意$q\in\mathcal{R}(\mathbb{X}_0)$

$$\begin{cases}\underline{\beta}_{\mathcal{V}}\left(|\boldsymbol{q}(t,k)|_{\boldsymbol{\mathcal{A}}}\right)\leqslant\mathcal{V}(\boldsymbol{q}(t,k))\leqslant\overline{\beta}_{\mathcal{V}}\left(|\boldsymbol{q}(t,k)|_{\mathcal{A}}\right)\\ \underline{\beta}_{\mathcal{V}_{ij}}\left(|\boldsymbol{q}_{ij}(t,k)|_{\boldsymbol{\mathcal{A}}}\right)\leqslant\mathcal{V}_{ij}(\boldsymbol{q}_{ij}(t,k))\leqslant\overline{\beta}_{\mathcal{V}_{ij}}\left(|\boldsymbol{q}_{ij}(t,k)|_{\mathcal{A}}\right)\end{cases}\tag{10-25}$$

结合(P2)、(P3)，式(10-14) 和式(10-16)，可知

$$\underline{c}_{\mathcal{V}_{ij}}|\boldsymbol{q}_{ij}|_{\boldsymbol{\mathcal{A}}}^2\leqslant\mathcal{V}_{ij}(\boldsymbol{q}_{ij})\tag{10-26}$$

其中，$\underline{c}_{\mathcal{V}_{ij}}$为正数，紧集$\boldsymbol{\mathcal{A}}=\{\boldsymbol{q}\in\mathbb{X}|\ \boldsymbol{\xi}=0,\boldsymbol{e}=0\}$。接下来考虑函数$\mathcal{V}$和$\{\mathcal{V}_{ij}\}_{(i,j)\in\mathcal{E}_{\text{cm}}}$在有/没有攻击的情形下是如何随着流动集和跳变集演化的。

当$\boldsymbol{q}\in D\cap\mathcal{R}(\mathbb{X}_0)$，接下来首先考虑$\{\mathcal{V}_{ij}\}_{(i,j)\in\mathcal{E}_{\text{cm}}}$的演化。$(\mathbf{J_a})$没有欺诈攻击$\left(s_{ij}\in\Theta^{ij}\right)$：对任意$(i,j)\in\mathcal{E}_{\text{cm}}$，基于式(10-12)、式(10-13)和式(10-18)，从定理 10.1(1c) 可知$\mathcal{V}_{ij}^+(\boldsymbol{q}_{ij})-\mathcal{V}_{ij}(\boldsymbol{q}_{ij})\leqslant-\left(\left|U_{l_{ij}}\right|+1\right)\gamma_{ij}W_{ij}^2\left(\delta_{ij}-\frac{c_{ij}^2}{\delta_{ij}}\right)+\tilde{\eta}_{ij}=0$，

进而 (P1) 和定理 10.1(1c) 表明$\mathcal{V}_i^+(\boldsymbol{q}_i)-\mathcal{V}_i(\boldsymbol{q}_i)\leqslant 0$，对任意$q\in D\cap\mathcal{R}(\mathbb{X}_0)$。

$(\mathbf{J_b})$有欺诈攻击$\left(s_{ij}\in\Xi^{ij}\right)$：基于式 (10-13)，其中$\boldsymbol{\xi}_{ij}^+=\boldsymbol{\xi}_{ij},\boldsymbol{e}_{ij}^+=\tilde{\mathcal{Q}}_{ij}(\boldsymbol{e}_{ij})$，$\phi_{ij}^+=\delta_{ij}^{-1}$以及$\eta_{ij}^+=\tilde{\eta}_{ij}$，由定理 10.1(1c) 可知$\mathcal{V}_{ij}^+(\boldsymbol{q}_{ij})-\mathcal{V}_{ij}(\boldsymbol{q}_{ij})\leqslant\left(\left|U_{li_j}\right|+1\right)\gamma_{ij}W_{ij}^2\left(-\delta_{ij}+\dfrac{\tilde{L}_{ij}^2}{\delta_{ij}}\right)+\tilde{\eta}_{ij}\leqslant 0$。因此可知

$$\mathcal{V}^+(\boldsymbol{q})\leqslant\mathcal{V}(\boldsymbol{q}) \tag{10-27}$$

对任意$\boldsymbol{q}\in D\cap\mathcal{R}(\mathbb{X}_0)$。

当$\boldsymbol{q}\in C\cap\mathcal{R}(\mathbb{X}_0)$，接下来考虑$\mathcal{V}$和$\mathcal{V}_{ij}$是如何演化的。$(\mathbf{F_a})$当$0\leqslant\tau_{ij}\leqslant\tau_{\text{miet}}^{ij}$，根据式 (10-18) 可知$\dot{\phi}_{ij}=-2L_{ij}\phi_{ij}-\gamma_{ij}\left(\lambda_{ij}\phi_{ij}^2+1\right)$。基于不等式$\Lambda: 2\gamma_{ij}\phi_{ij}W_{ij}H_{ij}\leqslant\gamma_{ij}^2\phi_{ij}^2W_{ij}^2+H_{ij}^2$，有

$$\begin{aligned}
&\left\langle\nabla\mathcal{V}(\boldsymbol{q}),\mathcal{F}(\boldsymbol{q},\boldsymbol{\omega})\right\rangle\\
\leqslant&\sum_{(i,j)\in\mathcal{E}_{cm}}\Big\{-\rho_{ij}V_{ij}(\boldsymbol{\xi}_{ij})-\varrho_{ij}\left(\left|\boldsymbol{\xi}_{ij}\right|\right)-H_{ij}^2(\boldsymbol{\xi}_{ij},\boldsymbol{e}_{ij},\boldsymbol{\omega}_{ij})\\
&+\left(\gamma_{ij}^2-\tilde{\rho}_{ij}\right)W_{ij}^2(\boldsymbol{e}_{ij})-\sum_{(i',i)\in Z_{l_{ij}}}\Big[\varrho_{i'i}\left(\left|\boldsymbol{\xi}_{ij}\right|\right)\\
&+\theta_{i'i}^2\left|\boldsymbol{\omega}_{i'i}\right|^2+\left(\gamma_{i'i}^2-\tilde{\rho}_{i'i}\right)W_{i'i}^2(\boldsymbol{e}_{i'i})-H_{i'i}^2(\boldsymbol{\xi}_{i'i},\boldsymbol{e}_{i'i})\Big]\\
&+\left(\left|U_{l_{ij}}\right|+1\right)\gamma_{ij}\left[-2L_{ij}\phi_{ij}-\gamma_{ij}\left(\lambda_{ij}\phi_{ij}^2+1\right)\right]W_{ij}^2\\
&+2\left(\left|U_{l_{ij}}\right|+1\right)\gamma_{ij}\phi_{ij}W_{ij}\left(L_{ij}W_{ij}+H_{ij}\right.\\
&\left.+\tilde{L}_{ij}W_{ij}\right)+\theta_{ij}^2\left|\omega_{ij}\right|^2+\Psi_{ij}\Big\}
\end{aligned} \tag{10-28}$$

$$\begin{aligned}
\leqslant&\sum_{(i,j)\in\mathcal{E}_{cm}}\Big\{-\rho_{ij}V_{ij}+\left(\left|U_{l_{ij}}\right|+1\right)\Big[-H_{ij}^2+\theta_{ij}^2\left|\boldsymbol{\omega}_{ij}\right|^2\\
&+\left(\gamma_{ij}^2-\tilde{\rho}_{ij}\right)W_{ij}^2-\varrho_{ij}\left(\left|\boldsymbol{\xi}_{ij}\right|\right)\Big]+\left(\left|U_{l_{ij}}\right|+1\right)\gamma_{ij}\\
&\times\left[-2L_{ij}\phi_{ij}-\gamma_{ij}\left(\lambda_{ij}\phi_{ij}^2+1\right)\right]W_{ij}^2+2\left(\left|U_{l_{ij}}\right|+1\right)\\
&\times\gamma_{ij}\phi_{ij}W_{ij}\left(L_{ij}W_{ij}+H_{ij}+\tilde{L}_{ij}W_{ij}\right)+\Psi_{ij}\Big\}\\
\underbrace{\leqslant}_{\Lambda}&\sum_{(i,j)\in\mathcal{E}_{cm}}\Big\{-\rho_{ij}V_{ij}+\left(\left|U_{l_{ij}}\right|+1\right)\Big(-\tilde{\rho}_{ij}W_{ij}^2\\
&+\theta_{ij}^2\left|\boldsymbol{\omega}_{ij}\right|^2+2\gamma_{ij}\phi_{ij}\tilde{L}_{ij}W_{ij}^2\Big)+\left(\left|U_{l_{ij}}\right|+1\right)
\end{aligned}$$

$$\times\left[-\varrho_{ij}\left(\left|\boldsymbol{\xi}_{ij}\right|\right)+\left(1-\lambda_{ij}\right)\gamma_{ij}^2\phi_{ij}^2W_{ij}^2\right]+\Psi_{ij}\Big\}$$

$$\leqslant\sum_{(i,j)\in\mathcal{E}_{cm}}\Big\{-\rho_{ij}V_{ij}+\left(\left|U_{l_{ij}}\right|+1\right)\left(-\tilde{\rho}_{ij}W_{ij}^2+\Psi_{ij}\right.$$

$$\left.+2\gamma_{ij}\phi_{ij}\tilde{L}_{ij}W_{ij}^2\right)+\left(\left|U_{l_{ij}}\right|+1\right)\theta_{ij}^2\left|\boldsymbol{\omega}_{ij}\right|^2$$

$$+\left(\left|U_{l_{ij}}\right|+1\right)\left[-\varrho_{ij}\left(\left|\boldsymbol{\xi}_{ij}\right|\right)+\left(1-\lambda_{ij}\right)\gamma_{ij}^2\phi_{ij}^2W_{ij}^2\right]\Big\}$$

可知

$$\left\langle\nabla\mathcal{V}(\boldsymbol{q}),\mathcal{F}(\boldsymbol{q},\boldsymbol{\omega})\right\rangle\leqslant\begin{cases}-\sum\limits_{(i,j)\in\mathcal{E}_{\mathrm{cm}}}\varpi_{ij}^0\mathcal{V}_{ij}+\sum\limits_{(i,j)\in\mathcal{E}_{\mathrm{cm}}}\left[\left(\left|U_{l_{ij}}\right|+1\right)\theta_{ij}^2\left|\boldsymbol{\omega}_{ij}\right|^2\right]\\ \qquad\tilde{L}_{ij}\equiv 0,\forall(i,j)\in\mathcal{E}_{\mathrm{cm}}\\ \sum\limits_{(i,j)\in\mathcal{E}_{\mathrm{cm}}}\varpi_{ij}^1\upsilon_{ij}+\sum\limits_{(i,j)\in\mathcal{E}_{\mathrm{cm}}}\left[\left(\left|U_{l_{ij}}\right|+1\right)\theta_{ij}^2\left|\boldsymbol{\omega}_{ij}\right|^2\right]\\ \qquad\tilde{L}_{ij}\neq 0,\exists(i,j)\in\mathcal{E}_{\mathrm{cm}}\end{cases}$$

(10-29)

其 中 ， $\varpi_{ij}^0=\min\left\{\rho_{ij},\dfrac{\hat{\rho}_{ij}\delta_{ij}}{\gamma_{ij}}\right\}$； $\varpi_{ij}^1=\dfrac{2\gamma_{ij}\tilde{L}_{ij}/\delta_{ij}-\tilde{\rho}_{ij}}{\gamma_{ij}\delta_{ij}}$； $\hat{\Psi}_{ij}=\left(\left|U_{l_{ij}}\right|+1\right)\left[\varrho_{ij}\left(\left|\boldsymbol{\xi}_{ij}\right|\right)+\left(\lambda_{ij}-1\right)\gamma_{ij}^2\phi_{ij}^2W_{ij}^2-\sigma_{ij}\left(\eta_{ij}\right)\right],\sigma_{ij}\left(\eta_{ij}\right)\geqslant\varpi_{ij}^0n_{ij}$。

$(\mathbf{F_b})$当$\tau_{ij}\geqslant\tau_{\mathrm{miet}}^{ij}$，由式 (10-18) 可知$\dot{\phi}_{ij}=0$，所以有

$$\left\langle\nabla\mathcal{V}(\boldsymbol{q}),\mathcal{F}(\boldsymbol{q},\boldsymbol{\omega})\right\rangle$$

$$\leqslant\sum_{(i,j)\in\mathcal{E}_{\mathrm{cm}}}\Big\{-\rho_{ij}V_{ij}\left(\boldsymbol{\xi}_{ij}\right)-\varrho_{ij}\left(\left|\boldsymbol{\xi}_{ij}\right|\right)-H_{ij}^2\left(\boldsymbol{\xi}_{ij},\boldsymbol{e}_{ij},\boldsymbol{\omega}_{ij}\right)+\left(\gamma_{ij}^2-\tilde{\rho}_{ij}\right)W_{ij}^2\left(\boldsymbol{e}_{ij}\right)$$

$$-\sum_{(i',i)\in Z_{l_{ij}}}\left[\rho_{i'i}\left(\left|\boldsymbol{\xi}_{ij}\right|\right)+\theta_{i'i}^2\left|\boldsymbol{\omega}_{i'i}\right|^2+\left(\gamma_{i'i}^2-\tilde{\rho}_{i'i}\right)W_{i'i}^2\left(\boldsymbol{e}_{i'i}\right)-H_{i'i}^2\left(\xi_{i'i},\boldsymbol{e}_{i'i}\right)\right]$$

$$+2\left(\left|U_{l_{ij}}\right|+1\right)\gamma_{ij}\phi_{ij}W_{ij}\left(L_{ij}W_{ij}+H_{ij}+\tilde{L}_{ij}W_{ij}\right)+\theta_{ij}^2\left|\boldsymbol{\omega}_{ij}\right|^2+\Psi_{ij}\Big\}$$

$$\leqslant\sum_{(i,j)\in\mathcal{E}_{\mathrm{cm}}}\Big\{-\rho_{ij}V_{ij}+\left(\left|U_{l_{ij}}\right|+1\right)\left(-\tilde{\rho}_{ij}W_{ij}^2+2\gamma_{ij}\phi_{ij}\tilde{L}_{ij}W_{ij}^2\right)+\left(\left|U_{l_{ij}}\right|+1\right)\theta_{ij}^2\left|\boldsymbol{\omega}_{ij}\right|^2$$

$$+\left(\left|U_{l_{ij}}\right|+1\right)\left[-\varrho_{ij}\left(\left|\boldsymbol{\xi}_{ij}\right|\right)+\left(\gamma_{ij}^2+2\gamma_{ij}\phi_{ij}L_{ij}+\gamma_{ij}^2\phi_{ij}^2\right)W_{ij}^2\right]+\Psi_{ij}\Big\}$$

(10-30)

因此，由式 (10-30) 可知

$$
\left\langle \nabla \mathcal{V}(\boldsymbol{q}), \mathcal{F}(\boldsymbol{q},\boldsymbol{\omega}) \right\rangle \leqslant \begin{cases} -\sum\limits_{(i,j)\in\varepsilon_{\mathrm{cm}}} \varpi_{ij}^{0} \boldsymbol{\mathcal{V}}_{ij} + \sum\limits_{(i,j)\in\varepsilon_{\mathrm{cm}}} \left\{ \left(\left| U_{l_{ij}} \right| + 1 \right) \theta_{ij}^{2} \left| \boldsymbol{\omega}_{ij} \right|^{2} \right\} \\ \qquad \tilde{L}_{ij} \equiv 0,\ \forall (i,j) \in \mathcal{E}_{\mathrm{cm}} \\ \sum\limits_{(i,j)\in\varepsilon_{\mathrm{cm}}} \varpi_{ij}^{1} \mathcal{V}_{ij} + \sum\limits_{(i,j)\in\varepsilon_{\mathrm{cm}}} \left\{ \left(\left| U_{l_{ij}} \right| + 1 \right) \theta_{ij}^{2} \left| \boldsymbol{\omega}_{ij} \right|^{2} \right\} \\ \qquad \tilde{L}_{ij} \neq 0,\ \exists (i,j) \in \mathcal{E}_{\mathrm{cm}} \end{cases}
$$

(10-31)

其中，$\varpi_{ij}^{0} = \min\left\{ \rho_{ij}, \dfrac{\tilde{\rho}_{ij}\delta_{ij}}{\gamma_{ij}} \right\}$；$\varpi_{ij}^{1} = \dfrac{2\gamma_{ij}\tilde{L}_{ij} / \delta_{ij} - \tilde{\rho}_{ij}}{\gamma_{ij}\delta_{ij}}$；$\tilde{\Psi}_{ij} = \left(\left| U_{l_{ij}} \right| + 1 \right) \Big[\varrho_{ij}\left(\left| \boldsymbol{\xi}_{ij} \right| \right) - \left(\gamma_{ij}^{2} + 2\gamma_{ij}\phi_{ij}L_{ij} + \gamma_{ij}^{2}\phi_{ij}^{2} \right) W_{ij}^{2} - \sigma_{ij}\left(\eta_{ij} \right) \Big]$；$\sigma_{ij}\left(\eta_{ij} \right) \geqslant \varpi_{ij}^{0}\eta_{ij}$。

实际上基于假设 10.2、假设 10.3，结合式 (10-24)、式 (10-29) 和式 (10-31) 以及定理 10.1 的条件可知，存在两列正常数$\left\{ \boldsymbol{\kappa}_{ij}^{0} \right\}_{(i,j)\in\mathcal{E}_{\mathrm{cm}}}$和$\left\{ \boldsymbol{\kappa}_{ij}^{1} \right\}_{(i,j)\in\mathcal{E}_{\mathrm{cm}}}$，使得对任意$(i,j) \in \mathcal{E}_{\mathrm{cm}}$，在欺诈攻击的情形下，有

$$
\left\langle \nabla \mathcal{V}_{ij}\left(\boldsymbol{q}_{ij} \right), \mathcal{F}_{ij}\left(\boldsymbol{q}_{ij}, \boldsymbol{\omega}_{ij} \right) \right\rangle \leqslant \begin{cases} -\kappa_{ij}^{0}\varpi_{0}\mathcal{V}_{ij}\left(\boldsymbol{q}_{ij} \right) + \left(\left| U_{l_{ij}} \right| + 1 \right) \theta_{ij}^{2} \left| \boldsymbol{\omega}_{ij} \right|^{2} & \tilde{L}_{ij} = 0 \\ \kappa_{ij}^{1}\varpi_{1}v_{ij}\left(\boldsymbol{q}_{ij} \right) + \left(\left| U_{l_{ij}} \right| + 1 \right) \theta_{ij}^{2} \left| \boldsymbol{\omega}_{ij} \right|^{2}, & \tilde{L}_{ij} \neq 0 \end{cases}
$$

(10-32)

由以上分析易知，基于事件驱动策略的混杂系统$\mathcal{H}_{\mathrm{dept}}$不会出现奇诺现象，因为最小传输间隔是个严格的正常数 [式 (10-19)]，而且结合式 (10-24)、式 (10-25)、式 (10-29) 和式 (10-31) 可知，混杂系统不存在有限逃逸时间，因此$\boldsymbol{\xi}$和$\boldsymbol{e}$是有界的，而且$\boldsymbol{\eta}$、$\boldsymbol{\tau}$、$\boldsymbol{s}$以及$\boldsymbol{m}$也不存在有限逃逸时间。基于 De 和 Postoyan[226]、Goebel 等人 [230] 的研究，定义 10.2 以及以上分析，可知系统$\mathcal{H}_{\mathrm{dept}}$是持续流动的，对任意初始条件$\mathbb{X}_0$。

根据式 (10-32)，接下来分析混杂系统$\mathcal{H}_{\mathrm{dept}}$ [式 (10-13)] 的向量$\boldsymbol{q} = \left(\boldsymbol{q}_{ij} \right)_{(i,j)\in\mathcal{E}_{\mathrm{cm}}}$中的元素$\boldsymbol{q}_{ij}$。结合 De 和 Postoyan[226] 以及 Dolk 等人 [83] 的研究，对任意给定的最大解$\boldsymbol{q}$，我们首先定义有 / 没有欺诈攻击下的关于q_{ij}的时间集，在任意给定的时间区间$\left[T_1, T_2 \right]$。首先对任意$(i,j) \in \mathcal{E}_{\mathrm{cm}}$，定义$\overline{\Theta}_{q_{ij}}^{ij}\left(T_1, T_2 \right) = \left\{ \tilde{t} \in \left(T_1, T_2 \right) \middle| \ \forall k \in \mathbb{Z}_{\geqslant 0}, \left(\tilde{t}, k \right) \in \operatorname{dom} \boldsymbol{q}_{ij}, \tilde{L}_{ij} = 0 \right\}$和$\overline{\Xi}_{q_{ij}}^{ij}\left(T_1, T_2 \right) = \left[T_1, T_2 \right] \setminus \overline{\Theta}_{q_{ij}}^{ij}\left(T_1, T_2 \right)$，那么$\overline{\Theta}_{q_{ij}}^{ij}\left(T_1, T_2 \right)$和$\overline{\Xi}_{q_{ij}}^{ij}\left(T_1, T_2 \right)$可以分别表示

为 $\overline{\Xi}_{q_{ij}}^{ij}(T_1,T_2)=\bigcup_{p_r^{ij}\in\mathbb{Z}_{\geqslant 0}}\tilde{Z}_{p_r^{ij}}^{ij}\cap[T_1,T_2]$ 和 $\overline{\Theta}_{q_{ij}}^{ij}(T_1,T_2)=\bigcup_{p_r^{ij}\in\mathbb{Z}_{\geqslant 0}}\tilde{W}_{p_r^{ij}-1}^{ij}\cap[T_1,T_2]$，其中 $p_r^{ij}\in\mathbb{Z}_{\geqslant 0}$，对任意 $k\in\mathbb{Z}_{\geqslant 0}$，$(i,j)\in\mathcal{E}_{\mathrm{cm}}$。另外，$\tilde{Z}_{p_r^{ij}}^{ij}$ 和 $\tilde{W}_{p_r^{ij}}^{ij}$ 定义为 $\tilde{Z}_{p_r^{ij}}^{ij}=\left[\zeta_{p_r^{ij}}^{ij},\zeta_{p_r^{ij}}^{ij}+v_{p_r^{ij}}^{ij}\right)$ 和 $\tilde{W}_{p_r^{ij}}^{ij}=\left[\zeta_{p_r^{ij}}^{ij}+v_{p_r^{ij}}^{ij},\zeta_{p_r^{ij}+1}^{ij}\right)$，其中 $v_{p_r^{ij}}^{ij}\geqslant 0$ 表示自 $\zeta_{p_i^{ij}}^{ij}$ 到下一次成功传输所经过的时间。由 $\upsilon_{p_r^{ij}}^{ij}=0$ 可知 $\tilde{Z}_{p_r^{ij}}^{ij}=\{\zeta_{p_r^{ij}}^{ij}\}$ 以及 $\tilde{W}_{p_r^{ij}}^{ij}=\left[\zeta_{p_r^{ij}}^{ij},\zeta_{p_r^{ij}+1}^{ij}\right)$，即欺诈攻击具有脉冲形式。另外，如果 $h_0^{ij}>0$，则有 $\zeta_0^{ij}=h_0^{ij}$ 以及 $\tilde{W}_{-1}^{ij}=\left[0,\zeta_0^{ij}\right)$，否则 $\tilde{W}_{-1}^{ij}=\varnothing$。基于假设 10.1 可知 $\left|\overline{\Xi}_{q_{ij}}^{ij}(T_1,T_2)\right|\leqslant\left|\Xi^{ij}(T_1,T_2)\right|+\left(1+n^{ij}(T_1,T_2)\right)\tau_{\mathrm{miet}}^{ij}$，这意味着欺诈攻击的时间集被 $\left(1+n_Z^{ij}(T_1,T_2)\right)\tau_{\mathrm{miet}}^{ij}$ 延长了。与原始区间 $\Xi^{ij}(T_1,T_2)$ 相比，有效的欺诈攻击时间间隔可以通过 $\tau_{\mathrm{miet}}^{ij}$ 趋近于 $\tau_{\mathrm{mati}}^{ij}$ [式 (10-19)] 进一步延长。基于假设 10.1，可以进一步得到

$$\left|\overline{\Xi}_{q_{ij}}^{ij}(T_1,T_2)\right|\leqslant\varsigma_*^{ij}+\frac{T_2-T_1}{T_*^{ij}},\tag{10-33}$$

其中 $\varsigma_*^{ij}=\varsigma^{ij}+v^{ij}\tau_{\mathrm{miet}}^{ij},T_*^{ij}=\dfrac{\tau_f^{ij}\tau_d^{ij}}{\tau_f^{ij}+\tau_d^{ij}\tau_{\mathrm{miet}}^{ij}}$。

接下来根据上述含义进行稳定分析和性能保证。基于 De 和 Postoyan[226]、De 和 Tesi[85] 以及 Dolk 等人 [83] 的研究，对任意 $(t,k)\in\left(\mathrm{dom}\,\boldsymbol{q}_{ij}\right)\cap\left(\tilde{W}_{p_r^{ij}}^{ij}\times\mathbb{Z}_{\geqslant 0}\right)$（没有欺诈攻击），$p_r^{ij}\in\mathbb{Z}_{\geqslant 0}\cup\{-1\}$，以及任意 $(t,k)\in\left(\mathrm{dom}\,\boldsymbol{q}_{ij}\right)\cap\left(\tilde{Z}_{p_r^{ij}}^{ij}\times\mathbb{Z}_{\geqslant 0}\right)$（有欺诈攻击），$p_r^{ij}\in\mathbb{Z}_{\geqslant 0}$，不等式 (10-27)、(10-32) 和 (10-33) 表明

$$\mathcal{V}_{ij}\left(q_{ij}(t,k)\right)\leqslant\begin{cases}\exp\left(-\kappa_{ij}^0\varpi_{ij}^0\left(t-\zeta_{p_r^{ij}}^{ij}-v_{p_r^{ij}}^{ij}\right)\right)\mathcal{V}_{ij}\left(\boldsymbol{q}_{ij}\left(\zeta_{p_r^{ij}}^{ij}+v_{p_r^{ij}}^{ij},k\right)\right)\\+\left(\left|U_{l_{ij}}\right|+1\right)\theta_{ij}^2\int_{\zeta_{p_r^{ij}}^{ij}+v_{p_r^{ij}}^{ij}}^{t}\exp\left(-\kappa_{ij}^0\varpi_{ij}^0(t-s)\right)\left|\omega_{ij}(s)\right|^2\mathrm{d}s\\\qquad\left(\tilde{L}_{ij}=0\right)\\\exp\left(\kappa_{ij}^1\varpi_{ij}^1\left(t-\zeta_{p_{k'}^{ij}}^{ij}\right)\right)\mathcal{V}_{ij}\left(\boldsymbol{q}_{ij}\left(\zeta_{p_r^{ij}}^{ij},k\right)\right)\\+\left(\left|U_{l_{ij}}\right|+1\right)\theta_{ij}^2\int_{\zeta_{p_r^{ij}}^{ij}}^{t}\exp\left(\kappa_{ij}^1\varpi_{ij}^1(t-s)\right)\left|\boldsymbol{\omega}_{ij}(s)\right|^2\mathrm{d}s\\\qquad\left(\tilde{L}_{ij}\neq 0\right)\end{cases}\tag{10-34}$$

作为连接有 / 没有欺诈攻击的桥梁，基于文献 [83] 中 Lemma 4，先给出以下引理。

引理 10.2

对任意$(t,k)\in \mathrm{dom}\,\boldsymbol{q}_{ij}$，有

$$\nu_{ij}\left(\boldsymbol{q}_{ij}(t,k)\right)\leqslant \mathcal{Y}_{ij}(0,t)\nu_{ij}\left(\boldsymbol{q}_{ij}(0,0)\right)+\left(\left|U_{l_{ij}}\right|+1\right)\theta_{ij}^{2}\int_{0}^{t}\mathcal{Y}_{ij}(s,t)\left|\boldsymbol{\omega}_{ij}(s)\right|^{2}\mathrm{d}s \tag{10-35}$$

其中，$\mathcal{Y}_{ij}(s,t)=\mathrm{e}^{-\kappa_{ij}^{0}\varpi_{ij}^{0}\left|\bar{\Theta}_{qij}(s,t)\right|}\mathrm{e}^{\kappa_{ij}^{1}\bar{w}_{ij}^{1}\left|\bar{\Xi}_{qij}(s,t)\right|}$。

根据式 (10-33) 和引理 10.2，有 $\mathcal{Y}_{ij}(T_1,T_2)\leqslant \mathrm{Im}_{ij}\mathrm{e}^{-\hat{\beta}_{ij}(T_2-T_1)}$，其中 $\mathrm{Im}_{ij}=\mathrm{e}^{\varsigma_{*}^{ij}\left(\kappa_{ij}^{0}\varpi_{ij}^{0}+\kappa_{ij}^{1}\varpi_{ij}^{1}\right)}$以及$\hat{\beta}_{ij}=\kappa_{ij}^{0}\varpi_{ij}^{0}-\dfrac{\kappa_{ij}^{0}\varpi_{ij}^{0}+\kappa_{ij}^{1}\varpi_{ij}^{1}}{T_{*}^{ij}}>0$。因此由式 (10-34) 和式 (10-35) 可知

$$\mathcal{V}_{ij}\left(\boldsymbol{q}_{ij}(t,k)\right)\leqslant\begin{cases}\mathrm{Im}_{ij}\mathrm{e}^{-\hat{\beta}_{ij}t}\mathcal{V}_{ij}\left(\boldsymbol{q}_{ij}(0,0)\right), & \boldsymbol{\omega}_{ij}\equiv 0,\forall(i,j)\in\mathcal{E}_{\mathrm{cm}}\\ \mathrm{Im}_{ij}\mathrm{e}^{-\hat{\beta}_{ij}t}\mathcal{V}_{ij}\left(\boldsymbol{q}_{ij}(0,0)\right)+\left[\left(\left|U_{l_{ij}}\right|+1\right)\mathrm{Im}_{ij}\theta_{ij}^{2}\right. & \\ \left.\times\int_{0}^{t}\mathrm{e}^{-\hat{\beta}_{ij}}\mathrm{d}s\right]\|\omega_{ij}\|_{\mathcal{L}_{\infty}}^{2}, & \boldsymbol{\omega}_{ij}\neq 0,\exists(i,j)\in\mathcal{E}_{\mathrm{cm}}\end{cases} \tag{10-36}$$

根据式 (10-36) 中第一个不等式，基于$\mathcal{V}_{ij}\left(\boldsymbol{q}_{ij}(0,0)\right)\leqslant\max\left(\hat{\bar{\alpha}}^{ij},\left(\left|U_{l_{ij}}\right|+1\right)\gamma_{ij}\dfrac{1}{\delta_{ij}}\hat{\bar{c}}_{W_{ij}}^{2}\right)\left|\boldsymbol{q}(0,0)\right|_{\mathcal{A}}^{2}$，其中正常数$\hat{\bar{\alpha}}^{ij}$和$\hat{\bar{c}}_{W_{ij}}$满足$\bar{\alpha}^{ij}\left(\left|\boldsymbol{\xi}_{ij}\right|\right)\leqslant\hat{\bar{\alpha}}^{ij}\left|\boldsymbol{\xi}_{ij}\right|$和$\bar{c}_{W_{ij}}^{2}\left(\left|\boldsymbol{e}_{ij}\right|\right)\leqslant\hat{\bar{c}}_{W_{ij}}^{2}\left|e_{ij}\right|$，不等式 (10-14)、(10-16) 和 (10-26) 表明

$$\left|\boldsymbol{q}_{ij}(t,k)\right|_{\mathcal{A}}\leqslant\Pi_{ij}^{0}\mathrm{e}^{-\left(\hat{\beta}_{ij}/2\right)t}\left|\boldsymbol{q}_{ij}(0,0)\right|_{\mathcal{A}} \tag{10-37}$$

其中，$\Pi_{ij}^{0}=\sqrt{\dfrac{\mathrm{Im}_{ij}\max\left(\hat{\bar{\alpha}}^{ij},\left(\left|U_{l_{ij}}\right|+1\right)\gamma_{ij}\dfrac{1}{\delta_{ij}}\hat{\bar{c}}_{W_{ij}}^{2}\right)}{\underline{c}_{\mathcal{V}_{ij}}}}$。基于式(10-36)中第二个不等式，$c_{z_{ij}}\left|\mathcal{O}_{ij}\left(\boldsymbol{\xi}_{ij}\right)\right|^{2}\leqslant V_{ij}\left(\boldsymbol{\xi}_{ij}\right)\leqslant\mathcal{V}_{ij}\left(\boldsymbol{q}_{ij}(t,k)\right)$表明

$$\| z_{ij} \|_{\mathcal{L}_\infty} \leqslant \Pi_{ij}^1 \left| \boldsymbol{q}_{ij}(0,0) \right|_{\mathcal{A}} + \hat{\Pi}_{ij} \| \omega_{ij} \|_{\mathcal{L}\infty} \tag{10-38}$$

其中，$\Pi_{ij}^1 = \sqrt{\dfrac{\mathrm{Im}_{ij} \max\left(\hat{\bar{\alpha}}^{ij}, \left(\left| U_{l_{ij}} \right| + 1 \right) \gamma_{ij} \dfrac{1}{\delta_{ij}} \hat{\bar{c}}_{W_{ij}}^2 \right)}{c_{z_{ij}}}}$；$\hat{\Pi}_{ij} = \theta_{ij} \sqrt{\dfrac{\left(\left| U_{l_{ij}} \right| + 1 \right) \mathrm{Im}_{ij}}{\hat{\beta}_{ij} c_{z_{ij}}}}$。

因此，基于定义 10.3、定义 10.4，当$\boldsymbol{\omega} \equiv 0$时，紧集$\boldsymbol{\mathcal{A}} = \{ \boldsymbol{q} \in \mathbb{X} | \boldsymbol{\xi} = 0, \boldsymbol{e} = 0 \}$是一致全局渐进稳定的；如果$\boldsymbol{\omega} \neq 0$，是$\mathcal{L}_\infty$稳定的，其中有限增益$\mathcal{L}_\infty$小于等于$\max_{(i,j) \in \varepsilon_{\mathrm{cm}}} \left\{ \theta_{ij} \sqrt{\dfrac{\left(\left| U_{ij} \right| + 1 \right) \mathrm{Im}_{ij}}{\hat{\beta}_{ij} c_{zij}}} \right\}$，$\mathrm{Im}_{ij} = \mathrm{e}^{\varsigma_*^{ij} \left(\kappa_{ij}^0 \varpi_{ij}^0 + \kappa_{ij}^1 \varpi_{ij}^1 \right)}$，$\hat{\beta}_{ij} = \kappa_{ij}^0 \varpi_{ij}^0 - \dfrac{\kappa_{ij}^0 \varpi_0 + \kappa_{ij}^1 \varpi_{ij}^1}{T_*^{ij}}, \varsigma_*^{ij} = \varsigma^{ij} + v^{ij} \tau_{\mathrm{miet}}^{ij}$以及$T_*^{ij} = \dfrac{\tau_j^{ij} \tau_d^{ij}}{\tau_f^{ij} + \tau_d^{ij} \tau_{\mathrm{miet}}^{ij}}$。至此证明了定理 10.1 的结果。 ■

情形 II：基于内邻近边的同步欺诈攻击，对任意$(i,j) \in \varepsilon_{\mathrm{cm}}$，即$\{ H_n^{ki} \}_{(k,i) \in Z_{l_{ij}}}$对于$(k,i) \in Z_{l_{ij}}$是相同的，其中$\chi_{ki} \neq 0$。

假设 10.4

对任意$i \in Z_j$和$(i,j) \in \mathcal{E}_{\mathrm{cm}}$，存在连续可微函数$V_j : \mathbb{R}^{2n|Z_j|} \to \mathbb{R}_{>0}$，函数$\underline{\alpha}^j$、$\overline{\alpha}^j$、$\varrho_{ij} \in \mathcal{K}_\infty$，正常数$c_{z_j}$、$\rho_j$、$\gamma_{ij}$和$\tilde{\rho}_{ij}$使得

i) 对任意$\xi_{ij} \in \mathbb{R}^{2n}$和输出$z_j = \left(z_{ij}(\boldsymbol{\xi}_{ij}) \right)_{i \in Z_j} = \mathcal{O}_j = \left(\mathcal{O}_{ij}(\boldsymbol{\xi}_{ij}) \right)_{i \in Z_j}$，满足

$$c_{z_j} \left| \mathcal{O}_j \right|^2 \leqslant V_j\left((\boldsymbol{\xi}_{ij})_{i \in Z_j} \right); \underline{\alpha}^j \left(\left| (\boldsymbol{\xi}_{ij})_{i \in Z_j} \right| \right) \leqslant V_j\left((\boldsymbol{\xi}_{ij})_{i \in Z_j} \right) \leqslant \overline{\alpha}^j \left(\left| (\boldsymbol{\xi}_{ij})_{i \in Z_j} \right| \right), \tag{10-39}$$

ii) 对任意$e_{ij} \in \mathbb{R}^{2n}$和几乎所有的$\xi_{ij} \in \mathbb{R}^{2n}$，有

$$\begin{aligned} &\left\langle \nabla V_j\left((\boldsymbol{\xi}_{ij})_{i \in Z_j} \right), \left(F_{ij}\left(\boldsymbol{\xi}_{\bar{Z}_{l_{ij}}}, \boldsymbol{e}_{\bar{Z}_{l_{ij}}}, \boldsymbol{\omega}_{ij} \right) \right)_{i \in Z_j} \right\rangle \\ &\leqslant -\rho_j V_j\left((\boldsymbol{\xi}_{ij})_{i \in Z_j} \right) - \sum_{i \in Z_j} \left[\varrho_{ij}\left(|\boldsymbol{\xi}_{ij}| \right) + \theta_{ij}^2 |\boldsymbol{\omega}_{ij}|^2 + \left(\gamma_{ij}^2 - \tilde{\rho}_{ij} \right) W_{ij}^2(\boldsymbol{e}_{ij}) - H_{ij}^2\left(\boldsymbol{\xi}_{ij}, \boldsymbol{e}_{ij}, \boldsymbol{\omega}_{ij} \right) \right], \end{aligned} \tag{10-40}$$

并且在$\tilde{L}_{ij} \neq 0$的情况下，有$\tilde{\rho}_{ij} < \gamma_{ij} \tilde{L}_{ij}$。

定理 10.2

混杂系统 (10-13) 满足条件 10.2.1 和假设 101、10.2 和 10.4。对任意 $i \in Z_j$ 和 $(i,j) \in \mathcal{E}_{\mathrm{cm}}$，假设以下条件满足：

(2a) 对于满足假设 10.1 的欺诈攻击：

$$\varpi_j^0 > \left(\frac{\max\limits_{i \in Z_j}\left(\tau_{\mathrm{miet}}^{ij}\right)}{\tau_f^j} + \frac{1}{\tau_d^j} \right)\left(\varpi_j^0 + \varpi_j^1\right) \tag{10-41}$$

其中，$\tau_f^j = \tau_f^{ij}$ 以及 $\tau_d^j = \tau_d^{ij}$ 满足假设 10.1(A1)，$\varpi_j^0 = \min\left\{\rho_j, \min_{i \in Z_j}\left(\frac{\delta_{ij}\tilde{\rho}_{ij}}{\gamma_{ij}}\right)\right\}$，以及 $\varpi_j^1 = \max_{i \in Z_j}\left(\frac{2\gamma_{ij}\frac{1}{\delta_{ij}}\tilde{L}_{ij} - \tilde{\rho}_{ij}}{\gamma_{ij}\delta_{ij}}\right)$，其中 $\tilde{L}_{ij} = \frac{\chi_{ij}}{\chi_{ij}+1}$ 满足假设 10.1(A2) 以及 $\delta_{ij} \leqslant \phi_{ij,\mathrm{miet}}$ 满足式 (10-21)。

(2b) 对于在式 (10-11) 中设计的参数满足假设 10.2 和假设 10.4 的边事件驱动策略：基于 $\sigma_{ij} \in \mathcal{K}_\infty$ 和 $\sigma_{ij}\left(\eta_{ij}\right) \geqslant \varpi_j^0\eta_{ij}$，有

$$\Psi_{ij} = \begin{cases} \hat{\Psi}_{ij}, & 0 \leqslant \tau_{ij} \leqslant \tau_{\mathrm{miet}}^{ij} \\ \tilde{\Psi}_{ij}, & \tau_{ij} \geqslant \tau_{\mathrm{miet}}^{ij} \end{cases} \tag{10-42}$$

其中，$\hat{\Psi}_{ij} = \varrho_{ij}\left(\left|\boldsymbol{\xi}_{ij}\right|\right) + \left(\lambda_{ij}-1\right)\gamma_{ij}^2 W_{ij}^2 \phi_{ij}^2 - \sigma_{ij}\left(\eta_{ij}\right)$；$\tilde{\Psi}_{ij} = \varrho_{ij}(\left|\boldsymbol{\xi}_{ij}\right|) - (\gamma_{ij}^2 + 2\gamma_{ij}\phi_{ij}L_{ij} + \gamma_{ij}^2\phi_{ij}^2)W_{ij}^2 - \sigma_{ij}\left(\eta_{ij}\right)$。

(2c) $\tilde{\eta}_{ij} = \gamma_{ij}W_{ij}^2\left(\delta_{ij} - \frac{c_{ij}^2}{\delta_{ij}}\right)$ 满足 $\delta_{ij} > c_{ij} \geqslant \tilde{L}_{ij}, \gamma_{ij}$、$W_{ij}$、$\tilde{L}_{ij}$ 满足假设 10.2，γ_{ij} 满足式 (10-18)，由式 (10-21) 可知 $\delta_{ij} \leqslant \phi_{ij,\mathrm{miet}}$。那么由定义 10.3 可知当 $\omega = 0$ 时，$\boldsymbol{\mathcal{A}} = \left\{\boldsymbol{q} \in \mathbb{X} \mid \boldsymbol{\xi} = 0, \boldsymbol{e} = 0\right\}$ 是一致全局渐近稳定的。紧集 $\boldsymbol{\mathcal{A}} = \left\{\boldsymbol{q} \in \mathbb{X} \mid \boldsymbol{\xi} = 0, \boldsymbol{e} = 0\right\}$ 是 $\mathcal{L}_\infty$ 稳定的，关于非零输入干扰 ω 和输出 z，而且增益 $\mathcal{L}_\infty$ 小于等于 $\max_{j \in \mathcal{N}}\left\{\left(\max_{i \in Z_j}\theta_{ij}\right)\sqrt{\frac{\mathrm{Im}_j}{\hat{\beta}_j c_{zj}}}\right\}$，其中 $\mathrm{Im}_j = e^{\varsigma_*^j\left(\varpi_j^0 + \varpi_j^1\right)}, \hat{\beta}_j = \varpi_j^0 - \frac{\varpi_j^0 + \varpi_j^1}{T_*^j}$，$\varsigma_*^j = \varsigma^j + \nu^j \max_{i \in Z_j}\left(\tau_{\mathrm{miet}}^{ij}\right)$，其中 $\varsigma^j = \varsigma^{ij}$

和$\nu^{j}=\nu^{ij}$满足假设 10.1，以及$T_{*}^{j}=\dfrac{\tau_{f}^{j}\tau_{d}^{j}}{\tau_{f}^{j}+\tau_{d}^{j}\max_{i\in Z_{j}}\left(\tau_{\text{miet}}^{ij}\right)}$。

证明：

基于假设 10.1、10.2 和 10.4，对于$\boldsymbol{q}=\left(\boldsymbol{q}_{ij}\right)_{(i,j)\in\mathcal{E}_{\text{cm}}}=\left(\left(\boldsymbol{q}_{ij}\right)_{(i)\in Z_{j}}\right)_{j\in\mathcal{N}}\in C\cup D$，其中$\left(\boldsymbol{q}_{ij}\right)_{(i)\in Z_{j}}=\boldsymbol{q}_{j}$，定义李雅普诺夫函数

$$\mathcal{V}_{j}\left(\boldsymbol{q}_{j}\right)=V_{j}\left(\left(\boldsymbol{\xi}_{ij}\right)_{i\in Z_{j}}\right)+\sum_{i\in Z_{j}}\left(\gamma_{ij}\phi_{ij}W_{ij}^{2}\left(\boldsymbol{e}_{ij}\right)+\eta_{ij}\right)$$

类似定理 10.1 的证明过程，基于定理 10.2(2a) ～ (2c)，有

$$\upsilon_{j}\left(\boldsymbol{q}_{j}\left(t,k\right)\right)\leqslant\begin{cases}\text{Im}_{j}e^{-\hat{\beta}_{j}t}\mathcal{V}_{j}\left(q_{j}\left(0,0\right)\right), & \left(\omega_{ij}\right)_{i\in Z_{j}}\equiv 0,\\ \text{Im}_{j}\text{e}^{-\hat{\beta}_{ij}t}\mathcal{V}_{j}\left(\boldsymbol{q}_{j}\left(0,0\right)\right)+\left(\text{Im}_{j}\int_{0}^{t}\text{e}^{-\hat{\beta}_{j}(t-s)}\text{d}s\right) & \\ \times\left(\max\limits_{i\in Z_{j}}\theta_{ij}^{2}\right)\|\left(\boldsymbol{\omega}_{ij}\right)_{i\in Z_{j}}\|_{\mathcal{L}_{\infty}}^{2}, & \omega_{ij}\neq 0,\exists i\in Z_{j}\end{cases}\tag{10-43}$$

根据式 (10-43) 中的第一个不等式，基于$\mathcal{V}_{j}\left(\boldsymbol{q}_{j}\left(0,0\right)\right)\leqslant\max_{i\in Z_{j}}\left(\hat{\alpha}^{j},\left|Z_{j}\right|\gamma_{ij}\dfrac{1}{\delta_{ij}}\hat{c}_{W_{ij}}^{2}\right)\left|\boldsymbol{q}_{j}\left(0,0\right)\right|_{\mathcal{A}}^{2}$，其中正常数$\hat{\alpha}^{ij}$和$\hat{c}_{W_{ij}}$分别满足$\overline{\alpha}^{j}\left(\left|\left(\boldsymbol{\xi}_{ij}\right)_{i\in Z_{j}}\right|\right)\leqslant\hat{\alpha}^{j}\left|\left(\boldsymbol{\xi}_{ij}\right)_{i\in Z_{j}}\right|$和$\overline{c}_{W_{ij}}^{2}\left(\left|\boldsymbol{e}_{ij}\right|\right)\leqslant\hat{c}_{W_{ij}}^{2}\left|\boldsymbol{e}_{ij}\right|$，类似式 (10-37)，有

$$\left|\boldsymbol{q}_{j}\left(t,k\right)\right|_{\mathcal{A}}\leqslant\Pi_{j}^{0}e^{-\left(\hat{\beta}_{j}/2\right)t}\left|\boldsymbol{q}_{j}\left(0,0\right)\right|_{\mathcal{A}}$$

其中，$\Pi_{j}^{0}=\sqrt{\dfrac{\text{Im}_{j}\max_{i\in Z_{j}}\left(\hat{\alpha},|Z_{j}|\gamma_{ij}\dfrac{1}{\delta_{ij}}\hat{c}_{W_{ij}}^{2}\right)}{\underline{c}_{\mathcal{V}_{j}}}}$。基于式(10-43)中的第二个不等式，不等式

$$c_{z_{j}}\left|\mathcal{O}_{j}\left(\left(\boldsymbol{\xi}_{ij}\right)_{i\in Z_{j}}\right)\right|^{2}\leqslant V_{j}\left(\left(\boldsymbol{\xi}_{ij}\right)_{i\in Z_{j}}\right)\leqslant\mathcal{V}_{j}\left(\boldsymbol{q}_{j}\left(t,k\right)\right)\text{表明}$$

$$\|\boldsymbol{z}_{j}\|_{\mathcal{L}_{\infty}}\leqslant\Pi_{j}^{1}\left|\boldsymbol{q}_{j}\left(0,0\right)\right|_{\mathcal{A}}+\hat{\Pi}_{j}\|\omega_{j}\|_{\mathcal{L}_{\infty}}$$

其中，$\Pi_j^1 = \sqrt{\dfrac{\mathrm{Im}_j \max_{i\in Z_j}\left(\hat{\bar{\alpha}}^j, |Z_j|\gamma_{ij}\dfrac{1}{\delta_{ij}}\hat{\bar{c}}_{W_{ij}}^2\right)}{c_{z_j}}}$以及$\hat{\Pi}_j = \max_{i\in Z_j}\{\theta_{ij}\}\sqrt{\dfrac{\mathrm{Im}_j}{\hat{\beta}_j c_{z_j}}}$。

基于定义 10.3、10.4，闭集$\boldsymbol{\mathcal{A}} = \{\boldsymbol{q} \in \mathbb{X} | \ \boldsymbol{\xi} = 0, \boldsymbol{e} = 0\}$是一致全局渐近稳定的，其中$\boldsymbol{\omega} \equiv 0$；是$\mathcal{L}_\infty$稳定的，关于非零输入干扰$\omega$，其中有限增益$\mathcal{L}_\infty$小于等于$\max_{j\in\mathcal{N}}\left\{\left(\max_{i\in Z_j}\theta_{ij}\right)\sqrt{\dfrac{\mathrm{Im}_j}{\hat{\beta}_j c_{z_j}}}\right\}$，其中$\mathrm{Im}_j = \mathrm{e}^{\varsigma_*^j\left(\varpi_j^0+\varpi_j^1\right)}$，$\hat{\beta}_j = \varpi_j^0 - \dfrac{\varpi_j^0 + \varpi_j^1}{T_*^j}, \varsigma_*^j = \varsigma^j + v^j \max_{i\in Z_j}\left(\tau_{\text{miet}}^{ij}\right)$以及$T_*^j = \dfrac{\tau_f^j \tau_d^j}{\tau_f^j + \tau_d^j \max_{i\in Z_j}\left(\tau_{\text{miet}}^{ij}\right)}$。

至证毕。 ■

情形Ⅲ：基于所有边的同步欺诈攻击，即对给定的$n \in \mathbb{Z}_{\geqslant 0}$，$\left\{H_n^{ij}\right\}_{(i,j)\in\mathcal{E}_{\text{cm}}}$对任意$(i,j) \in \mathcal{E}_{\text{cm}}$都是相同的，其中$\chi_{ij} \neq 0$。

假设 10.5

对任意$(i,j) \in \mathcal{E}_{\text{cm}}$，存在连续可微函数$V: \mathbb{R}^{2n|\mathcal{E}_{\text{cm}}|} \to \mathbb{R}_{>0}$、函数$\underline{\alpha}$、$\overline{\alpha}$、$\varrho_{ij} \in \mathcal{K}_\infty$、正常数$c_z$、$\rho$、$\gamma_{ij}$和$\tilde{\rho}_{ij}$，使得

i) 对任意$\boldsymbol{\xi}_{ij} \in \mathbb{R}^{2n}$和输出$\boldsymbol{z} = \left(\boldsymbol{z}_{ij}\left(\boldsymbol{\xi}_{ij}\right)\right)_{(i,j)\in\mathcal{E}_{\text{cm}}} = \mathcal{O} = \left(\mathcal{O}_{ij}\left(\boldsymbol{\xi}_{ij}\right)\right)_{(i,j)\in\mathcal{E}_{\text{cm}}}$，有

$$
\begin{gathered}
c_z |\mathcal{O}|^2 \leqslant V\left(\left(\boldsymbol{\xi}_{ij}\right)_{(i,j)\in\mathcal{E}_{\text{cm}}}\right), \\
\underline{\alpha}\left(\left|\left(\boldsymbol{\xi}_{ij}\right)_{(i,j)\in\mathcal{E}_{\text{cm}}}\right|\right) \leqslant V\left(\left(\boldsymbol{\xi}_{ij}\right)_{(i,j)\in\mathcal{E}_{\text{cm}}}\right) \leqslant \overline{\alpha}\left(\left|\left(\boldsymbol{\xi}_{ij}\right)_{(i,j)\in\mathcal{E}_{\text{cm}}}\right|\right)
\end{gathered} \tag{10-44}
$$

ii) 对任意$\boldsymbol{e}_{ij} \in \mathbb{R}^{2n}$和几乎所有的$\boldsymbol{\xi}_{ij} \in \mathbb{R}^{2n}$，有

$$
\begin{aligned}
&\left\langle \nabla V\left(\left(\boldsymbol{\xi}_{ij}\right)_{(i,j)\in\mathcal{E}_{\text{cm}}}\right), \left(F_{ij}\left(\boldsymbol{\xi}_{\bar{Z}l_{ij}}, \boldsymbol{e}_{\bar{Z}l_{ij}}, \boldsymbol{\omega}_{ij}\right)\right)_{(i,j)\in\mathcal{E}_{\text{cm}}}\right\rangle \\
&\leqslant -\rho V\left(\left(\boldsymbol{\xi}_{ij}\right)_{(i,j)\in\mathcal{E}_{\text{cm}}}\right) - \sum_{(i,j)\in\mathcal{E}_{\text{cm}}}\left[\varrho_{ij}\left(\left|\boldsymbol{\xi}_{ij}\right|\right) + \theta_{ij}^2\left|\boldsymbol{\omega}_{ij}\right|^2\right. \\
&\quad \left. + \left(\gamma_{ij}^2 - \tilde{\rho}_{ij}\right)W_{ij}^2\left(\boldsymbol{e}_{ij}\right) - H_{ij}^2\left(\boldsymbol{\xi}_{ij}, \boldsymbol{e}_{ij}, \boldsymbol{\omega}_{ij}\right)\right],
\end{aligned} \tag{10-45}
$$

并且在$\tilde{L}_{ij} \neq 0$的情况下，有$\tilde{\rho}_{ij} < \gamma_{ij}\tilde{L}_{ij}$。

定理 10.3

混杂系统 (10-13) 满足条件 10.2.1 和假设 10.1、10.2、10.5。对任意 $(i,j)\in\mathcal{E}_{\mathrm{cm}}$假设满足以下条件：

(3a) 对任意满足假设 10.1 的欺诈攻击，有

$$\varpi_0>\left(\frac{\max_{(i,j)\in\mathcal{E}_{\mathrm{cm}}}\left\{\tau_{\mathrm{miet}}^{ij}\right\}}{\tau_f}+\frac{1}{\tau_d}\right)(\varpi_0+\varpi_1) \tag{10-46}$$

其中，$\tau_f=\tau_f^{ij}$和$\tau_d=\tau_d^{ij}$满足假设 10.1(A1)；$\varpi_0=\min\left\{\rho,\min_{(i,j)\in\varepsilon_{\mathrm{cm}}}\left(\frac{\delta_{ij}\bar{\rho}_{ij}}{\gamma_{ij}}\right)\right\}$；$\varpi_1=\max_{(i,j)\in\varepsilon_{\mathrm{cm}}}\left(\frac{2\gamma_{ij}\frac{1}{\delta_{ij}}\tilde{L}_{ij}-\tilde{\rho}_{ij}}{\gamma_{ij}\delta_{ij}}\right)$；$\tilde{L}_{ij}=\frac{\chi_{ij}}{\chi_{ij}+1}$满足假设10.1(A2)；$\delta_{ij}\leqslant\phi_{ij,\mathrm{miet}}$满足式 (10-21)。

(3b) 对于式 (10-11) 中设计的参数满足假设 10.2 和 10.5 的边事件驱动策略，基于$\sigma_{ij}\in\mathcal{K}_\infty$和$\sigma_{ij}(\eta_{ij})\geqslant\varpi_0\eta_{ij}$，有

$$\varPsi_{ij}=\begin{cases}\hat{\varPsi}_{ij}, & 0\leqslant\tau_{ij}\leqslant\tau_{\mathrm{miet}}^{ij}\\ \tilde{\varPsi}_{ij}, & \tau_{ij}\geqslant\tau_{\mathrm{miet}}^{ij}\end{cases} \tag{10-47}$$

其中，$\hat{\varPsi}_{ij}=\varrho_{ij}(|\boldsymbol{\xi}_{ij}|)+(\lambda_{ij}-1)\gamma_{ij}^2W_{ij}^2\phi_{ij}^2-\sigma_{ij}(\eta_{ij})$；$\tilde{\varPsi}_{ij}=\varrho_{ij}(|\boldsymbol{\xi}_{ij}|)-(\gamma_{ij}^2+2\gamma_{ij}\phi_{ij}L_{ij}+\gamma_{ij}^2\phi_{ij}^2)W_{ij}^2-\sigma_{ij}(\eta_{ij})$。

(3c) $\tilde{\eta}_{ij}=\gamma_{ij}W_{ij}^2\left(\delta_{ij}-\frac{c_{ij}^2}{\delta_{ij}}\right)$满足$\delta_{ij}>c_{ij}\geqslant\tilde{L}_{ij},\gamma_{ij}$、$W_{ij}$和$\tilde{L}_{ij}$满足假设 10.2，$\gamma_{ij}$满足式 (10-18)，以及$\delta_{ij}\leqslant\phi_{ij,\mathrm{miet}}$。

那么由定义 10.3 可知，当$\omega=0$时，紧集$\boldsymbol{\mathcal{A}}=\{\boldsymbol{q}\in\mathbb{X}|\ \boldsymbol{\xi}=0,\boldsymbol{e}=0\}$是一致全局渐近稳定的。紧集$\boldsymbol{\mathcal{A}}=\{\boldsymbol{q}\in\mathbb{X}|\ \boldsymbol{\xi}=0,\boldsymbol{e}=0\}$是$\mathcal{L}_\infty$稳定的，关于非零输入干扰$\boldsymbol{\omega}$和输出$\boldsymbol{z}$，而且有限增益$\mathcal{L}_\infty$小于等于$\max_{(i,j)\in\varepsilon_{\mathrm{cm}}}\{\theta_{ij}\}\sqrt{\frac{\mathrm{Im}}{\hat{\beta}c_z}}$，其中$\mathrm{Im}=\mathrm{e}^{\varsigma_*(\varpi_0+\varpi_1)},\hat{\beta}=\varpi_0-\frac{\varpi_0+\varpi_1}{T_*},\varsigma_*=\varsigma+v\max_{(i,j)\in\varepsilon_{\mathrm{cm}}}\left\{\tau_{\mathrm{miet}}^{ij}\right\}$，其中$\varsigma=\varsigma^{ij}$和$v=v^{ij}$满足假设 10.1，以及$T_*=\frac{\tau_f\tau_d}{\tau_f+\tau_d\max_{(i,j)\in\varepsilon_{\mathrm{cm}}}\left\{\tau_{\mathrm{miet}}^{ij}\right\}}$。

证明：

定理 10.3 的证明与定理 10.1 和定理 10.2 的证明相似，此处略去。■

注释 10.6 有向图G的条件 10.2.1 假设是包含多个领导者的编队包围控制的先决条件[269]。在条件 10.2.1 中的关于通信拓扑的假设，做以下分析：①根据式 (10-37)，$\hat{\beta}_{ij} > 0$是为了保证追踪误差系统收敛（即实现编队包围控制）。②根据τ_f^{ij}，正参数$\hat{\beta}_{ij}$反映了收敛率，并且负相关于欺诈攻击的次数 / 时长 / 幅度，其中τ_d^{ij}和χ_{ij}满足假设 10.1 的条件，并且正参数$\hat{\beta}_{ij}$也负相关于最小传输间隔$\tau_{\mathrm{miet}}^{ij}$。也就是说，欺诈攻击的次数 / 时长 / 幅度越大或者传输间隔越大（即事件驱动的频率越低），参数$\hat{\beta}_{ij}$越小，因此追踪误差系统需要更多的时间收敛到一个更大的界内，如 Ge 等人[271]和 Ding 等人[272]研究中的分析。

注释 10.7 在**情形** Ⅰ 中，任何边都会在任何时候被攻击，即异步攻击，因此**情形** Ⅰ 是最一般的，包含**情形** Ⅱ 和**情形** Ⅲ，以及l_{ij}的外邻居集中的边被攻击的情况。进一步分析**情形** Ⅱ、**情形** Ⅲ 也是为了展示如何在假设 10.4、10.5 中降低条件的保守性。由于空间限制，我们没有进一步考虑l_{ij}的外邻居被攻击的情形，因为在这种情形中，受到影响的是节点$k\left(k \in U_j\right)$，通过$k \in Z_p$对智能体p进行分析，以确保子系统p是输入状态稳定的，这是对l_{ij}的外邻居被攻击情况做进一步分析的合理方法，但是对p子系统的稳定性做出比假设 10.3 更温和的条件并不是一件容易的工作，这在今后会做进一步的研究。

10.2.3 算法验证与分析

此次数值仿真将弹性混杂事件驱动策略用于在三维空间飞行的垂直起降无人机时变编队跟踪，其中$\mathcal{N} = \{1,2,\cdots,10\}, F = \{1,2,\cdots,6\}$以及$E = \{7,8,9,10\}$。令 6 个追随者的状态保持时变编队，同时追踪着 4 个领导者组成凸多边形。追随者和领导者的位置状态动力学模型可以分别表示为

$$\dot{x}_i = h_i + \omega_i, i \in F;\ \dot{x}_i = \omega_i, i \in E, \tag{10-48}$$

其中，$x_i=\left(x_{i1}\ x_{i2}\ x_{i3}\right)^{\mathrm{T}}\in\mathbb{R}^3$表示第 i 个无人机的状态；输入干扰$\omega_i=0.1i\sin t$；控制输入$h_i=\sum_{j\in F}a_{ij}\left(x_i-\hat{x}_j-r_i+\hat{r}_j\right)+\sum_{j\in E}a_{ij}\left(x_i-r_i+\hat{x}_j\right)$，其中$\hat{x}_j$和$\hat{r}_j$是采样值，$r_i\left(t\right)=\left(15\sin\left(t+\frac{i-1}{3}\pi\right),-15\cos\left(t+\frac{i-1}{3}\pi\right),15\cos\left(t+\frac{i-1}{3}\pi\right)\right)^{\mathrm{T}},i\in F$，即在三维空间中的同一个平面上，6 个跟随者一边形成六边形编队同时追踪着 4 个领导者的凸多边形。

为了简化数值仿真同时不失一般性，这里只考虑第一种情况（情形Ⅰ），而且有向互连图 G 满足条件 10.2.1，其拉普拉斯矩阵为$\boldsymbol{L}=\begin{bmatrix}\boldsymbol{L}_1 & \boldsymbol{L}_2\\ 0 & 0\end{bmatrix}_{N\times N}$。矩阵$\boldsymbol{L}_1$的六行依次为：[2 0 −1 0 0 0], [0 2−1 0 0 0], [−1 0 4 −1 0 0], [0 0 −1 3 −1 −1], [0 0 0 0 5 −1]，[0 0 0 0 −1 2]。矩阵 $\boldsymbol{L}_2$ 的六行依次为：$[-1\,0\,0\,0][0-1\,0\,0],[0-1-1\,0],[0\,0\,0\,0],[-1-1-1-1],[0\,0\,0-1]$。

令 $\xi_i=x_i-r_i-\sum_{d\in E}\alpha_d x_d$，那么有 $\boldsymbol{\xi}_{ij}=\left(\xi_i,\xi_j\right)$ 以及 $\boldsymbol{e}_{ij}=\hat{\boldsymbol{\xi}}_{ij}-\boldsymbol{\xi}_{ij}=\left(\hat{\xi}_i-\xi_i,\hat{\xi}_j-\xi_j\right)$，其中$\boldsymbol{e}_{ij}^{+}=\tilde{\mathcal{Q}}_{ij}\left(\boldsymbol{e}_{ij}\right)=\left(\mathcal{Q}_i\left(\xi_i\right),\mathcal{Q}_j\left(\xi_j\right)\right)$满足$\mathcal{Q}_\alpha\left(\cdot\right)=\sin\left(\cdot\right),\alpha=\{i,j\}$。由假设 10.1 可知$\tilde{L}_{ij}=\frac{\chi_{ij}}{\chi_{ij}+1}=\frac{1}{2}$，对任意$(i,j)\in\Lambda$。令$W_{ij}=\left|e_{ij}\right|$。令 $T_1=0,T_2=100$，假设 Λ 中的边在 [0, 100] 中被攻击，其中$\Lambda=\{l_{31},l_{54},l_{83},l_{10,6},l_{71},l_{32}\}$。为简单起见，每条边的某些参数是相同的，这可以很容易地扩展到它们彼此完全不同的情况。令$\psi_{ij}\left(s\right)=\tanh s$，其中$\lambda_{ij}=1,\varrho_{ij}\left(\left|\boldsymbol{\xi}_{ij}\right|\right)=\left|\boldsymbol{\xi}_{ij}\right|^2$以及$c_{z_{ij}}=1$。基于 Carnevale 等人[267] 的研究可知假设 10.2、10.3 成立，其中，$L_{ij}=1,c_{ij}=\frac{5}{8}$以及$\gamma_{ij}=1$。根据式 (10-19)，有 $\tau_{\text{mati}}^{ij}=\frac{1}{7}=0.142857$，其中，$\delta_{ij}=\frac{3}{4}$，所以可以选择 $\tau_{\text{miet}}^{ij}=\frac{3}{4}\tau_{\text{mati}}^{ij}=\frac{3}{28}=0.107143$。基于式 (10-22)，可知 $\varpi_{ij}^0=\frac{1}{4}$以及 $\varpi_{ij}^1=\frac{4}{3}$。令 $\hat{\beta}_{ij}=\frac{2}{3}\kappa_{ij}^0\varpi_{ij}^0$，其中 $\kappa_{ij}^0=\kappa_{ij}^1=1$，并选择 $v^{ij}=0$ 和 $\varsigma^{ij}=10$，那么有 $\frac{\tau_{\text{miet}}^{ij}}{\tau_f^{ij}}+\frac{1}{\tau_d^{ij}}\leqslant\frac{1}{3}\times\frac{\kappa_{ij}^0\varpi_{ij}^0}{\kappa_{ij}^0\varpi_{ij}^0+\kappa_{ij}^1\varpi_{ij}^1}=0.052632$ 满足式 (10-22)，因此增益 $\mathcal{L}_\infty$ 小于等于 4.894088。另外根据 $\frac{\tau_{\text{miet}}^{ij}}{\tau_f^{ij}}+\frac{1}{\tau_d^{ij}}\leqslant 0.052632$ 和假设 10.1，可以选择 $\tau_f^{ij}=5$ 和 $\tau_d^{ij}=42$，即有

$\frac{\tau_{\text{miet}}^{ij}}{\tau_f^{ij}}+\frac{1}{\tau_d^{ij}}=0.052380<0.052632$。因此假设 10.1 表明$n^{ij}\left[0,100\right]\leqslant 20$以及$\left|\Xi^{ij}\left(0,100\right)\right|\leqslant 12.380952$。

定义初始条件$p_1\left(0\right)=\left(14,0,2\right)$, $p_2\left(0\right)=\left(10,-1,2\right)$, $p_3\left(0\right)=\left(6,0,-2\right)$, $p_4\left(0\right)=\left(9,-4,1\right)$, $p_5\left(0\right)=\left(1,0,1\right)$, $p_6\left(0\right)=\left(3,1,4\right)$, $p_7\left(0\right)=\left(1,5,-2\right)$, $p_8\left(0\right)=\left(6,-2,4\right)$, $p_9\left(0\right)=\left(3,6,-1\right)$, $p_{10}\left(0\right)=\left(1,4,-3\right)$以及$v_i\left(0\right)=0$，其中$i\in\mathcal{N}$。因此有图 10-3(a)、(b) 展示了 6 个跟随者的状态快照图，其中“◇”表示追随者而“○”表示领导者；图 10-3(c) 展示了受攻击的 6 条边的触发时刻，其中受攻击的 6 条边分别为$l_{31},l_{54},l_{83},l_{10,6},l_{71}$和$l_{32}$，而图 10-3(d) ～ (f) 则展示了这 6 条边的受攻击情况，其中垂直条纹表示欺骗攻击的时间间隔。表 10-1 展示了每一条边的触发次数。图 10-4 描述了收敛误差在有 / 没有事件触发时候的状态轨迹，对任意$i\in F$。结合表 10-1 以及图 10-4 中的 $2.21-0.9926=1.2174$, $2.207-0.9926=1.2144$ 以及 $2.207-0.8326=1.3744$ 可知，所设计的混杂事件驱动策略在几乎不影响收敛性能的情况下，有效地节省了通信资源进而验证其有效性。

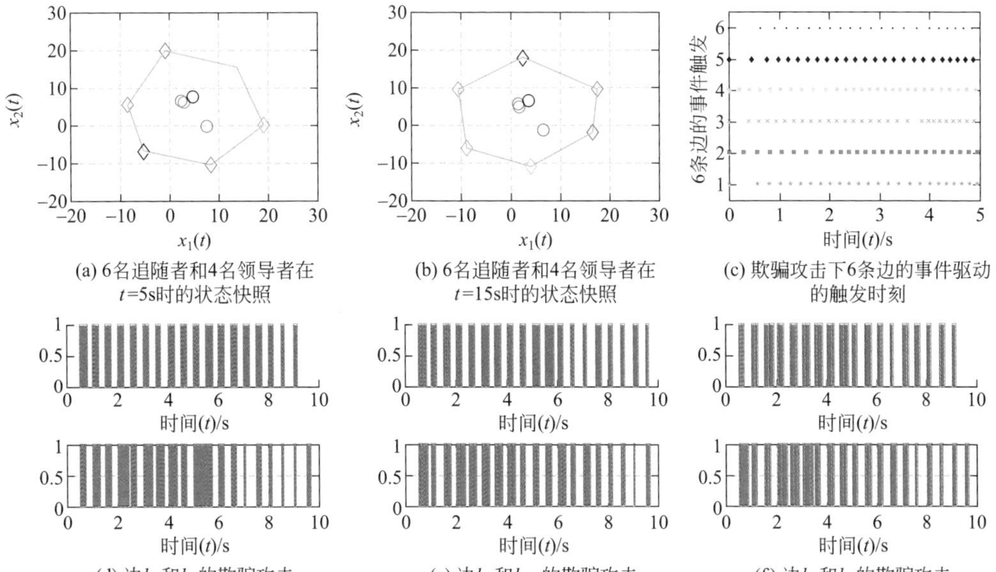

(a) 6名追随者和4名领导者在t=5s时的状态快照

(b) 6名追随者和4名领导者在t=15s时的状态快照

(c) 欺骗攻击下6条边的事件驱动的触发时刻

(d) 边l_{31}和l_{54}的欺骗攻击

(e) 边l_{83}和$l_{10,6}$的欺骗攻击

(f) 边l_{71}和l_{32}的欺骗攻击

图 10-3　垂直起降无人机的编队追踪图

表10-1　混杂事件驱动下的每条边在区间$t\in[0,20]$的驱动次数

边 l_{ij}	l_{31}	l_{71}	l_{32}	l_{82}	l_{13}	l_{43}	l_{83}	l_{93}
驱动次数	128	129	114	117	161	157	157	157
边 l_{ij}	l_{34}	l_{54}	l_{64}	l_{65}	l_{75}	l_{85}	l_{56}	$l_{10,6}$
驱动次数	131	129	129	138	142	142	138	143

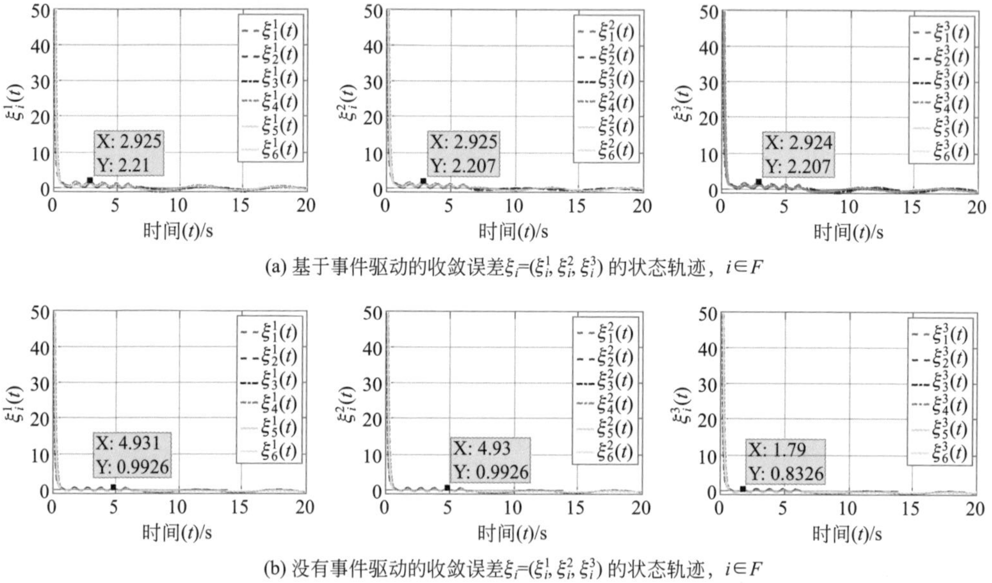

(a) 基于事件驱动的收敛误差$\xi_i=(\xi_i^1,\xi_i^2,\xi_i^3)$的状态轨迹，$i\in F$

(b) 没有事件驱动的收敛误差$\xi_i=(\xi_i^1,\xi_i^2,\xi_i^3)$的状态轨迹，$i\in F$

图10-4　垂直起降无人机的编队状态轨迹的收敛图

10.3
拒绝服务攻击下网络化多智能体系统的编队控制

令$G=(\mathcal{N},\mathcal{E}_{\rm cm},\boldsymbol{\mathcal{A}})$为一个加权图，其中节点集$\mathcal{N}=\{1,2,\cdots,N\}$，边集$\mathcal{E}_{\rm cm}\subseteq\mathcal{N}\times\mathcal{N}$，以及加权邻接矩阵$\boldsymbol{\mathcal{A}}=(a_{ij})_{\mathcal{N}\times\mathcal{N}}$。有向边$(j,i)\in\mathcal{E}_{\rm cm}$表示智能体$i$可以接收到智能体$j$广播的信息。加权矩阵$\boldsymbol{\mathcal{A}}$的元素满足：$a_{ij}>0\Leftrightarrow(j,i)\in\mathcal{E}_{\rm cm};a_{ij}=0\Leftrightarrow(j,i)\notin\mathcal{E}_{\rm cm};a_{ii}=0$，$\forall i\in\mathcal{N}$。拉普拉斯矩阵$\boldsymbol{L}=(l_{ij})_{N\times N}$的元素满足：$l_{ij}=-a_{ij}$当$i\neq j;l_{ii}=-\sum\limits_{j=1,j\neq i}^{N}l_{ij}$，对任意$i,j\in\mathcal{N}$。有

向图中，如果任意两个不同的节点之间存在有向路径[280-281]，则称其为强连通图。定义：①$Z_i=\left\{j\in\mathcal{N}|\left(j,i\right)\in\mathcal{E}_{\mathrm{cm}}\right\}$为智能体 i 的所有内邻居，智能体 i 可以接收到从这些内邻居广播的信息；②$U_i=\left\{j\in\mathcal{N}|\left(i,j\right)\in\mathcal{E}_{\mathrm{cm}}\right\}$为智能体 i 的外邻居，这些外邻居可以收到智能体 i 广播的信息。定义$\bar{Z}_i=Z_i\cup\{i\}$，其中$\left|\bar{Z}_i\right|\left(\left|Z_i\right|\right)$表示集合$\bar{Z}_i\left(Z_i\right)$的基数（集合中元素的个数），集合 U_i 也有类似的定义。

10.3.1 攻击下的分布式事件驱动策略

本节研究具有 N 个节点的非线性多智能体的编队控制问题，其中节点之间的有向通信拓扑是强连通的。令$\left\{t_k^i\right\}_{k\in\mathbb{Z}_{\geqslant 0}}$表示一系列传输时刻，其中$0\leqslant t_0^i<\cdots<t_k^i<\cdots$。在每一个传输时刻，智能体 i 会根据当前自身状态采样值和通过网络接收到的 Z_i 中智能体的最新广播值更新控制器，与此同时，智能体 i 会将当前自身状态通过网络广播给 U_i 中的多智能体。本章节假设在采样、传输和广播数据之间没有延迟，即传输时刻t_k^i也是采样时刻、更新时刻或广播时刻。

智能体 i 在输入扰动下的状态方程可以描述为：

$$\dot{\boldsymbol{x}}_i\left(t\right)=\boldsymbol{f}_{x_i}\left(\boldsymbol{x}_i\left(t\right),\boldsymbol{u}_i\left(t\right),\boldsymbol{\omega}_i\left(t\right)\right),\boldsymbol{u}_i\left(t\right)=\boldsymbol{g}_{u_i}\left(\hat{x}_{\bar{Z}_i}\left(t\right)\right),\boldsymbol{x}_i\left(t_0\right)=\boldsymbol{x}_{i0},t\in\left[t_k^i,t_{k+1}^i\right),\tag{10-49}$$

其中，$x_i\in\mathbb{R}^n$表示智能体i的状态；$u_i\in\mathbb{R}^m$表示控制输入；$\omega_i\in\mathbb{R}^l$表示$\mathcal{L}_\infty$中的外部干扰；局部利普希茨函数$\boldsymbol{f}_{x_i}:\mathbb{R}^{n\left|Z_i\right|}\times\mathbb{R}^n\times\mathbb{R}^m\times\mathbb{R}^l\to\mathbb{R}^n$满足$\boldsymbol{f}_{x_i}\left(0,0,0,0\right)=0$；函数$\boldsymbol{g}_{u_i}:\mathbb{R}^{n\left|\bar{Z}_i\right|}\to\mathbb{R}^m$满足$\boldsymbol{g}_{u_i}\left(0\right)=0$。另外，$\hat{\boldsymbol{x}}_{\bar{Z}_i}=\left\{\hat{x}_i\right\}\cup\hat{\boldsymbol{x}}_{Z_i}$，其中$\hat{\boldsymbol{x}}_{Z_i}=\left\{\hat{x}_{i_1},\cdots,\hat{x}_{i_{\left|Z_i\right|}}\right\}$表示$Z_i$中的智能体的最新广播值。

本节中智能体 i 的控制器$\boldsymbol{u}_i\left(t\right)$依赖$\hat{\boldsymbol{x}}_i\left(t\right)$而非$\boldsymbol{x}_i\left(t\right)$，这样做的目的是避免控制器根据$\boldsymbol{x}_i\left(t\right)$连续进行计算更新，其中$\hat{\boldsymbol{x}}_i\left(t\right)$靠零阶保持器在$t\in\left[t_k^i,t_{k+1}^i\right)$之间保持常数，即$\hat{\boldsymbol{x}}_i\left(t\right)=\boldsymbol{x}_i\left(t_k^i\right)$。在下一个更新时刻$t_{k+1}^i,\hat{\boldsymbol{x}}_i\left(t\right)$被重置为最新采样值$\boldsymbol{x}_i\left(t_{k+1}^i\right)$，即$\hat{\boldsymbol{x}}_i^+\left(t_{k+1}^i\right)=\boldsymbol{x}_i\left(t_{k+1}^i\right)$。与此同时，智能体$j\in U_i$会接收到智能体 i 的最新广播值$\hat{\boldsymbol{x}}_i\left(t_{k+1}^i\right)$并将其储存起来，以便在智能体 j 的传输时刻到来的时候用于更新其控制器 $\boldsymbol{u}_j$。因此，如果定义$\boldsymbol{e}_{x_i}\left(t\right)\overset{\text{def}}{=}\hat{\boldsymbol{x}}_i\left(t\right)-\boldsymbol{x}_i\left(t\right)$，那么在智能体 i 的更新时刻t_{k+1}^i，可以得到$\boldsymbol{e}_{x_i}^+\left(t_{k+1}^i\right)=0$，否则$\boldsymbol{e}_{x_i}^+\left(t_{k+1}^i\right)=\boldsymbol{e}_{x_i}\left(t_{k+1}^i\right)$。

总结起来就是，在每一个更新时刻t_{k+1}^i，控制器$\boldsymbol{u}_i$会基于当前自身状态采样值$\boldsymbol{x}_i\left(t_{k+1}^i\right)$和储存着的$\hat{\boldsymbol{x}}_{Z_i}$广播过来的最新更新值进行计算更新，并且利用零阶保持器保持该值直到下一个更新时刻。

不失一般性，令索引为 1 的智能体为参考者，进而可以定义$\boldsymbol{d}_{x_i}\in\mathbb{R}^n$为智能体 i 与参考者 1 之间的期望的相对位置向量，其中$\boldsymbol{d}_{x_1}=0$。接下来我们给出网络化多智能体 [式 (10-49)] 的编队控制定义。

定义 10.5[282]

如果对于任意给定的有界初始状态，成立$\lim\limits_{t\to\infty}\left(\boldsymbol{x}_i(t)-\boldsymbol{x}_1(t)-\boldsymbol{d}_{x_i}\right)=0$，其中$\boldsymbol{d}_{x_1}=0,\forall i\in\mathcal{N}$，那么称多智能体系统 (10-49) 实现了编队追踪控制。

本节考虑在任意的时间段内，最多有一个智能体受到拒绝服务攻击的情形，其中攻击可以发生在某个智能体的发送端或者接收端。当拒绝服务攻击发生在智能体 i 的接收端时，定义$\left\{H_{n_Z}^i\right\}_{n_Z\in\mathbb{Z}_{\geqslant 0}}$为拒绝服务攻击发生的一系列时间段，即$H_{n_Z}^i=\left\{h_{n_Z}^i\right\}\cup\left[h_{n_Z}^i,h_{n_Z}^i+\tau_{n_Z}^i\right)$，其中$h_{n_Z}^i\leqslant h_{n_Z}^i+\tau_{n_Z}^i<h_{(n_Z+1)}^i$[85]。因此对于任意时间区间$\left[T_1,T_2\right]$，其中$0\leqslant T_1\leqslant T_2$，拒绝服务攻击在智能体 i 的接收端发生的时间集合可以表示为：$\varXi_Z^i\left(T_1,T_2\right)=\left(\bigcup_{n_Z\in\mathbb{Z}_{\geqslant 0}}H_{n_Z}^i\right)\cap\left[T_1,T_2\right]$。相应地，有$\varTheta_Z^i\left(T_1,T_2\right)=\left[T_1,T_2\right]\setminus\varXi_Z^i\left(T_1,T_2\right)$，表示在时间$\left[T_1,T_2\right]$内智能体 i 的接收端没有拒绝服务攻击的时间集合。对于拒绝服务攻击发生在智能体 i 的发送端的情形，也有类似相应的定义：$\varXi_U^i\left(T_1,T_2\right)=\left(\bigcup_{n_U\in\mathbb{Z}_{\geqslant 0}}H_{n_U}^i\right)\cap\left[T_1,T_2\right]$ 和 $\varTheta_U^i\left(T_1,T_2\right)=\left[T_1,T_2\right]\setminus\varXi_U^i\left(T_1,T_2\right)$。令$\varTheta_Z^i\overset{\text{def}}{=}\varTheta_Z^i\left(T_1,T_2\right)$，$\varTheta_U^i\overset{\text{def}}{=}\varTheta_U^i\left(T_1,T_2\right),\varXi_Z^i\overset{\text{def}}{=}\varXi_Z^i\left(T_1,T_2\right)$和$\varXi_U^i\overset{\text{def}}{=}\varXi_U^i\left(T_1,T_2\right)$。定义$\left|\varXi_Z^i\right|$为时间区间$\varXi_Z^i$的长度，$\left|\varXi_U^i\right|$为$\varXi_U^i$的长度，$\left|\varTheta_Z^i\right|$为$\varTheta_Z^i$的长度，$\left|\varTheta_U^i\right|$为$\varTheta_U^i$的长度。

本节考虑$j\in U_i$的以下情形：

① 如果智能体 i 的接收端没有受到拒绝服务攻击，那么控制器$\boldsymbol{u}_i(t)$可以基于最新广播值$\hat{\boldsymbol{x}}_{Z_i}$和当前采样值$\boldsymbol{x}_i\left(t_k^i\right)$进行更新，其中$\hat{\boldsymbol{x}}_i(t)=x_i\left(t_k^i\right)$。a. 如果智能体 i 的发送端没有受到攻击，那么智能体 i 可以成功广播采样值$\hat{\boldsymbol{x}}_i\left(t_k^i\right)$给外邻居$j\in U_i$，进而当智能体 j 的更新时刻到来的时候，智能体 j 会根据内邻居 i 的广播值$\hat{\boldsymbol{x}}_i\left(t_k^i\right)$和其他内邻居最新广播值$\hat{\boldsymbol{x}}_{\bar{Z}_j\setminus\{i\}}$更新控制器$\boldsymbol{u}_j$;b.如果智能体 i 的发送端受到攻击，那么智能体 i 不能成功广

播采样值$\hat{\boldsymbol{x}}_i\left(t_k^i\right)$给$j\in U_i$，但是当智能体$j$的更新时刻到来时，智能体$j$依然会根据最新广播值$\hat{\boldsymbol{x}}_{\tilde{Z}_j}$来更新自己的控制器［不同于a，这里的$j$的控制器的更新依赖的不是$i$的最新更新值，因为此值不能被传播过来，即$\hat{\boldsymbol{x}}_i\left(t_k^i\right)\notin\hat{x}_{\bar{Z}_j}$］。因此，不管$i$的发送端是否被攻击，一旦$j$的更新时刻到来，外邻居$j$的控制器都可以更新并广播更新值。

② 如果智能体i的接收端受到拒绝服务攻击，那么控制器$u_i\left(t\right)$将不能更新也不会广播当前值给外邻居j。鉴于此，如果智能体i被攻击，可以总结为以下几种情形：只要i的接收端不被攻击（不管i的发送端是否被攻击），那么i的控制器都能更新；但是如果i的发送端在更新时刻t_k^i被攻击，那么i的更新值不能被广播，同时外邻居$j\in U_i$会在j的更新时刻更新，基于j的内邻居的最新广播值$\hat{\boldsymbol{x}}_{\bar{Z}_j}\left(\hat{\boldsymbol{x}}_i\left(t_k^i\right)\notin\hat{\boldsymbol{x}}_{\bar{Z}_j}\right)$，更新控制器$u_j$。

在以下的分析中，对于受到攻击的智能体i，本章节主要考虑时间区间Θ_Z^i和Ξ_Z^i，而不是Θ_U^i和Ξ_U^i。基于上边①和②中的分析，定义布尔逻辑变量$m_i\left(t\right)\in\left\{0,1\right\}$如下：$m_i\left(t\right)=0$表示接收端没有受到拒绝服务攻击，这跟上面①中时间区间Θ_Z^i代表的含义是一致的（即接收端没有受到攻击），所以可以成功传输并更新；$m_i\left(t\right)=1$则表示i的接收端受到攻击，这跟上面②中时间区间Ξ_Z^i代表的含义是一致的（即接收端受到攻击），所以不能传输，更不能更新。令$n_Z^i\left(T_1,T_2\right)$表示在时间段$\left[T_1,T_2\right)$内$i$的接收端受到拒绝服务攻击的次数。正如De和Tesi[85]、Feng和Tesi[236]所述，要想设计出合适的针对攻击的弹性策略，需要对拒绝服务攻击发生的频率和总长度$\left|\Xi_Z^i\right|$进行限制。

假设 10.6

对任意$i\in\mathcal{N}$和$T_2\geqslant T_1$，存在常数$v_Z^i\geqslant 0,\tau_{f_Z}^i>\varDelta>0,\varsigma_Z^i\geqslant 0$和$\tau_{d_Z}^i>1$，满足$n_Z^i\left(T_1,T_2\right)\leqslant v_Z^i+\left(T_2-T_1\right)/\tau_{f_Z}^i$和$\left|\Xi_Z^i\left(T_1,T_2\right)\right|\leqslant\varsigma_Z^i+\left(T_2-T_1\right)/\tau_{d_Z}^i$。

注释 10.8 类似于Feng等人[236]的研究，$\tau_{f_Z}^i$可以看成是连续两次拒绝服务攻击之间的平均驻留时间，而v_Z^i表示振动界限。假设10.6表明拒绝服务攻击的所有时间不多于$1/\tau_{d_Z}^i$的某倍数。具体地说，如果$T_1=T_2$而且$n_Z^i\left(T_1,T_2\right)=1$，那么参数$v_Z^i$是使得

不等式$1=n_Z^i(T_1,T_2)\leqslant v_Z^i+(T_2-T_1)/\tau_{f_Z}^i=v_Z^i$成立的必要条件。类似地，$\varsigma_Z^i$对于调节拒绝服务攻击的时间长度起着重要作用，也保证假设 10.6 中最后一个不等式成立的可行性，比如，如果$\left|\Xi_Z^i\left(h_{n_Z}^i,h_{n_Z}^i+\tau_{n_Z}^i\right)\right|=\tau_{n_Z}^i$，那么$\tau_{n_Z}^i>\tau_{nZ}^i/\tau_{d_Z}^i$成立，因为$\tau_{d_Z}^i>1$。

对于任意智能体 i，我们考虑如下的事件驱动策略

$$t_{k+1}^i=\inf\left\{t:t\geqslant t_k^i+\tau_{\text{miet}}^{i,m_i}\mid\ \eta_i(t)<0,k\in\mathbb{Z}_{\geqslant 0},t_0^i=0\right\}\tag{10-50}$$

其中，$\tau_{\text{miet}}^{i,m_i}$表示最小传输间隔，并且满足$0<\tau_{\text{miet}}^{i,1}\leqslant\tau_{\text{miet}}^{i,0}$和$\max_{i\in\mathcal{N}}\left\{\tau_{\text{miet}}^{i,0},\tau_{\text{miet}}^{i,1}\right\}\leqslant\underline{\Delta}$，其中$\underline{\Delta}$满足假设10.6。根据本节中的分析，$m_i\in\{0,1\}$的含义表示如下：$m_i=0$表示一旦不等式$\eta_i(t)<0$成立，那么传输尝试$t_{k+1}^i$会在最近（前）一次传输时刻$t_k^i$之后的$\tau_{\text{miet}}^{i,0}$时间后到达并成功传输；而$m_i=1$则表示尽管不等式$\eta_i(t)<0$成立，传输尝试也会在$\tau_{\text{miet}}^{i,0}$时间后到达，但是数据并不能成功传输，因而$i$的控制器不会更新，所以不能广播新数据给外邻居。事件驱动函数$\eta_i(t)$的定义如下：

$$\begin{cases}\dot{\eta}_i(t)=\Psi_i\left(m_i(t),\boldsymbol{o}_i(t),\eta_i(t)\right),\ t\in\left[t_k^i,t_{k+1}^i\right)\\ \eta_i^+\left(t_k^i\right)=\begin{cases}\tilde{\eta}_i\left(t_k^i\right), & t_k^i\in\Theta_Z^i\\ \eta_i\left(t_k^i\right), & t_k^i\in\Xi_Z^i\end{cases}\end{cases}\tag{10-51}$$

其中，$\boldsymbol{o}_i=(\boldsymbol{e}_i,\tau_i,\phi_i)\in\mathbb{R}^{2n}\times\mathbb{R}_{\geqslant 0}\times\left[\lambda_i,\lambda_i^{-1}\right]$代表局部变量；$\lambda_i\in(0,1)$。网络误差$\boldsymbol{e}_i$，时间变量$\tau_i$，以及函数$\phi_i$、$\Psi_i$和$\tilde{\eta}_i$会在接下来的章节给出具体描述。因此本节的目标是设计合适的事件驱动策略 (10-50) 进而产生合适的传输时刻来更新控制器，使得多智能体系统 (10-49) 的编队控制对拒绝服务攻击具有弹性作用。

注释 10.9　正如 De 和 Tesi[85] 所分析的那样，条件$\max_{i\in\mathcal{N}}\{\tau_{\text{miet}}^{i,0},\tau_{\text{miet}}^{i,1}\}\leqslant\underline{\Delta}$能有效防止拒绝服务攻击恰好发生在所有传输瞬间。条件$0<\tau_{\text{miet}}^{i,1}\leqslant\tau_{\text{miet}}^{i,0}$可以排除奇诺现象，同时也表明相比

于$m_i=0$的情况，$m_i=1$时的传输尝试需要更频繁地进行，以便事件驱动策略尽可能快地得知攻击何时结束，进而及时传输数据[226,229]。本章节中，我们使用确认方案，即事件驱动策略(10-50)随时都知道控制器端数据包是否被成功接收[83]。此外，所有智能体的确认方案都是互通的，即一旦智能体i的确认方案意识到i的接收端受到攻击，则通知其他智能体（未受到攻击）的确认方案，则所有智能体$i\in\mathcal{N}$都进入$m_i=1$的模式。值得说明的是所有智能体不会一直处在$m_i=1$的模式，因为本章节考虑的是在任意时间段内，最多只有一个智能体受到攻击的情形。

对任意$i\in\mathcal{N}$，定义$\boldsymbol{\xi}_i=\boldsymbol{x}_i-\boldsymbol{x}_1-\boldsymbol{d}_{x_i}$和$\boldsymbol{e}_i=\left(\boldsymbol{e}_{\xi_i},\boldsymbol{e}_{x_i}\right)$，其中$\boldsymbol{e}_{\xi_i}=\hat{\boldsymbol{\xi}}_i-\boldsymbol{\xi}_i=\left(\hat{\boldsymbol{x}}_i-\boldsymbol{x}_i\right)-\left(\hat{\boldsymbol{x}}_1-\boldsymbol{x}_1\right)$和$\boldsymbol{e}_{x_i}=\hat{\boldsymbol{x}}_i-\boldsymbol{x}_i$。多智能体系统(10-49)的脉冲模型可以描述为

$$\begin{cases}\dot{\boldsymbol{\xi}}_i=f_i\left(\boldsymbol{\xi}_{\bar{Z}_i},\boldsymbol{e}_{\bar{Z}_i},\boldsymbol{x}_{\bar{Z}_i},\boldsymbol{\omega}_i,\boldsymbol{x}_1,\boldsymbol{\omega}_1\right), & t\in\left[t_k^i,t_{k+1}^i\right)\\ \dot{\boldsymbol{e}}_i=g_i\left(\boldsymbol{\xi}_{\bar{Z}_i},\boldsymbol{e}_{\bar{Z}_i},\boldsymbol{x}_{\bar{Z}_i},\boldsymbol{\omega}_i,\boldsymbol{x}_1,\boldsymbol{\omega}_1\right), & t\in\left[t_k^i,t_{k+1}^i\right)\\ \dot{\eta}_i=\varPsi_i, & t\in\left[t_k^i,t_{k+1}^i\right)\\ \dot{\phi}_i=\varPhi_i, & t\in\left[t_k^i,t_{k+1}^i\right)\end{cases} \tag{10-52}$$

$$\begin{cases}\boldsymbol{\xi}_i^+\left(t_k^i\right)=\boldsymbol{\xi}_i\left(t_k^i\right),t_k^i\in\varTheta_Z^i,\\ \boldsymbol{e}_i^+\left(t_k^i\right)=0, \quad t_k^i\in\varTheta_Z^i,\\ \eta_i^+\left(t_k^i\right)=\tilde{\eta}_i, \quad t_k^i\in\varTheta_Z^i,\\ \phi_i^+\left(t_k^i\right)=\lambda_i^{-1}, \ t_k^i\in\varTheta_Z^i,\end{cases}\begin{cases}\boldsymbol{\xi}_i^+\left(t_k^i\right)=\boldsymbol{\xi}_i\left(t_k^i\right),t_k^i\in\varXi_Z^i\\ \boldsymbol{e}_i^+\left(t_k^i\right)=\boldsymbol{e}_i\left(t_k^i\right),t_k^i\in\varXi_Z^i\\ \eta_i^+\left(t_k^i\right)=n_i\left(t_k^i\right),t_k^i\in\varXi_Z^i\\ \phi_i^+\left(t_k^i\right)=\phi_i\left(t_k^i\right),t_k^i\in\varXi_Z^i\end{cases}$$

其中，$\boldsymbol{\xi}_{\bar{Z}_i}=\left\{\boldsymbol{\xi}_j\right\}_{j\in\bar{Z}_i},\boldsymbol{e}_{\bar{Z}_i}=\left\{\boldsymbol{e}_j\right\}_{j\in\bar{Z}_i}$和$\boldsymbol{x}_{\bar{Z}_i}=\left\{\boldsymbol{x}_j\right\}_{j\in\bar{Z}_i}$。令$\boldsymbol{z}_i\left(\boldsymbol{\xi}_i\right)=\mathcal{Q}_i\left(\boldsymbol{\xi}_i\right)$表示第$i$个智能体的输出。连续函数$f_i$和$g_i$的表达式可以从式(10-49)得到，后边会详细介绍$\varPhi_i$。为了不失一般性，脉冲系统(10-52)中的$\boldsymbol{f}_i$和$\boldsymbol{g}_i$依赖$\boldsymbol{\xi}_{\bar{Z}_i}$、$\boldsymbol{e}_{\bar{Z}_i}$、$\boldsymbol{x}_{\bar{Z}_i}$、$\boldsymbol{\omega}_i$、$\boldsymbol{x}_1$和$\boldsymbol{\omega}_1$，尽管它们可能只依赖于这些向量变量中的部分变量（如Postoyan等人[84]研究中的情况）。

令变量s_i表示整体时间，变量τ_i则表示从最近的传输尝试开始经过

的时间。定义$\boldsymbol{q}=(\boldsymbol{\xi},\boldsymbol{e},\boldsymbol{\eta},\boldsymbol{\phi},\boldsymbol{\tau},\boldsymbol{s},\boldsymbol{m})\in\mathbb{X}$，其中$\mathbb{X}=(\mathbb{R}^n)^N\times(\mathbb{R}^{2n})^N\times\mathbb{R}_{\geqslant0}^N\times[\lambda_i,\lambda_i^{-1}]^N\times\mathbb{R}_{\geqslant0}^N\times\mathbb{R}_{\geqslant0}^N\times\{0,1\}^N,\lambda_i\in(0,1)$。另外，令$q=(q_1,q_2,\cdots,q_N)$，其中$\boldsymbol{q}_i=(\boldsymbol{\xi}_i,\boldsymbol{e}_i,\eta_i,\phi_i,\tau_i,s_i,m_i),i\in\mathcal{N}$和$\boldsymbol{\iota}=(\iota_1,\cdots,\iota_N)$，$\boldsymbol{\iota}\in\{\boldsymbol{\xi},\boldsymbol{e},\boldsymbol{\eta},\boldsymbol{\phi},\boldsymbol{\tau},\boldsymbol{s},\boldsymbol{m},\boldsymbol{\omega},\boldsymbol{z}\}$。结合式 (10-49)、式 (10-51) 和式 (10-52)，根据 Goebel 等人 [230] 研究中的混杂模型，可以建立本章节的混杂模型$\mathcal{H}_{\text{dos}}$如下：

$$\mathcal{H}_{\text{dos}}:\begin{cases}\dot{\boldsymbol{q}}=F(\boldsymbol{q},\boldsymbol{\omega}), & \boldsymbol{q}\in C\\ \boldsymbol{q}^+\in G_{\text{dos}}(\boldsymbol{q}), & \boldsymbol{q}\in D\end{cases}\tag{10-53}$$

其中，$C=\{\boldsymbol{q}\in\mathbb{X}|\ \tau_i\leqslant\tau_{\text{miet}}^{i,m_i}或\eta_i\geqslant0,i\in\mathcal{N}\},D=\{\boldsymbol{q}\in\mathbb{X}|\ \tau_i\geqslant\tau_{\text{miet}}^{i,m_i}且\eta_i\leqslant0,i\in\mathcal{N}\}$，$G_{\text{dos}}(\boldsymbol{q})=G_0(\boldsymbol{q})\cup G_{1Z}(\boldsymbol{q})$。映射$F(\boldsymbol{q},\boldsymbol{\omega})$和$G_{\text{dos}}(\boldsymbol{q})$的表达式如下：

$$\begin{aligned}F(\boldsymbol{q},\boldsymbol{\omega})&=(f(\boldsymbol{\xi},\boldsymbol{e},\boldsymbol{\omega}),g(\boldsymbol{\xi},\boldsymbol{e},\boldsymbol{\omega}),\Psi(\boldsymbol{m},\boldsymbol{o},\boldsymbol{\eta}),f_\phi(\boldsymbol{\phi}),1,1,0)\\ G_0(q)&=\{G_i:G_i=(\boldsymbol{\xi}_1,\cdots,\boldsymbol{\xi}_N,\boldsymbol{e}_1,\cdots,\boldsymbol{e}_{i-1},0,\boldsymbol{e}_{i+1},\cdots,\boldsymbol{e}_N,\eta_1,\cdots,\\ &\quad\eta_{i-1},\tilde{\eta}_i,\eta_{i+1},\cdots,\eta_N,\phi_1,\cdots,\phi_{i-1},\lambda_i^{-1},\phi_{i+1},\cdots,\phi_N\\ &\quad\tau_1,\cdots,\tau_{i-1},0,\tau_{i+1},\cdots,\tau_N,s_1,\cdots,s_N,0,\cdots,0),s_i\in\Theta_Z^i,i\in\mathcal{N}\}\\ G_{1Z}(\boldsymbol{q})&=\{G_i:G_i=(\boldsymbol{\xi}_1,\cdots,\boldsymbol{\xi}_N,\boldsymbol{e}_1,\cdots,\boldsymbol{e}_i,\cdots,\boldsymbol{e}_N,\eta_1,\cdots,\eta_i,\cdots,\eta_N,\phi_1,\cdots,\\ &\quad\phi_i,\cdots,\phi_N,\tau_1,\cdots,\tau_i,\cdots,\tau_N,s_1,\cdots,s_N,1,\cdots,1),s_i\in\Xi_Z^i,i\in\mathcal{N}\}\end{aligned}$$

其中，$\boldsymbol{o}=(o_i)_{i\in\mathcal{N}},\boldsymbol{f}=(f_i)_{i\in\mathcal{N}},\boldsymbol{g}=(g_i)_{i\in\mathcal{N}},\boldsymbol{F}=(F_i)_{i\in\mathcal{N}}$，而$\boldsymbol{f}_\phi(\phi)=(\Phi_i)_{i\in\mathcal{N}}$会在后文中给出详细信息。

定义 10.6

混杂系统$\mathcal{H}_{\text{dos}}$ [式 (10-53)] 关于闭集$\boldsymbol{\mathcal{A}}=\{\boldsymbol{\xi}\in\mathbb{X}|\ \boldsymbol{\xi}=0,\boldsymbol{e}=0\}$是从干扰$\omega$到输出 $\boldsymbol{z}$ 上$\mathcal{L}_\infty$稳定的，其中增益小于等于$\hat{\theta}$，如果满足$\xi(0,0)\in\mathbb{X}_0$的所有最大解$\boldsymbol{\xi}$在 t 方向有无界的时间域，那么存在函数$\beta\in\mathcal{K}_\infty$，使得对于任意初始条件$\boldsymbol{\xi}(0,0)\in\mathbb{X}_0$和外部干扰$\omega\in\mathcal{L}_\infty$满足

$$|\boldsymbol{z}(t,k)|_{\mathcal{L}_\infty}\leqslant\beta(|\boldsymbol{\xi}(0,0)|_{\boldsymbol{\mathcal{A}}})+\hat{\theta}\|\boldsymbol{\omega}\|_{\mathcal{L}_\infty},(t,k)\in\text{dom}\ \xi$$

其中，$\mathbb{X}_0=\{\boldsymbol{q}\in\mathbb{X}|\ \tau_i\geqslant\tau_{\text{miet}}^{i,0};\ \eta_i=s_i=0,\phi_i=\phi_{i,\text{miet}}\}$。

注释 10.10 拒绝服务攻击所造成的影响类似于数据包丢失，信道访问限制或故障，即最新数据不能被成功传输来更新控制器 [264,283]，

尽管用于处理这些问题的方法很类似[284]，但是它们发生的机理不同。网络化控制系统存在固有的数据包丢失和信道约束，而拒绝服务攻击是由具有一定目的的外部的恶意攻击者发起的，而且在网络化控制系统的研究中，经常存在由真正的数据包丢失而引起的通信故障[283]，由拒绝服务攻击引起的情况可能有不同的配置[236]，而且不需要遵循任何概率分布。比如在 Wang 等人[264]和 Dolk 等人[283]的研究中，连续丢失数据包的数目可以被一个变量精确地捕获和跟踪，而在本节中，我们只是在式 (10-53) 中根据拒绝服务攻击的频率和时间的界限建立了拒绝服务攻击模型[85]，而这并不需要监控连续丢失数据包的确切数量。

10.3.2　多智能体系统的编队控制分析

为了建立事件驱动策略 (10-50) 中的分布式触发条件，我们首先给出网络多智能体 [式 (10-52)] 的稳定性和性能条件来约束触发条件中的参数。

假设 10.7

对任意$i\in\mathcal{N}$，存在局部利普希茨函数$W_i:\mathbb{R}^{2n}\to\mathbb{R}_{\geqslant 0}$，连续有界函数$H_i:\mathbb{R}^n\times\mathbb{R}^l\to\mathbb{R}$，函数 $\underline{c}_{W,i}$、$\overline{c}_{W,i}\in\mathcal{K}_\infty$和常数$L_i\geqslant 0$，使得：①对任意$e_i\in\mathbb{R}^{2n}$，有 $\underline{c}_{W,i}\left(\left|\boldsymbol{e}_i\right|\right)\leqslant W_i\left(\boldsymbol{e}_i\right)\leqslant\overline{c}_{W,i}\left(\left|\boldsymbol{e}_i\right|\right)$；②对任意$\boldsymbol{\xi}_i\in\mathbb{R}^n$和几乎所有$e_i\in\mathbb{R}^{2n}$，有$\left\langle\dfrac{\partial W_i\left(\boldsymbol{e}_i\right)}{\partial e_i},g_i\left(\xi_{\overline{Z}_i},e_{\overline{Z}_i},\omega_i\right)\right\rangle\leqslant L_iW_i\left(\boldsymbol{e}_i\right)+H_i\left(\boldsymbol{\xi}_i,\boldsymbol{\omega}_i\right)$。

假设 10.8

对任意$i\in\mathcal{N}$，存在局部利普希茨函数$V_i:\mathbb{R}^n\to\mathbb{R}_{>0}$，函数 $\underline{\alpha}^i$、$\overline{\alpha}^i$、$\sigma_i\in\mathcal{K}_\infty$，正常数$c_{z_i}$、$\rho_{V_i}$、$\theta_i$、$\gamma_i$和$\rho_{W_i}$满足$\rho_{W_i}\leqslant\gamma_i^2$，使得：①对任意$\xi_i\in\mathbb{R}^n$，有$c_{z_i}\left|\mathcal{Q}_i\left(\boldsymbol{\xi}_i\right)\right|^2\leqslant V_i\left(\boldsymbol{\xi}_i\right)$；②对任意$\boldsymbol{\xi}_i\in\mathbb{R}^n$，有 $\underline{\alpha}^i\left(\sum_{j\in\overline{Z}_i}\left|\boldsymbol{\xi}_j\right|\right)\leqslant V_i\left(\boldsymbol{\xi}_i\right)\leqslant\overline{\alpha}^i\left(\sum_{j\in\overline{Z}_i}\left|\boldsymbol{\xi}_j\right|\right)$；③对任意$\boldsymbol{e}_i\in\mathbb{R}^{2n}$和几乎所有$\boldsymbol{\xi}_i\in\mathbb{R}^n$，有$\left\langle\nabla V_i\left(\boldsymbol{\xi}_i\right),f_i\left(\boldsymbol{\xi}_i,\boldsymbol{e}_i,\boldsymbol{\omega}_i\right)\right\rangle\leqslant-\rho_{V_i}V_i\left(\boldsymbol{\xi}_i\right)-H_i^2\left(\boldsymbol{\xi}_i,\boldsymbol{\omega}_i\right)-\sum_{j\in Z_i}H_j^2\left(\boldsymbol{\xi}_i,\boldsymbol{\omega}_i\right)+\left(\gamma_i^2-\rho_{W_i}\right)$

$$W_i^2(\boldsymbol{e}_i)+\sum_{j\in Z_i}\left(\gamma_j^2-\rho_{W_j}\right)W_j^2(\boldsymbol{e}_j)-\sigma_i(W_i)-\sum_{j\in Z_i}\sigma_j(W_j)+\theta_i^2|\boldsymbol{\omega}_i|^2。$$

正如 Dolk 等人 [250] 研究中所描述的，假设 10.7 是关于子系统$\boldsymbol{e}_i$的稳定性，而假设 10.8 是为了保证子系统$\boldsymbol{\xi}$从$\left(W_i,(W_j)_{j\in Z_i}\right)$到$\left(H_i,(H_j)_{j\in Z_i}\right)$的$\mathcal{L}_2$稳定。接下来我们给出式 (10-52) 中函数$\varPhi_i$的表达式：

$$\varPhi_i=\begin{cases}-2L_i\phi_i-\gamma_i\left\{(m_i\epsilon_i+1)\phi_i^2+\left[m_i(\hat{\gamma}_i-\gamma_i)+\gamma_i\right]/\gamma_i\right\},\tau_i\in\left[0,\tau_{\text{miet}}^{i,m_i}\right]\\ 0, \qquad \tau_i>\tau_{\text{miet}}^{i,m_i}\end{cases} \tag{10-54}$$

其中，$\epsilon_i\geqslant 0$和$\hat{\gamma}_i\geqslant\gamma_i$，常数$L_i$和$\gamma_i$分别满足假设10.7和假设10.8，而时间常数$\tau_{\text{miet}}^{i,1}$和$\tau_{\text{miet}}^{i,0}$可以根据下边的公式取值[250]：

$$\tau_{\text{mati}}^{i,m_i}=\begin{cases}\dfrac{1}{L_ir_i}\arctan(\varGamma_i), & \varLambda_i>L_i\\ \tilde{\varGamma}_i, & \varLambda_i=L_i\\ \dfrac{1}{L_ir_i}\operatorname{arctanh}(\varGamma_i), & \varLambda_i<L_i\end{cases} \tag{10-55}$$

其中，$\lambda_i\in(0,1)$来自式(10-51)，$r_i=\sqrt{\left|\varLambda_i^2/L_i^2-1\right|}$，$\varLambda_i=\sqrt{\gamma_i(1+m_i\epsilon_i)\left[m_i\hat{\gamma}_i+\gamma_i(1-m_i)\right]}$，$\tilde{\varGamma}_i=1/\left[\lambda_i\gamma_i(1+m_i\epsilon_i)+L_i\right]-\lambda_i/\left[\lambda_iL_i+\gamma_i(1+m_i\epsilon_i)\right]$，$\varGamma_i=r_i(1-\lambda_i)/\left(\dfrac{\lambda_i}{1+\lambda_i}\left(\left(\gamma_i(1+m_i\epsilon_i)+m_i\hat{\gamma}_i+\gamma_i(1-m_i)\right)/L_i-2\right)+1+\lambda_i\right)$。

基于 Postoyan 等人 [84] 的研究，接下来我们给出关于式 (10-55) 的引理。

引理 10.3

假设$\tau_{\text{mati}}^{i,m_i}$满足式 (10-55)，那么以下方程

$$\dot{\tilde{\phi}}_i=-2L_i\tilde{\phi}_i-\gamma_i\left[(m_i\epsilon_i+1)\tilde{\phi}_i^2+m_i\frac{\hat{\gamma}_i}{\gamma_i}+1-m_i\right] \tag{10-56}$$

其中，$\tilde{\phi}_i(0)=\lambda_i^{-1}$，它的解满足$\tilde{\phi}_i(t)\in\left[\lambda_i,\lambda_i^{-1}\right]$，其中$t\in\left[0,\tau_{\text{mati}}^{i,m_i}\right]$，$\tilde{\phi}_i\left(\tau_{\text{mati}}^{i,m_i}\right)=\lambda_i$。

定义$\phi_{i,\text{miet}}=\tilde{\phi}_i\left(\tau_{\text{miet}}^{i,0}\right)$，其中$\tilde{\phi}_i$来自式 (10-56) 并且满足$\tilde{\phi}_i(0)=\lambda_i^{-1}$。基于式 (10-55) 和引理 10.3，我们可以根据如下不等式选择常数$\tau_{\text{miet}}^{i,1}$和$\tau_{\text{miet}}^{i,0}$：

$$\tau_{\text{miet}}^{i,1}\leqslant\tau_{\text{mati}}^{i,1}\leqslant\tau_{\text{mati}}^{i,0}\text{和}\tau_{\text{miet}}^{i,1}\leqslant\tau_{\text{miet}}^{i,0}\leqslant\tau_{\text{mati}}^{i,0}$$

注释 10.11 本节的事件触发策略的设计方法是基于 Dolk 等人[83]的研究。与之相比，本章节额外考虑两个非负参数ϵ_i和$\hat{\gamma}_i$，其中$\hat{\gamma}_i \geqslant \gamma_i$。非负参数$\epsilon_i$和$\hat{\gamma}_i$可以帮助确定并灵活调整拒绝服务攻击发生时的传输尝试：当$m_i = 1$时，ϵ_i或者$\hat{\gamma}_i$的值越大，那么传输间隔就越小（参见数值仿真中的表 10-2）。

接下来给出混杂系统 (10-53) 的$\mathcal{L}_\infty$稳定性。

定理 10.4

考虑混杂系统 (10-53) 具有强连通的有向拓扑结构，并且假设 10.6 ～假设 10.8 成立。参数$\tau_{\text{miet}}^{i,1}, \tau_{\text{miet}}^{i,0}$满足$\tau_{\text{miet}}^{i,1} \leqslant \tau_{\text{mati}}^{i,1} \leqslant \tau_{\text{mati}}^{i,0}$和$\tau_{\text{miet}}^{i,1} \leqslant \tau_{\text{miet}}^{i,0} \leqslant \tau_{\text{mati}}^{i,0}$。假设以下条件成立：

(4a) 存在两个正常数系列$\left\{\kappa_i^0\right\}_{i\in\mathcal{N}}$和$\left\{\kappa_i^1\right\}_{i\in\mathcal{N}}$使得

$$\kappa_i^0 \varpi_i^0 > \left(\tau_{\text{miet}}^{i,1} / \tau_{f_Z}^i + 1 / \tau_{d_Z}^i\right)\left(\kappa_i^0 \varpi_i^0 + \kappa_i^1 \varpi_i^1\right) \tag{10-57}$$

其中，$\varpi_i^0 = \min\left\{\rho_{V_i}, \lambda_i \rho_{W_i} / \gamma_i\right\}$；$\varpi_i^1 = \left(\hat{\gamma}_i - \rho_{W_i}\right) / \left(\gamma_i \phi_{i,\text{miet}}\right)$；$\bar{\gamma}_i = \gamma_i\left[2\phi_{i,\text{miet}} L_i + \gamma_i\left(1 + \phi_{i,\text{miet}}^2\right)\right]$。

(4b) 存在函数$\delta_{\eta_i} \in \mathcal{K}_\infty$，其中$\delta_{\eta_i}\left(\eta_i\right) \geqslant \varpi_i^0 \eta_i$和$\bar{\gamma}_i$来自上文的 (4a)，使得

$$\Psi_i = \begin{cases} -\delta_{\eta_i}\left(\eta_i\right), & 0 \leqslant \tau_i \leqslant \tau_{\text{miet}}^{i,0} \\ \tilde{\Psi}_i, & \tau_i > \tau_{\text{miet}}^{i,0} \\ 0, & 0 \leqslant \tau_i \leqslant \tau_{\text{miet}}^{i,1} \\ -1, & \tau_i > \tau_{\text{miet}}^{i,1} \end{cases} \tag{10-58}$$

其中，$\tilde{\Psi}_i = -\left(\left|U_i\right| + 1\right)\bar{\gamma}_i W_i^2\left(\boldsymbol{e}_i\right) - \delta_{\eta_i}\left(\eta_i\right)$。

(4c) 函数$\tilde{\eta}_i$满足$\tilde{\eta}_i = \gamma_i \phi_{i,\text{miet}} W_i^2\left(\boldsymbol{e}_i\right)^2$。

那么闭集合$\boldsymbol{\mathcal{A}} = \left\{\boldsymbol{\xi} \in \mathbb{X} \mid \boldsymbol{\xi} = 0, \boldsymbol{e} = 0\right\}$是$\mathcal{L}_\infty$稳定的，并且有限的$\mathcal{L}_\infty$增益小于等于$\max_{i\in\mathcal{N}}\left\{\theta_i \sqrt{\text{Im}_i / \left(\hat{\beta}_i c_{z_i}\right)}\right\}$，其中，$\text{Im}_i = \mathrm{e}^{\varsigma_*^i\left(\kappa_i^0 \varpi_i^0 + \kappa_i^1 \varpi_i^1\right)}, \hat{\beta}_i = \kappa_i^0 \varpi_i^0 - \left(\kappa_i^0 \varpi_i^0 + \kappa_i^1 \varpi_i^1\right) / T_*^i, \varsigma_*^i = \varsigma_Z^i + v_Z^i \tau_{\text{miet}}^{i,1}$，以及$T_*^i = \tau_{f_Z}^i \tau_{d_Z}^i / \left(\tau_{f_Z}^i + \tau_{d_Z}^i \tau_{\text{miet}}^{i,1}\right)$。

证明：

为了方便起见，在不引起混淆的情况下，接下来对定理 10.4 的证明会简化一些变量的符号，比如$V_i \overset{\text{def}}{=} V_i\left(\boldsymbol{\xi}_i\right), W_i \overset{\text{def}}{=} W_i\left(\boldsymbol{e}_i\right), H_i \overset{\text{def}}{=} H_i\left(\boldsymbol{\xi}_i, \boldsymbol{\omega}_i\right)$，

$\sigma_i \overset{\text{def}}{=} \sigma_i(W_i), \Psi_i \overset{\text{def}}{=} \Psi_i(m_i(t), \boldsymbol{o}_i(t), \eta_i(t))$，等等。令$\mathcal{R}(\mathbb{X}_0)$表示混杂系统$\mathcal{H}_{\text{dos}}$的可达状态，其中$\boldsymbol{q}(0,0) \in \mathbb{X}_0$，初始状态$\mathbb{X}_0$参见定义 10.6。由此可以得到，对任意$\boldsymbol{q} \in \mathcal{R}(\mathbb{X}_0)$，有：①$\tau_i \geqslant \tau_{\text{miet}}^{i,0} \Leftrightarrow \phi_i = \phi_{i,\text{miet}}$；②$\lambda_i^{-1} \geqslant \phi_i \geqslant \phi_{i,\text{miet}}$；③$\eta_i \geqslant 0$。

基于假设 10.7、10.8 和以上分析，为$q \in C \cup D$定义李雅普诺夫函数：

$$\mathcal{V}(\boldsymbol{q}) = \sum_{i=1}^{N} \mathcal{V}_i(\boldsymbol{q}_i)，\ \mathcal{V}_i(\boldsymbol{q}_i) = \mathcal{V}_i + (|U_i| + 1)\gamma_i \phi_i W_i^2 + \eta_i$$

基于假设10.7、10.8，可知对任意$i \in \mathcal{N}$，存在函数$\underline{\beta}_{\mathcal{V}_i}, \overline{\beta}_{\mathcal{V}_i} \in \mathcal{K}_\infty$使得对任意$\boldsymbol{q} = (\boldsymbol{q}_1, \boldsymbol{q}_2, \cdots, \boldsymbol{q}_N) \in \mathcal{R}(\mathbb{X}_0)$，有$\underline{\beta}_{\mathcal{V}_i}(|\boldsymbol{q}_i(t,k)|_{\mathcal{A}}) \leqslant \mathcal{V}_i(\boldsymbol{q}_i(t,k)) \leqslant \overline{\beta}_{\mathcal{V}_i}(|\boldsymbol{q}_i(t,k)|_{\mathcal{A}})$。为了研究混杂系统 (10-53) 的稳定性，接下来我们分析拒绝服务攻击下的函数$\mathcal{V}$和$\{\mathcal{V}_i\}_{i \in \mathcal{N}}$的状态变换。

现在讨论函数$\{\mathcal{V}_i\}_{i \in \mathcal{N}}$在$q \in C \cup D$中是如何演化的。当$\boldsymbol{q} \in D \cap \mathcal{R}(\mathbb{X}_0)$：

$(J_a)\, m_i = 0$：对任意$i \in \mathcal{N}$，基于式 (10-52) 和式 (10-54)，如果$\boldsymbol{q} \in \mathcal{R}(\mathbb{X}_0)$，有$\mathcal{V}_i^+ - \mathcal{V}_i = -(|U_i| + 1)\gamma_i \phi_i W_i^2 + \tilde{\eta}_i$。由定理 10.4(4c) 可知$\mathcal{V}_i^+(\boldsymbol{q}_i) - \mathcal{V}_i(\boldsymbol{q}_i) = 0$。

$(J_b)\, m_i = 1$：类似地，对任意$i \in \mathcal{N}$，基于式 (10-52) 有$\mathcal{V}_i^+(\boldsymbol{q}_i) - \mathcal{V}_i(\boldsymbol{q}_i) = 0$，其中$\boldsymbol{\xi}_i^+ = \boldsymbol{\xi}_i, \boldsymbol{e}_i^+ = \boldsymbol{e}_i, \phi_i^+ = \phi_i, \eta_i^+ = \eta_i$，因此有$\mathcal{V}^+(\boldsymbol{q}) - \mathcal{V}(\boldsymbol{q}) = 0$，当$\boldsymbol{q} \in C \cap \mathcal{R}(\mathbb{X}_0)$。

$(F_a)\, m_i = 0$：当$\tau_i \leqslant \tau_{\text{miet}}^{i,0}$，结合假设 10.7、10.8 和式 (10-54)，对任意$i \in \mathcal{N}$，有

$$\begin{aligned}
&\mathcal{V}^\circ(\boldsymbol{q}; F(\boldsymbol{q}, \boldsymbol{\omega})) \\
&\leqslant \sum_{i \in \mathcal{N}} \Big[-\rho_{V_i} V_i - (|U_i| + 1) H_i^2 + (|U_i| + 1)(\gamma_i^2 - \rho_{W_i}) W_i^2 \\
&\quad + \Psi_i - (|U_i| + 1)\sigma_i \Big] + \sum_{i \in \mathcal{N}} (|U_i| + 1) \{ \gamma_i [-2L_i \phi_i \\
&\quad - \gamma_i(\phi_i^2 + 1)] W_i^2 + 2\gamma_i \phi_i W_i (L_i W_i + H_i) \Big] \\
&\leqslant \sum_{i \in \mathcal{N}} \Big[-\rho_{V_i} V_i + (|U_i| + 1)(-\rho_{W_i} W_i^2 - \sigma_i) \\
&\quad - (|U_i| + 1)\hat{M}_i(q) + \Psi_i \Big]
\end{aligned}$$

其中，$\hat{M}_i(q) = (H_i + \gamma_i \phi_i W_i)^2$。当$\tau_i \geqslant \tau_{\text{miet}}^{i,0}$，结合假设10.7、10.8和式(10-54)，

类似地有

$$\begin{aligned}&\mathcal{V}^{\circ}\left(\boldsymbol{q};F\left(\boldsymbol{q},\boldsymbol{\omega}\right)\right)\\&\leqslant\sum_{i\in\mathcal{N}}\left[-\rho_{V_i}V_i+\left(\left|U_i\right|+1\right)\left(-\rho_{W_i}W_i^2-\sigma_i\left(W_i\right)\right)-\left(\left|U_i\right|+1\right)\tilde{M}_i\left(q\right)+\varPsi_i\right]\end{aligned}$$

其中，$\tilde{M}_i\left(q\right)=H_i^2-2\gamma_i\phi_iW_iH_i-\left(\gamma_i^2+2\gamma_i\phi_iL_i\right)W_i^2$。

$\left(F_b\right)m_i=1$：当$\tau_i\leqslant\tau_{\text{miet}}^{i,1}$，基于不等式$\varLambda:2\gamma_i\phi_iW_iH_i\leqslant\left(\epsilon_i+1\right)\gamma_i^2\phi_i^2W_i^2+H_i^2/\left(\epsilon_i+1\right)$，其中$\epsilon_i>0$定义于式(10-54)，$i\in\mathcal{N}$，以及式(10-54)和式(10-58)，可以得到

$$\begin{aligned}&\mathcal{V}^{\circ}\left(\boldsymbol{q};F\left(\boldsymbol{q},\boldsymbol{\omega}\right)\right)\\&\leqslant\sum_{i\in\mathcal{N}}\Big(-\rho_{V_i}V_i-\left(\left|U_i\right|+1\right)H_i^2+\left(\left|U_i\right|+1\right)\left(\gamma_i^2-\rho_{W_i}\right)\\&\times W_i^2-\left(\left|U_i\right|+1\right)\times\sigma_i\left(W_i\right)\Big)+\sum_{i\in\mathcal{N}}\Big(\left|U_i\right|+1\Big)\Big(\gamma_i\Big(-2L_i\phi_i\\&-\gamma_i((\epsilon_i+1)\phi_i^2+\frac{\hat{\gamma}_i}{\gamma_i}))W_i^2+2\gamma_i\phi_iW_i\left(L_iW_i+H_i\right)\Big)\\&\leqslant-\sum_{i\in\mathcal{N}}\left(\rho_{V_i}V_i-\left(\left|U_i\right|+1\right)\rho_{W_i}W_i^2\right)\end{aligned}$$

如果$\tau_i>\tau_{\text{miet}}^{i,1}$，根据式(10-54)可知$\dot{\phi}_i=0$。基于不等式$2\gamma_i\phi_iW_iH_i\leqslant\gamma_i^2\phi_i^2W_i^2+H_i^2$，可以得到$\mathcal{V}^{\circ}\left(q;F\left(q,\omega\right)\right)\leqslant\sum_{i\in\mathcal{N}}\left(\left|U_i\right|+1\right)\left(\bar{\gamma}_i-\rho_{W_i}\right)W_i^2$，其中$\bar{\gamma}_i=\gamma_i\left[2\phi_{i,\text{miet}}L_i+\gamma_i\left(1+\phi_{i,\text{miet}}^2\right)\right]$，进而可以得到

$$\begin{cases}\sum_{i=1}^{N}\mathcal{V}_i^{\circ}\left(\boldsymbol{q}_i;F_i\left(\boldsymbol{q}_i,\boldsymbol{\omega}_i\right)\right)\leqslant-\sum_{i=1}^{N}\varpi_i^0\mathcal{V}_i\left(q_i\right),\quad m_i=0\\\sum_{i=1}^{N}\mathcal{V}_i^{\circ}\left(\boldsymbol{q}_i;F_i\left(\boldsymbol{q}_i,\boldsymbol{\omega}_i\right)\right)\leqslant\sum_{i=1}^{N}\varpi_i^1\mathcal{V}_i\left(q_i\right),\quad m_i=1\end{cases}\tag{10-59}$$

其中，$\varpi_i^0=\min\left\{\rho_{V_i},\frac{\lambda_i\rho_{W_i}}{\gamma_i}\right\}>0;\ \varpi_i^1=\frac{\hat{\gamma}_i-\rho_{W_i}}{\gamma_i\phi_{i,\text{miet}}}>0$。

结合假设10.7、10.8和式(10-59)，以及Abdessameud等人[280]的研究可知，如果有向图是强连通的，那么$\left(\xi_i,e_i\right)$在$m_i=0$时，任意两个连续传输间隔之间的演化是有上界的，也就是说$\upsilon_i^0(q_i)$是有界的，所以$\mathcal{V}$是递减的。因此，当m_i=0的时候，存在正常数系列$\left\{\kappa_i^0\right\}_{i\in\mathcal{N}}$使得

$\mathcal{V}_i^0\left(\boldsymbol{q}_i;f_i\left(\boldsymbol{q}_i,\boldsymbol{\omega}_i\right)\right)\leqslant-\kappa_i^0\varpi_i^0\mathcal{V}_i\left(\tilde{\boldsymbol{q}}_i\right)$；而当$m_i=1$的时候，存在正常数系列$\left\{\kappa_i^1\right\}_{i\in\mathcal{N}}$使得$\mathcal{V}_i^\circ\left(q_i;\boldsymbol{f}_i\left(\boldsymbol{q}_i,\boldsymbol{\omega}_i\right)\right)\leqslant\kappa_i^1\varpi_i^1\mathcal{V}_i\left(\tilde{\boldsymbol{q}}_i\right)$。因此，有

$$\mathcal{V}_i^\circ\left(\boldsymbol{q}_i;f_i\left(\boldsymbol{q}_i,\boldsymbol{\omega}_i\right)\right)\leqslant-\kappa_i^{m_i}\varpi_i^{m_i}\mathcal{V}_i\left(\tilde{\boldsymbol{q}}_i\right) \tag{10-60}$$

其中，$m_i\in\{0,1\}$。

实际上，混杂系统$\mathcal{H}_{\text{dos}}$没有奇诺解，因为最小传输间隔$\tau_{\text{miet}}^{i,m_i}$严格正，而且上述关于函数$\{\mathcal{V}_i\}_{i\in\mathcal{N}}$在$\boldsymbol{q}\in C\cup D$中的演化分析也排除了系统有限的逃逸时间，即$\boldsymbol{\xi}$和$\boldsymbol{e}$是有界的，并且变量$\boldsymbol{\eta}$、$\boldsymbol{\tau}$、$\boldsymbol{s}$和$\boldsymbol{m}$的轨迹不显示有限的逃脱时间。基于 Dolk 等人 [83] 的研究和以上分析可知，混杂系统$\mathcal{H}_{\text{dos}}$的所有的最大解$\boldsymbol{\xi}$满足$\boldsymbol{\xi}(0,0)\in\mathbb{X}_0$在 t 方向有无界的时间域。

根据式 (10-60)，接下来我们考虑混杂系统 (10-53) 的$\boldsymbol{q}_i$。对于给定的最大解$\boldsymbol{q}$，在一个给定的时间区间$[T_1,T_2]$，我们首先为每个$\boldsymbol{q}_i$定义关于$m_i=0$和$m_i=1$的时间集合。对于$i\in\mathcal{N}$，定义$\tilde{\varTheta}_{q_i}^i\left(T_1,T_2\right)=\left\{\tilde{t}\in\left(T_1,T_2\right)\mid m_i\left(\tilde{t}\right)=0,\forall k\in\mathbb{Z}_{\geqslant0},\left(\tilde{t},k\right)\in\operatorname{dom}\boldsymbol{q}_i\right\}$和$\tilde{\varXi}_{q_i}^i\left(T_1,T_2\right)=\left[T_1,T_2\right]\setminus\tilde{\varTheta}_{q_i}^i\left(T_1,T_2\right)$，其中$\tilde{\varTheta}_{q_i}^i\left(T_1,T_2\right)$和$\tilde{\varXi}_{q_i}^i\left(T_1,T_2\right)$可以分别表示为$\tilde{\varXi}_{q_i}^i\left(T_1,T_2\right)=\bigcup_{k_t^i\in\mathbb{Z}_{\geqslant0}}\tilde{Z}_{k_t^i}^i\cap\left[T_1,T_2\right]$和$\tilde{\varTheta}_{q_i}^i\left(T_1,T_2\right)=\bigcup_{k_t^i\in\mathbb{Z}_{\geqslant0}}\tilde{W}_{k_t^i-1}^i\cap\left[T_1,T_2\right]$，其中$k_t^i\in\mathbb{Z}_{\geqslant0},i\in\mathcal{N}$。另外，$\tilde{Z}_{k_t^i}^i$和$\tilde{W}_{k_t^i}^i$分别定义为$\tilde{Z}_{k_t^i}^i=\left[\zeta_{k_t^i}^i,\zeta_{k_t^i}^i+v_{k_t^i}^i\right)$，以及$\tilde{W}_{k_t^i}^i=\left[\zeta_{k_t^i}^i+v_{k_t^i}^i,\zeta_{k_t^i+1}^i\right)$，其中$v_{k_t^i}^i\geqslant0$表示从$\zeta_{k_t^i}^i$到下一次成功传输尝试时刻所经过的时间。如果$v_{k_t^i}^i=0$，那么$\tilde{Z}_{k_t^i}^i=\left\{\zeta_{k_t^i}^i\right\}$以及$\tilde{W}_{k_t^i}^i=\left[\zeta_{k_t^i}^i,\zeta_{k_t^i+1}^i\right)$，即拒绝服务攻击是脉冲形式。另外$\zeta_0^i=h_0^i$和$\tilde{W}_{-1}^i=\left[0,\zeta_0^i\right)$成立，如果$h_0^i>0$；否则$\tilde{W}_{-1}^i=\varnothing$。

结合假设 10.6，有$\left|\tilde{\varXi}_{q_i}^i\left(T_1,T_2\right)\right|\leqslant\left|\varXi_Z^i\left(T_1,T_2\right)\right|+n_Z^i\left(T_1,T_2\right)\tau_{\text{miet}}^{i,1}$，即拒绝服务攻击的时间集合被$n_Z^i\left(T_1,T_2\right)\tau_{\text{miet}}^{i,1}$延长了。基于假设 10.6，有$\left|\tilde{\varXi}_{q_i}^i\left(T_1,T_2\right)\right|\leqslant\varsigma_*^i+\left(T_2-T_1\right)/T_*^i$，其中$\varsigma_*^i=\varsigma_Z^i+v_Z^i\tau_{\text{miet}}^{i,1}$，$T_*^i=\tau_{f_Z}^i\tau_{d_Z}^i/\left(\tau_{f_Z}^i+\tau_{d_Z}^i\tau_{\text{miet}}^{i,1}\right)$。接下来，基于以上分析可以得到混杂系统的稳定性结果。基于 De Persis 和 Tesi [85]、Dolk 等人 [83] 的研究，对任意$(t,k)\in\operatorname{dom}q_i$，可以由式 (10-60) 知

$$\mathcal{V}_i\left(q_i\left(t,k\right)\right)\leqslant\text{Im}_i\text{e}^{-\hat{\beta}_it}\mathcal{V}_i\left(q_i\left(0,0\right)\right)+\theta_i^2\int_0^t\text{Im}_i\text{e}^{-\hat{\beta}_i(t-s)}\text{d}s\parallel\omega_i\parallel_{\mathcal{L}_\infty}$$

其中，$\text{Im}_i=\text{e}^{\varsigma_{*i}\left(\kappa_i^0\varpi_i^0+\kappa_i^1\varpi_i^1\right)}$；$\hat{\beta}_i=\kappa_i^0\varpi_i^0-\left(\kappa_i^0\varpi_i^0+\kappa_i^1\varpi_i^1\right)/T_*^i$。由$\overline{\beta}_{\mathcal{V}_i}\left(\left|\boldsymbol{q}_i\right|_{\mathcal{A}}\right)\geqslant\mathcal{V}_i\left(\boldsymbol{q}_i\right)\geqslant V_i\left(\boldsymbol{\xi}_i\right)\geqslant c_{z_i}\left|\boldsymbol{z}_i\left(t,k\right)\right|^2$，可知

$$\| z_i \|_{\mathcal{L}_\infty} \leqslant \sqrt{\frac{\mathrm{Im}_i}{c_{z_i}} \overline{\beta}_{\mathcal{V}_i}\left(\left|\boldsymbol{q}_i(0,0)\right|_{\mathcal{A}}\right)} + \theta_i \sqrt{\frac{\mathrm{Im}_i}{\hat{\beta}_i c_{z_i}}} \| \boldsymbol{\omega}_i \|_{\mathcal{L}_\infty}$$

即有

$$\| \boldsymbol{z} \|_{\mathcal{L}_\infty} \leqslant \boldsymbol{\beta}_{\mathcal{V}}\left(\left|\boldsymbol{q}(0,0)\right|_{\mathcal{A}}\right) + \max_{i\in\mathcal{N}}\left\{\theta_i \sqrt{\mathrm{Im}_i / \left(\hat{\beta}_i c_{z_i}\right)}\right\} \| \boldsymbol{\omega} \|_{\mathcal{L}_\infty}$$

其中

$$\boldsymbol{\beta}_{\mathcal{V}} : \left(\left|\boldsymbol{q}(0,0)\right|_{\mathcal{A}}\right) \to \sqrt{\sum_{i\in\mathcal{N}} \mathrm{Im}_i / c_{z_i} \overline{\beta}_{\mathcal{V}_i}\left(\left|\boldsymbol{q}_i(0,0)\right|_{\mathcal{A}}\right)}$$

因此，闭集$\boldsymbol{\mathcal{A}}$是$\mathcal{L}_\infty$稳定性的，其中$\mathcal{L}_\infty$增益为$\max_{i\in\mathcal{N}}\left\{\theta_i \sqrt{\mathrm{Im}_i / \left(\hat{\beta}_i c_{z_i}\right)}\right\}$。至此证明了定理10.4的结论。 ■

注释 10.12 正如式(10–50)所示，驱动函数$\boldsymbol{\Psi}_i$会被连续监测并且会在$\tau_i = \tau_{\text{miet}}^{i,m_i}$之后的某时刻($\eta_i(t) < 0$的时候)被触发。类似于周期事件驱动，$\tau_{\text{miet}}^{i,m_i}$[式(10–50)]可以帮助排除奇诺现象，也可以减少资源的消耗。实际上，本章节的事件驱动函数会在最小触发间隔$\tau_i = \tau_{\text{miet}}^{i,m_i}$之后及时地根据需要进行触发，即可以潜在地增加触发频率，从而使得智能体可以进行更多的通信，也就是说跟周期触发相比，本章节的事件驱动策略(10–50)至少不会错过某些触发时刻，即本章节的事件驱动策略会在资源利用和触发通信频率之间更好地进行权衡。

注释 10.13 事实上，如果不了解拒绝服务攻击的频率和持续时间，就没法研究拒绝服务攻击的影响。因此在假设10.6中，我们给出可根据参数$v_Z^i \geqslant 0$和$\tau_{f_Z}^i > \varDelta > 0$进行调节的拒绝服务攻击的频率和持续时间的上界。另一方面，根据拒绝服务攻击的参数，我们可以调节$\tau_{\text{miet}}^{i,m_i}$，使得$\max_{i\in\mathcal{N}}\left\{\tau_{\text{miet}}^{i,0}\tau_{\text{miet}}^{i,1}\right\} \leqslant \varDelta$，为的是确保所研究的系统能通过设计合适的控制策略产生合适的更新时刻进而保持所需的性能，如定理10.4的式(10–58)根据$m_i = 0$和$m_i = 1$可以选择不同的事件触发条件。本章节中我们采用确认机制，这样可以通过$m_i=0$和$m_i=1$来确定拒绝服务攻击的开始和结束时刻，而作为一个重要而有趣的课题，探索更有效的攻击检测机制在理论上具有挑战性和实践难度，这将是我们未来的研究方向。

注释 10.14　现在分析参数$\{\kappa_i^0\}_{i\in\mathcal{N}}$和$\{\kappa_i^1\}_{i\in\mathcal{N}}$是如何影响$\mathcal{L}_\infty$增益。这里以$\kappa_i^0$为例详细解释一下，类似地可以得到$\kappa_i^1$的情形。首先，定义一个新函数$\Gamma_i(\kappa_i^0)=\mathrm{Im}_i/\hat{\beta}_i$，其中$\mathrm{Im}_i=e^{\varsigma_*^i(\kappa_i^0\varpi_i^0+\kappa_i^1\varpi_i^1)}$，$\hat{\beta}_i=\kappa_i^0\varpi_i^0-(\kappa_i^0\varpi_i^0+\kappa_i^1\varpi_i^1)/T_*^i>0$。接着定义$\Gamma_i(\kappa_i^0)=e^{\bar{a}_i\kappa_i^0+\bar{c}_i}/(\underline{a}_i\kappa_i^0+\underline{c}_i)$，其中$\kappa_i^0\in(-\underline{c}_i/\underline{a}_i,+\infty)$，常数变量$\bar{a}_i=\varsigma_*^i\varpi_i^0,\bar{c}_i=\varsigma_*^i\kappa_i^1\varpi_i^1,\underline{a}_i=\varpi_i^0(1-1/T_*^i)$以及$\underline{c}_i=-\kappa_i^1\varpi_i^1/T_*^i$。根据式(10-57)可知，$\Gamma_i(\kappa_i^0)$在$\kappa_i^0\in(-\underline{c}_i/\underline{a}_i,+\infty)$上是严格正的。接着可知$\frac{\mathrm{d}}{\mathrm{d}\kappa_i^0}\Gamma_i(\kappa_i^0)=\left[\bar{a}_i(\underline{a}_i\kappa_i^0+\underline{c}_i)-\underline{a}_i\right]e^{\bar{a}_i\kappa_i^0+\bar{c}_i}/(\underline{a}_i\kappa_i^0+\underline{c}_i)^2$，也就是说$\kappa_i^0=(\underline{a}_i-\bar{a}_i\underline{c}_i)/\bar{a}_i\underline{a}_i$是一个极值点，而且是最小值点。也就是说，函数$\Gamma_i(\kappa_i^0)$是关于$\kappa_i^0$的增函数，其中$\kappa_i^0\geqslant(\underline{a}_i-\bar{a}_i\underline{c}_i)/\bar{a}_i\underline{a}_i$；而函数$\Gamma_i(\kappa_i^0)$是关于$\kappa_i^0$的减函数，其中$\kappa_i^0\leqslant(\underline{a}_i-\bar{a}_i\underline{c}_i)/\bar{a}_i\underline{a}_i$。如果$\kappa_i^0=(\underline{a}_i-\bar{a}_i\underline{c}_i)/\bar{a}_i\underline{a}_i$满足式(10-60)，那么该值就是最优值，使系统的$\mathcal{L}_\infty$增益较小，因此，所选的参数值κ_i^0应该尽可能地接近$(\underline{a}_i-\bar{a}_i\underline{c}_i)/\bar{a}_i\underline{a}_i$。

10.3.3 算法验证与分析

为了验证本章节结果的有效性，本部分的数值仿真考虑了一个基准问题，即六架小型四旋翼原型的编队控制问题，而且该问题是基于Abdessameud等人[65]书中的例子而来的。拉普拉斯矩阵$\boldsymbol{L}=(l_{ij})_{6\times6}$的六行分别定义为：[1 0 0 0 0 −1], [−1 2 0 0 0 −1], [0 −1 2 −1 0 0], [0 0 −1 2 0 −1], [0 0 0 −1 1 0], [0 0 0 0 −1 1]，也就是说有向拓扑是强连通的。根据Abdessameud等人[65]的研究，这里所考虑的小型四旋翼原型可以建模为：

$\ddot{\boldsymbol{p}}_i = \mathcal{F}_i - \frac{u_i}{m_i} f_i\left(\boldsymbol{R}_i, \boldsymbol{R}_i^d\right), f_i\left(\boldsymbol{R}_i, \boldsymbol{R}_i\right) = \left(\boldsymbol{R}_i - \boldsymbol{R}_i^d\right)^{\mathrm{T}} \mathbf{e}_{3i}$ 和 $\mathcal{F}_i = g\mathbf{e}_{3i} - \frac{u_i}{m_i}\boldsymbol{R}_i^d \mathbf{e}_{3i}$，其中，$\mathbf{e}_{3i} = \left(0\ 0\ 1\right)^{\mathrm{T}}$。令$\mathbb{SO}(3)$表示旋转矩阵的特殊正交群，$\boldsymbol{R}_i, \boldsymbol{R}_i^d \in \mathbb{SO}(3)$则分别表示旋转矩阵和目标编队所需的旋转矩阵。为了简化数值仿真，我们主要考虑外环（即位置环），即假设$\boldsymbol{R}_i = \boldsymbol{R}_i^d$已成立[285]，而且根据 Naldi 等人[285]中 Appendix C 的研究中目标欧拉角的计算方式，我们可以得到$\boldsymbol{R}_i^d$。令$\mathcal{P}_i(t) = \left(\boldsymbol{p}_i(t) - \tilde{\boldsymbol{\delta}}_i(t), \dot{\boldsymbol{p}}_i(t) - \dot{\tilde{\boldsymbol{\delta}}}_i(t)\right), \tilde{\boldsymbol{\delta}}_i(t) = \boldsymbol{p}^*(t) + \boldsymbol{\delta}_i(t)$，其中$\boldsymbol{p}^*(t)$表示参照位置，向量$\boldsymbol{\delta}_i(t)$代表智能体 i 和参照点$\boldsymbol{p}^*(t)$之间的相对位置。

基于 Abdessameud 等人[65]的研究，这里考虑位置的推力控制策略为：$-\frac{u_i}{m_i}\boldsymbol{R}_i\left(\boldsymbol{q}_i^d\right)\mathbf{e}_{3i} = \left[\ddot{\tilde{\delta}}_i - \kappa_i^d \chi\left(\boldsymbol{e}_i - \boldsymbol{\psi}_i\right) - \kappa_i^p \chi\left(\boldsymbol{e}_i\right) - L_i^\zeta \mathbf{h}\left(\boldsymbol{\zeta}_i\right)\left(L_i^\zeta \chi\left(\boldsymbol{\zeta}_i\right) + L_i^\xi \chi\left(\boldsymbol{\xi}_i\right)\right)\right]$ $-g\mathbf{e}_{3i}$，其中 $\dot{\boldsymbol{\psi}}_i = \kappa_i^\psi\left(\boldsymbol{e}_i - \boldsymbol{\psi}_i\right)$，$\dot{\boldsymbol{r}}_i = \dot{\tilde{\boldsymbol{\delta}}}_i - L_i^\zeta \chi\left(\boldsymbol{\zeta}_i\right), \dot{\boldsymbol{\zeta}}_i = -L_i^\zeta \chi\left(\boldsymbol{\zeta}_i\right) + L_i^\xi \chi\left(\boldsymbol{\xi}_i\right)$，$\dot{\boldsymbol{\xi}}_i = -L_i^\xi \chi\left(\boldsymbol{\xi}_i\right) + \sum_{j \in \mathcal{N}} b_{ij}\left(\boldsymbol{r}_i - \boldsymbol{r}_j - \boldsymbol{\delta}_{ij}\right)$。对角矩阵$\mathbf{h}(r) = \mathrm{diag}\left(\partial\sigma\left(r^x\right)/\partial r^x, \partial\sigma\left(r^y\right)/\partial r^y, \partial\sigma\left(r^z\right)/\partial r^z\right)$满足$\|\mathbf{h}(r)\| \leqslant \sigma_{\mathbf{h}}$，其中$\sigma_{\mathbf{h}} = 1$。选择饱和函数$\sigma(s) = \tanh(s)$，其中$\chi(r) = \mathrm{col}\left[\sigma\left(r^k\right)\right]\left(k \in \{x, y, z\}\right)$，正如 Abdessameud 等人在文献 [65] 中 Section 2.1.3 研究中的定义，并且$\sigma_b = 1$。选择控制参数$L_i^\zeta = 0.5$, $L_i^\xi = 1, \kappa_i^d = 150, \kappa_i^p = 90$以及$\kappa_i^\psi = 3$，并且令$L_i = 9.81$。接下来我们考虑$\hat{\gamma}_i = 10$和$\epsilon_i = 0.1$的情形。选择$\rho_{V_i} = 0.8$，$\rho_{W_i} = 4.0561, \gamma_i = 4.0561, c_{z_i} = 1$和$\lambda_i = 0.8$，因此有$\varpi_i^0 = 0.8$和$\varpi_i^1 = 1.875$。基于式(10-55)，选择$\tau_{\mathrm{miet}}^{i,0} = 0.015656$和$\tau_{\mathrm{miet}}^{i,1} = 0.012743$。对于$v_Z^i$和$\varsigma_Z^i$，其中$T_1 = 7$和$T_2 = 17$，选择$v_Z^i = 5, \varsigma_Z^i = 0$，那么$\tau_{f_Z}^i$和$\tau_{d_Z}^i$满足$\kappa_i^0\varpi_i^0 > \left(\tau_{\mathrm{miet}}^{i,1}/\tau_{f_Z}^i + 1/\tau_{d_Z}^i\right)\left(\kappa_i^0\varpi_i^0 + \kappa_i^1\varpi_i^1\right)$，正如式(10-57)所示的那样，其中$\left(\kappa_i^0, \kappa_i^1\right) = (0.1, 1)$，即$\left(\tau_{\mathrm{miet}}^{i,1}/\tau_{f_Z}^i + 1/\tau_{d_Z}^i\right) < 0.040921$。令$\boldsymbol{p}^* = (10\cos(0.1t + 2), 10\cos(0.2t + 2.4), t)$ 和 $\boldsymbol{\delta}_i(t) = \left(10\sin\left(t + \frac{i-1}{3}\pi\right), -10\cos\left(t + \frac{i-1}{3}\pi\right), 10\cos\left(t + \frac{i-1}{3}\pi\right)\right)$。在此基础上，图 10-5 和图 10-6 分别给出第一个小型四旋翼的轨迹三维图以及六个小型四旋翼的向量轨迹的第一维跟踪误差。在图 10-6 中，垂直的灰色条纹表示拒绝服务攻击的攻击时间段，其中$\mathcal{P}_i = \left(\mathcal{P}_i^x, \mathcal{P}_i^y, \mathcal{P}_i^z\right), i \in \mathcal{N}$。表 10-2 展示了拒绝服务攻击下的平均传输时间，其中$\tau_{\mathrm{miet}}^{i,m_i} = \tau_{\mathrm{mati}}^{i,m_i}, \hat{\gamma}_i > \gamma_i = 4.0561$以及$\epsilon_i > 0$，即

$\tau_{\text{miet}}^{i,1} < \tau_{\text{miet}}^{i,0} = 0.015656$。平均传输次数是 160，如果$\epsilon_i = 0$，$\hat{\gamma}_i = \gamma_i$，即 $\tau_{\text{miet}}^{i,1} = \tau_{\text{miet}}^{i,0} = 0.015656$。表 10-2 和图 10-7 中给出的关于系统性能的两个参数的序列则显示了注释 10.11 中所描述的关于$\hat{\gamma}_i$和ϵ_i的灵活性。

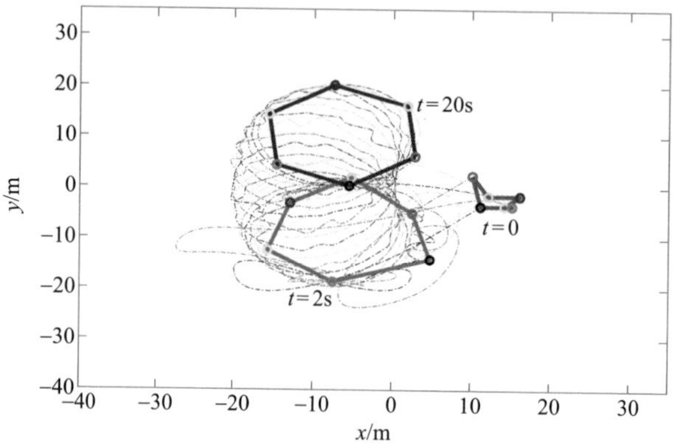

图 10-5　六个小型四旋翼的位置轨迹

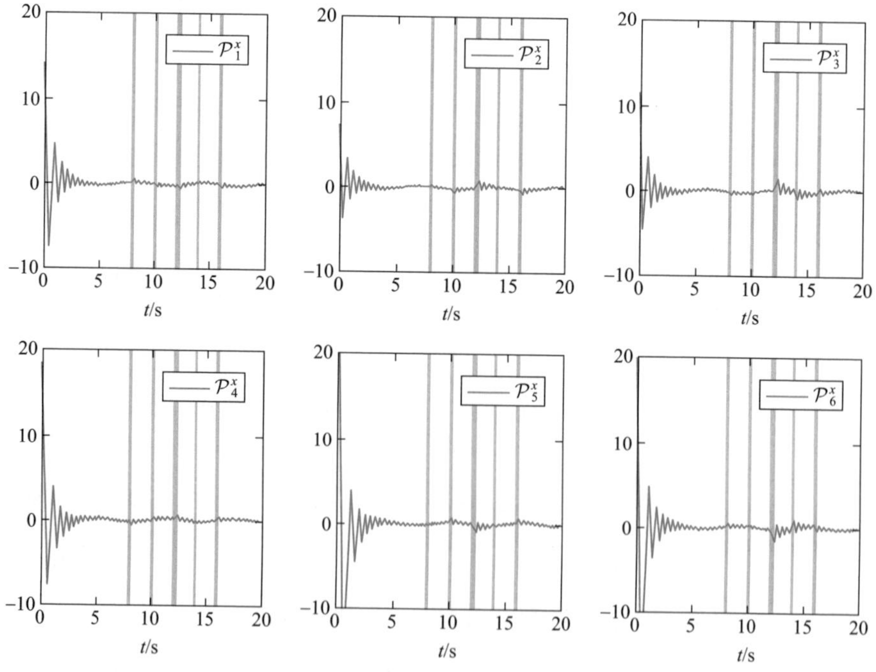

图 10-6　六个小型四旋翼轨迹向量的第一维跟踪误差

表10-2 $\hat{\gamma}_i$和ϵ_i的不同取值以及拒绝服务攻击下的平均传输时间的不同情形

$\hat{\gamma}_i / \epsilon_i$	$\tau_{\text{miet}}^{i,1}$	平均传输次数	γ_i / ϵ_i	$\tau_{\text{miet}}^{i,1}$	平均传输次数
5/0.1	0.014727	170	10/0.001	0.012897	194
7/0.1	0.013968	179	10/0.005	0.012891	194
8/0.1	0.013534	185	10/0.010	0.012882	194
10/0.1	0.012743	196	10/0.050	0.01282	195
12/0.1	0.01204	208	10/0.100	0.012743	196
14/0.1	0.011411	219	10/0.500	0.012155	206
16/0.1	0.010845	231	10/0.800	0.011749	213
18/0.1	0.010333	242	10/1.300	0.011130	225
20/0.1	0.009868	253	10/1.500	0.010901	229

注：表中$\tau_{\text{miet}}^{i,m_i} = \tau_{\text{mati}}^{i,m_i}, \hat{\gamma}_i > \gamma_i = 4.0561$以及$\epsilon_i > 0$，即$\tau_{\text{miet}}^{i,1} < \tau_{\text{miet}}^{i,0} = 0.015656$。平均传输次数是 160，如果$\epsilon_i = 0$和$\hat{\gamma}_i = \gamma_i$，即$\tau_{\text{miet}}^{i,1} = \tau_{\text{miet}}^{i,0} = 0.015656$。

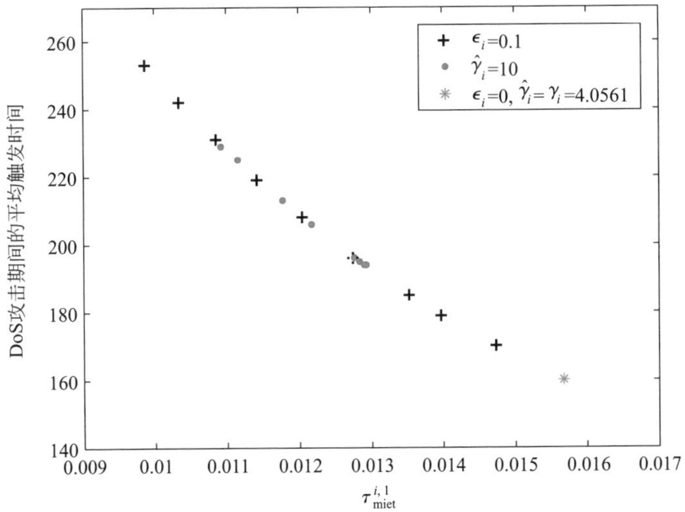

图 10-7 对应表 10-2 的拒绝服务攻击期间的平均传输次数

Digital Wave
Advanced Technology of Industrial Internet

Control of
Autonomous Intelligent Systems

自主智能系统控制

第 11 章

网络化状态估计的安全性分析以及攻击检测

11.1 概述

网络物理系统（CPS）是指物理过程与网络组件紧密集成的系统。CPS 具有如控制、计算和通信之类的诸多功能，已逐渐应用于各种关键基础设施，包括机器人系统、智能电网和环境监控。但是，通过入侵未受保护的无线传输网络，网络攻击也给 CPS 带来了潜在的安全隐患以及新风险。这可能会造成巨大的损害，例如经济损失、设施损失甚至巨大的生命损失。因此，安全问题的研究是保证 CPS 正常运行的基本要求，并且在近几年引起了极大的关注。

在远端状态估计系统，文献 [90-91] 提出了一种可以屏蔽传统χ^2检测器的线性中间人攻击。因此，为了解决该安全隐患，本节提出了一种基于伪随机数的安全数据传输模块，从而保证了远端估计器的估计性能以及提高了攻击的检测率。此外，针对分布式传感网络中存在的隐秘攻击，本章节分析了分布式滤波算法在攻击下仍能保证均方稳定性的充分条件。

11.2 抵御线性中间人攻击的安全远程状态估计

11.2.1 系统和攻击模型

考虑以下离散时间线性时不变（LTI）过程，可以将其建模为：

$$\boldsymbol{x}_{k+1}=\boldsymbol{A}\boldsymbol{x}_k+\boldsymbol{w}_k \tag{11-1}$$

$$\boldsymbol{y}_k=\boldsymbol{C}\boldsymbol{x}_k+\boldsymbol{v}_k \tag{11-2}$$

其中，$k\in\mathbb{N}$是时间指数；$\boldsymbol{x}_k\in\mathbb{R}^n$是过程状态；$\boldsymbol{y}_k\in\mathbb{R}^m$是传感器的测量值；$\boldsymbol{w}_k\in\mathbb{R}^n$和$\boldsymbol{v}_k\in\mathbb{R}^m$是相应的过程噪声以及测量噪声。它们都是独立同

分布的零均值高斯白噪声，其协方差分别为$\boldsymbol{Q} \geqslant 0$和$\boldsymbol{R} > 0$。$\boldsymbol{w}_k$和$\boldsymbol{v}_k$互不相关。此外，过程的初状态$\boldsymbol{x}_0$是零均值、方差为$\boldsymbol{\varPi}_0$的高斯白噪声，并且与$\boldsymbol{w}_k$和$\boldsymbol{v}_k$不相关。$\boldsymbol{A}$和$\boldsymbol{C}$分别是系统矩阵和测量矩阵。$(\boldsymbol{A},\boldsymbol{C})$是可检测的，$\left(\boldsymbol{A},\sqrt{\boldsymbol{Q}}\right)$是可镇定的。

基于测量方程 (11-2) 中的测量值$\boldsymbol{y}_k$，智能传感器采用标准卡尔曼滤波器对系统状态进行局部估计：

$$\hat{\boldsymbol{x}}_k^- = \boldsymbol{A}\hat{\boldsymbol{x}}_{k-1} \tag{11-3a}$$

$$\boldsymbol{P}_k^- = \boldsymbol{A}\boldsymbol{P}_{k-1}\boldsymbol{A}^{\mathrm{T}} + \boldsymbol{Q} \tag{11-3b}$$

$$\boldsymbol{K}_k = \boldsymbol{P}_k^-\boldsymbol{C}^{\mathrm{T}}\left(\boldsymbol{C}\boldsymbol{P}_k^-\boldsymbol{C}^{\mathrm{T}} + \boldsymbol{R}\right)^{-1} \tag{11-3c}$$

$$\hat{\boldsymbol{x}}_k = \hat{\boldsymbol{x}}_k^- + \boldsymbol{K}_k\left(\boldsymbol{y}_k - \boldsymbol{C}\hat{\boldsymbol{x}}_k^-\right) \tag{11-3d}$$

$$\boldsymbol{P}_k = \left(\boldsymbol{I} - \boldsymbol{K}_k\boldsymbol{C}\right)\boldsymbol{P}_k^- \tag{11-3e}$$

其中，$\hat{\boldsymbol{x}}_k^-$和$\hat{\boldsymbol{x}}_k$是状态$\boldsymbol{x}_k$的先验和后验最小均方误差（MMSE）估计；$\boldsymbol{P}_k^-$和$\boldsymbol{P}_k$是相应的估计误差协方差。卡尔曼滤波的递推迭代从$\hat{\boldsymbol{x}}_0 = 0$和$\boldsymbol{P}_0 = \boldsymbol{\varPi}_0 \geqslant 0$开始。然后，智能传感器就可以计算得到需要传输到远端的新息$\boldsymbol{z}_k$：

$$\boldsymbol{z}_k = \boldsymbol{y}_k - \boldsymbol{C}\hat{\boldsymbol{x}}_k^-,$$

其中$\boldsymbol{z}_k$为零均值，协方差为$\boldsymbol{\varSigma}_z \overset{\mathrm{def}}{=} \mathbb{E}\left[\boldsymbol{z}_k\boldsymbol{z}_k^{\mathrm{T}}\right] = \boldsymbol{C}\boldsymbol{P}_k^-\boldsymbol{C}^{\mathrm{T}} + \boldsymbol{R}$的独立同分布的高斯随机变量。众所周知，卡尔曼滤波器可以从任何初始条件以指数方式快速进入稳态[136]。因此，我们假设滤波器的迭代从稳态开始，即$\boldsymbol{\varPi}_0 = \lim\limits_{k\to\infty}\boldsymbol{P}_k^- = \overline{\boldsymbol{P}}$。在该假设下，除了$\boldsymbol{\varSigma}_z$之外，卡尔曼滤波器增益也是一个固定值：

$$\boldsymbol{K} = \overline{\boldsymbol{P}}\boldsymbol{C}^{\mathrm{T}}\left(\boldsymbol{C}\overline{\boldsymbol{P}}\boldsymbol{C}^{\mathrm{T}} + \boldsymbol{R}\right)^{-1}, \tag{11-4}$$

借助智能传感器通过无线网络发送的新息$\boldsymbol{z}_k$，远程估计器也可以利用卡尔曼滤波器[式(11-3a)～式(11-3e)]来监测过程的状态$\boldsymbol{x}_k$。

为了实现攻击检测以及过程监测，远端配置了基于残差的χ^2检测器。基于新息$\boldsymbol{z}_k$的统计特性，采用以下假设检验作为检测标准：

$$\boldsymbol{\xi}_k \overset{\mathrm{def}}{=} \sum_{i=k-J+1}^{k}\boldsymbol{z}_i^{\mathrm{T}}\boldsymbol{\varSigma}_z^{-1}z_i \underset{H_1}{\overset{H_0}{\lessgtr}} \zeta \tag{11-5}$$

其中，零假设H_0表示数据是可信的，而H_1则恰好相反，J和ζ分别表示检测器的窗口大小和阈值，ξ_k是归一化平方残差的和，并且服从自由度为mJ的

χ^2分布。ζ由置信系数P_ζ决定。因此，误报率为$\left(1-P_\zeta\right)$。当时间窗口$J\neq 1$，ξ_k不仅取决于当前接收到的数据，还取决于之前的数据。为了更清晰地反映攻击、检测器和估计器之间的一一对应关系，本章节设置$J=1$。需要注意的是，对于$J\neq 1$这种更一般的情况，分析方法和结果是相同的。

中间人攻击可以拦截通过无线网络的消息，然后注入伪造的数据[289]。特别地，文献 [90,91] 提出了一种基于新息的线性中间人攻击，其可以在不被 χ^2 检测器检测到的情况下修改传输的数据。由于被这种攻击篡改的数据可以保留与原始数据相同的统计特性，χ^2 检测器等基于统计的检测器无法检测到这种攻击。由于现有文献缺乏有效的防御方法，本节旨在解决由线性中间人攻击引起的安全问题，其攻击模型定义为：

$$\tilde{\boldsymbol{z}}_k=\boldsymbol{T}_k\boldsymbol{z}_k+\boldsymbol{b}_k \tag{11-6}$$

其中，$\tilde{\boldsymbol{z}}_k$是被攻击者篡改的新息；$\boldsymbol{T}_k\in\mathbb{R}^{m\times m}$是任意矩阵；$\boldsymbol{b}_k$是零均值、协方差为$\boldsymbol{\Sigma}_b$的高斯随机变量，与$\boldsymbol{z}_k$不相关。因此$\tilde{\boldsymbol{z}}_k$也遵循零均值、协方差为$\boldsymbol{T}_k\boldsymbol{\Sigma}_z\boldsymbol{T}_k^{\mathrm{T}}+\boldsymbol{\Sigma}_b$的正态分布。根据攻击是否存在，$\chi^2$检测器的报警率可以分为检测率和误报率。需要注意的是，检测率等于1减去漏报率。如果$\tilde{\boldsymbol{z}}_k$与原始的新息$\boldsymbol{z}_k$具有相同的统计特征，则攻击下的检测率[式(11-6)]，等于不存在攻击时的误报率$(1-P_\zeta)$[90]。换句话说，式(11-6)中的攻击应该满足以下可行性约束以保持隐蔽：

$$\boldsymbol{T}_k\boldsymbol{\Sigma}_z\boldsymbol{T}_k^{\mathrm{T}}+\boldsymbol{\Sigma}_b=\boldsymbol{\Sigma}_z \tag{11-7}$$

在所有满足条件(11-7)的攻击策略中，文献[90]进一步推导出了最优线性中间人攻击（OLMA）策略。

引理 11.1[90]

最优线性中间人攻击策略是$\boldsymbol{T}_k=-\boldsymbol{I}$以及$\boldsymbol{b}_k=0$，其可以在不被检测的条件下最大化远端估计器的估计误差协方差。相应的估计误差协方差的递推公式为：$\boldsymbol{P}_k=\boldsymbol{P}_k^-+3\overline{\boldsymbol{P}}\boldsymbol{C}^{\mathrm{T}}\boldsymbol{\Sigma}_z^{-1}\boldsymbol{C}\overline{\boldsymbol{P}}$。

11.2.2 数据传输的安全模块

基于原有的系统架构，我们进一步增加了如图 11-1 所示的数据传输的安全模块。图中，蓝色虚线框标出的系统块位于通信通道的两侧，是所谓的针对线性中间人攻击的数据传输安全模块。在发送端，我们构建了一个数据预处理子模块，其结构如下所示：

$$\boldsymbol{f}_k = a\boldsymbol{z}_k + \boldsymbol{m}_k \tag{11-8}$$

其中，$a \in \mathbb{R}$是一个常数因子；$\boldsymbol{m}_k \in \mathbb{R}^m$是作为水印的高斯随机变量，其均值为0，协方差为$\Sigma_m$，且与$\boldsymbol{z}_k$独立，$\boldsymbol{f}_k \in \mathbb{R}^m$是式(11-8)的输出。值得注意的是，通过无线网络传输的数据是$\boldsymbol{f}_k$，而不是真正的新息$\boldsymbol{z}_k$。在所提出的框架下，线性攻击[式(11-6)]可以改写为：

$$\widetilde{\boldsymbol{f}}_k = \boldsymbol{T}_k \boldsymbol{f}_k + \boldsymbol{b}_k \tag{11-9}$$

其中，$\tilde{\boldsymbol{f}}_k$是被攻击篡改的数据。为了在没有攻击的情况下从$\tilde{\boldsymbol{f}}_k$中恢复$\boldsymbol{z}_k$，我们将以下数据恢复子模块放置在远程端：

$$\boldsymbol{z}_k^r = \frac{\tilde{\boldsymbol{f}}_k - \boldsymbol{m}_k}{a} \tag{11-10}$$

其中，$\boldsymbol{z}_k^r$是式(11-10)的输出。通过将式(11-8)和式(11-9)代入式(11-10)，$\boldsymbol{z}_k^r$可以表示为：

$$\boldsymbol{z}_k^r = \boldsymbol{T}_k \boldsymbol{z}_k + \frac{1}{a}\boldsymbol{b}_k + \frac{1}{a}\left(\boldsymbol{T}_k - \boldsymbol{I}\right)\boldsymbol{m}_k \tag{11-11}$$

显然，$\boldsymbol{z}_k^r$也遵循零均值高斯分布，其协方差$\boldsymbol{\Sigma}_{\tilde{z}}$给出如下：

$$\boldsymbol{\Sigma}_{\tilde{z}} = \boldsymbol{T}_k \boldsymbol{\Sigma}_z \boldsymbol{T}_k^{\mathrm{T}} + \frac{1}{a^2}\boldsymbol{\Sigma}_b + \frac{1}{a^2}\left(\boldsymbol{T}_k - \boldsymbol{I}\right)\boldsymbol{\Sigma}_m \left(\boldsymbol{T}_k - \boldsymbol{I}\right)^{\mathrm{T}} \tag{11-12}$$

根据式(11-11)以及式(11-12)，在不存在攻击的时候，即$\boldsymbol{T}_k = \boldsymbol{I}, \boldsymbol{b}_k = 0$，远程端可以通过消除$\boldsymbol{z}_k^r$中附加的水印将$\boldsymbol{z}_k^r$恢复为$\boldsymbol{z}_k$。反之，$\boldsymbol{z}_k^r$会被附加上水印，从而使得$\chi^2$检测器实现攻击的检测。需要注意的是，由于新息$\boldsymbol{z}_k$可以在没有攻击的情况下被安全模块完全恢复，$\chi^2$检测器的误报率不受到安全模块的影响，仍然等于原始的误报率，即$(1-\boldsymbol{P}_\zeta)$。

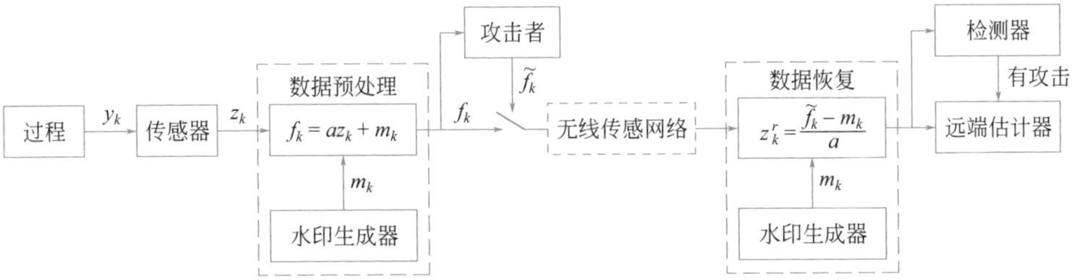

图 11-1 整个系统的框图

注释 11.1 为了使得 χ^2 检测器可以检测出文献 [90] 中的攻击，我们以式 (11-8) 和式 (11-10) 这种特定的形式构建了安全模块。对于更加复杂的攻击模型，我们可以基于本节提出的方法进一步设计安全模块的结构，并且结合其他更加强大的检测工具[287-288]实现攻击的检测。需要说明的是，本节只关注如图 11-1 所示中间人攻击。由于传感器在安全模块的保护范畴之外，因此所提出的方法无法抵御破坏测量值$\boldsymbol{y}_k$的攻击。

注释 11.2 与文献 [286,289] 中的攻击检测方法相比，我们所提出的方法的优点之一是可以保证在没有攻击的情况时的估计性能。这是因为我们所提出的方法可以完全消除水印，从而恢复原始的新息。

注释 11.3 当式 (11-11) 中的$\boldsymbol{m}_k$为常数$\overline{m}$时，z_k^r的均值为$\frac{1}{a}(\boldsymbol{T}_k-\boldsymbol{I})\overline{m}$，协方差为$\boldsymbol{T}_k\boldsymbol{\Sigma}_z\boldsymbol{T}_k^{\mathrm{T}}+\frac{1}{a^2}\boldsymbol{\Sigma}_b$。因此，在任何的线性攻击下，即$\boldsymbol{T}_k\neq\boldsymbol{I}$或$\boldsymbol{b}_k\neq 0, z_k^r$与新息$\boldsymbol{z}_k$之间至少有一个统计特性不相同。这意味着常数似乎也可以用作水印。然而，在这种情况下，传输数据$\boldsymbol{f}_k$的均值和协方差分别为$\overline{m}$和$a^2\boldsymbol{\Sigma}_z$。通过在无线网络中长期窃听$\boldsymbol{f}_k$，具有$\boldsymbol{\Sigma}_z$先验知识的攻击者可以估计出 a 和$\overline{m}$的准确值，并且计算出$\boldsymbol{z}_k$。因此，使用这种水印的安全模块可能会被攻击者识破，从而完全失去效用。这也是我们使用随机数作为水印$\boldsymbol{m}_k$的主要原因。

为了保证我们的方法的可行性，还需要解决两个关键问题：

① 如何使式 (11-8) 和式 (11-10) 中的水印序列在同一时刻步保持一致？

② 攻击者能否从截获的数据$\boldsymbol{f}_k$中推断出真正的新息$\boldsymbol{z}_k$？

第一个问题旨在确保在没有攻击的情况下可以完全消除水印，第二个问题确保我们的方法的有效性。相应的答案如下：

① 受到信息科学中研究的伪随机数的启发[290]，我们将满足高斯分布的伪随机序列作为水印m_k。由于相同的种子会生成唯一且确定的序列，因此我们可以将任意常数作为种子，并利用一些现有的密钥分发和管理技术[291]预先将其写入两个水印生成器的板载芯片中，这样双方的水印序列就可以保持一致。另一方面，由于攻击者不知道种子的信息，伪随机序列对于攻击者而言仍然是随机且未知的。当网络中存在时延的情况下，式(11-10)生成的水印可能与传输数据中的水印不匹配。然而，基于数据包上的时间戳[292]，我们仍然可以解决这个问题。

② 式(11-8)和式(11-10)可以视为加密以及解密的过程，而伪随机序列的种子可以视为密匙。事实上，伪随机数在密码学[293]中被广泛使用。并且，通过采用强密码学的伪随机数生成算法，我们仍然可以保证水印序列的隐蔽性。另外，由于传输数据f_k中z_k的随机性，攻击者无法从f_k中提取出m_k。

11.2.3 不同信息泄露场景下的检测性能与估计性能分析

攻击者需要基于其所获得的系统知识来设计攻击策略。对于攻击者而言，系统信息泄露的程度可分为以下三种情况：

① 攻击者不知道安全模块的存在；

② 攻击者知道式(11-8)和式(11-10)的结构，但是不知道水印m_k和a的具体信息；

③ 攻击者既知道安全模块的模型结构，还知道水印信息m_k和a的准确值。

针对上述三种情况，我们接下来主要分析安全模块［式(11-8)和式(11-10)］中的水印参数a和$\boldsymbol{\Sigma}_m$的选择对远端估计器的估计性能以及检测器的检测性能的影响。此外，我们推导出最优的水印参数a和$\boldsymbol{\Sigma}_m$，实现线性攻击下估计误差的最小化。

场景一：无信息泄露

我们假设攻击者不知道安全模块的存在，并且仍然在引理11.1中原始的最优线性欺诈攻击（OLMA）。不失一般性，$\boldsymbol{m}_k$的协方差设计为

$$\boldsymbol{\Sigma}_m = \eta \boldsymbol{\Sigma}_z \tag{11-13}$$

其中，$\eta \in \mathbb{R}$是一个非负标量。我们将在本文的其余部分采用这样的水印$\boldsymbol{m}_k$。在OLMA下，$\boldsymbol{z}_k^r$和$\boldsymbol{\Sigma}_{\tilde{z}}$可以改写为

$$\boldsymbol{z}_k^r = -\boldsymbol{z}_k - \frac{2}{a}\boldsymbol{m}_k, \boldsymbol{\Sigma}_{\tilde{z}} = \left(1 + \frac{4\eta}{a^2}\right)\boldsymbol{\Sigma}_z \tag{11-14}$$

其中，$\boldsymbol{\Sigma}_{\tilde{z}}$的推导基于式(11-13)。通过配备安全模块，OLMA下χ^2检测器的检测率如下。

定理 11.1

对于图 11-1 中的系统，χ^2 检测器 [式 (11-5)] 在 OLMA 下对$\boldsymbol{z}_k^r$的检测率为

$$F\left(\tilde{\zeta}, m\right) \stackrel{\text{def}}{=} \int_{\tilde{\xi}}^{\infty} \frac{1}{2^{m/2}\Gamma\left(m/2\right)} e^{-x/2} x^{\frac{m}{2}-1} \mathrm{d}x,$$

其随着$\tilde{\zeta} \stackrel{\text{def}}{=} \dfrac{\zeta}{1+4\eta/a^2}$单调递增。此外，当式(11-8)和式(11-10)的参数选择为$\eta/a^2 \to \infty$时，检测率等于100%。

证明：

由于$\boldsymbol{\Sigma}_z$的正定性，我们可以将$\boldsymbol{\Sigma}_z$的特征值分解为$\boldsymbol{\Sigma}_z = \boldsymbol{U}_1^{'}\boldsymbol{\Lambda}_1\boldsymbol{U}_1$，其中$\boldsymbol{U}_1$是酉矩阵，$\boldsymbol{\Lambda}_1 \stackrel{\text{def}}{=} \mathrm{diag}\left(\lambda_1, \cdots, \lambda_m\right)$是包含$\boldsymbol{\Sigma}_z$特征值$\left\{\lambda_1, \cdots, \lambda_m\right\}$的对角矩阵。那么$\boldsymbol{z}_k^r$可以转化为$\boldsymbol{\epsilon}_{k1} \stackrel{\text{def}}{=} \boldsymbol{\Lambda}_2^{-\frac{1}{2}}\boldsymbol{U}_1\boldsymbol{z}_k^r$，其中$\boldsymbol{\Lambda}_2 = \left(1+4\eta/a^2\right)\boldsymbol{\Lambda}_1$和$\boldsymbol{\epsilon}_{k1}$遵循标准的零均值高斯分布，其协方差是单位矩阵。因此，式 (11-5) 的左侧可以改写为

$$\begin{aligned}
\boldsymbol{\xi}_k &= \left(\boldsymbol{z}_k^r\right)^{\mathrm{T}} \boldsymbol{\Sigma}_z^{-1} \boldsymbol{z}_k^r \\
&= \left(\boldsymbol{z}_k^r\right)^{\mathrm{T}} \boldsymbol{U}_1^{\mathrm{T}} \boldsymbol{\Lambda}_1^{-1} \boldsymbol{U}_1 \boldsymbol{z}_k^r \\
&= \left(1+\frac{4\eta}{a^2}\right)\left[\left(\boldsymbol{z}_k^r\right)^{\mathrm{T}} \boldsymbol{U}_1^{\mathrm{T}} \boldsymbol{\Lambda}_2^{-\frac{1}{2}}\right]\left[\boldsymbol{\Lambda}_2^{-\frac{1}{2}} \boldsymbol{U}_1 \boldsymbol{z}_k^r\right] \\
&= \left(1+\frac{4\eta}{a^2}\right)\boldsymbol{\epsilon}_{k1}^{\mathrm{T}}\boldsymbol{\epsilon}_{k1} \\
&= \left(1+\frac{4\eta}{a^2}\right)\sum_{i=1}^{m}\left(\epsilon_{k1}^{i}\right)^2
\end{aligned} \tag{11-15}$$

其中，ϵ_{k1}^{i}是ϵ_{k1}的j元素。根据ϵ_{k1}的统计性质，$\sum_{i=1}^{m}\left(\epsilon_{k1}^{i}\right)^{2}$服从自由度为$m$的$\chi^2$分布。因此触发警报的概率可以表示为

$$\mathbb{P}\left[\xi_k>\zeta\right]=\mathbb{P}\left[\sum_{i=1}^{m}\left(\epsilon_{k1}^{i}\right)^{2}>\tilde{\zeta}\right]$$

$$=\int_{\tilde{\xi}}^{\infty}\frac{1}{2^{m/2}\Gamma\left(m/2\right)}\mathrm{e}^{-x/2}x^{\frac{m}{2}-1}\mathrm{d}x\overset{\text{def}}{=}F\left(\tilde{\zeta},m\right)$$

它取决于η/a^2。因此，可以推导出$\lim\limits_{\eta/a^2\to\infty}F\left(\tilde{\zeta},m\right)=100\%$。这表明当$\eta/a^2\to\infty$时，$\chi^2$检测器完全可以检测到OLMA。■

然后，我们分析了OLMA下估计性能与水印参数a和η的关系。将γ_k定义为χ^2检测器在每个时刻的警报，即

$$\gamma_k=\begin{cases}0, & \left(z_k^r\right)^{\mathrm{T}}\boldsymbol{\Sigma}_z^{-1}z_k^r>\zeta\\1, & \text{其他}\end{cases}\tag{11-16}$$

远程估计器根据γ_k决定如何估计状态。当$\gamma_k=1$时，远程估计器决定融合z_k^r。否则，它放弃z_k^r并用一步预测$\tilde{x}_k^-$替换状态估计。因此，状态估计的递归可以表示为

$$\tilde{\boldsymbol{x}}_k^-=\boldsymbol{A}\tilde{\boldsymbol{x}}_{k-1}\tag{11-17}$$

$$\tilde{\boldsymbol{x}}_k=\tilde{\boldsymbol{x}}_k^-+\gamma_k\boldsymbol{K}z_k^r\tag{11-18}$$

基于$\tilde{\boldsymbol{x}}_0=\hat{\boldsymbol{x}}_0$的稳态假设，式(11-18)可以递归地扩展为

$$\tilde{\boldsymbol{x}}_k=\boldsymbol{A}^k\hat{\boldsymbol{x}}_0+\sum_{i=1}^{k}\gamma_i\boldsymbol{A}^{k-i}\boldsymbol{K}z_i^r\tag{11-19}$$

这表明$\tilde{\boldsymbol{x}}_k$依赖于$\boldsymbol{I}_k\overset{\text{def}}{=}\left\{\gamma_k,\gamma_k z_k\right\}$。首先，我们给出如下引理。

引理 11.2

令$x\in\mathbb{R}$是一个零均值高斯变量，其协方差为$\mathbb{E}\left[x^2\right]=1$。令$\varDelta$是一个非负常数。则：(A1) $\mathbb{E}\left[x^2\mid|x|\leqslant\varDelta\right]$随着$\varDelta$单调递增；(A2) 如果$\varDelta>0$，则有$\mathbb{E}\left[x^2\mid|x|\leqslant\varDelta\right]<\varDelta^2$。

证明：

基于条件期望的定义，我们可以将$\mathbb{E}\left[x^2\mid|x|\leqslant\varDelta\right]$写成

$$\mathbb{E}\left[x^2 \mid |x| \leqslant \Delta\right]=\frac{\int_{-\Delta}^{\Delta}\frac{t^2}{\sqrt{2\pi}}\mathrm{e}^{-\frac{t^2}{2}}\mathrm{d}t}{\int_{-\Delta}^{\Delta}\frac{1}{\sqrt{2\pi}}\mathrm{e}^{-\frac{t^2}{2}}\mathrm{d}t}=\frac{\int_{0}^{\Delta}\frac{t^2}{\sqrt{2\pi}}\mathrm{e}^{-\frac{t^2}{2}}\mathrm{d}t}{\int_{0}^{\Delta}\frac{1}{\sqrt{2\pi}}\mathrm{e}^{-\frac{t^2}{2}}\mathrm{d}t}$$

然后，我们对$\mathbb{E}\left[x^2 \mid |x| \leqslant \Delta\right]$求导，并且得到

$$\frac{\mathrm{d}\left(\mathbb{E}\left[x^2 \mid |x| \leqslant \Delta\right]\right)}{\mathrm{d}\Delta}$$

$$=\frac{\frac{\Delta^2}{\sqrt{2\pi}}\mathrm{e}^{-\frac{\Delta^2}{2}}\int_{0}^{\Delta}\frac{1}{\sqrt{2\pi}}\mathrm{e}^{-\frac{t^2}{2}}\mathrm{d}t-\frac{1}{\sqrt{2\pi}}\mathrm{e}^{-\frac{\Delta^2}{2}}\int_{0}^{\Delta}\frac{t^2}{\sqrt{2\pi}}\mathrm{e}^{-\frac{t^2}{2}}\mathrm{d}t}{\left(\int_{0}^{\Delta}\frac{1}{\sqrt{2\pi}}\mathrm{e}^{-\frac{t^2}{2}}\mathrm{d}t\right)^2}$$

$$=\frac{\frac{1}{\sqrt{2\pi}}\mathrm{e}^{-\frac{\Delta^2}{2}}}{\left(\int_{0}^{\Delta}\frac{1}{\sqrt{2\pi}}\mathrm{e}^{-\frac{t^2}{2}}\mathrm{d}t\right)^2}\int_{0}^{\Delta}\frac{\Delta^2-t^2}{\sqrt{2\pi}}\mathrm{e}^{-\frac{t^2}{2}}\mathrm{d}t>0,$$

这完成了(A1)的证明。由于对$\forall t\in[0,\Delta]$，有$\Delta^2\geqslant t^2\geqslant 0$，我们可以证明

$$\mathbb{E}\left[x^2 \mid |x| \leqslant \Delta\right]=\frac{\int_{0}^{\Delta}\frac{t^2}{\sqrt{2\pi}}\mathrm{e}^{-\frac{t^2}{2}}\mathrm{d}t}{\int_{0}^{\Delta}\frac{1}{\sqrt{2\pi}}\mathrm{e}^{-\frac{t^2}{2}}\mathrm{d}t}<\frac{\int_{0}^{\Delta}\frac{\Delta^2}{\sqrt{2\pi}}\mathrm{e}^{-\frac{t^2}{2}}\mathrm{d}t}{\int_{0}^{\Delta}\frac{1}{\sqrt{2\pi}}\mathrm{e}^{-\frac{t^2}{2}}\mathrm{d}t}=\Delta^2$$

基于上述引理，不同γ_k条件下的估计误差协方差推导如下。 ■

引理 11.3

对于图 11-1 中 OLMA 下的系统，$\gamma_k=1$条件下的估计误差协方差$\tilde{P}_k^1$，有界于

$$\tilde{\boldsymbol{P}}_k^-+2\overline{\boldsymbol{P}}\boldsymbol{C}^{\mathrm{T}}\boldsymbol{\Sigma}\boldsymbol{C}\overline{\boldsymbol{P}}\leqslant\tilde{\boldsymbol{P}}_k^1\leqslant\tilde{\boldsymbol{P}}_k^-+(2+\zeta)\overline{\boldsymbol{P}}\boldsymbol{C}^{\mathrm{T}}\boldsymbol{\Sigma}\boldsymbol{C}\overline{\boldsymbol{P}} \tag{11-20}$$

其中，$\tilde{\boldsymbol{P}}_k^-=\boldsymbol{A}\tilde{\boldsymbol{P}}_{k-1}\boldsymbol{A}'+\boldsymbol{Q}$。在$\gamma_k=0$的条件下估计误差协方差$\tilde{\boldsymbol{P}}_k^2$等于$\tilde{\boldsymbol{P}}_k^-$。

证明：

我们首先推导$\gamma_k=1$时的情况，并得到❶

❶ 在本节中，$(x)(x)'$等形式缩写成“$(x)(\bullet)'$”。即该点表示前面所对应的x。

$$
\begin{aligned}
\tilde{\boldsymbol{P}}_k^1 &= \mathbb{E}\left[\left(\boldsymbol{x}_k - \tilde{\boldsymbol{x}}_k\right)\left(\boldsymbol{x}_k - \tilde{\boldsymbol{x}}_k\right)^{\mathrm{T}} \mid \boldsymbol{I}_{k-1}, \gamma_k = 1\right] \\
&= \mathbb{E}\left[\left(\boldsymbol{x}_k - \tilde{\boldsymbol{x}}_k^- - \boldsymbol{K}\boldsymbol{z}_k^r\right)(\bullet)^{\mathrm{T}} \mid \boldsymbol{I}_{k-1}, \gamma_k = 1\right] \\
&= \mathbb{E}\left[\left(\boldsymbol{x}_k - \tilde{\boldsymbol{x}}_k^-\right)\left(\boldsymbol{x}_k - \tilde{\boldsymbol{x}}_k^-\right)^{\mathrm{T}} \mid \boldsymbol{I}_{k-1}, \gamma_k = 1\right] \\
&\quad + \boldsymbol{K}\mathbb{E}\left[\boldsymbol{z}_k^r\left(\boldsymbol{z}_k^r\right)^{\mathrm{T}} \mid \boldsymbol{I}_{k-1}, \gamma_k = 1\right]\boldsymbol{K}' \\
&\quad - \mathbb{E}\left[\left(\boldsymbol{x}_k - \tilde{\boldsymbol{x}}_k^-\right)\left(\boldsymbol{z}_k^r\right)^{\mathrm{T}} \mid \boldsymbol{I}_{k-1}, \gamma_k = 1\right]\boldsymbol{K}' \\
&\quad - \boldsymbol{K}\mathbb{E}\left[\boldsymbol{z}_k^r\left(\boldsymbol{x}_k - \tilde{\boldsymbol{x}}_k^-\right)^{\mathrm{T}} \mid \boldsymbol{I}_{k-1}, \gamma_k = 1\right],
\end{aligned} \tag{11-21}
$$

它由四个需要计算的项组成。对于式(11-21)的第一项，我们可以得到

$$
\begin{aligned}
&\mathbb{E}\left[\left(\boldsymbol{x}_k - \tilde{\boldsymbol{x}}_k^-\right)\left(\boldsymbol{x}_k - \tilde{\boldsymbol{x}}_k^-\right)^{\mathrm{T}} \mid \boldsymbol{I}_{k-1}, \gamma_k = 1\right] \\
=&\mathbb{E}\left[\left(\boldsymbol{A}\boldsymbol{x}_{k-1} + \boldsymbol{w}_{k-1} - \boldsymbol{A}\tilde{\boldsymbol{x}}_{k-1}\right)(\bullet)^{\mathrm{T}} \mid \boldsymbol{I}_{k-1}, \gamma_k = 1\right] \\
=&\boldsymbol{A}\mathbb{E}\left[\left(\boldsymbol{x}_{k-1} - \tilde{\boldsymbol{x}}_{k-1}\right)\left(\boldsymbol{x}_{k-1} - \tilde{\boldsymbol{x}}_{k-1}\right)^{\mathrm{T}} \mid \boldsymbol{I}_{k-1}, \gamma_k = 1\right]\boldsymbol{A}^{\mathrm{T}} \\
&\quad + \mathbb{E}\left[\boldsymbol{w}_{k-1}\boldsymbol{w}_{k-1}^{\mathrm{T}} \mid \boldsymbol{I}_{k-1}, \gamma_k = 1\right] \\
\overset{u}{=}&\boldsymbol{A}\mathbb{E}\left[\left(\boldsymbol{x}_{k-1} - \tilde{\boldsymbol{x}}_{k-1}\right)(\bullet)^{\mathrm{T}} \mid \boldsymbol{I}_{k-1}\right]\boldsymbol{A}^{\mathrm{T}} + \mathbb{E}\left[\boldsymbol{w}_{k-1}\boldsymbol{w}_{k-1}^{\mathrm{T}}\right] \\
=&\boldsymbol{A}\tilde{\boldsymbol{P}}_{k-1}\boldsymbol{A}^{\mathrm{T}} + \boldsymbol{Q},
\end{aligned} \tag{11-22}
$$

下文注释11.4给出了如何推导出方程(u)的具体解释。由于$\boldsymbol{z}_i^r$和$\boldsymbol{z}_j^r$之间的独立性，$\forall i \neq j$，式(11-21)的第二项可以改写为

$$
\begin{aligned}
&\boldsymbol{K}\mathbb{E}\left[\boldsymbol{z}_k^r\left(\boldsymbol{z}_k^r\right)^{\mathrm{T}} \mid \gamma_k = 1\right]\boldsymbol{K}^{\mathrm{T}} \\
=&\boldsymbol{K}\mathbb{E}\left[\left(\boldsymbol{U}_1^{\mathrm{T}}\boldsymbol{\Lambda}_1^{\frac{1}{2}}\boldsymbol{\Lambda}_1^{-\frac{1}{2}}\boldsymbol{U}_1\boldsymbol{z}_k^r\right)(\bullet)^{\mathrm{T}} \mid \gamma_k = 1\right]\boldsymbol{K}^{\mathrm{T}} \\
=&\left(1 + \frac{4\eta}{a^2}\right)\boldsymbol{K}\mathbb{E}\left[\left(\boldsymbol{U}_1^{\mathrm{T}}\boldsymbol{\Lambda}_1^{\frac{1}{2}}\boldsymbol{\Lambda}_2^{-\frac{1}{2}}\boldsymbol{U}_1\boldsymbol{z}_k^r\right)(\bullet)^{\mathrm{T}} \mid \gamma_k = 1\right]\boldsymbol{K}^{\mathrm{T}} \\
=&\left(1 + \frac{4\eta}{a^2}\right)\boldsymbol{K}\boldsymbol{U}_1^{\mathrm{T}}\boldsymbol{\Lambda}_1^{\frac{1}{2}}\mathbb{E}\left[\boldsymbol{\epsilon}_{k1}\boldsymbol{\epsilon}_{k1}^{\mathrm{T}} \mid \boldsymbol{\epsilon}_{k1}^{\mathrm{T}}\boldsymbol{\epsilon}_{k1} \leqslant \tilde{\zeta}\right]\boldsymbol{\Lambda}_1^{\frac{1}{2}}\boldsymbol{U}_1\boldsymbol{K}^{\mathrm{T}} \\
=&\left(1 + \frac{4\eta}{a^2}\right)\boldsymbol{K}\boldsymbol{U}_1^{\mathrm{T}}\boldsymbol{\Lambda}_1^{\frac{1}{2}}\mathrm{diag}\left(\varPhi_1, \cdots, \varPhi_m\right)\boldsymbol{\Lambda}_1^{\frac{1}{2}}\boldsymbol{U}_1\boldsymbol{K}^{\mathrm{T}}
\end{aligned} \tag{11-23}
$$

其中，$\Phi_i \overset{\text{def}}{=} \mathbb{E}\left[\left(\epsilon_{k1}^i\right)^2 | \sum_{i=1}^{m}\left(\epsilon_{k1}^i\right)^2 \leqslant \tilde{\zeta}\right]$。由于$0 \leqslant \left(\epsilon_{k1}^i\right)^2 \leqslant \tilde{\zeta} - \sum_{j=1, j\neq i}^{m}\left(\epsilon_{k1}^j\right)^2 \leqslant \tilde{\zeta}$和引理11.2，我们可以得到$0 = \mathbb{E}\left[\left(\epsilon_{k1}^i\right)^2 | \left(\epsilon_{k1}^i\right)^2 = 0\right] \leqslant \Phi_i \leqslant \mathbb{E}\left[\left(\epsilon_{k1}^i\right)^2 | \left(\epsilon_{k1}^i\right)^2 \leqslant \tilde{\zeta}\right] < \tilde{\zeta}$。那么我们可以得到式(11-23)的上界，即$\left(1+\frac{4\eta}{a^2}\right)\boldsymbol{K}\boldsymbol{U}_1^{\mathrm{T}}\boldsymbol{\Lambda}_1^{\frac{1}{2}}\left(\tilde{\zeta}\boldsymbol{I}\right)\boldsymbol{\Lambda}_1^{\frac{1}{2}}\boldsymbol{U}_1\boldsymbol{K}^{\mathrm{T}} = \zeta\boldsymbol{K}\boldsymbol{\Sigma}_z\boldsymbol{K}^{\mathrm{T}}$。类似地，可以计算出式(11-23)的下界等于0。由于卡尔曼增益$\boldsymbol{K}$可能不是满秩矩阵，我们可以得到

$$0 \leqslant \boldsymbol{K}\mathbb{E}\left[\boldsymbol{z}_k^r\left(\boldsymbol{z}_k^r\right)^{\mathrm{T}} | \ \gamma_k = 1\right]\boldsymbol{K}^{\mathrm{T}} \leqslant \zeta\boldsymbol{K}\boldsymbol{\Sigma}_z\boldsymbol{K}^{\mathrm{T}}. \tag{11-24}$$

基于式(11-1)、式(11-17)和式(11-19)，我们可以得到

$$\begin{aligned}&\boldsymbol{x}_k - \tilde{\boldsymbol{x}}_k^- \\ &= \boldsymbol{A}^k\left(\boldsymbol{x}_0 - \hat{\boldsymbol{x}}_0\right) + \sum_{i=0}^{k-1}\boldsymbol{A}^i\boldsymbol{w}_{k-1-i} - \sum_{i=1}^{k-1}\gamma_i\boldsymbol{A}^{k-i}\boldsymbol{K}\boldsymbol{z}_i^r.\end{aligned} \tag{11-25}$$

由于$\mathbb{E}\left[\boldsymbol{z}_i^r\left(\boldsymbol{z}_j^r\right)^{\mathrm{T}}\right] = 0, \forall i \neq j$，并且$\boldsymbol{m}_k$独立于式(11-25)的所有项，所以式(11-21)的倒数第二项是

$$\begin{aligned}&\mathbb{E}\left[\left(\boldsymbol{x}_k - \tilde{\boldsymbol{x}}_k^-\right)\left(\boldsymbol{z}_k^r\right)^{\mathrm{T}} | \ \boldsymbol{I}_{k-1}, \gamma_k = 1\right]\boldsymbol{K}^{\mathrm{T}} \\ &= \mathbb{E}\left[\left\{\boldsymbol{A}^k\left(\boldsymbol{x}_0 - \hat{\boldsymbol{x}}_0\right) + \sum_{i=0}^{k-1}\boldsymbol{A}^i\boldsymbol{w}_{k-1-i}\right\}\left\{\boldsymbol{z}_k^r\right\}^{\mathrm{T}} | \ \boldsymbol{I}_{k-1}, \gamma_k = 1\right]\boldsymbol{K}^{\mathrm{T}} \\ &= \mathbb{E}\left[\left\{\boldsymbol{A}^k\left(\boldsymbol{x}_0 - \hat{\boldsymbol{x}}_0\right) + \sum_{i=0}^{k-1}\boldsymbol{A}^i\boldsymbol{w}_{k-1-i}\right\}\left\{-\boldsymbol{z}_k - \frac{2}{a}\boldsymbol{m}_k\right\}^{\mathrm{T}} | \ \right]\boldsymbol{K}^{\mathrm{T}} \\ &= \mathbb{E}\left[\left\{\boldsymbol{A}^k\left(\boldsymbol{x}_0 - \hat{\boldsymbol{x}}_0\right) + \sum_{i=0}^{k-1}\boldsymbol{A}^i\boldsymbol{w}_{k-1-i}\right\}\left\{-\boldsymbol{z}_k\right\}^{\mathrm{T}} | \ \right]\boldsymbol{K}^{\mathrm{T}} \\ &\overset{v}{=} -\mathbb{E}\left[\left\{\boldsymbol{A}^k\left(\boldsymbol{x}_0 - \hat{\boldsymbol{x}}_0\right) + \sum_{i=0}^{k-1}\boldsymbol{A}^i\boldsymbol{w}_{k-1-i}\right\}\boldsymbol{z}_k^{\mathrm{T}}\right]\boldsymbol{K}^{\mathrm{T}} \\ &= -\overline{\boldsymbol{P}}\boldsymbol{C}^{\mathrm{T}}\boldsymbol{K}^{\mathrm{T}},\end{aligned} \tag{11-26}$$

其中(v)的推导与(u)类似，式(11-26)的证明与文献[90]中的定理1相同，故省略。类似地，我们可以得到

$$\boldsymbol{K}\mathbb{E}\left[\boldsymbol{z}_k^r\left(\boldsymbol{x}_k - \tilde{\boldsymbol{x}}_k^-\right)^{\mathrm{T}} | \ \boldsymbol{I}_{k-1}, \gamma_k = 1\right] = -\boldsymbol{K}\boldsymbol{C}\overline{\boldsymbol{P}}. \tag{11-27}$$

将式(11-22)、式(11-24)、式(11-26)和式(11-27)代入式(11-21)，误差协方差的下界$\tilde{\boldsymbol{P}}_k^1$以式(11-20)的形式给出。同理，我们可以得到$\tilde{\boldsymbol{P}}_k^2 = \tilde{\boldsymbol{P}}_k^-$。 ■

注释 11.4　新息$\boldsymbol{z}_k$不仅仅可以用于估计当前的状态，还可以用于更新以前的状态估计，而这也是卡尔曼滤波是所谓的平滑过程[136]。根据式(11-14)和式(11-16)，γ_k是由$\boldsymbol{m}_k$和$\boldsymbol{z}_k$组成的。因此，γ_k似乎可以提供$\boldsymbol{z}_k$的一些信息，从而影响式(11-22)和式(11-26)中的条件期望。然而，下文中的引理11.4表明γ_k无法提供关于$\boldsymbol{z}_k$的任何信息。这是推导(u)和(v)的理论原因。

引理 11.4

对于图11-1中在OLMA攻击下的系统，有

$$f_{z|\gamma}\left(\boldsymbol{z}_k = z|\ \gamma_k\right) = f_z\left(\boldsymbol{z}_k = z\right) \tag{11-28}$$

其中，$f_z\left(\boldsymbol{z}_k = z\right)$表示$\boldsymbol{z}_k$的概率密度函数（pdf），并且$f_{z|\gamma}\left(\boldsymbol{z}_k = z|\ \gamma_k\right)$是$\boldsymbol{z}_k$在$\gamma_k$条件下的条件概率密度函数。

证明：

首先考虑$\gamma_k = 1$的情况。注意到$\left(\boldsymbol{z}_k^r\right)^{\mathrm{T}} \boldsymbol{\Sigma}_z^{-1} \boldsymbol{z}_k^r$可以简化为

$$\begin{aligned}
&\left(\boldsymbol{z}_k^r\right)^{\mathrm{T}} \boldsymbol{U}_1^{\mathrm{T}} \boldsymbol{\Lambda}_1^{-1} \boldsymbol{U}_1 \boldsymbol{z}_k^r \\
&=\left(-\boldsymbol{z}_k - \frac{2}{a}\boldsymbol{m}_k\right)^{\mathrm{T}} \boldsymbol{U}_1^{\mathrm{T}} \boldsymbol{\Lambda}_1^{-\frac{1}{2}} \boldsymbol{\Lambda}_1^{-\frac{1}{2}} \boldsymbol{U}_1 \left(-\boldsymbol{z}_k - \frac{2}{a}\boldsymbol{m}_k\right) \\
&=\left(\boldsymbol{\Lambda}_1^{-\frac{1}{2}} \boldsymbol{U}_1 \boldsymbol{z}_k + \frac{2}{a}\boldsymbol{\Lambda}_1^{-\frac{1}{2}} \boldsymbol{U}_1 \boldsymbol{m}_k\right)^{\mathrm{T}} \left(\boldsymbol{\Lambda}_1^{-\frac{1}{2}} \boldsymbol{U}_1 \boldsymbol{z}_k + \frac{2}{a}\boldsymbol{\Lambda}_1^{-\frac{1}{2}} \boldsymbol{U}_1 \boldsymbol{m}_k\right) \\
&=\left(\boldsymbol{\epsilon}_{k5} + \boldsymbol{\epsilon}_{k6}\right)^{\mathrm{T}} \left(\boldsymbol{\epsilon}_{k5} + \boldsymbol{\epsilon}_{k6}\right) = \sum_{i=1}^{m} \left(\epsilon_{k5}^i + \epsilon_{k6}^i\right)^2
\end{aligned}$$

其中，$\boldsymbol{\epsilon}_{k5} \overset{\text{def}}{=} \boldsymbol{\Lambda}_1^{-\frac{1}{2}} \boldsymbol{U}_1 \boldsymbol{z}_k, \boldsymbol{\epsilon}_{k6} \overset{\text{def}}{=} \frac{2}{a}\boldsymbol{\Lambda}_1^{-\frac{1}{2}} \boldsymbol{U}_1 \boldsymbol{m}_k$分别是具有协方差$\boldsymbol{I}$以及$\frac{4\eta}{a^2}\boldsymbol{I}$的零均值高斯变量，且它们相互独立，而$\epsilon_{k5}^i$和$\epsilon_{k6}^i$分别是$\boldsymbol{\epsilon}_{k5}$和$\boldsymbol{\epsilon}_{k6}$的第$i^{th}$个元素。对于满足条件$\sum_{i=1}^{m} \zeta_i = \zeta$的一系列正常数$\{\zeta_i, i = 1, \cdots, m\}$，条件$\gamma_k = 1$可以转

换为$\left(\epsilon_{k5}^{i}+\epsilon_{k6}^{i}\right)^{2}<\zeta_{i}, i=1,\cdots,m$。然后，我们可以得出

$$
\begin{aligned}
& f_{z|\gamma}\left(\boldsymbol{z}_{k}=z \mid \gamma_{k}=1\right) \\
& =f_{z|\gamma}\left(\boldsymbol{\epsilon}_{k5}=\boldsymbol{\Lambda}_{1}^{-\frac{1}{2}} \boldsymbol{U}_{1} \boldsymbol{z} \mid \gamma_{k}=1\right) \\
& =f_{z|\gamma}\left(\boldsymbol{\epsilon}_{k5}=\tilde{\boldsymbol{\epsilon}} \mid\left(\epsilon_{k5}^{i}+\epsilon_{k6}^{i}\right)^{2}<\zeta_{i}, i=1,\cdots,m\right) \\
& =\prod_{i=1}^{m} f_{z|\gamma}\left(\epsilon_{k5}^{i}=(\tilde{\epsilon})^{i} \mid\left(\epsilon_{k5}^{i}+\epsilon_{k6}^{i}\right)^{2}<\zeta_{i}\right) \\
& =\prod_{i=1}^{m} \frac{f_{z,\gamma}\left(\epsilon_{k5}^{i}=(\tilde{\epsilon})^{i},\left(\epsilon_{k5}^{i}+\epsilon_{k6}^{i}\right)^{2}<\zeta_{i}\right)}{f_{z,\gamma}\left(\left(\epsilon_{k5}^{i}+\epsilon_{k6}^{i}\right)^{2}<\zeta_{i}\right)} \\
& =\prod_{i=1}^{m} \frac{\int_{\sqrt{\zeta_{i}}-(\tilde{\epsilon})^{i}}^{-\sqrt{\zeta_{i}}-(\tilde{\epsilon})^{i}} f_{z,\gamma}\left(\epsilon_{k5}^{i}=(\tilde{\epsilon})^{i}, \epsilon_{k6}^{i}\right) \mathrm{d} \epsilon_{k6}^{i}}{\int_{-\infty}^{\infty} \int_{\sqrt{\zeta_{i}}-(\tilde{\epsilon})^{i}}^{-\sqrt{\zeta_{i}}-(\tilde{\epsilon})^{i}} f_{z,\gamma}\left(\epsilon_{k5}^{i}, \epsilon_{k6}^{i}\right) \mathrm{d} \epsilon_{k5}^{i} \mathrm{d} \epsilon_{k6}^{i}} \\
& =\prod_{i=1}^{m} \frac{f_{z}\left(\epsilon_{k5}^{i}=(\tilde{\epsilon})^{i}\right)\left[\int_{\sqrt{\zeta_{i}}-(\tilde{\epsilon})^{i}}^{-\sqrt{\zeta_{i}}-(\tilde{\epsilon})^{i}} f_{\gamma}\left(\epsilon_{k6}^{i}\right) \mathrm{d} \epsilon_{k6}^{i}\right]}{\left[\int_{-\infty}^{\infty} f_{z}\left(\epsilon_{k5}^{i}\right) \mathrm{d} \epsilon_{k5}^{i}\right]\left[\int_{\sqrt{\zeta_{i}}-(\tilde{\epsilon})^{i}}^{-\sqrt{\zeta_{i}}-(\tilde{\epsilon})^{i}} f_{\gamma}\left(\epsilon_{k6}^{i}\right) \mathrm{d} \epsilon_{k6}^{i}\right]} \\
& =\prod_{i=1}^{m} f_{z}\left(\epsilon_{k5}^{i}=(\tilde{\epsilon})^{i}\right)=f_{z}\left(\boldsymbol{\epsilon}_{k5}=\boldsymbol{\Lambda}_{1}^{-\frac{1}{2}} \boldsymbol{U}_{1} \boldsymbol{z}\right)=f_{z}\left(\boldsymbol{z}_{k}=z\right)
\end{aligned}
$$

其中，$\tilde{\boldsymbol{\epsilon}} \stackrel{\text{def}}{=} \boldsymbol{\Lambda}_{1}^{-\frac{1}{2}} \boldsymbol{U}_{1} \boldsymbol{z}$，并且$(\tilde{\epsilon})^{i}$是$\tilde{\boldsymbol{\epsilon}}$的第$j^{th}$个元素，$f_{z}\left(\epsilon_{k5}^{i}\right)$和$f_{\gamma}\left(\epsilon_{k6}^{i}\right)$分别是$\epsilon_{k5}^{i}$和$\epsilon_{k6}^{i}$的概率密度函数，$f_{z,\gamma}\left(\epsilon_{k5}^{i}, \epsilon_{k6}^{i}\right)$是$\epsilon_{k5}^{i}$和$\epsilon_{k6}^{i}$的联合概率密度函数。由于$\epsilon_{k5}^{i}$和$\epsilon_{k6}^{i}$之间的无关性，$f_{z,\gamma}\left(\epsilon_{k5}^{i}, \epsilon_{k6}^{i}\right)$可以分解为$f_{z}\left(\epsilon_{k5}^{i}\right)$和$f_{\gamma}\left(\epsilon_{k6}^{i}\right)$。类似地，我们可以推导出$f_{z|\gamma}\left(\boldsymbol{z}_{k}=z \mid \gamma_{k}=0\right)=f_{z}\left(\boldsymbol{z}_{k}=z\right)$。这样我们就完成了式(11-28)的证明。 ■

基于引理 11.3，推导出 OLMA 下期望估计误差协方差。

定理 11.2

对于图 11-1 中的系统，OLMA 下远程估计器的最小期望估计误差协方差为

$$
\lim_{\frac{\eta}{a^{2}} \rightarrow \infty} \mathbb{E}\left[\tilde{\boldsymbol{P}}_{k} \mid \boldsymbol{I}_{k-1}\right]=\tilde{P}_{k}^{-} \tag{11-29}
$$

当安全模块式 (11-8) 和式 (11-10) 中的参数满足$\eta / a^2 \to \infty$时，$\tilde{\zeta} \to 0$。

证明：

根据式 (11-15) 以及式 (11-16)，警报γ_k取决于χ^2变量$\sum_{i=1}^{m}\left(\epsilon_{k1}^{i}\right)^2$，其概率密度函数 (pdf) 定义为$f_{\chi^2}\left(t\right)$。结合引理 11.3 中的结果，我们可以得到$\mathbb{E}\left[\tilde{\boldsymbol{P}}_k | \boldsymbol{I}_{k-1}\right]$：

$$
\begin{aligned}
&\mathbb{E}\left[\tilde{\boldsymbol{P}}_k | \boldsymbol{I}_{k-1}\right] = \int_0^{\infty} \tilde{\boldsymbol{P}}_k f_{\chi^2}\left(t\right)\mathrm{d}t \\
&= \int_0^{\tilde{\xi}} \tilde{\boldsymbol{P}}_k^1 f_{\chi^2}\left(t\right)\mathrm{d}t + \int_{\tilde{\xi}}^{\infty} \tilde{\boldsymbol{P}}_k^2 f_{\chi^2}\left(t\right)\mathrm{d}t \\
&\geqslant \int_0^{\xi}\left[\tilde{\boldsymbol{P}}_k^- + 2\overline{\boldsymbol{P}}\boldsymbol{C}'\boldsymbol{\Sigma}\boldsymbol{C}\overline{\boldsymbol{P}}\right] f_{\chi^2}\left(t\right)\mathrm{d}t + \int_{\tilde{\xi}}^{\infty} \tilde{\boldsymbol{P}}_k^- f_{\chi^2}\left(t\right)\mathrm{d}t \\
&= \tilde{\boldsymbol{P}}_k^- + \left[2\int_0^{\tilde{\zeta}} f_{\chi^2}\left(t\right)\mathrm{d}t\right]\overline{\boldsymbol{P}}\boldsymbol{C}^{\mathrm{T}}\boldsymbol{\Sigma}\boldsymbol{C}\overline{\boldsymbol{P}}
\end{aligned}
\tag{11-30}
$$

其极限为$\tilde{\boldsymbol{P}}_k^-$和$\tilde{\zeta} \to 0$。类似地，可以得到$\mathbb{E}\left[\tilde{\boldsymbol{P}}_k | \boldsymbol{I}_{k-1}\right]$上界的极限，其与下界的极限相同。利用夹逼定理 [294]，可以推导出式 (11-29) 中的结果。由于矩阵$\overline{\boldsymbol{P}}\boldsymbol{C}^{\mathrm{T}}\boldsymbol{\Sigma}\boldsymbol{C}\overline{\boldsymbol{P}}$的正定性，可以通过式 (11-30) 得到$\mathbb{E}\left[\tilde{\boldsymbol{P}}_k | \boldsymbol{I}_{k-1}\right] \geqslant \tilde{\boldsymbol{P}}_k^-$。因此，当$\tilde{\zeta} \to 0$时，预期的估计误差协方差最小。 ■

注释 11.5 当远程估计器配备随机χ^2检测器时，文献 [289] 证明在 OLMA 下期望的估计误差协方差为$\tilde{\boldsymbol{P}}_k^- + \frac{9}{4}\tilde{\boldsymbol{P}}_k^- \boldsymbol{C}^{\mathrm{T}}\boldsymbol{\Sigma}\boldsymbol{C}\tilde{\boldsymbol{P}}_k^-$，该结果比式 (11-29) 大。对于没有攻击的情况，安全模块不会牺牲估计性能。然而，随机χ^2检测器仍然随机地改变其阈值。因此，和随机χ^2检测器对比，我们的方法更能在 OLMA 下保护远端估计器的估计性能。我们在 4.6 节中通过数值模拟进一步比较了两个检测器。

当远端缺少攻击者信息时，它只能在检测器报警时丢弃可疑数据。然而，当 OLMA 的攻击模型已知时，远程估计器可以通过将假数据$\boldsymbol{z}_k^r$修

正为z_k来进一步提高估计性能。数据校正可以表示为

$$z_k^{rc} = -z_k^r - \frac{2}{a} m_k \tag{11-31}$$

其中，z_k^{rc}是式(11-31)的输出，并且在OLMA发生时等于z_k。需要注意的是，是否触发数据校正[式(11-31)]取决于检测器的决策。如果χ^2探测器的警报被触发，远程估计器将利用式(11-31)中的更正步骤来修改可疑数据。否则，它会直接融合接收到的数据。然而，由于误报率$(1-P_\zeta)$的存在，χ^2检测器可能在不存在攻击的时候错误地将新息z_k视为z_k^r。我们可以通过以下贝叶斯规则来描述警报和攻击之间的关系。

$$\mathbb{P}[\alpha|\ \beta] = \frac{\mathbb{P}[\beta|\ \alpha]\mathbb{P}[\alpha]}{\mathbb{P}[\beta|\ \alpha]\mathbb{P}[\alpha] + \mathbb{P}[\beta|\overline{\alpha}]\mathbb{P}[\overline{\alpha}]} = \frac{\mathbb{P}[\alpha]}{\mathbb{P}[\alpha] + (1-P_\zeta)(1-\mathbb{P}[\alpha])} \tag{11-32}$$

其中，α表示存在攻击，$\overline{\alpha}$与之相反；β表示触发警报。由于OLMA可能会间歇性发生，即$\mathbb{P}[\alpha] \neq 100\%$，报警并不意味着OLMA以100%的概率存在。在没有攻击的情况下，式(11-31)可能会错误地将z_k修改为

$$\tilde{z}_k^{rc} = -z_k - \frac{2}{a} m_k, \tag{11-33}$$

其中，$\tilde{z}_k^{rc}$是校正错误的数据，正好等于z_k^r。受到定理11.1的启发，校正错误的数据不可能通过χ^2检测。因此，解决方案是再次运行χ^2检测器来校验$\tilde{z}_k^{rc}$。对于由式(11-31)修改的数据，远程估计器只会接收这些通过χ^2检测的数据。通过这种方式，我们可以避免当检测器误报时数据校正[式(11-31)]错误地将z_k修改为z_k^r的情况，从而将估计性能进一步提高到不存在攻击时的估计性能。

场景二：部分信息泄露

使用原本的可行性约束 [式 (11-7)]，攻击者无法欺骗图 11-1 中配置了安全模块 [式 (11-8) 和式 (11-10)] 的系统。此外，当攻击者只有部分的信息时，也无法重新构建有效的可行性约束。因此我们假设攻击者以式 (11-6) 的形式发起任意线性攻击，并分析这种攻击下的检测性能。对于 χ^2 检测器，它满足如下性质。

引理 11.5

假设有两个随机序列$\{z_{k3}\}$和$\{z_{k4}\}$，其协方差分别为$\boldsymbol{\Sigma}_{z3}$和$\boldsymbol{\Sigma}_{z4}$。如果$\boldsymbol{\Sigma}_{z3} > \boldsymbol{\Sigma}_{z4}$，则$\mathbb{E}[\xi_{k3}] > \mathbb{E}[\xi_{k4}]$，其中$\xi_{k3}$和$\xi_{k4}$由式 (11-5) 计算得出。

证明：

类似于定理 11.1，利用特征值分解得到$\boldsymbol{\Sigma}_{z3} = \boldsymbol{U}_3^{\mathrm{T}}\boldsymbol{\Lambda}_3\boldsymbol{U}_3$和$\boldsymbol{\Sigma}_{z4} = \boldsymbol{U}_4^{\mathrm{T}}\boldsymbol{\Lambda}_4\boldsymbol{U}_4$，并将$z_{k3}$和$z_{k4}$转化为$\epsilon_{k3} \overset{\text{def}}{=} \boldsymbol{\Lambda}_3^{-\frac{1}{2}}\boldsymbol{U}_3 z_{k3}, \epsilon_{k4} \overset{\text{def}}{=} \boldsymbol{\Lambda}_4^{-\frac{1}{2}}\boldsymbol{U}_4 z_{k4}$，其中$\epsilon_{k3},\epsilon_{k4}$都是协方差为单位矩阵的标准高斯变量。然后，计算$\xi_{k3}$的期望值。

$$
\begin{aligned}
\mathbb{E}[\xi_{k3}] &= \mathbb{E}\left[z_{k3}^{\mathrm{T}}\boldsymbol{U}_3^{\mathrm{T}}\boldsymbol{\Lambda}_3^{-\frac{1}{2}}\boldsymbol{\Lambda}_3^{\frac{1}{2}}\boldsymbol{U}_3\boldsymbol{\Sigma}_z^{-1}\boldsymbol{U}_3^{\mathrm{T}}\boldsymbol{\Lambda}_3^{\frac{1}{2}}\boldsymbol{\Lambda}_3^{-\frac{1}{2}}\boldsymbol{U}_3 z_{k3} \right] \\
&= \mathbb{E}\left[\epsilon_{k3}^{\mathrm{T}}\left(\boldsymbol{\Lambda}_3^{\frac{1}{2}}\boldsymbol{U}_3\boldsymbol{\Sigma}_z^{-1}\boldsymbol{U}_3^{\mathrm{T}}\boldsymbol{\Lambda}_3^{\frac{1}{2}} \right)\epsilon_{k3} \right] \\
&= \mathrm{Tr}\left\{ \boldsymbol{\Lambda}_3^{\frac{1}{2}}\boldsymbol{U}_3\boldsymbol{\Sigma}_z^{-1}\boldsymbol{U}_3^{\mathrm{T}}\boldsymbol{\Lambda}_3^{\frac{1}{2}} \right\} \\
&= \mathrm{Tr}\left\{ \boldsymbol{\Sigma}_z^{-1}\boldsymbol{\Sigma}_{z3} \right\} \\
&= \mathrm{Tr}\left\{ \boldsymbol{\Lambda}_1^{-\frac{1}{2}}\boldsymbol{U}_1\boldsymbol{\Sigma}_{z3}\boldsymbol{U}_1^{\mathrm{T}}\boldsymbol{\Lambda}_1^{-\frac{1}{2}} \right\}
\end{aligned}
$$

类似地，可以得到$\mathbb{E}[\xi_{k4}] = \mathrm{Tr}\left\{ \boldsymbol{\Lambda}_1^{-\frac{1}{2}}\boldsymbol{U}_1\boldsymbol{\Sigma}_{z4}\boldsymbol{U}_1^{\mathrm{T}}\boldsymbol{\Lambda}_1^{-\frac{1}{2}} \right\}$。基于$\boldsymbol{\Sigma}_{z3} > \boldsymbol{\Sigma}_{z4}$，可以证明$\boldsymbol{\Lambda}_1^{-\frac{1}{2}}\boldsymbol{U}_1(\boldsymbol{\Sigma}_{z3} - \boldsymbol{\Sigma}_{z4})\boldsymbol{U}_1^{\mathrm{T}}\boldsymbol{\Lambda}_1^{-\frac{1}{2}}$是正定矩阵，因此，可以获得$\mathbb{E}[\xi_{k3}] > \mathbb{E}[\xi_{k4}]$。

■

将式(11-13)代入式(11-12)中，$\boldsymbol{\Sigma}_{\tilde{z}}$可以改写为$\boldsymbol{\Sigma}_{\tilde{z}} = \boldsymbol{T}_k\boldsymbol{\Sigma}_z\boldsymbol{T}_k^{\mathrm{T}} + 1/a^2\boldsymbol{\Sigma}_b + \eta/a^2(\boldsymbol{T}_k - \boldsymbol{I})\boldsymbol{\Sigma}_z(\boldsymbol{T}_k - \boldsymbol{I})^{\mathrm{T}}$，其中$1/a^2$和$\eta/a^2$可以看作是可以在攻击存在的情况下的放大因子。当安全模块的参数设置为

$$
a \to 0, \frac{\eta}{a^2} \to \infty, \tag{11-34}
$$

$\boldsymbol{\Sigma}_{\tilde{z}}$在式(11-6)中几乎所有的线性攻击下都可以放大到无穷大（除了$\boldsymbol{T}_k \to \boldsymbol{I}, \boldsymbol{\Sigma}_b \to 0$)。根据引理11.5，可以得到$\mathbb{E}[\xi_k | \alpha] \to \infty$。由于没有攻击时的$\xi_k$是一个自由度为$m$的$\chi^2$变量，可以得到$\mathbb{E}[\xi_k | \overline{\alpha}] = m$。因此，在参数设置为式(11-34)的情况下，$\chi^2$检测器可以显著地检测出几乎所有线性攻击。

注释 11.6

当攻击者细微地篡改数据时，即$\boldsymbol{T}_k \to \boldsymbol{I}, \boldsymbol{\Sigma}_b \to 0$，我们是无法确定$\boldsymbol{\Sigma}_{\bar{z}}$是否可以放大的。因此，参数为式 (11-34) 的χ^2检测器可能无法显著区分$\mathbb{E}\left[\xi_k | \alpha\right]$和$\mathbb{E}\left[\xi_k | \bar{\alpha}\right]$，从而导致攻击检测的失败。需要注意的是，对于未配备安全模块的原始系统，这种攻击几乎无法修改传输的数据，对性能下降几乎没有影响。因此，解决方案是通过切断安全模块将系统切换到原始系统。这取决于攻击者和系统之间的认知程度，并且可以在博弈论框架中进一步建模。

场景三：信息完全泄露

当线性中间人攻击具有整个系统模型的知识时，它可以利用 a 和 η 的信息重新构建攻击策略$\boldsymbol{T}_k$和$\boldsymbol{\Sigma}_b$，并且通过满足以下的可行性约束从而避免被χ^2检测器检测：

$$\boldsymbol{T}_k \boldsymbol{\Sigma}_z \boldsymbol{T}_k^{\mathrm{T}} + \frac{1}{a^2}\boldsymbol{\Sigma}_b + \frac{\eta}{a^2}\left(\boldsymbol{T}_k - \boldsymbol{I}\right)\boldsymbol{\Sigma}_z\left(\boldsymbol{T}_k - \boldsymbol{I}\right)^{\mathrm{T}} = \boldsymbol{\Sigma}_z, \tag{11-35}$$

在该可行性条件下，$\boldsymbol{z}_k^r$与$\boldsymbol{z}_k$具有相同的统计特性，因此χ^2检测器的检测率为$(1-P_\zeta)$。由于χ^2检测器无法检测到满足条件(11-35)的攻击，这似乎意味着安全模块是无效的。然而，通过合适地选择水印参数a和η，我们可以证明这种攻击对估计性能下降没有影响。特别地，我们将在这种攻击下估计误差协方差表示为$\tilde{\boldsymbol{P}}_k$。

引理 11.6

当线性攻击 [式 (11-9)] 满足可行性约束条件 (11-35) 时，远程估计器在这种攻击下的估计误差协方差的递归公式如下：

$$\tilde{\boldsymbol{P}}_k = \boldsymbol{A}\tilde{\boldsymbol{P}}_{k-1}\boldsymbol{A}^{\mathrm{T}} + \boldsymbol{Q} + \overline{\boldsymbol{P}}\boldsymbol{C}^{\mathrm{T}}\left(\boldsymbol{\Sigma} - \boldsymbol{T}_k^{\mathrm{T}}\boldsymbol{\Sigma} - \boldsymbol{\Sigma}\boldsymbol{T}_k\right)\boldsymbol{C}\overline{\boldsymbol{P}}, \tag{11-36}$$

其中，$\boldsymbol{\Sigma} = \left(\boldsymbol{C}\overline{\boldsymbol{P}}\boldsymbol{C}' + \boldsymbol{R}\right)^{-1}$。

证明：

首先将式 (11-11) 的表达式改写为

$$\boldsymbol{z}_k^r = \boldsymbol{T}_k \boldsymbol{z}_k + \boldsymbol{n}_k, \tag{11-37}$$

其中，$\boldsymbol{n}_k = \frac{1}{a}\boldsymbol{b}_k + \frac{1}{a}\left(\boldsymbol{T}_k - \boldsymbol{I}\right)\boldsymbol{m}_k$是零均值高斯变量，且与$\boldsymbol{z}_k$独立无关。由于

$\boldsymbol{z}_k^r$在式(11-37)中的结构与文献[90]中的相同，因此后续证明与文献[90]中的类似，故此省略。 ■

为了分析满足条件 (11-35) 的所有线性攻击下的估计性能界限，需要推导出最优的$\boldsymbol{T}_k^*$、$\boldsymbol{b}_k^*$及其对应的估计误差协方差$\tilde{\boldsymbol{P}}_k^*$，具体如下。

引理 11.7

对于图 11-1 中配置了安全模块 [式 (11-8) 和式 (11-10)] 的系统，式 (11-9) 中的最优攻击为

$$\boldsymbol{T}_k^* = \frac{\eta / a^2 - 1}{\eta / a^2 + 1}\boldsymbol{I}，\ \boldsymbol{b}_k^* = 0 \tag{11-38}$$

其导致了最大的估计误差协方差：

$$\tilde{\boldsymbol{P}}_k^* = \boldsymbol{A}\tilde{\boldsymbol{P}}_{k-1}\boldsymbol{A}^{\mathrm{T}} + \boldsymbol{Q} + \frac{3 - \eta / a^2}{\eta / a^2 + 1}\overline{\boldsymbol{P}}\boldsymbol{C}'\boldsymbol{\Sigma}\boldsymbol{C}\overline{\boldsymbol{P}} \tag{11-39}$$

证明：

与文献 [90] 类似，我们利用最优攻击和最优估计之间的对偶关系推导出最优攻击策略。根据 LTI 过程中状态估计的 Kalman 滤波器的最优性 [136]，估计误差协方差在没有攻击的情况下是最小的。因此，对于任何攻击$\boldsymbol{T}_{k2} = \boldsymbol{I} + \boldsymbol{M}$，其中 $\boldsymbol{M}$ 是满足如下约束的任意矩阵：

$$(\boldsymbol{I} + \boldsymbol{M})\boldsymbol{\Sigma}_z(\boldsymbol{I} + \boldsymbol{M})^{\mathrm{T}} + \frac{\eta}{a^2}\boldsymbol{M}\boldsymbol{\Sigma}_z\boldsymbol{M}' \leqslant \boldsymbol{\Sigma}_z \tag{11-40}$$

可以得到以下不等式

$$\begin{aligned}&\overline{\boldsymbol{P}}\boldsymbol{C}^{\mathrm{T}}\left[\left(\boldsymbol{\Sigma} - \boldsymbol{T}_{k2}^{\mathrm{T}}\boldsymbol{\Sigma} - \boldsymbol{\Sigma}\boldsymbol{T}_{k2}\right) - \left(\boldsymbol{\Sigma} - \boldsymbol{T}_{k1}^{\mathrm{T}}\boldsymbol{\Sigma} - \boldsymbol{\Sigma}\boldsymbol{T}_{k1}\right)\right]\boldsymbol{C}\overline{\boldsymbol{P}} \\ &= \overline{\boldsymbol{P}}\boldsymbol{C}^{\mathrm{T}}\left[-\boldsymbol{M}^{\mathrm{T}}\boldsymbol{\Sigma} - \boldsymbol{\Sigma}\boldsymbol{M}\right]\boldsymbol{C}\overline{\boldsymbol{P}} \geqslant 0\end{aligned} \tag{11-41}$$

假设可以找到形式为$\boldsymbol{T}_{k3} = \rho\boldsymbol{I}$的攻击策略，其中$\rho \in \mathbb{R}$，对于任何$\boldsymbol{T}_{k4} = \rho\boldsymbol{I} - \boldsymbol{M}$，其中 $\boldsymbol{M}$ 是满足如下约束的任意矩阵：

$$\begin{aligned}&(\rho\boldsymbol{I} - \boldsymbol{M})\boldsymbol{\Sigma}_z(\rho\boldsymbol{I} - \boldsymbol{M})^{\mathrm{T}} \\ &+ \frac{\eta}{a^2}(\rho\boldsymbol{I} - \boldsymbol{M} - \boldsymbol{I})\boldsymbol{\Sigma}_z(\rho\boldsymbol{I} - \boldsymbol{M} - \boldsymbol{I})^{\mathrm{T}} \leqslant \boldsymbol{\Sigma}_z\end{aligned} \tag{11-42}$$

可以得到

$$\begin{aligned}&\overline{\boldsymbol{P}}\boldsymbol{C}^{\mathrm{T}}\left[\left(\boldsymbol{\Sigma} - \boldsymbol{T}_{k3}^{\mathrm{T}}\boldsymbol{\Sigma} - \boldsymbol{\Sigma}\boldsymbol{T}_{k3}\right) - \left(\boldsymbol{\Sigma} - \boldsymbol{T}_{k4}^{\mathrm{T}}\boldsymbol{\Sigma} - \boldsymbol{\Sigma}\boldsymbol{T}_{k4}\right)\right]\boldsymbol{C}\overline{\boldsymbol{P}} \\ &= \overline{\boldsymbol{P}}\boldsymbol{C}^{\mathrm{T}}\left[-\boldsymbol{M}^{\mathrm{T}}\boldsymbol{\Sigma} - \boldsymbol{\Sigma}\boldsymbol{M}\right]\boldsymbol{C}\overline{\boldsymbol{P}} \geqslant 0\end{aligned} \tag{11-43}$$

其推导是基于式 (11-41) 的，这意味着$\boldsymbol{T}_{k3}$是最优的攻击策略，因为它对应的估计误差协方差比任何攻击$\boldsymbol{T}_{k4}$下的都要大。因此，我们可以将式 (11-40) 和式 (11-42) 的左侧简化为

$$\left(1+\frac{\eta}{a^2}\right)\boldsymbol{M}\boldsymbol{\Sigma}_z\boldsymbol{M}^{\mathrm{T}}+\boldsymbol{\Sigma}_z+\boldsymbol{M}\boldsymbol{\Sigma}_z+\boldsymbol{\Sigma}_z\boldsymbol{M}^{\mathrm{T}} \tag{11-44}$$

$$\left(1+\frac{\eta}{a^2}\right)\boldsymbol{M}\boldsymbol{\Sigma}_z\boldsymbol{M}^{\mathrm{T}}+\left[\rho^2+\frac{\eta}{a^2}(\rho-1)^2\right]\boldsymbol{\Sigma}_z+ \\ \left[-\rho-\frac{\eta}{a^2}(\rho-1)\right]\boldsymbol{\Sigma}_z\boldsymbol{M}^{\mathrm{T}}+\left[-\rho-\frac{\eta}{a^2}(\rho-1)\right]\boldsymbol{M}\boldsymbol{\Sigma}_z \tag{11-45}$$

基于最优攻击和最优估计之间的一一对应关系，$\boldsymbol{M}$应该同时满足式(11-40)和式(11-42）。因此式(11-44)和式(11-45)的系数应该相同，即

$$\begin{cases}\rho^2+\eta/a^2(\rho-1)^2=1\\ -\rho-\eta/a^2(\rho-1)=1\end{cases} \tag{11-46}$$

从式(11-46)可以得到$\rho=\dfrac{\eta/a^2-1}{\eta/a^2+1}$，这意味着最优攻击以$\boldsymbol{T}_k^*=\rho\boldsymbol{I}$的形式存在。然后将$\boldsymbol{T}_k^*$和式(11-13)代入约束条件(11-35)，得到$\boldsymbol{\Sigma}_b=0$，即$\boldsymbol{b}_k^*=0$。最后，通过将$\boldsymbol{T}_k^*$代入式(11-36)，可以得到$\tilde{\boldsymbol{P}}_k^*$。 ■

注释 11.7　当$\eta=0$和$a=1$时，最优线性攻击策略为$\boldsymbol{T}_k^*=-\boldsymbol{I}$和$\boldsymbol{b}_k^*=0$，这意味着文献 [90] 中的定理 2 是引理 11.7 的特例。

引理 11.7 表明攻击效果的上限受到η和a的限制。在满足条件 (11-35) 的线性攻击下，最小化估计误差协方差的最优水印参数如下。

定理 11.3

对于图 11-1 中的系统，当式 (11-8) 和式 (11-10) 的参数选择为$\eta/a^2\to\infty$时，在任何满足可行性约束条件 (11-35) 的线性中间人攻击 [式 (11-9)] 下，估计误差协方差趋向于

$$\tilde{\boldsymbol{P}}_k=\boldsymbol{A}\tilde{\boldsymbol{P}}_{k-1}\boldsymbol{A}^{\mathrm{T}}+\boldsymbol{Q}-\overline{\boldsymbol{P}}\boldsymbol{C}^{\mathrm{T}}\boldsymbol{\Sigma}\boldsymbol{C}\overline{\boldsymbol{P}},$$

其等于没有攻击时的估计误差协方差。

证明：

式 (11-36) 中的估计误差协方差$\tilde{\boldsymbol{P}}_k$可以改写为$\boldsymbol{A}\tilde{\boldsymbol{P}}_{k-1}\boldsymbol{A}^{\mathrm{T}}+\boldsymbol{Q}+\boldsymbol{\Sigma}_w$，其中$\boldsymbol{\Sigma}_w=\bar{\boldsymbol{P}}\boldsymbol{C}^{\mathrm{T}}\left(\boldsymbol{\Sigma}-\boldsymbol{T}_k^{\mathrm{T}}\boldsymbol{\Sigma}-\boldsymbol{\Sigma}\boldsymbol{T}_k\right)\boldsymbol{C}\bar{\boldsymbol{P}}$。需要注意的是，$\boldsymbol{T}_k=\boldsymbol{I},\boldsymbol{b}_k=0$产生最优估计，这也是式 (11-36) 的下界。结合引理 11.7 中的结果，我们可以得到

$$-\bar{\boldsymbol{P}}\boldsymbol{C}^{\mathrm{T}}\boldsymbol{\Sigma}\boldsymbol{C}\bar{\boldsymbol{P}}\leqslant\boldsymbol{\Sigma}_w\leqslant\frac{3-\eta/a^2}{\eta/a^2+1}\bar{\boldsymbol{P}}\boldsymbol{C}^{\mathrm{T}}\boldsymbol{\Sigma}\boldsymbol{C}\bar{\boldsymbol{P}} \tag{11-47}$$

由于$\eta/a^2\to\infty$，式(11-47)的上界趋于下界，即$\frac{3-\eta/a^2}{\eta/a^2+1}\bar{\boldsymbol{P}}\boldsymbol{C}^{\mathrm{T}}\boldsymbol{\Sigma}\boldsymbol{C}\bar{\boldsymbol{P}}\to-\bar{\boldsymbol{P}}\boldsymbol{C}^{\mathrm{T}}\boldsymbol{\Sigma}\boldsymbol{C}\bar{\boldsymbol{P}}$。 ■

注释 11.8 虽然$\eta/a^2\to\infty$是安全模块的最优水印参数，但是它在实际应用中没有必要选为无穷大。如 4.6 节所示，当η/a^2较大时，我们在仿真中也可以得到满意的结果。需要说明的是，我们从保护数据完整性的角度推导出$\eta/a^2\to\infty$，然而这似乎不利于保护数据的机密性。因为在这种情况下，伪随机序列几乎暴露给攻击者。因此，η/a^2的值的选择应该是保密性和完整性之间的权衡，而这是我们未来综合评估安全性的工作之一。

扩展到检测其他攻击

首先，考虑虚假数据注入攻击，其攻击模型为

$$\tilde{\boldsymbol{f}}_k=\boldsymbol{f}_k+\boldsymbol{l}_k \tag{11-48}$$

其中，$\boldsymbol{l}_k\in\mathbb{R}^m$是注入到传输数据$\boldsymbol{f}_k$中的任意虚假数据。将式(11-8)和式(11-48)代入式(11-10)，χ^2检测器检测到的数据为$\boldsymbol{z}_k^r=\boldsymbol{z}_k+\frac{1}{a}\boldsymbol{l}_k$。式(11-5)的左侧可以表示

$$\begin{aligned}\mathbb{E}\left[\boldsymbol{\xi}_k|\ \alpha\right]&=\mathbb{E}\left[\left(\boldsymbol{z}_k+\frac{1}{a}\boldsymbol{l}_k\right)^{\mathrm{T}}\boldsymbol{\Sigma}_z^{-1}\left(\boldsymbol{z}_k+\frac{1}{a}\boldsymbol{l}_k\right)\right]\\&=\mathbb{E}\left[\boldsymbol{z}_k^{\mathrm{T}}\boldsymbol{\Sigma}_z^{-1}\boldsymbol{z}_k+\frac{\boldsymbol{l}_k^{\mathrm{T}}\boldsymbol{\Sigma}_z^{-1}\boldsymbol{l}_k}{a^2}+\frac{\boldsymbol{z}_k^{\mathrm{T}}\boldsymbol{\Sigma}_z^{-1}\boldsymbol{l}_k}{a}+\frac{\boldsymbol{l}_k^{\mathrm{T}}\boldsymbol{\Sigma}_z^{-1}\boldsymbol{z}_k}{a}\right]\\&=m+\frac{1}{a}\mathbb{E}\left[\left(\frac{1}{a}\boldsymbol{l}_k+2\boldsymbol{z}_k\right)^{\mathrm{T}}\boldsymbol{\Sigma}_z^{-1}\boldsymbol{l}_k\right]\end{aligned} \tag{11-49}$$

当安全模块选择式(11-34)中的参数时，攻击需要满足$\boldsymbol{l}_k = -2a\boldsymbol{z}_k$以避免被检测。否则，式(11-49)将被放大到无穷大。然而，由于新息$\boldsymbol{z}_k$是被水印$\boldsymbol{m}_k$加密的，攻击者无法获得$\boldsymbol{z}_k$的准确信息，因此，攻击者无法成功地注入虚假数据。

其次，我们考虑重放攻击。重放攻击可以记录先前传输的数据$\boldsymbol{f}_k$，并在之后重放这些数据 [86]，它可以表述为$\widetilde{\boldsymbol{f}_k} = \boldsymbol{f}_{k-\tau}$，其中$\tau \in \mathbb{N}$。由于重放攻击的效果类似于网络延迟，我们利用时间戳来区分它们。附加时间戳后，传输的数据包变为$\{\boldsymbol{f}_k, k\}$。以下定理分析了$\boldsymbol{f}_k$与其时间戳 k 错误匹配的情况。

定理 11.4

对于图 11-1 中的系统，如果$\boldsymbol{f}_k$与其时间戳 k 不匹配，则检测器的报警率为

$$F\left(\frac{\zeta}{1+2\eta / a^2}, m\right) \tag{11-50}$$

当式 (11-8) 和式 (11-10) 的参数选择为$\eta / a^2 \to \infty$时，χ^2检测器以概率 100% 触发报警。

证明：

假设附加到传输数据$\boldsymbol{f}_k$的时间戳是$k-\tau$。式 (11-10) 生成$\boldsymbol{m}_{k-\tau}$作为水印，这导致$\boldsymbol{z}_k^r$为

$$\boldsymbol{z}_k^r = \boldsymbol{z}_k - \frac{1}{a}\left(\boldsymbol{m}_k - \boldsymbol{m}_{k-\tau}\right) \tag{11-51}$$

由于$\boldsymbol{m}_k$是一个高斯变量，式(11-51)的协方差可以被推导为$\boldsymbol{\Sigma}_{\tilde{z}} = \left(1+\frac{2\eta}{a^2}\right)\sum_z$。与定理11.1类似，我们可以得到如式(11-50)所示的报警率。当$\eta / a^2 \to \infty$时，检测器一定会触发警报。■

受定理 11.4 的启发，我们可以应用时间戳和χ^2检测器来防御重放攻击。假设攻击者可以识别数据包$\{\boldsymbol{f}_k, k\}$的结构。基于定理 11.4，当攻击者只是修改了$\boldsymbol{f}_k$而不改变时间戳时，即$\{\boldsymbol{f}_{k-\tau}, k\}$，$\chi^2$检测器一定会触发报警。否则，如果攻击者将数据包修改为$\{\boldsymbol{f}_{k-\tau}, k-\tau\}$，则时间戳$k-\tau$将

再次出现。因此，可以很容易检测到此类攻击。

11.2.4 仿真示例 1

在本节中，给出一些数值仿真来证明所推导的结果。为了说明安全模块对线性中间人攻击的效果，采用与文献 [90] 中相同的模型。参数为 $A=0.8, C=1.2, Q=1, R=1$。一般来说，χ^2 检测器的窗口大小 $\boldsymbol{J}$ 是由我们的实践经验决定的。在 $\boldsymbol{J}$ 固定的情况下，阈值 ζ 可以根据预先确定的置信系数从 χ^2 表中选择。将其窗口大小设置为 $\boldsymbol{J}=1$，并将其置信系数设置为 95%，那么阈值是 $\zeta=3.841$。进行 10000 次蒙特卡罗模拟，得到以下仿真结果。

11.2.5 场景 I 的仿真结果

在 OLMA 下，将文献 [289] 中的随机 χ^2 检测器（表示为 D1) 与配备安全模块的 χ^2 检测器进行比较，其中 $\eta / a^2=9999$（表示为 D2)。在图 11-2 中，攻击者仅在时间段 [60,80] 内发起 OLMA 攻击，而两个检测器 D1 和 D2 同时在时间段 [20,40] 以及 [60,80] 内开启攻击检测。可以看出，无论 OLMA 是否发生，D1 的估计误差协方差总是大于D2，这意味着 D2 可以更好地保护估计性能。除此之外，D2 比 D1 更有效地检测出 OLMA 攻击。在图 11-2 中，检测器 D1 在时间段 [20,40] 内的报警率几乎和时间段 [60,80] 内的一样，这意味着 D1 无法识别出 OLMA 攻击的存在性。与之形成鲜明对比的是，检测器 D2 可以以 100% 的概率检测到 OLMA 攻击，且其误报率在时间段 [20,40] 内低于检测器 D1。当 OLMA 的攻击模式已知时，检测器 D2 可以进一步结合式 (11-31) 中的数据校正步骤（表示为 D3），以提升时间段 [60,80] 内的估计性能。请注意，在时间段 [20,40] 内，式 (11-31) 会以误报率 5% 错误地将 $\boldsymbol{z}_k$ 修改为式 (11-33) 中的假数据 $\tilde{\boldsymbol{z}}_k^{rc}$。然而，通过再次使用 χ^2 检测器检查 $\tilde{\boldsymbol{z}}_k^{rc}$，我们可以避免上述情况，使得 D3 可以在 [20,40] 期间保持与 D2 相同的估计性能。

此外，我们进一步揭示了 η / a^2 的选择与图 11-3 中的检测性能之间的关系。随着η/a^2的增加，从图 11-3 可以看出，OLMA 的实际检测率是单调递增的，与定理 11.1 中所推导的结果一致。和注释 11.8 中所说的一样，这说明了η/a^2在实际应用中只要相对较大，检测器 D2 或 D3 在 OLMA 下就能得到满意的检测性能。此外，因为可以在没有攻击的情况下消除水印，所以检测器 D2 或 D3 的误报率和初始的χ^2检测器的误报率一致，即为 5%，而和η/a^2的选择无关。这说明η/a^2的选择不是误报率和检测率之间的权衡。

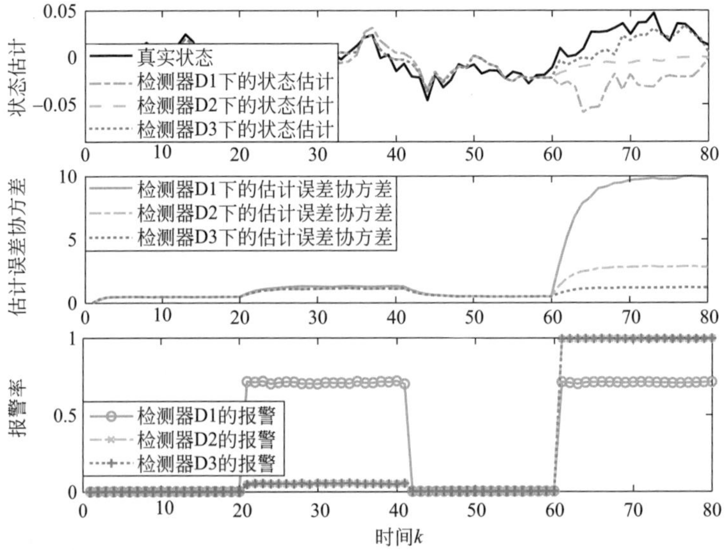

图 11-2　检测器 D1、D2 和 D3 在 OLMA 下的估计性能和检测性能的比较

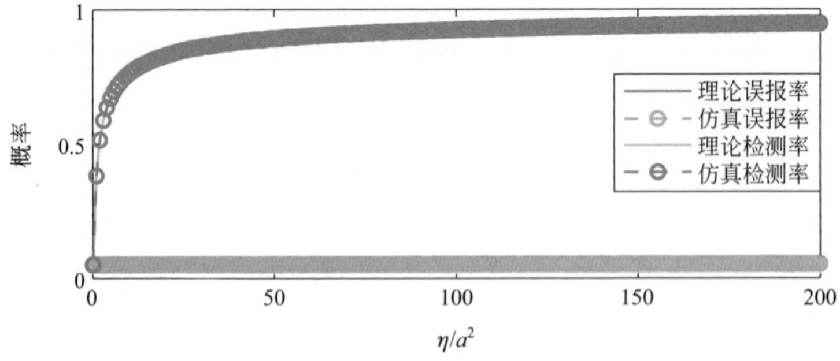

图 11-3　在 OLMA 以及不同的η/a^2下，D2 或 D3 的检测率以及误报率

在图 11-4 中，我们采用文献 [287] 中的累积和（CUSUM）检测器和多元指数加权移动平均（MEWMA）检测器作为基准来比较所提出方法的检测性能。所有检测器在时间段 [20,40] 和 [60,80] 内同时开启，而攻击者仅在时间段 [60,80] 内发起攻击。如果没有使用本节所提出的水印方法，所有检测器都无法检测到 OLMA 攻击，因为它们的检测率都非常低，并且在时间段 [60,80] 内大约接近 5%。与此相反的是，通过利用所提出的水印方法，上述检测器可以以接近 100% 的概率有效地检测到攻击。这反映了所提出的水印方法不仅可以帮助 χ^2 检测器，还可以帮助其他基于统计的检测器检测 OLMA 攻击。

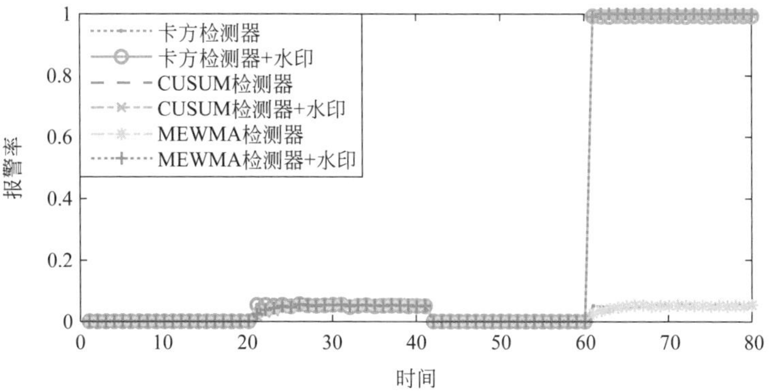

图 11-4　不同检测器之间检测性能的比较

11.2.6　场景 II 的仿真结果

在图 11-5 中，我们在攻击策略 [式 (11-9)] 中任意地选择了三种不同的线性攻击，分别为 $\boldsymbol{T}_{k4}=0,\boldsymbol{\Sigma}_{b4}=3\boldsymbol{\Sigma}_z$ 和 $\boldsymbol{T}_{k5}=\dfrac{1}{2}\boldsymbol{I},\boldsymbol{\Sigma}_{b5}=2\boldsymbol{\Sigma}_z$，以及 $\boldsymbol{T}_{k6}=\dfrac{3}{4}\boldsymbol{I},\boldsymbol{\Sigma}_{b6}=\boldsymbol{\Sigma}_z$。此外，所有的攻击都发生在时间段 [100,150] 内。通过将 ξ_k 的值与 χ^2 检测器的阈值进行比较，图 11-5 直观地反映了 D2 的检测性能。注意到当 $\boldsymbol{T}_k\to\boldsymbol{I},\boldsymbol{\Sigma}_b\to 0$ 时，ξ_k 的值会变小。尽管如此，式 (11-9) 中的大多数线性攻击都可以被有效地检测到，因为这些攻击下 ξ_k 都比阈值大 10^5 倍。

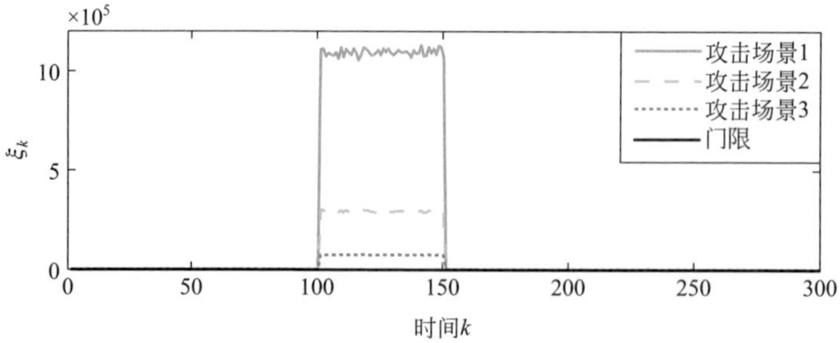

图 11-5　利用ξ_k的值来检测发生在时间段 [100,150] 内的三种不同的线性攻击

11.2.7　场景 III 的仿真结果

在信息泄露的情况下，攻击者可以构建满足约束条件 (11-35) 的线性中间人攻击。在图 11-6 中的时间段 [0,20]、[40,60] 以及 [80,100] 内，系统在正

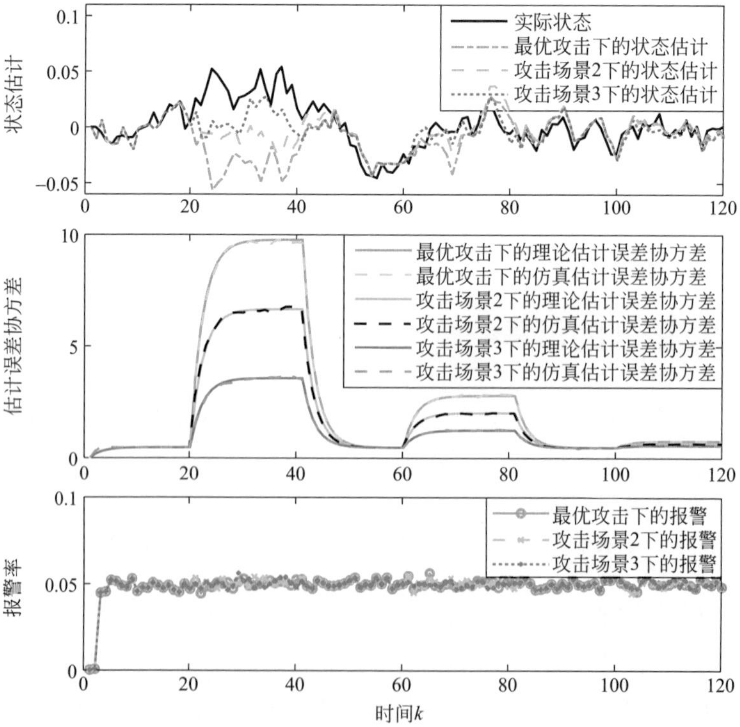

图 11-6　图 11-1 中系统在三种不同的线性中间人攻击下的估计和检测性能

常情况下运行，使得远程估计器进入稳定状态。然而，在余下的时间段内，无线传感网络受到了三种不同的线性中间人攻击。除了式 (11-38) 中的最优策略，其他两种攻击策略设置为$\boldsymbol{T}_{k2}=\dfrac{3\eta/a^2-1}{3\eta/a^2+3}\boldsymbol{I},\boldsymbol{\Sigma}_{b2}=\dfrac{8a^2}{9+9\eta/a^2}\boldsymbol{\Sigma}_z$以及$\boldsymbol{T}_{k3}=\dfrac{3\eta/a^2+1}{3\eta/a^2+3}\boldsymbol{I},\boldsymbol{\Sigma}_{b3}=\dfrac{8a^2}{9+9\eta/a^2}\boldsymbol{\Sigma}_z$。注意到，这三种攻击都满足约束条件 (11-35)，从而具备不可检测性，即在图 11-6 中它们的检测率等于误报率的 5%。为了分析攻击效果与η/a^2的关系，我们将时间段 [20,40]、[60,80] 和 [100,120] 内的参数分别设置为$\{\eta_1 a_1=0,a_1=1\}$、$\{\eta_2=3,a_2=1\}$和$\{\eta_3=30,a_3=1\}$。由于式 (11-38) 中的攻击始终可以导致最大的估计误差协方差，我们可以验证引理 11.7 中的结果。当η/a^2变大时，三种攻击下的估计误差协方差逐渐接近于没有攻击下的估计协方差，因此我们验证了定理 11.3 中的结果。

11.2.8 扩展：检测重放攻击

针对重放攻击的检测性能如图 11-7 所示。攻击者记录时间段 [0,20] 内的数据，并在时间段 [20,40] 和 [60,80] 内用两种不同的策略向远程端

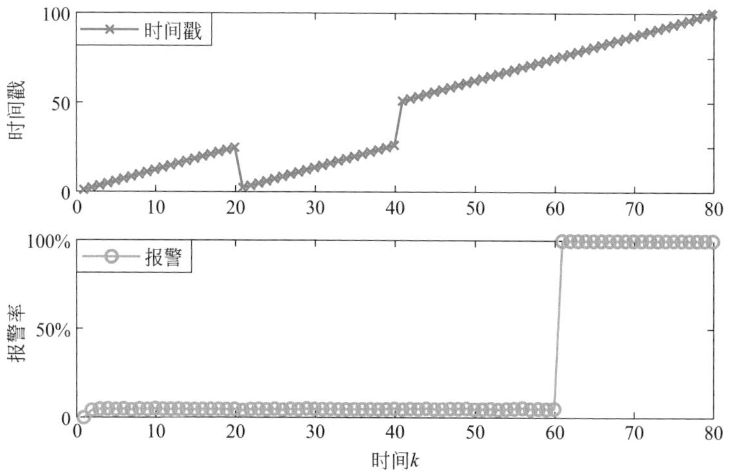

图 11-7　利用时间戳和检测器 D2 来检测发生在 [20,40] 和 [60,80] 期间的重放攻击

重放之前的数据。对于前一种，攻击者直接重放整个数据包。由于时间戳重复出现，攻击很容易被检测到。对于后一种，攻击者首先将记录数据的时间戳修改为当前时间戳，然后重放数据。通过使用检测器 D2，这种攻击以 100% 的概率触发警报。

11.3 隐秘攻击下分布式状态估计的收敛性分析

11.3.1 系统和攻击模型

考虑如下的线性离散时不变系统：

$$\boldsymbol{x}(k+1)=\boldsymbol{A}\boldsymbol{x}(k)+\boldsymbol{w}(k) \tag{11-52}$$

其中，$\boldsymbol{A}$是系统矩阵；$k\in\mathbb{N}$是时间序列；$\boldsymbol{x}(k)\in\mathbb{R}^n$表示系统的真实状态；过程噪声$\boldsymbol{w}(k)$以及状态初始值$\boldsymbol{x}(0)$是独立零均值高斯变量，其协方差分别为$\boldsymbol{Q}>0$以及$\boldsymbol{\Pi}_0\geqslant 0$。在分布式传感网络中，共有$N$个用于共同监测系统的状态$\boldsymbol{x}(k)$。第$i$个传感器的观测方程如下所示：

$$\boldsymbol{y}_i(k)=\boldsymbol{C}_i\boldsymbol{x}(k)+\boldsymbol{v}_i(k) \tag{11-53}$$

其中，$\boldsymbol{y}_i(k)\in\mathbb{R}^{m_i}$表示测量值；$\boldsymbol{v}_i(k)$是零均值独立同分布的高斯测量噪声，其协方差为$\boldsymbol{R}_i>0$。注意$\boldsymbol{v}_i(k)$和$\boldsymbol{x}(0)$以及$\boldsymbol{w}(k)$独立无关，并且与$\boldsymbol{v}_j(k)$也独立无关，$\forall i\neq j$。无线传感器网络的通信拓扑建模为有向图$G=(V,E)$，其中节点$V=\{1,2,\cdots,N\}$以及边$E\subset V\times V$分别表示传感器集合以及通信链路。如果存在一条链路$(i,j)\subset E$，这就意味着节点$j$可以将其信息传输给节点$i$。鉴于此，节点$i$的入邻居被定义为$N_i=\{j:(i,j)\subset E\}$，且其维度被定义为$d_i=|N_i|$。类似地，$\overline{N}_i=\{j:(j,i)\subset E\}$表示节点$i$的出邻居。图被假设是强连通的。为了描述传感网络的物理拓扑，如图11-8所示，其中，蓝色虚线表示攻击者拦截了从节点3传输到节点2的数据，并将其篡改为错误虚假数据。我们将图G的拉普拉斯矩阵定义为$\boldsymbol{L}=[l_{ij}]$，其中，

$$l_{ij}=\begin{cases}-1, & 如果(i,j)\subset E, i\neq j,\\ -\sum\limits_{j\in N_i} l_{ij}, & 如果 i=j,\\ 0, & 其他\end{cases}$$

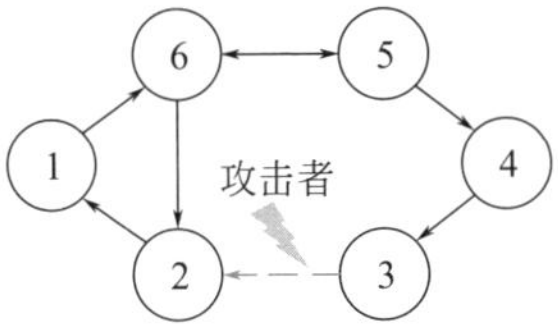

图 11-8　攻击下的传感网络

对于线性高斯系统 [式 (11-52) 以及式 (11-53)]，我们需要选择一种分布式滤波算法来估计其系统状态，而不是应用估计无噪声系统状态的分布式观测器 [295]。在现有文献中，分布式滤波要求交换的信息包括信息矩阵以及信息矢量两种信息 [296]。然而，这些信息可能同时被攻击者所篡改，这使得更加难以设计相应的攻击检测方法。因此，本节聚焦于仅需交互局部状态估计的分布式一致性滤波。对于第 i 个节点，其局部滤波算法如下所示：

$$\hat{\boldsymbol{x}}_i(k+1)=\boldsymbol{A}\hat{\boldsymbol{x}}_i(k)+\boldsymbol{K}_i(k)\left[\boldsymbol{y}_i(k)-\boldsymbol{C}_i\hat{\boldsymbol{x}}_i(k)\right]-\varepsilon\boldsymbol{A}\sum_{j\in N_i}\left[\hat{\boldsymbol{x}}_i(k)-\hat{\boldsymbol{x}}_j(k)\right] \tag{11-54}$$

其中，$\hat{\boldsymbol{x}}_i(k)$表示在不存在网络攻击时，第$i$个节点在第$k$个时刻的局部状态估计值，$\boldsymbol{K}_i(k)$以及$\varepsilon$分别是估计器的卡尔曼增益以及一致性增益。$\varepsilon$在$(0,1/\boldsymbol{\Delta})$的范围中取值，其中$\boldsymbol{\Delta}=\max_i d_i$。对于式(11-54)，其最后一项对每个节点的状态估计的偏差实现了补偿，从而使得传感网络达到一致。

在不受保护的无线通道中，传输的数据可能被恶意攻击者所篡改。在单传感器网络和集中式多传感器网络中，文献 [289,297] 利用传输新息$\boldsymbol{z}_i(k)\overset{\text{def}}{=}\boldsymbol{y}_i(k)-\boldsymbol{C}_i\hat{\boldsymbol{x}}_i(k)$的统计特性来诊断攻击所引起的异常。然而，在分布式传感网络，第 i 个节点所接收的数据是其邻居节点的局部状态估计$\hat{\boldsymbol{x}}_j$，而不是其新息。因此，这引起了新的挑战，也就是如何利用状态估计值来检测攻击。如图 11-9 所示，在多智能体等诸多实际应用中，装载的传感器配置于节点内部，并且和节点其余部件通过有线连接。图中，所设计的检测器用以检测邻居节点所发送过来的数据是否被攻击者篡改。由于本章所聚焦的是篡改无线传感数据的网络攻击，而不是直接入侵节点内部的传感器攻击，因此，我们假设测量数据$\boldsymbol{y}_i(k)$对于第 i 个节点本身是可靠的，从而可以被第 i 个节点用作攻击检测的基准。接下来，我们为每一个传感器提出了一种防御方法来抵抗恶意攻击：

$$\gamma_{ij}(k)=\begin{cases}1, & \text{如果}\zeta_{ij}(k)>\delta_{ij},\\ 0, & \text{其他},\end{cases}$$

其中

$$\begin{aligned}\zeta_{ij}(k)&=\left(\boldsymbol{y}_i(k)-\boldsymbol{C}_i\tilde{\boldsymbol{x}}_j^a(k)\right)^{\mathrm{T}}Z_i^{-1}\left(\boldsymbol{y}_i(k)-\boldsymbol{C}_i\tilde{\boldsymbol{x}}_j^a(k)\right)\\ Z_i&=3\|\boldsymbol{R}_i\|\end{aligned}\tag{11-55}$$

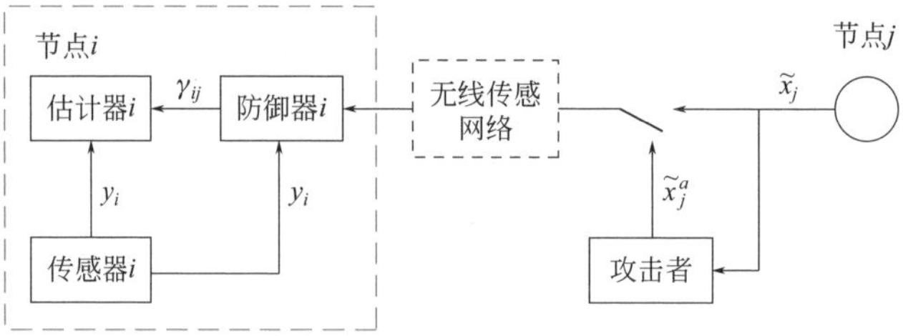

图 11-9　每个节点的内部结构

由于无线传感网络可能受到图11-9中所示的网络攻击，$\tilde{\boldsymbol{x}}_j^a(k)$表示第$i$个节点从其邻居节点$j\in N_i$所接收到的被攻击者篡改后的数据。$\gamma_{ij}(k)$表示第$i$个节点的防御器是否会对$\tilde{\boldsymbol{x}}_j^a(k)$触发报警。由于$\gamma_{ij}(k)$取决于预设的门限值$\delta_{ij}$，因此，我们需要合理地选择门限值来平衡误报率以及防御器的检测精度。Z_i是一个用以方便调节δ_{ij}的参数，其选择基于工程实践中故障诊断所常用的拉依达准则（即3δ准则）。为了阐述防御器的机理，我们求$\zeta_{ij}(k)$的期望值：

$$\begin{aligned}\mathbb{E}\left[\zeta_{ij}(k)\right]&=Z_i^{-1}\mathrm{tr}\left(\mathbb{E}\left[\left(\boldsymbol{y}_i(k)-\boldsymbol{C}_i\tilde{\boldsymbol{x}}_j^a(k)\right)\left(\boldsymbol{y}_i(k)-\boldsymbol{C}_i\tilde{\boldsymbol{x}}_j^a(k)\right)^{\mathrm{T}}\right]\right)\\ &=Z_i^{-1}\mathrm{tr}\left(\boldsymbol{C}_i^{\mathrm{T}}\tilde{\boldsymbol{P}}_j^a(k)\boldsymbol{C}_i+\boldsymbol{R}_i\right)\end{aligned}\tag{11-56}$$

其中，$\tilde{\boldsymbol{P}}_j^a(k) \stackrel{\text{def}}{=} \mathbb{E}\left[\left(\tilde{\boldsymbol{x}}_j^a(k)-\boldsymbol{x}(k)\right)\left(\tilde{\boldsymbol{x}}_j^a(k)-\boldsymbol{x}(k)\right)^{\mathrm{T}}\right]$表示$\tilde{\boldsymbol{x}}_j^a(k)$的估计误差协方差。对于$\hat{\tilde{\boldsymbol{P}}}_j^a(k) > \check{\tilde{\boldsymbol{P}}}_j^a(k)$，可以导出$\mathbb{E}\left[\hat{\zeta}_{ij}(k)\right]-\mathbb{E}\left[\check{\zeta}_{ij}(k)\right] \geqslant 0$，其中$\mathbb{E}\left[\hat{\zeta}_{ij}(k)\right]$和$\mathbb{E}\left[\check{\zeta}_{ij}(k)\right]$分别对应于$\hat{\tilde{\boldsymbol{P}}}_j^a(k)$以及$\hat{\tilde{\boldsymbol{P}}}_j^a(k)$。也就是说，和$\check{\tilde{\boldsymbol{P}}}_j^a(k)$相比，$\hat{\tilde{\boldsymbol{P}}}_j^a(k)$导致$\mathbb{E}\left[\zeta_{ij}(k)\right]$上升得更大，从而使得防御机制更容易触发报警。因此，该检测方法是基于$\tilde{\boldsymbol{P}}_j^a(k)$来检测攻击的。

注释 11.9　在“f- 局部”攻击者的假定下，每个节点可以直接从它所接收的数据中丢弃f个最大的以及f个最小的数据，从而对抗分布式传感网络中的攻击 [298]。这里所提出的防御方法基于传输数据的估计误差协方差，因此可以更具选择性地丢弃偏离真实状态$\boldsymbol{x}_k$的数据。另外，式 (11-55) 中的防御机制直接利用传输的状态估计值进行攻击的检测，而不是依赖于额外所附加的检验数据。因此，防御机制 [式 (11-55)] 不需要遵守系统状态的维度和测量值的维度相等的要求，即$n \neq m_i, \forall i = 1,\cdots,N$。因此，它比文献 [299] 中的保护器更具一般性。

注释 11.10　根据式 (11-56)，δ_{ij}的选择取决于测量矩阵$\boldsymbol{C}_i$、测量噪声的方差$\boldsymbol{R}_i$、参数Z_i以及估计误差协方差$\tilde{\boldsymbol{P}}_j^a(k)$。然而，$\tilde{\boldsymbol{P}}_j^a(k)$是第$j$个的节点的局部信息，而对于第$i$个的节点是未知的。

因此，由于信息缺失，我们很难为分布式传感网络中的防御器 [式 (11-55)] 推导出δ_{ij}的准确值。需要注意的是，门限值δ_{ij}过小，会降低分布式滤波的收敛速率，甚至会破坏传感网络的一致性。然而，在门限值δ_{ij}选择过大时，防御器会完全失去其抵御攻击的效果，分布式滤波可能在网络攻击的影响下发散。因此，门限值δ_{ij}的选择是安全性以及一致性之间的均衡。在实际应用中，需要将δ_{ij}调节至合适的临界值，使得防御器的误报率相对较低，从而保证正常滤波系统的稳定性。

尽管式 (11-55) 中的防御器可以检测部分恶意攻击，但是隐秘攻击者仍能设计其攻击策略达到欺诈防御器的效果。假设攻击者拥有完整的系统模型知识，甚至是预设门限δ_{ij}的准确值。为了避免被防御器检测，攻击者需要保证其篡改的数据$\tilde{\boldsymbol{x}}_j^a(k)$不会触发防御器的报警，即$\zeta_{ij}(k) \leqslant \delta_{ij}, \forall k \in N$。通过对上述不等式两边求期望，有$\mathbb{E}\left[\zeta_{ij}(k)\right] \leqslant \delta_{ij}, \forall k \in N$。根据式 (11-56) 中$\mathbb{E}\left[\zeta_{ij}(k)\right]$和$\tilde{\boldsymbol{P}}_j^a(k)$之间的相互关系，这意味着隐秘攻击者需要精密地设计其攻击策略，从而满足以下条件：

$$\tilde{\boldsymbol{P}}_j^a(k) \leqslant \left(\boldsymbol{P}_j^i\right)^* \tag{11-57}$$

其中，$\left(\boldsymbol{P}_j^i\right)^*$表示第$i$个节点的防御器对$\tilde{\boldsymbol{x}}_j^a(k)$的最大容忍度。此时，$\mathbb{E}\left[\zeta_{ij}(k)\right]$将达到其最大值，即门限值$\delta_{ij}$。根据式(11-56)中的防御机制，攻击者需要获得实际状态值$x(k)$的近似甚至是准确值来估算出$\tilde{\boldsymbol{x}}_j^a(k)$和$\boldsymbol{x}(k)$之间的偏离程度，从而保证隐秘性条件(11-57)成立。如文献[300]所述，真实状态值$\boldsymbol{x}(k)$也是一类攻击者可以获取的信息，因此，基于真实状态值$\boldsymbol{x}(k)$，我们提出了如下两种不同攻击场景下的隐秘攻击策略。

（1）通信资源充足的隐秘攻击

在实际应用中，攻击者需要消耗其能量以及资源来破坏无线传感网络。在第一种场景下，我们假设攻击者可以在任何时刻，通过任意的通信链路对传输数据进行篡改。受到远端状态估计系统中的线性欺诈攻击模型 [301] 的启发，我们对分布式传感网络提出了如下的隐秘攻击模型：

$$\tilde{\boldsymbol{x}}_j^a(k) = \boldsymbol{x}(k) + \boldsymbol{T}_{ij}(k)\left(\tilde{\boldsymbol{x}}_j(k) - \boldsymbol{x}(k)\right) + \boldsymbol{b}_{ij}(k) \tag{11-58}$$

其中，$\tilde{\boldsymbol{x}}_j(k)$表示在攻击存在的情况下，第$j$个节点的局部状态估计。区分于$\hat{\boldsymbol{x}}_j(k)$,$\tilde{\boldsymbol{x}}_j(k)$进一步强调了攻击对状态估计的影响。因此，$\tilde{\boldsymbol{x}}_j(k)$是$\hat{\boldsymbol{x}}_j(k)$更具一般性的表述方式。$\boldsymbol{T}_{ij}(k) \in \mathbb{R}^{n \times n}$是一个任意的矩阵，$\boldsymbol{b}_{ij}(k) \in \mathbb{R}^n$是一个与$\boldsymbol{x}(k)$以及$\tilde{\boldsymbol{x}}_j(k)$独立无关的高斯随机变量，其方差为$\boldsymbol{\Sigma}_{ij}^b$。注意到第$j$个节点会将局部状态估计值$\tilde{\boldsymbol{x}}_j(k)$广播给其邻居节点$i, \forall i \in \overline{N}_j$。如果在第$k$个时刻，攻击者并未篡改从第$j$个节点传输给第$i$个节点的传输数据，即$\boldsymbol{T}_{ij}(k) = \boldsymbol{I}, \boldsymbol{b}_{ij}(k) = 0$，则$\tilde{\boldsymbol{x}}_j^a(k)$等于$\tilde{\boldsymbol{x}}_j(k)$。否则，根据攻击模型(11-58)，$\tilde{\boldsymbol{x}}_j^a(k)$将偏离$\tilde{\boldsymbol{x}}_j(k)$。在这种情况下，隐秘性条件(11-57)可以重写成：

$$\tilde{\boldsymbol{P}}_j^a(k)=\boldsymbol{T}_{ij}(k)\tilde{\boldsymbol{P}}_j(k)\boldsymbol{T}_{ij}(k)^{\mathrm{T}}+\boldsymbol{\Sigma}_{ij}^b\leqslant(\boldsymbol{P}_j^i)^* \tag{11-59}$$

其中，$\tilde{\boldsymbol{P}}_j(k)$是$\tilde{\boldsymbol{x}}_j(k)$的估计误差协方差。当$\tilde{\boldsymbol{P}}_j(k)>(\boldsymbol{P}_j^i)^*$时，攻击者[式(11-58)]可以通过调节$\boldsymbol{T}_{ij}(k)$使得$\tilde{\boldsymbol{P}}_j^a(k)$满足式(11-59)中的隐秘性条件。否则，攻击者可以调节$\boldsymbol{T}_{ij}(k)$来增大$\tilde{\boldsymbol{P}}_j^a(k)$从而破坏第$i$个节点的估计性能。

注释 11.11 当$\boldsymbol{T}_{ij}(k)=\boldsymbol{I}$,文献 [299] 中所研究的错误数据注入攻击是所提出的攻击模型 (11-58) 的一个特例。另外，和文献 [299] 相比，式 (11-58) 中所提出的攻击具有更大的自由度，来调节其策略$\boldsymbol{T}_{ij}(k)$以实现隐身。

（2）通信资源受限的隐秘攻击

对于通信资源受限的攻击者，主要有两种方式建立攻击模型。第一种是假设在有限的时间域 T 内，攻击者可以进行 t 次有限数目的攻击，$t\leqslant T$ [302]。而第二种是以一定的概率随机地攻击无线信道 [289]。本节采用了后者来描述攻击，因为攻击的随机性使得隐秘攻击更加难以被检测。我们用一个伯努利分布的二进制变量$\alpha_{ij}(k)$来表示攻击行为。$\alpha_{ij}(k)=1$表示攻击者篡改通信边(i,j)的传输数据。定义攻击者在各条通信链路上的概率为$\mathbb{P}[\alpha_{ij}(k)=1]=\rho_{ij}$。因此，这种情况下，隐秘攻击模型可以写成

$$\tilde{\boldsymbol{x}}_j^a(k)=\alpha_{ij}(k)\left[\boldsymbol{x}(k)+\boldsymbol{T}_{ij}(k)\left(\tilde{\boldsymbol{x}}_j(k)-\boldsymbol{x}(k)\right)+\boldsymbol{b}_{ij}(k)\right]+\left(1-\alpha_{ij}(k)\right)\tilde{\boldsymbol{x}}_j(k) \tag{11-60}$$

这意味着当$\alpha_{ij}(k)=0$时，攻击者无法拦截通过链路(i,j)传输给第i个传感器的数据$\tilde{\boldsymbol{x}}_j(k)$，因此，为了保证防御机制不会报警，隐秘攻击者在这种情况下应该满足以下约束：

$$\tilde{\boldsymbol{P}}_j^a(\alpha_{ij}(k)=0)=\tilde{\boldsymbol{P}}_j(k)\leqslant(\boldsymbol{P}_j^i)^* \tag{11-61}$$

$$\tilde{\boldsymbol{P}}_j^a(\alpha_{ij}(k)=1)=\boldsymbol{T}_{ij}(k)\tilde{\boldsymbol{P}}_j(k)\boldsymbol{T}_{ij}(k)^{\mathrm{T}}+\boldsymbol{\Sigma}_{ij}^b\leqslant(\boldsymbol{P}_j^i)^* \tag{11-62}$$

和第一种攻击场景相比，通信资源受限的攻击者需要式(11-60)等来更严格地设计其攻击策略以欺诈防御器。需要注意的是，第j个传感器需要将其局部状态估计值$\tilde{\boldsymbol{x}}_j(k)$广播给所有的邻居节点$i\in\overline{N}_j$。为了屏蔽这些邻

居节点相应的防御器，攻击者需要保证隐秘条件(11-61)对于节点$\forall i \in \overline{N}_j$都是满足的。也就是说，在这种隐秘攻击下，每个节点的估计误差协方差$\forall i \in \overline{N}_j$存在上界：

$$\tilde{\boldsymbol{P}}_j(k) \leqslant \min_{i \in \overline{N}_j}\left(\boldsymbol{P}_j^i\right)^* \tag{11-63}$$

这体现出防御器[式(11-55)]能大大地限制隐秘攻击对估计性能所造成的破坏。不仅仅局限于式(11-60)所给定的攻击模型，防御器对任意资源受限的隐秘攻击都具备式(11-63)类似的约束。

11.3.2 资源充足的隐秘攻击下的分布式一致性估计

在本节中，我们提出式 (11-58) 中最优的隐秘攻击的充分条件。这种最优的隐秘攻击可以在欺诈防御器 [式 (11-55)] 的同时，最大化地破坏分布式滤波 [式 (11-54)] 的估计性能。此外，分析了在状态估计因攻击而发散的这种极端情况下，分布式滤波的弹性。

考虑到屏蔽防御器的隐秘攻击的存在，攻击下的分布式估计算法 (11-54) 可以写为：

$$\tilde{\boldsymbol{x}}_i(k+1) = \boldsymbol{A}\tilde{\boldsymbol{x}}_i(k) + \boldsymbol{K}_i(k)\left[\boldsymbol{y}_i(k) - \boldsymbol{C}_i\tilde{\boldsymbol{x}}_i(k)\right] - \varepsilon\boldsymbol{A}\sum_{j \in N_i}\left[\tilde{\boldsymbol{x}}_i(k) - \tilde{\boldsymbol{x}}_j^a(k)\right] \tag{11-64}$$

其中，$\tilde{\boldsymbol{x}}_i(0) = \hat{\boldsymbol{x}}_i(0)$。通过式(11-52)与式(11-64)之间相减，估计误差如下所示：

$$\begin{aligned}\tilde{\boldsymbol{e}}_i(k+1) = &\left(\boldsymbol{A} - \boldsymbol{K}_i(k)\boldsymbol{C}_i\right)\tilde{\boldsymbol{e}}_i(k) + \boldsymbol{K}_i(k)\boldsymbol{v}_i(k) - \boldsymbol{w}(k) \\ &- \varepsilon\boldsymbol{A}\sum_{j \in N_i}\left(\tilde{\boldsymbol{e}}_i(k) - \boldsymbol{T}_{ij}\tilde{\boldsymbol{e}}_j(k) - \boldsymbol{b}_{ij}(k)\right)\end{aligned} \tag{11-65}$$

其中，$\tilde{\boldsymbol{e}}_i(k) \overset{\text{def}}{=} \tilde{\boldsymbol{x}}_i(k) - \boldsymbol{x}(k)$。定义

$$\tilde{\boldsymbol{e}}(k) \overset{\text{def}}{=} \left[\tilde{\boldsymbol{e}}_1^{\mathrm{T}}(k), \cdots, \tilde{\boldsymbol{e}}_i^{\mathrm{T}}(k), \cdots, \tilde{\boldsymbol{e}}_N^{\mathrm{T}}(k)\right]^{\mathrm{T}}$$

$$\boldsymbol{v} \overset{\text{def}}{=} \left[\boldsymbol{v}_1^{\mathrm{T}}(k), \cdots, \boldsymbol{v}_i^{\mathrm{T}}(k), \cdots, \boldsymbol{v}_N^{\mathrm{T}}(k)\right]^{\mathrm{T}}$$

$$\boldsymbol{B} \overset{\text{def}}{=} \left[\boldsymbol{b}_1^{\mathrm{T}}(k), \cdots, \boldsymbol{b}_i^{\mathrm{T}}(k), \cdots, \boldsymbol{b}_N^{\mathrm{T}}(k)\right]^{\mathrm{T}}$$

其中，$\boldsymbol{b}_i(k) \overset{\text{def}}{=} \left[\boldsymbol{b}_{i1}^{\mathrm{T}}(k), \cdots, \boldsymbol{b}_{iN}^{\mathrm{T}}(k)\right]^{\mathrm{T}}$。我们进一步定义$\boldsymbol{H}$为图$G$的伴随矩

阵，并且定义$\boldsymbol{H}_i$为$\boldsymbol{H}$的第i列。通过将所有的$\tilde{\boldsymbol{e}}_i(k)$堆叠成$\tilde{\boldsymbol{e}}(k)$，我们有

$$\tilde{\boldsymbol{e}}(k+1)=\left(\boldsymbol{\Gamma}_1(k)+\varepsilon\boldsymbol{H}\otimes\boldsymbol{A}\boldsymbol{T}\right)\tilde{\boldsymbol{e}}(k)+\boldsymbol{\Psi}_1(k) \tag{11-66}$$

其中

$$\boldsymbol{\Gamma}_1(k)\overset{\text{def}}{=}\left(\boldsymbol{I}_N-\varepsilon\boldsymbol{L}\right)\otimes\boldsymbol{A}-\operatorname{diag}\left(\boldsymbol{K}_i(k)\boldsymbol{C}_i\right)-\varepsilon\boldsymbol{H}\otimes\boldsymbol{A}\left(\boldsymbol{1}_N\boldsymbol{1}_N^{\mathrm{T}}\right)\otimes\boldsymbol{I}_n,$$

$$\boldsymbol{\Psi}_1(k)\overset{\text{def}}{=}\operatorname{diag}\left(\boldsymbol{K}_i(k)\right)\boldsymbol{v}-\boldsymbol{1}_N\otimes\boldsymbol{w}(k)+\varepsilon\left(\operatorname{diag}\left(\boldsymbol{H}_i\right)\otimes\boldsymbol{A}\right)\boldsymbol{B}$$

$\boldsymbol{T}\in\mathbb{R}^{Nn\times Nn}$表示第$i$行、第$j$列的元素为$T_{ij}$的对角块矩阵。根据式(11-66)，$\tilde{\boldsymbol{e}}(k+1)$的估计误差协方差为

$$\begin{aligned}\tilde{\boldsymbol{P}}(k+1)=&\left(\boldsymbol{\Gamma}_1(k)+\varepsilon\boldsymbol{H}\otimes\boldsymbol{A}\boldsymbol{T}\right)\tilde{\boldsymbol{P}}(k)(\cdot)^{\mathrm{T}}+\left(\boldsymbol{1}_N\boldsymbol{1}_N^{\mathrm{T}}\right)\otimes\boldsymbol{Q}\\&+\operatorname{diag}\left(\boldsymbol{K}_i(k)\right)\boldsymbol{R}(\cdot)^{\mathbf{T}}+\varepsilon^2\left(\operatorname{diag}\left(\boldsymbol{H}_i\right)\otimes A\right)\boldsymbol{\Sigma}_B(\cdot)^{\mathrm{T}}\end{aligned}$$

其中，$\boldsymbol{R}$和$\boldsymbol{\Sigma}_B$分别表示$\boldsymbol{v}$和$\boldsymbol{B}$的方差。

考虑到最优隐秘攻击 [式 (11-58)] 可以对分布式滤波 [式 (11-54)] 的估计性能造成最大程度的破坏，我们在下面的引理中提出了最优隐秘攻击策略的充分条件。

引理 11.8

在防御器 [式 (11-55)] 的保护下，如果$\boldsymbol{T}^*$满足隐秘条件 (11-59) 以及以下约束条件：

$$\begin{aligned}&\left(\boldsymbol{H}\otimes\boldsymbol{A}\right)^{\mathrm{T}}\boldsymbol{\Gamma}_1(k)\tilde{\boldsymbol{P}}(k)=\\&\varepsilon\left[\sum_{i=1}^{N}\sum_{j=1}^{N}\left(\boldsymbol{\Delta}_{ij}^{\mathrm{T}}\boldsymbol{\Delta}_{ij}\right)\boldsymbol{T}^*\left(\boldsymbol{\Omega}_j\tilde{\boldsymbol{P}}(k)\boldsymbol{\Omega}_j^{\mathrm{T}}\right)-\left(\boldsymbol{H}\otimes\boldsymbol{A}\right)^{\mathrm{T}}\left(\boldsymbol{H}\otimes\boldsymbol{A}\right)\boldsymbol{T}^*\tilde{\boldsymbol{P}}(k)\right]\end{aligned}$$

其中，$\boldsymbol{\Delta}_{ij}\overset{\text{def}}{=}\left(\operatorname{diag}\left(\boldsymbol{H}_i\right)\left(\boldsymbol{\theta}_i\boldsymbol{\theta}_i^{\mathrm{T}}\right)\otimes\boldsymbol{\theta}_j\right)\otimes\boldsymbol{A},\boldsymbol{\Omega}_j\overset{\text{def}}{=}\left(\boldsymbol{\theta}_j\boldsymbol{\theta}_j^{\mathrm{T}}\right)\otimes\boldsymbol{I}_n$，则$\boldsymbol{T}^*$是式(11-58)中的最优隐秘攻击。该攻击使分布式滤波[式(11-54)]的估计误差协方差$\tilde{\boldsymbol{P}}_{T^*}(k)$最大。

证明：

基于b_{ij}之间的独立无关性以及隐秘性条件 (11-59)，我们有

$$\begin{aligned}\boldsymbol{\Sigma}_B=\operatorname{diag}\left(\boldsymbol{\Sigma}_{ij}^b\right)&\leqslant\operatorname{diag}\left(\left(\boldsymbol{P}_j^i\right)^*-\boldsymbol{T}_{ij}(k)\tilde{\boldsymbol{P}}_j(k)\boldsymbol{T}_{ij}(k)^{\mathrm{T}}\right)\\&=\boldsymbol{P}^*-\operatorname{diag}\left(\boldsymbol{T}_{ij}(k)\tilde{\boldsymbol{P}}_j(k)\boldsymbol{T}_{ij}(k)^{\mathrm{T}}\right)\end{aligned}$$

其中，$\boldsymbol{P}^* \stackrel{\text{def}}{=} \text{diag}\left(\left(\boldsymbol{P}_j^i\right)^*\right)$。由于$\boldsymbol{\Sigma}_B$的正定性，应该取其上界来最大化$\tilde{\boldsymbol{P}}(k)$。因此，我们有

$$
\begin{aligned}
&\max_{\boldsymbol{T},\boldsymbol{\Sigma}_B} \tilde{\boldsymbol{P}}(k+1) \\
&= \max_{\boldsymbol{T},\boldsymbol{\Sigma}_B} \Big[\left(\boldsymbol{\Gamma}_1(k) + \varepsilon \boldsymbol{H} \otimes \boldsymbol{A}\boldsymbol{T}\right) \tilde{\boldsymbol{P}}(k)(\cdot)^{\mathrm{T}} + \left(\boldsymbol{1}_N \boldsymbol{1}_N^{\mathrm{T}}\right) \otimes \boldsymbol{Q} \\
&\qquad + \text{diag}\left(\boldsymbol{K}_i(k)\right) \boldsymbol{R}(\cdot)^{\mathrm{T}} + \varepsilon^2 \left(\text{diag}\left(\boldsymbol{H}_i\right) \otimes \boldsymbol{A}\right) \boldsymbol{\Sigma}_B (\cdot)^{\mathrm{T}} \Big] \\
&= \max_{\boldsymbol{T}} \Big[-\varepsilon^2 \left(\text{diag}\left(\boldsymbol{H}_i\right) \otimes \boldsymbol{A}\right) \text{diag}\left(\boldsymbol{T}_{ij}(k) \tilde{\boldsymbol{P}}_j(k) \boldsymbol{T}_{ij}(k)^{\mathrm{T}}\right)(\cdot)^{\mathrm{T}} \\
&\qquad + \left(\boldsymbol{\Gamma}_1(k) + \varepsilon \boldsymbol{H} \otimes \boldsymbol{A}\boldsymbol{T}\right) \tilde{\boldsymbol{P}}(k)(\cdot)^{\mathrm{T}} \Big] \\
&\qquad + \left(\boldsymbol{1}_N \boldsymbol{1}_N^{\mathrm{T}}\right) \otimes \boldsymbol{Q} + \text{diag}\left(\boldsymbol{K}_i(k)\right) \boldsymbol{R}(\cdot)^{\mathrm{T}} \\
&\qquad + \varepsilon^2 \left(\text{diag}\left(\boldsymbol{H}_i\right) \otimes \boldsymbol{A}\right) \boldsymbol{P}^* (\cdot)^{\mathrm{T}}
\end{aligned} \tag{11-67}
$$

将下列等式代入式 (11-67) 中，

$$
\text{diag}\left(\boldsymbol{T}_{ij}(k) \tilde{\boldsymbol{P}}_j(k) \boldsymbol{T}_{ij}(k)^{\mathrm{T}}\right) = \sum_{i=1}^{N} \sum_{j=1}^{N} \Big[\left(\left(\boldsymbol{\theta}_i \boldsymbol{\theta}_i^{\mathrm{T}}\right) \otimes \boldsymbol{\theta}_j \otimes \boldsymbol{I}_n\right) \boldsymbol{T} \left(\boldsymbol{\theta}_j \boldsymbol{\theta}_j^{\mathrm{T}}\right) \otimes \boldsymbol{I}_n \Big] \tilde{\boldsymbol{P}}(k) [\cdot]^{\mathrm{T}}
$$

则$\tilde{\boldsymbol{P}}(k)$可以被视为是一个关于$\boldsymbol{T}$的函数。因此，最优攻击$\boldsymbol{T}^*$是下列方程的解：

$$
\frac{\partial \text{tr}\left(\tilde{\boldsymbol{P}}(k)\right)}{\partial \boldsymbol{T}} = 0.
$$

根据矩阵运算，下面等式成立：

$$
\frac{\partial\left(\boldsymbol{A}\boldsymbol{X}\boldsymbol{B}\right)}{\partial \boldsymbol{X}} = \boldsymbol{A}^{\mathrm{T}} \boldsymbol{B}^{\mathrm{T}}, \frac{\partial\left(\boldsymbol{A}\boldsymbol{X}\boldsymbol{B}\boldsymbol{X}^{\mathrm{T}}\boldsymbol{A}^{\mathrm{T}}\right)}{\partial \boldsymbol{X}} = \boldsymbol{A}^{\mathrm{T}} \boldsymbol{A}\boldsymbol{X}\left(\boldsymbol{B}^{\mathrm{T}} + \boldsymbol{B}\right)
$$

其中，$\boldsymbol{A},\boldsymbol{B}$以及$\boldsymbol{X}$是任意矩阵。然后，我们有

$$
\begin{aligned}
&\frac{\partial \text{tr}\left(\tilde{\boldsymbol{P}}(k+1)\right)}{\partial \boldsymbol{T}} \\
&= 2\varepsilon \left(\boldsymbol{H} \otimes \boldsymbol{A}\right)^{\mathrm{T}} \boldsymbol{\Gamma}_1(k) \tilde{\boldsymbol{P}}(k) + 2\varepsilon^2 \left(\boldsymbol{H} \otimes \boldsymbol{A}\right)^{\mathrm{T}} \left(\boldsymbol{H} \otimes \boldsymbol{A}\right) \boldsymbol{T} \tilde{\boldsymbol{P}}(k) \\
&-2\varepsilon^2 \sum_{i=1}^{N} \sum_{j=1}^{N} \left(\boldsymbol{\Delta}_{ij}^{\mathrm{T}} \boldsymbol{\Delta}_{ij}\right) \boldsymbol{T} \left(\boldsymbol{\Omega}_j \tilde{\boldsymbol{P}}(k) \boldsymbol{\Omega}_j^{\mathrm{T}}\right) = 0
\end{aligned}
$$

如果推导出的结果进一步满足隐秘性条件(11-59)，则它即是最大化估计误差协方差的最优隐秘攻击。 ■

基于引理 11.8，可以得到在隐秘攻击 [式 (11-58)] 下分布式滤波 [式 (11-54)] 的估计误差的上界，即$\tilde{\boldsymbol{P}}(k+1) \leqslant \tilde{\boldsymbol{P}}_{T^*}(k+1)$，进一步考虑估计误差协方差因攻击而发散的这种最坏的情况，即$\mathrm{tr}\left(\tilde{\boldsymbol{P}}_i(k)\right) \to \infty, i=1,\cdots,N$。在这种情况下，攻击者为了满足式 (11-59) 中的隐秘性条件，所制定的攻击策略为：

$$\boldsymbol{T}_{ij}(k) \to 0, \boldsymbol{\Sigma}_{ij}^b \leqslant \left(\boldsymbol{P}_j^i\right)^*,\ \forall (i,j) \subset E \tag{11-68}$$

当$\tilde{\boldsymbol{P}}_i(k)$发散时，攻击者将一直持续这种攻击策略来保持隐身。因此，在这种极端情况下，分布式滤波的弹性取决于估计误差协方差[式(11-65)]是否满足以下条件：

$$\begin{aligned}\tilde{\boldsymbol{P}}_i(k+1) = &\left[\left(1-\varepsilon d_i\right)\boldsymbol{A} - \boldsymbol{K}_i(k)\boldsymbol{C}_i\right]\tilde{\boldsymbol{P}}_i(k)[\cdot]^{\mathrm{T}} + \boldsymbol{Q} \\ &+ \boldsymbol{K}_i(k)\boldsymbol{R}_i(\cdot)^{\mathrm{T}} + \varepsilon^2 \boldsymbol{A}\sum_{j\in N_i}\left(\boldsymbol{P}_j^i\right)^* \boldsymbol{A}^{\mathrm{T}} \leqslant \tilde{\boldsymbol{P}}_i(k)\end{aligned}$$

注释 11.12 在防御器 [式 (11-55)] 的保护下，通过仿真进一步对比了分布式传感网络和单传感器系统在隐秘攻击下状态估计的弹性。仿真结果表明，如果合理地选择门限值δ_{ij}，在隐秘攻击下，分布式传感网络的估计性能优于单传感器系统。

11.3.3 资源受限的隐秘攻击下的分布式一致性估计

本节首先推导出分布式滤波 [式 (11-54)] 的最优卡尔曼增益，从而最小化隐秘攻击 [式 (11-60)] 下的估计误差协方差。然后，分析在攻击下分布式滤波 [式 (11-54)] 的收敛性。最后，提出一种构造隐秘攻击的方法。

首先引入如下两个合理的假设。

假设 11.1

图 G 是强连通的。

假设 11.2

$\left\{\boldsymbol{A},\boldsymbol{Q}^{1/2}\right\}$是可镇定的。

在隐秘攻击 [式 (11-60)] 下，第 i 个节点的状态估计的误差满足如下递推方程：

$$
\begin{aligned}
\tilde{\boldsymbol{e}}_i(k+1)=&\left(\boldsymbol{A}-\boldsymbol{K}_i(k)\boldsymbol{C}_i\right)\tilde{\boldsymbol{e}}_i(k)-\varepsilon\boldsymbol{A}\sum_{j\in N_i}\left(\tilde{\boldsymbol{e}}_i(k)-\tilde{\boldsymbol{e}}_j(k)\right)\\
&-\varepsilon\boldsymbol{A}\sum_{j\in N_i}\alpha_{ij}(k)\left[\tilde{\boldsymbol{e}}_j(k)-\left(\boldsymbol{T}_{ij}(k)+\boldsymbol{b}_{ij}\right)\right]\\
&+\boldsymbol{K}_i(k)\boldsymbol{v}_i(k)-\boldsymbol{w}(k)
\end{aligned}
\tag{11-69}
$$

为了最小化估计误差协方差，我们在隐秘攻击 [式 (11-60)] 下，为分布式滤波推导出了最优卡尔曼增益。

定理 11.5

在隐秘攻击 [式 (11-60)] 下，最小化式 (11-69) 的估计误差协方差 $\tilde{\boldsymbol{P}}_i(k)$的最优卡尔曼增益为

$$
\begin{aligned}
\boldsymbol{K}_i^*(k)=&\boldsymbol{A}[\tilde{\boldsymbol{P}}_i(k)+\varepsilon\sum_{j\in N_i}\left(\boldsymbol{M}_{ij}\tilde{\boldsymbol{P}}_{ij}(k)\right.\\
&\left.-\tilde{\boldsymbol{P}}_i(k)\right)]\boldsymbol{C}_i^{\mathrm{T}}\left(\boldsymbol{C}_i\tilde{\boldsymbol{P}}_i(k)\boldsymbol{C}_i^{\mathrm{T}}+\boldsymbol{R}_i\right)^{-1}
\end{aligned}
\tag{11-70}
$$

其中，$\boldsymbol{M}_{ij}\overset{\text{def}}{=}1-\alpha_{ij}(k)\left(\boldsymbol{I}_n-\boldsymbol{T}_{ij}(k)\right)$。

证明：

通过式 (11-69)，我们可以将其估计误差协方差写为

$$
\begin{aligned}
&\tilde{\boldsymbol{P}}_i(k+1)\\
&=\mathbb{E}\left[\tilde{\boldsymbol{e}}_i(k+1)\tilde{\boldsymbol{e}}_i(k+1)^{\mathrm{T}}\right]\\
&=\left(\boldsymbol{A}-\boldsymbol{K}_i(k)\boldsymbol{C}_i\right)\tilde{\boldsymbol{P}}_i(k)(\cdot)^{\mathrm{T}}+\boldsymbol{K}_i(k)\boldsymbol{R}_i\boldsymbol{K}_i^{\mathrm{T}}(k)\\
&+\varepsilon\left(\boldsymbol{A}-\boldsymbol{K}_i(k)\boldsymbol{C}_i\right)\sum_{j\in N_i}\left(\boldsymbol{M}_{ij}\tilde{\boldsymbol{P}}_{ij}(k)-\tilde{\boldsymbol{P}}_i(k)\right)\boldsymbol{A}^{\mathrm{T}}\\
&+\varepsilon\boldsymbol{A}\sum_{j\in N_i}\left(\boldsymbol{M}_{ij}\tilde{\boldsymbol{P}}_{ij}(k)-\tilde{\boldsymbol{P}}_i(k)\right)^{\mathrm{T}}\left(\boldsymbol{A}-\boldsymbol{K}_i(k)\boldsymbol{C}_i\right)^{\mathrm{T}}\\
&+\varepsilon^2\boldsymbol{A}\sum_{r\in N_i}\sum_{s\in N_i}\left(\tilde{\boldsymbol{P}}_i(k)+\boldsymbol{M}_{is}\tilde{\boldsymbol{P}}_{rs}(k)\boldsymbol{M}_{ir}^{\mathrm{T}}-\boldsymbol{M}_{is}\tilde{\boldsymbol{P}}_{si}(k)\right.\\
&\left.-\tilde{\boldsymbol{P}}_{ir}(k)\boldsymbol{M}_{ir}^{\mathrm{T}}\right)\boldsymbol{A}^{\mathrm{T}}+\boldsymbol{Q}+\varepsilon^2\boldsymbol{A}\sum_{j\in N_i}\left(\alpha_{ij}^2(k)\boldsymbol{\Sigma}_{ij}^b\right)\boldsymbol{A}^{\mathrm{T}}
\end{aligned}
$$

类似于引理 11.8 的证明，基于以下矩阵运算的性质，

$$\frac{\partial(\boldsymbol{AX})}{\partial \boldsymbol{X}}=\boldsymbol{A}^{\mathrm{T}}, \frac{\partial\left(\boldsymbol{X}^{\mathrm{T}} \boldsymbol{AX}\right)}{\partial \boldsymbol{X}}=\left(\boldsymbol{A}+\boldsymbol{A}^{\mathrm{T}}\right) \boldsymbol{X}$$

我们有

$$\begin{aligned} \frac{\partial \operatorname{tr}\left(\tilde{\boldsymbol{P}}_i(k+1)\right)}{\partial \boldsymbol{K}_i(k)}= & 2 \boldsymbol{K}_i(k)\left(\boldsymbol{C}_i \tilde{\boldsymbol{P}}_i(k) \boldsymbol{C}_i^{\mathrm{T}}+\boldsymbol{R}_i\right)-2 \boldsymbol{A} \tilde{\boldsymbol{P}}_i(k) \boldsymbol{C}_i^{\mathrm{T}} \\ & -2 \varepsilon \boldsymbol{A} \sum_{j \in N_i}\left(\boldsymbol{M}_{ij} \tilde{\boldsymbol{P}}_{ij}(k)-\tilde{\boldsymbol{P}}_i(k)\right) \boldsymbol{C}_i^{\mathrm{T}}=0 \end{aligned}$$

从中可以导出最优卡尔曼增益[式(11-70)]。■

注释 11.13 最优卡尔曼增益 [式 (11-70)] 取决于隐秘攻击策略 [式 (11-60)]，而该攻击策略又受到了防御器 [式 (11-55)] 的约束。另外，在不存在攻击的时候，即$\boldsymbol{M}_{ij}=\boldsymbol{I}$，式 (11-70) 可以化简为文献 [303] 所给出的形式。因此，文献 [303] 中的最优卡尔曼增益是式 (11-70) 的特例。

注意到式(11-70)中的最优卡尔曼增益$\boldsymbol{K}_i^*(k)$取决于攻击策略。然而，这是估计器不可获得的信息，因为隐秘攻击者可以欺诈式 (11-55) 中的防御器。即便如此，当估计器拥有$\boldsymbol{M}_{ij}$相关的先验知识时，可以通过调节其卡尔曼增益来提升其在隐秘攻击下的弹性以及鲁棒性。

推论 11.1

假设分布式滤波对于隐秘攻击 [式 (11-60)] 的先验知识为 $\mathcal{Y}$。最小化估计误差协方差$\mathbb{E}\left[\tilde{\boldsymbol{P}}_i(k)\right]$的最优卡尔曼增益为

$$\overline{\boldsymbol{K}}_i^*(k)=\boldsymbol{A}\left[\tilde{\boldsymbol{P}}_i(k)+\varepsilon \sum_{j \in N_i}\left(\overline{\boldsymbol{M}}_{ij} \tilde{\boldsymbol{P}}_{ij}(k)-\tilde{\boldsymbol{P}}_i(k)\right)\right] \boldsymbol{C}_i^{\mathrm{T}}\left(\boldsymbol{C}_i \tilde{\boldsymbol{P}}_i(k) \boldsymbol{C}_i^{\mathrm{T}}+\boldsymbol{R}_i\right)^{-1} \tag{11-71}$$

其中，$\overline{\boldsymbol{M}}_{ij} \stackrel{\text{def}}{=} 1-\bar{\rho}_{ij}\left(\boldsymbol{I}_n-\overline{\boldsymbol{T}}_{ij}(k)\right), \bar{\rho}_{ij} \stackrel{\text{def}}{=} \mathbb{E}\left[\rho_{ij} \mid \mathcal{Y}\right], \overline{\boldsymbol{T}}_{ij}(k) \stackrel{\text{def}}{=} \mathbb{E}\left[\boldsymbol{T}_{ij} \mid \mathcal{Y}\right]$。

证明：

该证明和定理 11.5 中的类似，因此省略。■

注释 11.14 如果$\bar{\rho}_{ij} \rightarrow \rho_{ij}, \overline{\boldsymbol{T}}_{ij}(k) \rightarrow \boldsymbol{T}_{ij}(k)$，则估计器可以最大程度地提高其估计性能。因此，一个有趣的研究课题是如何获得 $\mathcal{Y}$ 的准确的先验知识。注意到，隐秘攻击需要满足式 (11-61)

以及式 (11-62) 的隐秘性条件，并且我们在本节中进一步导出了攻击者要保证隐身的充分条件。相应地，可以从中抽取出一些先验知识来设计滤波器的增益。

接下来，分析估计器在隐秘攻击 [式 (11-60)] 下的收敛性。如图 11-9 所示，由于随机的攻击行为$\alpha_{ij}(k)$，我们进一步定义一个矩阵$\overline{\boldsymbol{L}}^a=\{\alpha_{ij}\}$来描述第 k 个时刻整个传感网络的攻击决策。定义$\overline{\boldsymbol{L}}_i^a$为$\overline{\boldsymbol{L}}^a$的第 i 行，则隐秘攻击 [式 (11-60)] 下的估计误差$\tilde{\boldsymbol{e}}(k)$为

$$\begin{aligned}\tilde{\boldsymbol{e}}(k+1)=&\left\{\boldsymbol{\Gamma}_2(k)-\varepsilon\left(\overline{\boldsymbol{L}}^a\otimes\boldsymbol{A}\right)\left[\left(\boldsymbol{1}_N\boldsymbol{1}_N^{\mathrm{T}}\right)\otimes\boldsymbol{I}_n-\boldsymbol{T}\right]\right\}\tilde{\boldsymbol{e}}(k)\\&+\boldsymbol{\Psi}_2(k)+\varepsilon\left(\operatorname{diag}\left(\overline{\boldsymbol{L}}_i^a\right)\otimes\boldsymbol{A}\right)\boldsymbol{B}\end{aligned}\tag{11-72}$$

其中，

$$\boldsymbol{\Gamma}_2(k)\overset{\text{def}}{=}(\boldsymbol{I}_N-\varepsilon\boldsymbol{L})\otimes\boldsymbol{A}-\operatorname{diag}\left(\boldsymbol{K}_i(k)\boldsymbol{C}_i\right)$$

$$\boldsymbol{\Psi}_2(k)\overset{\text{def}}{=}\operatorname{diag}\left(\boldsymbol{K}_i(k)\right)\boldsymbol{v}-\boldsymbol{1}_N\otimes\boldsymbol{w}(k)$$

由于式 (11-72) 中估计误差$\tilde{\boldsymbol{e}}(k)$包含随机矩阵$\overline{\boldsymbol{L}}^a$，我们很难直接分析其收敛性。因此，首先推导除了估计误差协方差的一个上界。注意到$\mathbb{E}\left[\overline{\boldsymbol{L}}^a\right]\overset{\text{def}}{=}\boldsymbol{\Lambda}$并且$\Lambda_{ij}=\rho_{ij}$。因此，若定义$\alpha_{ij}$的方差为$\sigma_{ij}^2=\rho_{ij}(1-\rho_{ij})$，则有$\mathbb{E}\left[\alpha_{ij}^2\right]=\rho_{ij}^2+\sigma_{ij}^2$。另外，定义$\mathbb{E}[\boldsymbol{\sigma}_i]\overset{\text{def}}{=}\overline{\boldsymbol{\Lambda}}_i$，其中$\boldsymbol{\sigma}_i\overset{\text{def}}{=}[\sigma_{i1},\cdots,\sigma_{iN}]$，以及定义$\mathbb{E}\left[\overline{\boldsymbol{L}}_i^a\right]\overset{\text{def}}{=}\boldsymbol{\Lambda}_i$。

将一个修正的代数黎卡提方程定义如下：

$$\begin{aligned}&\tilde{\boldsymbol{P}}^t(k+1)\\&=\left[\boldsymbol{\Gamma}_2(k)-\varepsilon(\boldsymbol{\Lambda}\otimes\boldsymbol{A})\boldsymbol{\Phi}\right]\tilde{\boldsymbol{P}}^t(k)\left[\boldsymbol{\Gamma}_2(k)-\varepsilon(\boldsymbol{\Lambda}\otimes\boldsymbol{A})\boldsymbol{\Phi}\right]^{\mathrm{T}}\\&\quad+\varepsilon^2\sum_{i=1}^N\sum_{j=1}^N\left[\sigma_{ij}\left(\boldsymbol{\theta}_i\boldsymbol{\theta}_j^{\mathrm{T}}\right)\otimes\boldsymbol{A}\boldsymbol{\Phi}\right]\tilde{\boldsymbol{P}}^t(k)[\cdot]^{\mathrm{T}}\\&\quad+\varepsilon^2\left(\operatorname{diag}(\boldsymbol{\Lambda}_i)\otimes\boldsymbol{A}\right)\boldsymbol{P}^*(\cdot)^{\mathrm{T}}+\varepsilon^2\left(\operatorname{diag}\left(\overline{\boldsymbol{\Lambda}}_i\right)\otimes\boldsymbol{A}\right)\boldsymbol{P}^*(\cdot)^{\mathrm{T}}\\&\quad+\left(\boldsymbol{1}_N\boldsymbol{1}_N^{\mathrm{T}}\right)\otimes\boldsymbol{Q}+\operatorname{diag}\left(\overline{\boldsymbol{K}}_i(k)\right)\boldsymbol{R}(\cdot)^{\mathrm{T}}\end{aligned}\tag{11-73}$$

其中，$\boldsymbol{\Phi}\overset{\text{def}}{=}\left(\boldsymbol{1}_N\boldsymbol{1}_N^{\mathrm{T}}\right)\otimes\boldsymbol{I}_n-\boldsymbol{T},\boldsymbol{P}^*\overset{\text{def}}{=}\operatorname{diag}\left(\left(\boldsymbol{P}_j^i\right)^*\right)$。类似于定理11.5的证明方法，我们可以得到当卡尔曼增益选择为式(11-71)中的最优值时，$\tilde{\boldsymbol{P}}^t(k)$是最小的。

引理 11.9

考虑如下修正的代数黎卡提方程：

$$
\begin{aligned}
\boldsymbol{\Xi}(k+1) \overset{\text{def}}{=} & \left[\boldsymbol{\Gamma}_2(k)-\varepsilon\left(\overline{\boldsymbol{L}}^a \otimes \boldsymbol{A}\right)\boldsymbol{\Phi}\right]\boldsymbol{\Xi}(k)[\cdot]^{\mathrm{T}} \\
& +\left(\boldsymbol{1}_N \boldsymbol{1}_N^{\mathrm{T}}\right)\otimes \boldsymbol{Q}+\operatorname{diag}\left(\boldsymbol{K}_i(k)\right)\boldsymbol{R}(\cdot)^{\mathrm{T}} \\
& +\varepsilon^2\left(\operatorname{diag}\left(\overline{\boldsymbol{L}}_i^a\right)\otimes \boldsymbol{A}\right)\boldsymbol{\Sigma}_B(\cdot)^{\mathrm{T}}
\end{aligned}
$$

其中，$\boldsymbol{K}_i(k)=\boldsymbol{K}_i^*(k)$。当$\boldsymbol{K}_i(k)=\overline{\boldsymbol{K}}_i^*(k)$时，对于任意的$\boldsymbol{\Xi}(0)=\tilde{\boldsymbol{P}}^t(0)\geqslant 0$，下列不等式成立：

$$
\mathbb{E}\left[\boldsymbol{\Xi}(k+1)\right]\leqslant \tilde{\boldsymbol{P}}^t(k+1), \forall k
$$

证明：

因为当$\boldsymbol{K}_i(k)=\boldsymbol{K}_i^*(k)$时，$\boldsymbol{\Xi}(k)$是最小的。假设$\mathbb{E}\left[\boldsymbol{\Xi}(k)\right]\leqslant \tilde{\boldsymbol{P}}^t(k)$，且初始条件为$\boldsymbol{\Xi}(0)=\tilde{\boldsymbol{P}}^t(0)$，我们有

$$
\begin{aligned}
& \mathbb{E}\left[\boldsymbol{\Xi}(k+1)\right] \\
& =\mathbb{E}\left\{\left[\boldsymbol{\Gamma}_2(k)-\varepsilon\left(\overline{\boldsymbol{L}}^a \otimes \boldsymbol{A}\right)\boldsymbol{\Phi}\right]\boldsymbol{\Xi}(k)(\cdot)^{\mathrm{T}}+\left(\boldsymbol{1}_N \boldsymbol{1}_N^{\mathrm{T}}\right)\otimes \boldsymbol{Q}\right. \\
& \quad \left.+\operatorname{diag}\left(\boldsymbol{K}_i^*(k)\right)\boldsymbol{R}(\cdot)^{\mathrm{T}}+\varepsilon^2\left(\operatorname{diag}\left(\overline{\boldsymbol{L}}_i^a\right)\otimes \boldsymbol{A}\right)\boldsymbol{\Sigma}_B(\cdot)^{\mathrm{T}}\right\} \\
& \leqslant \mathbb{E}\left\{\left[\boldsymbol{\Gamma}_2(k)-\varepsilon\left(\overline{\boldsymbol{L}}^a \otimes \boldsymbol{A}\right)\boldsymbol{\Phi}\right]\tilde{\boldsymbol{P}}^t(k)(\cdot)^{\mathrm{T}}+\left(\boldsymbol{1}_N \boldsymbol{1}_N^{\mathrm{T}}\right)\otimes \boldsymbol{Q}\right. \\
& \quad \left.+\operatorname{diag}\left(\boldsymbol{K}_i^*(k)\right)\boldsymbol{R}(\cdot)^{\mathrm{T}}+\varepsilon^2\left(\operatorname{diag}\left(\overline{\boldsymbol{L}}_i^a\right)\otimes \boldsymbol{A}\right)\boldsymbol{\Sigma}_B(\cdot)^{\mathrm{T}}\right\} \\
& \leqslant \mathbb{E}\left\{\left[\boldsymbol{\Gamma}_2(k)-\varepsilon\left(\overline{\boldsymbol{L}}^a \otimes \boldsymbol{A}\right)\boldsymbol{\Phi}\right]\tilde{\boldsymbol{P}}^t(k)(\cdot)^{\mathrm{T}}+\left(\boldsymbol{1}_N \boldsymbol{1}_N^{\mathrm{T}}\right)\otimes \boldsymbol{Q}\right. \\
& \quad +\operatorname{diag}\left(\boldsymbol{K}_i^*(k)\right)\boldsymbol{R}(\cdot)^{\mathrm{T}}+\varepsilon^2\left(\operatorname{diag}\left(\overline{\boldsymbol{L}}_i^a\right)\otimes \boldsymbol{A}\right)\boldsymbol{P}^*(\cdot)^{\mathrm{T}} \\
& \quad \left.-\varepsilon^2\left(\operatorname{diag}\left(\overline{\boldsymbol{L}}_i^a\right)\otimes \boldsymbol{A}\right)\operatorname{diag}\left(\boldsymbol{T}_{ij}(k)\tilde{\boldsymbol{P}}_j^t(k)\boldsymbol{T}_{ij}(k)^{\mathrm{T}}\right)(\cdot)^{\mathrm{T}}\right\} \\
& \leqslant \mathbb{E}\left\{\left[\boldsymbol{\Gamma}_2(k)-\varepsilon\left(\overline{\boldsymbol{L}}^a \otimes \boldsymbol{A}\right)\boldsymbol{\Phi}\right]\tilde{\boldsymbol{P}}^t(k)(\cdot)^{\mathrm{T}}+\left(\boldsymbol{1}_N \boldsymbol{1}_N^{\mathrm{T}}\right)\otimes \boldsymbol{Q}\right. \\
& \quad \left.+\operatorname{diag}\left(\boldsymbol{K}_i^*(k)\right)\boldsymbol{R}(\cdot)^{\mathrm{T}}+\varepsilon^2\left(\operatorname{diag}\left(\overline{\boldsymbol{L}}_i^a\right)\otimes \boldsymbol{A}\right)\boldsymbol{P}^*(\cdot)^{\mathrm{T}}\right\} \\
& =\left[\boldsymbol{\Gamma}_2(k)-\varepsilon(\boldsymbol{\Lambda}\otimes \boldsymbol{A})\boldsymbol{\Phi}\right]\tilde{\boldsymbol{P}}^t(k)\left[\boldsymbol{\Gamma}_2(k)-\varepsilon(\boldsymbol{\Lambda}\otimes \boldsymbol{A})\boldsymbol{\Phi}\right]^{\mathrm{T}}
\end{aligned}
$$

$$
\begin{aligned}
&+\varepsilon^{2}\sum_{i=1}^{N}\sum_{j=1}^{N}\left[\sigma_{ij}\left(\boldsymbol{\theta}_{i}\boldsymbol{\theta}_{j}^{\mathrm{T}}\right)\otimes \boldsymbol{A}\boldsymbol{\Phi}\right]\tilde{\boldsymbol{P}}^{t}\left(k\right)[\cdot]^{\mathrm{T}}\\
&+\varepsilon^{2}\left(\operatorname{diag}\left(\boldsymbol{\Lambda}_{i}\right)\otimes \boldsymbol{A}\right)\boldsymbol{P}^{*}\left(\cdot\right)^{\mathrm{T}}+\varepsilon^{2}\left(\operatorname{diag}\left(\overline{\boldsymbol{\Lambda}}_{i}\right)\otimes \boldsymbol{A}\right)\boldsymbol{P}^{*}\left(\cdot\right)^{\mathrm{T}}\\
&+\left(\boldsymbol{I}_{N}\boldsymbol{I}_{N}^{\mathrm{T}}\right)\otimes \boldsymbol{Q}+\operatorname{diag}\left(\overline{\boldsymbol{K}}_{i}\left(k\right)\right)\boldsymbol{R}\left(\cdot\right)^{\mathrm{T}}\\
=&\tilde{\boldsymbol{P}}^{t}\left(k+1\right)
\end{aligned}
$$

证毕。 ■

引理 11.9 给出了估计误差协方差$\tilde{\boldsymbol{P}}(k)$的一个上界，即$\tilde{\boldsymbol{P}}^{t}(k)$。因此，若上界$\tilde{\boldsymbol{P}}^{t}(k)$随着时间的递推而收敛，则$\tilde{\boldsymbol{P}}(k)$同时也一定收敛。我们需要如下假定条件来分析$\tilde{\boldsymbol{P}}^{t}(k)$的收敛性。

假设 11.3

对于任意的$\boldsymbol{\chi}(0)$，假设存在一个定常矩阵$\boldsymbol{K}_{i}^{c}$，使得下列离散时间随机系统均方稳定：

$$
\boldsymbol{\chi}\left(k+1\right)=\left[\boldsymbol{\Gamma}_{2}\left(k\right)-\varepsilon\left(\boldsymbol{\Lambda}\otimes \boldsymbol{A}\right)\boldsymbol{\Phi}+\varepsilon\sum_{i=1}^{N}\sum_{j=1}^{N}\sigma_{ij}\left(\boldsymbol{\theta}_{i}\boldsymbol{\theta}_{j}^{\mathrm{T}}\right)\otimes \boldsymbol{A}\boldsymbol{\Phi}\boldsymbol{\tau}_{ij}\left(k\right)\right]\boldsymbol{\chi}\left(k\right) \tag{11-74}
$$

其中，$\boldsymbol{\chi}(0)$与过程状态$\boldsymbol{\tau}_{ij}(k)$独立无关，$\boldsymbol{\tau}_{ij}(k)$是一个独立同分布的随机变量，其均值为$\mathbb{E}\left[\boldsymbol{\tau}_{ij}(k)\right]=0$，其方差为$\mathbb{E}\left[\boldsymbol{\tau}_{ij}(k)\boldsymbol{\tau}_{ij}(k)^{\mathrm{T}}\right]=1$。

定理 11.6

在假设 11.1 ～假设 11.3 满足时，对于任意的非负定对称矩阵$\tilde{\boldsymbol{P}}^{t}(0)$，如果$\left\{\left(\boldsymbol{I}_{N}-\varepsilon\boldsymbol{L}\right)\otimes \boldsymbol{A},\operatorname{diag}\left(\boldsymbol{C}_{i}\right)\right\}$是可检测的，则在隐秘攻击 [式 (11-60)] 下，估计误差协方差$\tilde{\boldsymbol{P}}(k)$在任意时刻都是有界的，且收敛到一个确定的$\overline{\boldsymbol{P}}>0$。

证明：

该证明和文献 [304] 中定理 2 中的证明类似，因此省略。证毕。 ■

注释 11.15 即使对于不存在攻击的分布式滤波 [式 (11-54)]，文献 [304] 也指出系统需要满足几个类似的假设来保证收敛性。唯一的区别如下：文献 [304] 中，假设系统$\boldsymbol{\chi}(k+1)=\boldsymbol{\Gamma}_{2}(k)\boldsymbol{\chi}(k)$是均方稳定的，而在本节的假设 11.3 中，要求式 (11-74) 中的离散时间随机系统是稳定的。注意到，当不存在攻击时，式(11-74)中的第三项将等于零矩阵。因此，式(11-74)是更一般的表述形式。

考虑到隐秘攻击需要满足隐秘性条件(11-61)和(11-62)，下面提供一个方法来设计其攻击策略。另外，通过分析隐秘攻击的设计方法，估计器也可以从中得到一些先验知识来设计其卡尔曼增益[式(11-71)]。接下来，我们给出了攻击者满足隐秘条件(11-57)的充分条件。

定理 11.7

对于给定的初始条件$\hat{\boldsymbol{P}}(0)>0$，如果式(11-60)中的攻击策略，即$\boldsymbol{\Lambda}$、$\boldsymbol{T}$满足下面的条件：

$$\boldsymbol{P}^* \geqslant \boldsymbol{I}_N \otimes \left\{\operatorname{diag}\left[\boldsymbol{\theta}_{1+jN}^{\mathrm{T}}\left(\boldsymbol{X}^k \operatorname{vec}\left\{\hat{\boldsymbol{P}}(0)\right\}+\sum_{i=0}^{k}\boldsymbol{X}^i\boldsymbol{Y}\right)\right]\right\}, \forall k$$

其中，$j=0,\cdots,(N-1)$，以及

$$\boldsymbol{X}=\left[\boldsymbol{\Gamma}_2(k)-\varepsilon(\boldsymbol{\Lambda}\otimes\boldsymbol{A})\boldsymbol{\Phi}\right]\otimes\left[\boldsymbol{\Gamma}_2(k)-\varepsilon(\boldsymbol{\Lambda}\otimes\boldsymbol{A})\boldsymbol{\Phi}\right]$$
$$+\varepsilon^2\sum_{i=1}^{N}\sum_{j=1}^{N}\left[\sigma_{ij}\left(\boldsymbol{\theta}_i\boldsymbol{\theta}_j^{\mathrm{T}}\right)\otimes\boldsymbol{A\Phi}\right]\otimes\left[\sigma_{ij}\left(\boldsymbol{\theta}_i\boldsymbol{\theta}_j^{\mathrm{T}}\right)\otimes\boldsymbol{A\Phi}\right]$$

$$\boldsymbol{Y}=\operatorname{vec}\left\{\varepsilon^2\left(\operatorname{diag}(\boldsymbol{\Lambda}_i)\otimes\boldsymbol{A}\right)\boldsymbol{P}^*(\cdot)^{\mathrm{T}}+\left(\boldsymbol{1}_N\boldsymbol{1}_N^{\mathrm{T}}\right)\otimes\boldsymbol{Q}\right.$$
$$\left.+\varepsilon^2\left(\operatorname{diag}(\overline{\boldsymbol{\Lambda}}_i)\otimes\boldsymbol{A}\right)\boldsymbol{P}^*(\cdot)^{\mathrm{T}}+\operatorname{diag}\left(\overline{\boldsymbol{K}}_i(k)\right)\boldsymbol{R}(\cdot)^{\mathrm{T}}\right\}$$

则攻击可以满足隐秘条件(11-61)和(11-62)来欺诈防御器[式(11-55)]。

证明：

通过将矩阵矢量化，我们可以将式(11-73)转换成列向量的形式：

$$\operatorname{vec}\left\{\tilde{\boldsymbol{P}}^t(k+1)\right\}$$
$$=\left[\boldsymbol{\Gamma}_2(k)-\varepsilon(\boldsymbol{\Lambda}\otimes\boldsymbol{A})\boldsymbol{\Phi}\right]\otimes\left[\boldsymbol{\Gamma}_2(k)-\varepsilon(\boldsymbol{\Lambda}\otimes\boldsymbol{A})\boldsymbol{\Phi}\right]\operatorname{vec}\left\{\tilde{\boldsymbol{P}}^t(k)\right\}$$
$$+\varepsilon^2\sum_{i=1}^{N}\sum_{j=1}^{N}\left[\sigma_{ij}\left(\boldsymbol{\theta}_i\boldsymbol{\theta}_j^{\mathrm{T}}\right)\otimes\boldsymbol{A\Phi}\right]\otimes\left[\sigma_{ij}\left(\boldsymbol{\theta}_i\boldsymbol{\theta}_j^{\mathrm{T}}\right)\otimes\boldsymbol{A\Phi}\right]$$
$$\operatorname{vec}\left\{\tilde{\boldsymbol{P}}^t(k)\right\}+\operatorname{vec}\left\{\varepsilon^2\left(\operatorname{diag}(\boldsymbol{\Lambda}_i)\otimes\boldsymbol{A}\right)\boldsymbol{P}^*(\cdot)^{\mathrm{T}}+\left(\boldsymbol{1}_N\boldsymbol{1}_N^{\mathrm{T}}\right)\otimes\boldsymbol{Q}\right.$$
$$\left.+\varepsilon^2\left(\operatorname{diag}(\overline{\boldsymbol{\Lambda}}_i)\otimes\boldsymbol{A}\right)\boldsymbol{P}^*(\cdot)^{\mathrm{T}}+\operatorname{diag}\left(\overline{\boldsymbol{K}}_i(k)\right)\boldsymbol{R}(\cdot)^{\mathrm{T}}\right\}$$
$$=\boldsymbol{X}\operatorname{vec}\left\{\tilde{\boldsymbol{P}}^t(k)\right\}+\boldsymbol{Y}$$

然后，因为$\tilde{\boldsymbol{P}}^t(0)=\tilde{\boldsymbol{P}}(0)=\hat{\boldsymbol{P}}(0)$，我们可以得到

$$\operatorname{vec}\left\{\tilde{\boldsymbol{P}}^t(k+1)\right\}=\boldsymbol{X}^k\operatorname{vec}\left\{\hat{\boldsymbol{P}}(0)\right\}+\sum_{i=0}^{k}\boldsymbol{X}^i\boldsymbol{Y}$$

如果攻击者想要欺诈第 i 个节点的防御器，则必须满足条件$\left(\boldsymbol{P}_j^i\right)^*\geqslant\tilde{\boldsymbol{P}}_j^t(k+1)$，其中$\tilde{\boldsymbol{P}}_j^t(k+1)$表示$\operatorname{vec}\left\{\tilde{\boldsymbol{P}}^t(k+1)\right\}$的第$\operatorname{vec}\left\{\tilde{\boldsymbol{P}}^t(k+1)\right\}$个元素。因此，为了欺诈所有节点的防御器，我们可以由此导出定理 11.7。证毕。 ■

基于定理 11.7，攻击者可以利用动态规划等数学工具来实现隐秘攻击策略的设计。

11.3.4 仿真示例 2

本节我们给出了一些数值仿真来验证理论分析的结果。我们聚焦于一个由 6 个节点构成的无线传感网络，其拓扑结构如图 11-8 所示。系统的各个参数如下所示：

$$\boldsymbol{A}=\begin{pmatrix}0.8&0&0&0.4&0\\0&-0.6&0.5&0&0\\0&0&-0.3&0&1\\0&0&0&-0.7&0.1\\0&0&0&0&1.01\end{pmatrix}$$

$$\boldsymbol{Q}=\begin{pmatrix}2&0&0&0&0\\0&1&0&0&0\\0&0&1&0&0\\0&0&0&2&0\\0&0&0&0&3\end{pmatrix},\quad \boldsymbol{R}_i=\begin{pmatrix}2\xi_i&0&0\\0&\xi_i&0\\0&0&3\xi_i\end{pmatrix}$$

$$\boldsymbol{C}_i=\begin{pmatrix}\eta_i&-\eta_i&2\eta_i&0&0\\-\eta_i&2\eta_i&0&3\eta_i&0\\2\eta_i&\eta_i&0&0&4\eta_i\end{pmatrix}$$

其中，对于任意的i、η_i和ξ_i在 (0,1] 的范围内取值，$\varepsilon=0.1$。仿真实例是通过 10000 次蒙特卡洛模拟得到的。

我们首先验证了防御器 [式 (11-55)] 对文献 [26] 中的错误数据注入

攻击的检测效果。该攻击在不考虑隐秘条件 (11-57) 的情况下，篡改了无线通信数据。具体地说，分布式滤波在时间段 [0,50] 内并未受到攻击。因此，每个节点的估计误差协方差能够进入稳态。而攻击者在时间段 [51,100] 内连续地向通信边 (3,4) 注入高斯随机噪声。将真实状态值$\boldsymbol{x}(k)$以及第 i 个节点的状态估计值$\tilde{\boldsymbol{x}}_i(k)$的第 s 个元素分别定义为$\boldsymbol{x}_s(k)$以及$\tilde{\boldsymbol{x}}_{s,i}(k)$，则$\tilde{\boldsymbol{x}}_{s,i}(k)$的估计误差协方差$\tilde{\boldsymbol{P}}_{s,i}(k)\overset{\text{def}}{=}\mathbb{E}\left[\left(\tilde{\boldsymbol{x}}_{s,i}(k)-\boldsymbol{x}_s(k)\right)\left(\tilde{\boldsymbol{x}}_{s,i}(k)-\boldsymbol{x}_s(k)\right)^{\mathrm{T}}\right]$的演变过程如图 11-10 所示。由于攻击者所篡改的是通信边 (3,4) 的传输数据，因此直接融合虚假数据进行状态估计的是第 3 个节点。这将导致攻击对第 3 个节点的估计性能破坏是最大的。随着第 3 个节点的信息向邻居传播，第 4

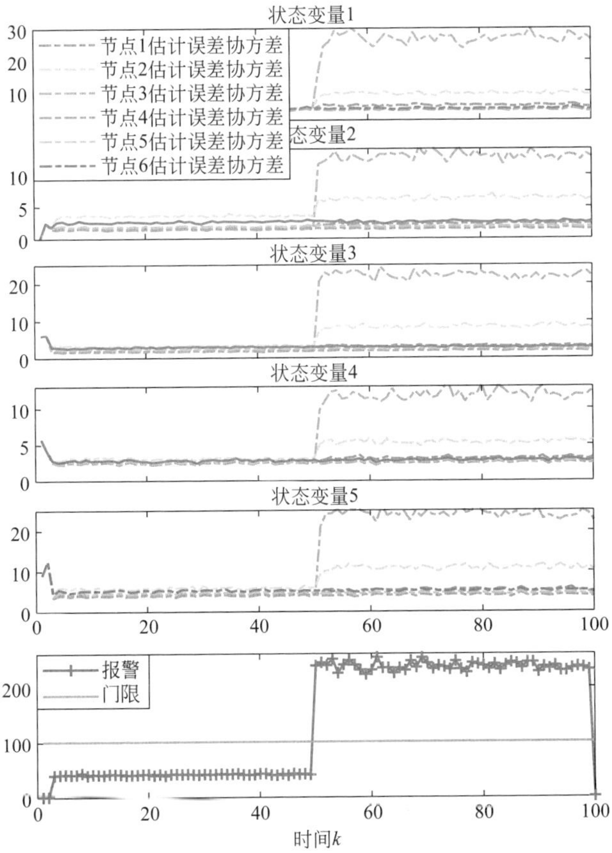

图 11-10　在错误数据注入攻击下，分布式滤波的估计性能，以及防御器抵御攻击的效果

个节点的估计误差协方差也因此而上升。然而，由于传感网络的弹性，攻击几乎对其余别的节点的估计性能没有影响。另外，我们以第 3 个节点的防御器为例，并将其门限值δ_{ij}设置为 100。在时间段 [0,50] 内，防御器的报警值$\zeta_{ij}\left(k\right)$低于该门限。这意味着防御器在该门限下的误报率相对较低。然而，在时间段 [51,100] 内，防御器可以有效地诊断出攻击所引起的异常。

当拥有足够的通信资源来破坏传感网络时，攻击者使用式 (11-68) 中的隐秘攻击策略来篡改所有无线信道的通信数据。当满足隐秘性条件 (11-57) 时，攻击者可以精密地设计其攻击策略以保证其隐秘性。如图 11-11 所示，在上述攻击下，第 3 个节点的防御器的报警值接近但是

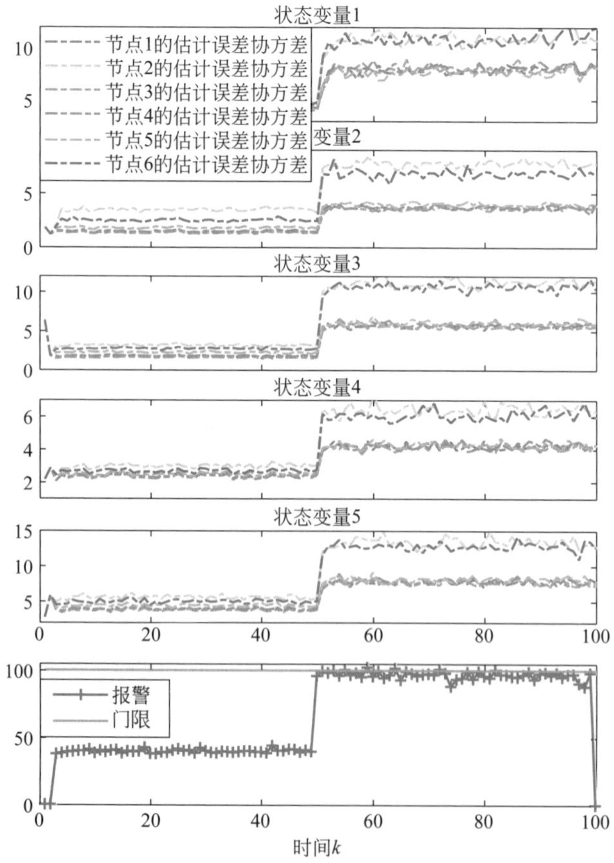

图 11-11 在资源充足的隐秘攻击下，分布式滤波的估计性能，以及防御器抵御攻击的效果

并未超过门限值。即便如此，由于传感网络的弹性，防御器仍然能有效地限制这种无法检测的攻击对估计性能的影响。在图 11-11 中，$\tilde{\boldsymbol{x}}_{s,i}(k)$的估计误差协方差仍然在一定有界的范围内波动。这意味着$\tilde{\boldsymbol{x}}_{s,i}(k)$仍然能有效地追踪真实状态$\boldsymbol{x}_s(k)$。另外，图 11-11 中的估计性能下降小于图 11-10。这意味着隐秘攻击是以减少其攻击效果来保证其隐秘性的。

对于通信资源受限的攻击，我们假设其采取定理 11.7 所设计的攻击策略来生成虚假的数据，并且以概率$\rho_{ij}=80\%$攻击每条通信链路。如图 11-12 所示，第 3 个节点的防御器的报警值仍然无法超过门限。和图 11-11 所不同的是，攻击下防御器的报警值将偏离预设的门限值。这

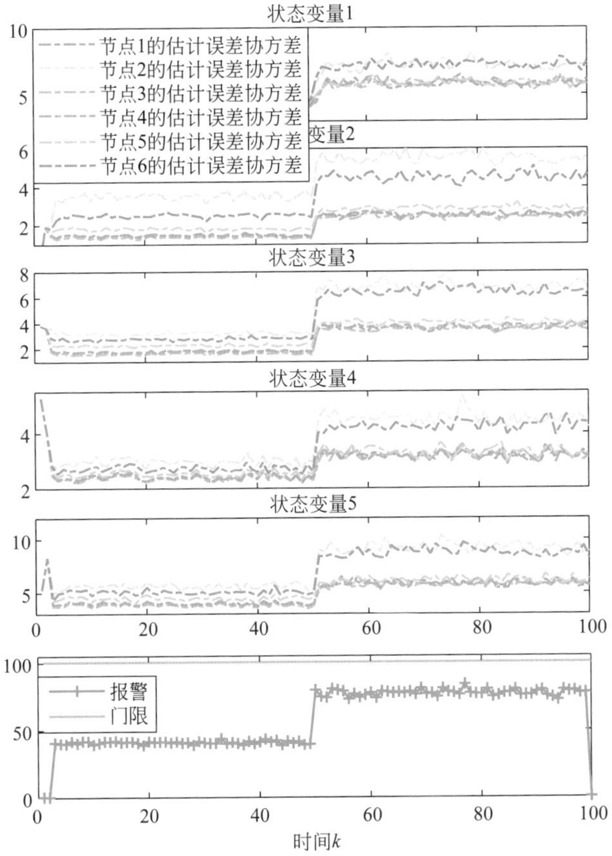

图 11-12　在资源受限的隐秘攻击下，分布式滤波的估计性能，以及防御器抵御攻击的效果

是因为攻击者在通信资源受限的情况下需要满足更为严苛的隐秘性条件 (11-61) 以及 (11-62)，来保证攻击无法被检测。另外，这种隐秘攻击对估计性能的影响很小，并且每个节点的状态估计值仍能收敛至一致。

在图 11-13 中，我们进一步对比了在隐秘攻击下，分布式传感网络与单传感器系统的状态估计的弹性。案例 1 表示在分布式传感网络中通信边 (3,4) 受到攻击；案例 2 表示在单传感器系统中通信边 (3,4) 受到攻击。在单传感器的场景下，第 4 个节点基于其测量值实现对状态的局部估计，并且将该状态估计值传输给第 3 个节点。假设在时间段 [51,100]，攻击者采用隐秘攻击策略 [式 (11-68)] 来篡改无线信道 (3,4) 的通信数据。第 3 个节点利用其防御器来检测恶意攻击。如图 11-13 所示，在两种情况下，时间段 [51,100] 内防御器的报警值均超过门限。这意味着攻击者可以在两种情况下都保持隐身。以真实状态 $\boldsymbol{x}(k)$ 的第 1 个元素的状态估计为例。在时间段 [1,50] 内，传感网络和单传感器的估计性能几乎相同。然而，存在攻击时，传感网络在状态估计上的弹性是优于单传感系统的。

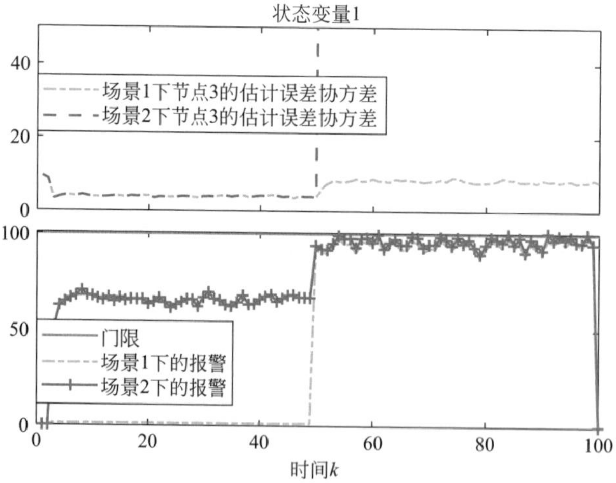

图 11-13 在隐秘攻击下，对比案例 1 和案例 2 的弹性

参考文献

[1] Yeong D J, Velasco-Hernandez G, Barry J, et al. Sensor and sensor fusion technology in autonomous vehicles: A review[J]. Sensors, 2021, 21(6): 2140.

[2] Chen L, Wang S, McDonald-Maier K, et al. Towards autonomous localization and mapping of auvs: a survey[J]. International Journal of Intelligent Unmanned Systems, 2013.

[3] 孙琦钰，赵超强，唐漾，等 . 基于无监督域自适应的计算机视觉任务研究进展 [J] 中国科学 : 技术科学 , 2022, 52(1): 29.

[4] Chen J, Sun J, Wang G. From unmanned systems to autonomous intelligent systems[J]. Engineering, 2022, 12(5): 16-19.

[5] Czubenko M, Kowalczuk Z, Ordys A. Autonomous driver based on an intelligent system of decision-making[J]. Cognitive computation, 2015, 7(5): 569-581.

[6] Wang J R, Hong Y, Wang J L, et al. Cooperative and competitive multi-agent systems: From optimization to games[J]. IEEE/CAA Journal of Automatica Sinica, 2022, 9(5): 763-783.

[7] Dutta V, Zielińska T. Cybersecurity of robotic systems: Leading challenges and robotic system design methodology[J]. Electronics, 2021, 10(22): 2850.

[8] He W, Xu W, Ge X, et al. Secure control of multi-agent systems against malicious attacks: A brief survey[J]. IEEE Transactions on Industrial Informatics, 2022, 18(6): 3595-3608.

[9] Zhang D, Feng G, Shi Y, et al. Physical safety and cyber security analysis of multi-agent systems: A survey of recent advances[J]. IEEE/CAA Journal of Automatica Sinica, 2021, 8(2): 319-333.

[10] Yang W, Chen G, Wang X, et al. Stochastic sensor activation for distributed state estimation over a sensor network[J]. Automatica, 2014, 50(8): 2070-2076.

[11] Das S, Moura J M. Distributed Kalman filtering with dynamic observations consensus[J]. IEEE Transactions on Signal Processing, 2015, 63(17): 4458-4473.

[12] Li W, Jia Y, Du J. Event-triggered Kalman consensus filter over sensor networks[J]. IET Control Theory & Applications, 2016, 10(1): 103-110.

[13] Liu Q, Wang Z, He X, et al. Event-based recursive distributed filtering over wireless sensor networks[J]. IEEE Transactions on Automatic Control, 2015, 60(9): 2470-2475.

[14] Battistelli G, Chisci L, Selvi D. A distributed Kalman filter with event-triggered communication and guaranteed stability[J]. Automatica, 2018, 93: 75-82.

[15] Muehlebach M, Trimpe S. Distributed event-based state estimation for networked systems: An LMI approach[J]. IEEE Transactions on Automatic Control, 2018, 63(1): 269-276.

[16] Lynch N A. Distributed algorithms[M]. San Francisco, CA: Morgan Kaufmann, 1997.

[17] Borkar V, Varaiya P P. Asymptotic Agreement in Distributed Estimation[J]. IEEE Transactions on Automatic Control, 1982, 27(3): 650-655.

[18] Tsitsiklis J N, Bertsekas D P, Athans M. Distributed Asynchronous Deterministic and Stochastic Gradient Optimization Algorithms[J]. IEEE Transactions on Automatic Control, 1986, 31(9): 803-812.

[19] Olfati-Saber R, Fax J A, Murray R M. Consensus and cooperation in networked multi-agent systems[J]. Proceedings of the IEEE, 2007, 95(1): 215-233.

[20] Cao Y, Yu W, Ren W, et al. An Overview of Recent Progress in the Study of Distributed Multi-Agent Coordination[J]. IEEE Transactions on Industrial Informatics, 2013, 9(1): 427-438.

[21] Qin J, Ma Q, Shi Y, et al. Recent advances in consensus of multi-agent systems: A brief survey[J]. IEEE Transactions on Industrial Electronics, 2017, 64(6): 4972-4983.

[22] Ren W. On consensus algorithms for double-integrator dynamics[J]. IEEE Transactions on Automatic Control, 2008, 53(6): 1503-1509.

[23] Tuna S E. Conditions for Synchronizability in Arrays of Coupled Linear Systems[J]. IEEE Transactions on Automatic Control, 2009, 54(10): 2416-2420.

[24] Lin Z, FRANCIS B, MAGGIORE M. State agreement for continuous-time coupled nonlinear systems[J]. SIAM Journal on Control and Optimization, 2007, 46(1): 288-307.

[25] Ren W. Consensus Tracking Under Directed Interaction Topologies :[J]. IEEE Transactions on Control Systems Technology, 2010, 18(1): 230-237.

[26] Parlangeli G, Valcher M E. Accelerating Consensus in High-Order Leader-Follower Networks[J]. IEEE Control

Systems Letters, 2018, 2(3): 381-386.
[27] Cao Y, Stuart D, Ren W, et al. Distributed containment control for multiple autonomous vehicles with double-integrator dynamics: Algorithms and experiments[J]. IEEE Transactions on Control Systems Technology, 2011, 19(4): 929-938.
[28] Chung S J, Ahsun U, Slotine J. Application of synchronization to formation flying spacecraft: Lagrangian approach[J]. Journal of Guidance Control & Dynamics, 2009, 32(2): 512-526.
[29] Nigam N, Bieniawski S, Kroo I, et al. Control of multiple uavs for persistent surveillance: Algorithm and flight test results[J]. IEEE Transactions on Control Systems Technology, 2012, 20(5): 1236-1251.
[30] Dong X, Yu B, Shi Z, et al. Time-varying formation control for unmanned aerial vehicles: Theories and applications[J]. IEEE Transactions on Control Systems Technology, 2015, 23(1): 340-348.
[31] Oh K K, Park M C, Ahn H S. A survey of multi-agent formation control[J]. Automatica, 2015, 53: 424-440.
[32] Ren W, Beard R W, Ella M A. Informtion consensus in Multivehicle Cooperative Control[J]. IEEE Control Systems Magazine, 2007, 27(2): 71-82.
[33] Ji M, Egerstedt M. Distributed coordination control of multiagent systems while preserving connectedness[J]. IEEE Transactions on Robotics, 2007, 23(4): 693-703.
[34] Bae Y B, Lim Y H, Ahn H S. Distributed Robust Adaptive Gradient Controller in Distance-based Formation Control with Exogenous Disturbance[J]. IEEE Transactions on Automatic Control, 2021, 66(6): 2868-2874.
[35] Zuo Z, Han Q L, Ning B, et al. An Overview of Recent Advances in Fixed-Time Cooperative Control of Multiagent Systems[J]. IEEE Transactions on Industrial Informatics, 2018, 14(6): 2322-2334.
[36] Qin J, Li M, Shi Y, et al. Optimal Synchronization Control of Multiagent Systems with Input Saturation via Off-Policy Reinforcement Learning[J]. IEEE Transactions on Neural Networks and Learning Systems, 2019, 30(1): 85-96.
[37] Bhat S P, Bernstein D S. Finite-time Stability of Continuous Autonomous Systems[J]. SIAM Journal on Control and Optimization, 2000, 38(3): 751-766.
[38] Zuo Z, Tie L. A new class of finite-time nonlinear consensus protocols for multi-agent systems[J]. International Journal of Control, 2014, 87(2): 363-370.
[39] Wang L, Xiao F. Finite-time consensus problems for networks of dynamic agents[J]. IEEE Transactions on Automatic Control, 2010, 55(4): 950-955.
[40] Xiao F, Wang L, Chen J, et al. Finite-time formation control for multi-agent systems[J]. Automatica, 2009, 45(11): 2605-2611.
[41] Cortés J. Finite-time convergent gradient flows with applications to network consensus[J]. Automatica, 2006, 42(11): 1993-2000.
[42] Cao Y, Ren W. Finite-time consensus for multi-agent networks with unknown inherent nonlinear dynamics[J]. Automatica, 2014, 50(10): 2648-2656.
[43] Song Y, Wang Y, Holloway J, et al. Time-varying feedback for regulation of normal-form nonlinear systems in prescribed finite time[J]. Automatica, 2017, 83: 243-251.
[44] Wang Y, Song Y. Leader-following control of high-order multi-agent systems under directed graphs: Pre-specified finite time approach[J]. Automatica, 2018, 87: 113-120.
[45] Ferrari-Trecate G, Galbusera L, Marciandi M P E, et al. Model predictive control schemes for consensus in multi-agent systems with single-and double-integrator dynamics[J]. IEEE Trans-actions on Automatic Control, 2009, 54(11): 2560-2572.
[46] 黄超，薄翠梅，郭伟，等 . 工业污水处理溶解氧的双模糊控制系统研究 [J]. 控制工程 , 2019, 26(2): 6.
[47] 郑晓斌 . 基于模糊滑模变结构的工业机械臂控制系统研究 [J]. 陕西理工大学学报：自然科学版 , 2019, 35(1): 6.
[48] 曾干文 . 工业锅炉燃烧系统模糊控制算法 [J]. 湘潭大学自然科学学报 , 1992(01): 113-117.
[49] 赵弘，杨晓丽 . 用于工业 CT 的电液位置控制系统模糊 PID 研究 [J]. 石油机械 , 2017, 45(10): 6.
[50] 王立红，关云鹤，吕丽 . 基于模糊 PID 的工业烤箱温度控制系统设计 [J]. 辽宁工业大学学报：自然科学版 , 2013, 33(3): 153-155.
[51] 黄金侠，李凤霞，韩华，等 . 模糊自整定 PID 在工业温度控制系统中的应用与研究 [J]. 长沙通信职业技术学院学报 , 2011, 10(4): 101-105.
[52] 宋绍剑，黄清宝，林小峰 . 基于工业以太网的专家控制系统 [J]. 自动化与仪器仪表 , 2004 (5): 57-58.
[53] 崔嵬，李燕，马玥 . 基于 LonWorks 的专家 PID 工业控制系统设计 [J]. 科技信息 , 2011(10):218-219.

[54] 谢鑫，邓通通，邱淑建，等 . 基于专家系统的工业炉节能控制系统研究 [J]. 工业炉 , 2018(1): 62-64.
[55] 朱清智，吴会敏 . 神经网络控制在工业机械臂的应用 [J]. 自动化与仪器仪表 , 2014(12): 115-116.
[56] 王绍亮，张乐，李耀，等 . 镀锌生产线神经网络控制系统的研发与工业应用 [J]. 冶金自动化 , 2017, 041(001): 13-17,23.
[57] 刘挺 . 基于计算机神经网络的工业电弧炉智能化控制系统设计 [J]. 工业加热 , 2020, 49(10): 26-28.
[58] 石乐义，徐兴华，刘祎豪，等 . 一种改进概率神经网络的工业控制系统安全态势评估方法 [J]. 信息网络安全 , 2021, 21(03): 15-25.
[59] 袁鹰，卫志农，陈洪涛，等 . 基于差分进化膜计算的 MMC-HVDC 控制参数优化 [J]. 电力系统及其自动化学报 , 2017, 29(007): 13-19.
[60] 衣文娟 . 进化计算的 PID 控制参数快速整定算法 [J]. 科技通报 , 2013(09): 126-129.
[61] 王国鑫 . 基于遗传算法的工业锅炉变偏置双交叉限幅燃烧控制研究 [D]. 西安科技大学 .
[62] Leonard N E, Fiorelli E. Virtual leaders, artificial potentials and coordinated control of groups[C]// Proceedings of the 40th IEEE Conference on Decision and Control. 2001, 3: 2968-2973.
[63] Lawton J R, Beard R W, Young B J. A decentralized approach to formation maneuvers[J]. IEEE transactions on robotics and automation, 2003, 19(6): 933-941.
[64] Lewis M A, Tan K H. High precision formation control of mobile robots using virtual structures[J]. Autonomous robots, 1997, 4(4): 387-403.
[65] Abdessameud A, Tayebi A. Motion coordination for vtol unmanned aerial vehicles: Attitude synchronisation and formation control[M]. London: Springer Science & Business Media, 2013.
[66] Koditschek D E, Rimon E. Robot navigation functions on manifolds with boundary[J]. Advances in applied mathematics, 1990, 11(4): 412-442.
[67] Tanner H G, Kumar A. Towards decentralization of multi-robot navigation functions[C]// Proceedings of the 2005 IEEE International Conference on Robotics and Automation. 2005: 4132-4137.
[68] Stipanović D M, Inalhan G, Teo R, et al. Decentralized overlapping control of a formation of unmanned aerial vehicles[J]. Automatica, 2004, 40(8): 1285-1296.
[69] Ogren P, Egerstedt M, Hu X. A control lyapunov function approach to multi-agent coordination[C]// Proceedings of the 40th IEEE Conference on Decision and Control. 2001, 2: 1150-1155.
[70] Cortes J, Martinez S, Karatas T, et al. Coverage control for mobile sensing networks[J]. IEEE Transactions on robotics and Automation, 2004, 20(2): 243-255.
[71] Ding D, Wang Z, Ho D W, et al. Observer-based event-triggering consensus control for multi-agent systems with lossy sensors and cyber-attacks[J]. IEEE transactions on cybernetics, 2016, 47(8): 1936-1947.
[72] Guerrero-Higueras Á M, DeCastro-Garcia N, Matellan V. Detection of cyber-attacks to indoor real time localization systems for autonomous robots[J]. Robotics and Autonomous Systems, 2018, 99: 75-83.
[73] Speyer J L, Fan C H, Banavar R N. Optimal stochastic estimation with exponential cost criteria[C]// Proceedings of the 31st IEEE Conference on Decision and Control (CDC). 1992, 2: 2293-2299.
[74] Ephraim Y, Merhav N. Hidden Markov processes[J]. IEEE Transactions on Information Theory, 2002, 48(6): 1518-1569.
[75] Dymarski P. Hidden markov models, theory and applications[M]. Rijeka, Croatia: InTech, 2011.
[76] Rabiner L R. A tutorial on hidden Markov models and selected applications in speech recognition[J]. Proceedings of the IEEE, 1989, 77(2): 257-286.
[77] Thunberg J, Goncalves J, Hu X. Consensus and formation control on SE(3) for switching topologies[J]. Automatica, 2016, 66: 109-121.
[78] Chaturvedi N A, Sanyal A K, McClamroch N H. Rigid-body attitude control[J]. IEEE Control Systems Magazine, 2011, 31(3): 30-51.
[79] Wei J, Zhang S, Adaldo A, et al. Finite-time attitude synchronization with distributed discon-tinuous protocols[J]. IEEE Transactions on Automatic Control, 2018, 63(10): 3608-3615.
[80] Tang Y, Xing X, Karimi H R, et al. Tracking control of networked multi-agent systems under new characterizations of impulses and its applications in robotic systems[J]. IEEE Transactions on Industrial Electronics, 2015, 63(2): 1299-1307.
[81] Tang Y, Gao H, Zhang W, et al. Leader-following consensus of a class of stochastic delayed multi-agent systems with partial mixed impulses[J]. Automatica, 2015, 53: 346-354.
[82] Heemels W M H, Teel A R, Van de Wouw N, et al. Networked control systems with communication constraints:

Tradeoffs between transmission intervals, delays and performance[J]. IEEE Transactions on Automatic control, 2010, 55(8): 1781-1796.
[83] Dolk V, Tesi P, De Persis C, et al. Event-triggered control systems under denial-of-service attacks[J]. IEEE Transactions on Control of Network Systems, 2016, 4(1): 93-105.
[84] Postoyan R, Van de Wouw N, Nešić D, et al. Tracking control for nonlinear networked control systems[J]. IEEE Transactions on Automatic Control, 2014, 59(6): 1539-1554.
[85] De Persis C, Tesi P. Input-to-state stabilizing control under denial-of-service[J]. IEEE Transactions on Automatic Control, 2015, 60(11): 2930-2944.
[86] Teixeira A, Pérez D, Sandberg H, et al. Attack models and scenarios for networked control systems[C]// Proceedings of the 1st international conference on High Confidence Networked Systems. ACM, 2012: 55-64.
[87] Li Y, Quevedo D E, Dey S, et al. Sinr-based dos attack on remote state estimation: A game-theoretic approach[J]. IEEE Transactions on Control of Network Systems, 2016, 4(3): 632-642.
[88] Mo Y, Sinopoli B. Secure control against replay attacks[C]// 2009 47th annual Allerton conference on communication, control, and computing (Allerton). IEEE, 2009: 911-918.
[89] Mo Y, Chabukswar R, Sinopoli B. Detecting integrity attacks on scada systems[J]. IEEE Transactions on Control Systems Technology, 2013, 22(4): 1396-1407.
[90] Guo Z, Shi D, Johansson K H, et al. Optimal linear cyber-attack on remote state estimation[J]. IEEE Transactions on Control of Network Systems, 2017, 4(1): 4-13.
[91] Guo Z, Shi D, Johansson K H, et al. Worst-case innovation-based integrity attacks with side information on remote state estimation[J]. IEEE Transactions on Control of Network Systems, 2019, 6(1): 48-59.
[92] Shi D, Elliott R J, Chen T. Event-based state estimation of discrete-state hidden Markov models[J]. Automatica, 2016, 65: 12-26.
[93] Chen W, Wang J, Shi D, et al. Event-based state estimation of hidden Markov models through a Gilbert-Elliott channel[J]. IEEE Transactions on Automatic Control, 2017, 62(7): 3626-3633.
[94] Dey S, Moore J B. Risk-sensitive filtering and smoothing via reference probability methods[J]. IEEE Transactions on Automatic Control, 1997, 42(11): 1587-1591.
[95] Gao H, Li X. Robust filtering for uncertain systems[M]. Cham, Switzerland: Springer, 2014.
[96] Tang Y, Gao H, Kurths J. Robust H_∞ self-triggered control of networked systems under packet dropouts[J]. IEEE Transactions on Cybernetics, 2016, 46(12): 3294-3305.
[97] Boel R K, James M R, Petersen I R. Robustness and risk-sensitive filtering[J]. IEEE Transactions on Automatic Control, 2002, 47(3): 451-461.
[98] Dey S, Moore J B. Risk-sensitive filtering and smoothing for hidden Markov models[J]. Systems Control Letters, 1995, 25(5): 361-366.
[99] Ramezani V R, Marcus S I, et al. Estimation of hidden Markov models: Risk-sensitive filter banks and qualitative analysis of their sample paths[J]. IEEE Transactions on Automatic Control, 2002, 47(12): 1999-2009.
[100] Kung E, Wu J, Shi D, et al. On the nonexistence of event-based triggers that preserve Gaussian state in presence of package-drop[C]// Proceedings of the 2017 American Control Conference (ACC). 2017: 1233-1237.
[101] Sijs J, Lazar M. Event based state estimation with time synchronous updates[J]. IEEE Transactions on Automatic Control, 2012, 57(10): 2650-2655.
[102] Wu J, Jia Q S, Johansson K H, et al. Event-based sensor data scheduling: Trade-off between communication rate and estimation quality[J]. IEEE Transactions on Automatic Control, 2013, 58(4): 1041-1046.
[103] Shi D, Chen T, Shi L. An event-triggered approach to state estimation with multiple point-and set-valued measurements[J]. Automatica, 2014, 50(6): 1641-1648.
[104] Han D, Mo Y, Wu J, et al. Stochastic event-triggered sensor schedule for remote state estimation[J]. IEEE Transactions on Automatic Control, 2015, 60(10): 2661-2675.
[105] Weerakkody S, Mo Y, Sinopoli B, et al. Multi-sensor scheduling for state estimation with event-based, stochastic triggers[J]. IEEE Transactions on Automatic Control, 2016, 61(9): 2695-2701.
[106] Huang J, Shi D, Chen T. Energy-based event-triggered state estimation for hidden Markov models[J]. Automatica, 2017, 79: 256-264.
[107] Elliott R J, Aggoun L, Moore J B. Hidden markov models: Estimation and control[M]. New York, NY, USA:

Springer, 1995.
[108] Cohen S N, Elliott R J. Stochastic calculus and applications[M]. 2 nd ed. New York, NY, USA: Springer, 2015.
[109] Levy B C. Principles of signal detection and parameter estimation[M]. New York, NY, USA: Springer, 2008.
[110] Miskowicz M. Send-on-delta concept: an event-based data reporting strategy[J]. Sensors, 2006, 6(1): 49-63.
[111]Tai A H, Ching W K, Chan L Y. Detection of machine failure: Hidden Markov model approach[J]. Computers & Industrial Engineering, 2009, 57(2): 608-619.
[112] Chen W, Shi D, Wang J, et al. Event-triggered state estimation: Experimental performance assessment and comparative study[J]. IEEE Transactions on Control Systems Technology, 2017, 25(5): 1865-1872.
[113] Petersen I R, James M R, Dupuis P. Minimax optimal control of stochastic uncertain systems with relative entropy constraints[J]. IEEE Transactions on Automatic Control, 2000, 45(3): 398-412.
[114] Yoon M G, Ugrinovskii V A, Petersen I R. Robust finite horizon minimax filtering for discrete-time stochastic uncertain systems[J]. Systems & Control Letters, 2004, 52(2): 99-112.
[115] Gray R M. Entropy and information theory[M]. 2nd ed. New York, NY, USA: Springer, 2011.
[116] Shi D, Chen T, Darouach M. Event-based state estimation of linear dynamic systems with unknown exogenous inputs[J]. Automatica, 2016, 69: 275-288.
[117] Trimpe S, Campi M C. On the choice of the event trigger in event-based estimation[C]//Proceedings of the 1st International Conference on Event-based Control, Communication, and Signal Processing. 2015: 1-8.
[118] Dai Pra P, Meneghini L, Runggaldier W J. Connections between stochastic control and dynamic games[J]. Mathematics of Control, Signals and Systems, 1996, 9(4): 303-326.
[119] Kumar P R, Varaiya P. Stochastic systems: Estimation, identification, and adaptive control[M]. Englewood Cliffs, NJ, USA: Prentice-Hall, 1986.
[120] Collings I B, James M R, Moore J B. An information-state approach to risk-sensitive tracking problems[J]. Journal of Mathematical Systems Estimation and Control, 1996, 6(3): 1-24.
[121] Shaiju A, Petersen I R. A formula for the optimal cost in the general discrete-time LEQG problem[J]. Automatica, 2009, 45(10): 2419-2426.
[122] Huang J, Shi D, Chen T. Robust event-triggered state estimation: A risk-sensitive approach[J]. Automatica, 2019, 99: 253-265.
[123] Bertsekas D P. Dynamic programming and optimal control: Vol. 1[M]. 3rd ed. Belmont, MA, USA: Athena scientific, 2005.
[124] Shi D, Chen T, Shi L. Event-triggered maximum likelihood state estimation[J]. Automatica, 2014, 50(1): 247-254.
[125] Xie L, Ugrinovskii V A, Petersen I R. A posteriori probability distances between finite-alphabet hidden Markov models[J]. IEEE Transactions on Information Theory, 2007, 53(2): 783-793.
[126] Sinopoli B, Schenato L, Franceschetti M, et al. Kalman filtering with intermittent observations[J]. IEEE Transactions on Automatic Control, 2004, 49(9): 1453-1464.
[127] Huang M, Dey S. Stability of Kalman filtering with Markovian packet losses[J]. Automatica, 2007, 43(4): 598-607.
[128] Xie L, Xie L. Stability of a random Riccati equation with Markovian binary switching[J]. IEEE Transactions on Automatic Control, 2008, 53(7): 1759-1764.
[129] You K, Fu M, Xie L. Mean square stability for Kalman filtering with Markovian packet losses[J]. Automatica, 2011, 47(12): 2647-2657.
[130] Sui T, You K, Fu M. Stability conditions for multi-sensor state estimation over a lossy network[J]. Automatica, 2015, 53: 1-9.
[131] Wu J, Shi G, Anderson B D, et al. Kalman filtering over Gilbert-Elliott channels: Stability conditions and critical curve[J]. IEEE Transactions on Automatic Control, 2018, 63(4): 1003-1017.
[132] Lin H, Lam J, Wang Z, et al. State estimation over non-acknowledgment networks with Markovian packet dropouts[J]. Automatica, 2019, 109(108484):1-10.
[133] Smith S C, Seiler P. Estimation with lossy measurements: Jump estimators for jump systems[J]. IEEE Transactions on Automatic Control, 2003, 48(12): 2163-2171.
[134] Fletcher A K, Rangan S, Goyal V K. Estimation from lossy sensor data: Jump linear modeling and Kalman filtering[C]// Proceedings of the 3rd International Symposium on Information Processing in Sensor Networks. 2004: 251-258.

[135] Harville D A. Matrix algebra from a statistician' s perspective[M]. New York, NY, USA: Springer, 1997.
[136] Anderson B D, Moore J B. Optimal filtering[M]. Englewood Cliffs, NJ, USA: Prentice-Hall, 1979.
[137] Schenato L, Sinopoli B, Franceschetti M, et al. Foundations of control and estimation over lossy networks[J]. Proceedings of the IEEE, 2007, 95(1): 163-187.
[138] Moon J, Başar T. Minimax estimation with intermittent observations[J]. Automatica, 2015, 62: 122-133.
[139] Costa O L V, Fragoso M D, Marques R P. Discrete-time Markov jump linear systems[M]. London, UK: Springer, 2005.
[140] Costa O L. Discrete-time coupled Riccati equations for systems with Markov switching parameters[J]. Journal of Mathematical Analysis and Applications, 1995, 194(1): 197-216.
[141] Shen L, Sun J, Wu Q. Observability and detectability of discrete-time stochastic systems with Markovian jump[J]. Systems & Control Letters, 2013, 62(1): 37-42.
[142] Costa O L V, Marques R P. Maximal and stabilizing Hermitian solutions for discrete-time coupled algebraic Riccati equations[J]. Mathematics of Control, Signals and Systems, 1999, 12 (2): 167-195.
[143] Ungureanu V M. Stabilizing solution for a discrete-time modified algebraic Riccati equation in infinite dimensions[J]. Discrete Dynamics in Nature and Society, 2015, 2015: 293930.
[144] Gu G. Discrete-time linear systems: Theory and design with applications[M]. New York, NY, USA: Springer, 2012.
[145] Wonham W. On pole assignment in multi-input controllable linear systems[J]. IEEE Transactions on Automatic Control, 1967, 12(6): 660-665.
[146] Gu G, Qiu L. Networked stabilization of multi-input systems with channel resource allocation[C]// Proceedings of the 17th International Federation of Automatic Control (IFAC) World Congress. 2008: 625-630.
[147] Gu G, Wan S, Qiu L. Networked stabilization for multi-input systems over quantized fading channels[J]. Automatica, 2015, 61: 1-8.
[148] Xiao N, Xie L, Qiu L. Feedback stabilization of discrete-time networked systems over fading channels[J]. IEEE Transactions on Automatic Control, 2012, 57(9): 2176-2189.
[149] Mo Y, Garone E, Sinopoli B. LQG control with Markovian packet loss[C]// Proceedings of the 2013 European Control Conference. 2013: 2380-2385.
[150] Löfberg J. YALMIP : A toolbox for modeling and optimization in MATLAB[C]// Proceedings of the 2004 IEEE International Symposium on Computer Aided Control Systems Design. 2004: 284-289.
[151] Singer R A. Estimating optimal tracking filter performance for manned maneuvering targets[J]. IEEE Transactions on Aerospace and Electronic Systems, 1970, 6(4): 473-483.
[152] Mei J, Ren W, Ma G. Distributed coordinated tracking with a dynamic leader for multiple euler-lagrange systems[J]. IEEE Transactions on Automatic Control, 2011, 56(6): 1415-1421.
[153] Mei J, Ren W, Ma G. Distributed containment control for Lagrangian networks with parametric uncertainties under a directed graph[J]. Automatica, 2012, 48(4): 653-659.
[154] Meng Z, Ren W, You Z. Distributed finite-time attitude containment control for multiple rigid bodies[J]. Automatica, 2010, 46(12): 2092-2099.
[155] Ghasemi M, Nersesov S G. Finite-time coordination in multiagent systems using sliding mode control approach[J]. Automatica, 2014, 50(4): 1209-1216.
[156] Pal A K, Kamal S, Nagar S K, et al. Design of controllers with arbitrary convergence time[J]. Automatica, 2020, 112: 108710.
[157] Shtessel Y B, Moreno J A, Fridman L M. Twisting sliding mode control with adaptation: Lyapunov design, methodology and application[J]. Automatica, 2017, 75: 229-235.
[158] Zou A M, de Ruiter A H, Kumar K D. Distributed finite-time velocity-free attitude coordination control for spacecraft formations[J]. Automatica, 2016, 67: 46-53.
[159] Gui H, De Ruiter A H. Global finite-time attitude consensus of leader-following spacecraft systems based on distributed observers[J]. Automatica, 2018, 91: 225-232.
[160] Lu W, Liu X, Chen T. A note on finite-time and fixed-time stability[J]. Neural Networks, 2016, 81: 11-15.
[161] Noreen G, Shambayati S, Piazzolla S, et al. Low cost deep space hybrid optical/RF communications architecture[C]// 2009 IEEE Aerospace conference. Big sky, MT, USA, 2009: 1-15.
[162] Hu B, Guan Z H, Fu M. Distributed event-driven control for finite-time consensus[J]. Automatica, 2019, 103: 88-95.

[163] Ren W. Distributed cooperative attitude synchronization and tracking for multiple rigid bodies[J]. IEEE Transactions on Control Systems Technology, 2010, 18(2): 383-392.
[164] Thunberg J, Hu X, Goncalves J. Local Lyapunov Functions for Consensus in Switching Non-linear Systems[J]. IEEE Transactions on Automatic Control, 2017, 62(12): 6466-6472.
[165] Cai H, Huang J. Leader-following attitude consensus of multiple rigid body systems by attitude feedback control[J]. Automatica, 2016, 69: 87-92.
[166] Tang Y, Zhang D, Jin X, et al. A resilient attitude tracking algorithm for mechanical systems[J]. IEEE/ASME Transactions on Mechatronics, 2019, 24(6): 2550-2561.
[167] Weng S, Yue D. Distributed event-triggered cooperative attitude control of multiple rigid bodies with leader-follower architecture[J]. International Journal of Systems Science, 2016, 47(3): 631-643.
[168] Thunberg J, Song W, Montijano E, et al. Distributed attitude synchronization control of multi-agent systems with switching topologies[J]. Automatica, 2014, 50(3): 832-840.
[169] Hardy G, Littlewood J, Pólya G. Inequalities[M]. Cambridge: The University Press, 1952.
[170] Pólya G. Über positive darstellung von polynomen[J]. Vierteljahrschrift Naturforschenden Gesellschaft, 1928, 73: 141-145.
[171] Bhat S, Bernstein D. Geometric homogeneity with applications to finite-time stability[J]. Mathematics of Control Signals and Systems, 2005, 17(2): 101-127.
[172] Zhang H, Lewis F L, Qu Z. Lyapunov, adaptive, and optimal design techniques for cooperative systems on directed communication graphs[J]. IEEE Transactions on Industrial Electronics, 2012, 59(7): 3026-3041.
[173] Khalil H. Nonlinear systems[M]. Upper Saddle River: Prentice Hall, 2002.
[174] Polyakov A. Nonlinear feedback design for fixed-time stabilization of linear control systems[J]. IEEE Transactions on Automatic Control, 2012, 57(8): 2106-2110.
[175] Jin X, Du W, He W, et al. Twisting-based finite-time consensus for Euler-Lagrange systems with an event-triggered strategy[J/OL]. IEEE Transactions on Network Science and Engineering, 2020, 7(3): 1007-1018.
[176] Cheng B, Li Z. Fully distributed event-triggered protocols for linear multiagent networks[J]. IEEE Transactions on Automatic Control, 2019, 64(4): 1655-1662.
[177] Zhu W, Zhou Q, Wang D. Consensus of linear multi-agent systems via adaptive event-based protocols[J]. Neurocomputing, 2018, 318: 175-181.
[178] Zhu W, Jiang Z, Feng G. Event-based consensus of multi-agent systems with general linear models[J]. Automatica, 2014, 50(2): 552-558.
[179] Li H, Liao X, Huang T, et al. Event-triggering sampling based leader-following consensus in second-order multi-agent systems[J]. IEEE Transactions on Automatic Control, 2015, 60(7): 1998-2003.
[180] Yi X, Liu K, Dimarogonas D V, et al. Dynamic Event-Triggered and Self-Triggered Control for Multi-agent Systems[J]. IEEE Transactions on Automatic Control, 2019, 64(8): 3300-3307.
[181] Lu W, Chen T. New approach to synchronization analysis of linearly coupled ordinary differential systems[J]. Physica D-nonlinear Phenomena, 2006, 213(2): 214-230.
[182] Sánchez T, Moreno J. A constructive lyapunov function design method for a class of homogeneous systems[C]// 53rd IEEE Conference on Decision and Control (CDC). 2014: 5500-5505.
[183] Gonzalez V, Sánchez T, Fridman L, et al. Design of continuous twisting algorithm[J]. Automatica, 2017, 80: 119-126.
[184] Liu X, Du C, Lu P, et al. Decentralised consensus for multiple lagrangian systems based on event-triggered strategy[J]. Int. J. Control, 2016, 89(6): 1111-1124.
[185] He W, Xu C, Han Q, et al. Finite-time $\mathcal{L}_2$ leader-follower consensus of networked euler-lagrange systems with external disturbances[J]. IEEE Transactions on Systems, Man, and Cybernetics: Systems, 2018, 48(11): 1920-1928.
[186] Du C, Liu X, Ren W, et al. Finite-time consensus for linear multiagent systems via event-triggered strategy without continuous communication[J]. IEEE Transactions on Control of Network Systems, 2020, 7(1): 19-29.
[187] Hadaegh F, Smith R. Control topologies for deep space formation flying spacecraft[J]. Journal of Guidance, Control and Dynamics, 2005, 28(1): 106-114.
[188] Magnis L, Petit N. Angular velocity nonlinear observer from single vector measurements[J]. IEEE Transactions on Automatic Control, 2016, 61(9): 2473-2483.
[189] Tayebi A. Unit quaternion-based output feedback for the attitude tracking problem[J]. IEEE Transactions on

Automatic Control, 2008, 53(6): 1516-1520.
[190] Zou A, De Ruiter A H, Kumar K D. Distributed finite-time velocity-free attitude coordination control for spacecraft formations[J]. Automatica, 2016, 67: 46-53.
[191] Frias A, De Ruiter A H, Kumar K D. Velocity-free spacecraft attitude stabilization using two control torques[J]. Automatica, 2019, 109: 108553.
[192] Gui H, Vukovich G. Finite-time output-feedback position and attitude tracking of a rigid body[J]. Automatica, 2016, 74: 270-278.
[193] Zou A, Fan Z. Fixed-time attitude tracking control for rigid spacecraft without angular velocity measurements[J]. IEEE Transactions on Industrial Electronics, 2020, 67(8): 6795-6805.
[194] Abdessameud A, Tayebi A. Attitude synchronization of a group of spacecraft without velocity measurements[J]. IEEE Transactions on Automatic Control, 2009, 54(11): 2642-2648.
[195] Abdessameud A, Tayebi A, Polushin I G. Attitude Synchronziation of Multiple Rigid Bodies With Communications Delays[J]. IEEE Transactions on Automatic Control, 2012, 57(9): 2405-2411.
[196] Mayhew C, Sanfelice R, Sheng J, et al. Quaternion-based hybrid feedback for robust global attitude synchronization[J]. IEEE Transactions on Automatic Control, 2012, 57(8): 2122-2127.
[197] Zou Y, Meng Z. Velocity-free leader-follower cooperative attitude tracking of multiple rigid bodies on SO(3)[J]. IEEE Transactions on Cybernetics, 2019, 49(12): 4078-4089.
[198] Tron R, Afsari B, Vidal R. Riemannian consensus for manifolds with bounded curvature[J]. IEEE Transactions on Automatic Control, 2013, 58(4): 921-934.
[199] Davarian F, Asmar S, Angert M, et al. Improving Small Satellite Communications and Tracking in Deep Space - A Review of the Existing Systems and Technologies with Recommendations for Improvement. Part II: Small Satellite Navigation, Proximity Links, and Communications Link Science[J/OL]. IEEE Aerospace and Electronic Systems Magazine, 2020, 35(7): 26-40.
[200] Xu C, Wu B, Cao X, et al. Distributed adaptive event-triggered control for attitude synchronization of multiple spacecraft[J]. Nonlinear Dynamics, 2019, 95: 2625-2638.
[201] Guo F, Zhang S. Event-triggered coordinated attitude method for chip satellite cluster[J]. Acta Astronautica, 2019, 160: 451-460.
[202] Wang C, Guo L, Wen C, et al. Event-triggered adaptive attitude tracking control for spacecraft with unknown actuator faults[J]. IEEE Transactions on Industrial Electronics, 2020, 67(3): 2241-2250.
[203] Tron R, Afsari B, Vidal R. Intrinsic consensus on SO(3) with almost-global convergence[C]// 51st IEEE Conference on Decision and Control. Maui, HI, USA, 2012: 2052-2058.
[204] Markdahl J, Thunberg J, Gonçalves J. Almost global consensus on the *n*-sphere[J]. IEEE Transactions on Automatic Control, 2018, 63(6): 1664-1675.
[205] Weng S, Yue D, Xie X, et al. Distributed event-triggered cooperative attitude control of multiple groups of rigid bodies on manifold SO(3) [J]. Information Sciences, 2016, 370-371: 636-649.
[206] Berneburg J, Nowzari C. Distributed dynamic event-triggered coordination with a designable minimum inter-event time[C]// 2019 American Control Conference (ACC). Philadelphia, PA, USA, 2019: 1424-1429.
[207] Jin X, Shi Y, Tang Y, et al. Event-triggered attitude consensus with absolute and relative attitude measurements[J]. Automatica, 2020, 122: 109245.
[208] Yi X, Lu W, Chen T. Pull-based distributed event-triggered circle formation control for multi-agent systems with directed topologies[J]. IEEE Transactions on Neural Networks and Learning Systems, 2017, 28(1): 71-79.
[209] Su Y, Huang J. Stability of a class of linear switching systems with applications to two consensus problems[J]. IEEE Transactions on Automatic Control, 2012, 57(6): 1420-1430.
[210] Shuster M. A survey of attitude representations[J]. The Journal of the Astronautical Sciences, 1993, 41(4): 439-517.
[211] Thunberg J, Markdahl J, Bernard F, et al. A lifting method for analyzing distributed synchronization on the unit sphere[J]. Automatica, 2018, 96: 253-258.
[212] Lageman C, Sun Z. Consensus on spheres: Convergence analysis and perturbation theory [C]// 55th IEEE Conference on Decision and Control (CDC). 2016: 19-24.
[213] Afsari B. Riemannian L_p center of mass: existence, uniqueness, and convexity[J]. Proceedings of the American Mathematical Society, 2011, 139(2): 655-673.
[214] Sutton R S, Barto A G. Reinforcement learning: An introduction[M]. Cambridge, MA, USA:MIT Press, 1998.

[215] Kelly R, Santibáñez V, Loría A. Control of robot manipulators in joint space[M]. London, UK: Springer-Verlag, 2005.
[216] Yang Y, Li Y, Yue D, et al. Adaptive event-triggered consensus control of a class of second-order nonlinear multiagent systems[J]. IEEE Transactions on Cybernetics, 2020, 50(12): 5010-5020.
[217] Murray J J, Cox C J, Lendaris G G, et al. Adaptive dynamic programming[J]. IEEE Transactions on Systems, Man and Cybernetics, Part C (Applications and Reviews), 2002, 32(2): 140-153.
[218] Zhang J, Zhang H, Feng T. Distributed optimal consensus control for nonlinear multiagent system with unknown dynamic[J]. IEEE Transactions on Neural Networks and Learning Systems, 2018, 29(8): 3339-3348.
[219] Bemporad A, Heemels M, Johansson M. Networked control systems[M]. Berlin: Springer, 2011.
[220] Dong L, Zhong X, Sun C, et al. Event-triggered adaptive dynamic programming for continuous-time systems with control constraints[J]. IEEE Transactions on Neural Networks and Learning Systems, 2017, 28(8): 1941-1952.
[221] Ioannou P, Fidan B. Adaptive control tutorial[M]. Society for Industrial and Applied Mathematics, 2006.
[222] Khalil H K. Nonlinear systems[M]. Upper Saddle River, NJ, USA: Prentice Hall, 2002.
[223] Jin X, Du W, He W, et al. Twisting-based finite-time consensus for euler-lagrange systems with an event-triggered strategy[J]. IEEE Transactions on Network Science and Engineering, 2020, 7(3): 1007-1018.
[224] Oh K K, Park M C, Ahn H S. A survey of multi-agent formation control[J]. Automatica, 2015, 53: 424-440.
[225] Chen H. Robust stabilization for a class of dynamic feedback uncertain nonholonomic mobile robots with input saturation[J]. International Journal of Control, Automation and Systems, 2014, 12(6): 1216-1224.
[226] De Persis C, Postoyan R. A lyapunov redesign of coordination algorithms for cyber-physical systems[J]. IEEE Transactions on Automatic Control, 2016, 62(2): 808-823.
[227] Teixeira A, Pérez D, Sandberg H, et al. Attack models and scenarios for networked control systems [C]// Proceedings of the 1st international conference on High Confidence Networked Systems. 2012: 55-64.
[228] Liu Y, Ning P, Reiter M K. False data injection attacks against state estimation in electric power grids[J]. ACM Transactions on Information and System Security (TISSEC), 2011, 14(1): 1-33.
[229] Senejohnny D, Tesi P, De Persis C. A jamming-resilient algorithm for self-triggered network coordination[J]. IEEE Transactions on Control of Network Systems, 2017, 5(3): 981-990.
[230] Goebel R, Sanfelice R G, Teel A R. Hybrid dynamical systems: modeling stability, and robustness[M]. Princeton, NJ: Princeton University Press, 2012.
[231] Jiang Z P, Nijmeijer H. Tracking control of mobile robots: A case study in backstepping[J]. Automatica, 1997, 33(7): 1393-1399.
[232] Do K D. Global output-feedback path-following control of unicycle-type mobile robots: A level curve approach[J]. Robotics and Autonomous Systems, 2015, 74: 229-242.
[233] Chen H, Zhang B, Zhao T, et al. Finite-time tracking control for extended nonholonomic chained-form systems with parametric uncertainty and external disturbance[J]. Journal of Vi- bration and Control, 2018, 24(1): 100-109.
[234] Krstic M, Kokotovic P V, Kanellakopoulos I. Nonlinear and adaptive control design[M]. New York: Wiley, 1995.
[235] Mayhew C G, Sanfelice R G, Teel A R. On path-lifting mechanisms and unwinding in quaternion-based attitude control[J]. IEEE Transactions on Automatic Control, 2012, 58(5): 1179-1191.
[236] Feng S, Tesi P. Resilient control under denial-of-service: Robust design[J]. Automatica, 2017, 79: 42-51.
[237] Mayhew C G, Sanfelice R G, Teel A R. Quaternion-based hybrid control for robust global attitude tracking[J]. IEEE Transactions on Automatic control, 2011, 56(11): 2555-2566.
[238] Wang Y, Karimi H R, Lam H K, et al. Notice of violation of ieee publication principles: An improved result on exponential stabilization of sampled-data fuzzy systems[J]. IEEE Transactions on Fuzzy Systems, 2018, 26(6): 3875-3883.
[239] Wang Y, Yang X, Yan H. Reliable fuzzy tracking control of near-space hypersonic vehicle using aperiodic measurement information[J]. IEEE Transactions on Industrial Electronics, 2019, 66 (12): 9439-9447.
[240] Wang Y, Karimi H R, Yan H. An adaptive event-triggered synchronization approach for chaotic lur' e systems subject to aperiodic sampled data[J]. IEEE Transaction on Circuits and Systems Ⅱ : Express Briefs, 2019, 66(3): 442-446.
[241] Du H, Li S. Attitude synchronization for flexible spacecraft with communication delays[J]. IEEE Transactions

on Automatic Control, 2016, 61(11): 3625-3630.

[242] Du H, Li S. Attitude synchronization control for a group of flexible spacecraft[J]. Automatica, 2014, 50(2): 646-651.

[243] Abdessameud A, Polushin I G, Tayebi A. Distributed coordination of dynamical multi-agent systems under directed graphs and constrained information exchange[J]. IEEE Transactions on Automatic Control, 2016, 62(4): 1668-1683.

[244] Berkane S, Tayebi A. Construction of synergistic potential functions on SO(3) with application to velocity-free hybrid attitude stabilization[J]. IEEE Transactions on Automatic Control, 2016, 62(1): 495-501.

[245] Berkane S, Abdessameud A, Tayebi A. Hybrid attitude and gyro-bias observer design on SO(3)[J]. IEEE Transactions on Automatic Control, 2017, 62(11): 6044-6050.

[246] Abdessameud A, Tayebi A. Global trajectory tracking control of vtol-uavs without linear velocity measurements[J]. Automatica, 2010, 46(6): 1053-1059.

[247] Li H, Liao X, Huang T, et al. Event-triggering sampling based leader-following consensus in second-order multi-agent systems[J]. IEEE Transactions on Automatic Control, 2014, 60(7): 1998-2003.

[248] Do K D. Formation tracking control of unicycle-type mobile robots with limited sensing ranges[J]. IEEE Transactions on Control Systems Technology, 2008, 16(3): 527-538.

[249] Postoyan R, Bragagnolo M C, Galbrun E, et al. Event-triggered tracking control of unicycle mobile robots[J]. Automatica, 2015, 52: 302-308.

[250] Dolk V, Borgers D P, Heemels W. Output-based and decentralized dynamic event-triggered control with guaranteed $\mathcal{L}_p$-gain performance and zeno-freeness[J]. IEEE Transactions on Automatic Control, 2016, 62(1): 34-49.

[251] Nesic D, Teel A R. Input-output stability properties of networked control systems[J]. IEEE Transactions on Automatic Control, 2004, 49(10): 1650-1667.

[252] Hespanha J P, Morse A S. Stability of switched systems with average dwell-time[C]// Proceedings of the 38th IEEE conference on decision and control (Cat. No. 99CH36304). IEEE, 1999: 2655-2660.

[253] Postoyan R, Tabuada P, Nesic D, et al. A framework for the event-triggered stabilization of nonlinear systems[J]. IEEE Transactions on Automatic Control, 2014, 60(4): 982-996.

[254] Isidori A. Nonlinear control systems[M]. London: Springer Verlag, London, 1999.

[255] Ding Z. Distributed adaptive consensus output regulation of network-connected heterogeneous unknown linear systems on directed graphs[J]. IEEE Transactions on Automatic Control, 2016, 62(9): 4683-4690.

[256] Zhang J, Shi P, Xia Y. Fuzzy delay compensation control for ts fuzzy systems over network[J]. IEEE Transactions on Cybernetics, 2012, 43(1): 259-268.

[257] Hespanha J P, Naghshtabrizi P, Xu Y. A survey of recent results in networked control systems[J]. Proceedings of the IEEE, 2007, 95(1): 138-162.

[258] Amin S, Litrico X, Sastry S S, et al. Stealthy deception attacks on water scada systems[C]// Proceedings of the 13th ACM international conference on Hybrid systems: Computation and Control. 2010: 161-170.

[259] Xie L, Mo Y, Sinopoli B. False data injection attacks in electricity markets[C]// 2010 First IEEE International Conference on Smart Grid Communications. IEEE, 2010: 226-231.

[260] Pang Z H, Liu G P. Design and implementation of secure networked predictive control systems under deception attacks[J]. IEEE Transactions on Control Systems Technology, 2011, 20(5): 1334-1342.

[261] Zhu M, Martinez S. Attack-resilient distributed formation control via online adaptation[C]// 2011 50th IEEE Conference on Decision and Control and European Control Conference. IEEE, 2011: 6624-6629.

[262] Zhang X M, Han Q L. Network-based H_∞ filtering using a logic jumping-like trigger[J]. Automatica, 2013, 49(5): 1428-1435.

[263] Meng D, Moore K L. Studies on resilient control through multiagent consensus networks subject to disturbances[J]. IEEE Transactions on Cybernetics, 2014, 44(11): 2050-2064.

[264] Wang X, Lemmon M D. Event-triggering in distributed networked control systems[J]. IEEE Transactions on Automatic Control, 2010, 56(3): 586-601.

[265] Zhang X M, Han Q L, Zhang B L. An overview and deep investigation on sampled-data-based event-triggered control and filtering for networked systems[J]. IEEE Transactions on Industrial Informatics, 2016, 13(1): 4-16.

[266] Du W, Leung S Y S, Tang Y, et al. Differential evolution with event-triggered impulsive control[J]. IEEE Transactions on Cybernetics, 2016, 47(1): 244-257.

[267] Carnevale D, Teel A R, Nesic D. A lyapunov proof of an improved maximum allowable transfer interval for networked control systems[J]. IEEE Transactions on Automatic Control, 2007, 52 (5): 892-897.
[268] Ding L, Han Q L, Zhang X M. Distributed secondary control for active power sharing and frequency regulation in islanded microgrids using an event-triggered communication mechanism[J]. IEEE Transactions on Industrial Informatics, 2018, 15(7): 3910-3922.
[269] Wu Y, Su H, Shi P, et al. Output synchronization of nonidentical linear multiagent systems[J]. IEEE Transactions on Cybernetics, 2015, 47(1): 130-141.
[270] Ge X, Han QL. Distributed formation control of networked multi-agent systems using a dynamic event-triggered communication mechanism[J]. IEEE Transactions on Industrial Electronics, 2017, 64(10): 8118-8127.
[271] Ge X, Han Q L, Ding D, et al. A survey on recent advances in distributed sampled-data cooperative control of multi-agent systems[J]. Neurocomputing, 2018, 275: 1684-1701.
[272] Ding L, Han Q L, Ge X, et al. An overview of recent advances in event-triggered consensus of multiagent systems[J]. IEEE Transactions on Cybernetics, 2017, 48(4): 1110-1123.
[273] Hu W, Liu L, Feng G. Consensus of linear multi-agent systems by distributed event-triggered strategy[J]. IEEE Transactions on Cybernetics, 2015, 46(1): 148-157.
[274] Ning B, Han Q L, Zuo Z. Distributed optimization for multiagent systems: An edge-based fixed-time consensus approach[J]. IEEE Transactions on Cybernetics, 2017, 49(1): 122-132.
[275] Ning B, Han Q L. Prescribed finite-time consensus tracking for multiagent systems with non-holonomic chained-form dynamics[J]. IEEE Transactions on Automatic Control, 2018, 64(4): 1686-1693.
[276] Ge X, Han Q L, Zhang X M. Achieving cluster formation of multi-agent systems under aperiodic sampling and communication delays[J]. IEEE Transactions on Industrial Electronics, 2017, 65 (4): 3417-3426.
[277] Ning B, Han Q L, Zuo Z, et al. Collective behaviors of mobile robots beyond the nearest neighbor rules with switching topology[J]. IEEE Transactions on Cybernetics, 2017, 48(5): 1577-1590.
[278] Zhang X M, Han Q L, Yu X. Survey on Recent advances in networked control systems[J]. IEEE Transactions on Industrial Informatics, 2015, 12(5): 1740-1752.
[279] Teel A R, Subbaraman A, Sferlazza A. Stability analysis for stochastic hybrid systems: A survey[J]. Automatica, 2014, 50(10): 2435-2456.
[280] Abdessameud A, Tayebi A, Polushin I G. Attitude synchronization of multiple rigid bodies with communication delays[J]. IEEE Transactions on Automatic Control, 2012, 57(9): 2405-2411.
[281] Jungnickel D, Jungnickel D. Graphs, networks and algorithms[M]. Berlin: Springer, 2005.
[282] Dong X, Hu G. Time-varying formation tracking for linear multiagent systems with multiple leaders[J]. IEEE Transactions on Automatic Control, 2017, 62(7): 3658-3664.
[283] Dolk V, Heemels M. Event-triggered control systems under packet losses[J]. Automatica, 2017, 80: 143-155.
[284] Mo Y, Kim T H J, Brancik K, et al. Cyber-physical security of a smart grid infrastructure[J]. Proceedings of the IEEE, 2011, 100(1): 195-209.
[285] Naldi R, Furci M, Sanfelice R G, et al. Robust global trajectory tracking for underactuated vtol aerial vehicles using inner-outer loop control paradigms[J]. IEEE Transactions on Automatic Control, 2016, 62(1): 97-112.
[286] Mo Y, Weerakkody S, Sinopoli B. Physical authentication of control systems: designing water-marked control inputs to detect counterfeit sensor outputs[J]. IEEE Control Systems Magazine, 2015, 35(1): 93-109.
[287] Porter M, Joshi A, Hespanho P, et al. Simulation and real-world evaluation of attack detection schemes[C]// 2019 American Control Conference (ACC). 2019: 551-558.
[288] Satchidanandan B, Kumar P R. Dynamic watermarking: Active defense of networked cyber-physical systems[J]. Proceedings of the IEEE, 2017, 105(2): 219-240.
[289] Li Y, Chen T. Stochastic detector against linear deception attacks on remote state estimation[C]// 55th IEEE Conference on Decision and Control (CDC). 2016: 6291-6296.
[290] Niederreiter H. Random number generation and quasi-monte carlo methods[M]. Philadelphia, Pa.: Society for Industrial and Applied Mathematics, 1992.
[291] Menezes A J, Katz J, Van Oorschot P C, et al. Handbook of applied cryptography[M]. Boca Raton: CRC Press Series on Discrete Mathematics and Its Applications (CRC press), 1996.
[292] Chen B, Ho D W, Hu G, et al. Secure fusion estimation for bandwidth constrained cyber-physical systems under replay attacks[J]. IEEE Transactions on Cybernetics, 2018, 48(6): 1862-1876.

[293] Blum M, Micali S. How to generate cryptographically strong sequences of pseudorandom bits[J]. SIAM journal on Computing, 1984, 13(4): 850-864.
[294] Knapp A W. Basic real analysis[M]. Boston, MA: Birkhäuser, 2005.
[295] Park S, Martins N C. Design of distributed lti observers for state omniscience[J]. IEEE Transactions on Automatic Control, 2017, 62(2): 561-576.
[296] Kamal A T, Farrell J A, Roy-Chowdhury A K. Information weighted consensus filters and their application in distributed camera networks[J]. IEEE Transactions on Automatic Control, 2013, 58(12): 3112-3125.
[297] Li Y, Shi L, Chen T. Detection against linear deception attacks on multi-sensor remote state estimation[J]. IEEE Transactions on Control of Network Systems, 2018, 5(3): 846-856.
[298] Mitra A, Sundaram S. Secure distributed state estimation of an lti system over time-varying networks and analog erasure channels[C]// 2018 Annual American Control Conference (ACC). 2018: 6578-6583.
[299] Yang W, Zhang Y, Chen G, et al. Distributed filtering under false data injection attacks[J]. Automatica, 2019, 102: 34-44.
[300] Bai C Z, Gupta V, Pasqualetti F. On Kalman filtering with compromised sensors: Attack stealthiness and performance bounds[J]. IEEE Transactions on Automatic Control, 2017, 62(12): 6641-6648.
[301] Guo Z, Shi D, Johansson K H, et al. Optimal linear cyber-attack on remote state estimation[J]. IEEE Transactions on Control of Network Systems, 2016, 4(1): 4-13.
[302] Zhang H, Cheng P, Shi L, et al. Optimal denial-of-service attack scheduling with energy constraint[J]. IEEE Transactions on Automatic Control, 2015, 60(11): 3023-3028.
[303] Olfati-Saber R. Kalman-consensus filter: Optimality, stability, and performance[C]// Proceedings of the 48h IEEE Conference on Decision and Control (CDC) held jointly with 2009 28th Chinese Control Conference. 2009: 7036-7042.
[304] Yang W, Yang C, Shi H, et al. Stochastic link activation for distributed filtering under sensor power constraint[J]. Automatica, 2017, 75: 109-118.